Advances in Composite Science and Technology (ACST 2019)

IOP Conference Series: Materials Science and Engineering
Volume 934

Moscow, Russia
20 - 21 November 2019

ISBN: 978-1-7138-4770-0
ISSN: 1757-8981

Printed from e-media with permission by:

Curran Associates, Inc.
57 Morehouse Lane
Red Hook, NY 12571

Some format issues inherent in the e-media version may also appear in this print version.

Institute of Physics
Dirac House, Temple Back
Bristol BS1 6BE UK

Phone: 44 1 17 929 7481
Fax: 44 1 17 920 0979

techtracking@iop.org

Additional copies of this publication are available from:

Curran Associates, Inc.
57 Morehouse Lane
Red Hook, NY 12571 USA
Phone: 845-758-0400
Fax: 845-758-2633
Email: curran@proceedings.com
Web: www.proceedings.com

TABLE OF CONTENTS

IOP Conf. Series: Materials Science and Engineering **934** (2020) 011001 doi:10.1088/1757-899X/934/1/011001

Preface

The 2nd international conference on composites **"Advances in Composite Science and Technologies"** was held in Bauman Moscow State Technical University from 20-21 November 2019. The conference is meant to be a unified international platform for the knowledge and experience exchange between representatives of the scientific community within the field of composite materials. The following sections covered the most important topics in the composite sciences field.

Session 1. "Smart" composite materials and composites for biomedical application

Session 2. Advanced metal matrix composites

Session 3. Physical and technological maintenance of the quality of composite materials products

Session 4. Composite materials for rocket and space applications

Session 5. Mechanics of composite materials and structures

Session 6. Inorganic composite materials

Session 7. Fiber surface modification methods

Session 8. Modelling in composite science

Session 9. High Temperature composites

Session 10. Russian engineering school (educational session)

The success of the conference is supported with more than 80 proceeding papers received by the Organizing Committee with 70 selected for publication after the reviewing process. We are grateful to all participants as well as to discussion leaders, members of Organizing and Scientific boards for taking part in the conference and we are looking forward to next year's event.

IOP Conf. Series: Materials Science and Engineering **934** (2020) 011002 doi:10.1088/1757-899X/934/1/011002

Peer review declaration

All papers published in this volume of *IOP Conference Series: Materials Science and Engineering* have been peer reviewed through processes administered by the Editors. Reviews were conducted by expert referees to the professional and scientific standards expected of a proceedings journal published by IOP Publishing.

- **Type of peer review:** Single-blind

- **Conference submission management system:** First author submits an article on one of the proposed forum topics (7 topics in total), then the head of the section checks the article, then the head of the section either sends it to the author for revision, or transfers the article to the reviewers, then the reviewer sends the conclusion back to the head of the section, the head of the section sends the conclusion to the editorial department, the editorial department sends the necessary information to the authors.

- **Number of submissions received:** 80

- **Number of submissions sent for review:** 80

- **Number of submissions accepted:** 72

- **Acceptance Rate (Number of Submissions Accepted / Number of Submissions Received X 100):** 90

- **Average number of reviews per paper:** 1-2

- **Total number of reviewers involved:** 2

- **Any additional info on review process:** We initially provide the forum participants with the specified topics (they are also sections), the conference participants prepare materials (articles and reports) in advance, so we achieve the maximum coefficient of approved articles.

- **Contact person for queries:** Jane Kim (jkim@emtc.ru)

IOP Conf. Series: Materials Science and Engineering **934** (2020) 012001 doi:10.1088/1757-899X/934/1/012001

Investigation of the effect of zinc borate on the physic mechanical properties of PVC plastic

T A Borukaev, A Kh Shaov, A M Kharaev and A S Borodulin

Kabardino-Balkarian State University, 360004, Nalchik, Chernishevskiy St., 173, Russian

E-mail: boruk-chemical@mail.ru

Abstract. Zinc borate crystalline hydrate of the composition 2ZnO•3B2O3•3,5H2O was obtained by the interaction of zinc oxide and boric acid in the presence of a seed. The composition of the obtained crystalhydrate of zinc borate was confirmed using IR spectroscopy and gravimetric analysis. It is shown at what temperatures the obtained zinc borate crystalline hydrate loses crystallization water and in what quantity. To evaluate the antipyretic properties, the obtained zinc borate crystalline hydrate was introduced into the polyvinylchloride plastic compound grade NGP30-32, replacing industrial zinc borate in the formulation. It was found that polyvinylchloride plastic containing the obtained zinc borate crystalline hydrate is not inferior in its thermal, fire-resistant and physic mechanical characteristics to a compound containing industrial zinc borate. At the same time, the main indicators of polyvinylchloride plastic containing laboratory zinc borate crystalline hydrate complies with the requirements for fireproof plastic compounds.

Keywords: zinc borate, synthesis, composition, polyvinylchloride, plastic compound, properties

1. Introduction

At present, it is impossible to imagine a modern industry without materials based on polyvinylchloride (PVC) [1, 2]. In terms of production and industrial applications, PVC is one of the most popular polymers [3, 4]. The fields of their application in modern technology are quite extensive: electrical engineering; household industry, medicine, etc. [5, 6]. However, the widespread use of PVC materials in industry, especially in the cable industry as insulation and sheathing of electrical wires, poses certain problems. In particular, PVC compounds are fire hazardous materials. At the same time, fires caused by ignition and burning of polymeric materials annually cause significant damage to various industries [7]. In this regard, the decrease in flammability and combustibility of PVC plastic compounds, the creation of fireproof compounds, is an important problem that needs solutions. To solve this problem - to reduce the flammability of PVC plastic compounds, modern technology uses various extinguishing agents - flame retardants (FR) [8-10], which are mainly of imported origin. In this regard, the urgent problem is the development of technology for domestic effective FR to reduce the combustibility of polymeric materials [11]. At the same time, it is important to use source materials of domestic origin in the process of obtaining flame retardants, which will make it possible to exclude dependence on imported raw materials.

The present work is devoted to the development of an effective method for the production of zinc borate (ZB) of a specific chemical composition (2ZnO•3B2O3•3,5H2O) and the study of its effect on the properties of PVC compound when replaced in the formulation of an industrial ZB.

2. Experimental

Zinc borate was prepared as follows. 36,25 g of boric acid were dissolved in 500 ml of distilled water. 0,625 g of zinc borate was added to the solution as a seed. Then, 47,85 g of zinc oxide and 108,75 g of boric acid were added to the solution. The reaction mixture was stirred for 90 min at a temperature of 60 °C. The temperature of the reaction mixture was raised to 90 °C and stirred for another 4 hours. The resulting product was filtered, washed with water and dried in vacuum at 105 °C. The yield of ZB was 95 %. The particle size is 1-3 mcm. The amount of crystallization water in zinc borate was determined by the gravimetric method, according to the procedure [12].

In our work, we used NGP30-32 industrial plastic compound manufactured in accordance with GOST 5960-72, which includes PVC, dioctyl phthalate, tribasic lead sulfide (TOCS), chalk, stearic acid, calcium stearate, zinc borate, as the base polyvinyl chloride (PVC) plastic compounds, antimony oxide, zinc oxide and aluminum hydroxide.

As an alternative to industrial zinc borate, zinc borate obtained in the laboratory was used in the work. At the same time, PVC plastic compounds were prepared in accordance with GOST using a two-speed mixer R600/HC2500 manufactured by “Diosna”, followed by extrusion of the mixture using a twin-screw extruder (145-170 °C), screw rotation speed 48 rpm.

Samples of PVC plastic compounds for testing were prepared by injection molding on a Ray-Ran Politest injection molding machine at a material cylinder temperature of 165-170 °C, a mold temperature of 60 °C and a closing pressure of 8 bar.

Before conducting physical and mechanical tests, the samples were conditioned in a standard atmosphere according to GOST 12423-66 for at least 16 hours.

The oxygen index (OI) was determined by the GOST 12.1.044-89 on the instrument of the company Noselab ast. Fire testing model Oxygen Index (Italy). The samples were bars with a length of 80-120 mm, wide 10±0,5 mm and thick 3-4 mm.

Thermogravimetric analysis (TGA) of the obtained ZB was performed using a TGA 4000 derivatograph from Perkin Elmer (USA). The measurement was carried out in a dynamic heating mode in the temperature range from 25 to 700 °C at a heating rate of 2,5 °C/min in an air atmosphere.

The thermo stability of PVC compounds was determined by the time of the induction period of the color change of the Congo-red indicator during the release of HCl during the aging of the compound (200 °C) according to [13].

The melt flow rate (MFR) was determined on an IIRT-A capillary viscometer (load - 10 kg, temperature 190 °C). The diameter of the capillary is 2 mm. The density of PVC compounds was determined according to GOST 15139-69 by hydrostatic weighing in distilled water at a temperature of (20±2) °C on injection samples.

Uniaxial tensile mechanical tests were performed on samples in the form of a double-sided blade with dimensions according to GOST 112-80. The tests were carried out on a universal testing machine Gotech Testing Machine CT-TCS 2000 manufactured in Taiwan at a temperature from -30 °C to +70 °C and a tensile speed of 10 mm/min.

3. Results and discussion

The use of ZB as a flame retardant makes it possible to obtain polymer composites of low combustibility with high operational characteristics [14, 15]. An important factor from a practical point of view is the low cost, the availability of the starting compounds used by us upon receipt of the ZB, and its safety from the point of view of environmental impact [16].

The composition of the synthesized ZB was studied using IR spectroscopy (Fig. 1). The data of IR spectroscopy showed that the main characteristic bands of the laboratory and industrial BC practically coincide. This suggests that the composition of the laboratory BC is 2ZnO•3B2O3•3,5H2O.

IOP Conf. Series: Materials Science and Engineering **934** (2020) 012001 doi:10.1088/1757-899X/934/1/012001

Zinc borate releases water at 330 - 450 °C. Crystal hydrate contains 12 - 15% of crystallization water. To confirm the proposed chemical formula, crystallization water was determined by the appropriate method [17]. The research results are shown in table 1, which confirm the amount of crystallization water corresponding to the above ZB formula.

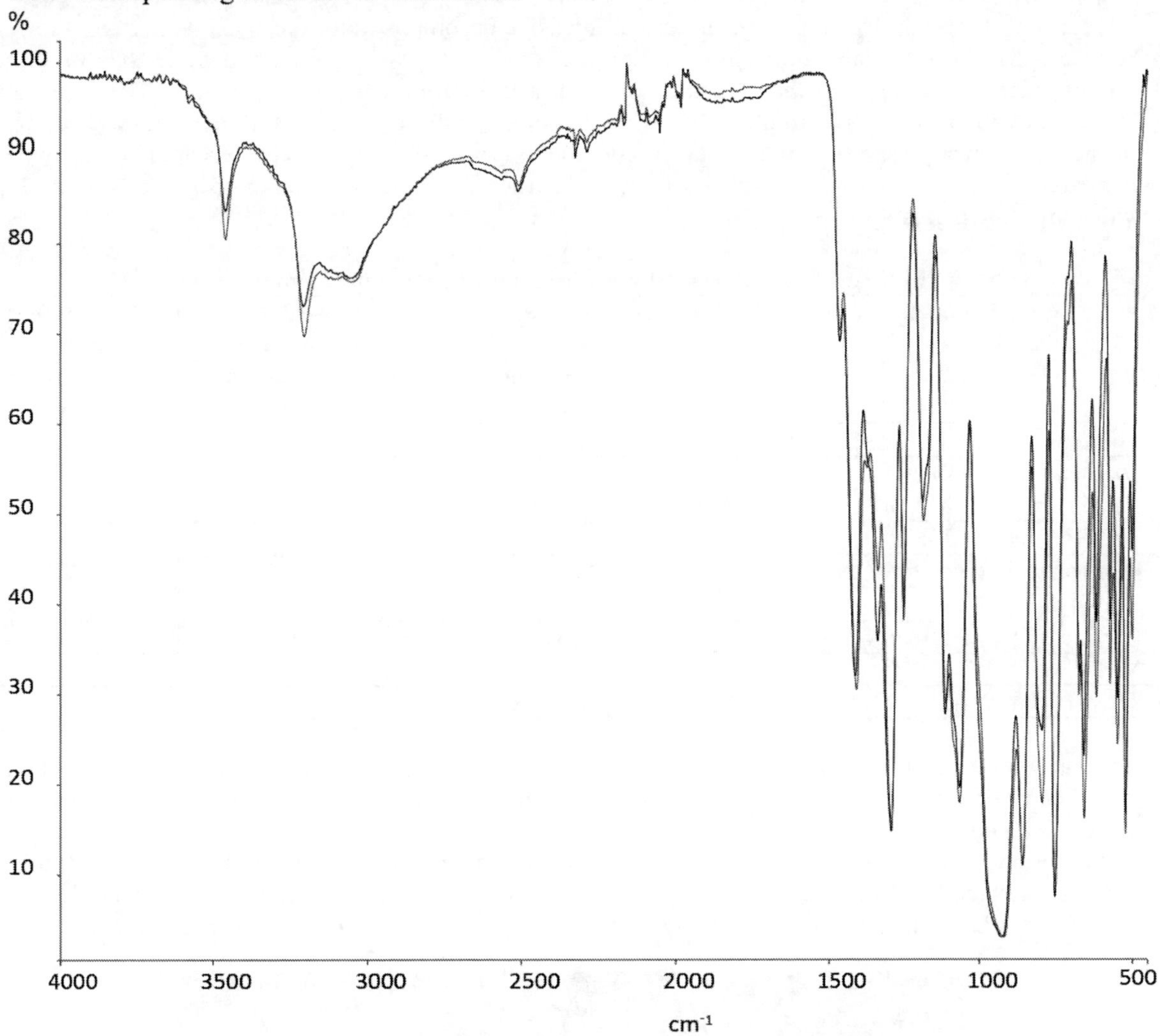

Fig. 1. IR spectrum of zinc borate (black - labor., red - industr.)

Table 1 Determination of crystallization water in ZB

The mass of the source component					
m_{T1} = 10,62915 g			m_{T2} = 10,62510 g		
m_{T1} (with substance) = 12,62915 g			m_{T2} (with substance) = 12,62510 g		
Mass after calcinations with the substance					
τ, min	m_{T1}	m_{T2}	Δm_1	Δm_2	t, °C
60	12,54520	12,56270	0,08395	0,0624	330
60	12,36040	12,35250	0,26875	0,27260	450

note: where, m t_1 - crucible mass 1; m_{t2} is the mass of the crucible 2; in 1 - industrial ZB;
in 2 - laboratory ZB.

Thermo gravimetric studies of the laboratory ZB show that the mass loss of the sample begins at a temperature of 250 °C, which is associated with the onset of loss of crystallization water. In this case, the maximum loss of crystallization water occurs at a temperature of 312 °C. A further increase in temperature leads to the decomposition of zinc borate. Therefore, the obtained ZB can be used for polymeric materials in which the temperature of intensive decomposition exceeds 250 °C.
To assess the effect of a laboratory ZB on the properties of PVC plastic compound grade NGP30-32 in our formulation, we replaced the industrial ZB with a laboratory sample. The resulting PVC plastic compound with a laboratory sample of ZB was examined for fire resistance, thermal stability, and the physic mechanical properties of the plastic compound were determined. The research results were compared with the characteristics of the base (industrial) PVC plastic compound grade NGP30-32, which are shown in table 2.

Table 2 Physic mechanical properties of PVC plastic compounds grade NGP 30-32

Parameters	NGP 30-32 with industrial ZB tested on injection molds	NGP30-32 with laboratory ZB laboratory tests on injection molds	NGP30-32 with industrial ZB tested on rolled samples
Thermal stability at 200 °C, min	120	113	120
Oxygen Index (OI), %	32	32	32
Tensile at stretching, MPa	12,7	12,1	14,5
Tensile strength, MPa	12,5	12,0	14,0
Elongation in tension, %	123,2	196,7	250
Density, g/cm^3	1,54	1,52	1,50
Hardness D scale, (max/relax)	53,3/39	51/40	-
Melt flow rate ($MFR^{190}{}_{10}$, g/10 min)	30,6	30,0	29,5
Thermal shrinkage, %	2,3	2,3	-

As tests showed, all three samples of PVC compound burn when in the flame zone (OI = 32 %, table 2). When removed from the flame zone, the combustion of the samples ceased almost immediately. In addition, the combustion process in the flame zone of all samples of PVC compound was accompanied by the formation of a small amount of smoke, it was minimal. Obviously, this behavior of all samples is due to the fact that the NGP30-32 PVC plastic compound is classified as fireproof (PP) material [18, 19], which, when exposed to an open flame, forms a durable flame-retardant coke (Fig. 2).

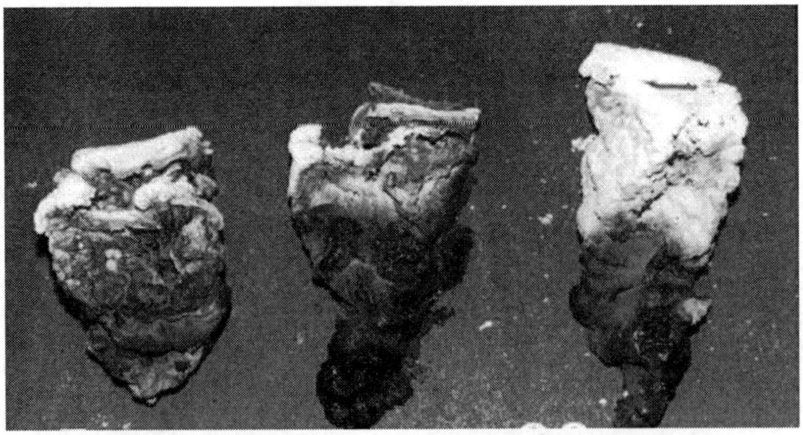

1 2 3

Fig. 2. State of samples after exposure to open flame:
1 - 5 min.; 2 – 10 min.; 3 – 30 min.

The presence of such coke helps to reduce flammability, reduce smoke and toxic carbon monoxide. The latter circumstance plays an important role in preventing the entry of harmful decomposition products of polymeric materials and flame retardants into the natural environment.
Studies of the thermal stability of PVC plastic compounds by thermo stating the obtained samples at a temperature of 200 °C showed that all samples of PVC plastic compound can withstand more than 100 min. However, the thermal stability of a sample of a plastic compound containing a laboratory ZB is slightly (7 min) lower than that of an industrial ZB, which is not significant. According to the requirements for non-combustible PVC plastic compound, the thermal stability of the compounds at 200 °C should not be less than 100 min [13]. In this regard, PVC plastic compound containing a laboratory ZB meets the requirements.
Another important indicator for PVC plastic compounds is manufacturability. The results of a study of the technological properties of PVC compounds according to the value of the melt flow rate (MFR) show that replacing an industrial ZB in a plastic compound with a laboratory one does not lead to a change in the process ability of the material. As can be seen from table 2, all three samples of PVC compound have almost the same MFI values, which show the same behavior of ZB particles of different origin in PVC compound.
In the work, it was important to determine whether the physic mechanical parameters of the PVC compound of the grade NGP30-32 change when the industrial ZB is replaced in the formulation with a laboratory one and to assess the degree of these changes. In particular, the density of PVC compound [20] used for insulation and sheathing of electrical cables is important. The mass of the cable and the amount of consumed insulating material per unit of product significantly depend on the density of the PVC compound. In this regard, it was of interest to determine the effect of replacing an industrial ZB on a laboratory sample. The results of determining the density of PVC plastic compounds are shown in table 2. As can be seen from the table, the replacement in the PVC compound formulation of an industrial BC with a laboratory sample does not lead to a noticeable change in the density of the material. It should be noted that the change in density occurs within acceptable values, i.e. in the area of 1,5 g/cm^3.
Preservation of the density of the material should be reflected in the hardness and mechanical properties of the PVC compound, i.e. the replacement in the PVC compound formulation of an industrial ZB with a laboratory sample should not have a significant effect on the mechanical properties of the compound in the direction of their deterioration (see table 2). As can be seen from the table, the strength properties, material hardness, tensile stress and elongation of PVC plastic compounds tested in the laboratory conditions of KBSU are almost the same. This shows that replacing an industrial ZB in a PVC compound with a laboratory sample does not lead to deterioration in the basic physical and mechanical properties of the compound.
However, comparing the relative elongation of these samples, it is found that PVC plastic compound containing ZB of laboratory origin is more deformable.

4. Conclusions

Thus, zinc borate crystalline hydrate with certain characteristics in terms of chemical composition and properties was obtained. Replacement in the formulation of PVC plastic compound of the NGP30-32 grade of industrial zinc borate with a laboratory sample does not lead to a deterioration in the basic physical, mechanical, thermal properties and fire resistance of the compound. The main indicators of PVC compound containing laboratory zinc borate meet the requirements for plastic compound grade NGP30-32.

References

[1] Fadina Yu I 2017 Analysis of the Russian polymer market and further ways of its development *Business education in the knowledge economy* **1** 99 - 101
[2] Grossman R F and Guzeeva V V 2009 Guide to the development of compositions based on PVC (SPb: Scientific Foundations and technologies) 614

IOP Conf. Series: Materials Science and Engineering **934** (2020) 012001 doi:10.1088/1757-899X/934/1/012001

[3] Wilkie C, Summers J and Daniels C 2007 Polyvinyl chloride (SPb: Profession) 728
[4] Khazova T N and Cherepova E B 2006 State and structure of the Russian market of polyvinyl chloride, *Polymer materials* **1** 24 - 31
[5] Borodulin A, Kalinnikov A, Tereshkov A, Kharaev A New polymeric binders for the production of composites. *Materials Today: Proceedings* 2019. 139-143.
[6] Borodulin A S, Kalinnikov A N, Bazheva R Ch, Kharaev A M, Beshtoev B Z Receipt and investigation of performance characteristics of super constructions polyesters *International Journal of Mechanical Engineering and Technology*. 2018. **9(13)**. 1117-1127.
[7] Knyazkov N A and Khayrullin R Z 2017 Features of the combustion process of polymeric materials, *Scientific Almanac* **3-5** 91-94
[8] Borodulin A.S. Polyester resins for production of goods from polymer composite materials by pressing methods *Polymer Science. Series D*. 2013. **6(3).** pp. 269-270.
[9] Borisov V A, Bedanokov A Yu, Karmokov A M, Mikitaev A K, Mikitaev M A and Turaev E R 2007. *Plastic. Masses* **5** 30-33.
[10] Nelyub V A, Borodulin A S Properties of epoxy materials used for production of glass-reinforced plastics by the winding method *Polymer Science. Series D*. 2018 **11(2).** pp. 147-153.
[11] Nelyub V A, Borodulin A S, Kobets L P, Malysheva G V. Thixotropy hysteresis and structure formation in elastomeric suspensions Inorganic Materials: Applied Research. 2018. 9(4). pp. 603-608.
[12] Sultigova Z Kh, Kitieva L I and Borukaev T A 2019 *Using zinc borate as effective flame retardant, Engineering Materials* **816** 129-133
[13] GOST 14041-91 (ISO 182 / 1-90) Plastics. Determination of the tendency to release hydrogen chloride and other acid products at high temperature in compositions and products based on vinyl chloride homopolymers and copolymers. Congo method is red.
[14] Matveev M D, Marakulin S D, Sertsova A N and Yurts E V 2015 The effect of micro- and nanoparticles of zinc borate on the combustibility of a composite material based on polymethylmethacrylate, *Advances in Chemistry and Chemical Technology* **6** 99-101
[15] Petrova N P, Ushmarin N F and Koltsov N I 2012 Improving the fire resistance of rubber based on BNK using combinations of trichloroethylphosphate with various flame retardants, *Bulletin of Kazan Technological University* **19** 94-97
[16] Hoang T Kh, Sertsova A A and Yurtov E V 2016 Methods of modifying zinc borate to create flame retardant composites, *Advances in Chemistry and Chemical Technology* **12** pp. 55-56.
[17] Sukhankina N V and Kozlova-Kozyrevskaya A L 2017 Analytical chemistry. Quantitative analysis: workshop (Minsk: BSPU) 96
[18] Ableev R I 2009 Actual problems in the design and manufacture of non-combustible polymer compounds for the cable industry, *Cable-news* **6-7** 64-69
[19] GOST 31565-2012. Cable products. Fire safety requirements.
[20] Guzeev V V 2012 Structure and properties of filled polyvinyl chloride (SPb: Scientific foundations and technologies) 426

Simulation of the kinetics of the curing process of polymer composite materials based on epoxy binders

Yangyang Chen[1], Pyi Phyo Maung[2], G V Malysheva [3] and A A Paduchin [4]

[1,2,3] Bauman Moscow State Technical University, Moscow, Russia Federation

[4] K.A. Kerimov School №1583, Moscow state budget educational institution, Moscow, Russia Federation

E-mail: [1]yangyangchen@mail.ru, [2]pyaephyo@mail.ru, [3]malyin@mail.ru, [4]a@inginirium.ru

Abstract. This article investigated the kinetics of the curing process of composite material based on epoxy binders under different conditions of heat transfer. A technique for determining the heat capacity and thermal conductivity depending on the degree of cure of epoxy binders was carried out. The kinetic parameters of the binders used to describe the exothermic effects of binders in the chemical reaction were determined. Taking into account the phase state changes and their exothermic effects of the binders during curing, the simulation of curing process of glass fiber reinforced plastics (GFRP) was carried out. Experimental results of determination of heat capacity, thermal conductivity and kinetic parameters of epoxy binders were used as initial inputs for modeling the curing process of polymer composite materials. The influence of the heat transfer coefficient on the temperature state of GFPR was investigated. It was found that the heat transfer condition strongly affects the temperature state of fiberglass during their curing.

1. Introduction

Modern glass fiber reinforced plastics (GFRP) and carbon fiber reinforced plastics (CFRP) are used as structural materials in the manufacture of products in mechanical engineering, aviation, aerospace and many other industries [1-3]. It is known that technological errors can be inherited, but for parts made of composites based on thermosetting binders, their quality is finally formed during the technological operation — curing [4-6].

The problem of studying the kinetics of curing processes is given much attention, especially in the development of the technology for forming large-sized and complex geometrical shaped products, in which the curing processes proceed unevenly, is solved [7-10]. In the technical literature [11-12], the study of the temperature field of products in the curing process of epoxy (and other thermosetting) materials only in the solid aggregate state is given. The researchers do not take into account the kinetics of changes in thermal properties during heating, which leads to a large error in the study of the temperature state of parts from GFRP during their curing. The curing of polymer composites is carried out, as a rule, in autoclaves or electric furnaces [13-15]. With different heating technologies during curing, the heat transfer coefficients vary from 5 to 200 W/(m2·K). For conditions of free air convection, the heat transfer coefficient is 5-25 W/(m2·K). In forced convection, the heat transfer

coefficient is 10-200 W/(m2·K). In the process of curing GFRP products, heat transfer conditions between the environment and the sample must be taken into account.
The aim of this work is to study the curing process of products taking into account the phase change and exothermic effects of the binders in the curing process under different convection conditions.

2. Determination of the thermophysical properties of resins

As an object of research used epoxy binder hot curing, consisting of an epoxy resin and a hardener based on aromatic amine.
In the process of curing composite materials based on epoxy binders, the thermophysical properties of the binders vary depending on the degree of cure. The heat capacity (Table 1) of the binders, depending on the curing degree, is determined by the differential calorimeter method. The thermal conductivity (Table 1) of the binders, depending on the degree of curing, is determined by the laser flash method. The feature of these tests was the use of a sample in a liquid state, which did not allow to withstand the required dimensions. In addition, the composition used is an optically transparent material for laser radiation. A special crucible was used to measure the thermal conductivity of the material in the liquid state of aggregation.

Table 1. Dependence of heat capacity and thermal conductivity on the degree of curing

Degree of curing, %	Cp, J/(kg/K)	λ, W/(m/K)
0	1973	0,08
5	1973	0,08
16	1921	0,14
28	1820	0,19
75	1406	0,22
87	1338	0,25
100	1300	0,27

It is established that with increasing the degree of curing the heat capacity increases and the thermal conductivity decreases.

3. Determination of cure kinetic parameters of resins

To study the thermal effect of the chemical reaction during the curing of the binders, a differential scanning calorimeter (DSC) was used. Using the DSC method, the curve of the thermal effect of epoxy binders was obtained for different heating rates of 1, 2, 3, 5 K / min (Figure.1). The amount of heat during the chemical reaction is determined by the peak area of the DSC curve.

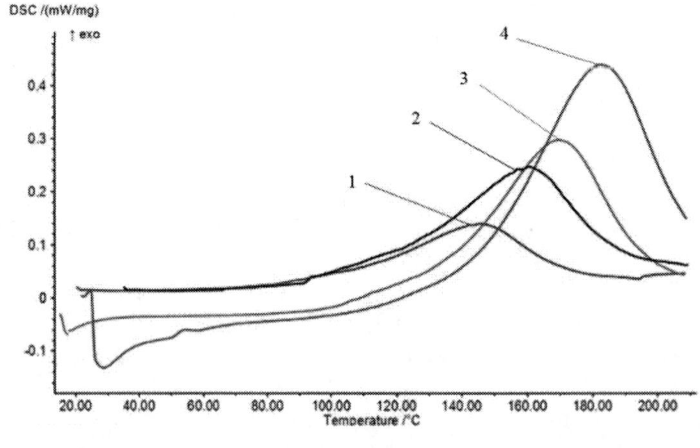

Figure 1. DSC curve at the heating rate of 1 K / min (1), 2 K / min (2) 3 K / min (3), 5 K / min (4).

The equation for describing the curing process of these binders is:

$$C\frac{d\beta}{dT} = A\exp\left(-\frac{E}{RT}\right)\beta^{m}(1-\beta)^{n} \tag{1}$$

where C – heating rate in the curing mode, K/s; β – degree of curing, %; m, n – the orders of the chemical reaction; A – frequency factor, s-1; E – activation energy, J/mol; T – absolute temperature, K; R – universal gas constant, J/(mol·K).

The activation energy for epoxy resin is defined by equation Kissinger-Akahira-Sunose (KAS)[16-18]:

$$\frac{d\left[\ln\left(\frac{\beta}{T_p^2}\right)\right]}{d\left(\frac{1}{T_p}\right)} = -\frac{E}{R} \tag{2}$$

where T_p –temperature at the peak of the DSC curve, K.
The kinetic parameters of the binders used were determined from the equations for describing the curing process (1) (2) and the results of DSC (Table 2).

Table 2. Curing kinetic parameters

Parameter	Value
Activation energy E, J/mol	61844
Frequency factor A, s^{-1}	$3.6 \cdot 10^4$
m	0.12
n	0.62

The obtained kinetic parameters will be used to determine the exothermic effects of binders during the curing process.

4. Simulation of the curing process

During the curing process of composite materials, the thermophysical and kinetic properties of the binders, the efficiency of convection in the heat transfer process affects the temperature state of the sample. To analyze heat transfer during curing, the equations [19, 20] were used:

$$\rho C_p(\beta)\frac{dT}{dt} = Q_\lambda + Q_\alpha + Q_v \tag{3}$$

$$Q_\lambda = \lambda_{xx}\frac{\partial^2 T}{\partial x^2} + 2\lambda_{xz}\frac{\partial^2 T}{\partial x \partial z} + \lambda_{zz}\frac{\partial^2 T}{\partial z^2} \tag{4}$$

$$Q_\alpha = \alpha(T_0 - T)\frac{S_h}{V} \tag{5}$$

$$Q_v = \rho H_r \frac{d\beta}{dt} \tag{6}$$

where ρ – density of GFRC, kg/m3; C_p – specific heat, J/(kg·K); T –absolute temperature, K; t- time, s; Q_λ –heat flux with thermal conductivity, W/m3; λxx, λxz, λzz – thermal conductivity of anisotropic material, W/(m·K); Qα –convective flow to the free surface of the sample, W/m3; α–heat transfer coefficient, W/(m2·K); T0 –the ambient temperature, K; Sh –the area of action of convective flow, m2; Qv –heat dissipation, W/m3; H_r – the total amount of heat released during the curing process, J /kg.

The ESI PAM-RTM software was used to simulate the curing process. All calculations were carried out on samples with a size of 25x25x25 mm. As the initial data for modeling, the above determined results of the heat capacity of thermal conductivity and kinetic parameters were used. In the curing mode rate of temperature rise is 2 K / min. The curing process (Figure 2, 3) was simulated taking into account the change in the phase state and their exothermic effects of the binders in the ideal case, under conditions of free air convection (heat transfer coefficient: 10 W/(m2·K)), with forced convection (heat transfer coefficient 20, 100 W/(m2·K)). The temperature state depending on the curing time for different convection coefficients was shown in figure 2. The table 3 shows the temperature gradient under different heat transfer conditions.

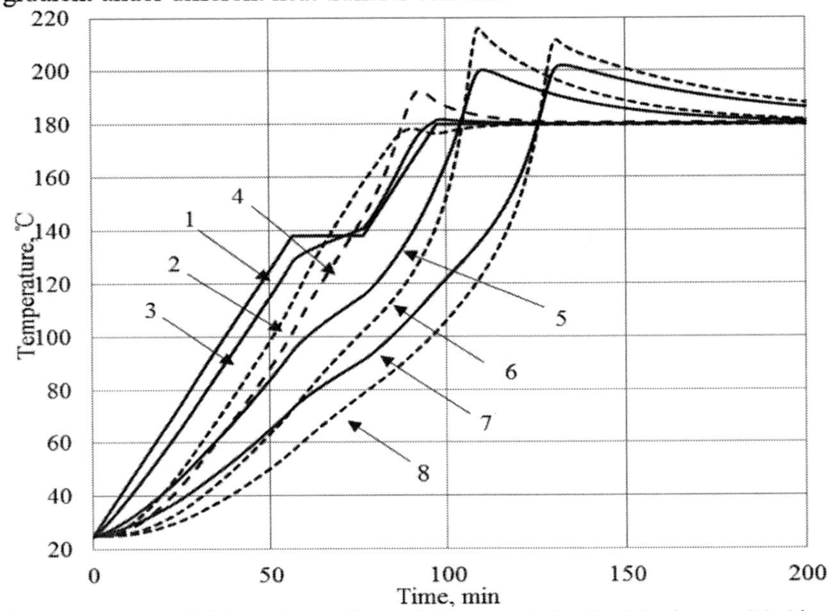

Figure 2. The temperature field on the surface of the sample in the ideal case (1) (the given curing mode) and when the convection coefficient is 100 (3), 20 (5), 10 (7); the temperature field in the center of the sample in the ideal case (2) and when the convection coefficient is 100 (4), 20 (6), 10 (8).

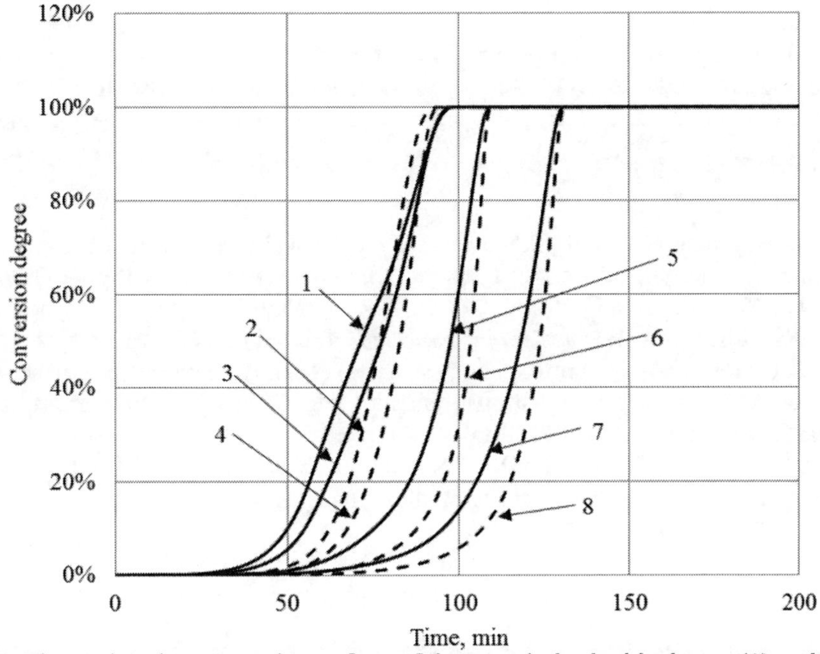

Figure 3. The curing degree on the surface of the sample in the ideal case (1) and when the convection coefficient is 100 (3), 20 (5), 10 (7); the degree of curing in the center of the sample in the ideal case (2) and when the convection coefficient is 100 (4), 20 (6), 10 (8).

Table 3. The temperature gradient under different heat transfer conditions

Option Number	Heat transfer coefficient ($W/m^2 \cdot K$)	Maximum temperature gradient (℃)
1	10	16,2
2	20	21,8
3	100	26,9
4	∞	27,5

It was found that in the first section of the curing process, the temperature on the surface of the sample is higher than the temperature in the center of the sample. Due to the exothermic effects of binders, in the third section of the curing process the gradient of the temperature at the surface and in the center of the sample decreases. At the temperature rise section of the curing mode there is a moment where the temperature inside the sample is higher than the temperature on the surface of the sample. The temperature of the sample reaches its maximum value in the fourth section. In this area, heat release strongly affects the temperature field of the sample. The exothermic effect of the binders can lead to overheat in the fourth section of the curing mode. When the heat transfer coefficient has a larger value (100 and ∞) at the end of the curing process, there is a place where the degree of cure on the surface is higher than the degree of cure in the center of the sample.

5. Conclusion

As the result of simulation, it was found that the increasing the values of the heat transfer coefficients, the difference in temperature state from the set mode is higher. With increasing heat transfer coefficient, the maximum temperature gradient on the surface and in the center of the sample does not increase very much. The value of the heat transfer coefficient strongly affects the temperature state of the sample, and not so much affects the temperature difference on the surface and in the center of the sample.

Acknowledgement
The research results were partially obtained within the framework of the Russian Foundation for Basic Research (RFBR) grant No. 18-29-19034 / 18 "Development of theoretical and technological foundations for the manufacture of high-strength carbon plastics with a set of new functional properties based on metallized fibers"

References
[1] Zorin V A, Baurova N I and Shakurova A M 2014 Investigation of the structure of an encapsulated anaerobic adhesive J *Polymer Science* Series D vol **7(4)** pp 303-305.
[2] Baurova N I, Zorin V A and Prikhodko V M 2016 Manifestation of a synergistic effect in technological heredity *Polymer Science Series D* vol. **9(2)** pp 209-211.
[3] Maung P P, Malysheva G V and Gusev S A 2016 A study of the effect of network angle of fabrics on kinetics of impregnation upon molding of articles made from carbon plastics *Polymer Science* Series D vol **9(4)** pp 407-410.
[4] Sergeev A.Y., Turusov R.A., Baurova N.I. 2016 Strength of the joint of an anisotropic composite and a cylindrical element of the exhaust system of road vehicles *Mechanics of Composite Materials* vol. **52(1)** pp. 89-98.
[5] Yangyang C, Malysheva G 2019 Method for Determining the Rational Regimes of Curing Products from Polymer Composite Materials *Materials Today: Proceedings* vol **11** pp 128-133.
[6] Nelyub V., Borodulin A., Kobets L., Malysheva G. 2017 Viscous hysteresis in filled siloxane binders *Polymer Science – Series D.* vol. **10(1)** pp. 19-22.
[7] Reznik S V, Timoshenko V P, Shulyakovsky AV and Denisov O V 2013 Thermal-vacuum tests of hollow composite rods intended for structures in space *Polymer Science* Series D vol. **6(3)** pp. 242-245.
[8] Reznik S V, Prosuntsov P V, Azarov A V 2015 Modeling of the temperature and stressed-strained states of the reflector of a mirror space antenna *Eng Phys Thermophy* vol **88(4)** pp 978–983.
[9] Prosuntsov P V, Reznik S V, Mikhaylovskii K V and Belenkov E S 2018 Multiscale modeling of the binder polymer composite materials heating using microwave radiation *Journal of Physics Conference* Series vol **1134(1)** p 012048.
[10] Reznik, S V, Timoshenko, V P, Prosuntsov, P V and Mial' L V 2014 Theoretical Principles of Determining the Longitudinal Thermal Conductivity of Thin-Walled Structural Elements from Composite Materials *Eng. Phys. Thermophy* vol **87(4)** pp 864–870.
[11] Pyi Phyo Maung, Malysheva G, Romanova I 2016 Optimization of the rheological properties of epoxy resins for glass and carbon reinforced plastics *IOP Journal Conferenceseries Materials scienceandengineering* vol **153(1)** p 6.
[12] Pyi Phyo Maung, Malysheva G and Tatanikov O 2016 Optimization of the curing process of a sandwich panel *IOP Journal Conference series Materials science and engineering* vol **153(1)** p 6.
[13] Borodulin A S, Marycheva A N, Malysheva G V 2015 Simulation of impregnation kinetics of fabric fillers in the production of fiberglass articles *Glass Physics and Chemistry* vol **41(6)** pp 660-664.
[14] Mantell S C 1995 Cure Kinetics and rheology models for ICI fiberite 977-3 and 977-2 thermosetting resins *Journal of reinforced plastics and composites* vol **14(8)** pp 847-865.
[15] Hickey C M D 2013 Cure kinetics and rheology characterisation and modelling of ambient temperature curing epoxy resins for resin infusion, VARTM and wet layup applications *Journal of materials science* vol **8** pp 690-701.
[16] Kamal M R and Sourour S 1973 Kinetics and thermal characterization of thermoset cure Polymer Engineering & Science 1973 vol **13(1)** pp 59-64.

[17] Kissinger H E 1957 Reaction kinetics in differential thermal analysis *Analytical chemistry*. vol **29(11).** pp 1702-1706.

[18] Akahira T and Sunose T T 1971 Joint Convention of Four Electrical Institutes *Chiba Institute of Technology* 1971 vol **16** pp 22-31.

[19] Liang J Z, Li F H 2007 Simulation of heat transfer in hollow-glass-bead-filled polypropylene composites by finite element method *Polymer Testing* vol **26(3)** pp 419-424.

[20] Sonmez F O and Hahn H T 1997 Modeling of heat transfer and crystallization in thermoplastic composite tape placement process *Journal of Thermoplastic Composite Materials* vol **10(3)** pp 198-240.

Assessment of technological resources for the production of composite products based on mathematical methods

A E Brom, M V Stoyanova, M V Yazev and S A Korolev

Bauman Moscow State Technical University, 2nd Baumanskaya street, Moscow, Russia

E-mail: abrom@yandex.ru

Abstracts. This article is devoted to the selection and justification of the use of the corresponding mathematical methods in the formation of a composite product production program and the assessment of the importance of provided technological resources. The choice of marginal profit as criteria of the optimality of the production program is proved. The formalized mathematical formulations of integer and dual tasks of linear programming are presented. The tasks of analyzing the annual release plan for a range of six products is solved. It is concluded that the usage of mathematical methods of linear programming and the theory of dual is successful decision in the field of development of composite production programs.

Key words: composite products, technological production resources, economic and mathematical methods.

1. Introduction

The current level of technological development necessitates the usage of materials that have special properties that are not inherent in traditional structural materials [10]. In this regard, in various areas of technology, composite materials are being actively introduced. [5]. Which leads to breakthrough growth of the composite industry. Thus, according to the estimates of the reputable statistical portal Statista, the global market for composite materials will grow in 2017 to 2024 from 76.71 to 130.83 billion US dollars [9].

High-intensity development of the industry requires effective management of production [4]. This task can be attributed to the field of engineering. One of the important sections of engineering is the development of recommendations to achieve the best result from the spent investments in the organization of the technological process of production [3].

The specificity of the manufacturing process for the production of composite products suggests the possibility of using versatile capital assets for the production of a wide range of products applicable in various engineering industries [1]. This raises the problem: what types of products are the most profitable for the company to produce taking into account the available technological resources and in the conditions of a limited available stock of working hours of capital assets?

The purpose of this study is the selection and justification of the use of the corresponding mathematical methods for the formation of the optimal production program for the composite industry enterprise.

2. Research methods

The task of finding the best distribution of limited technological resources between manufactured products is the task of integer linear programming. In order to build a mathematical model of this task, it is necessary to choose criteria of optimality [8]. It is appropriate to use marginal profit as criteria of optimality, since its specific value does not depend on the volume of output and allows to bring out an objective function that is linear in volume of output [7]. At the same time, the maximum of margin profit and the maximum of operating profit - a key indicator of the effectiveness of the main production activities of the enterprise [6], are achieved at the same point in the space of acceptable release plans [2].

The limitations are, on the one hand, the available working time funds for each type of main technical equipment (capital assets) and production plans for each type of product on the other.

The total system of conditions is as follows:

$$\begin{cases} \sum_{j=1}^{n} X_j \cdot a_{i,j} \le b_i, \ \forall i \in [1;m] \\ X_j \ge B_j, \ \forall j \in J^1 \\ X_j = B_j, \ \forall j \in J^2 \\ X_j \in Z \\ \sum_{j=1}^{n} X_j \cdot p_j \to \max \end{cases} \tag{1}$$

Where j – sequence number of product; i – sequence number of technical equipment; $a_{i,j}$ – time soent on processing the j- product on i-type of technical equipment; b_i – Fund of working hours of the i-type of technical equipment; p_j – marginal profit from the sale of the j-product; B_j – Production plan for type j product; J^1 – set of products for which it is permissible to overfulfill the plan, for the reason that they can be sold on the open market; J^2 – set of products for which exact fulfillment of the plan is required, that is, they cannot be sold on the open market and are available only on pre-order; X_j – the number of manufactured products of type j.

The solution to this problem is the vector of optimal volume with fixed current restrictions [12]. To further improvement of the proposed solution, it is necessary to consider the effect of restrictions on the objective function. It is required to determine the deficiency of resources and the criticality of product release plans. A quantitative assessment of these parameters will make it possible to draw conclusions about possible ways to increase profits by revising obligations to supply certain types of products or by attracting additional equipment working time.

This analysis is performed by solving the dual linear programming task, in the following statement:

$$\begin{cases} \sum_{i=1}^{m} y_i \cdot a_{ij} - V_j \ge p_j, \ \forall j \in [1;n] \\ y_i \ge 0, \ \forall i \in [1;m] \\ V_j \ge 0, \ \forall j \in J^1 \\ \sum_{i=1}^{m} y_i b_i - \sum_{j=1}^{n} V_j B_j \to \min \end{cases} \tag{2}$$

Where: y_i – dual resource rating; V_j – dual product rating.

An economically dual resource rating can be interpreted as the rental cost of an hour of operation of this type of technical equipment. Thus, if an enterprise has the opportunity, in one way or another, to increase up to some point the available fund of working time of technical equipment with unit costs less than the value of the dual raring of the hours of operation of this technical equipment, then you it is feasible to take this opportunity. The result of making the described management decision will be an increase in the total marginal profit by

$$\sum_{i=1}^{m} (y_i - r_i) \cdot \Delta b_i \tag{3}$$

Where r_i – the rental cost of an hour of operation of i type of technical equipment; Δb_i – increase in available working time of i type of technical equipment.

The positive dual rating of a product can be interpreted as the amount of a penalty for failure to fulfill a delivery plan per one unit of a product. If the enterprise has the opportunity to refuse to release a certain number of products included in the plan and also having a positive dual rating, while suffering specific losses associated with the payment of the penalty, lower than the dual rating, then this opportunity should

IOP Conf. Series: Materials Science and Engineering **934** (2020) 012003 doi:10.1088/1757-899X/934/1/012003

be taken. The negative dual rating of a product can be interpreted as the amount of discounts provided in conditions of price discrimination that a manufacturer can offer to increase demand for this type of product. If it is possible to sell an additional unit of products that have a negative dual rating, lowering the price by an amount less than the absolute value of the dual rating, then this move will bring additional profit [11].

The result of making the described management decisions will be an increase in the total marginal profit by:

$$\sum_{j=1}^{n} P_j \quad (4)$$

Where P_j – profit margin increase due to changes in the plan for the j product type, calculating as follows:

$$P_j=\begin{cases} -(V_j-l_j)\cdot\Delta B_j, & V_j>0 \\ -(V_j+s_j)\cdot\Delta B_j, & V_j<0 \\ 0, & V_j=0 \end{cases} \quad (5)$$

Where l_j – the amount of the penalty paid for the refusal to supply one unit of the j product type; ΔB_j – change of the planned production volume for the j product type; s_j – discount provided for an additional unit of the j product type.

For a relatively small nomenclature of production, the numerical solution of the two tasks formulated above can be implemented in the MS Excel using the Solver Add-in. Also, this tool provides an opportunity to assess the boundaries in which the criticality of resources does not change and the linearity of profits by changing restrictions is keeping the same. Described ratings can be extracted from the Sensitivity Report. [13].

3. Results

In this study, we analyzed the annual plan for the production of a product range consisting of six positions: a helicopter blade, a covering for the main beam of a helicopter tail rotor, a tactical helmet, and a fender of the car, hood of the car and body panel of a car, taking into consideration that the company operates in one shift. In view of the specifics of the products, the products «Helicopter blade» a «Covering for the main beam of a helicopter tail rotor» are made on pre-order, and therefore the volume of products is strictly equal to the production plan. Other types of products are produced in the amount not less than planned. The following solution is obtained for the direct integer linear programming task.

Table 1. Solution direct integer linear programming task

Technical equipment type	Products						Fund of working hours of technical equipment (hours/year)	Total technical equipment runtime
	Helicopter blade	Covering for the main beam of a helicopter tail rotor	Tactical helmet	Fender of the car	Hood of the car	Body panel of the car		
Vacuum pump	0,5	4	0,5	3	4	2	1682,15	1681,00
Oven	0	2,5	2,5	2,5	2,5	0	1781,10	882,50
Autoclave	3	0	0	0	0	2,5	1781,10	1780,00
Milling machine	1,2	1,2	0,3	0,5	0,7	1,5	1721,73	1 104,90
Vertical drilling machine	0,3	0,5	0,2	0,1	0,3	0,3	1840,47	286,90
Grinding machine	1	0,2	0,5	0	0	2	1840,47	1 125,00
Spray booth	2	2	0,5	1	1	0	1840,47	904,00

Infrared lamp 1	1	1	1	1	1	0	1880,05	633,00
Infrared lamp 2	1	1	0	0	0	0	1880,05	350,00
Infrared lamp 3	1	1	0	0	0	0	1880,05	350,00
Infrared lamp 4	1	1	0	0	0	0	1880,05	350,00
Release plan	280	70	100	70	55	70		
Production volume	280	70	158	70	55	376		
Marginal profit per unit	77 400	54 180	10 000	20 000	50 000	80 000		
Total marginal profit	21 672 000	3 792 600	1 580 000	1 400 000	2 750 000	30 080 000	61 274 600	

The optimal production plan with fixed restrictions is presented in the line «Production volume». The total marginal profit provided by this release plan is presented in the last cell of the «Total marginal profit» line.
To further improvement of the proposed production plan, it is necessary to consider the solution of the dual task, reflecting the influence of restrictions on the objective function.

Table 2. Solution of dual task

Technical equipment type	Products						Fund of working hours of technical equipment (hours/year)	Shadow price
	Helicopter blade	Covering for the main beam of a helicopter tail rotor	Tactical helmet	Fender of the car	Hood of the car	Body panel of the car		
Vacuum pump	0,5	4	0,5	3	4	2	1682,15	20 000
Oven	0	2,5	2,5	2,5	2,5	0	1781,10	0
Autoclave	3	0	0	0	0	2,5	1781,10	16 000
Milling machine	1,2	1,2	0,3	0,5	0,7	1,5	1721,73	0
Vertical drilling machine	0,3	0,5	0,2	0,1	0,3	0,3	1840,47	0
Grinding machine	1	0,2	0,5	0	0	2	1840,47	0
Spray booth	2	2	0,5	1	1	0	1840,47	0
Infrared lamp 1	1	1	1	1	1	0	1880,05	0
Infrared lamp 2	1	1	0	0	0	0	1880,05	0
Infrared lamp 3	1	1	0	0	0	0	1880,05	0
Infrared lamp 4	1	1	0	0	0	0	1880,05	0
Release plan	280	70	100	70	55	70		
Reduced Cost	-19 400	25 820	0	40 000	30 000	0		
Marginal profit per unit	77 400	54 180	10 000	20 000	50 000	80 000		
Total marginal profit							61 315 200	

According to the calculation results, it was determined that the «Vacuum Pump» and the «Autoclave» are lacking resources, the dual ratings of which are greater than 0. Each additional hour of operation of these types of technical equipment will bring an additional 20,000 and 16,000 rubles. into the objective function. Products with dual ratings <0 ("Helicopter Blade") are profitable for production, but are produced strictly according to plan. An increase in the output plan for these products would increase the value of the objective function by the absolute value of the dual valuation of 19,400 rubles. The types of products which have received dual ratings = 0 («tactical helmet», «body panel») are produced in accordance with the optimal plan; Products with dual ratings> 0 («Covering for the main beam of a helicopter tail rotor», «fender», «hood of the car») are not profitable for production, thus reducing the plan per unit of this type of products will lead to an increase in the value of the objective function by the absolute value of the corresponding dual assessment (25,820 rubles for the («Covering for the main beam of a helicopter tail rotor»; 40,000 rubles for the «fender» and 30,000 rubles for the «hood of the car»)

The boundaries within which the calculated dual ratings are valid can be extracted from the Sensitivity report.

Table 3. The Sensitivity Report

Products dual ratings						
Cell	Name	Final Value	Reduced Cost $-V_j$	Objective Coefficient	Allowable Increase	Allowable Decrease
B15	Helicopter blade	280	19400	77400	19400	∞
C15	Covering for the main beam of a helicopter tail rotor	70	-25820	54180	25820	∞
D15	Tactical helmet	158,54	0	10000	5105,26	3227,50
E15	Fender of the car	70	-40000	20000	40000	∞
F15	Hood of the car	55	-30000	50000	30000	∞
G15	Body panel of the car	376,44	0	80000	∞	16166,67
Technical equipment dual ratings						
Cell	Name	Final Value	Shadow price y_i	Constraint R.H. Side	Allowable Increase	Allowable Decrease
I3	Vacuum pump	1682,15	20000	1682,15	179,45	29,27
I4	Oven	883,85	0	1781,10	∞	897,25
I5	Autoclave	1781,1	16000	1781,10	36,59	224,31
I6	Milling machine	1105,722	0	1721,73	∞	616,01
I7	Vertical drilling machine	287,14	0	1840,47	∞	1553,33
I8	Grinding machine	1126,15	0	1840,47	∞	714,32
I9	Spray booth	904,27	0	1840,47	∞	936,20
I10	Infrared lamp 1	633,54	0	1880,05	∞	1246,51
I11	Infrared lamp 2	350,00	0	1880,05	∞	1530,05
I12	Infrared lamp 3	350,00	0	1880,05	∞	1530,05
I13	Infrared lamp 4	350,00	0	1880,05	∞	1530,05

The boundaries in which dual ratings are keeping the same are contained in the columns «Allowable Increase» and «Allowable Decrease».

4. Conclusion

According to the results of the study, it is clear that the mathematical methods of integer linear programming and duality theory can be effectively used in solving engineering issues during the process of the formation of a composite production program and assessing the significance of available technological resources.

References

[1] Baurova, N I 2016 The use of polymer composite materials in the manufacture and repair of machines: textbook. allowance / Baurova N I, Zorin V A . – Moscow, MADI, 264 p.

[2] Brom A E and Popova M V 2015 Engineering services price formation approach based on marginal profit concept. *Economic sciences*, **10 (131)** 97–100.

[3] Brom A E and Stoyanova M V Engineering Rent In Machine-Building: Specificity, Problems, Possibilities of Assessment. *Bulletin of the Moscow Region State University. Series: Economics*, **4** 48-55.

[4] Brom A E and Stoyanova M V 2017 The Choice Algorithm of the Engineering Company Based on the Assessment of Its Competences With Use of the Method of Analytical Hierarchy Process Development. *Azimuth of Scientific Research: Economics and Administration*, **6-3(20)** 75-77.

[5] Brom A, Gorlacheva E and Pekshev I 2017 The perspectives of additive manufacturing technologies implementation in mass production of short range unmanned aerial vehicle. *2017 International Conference on Mechanical, System and Control Engineering (ICMSC)*, 27-30. DOI: 10.1109/ICMSC.2017.7959436

[6] Ciemleja G and Lace N 2011 The Model of Sustainable Performance of Small and Medium-sized Enterprise. *Engineering Economics,* **22(5)** 501-509. DOI: 10.5755/j01.ee.22.5.968

[7] Gale D 1960 The Theory of Linear Economic Models. New York, McGraw-Hill 352

[8] Gorlacheva E N, Gudkov A G, Omelchenko, I N, Drogovoz P A and Koznov D V 2018 Knowledge Management Capability Impact on Enterprise Performance in Russian High-Tech Sector. 2018 IEEE International Conference on Engineering, Technology and Innovation (ICE/ITMC),1-9. DOI: 10.1109/ICE.2018.8436316

[9] Market value of composite materials worldwide from 2015 to 2024 (in billion U.S. dollars). [Electronic resource] (Updated: 11.10.2019) URL: https://www.statista.com/statistics/944471/global-market-value-of-composites/ (Date of the request: 15.11.2019).

[10] Nelyub V A and Borodulin A S 2018 Properties of Epoxy Materials Used for Production of Glass-Reinforced Plastics by the Winding Method. *Polymer science. Series D*,**11-2** 147-153. DOI: 10.1134/S1995421218020132

[11] Orlov A I 2003 Management. Textbook. Moscow, Publishing House "Emerald",298 p.

[12] Schofield N 2014 Mathematical Methods in Economics and Social Choice. Berlin, Springer, 262

[13] Vukolov E A 2012 Fundamentals of statistical analysis. Workshop on statistical methods and operations research using STATISTIS and EXCEL packages. Moscow, Forum, 464

Composite materialis based on propylene and modified Na +-montmorillonite N, N- diallylacrilamide

M B Begieva, R Ch Bazheva and A M Kharaev and A S Borodulin

Kabardino-Balkar State University Chernishevsky St.,173, Nalchik,360004

Email: bsk@kbsu.ru

Abstract: The monomer N, N-diallylacrylamide was synthesized by alkylation of acrylamide. The IR spectra of acrylamide and N, N-diallylacrylamide are analyzed. The possibility of obtaining alkyl acrylamide is shown.Modified organoclay is obtained on the basis of montmorillonite of the Na + - form and acrylamide, a water-soluble monomer. The structures and properties of the synthesized compounds were investigated. Composites based on polypropylene modified montmorillonite prepared by mixing in a twin screw melt extruder of the company "JiangsuXindaScience & Technology". The structure of the composites was investigated using X-ray diffraction and scanning electron microscopy. It is shown that, compared with the unfilled polypropylene, modulus, toughness, melt flow index composites increases, and the time of combustion and the flame propagation velocity is decreased. Shown, that with the introduction of 3-5 wt.% Of the modified Na + - montmorillonite + N, N-diallylacrylamide into the PP, a decrease in the flame propagation rate by 74-79% is observed. The slowing down of the burning rate of the composite is associated with a limitation of the diffusion of atmospheric oxygen into the material due to the presence of a barrier layer created by plane of montmorillonite particles. The presented data are consistent with the results previously obtained in.

Keywords: N,N-diallilacrilamide, composite, polypropylene, modified Na+ - montmorillonite, polymer, organoclay, elastic modulus

1. Introduction

Currently, there is an increased interest of researchers in obtaining new polymeric nanocomposite materials with improved physicochemical and physic mechanical properties based on polymers and layered silicate materials, in particular, montmorillonite (MMT). The poor compatibility of organic and inorganic components complicates the creation of nanocomposites, as well as the need to select the optimal conditions for dispersion of the silicate filler. Earlier there was an attempt to solve this problem by modifying clay by introducing surfactants into its structure; in particular, quaternary ammonium compounds with different lengths of the N-alkyl substituent [1-4]. Such organoclay makes it possible to destroy agglomerates formed because of the adhesion of individual crystallites of layer silicate. Surfactant adsorption increases the hydrophobicity of clay, thereby improving the penetration of polymer macromolecules into interplanar spaces. An analysis of the literature data in [5–9] devoted to the problem of obtaining nanocomposites based on various types of polar polymers (polyesters, polyamides, and others) and montmorillonite modified quaternary ammonium bases shows the possibility of successful acquire of intercalated and exfoliated structures using polyolefin for receiving nanocomposites, which are more modest [10, 11]. In this regard, our work aimed to obtain a modified sodium montmorillonite using N, N-diallylacrilamide, structurally different from the previously studied types of quaternary ammonium bases [12], and to study the effect of layered silicates on the properties and structure of polypropylene-based nanocomposites.

2. Experimental part

The studies were held on a homopolymer of PPG 1035-08 polypropylene (hereinafter referred to as PPS), tradename Stavrolen, with a melt flow index (MFI) of 3 g / 10 min.

According to procedure, alkylation of acrilamide with allyl chloride (allyl bromide) obtained an N, N-diallylacrilamide. [12,13].

The activated Na + form of montmorillonite (Na +- MMT) was prepared from natural clay from the Herpegezh deposit (Russia, Kabardino-Balkarian Republic); according to procedure [14], the cation exchange capacity of the clay was 95 meq/100 gr of clay. Absorption bands in the IR spectra characteristic of calcite controlled the presence of Ca2 + cations.

Adsorption on Na + -MMT of a 30% aqueous solution of AIBN obtained Modified Na + -MMT-AIBN.

Composites were prepared by adding 3-7 wt.% modified organoclay to the PP polymer. Dry component mixtures were prepared, afterward, the melt mixing have been conductedon a Jiangsu Xinda Science & Technology twin-screw extruder with a screw diameter of 20 mm at a temperature of 210-230 ° C and a screw rotation speed of 120-150 rpm. Then the extrudate was granulated and used to make the appropriate samples for research.

The molding injection obtained test samples on a Ray-Ran Polestest machine at a material cylinder temperature of 230-240 ° C, a mold temperature of 100 ° C and a shut-off pressure of 8 bars.

Physical and mechanical tests were held according to these methods: melt flow rate (MFR) - government standard11645-73; Izod impact strength - government standard19109-84; modulus of elasticity, ultimate strength, and elongation at the break on specimens in the form of a double-sided blade with dimensions according to government standard 112 62-80. These tests were held on a universal testing machine Gotech Testing Machine CT-TCS 2000, production Germany, at a temperature of 293 K and a strain rate of~ 2×10-3 c-1.

IR spectra of polypropylene mixtures were obtained using a Perkin Elmer FT-IR Spectrometer-Spectrum TWO instrument.

Shore microhardness measurements according to Shore (scale D) were performed according to government standard 24 621-91 on a Hildebrand hardness tester and after 1 s (maximum HB value) and 15 s (HB value after voltage relaxation) of the sample under load. At least five microhardness measurements were performed in different places of the sample surface at a distance of at least 6 mm from the point of the previous measurement. The arithmetic average of five measurements was taken as the test result at least. The samples for measuring HB had a cylindrical shape with a diameter of 40 mm and a height of 5 mm.

IR spectra were recorded on a Spekord M-82 spectrophotometer in the region of 4000-400 cm-1. A finely ground polymer sample was mixed with KBr powder and pressed under high pressure to transparent thin disks. The thickness of the pressed disk is 1 mm, the diameter is 12 mm, and the transmission is 95%. The weighed portions in all cases were the same; it allowed us to compare the results of spectral studies (as well as recorded on a Perkin Elmer Spektrum Two in the region of 4000-400 cm-1).

The X-ray diffraction (XRD) method studied the structure of the obtained composites using a D2 PHASER compact powder diffractometer and SEM Vega 3 TESCAN (Czech Republic) with an X-MAX X-ray microanalysis detector OXFORD (Great Britain).

IOP Conf. Series: Materials Science and Engineering **934** (2020) 012004 doi:10.1088/1757-899X/934/1/012004

3. Results and discussion

It was authenticated previously that the use of carcinogenic modifier with at least 16–18 CH2 groups in the main fat-aliphatic chain [3] at an optimal concentration of 25-75% wt.% of cation exchange capacity (CFU) clay is necessary for the formation of intercalated and exfoliated organoclay structures.

At the same time, it was possible to increase the elastic modulus of PE (by 50%), the ultimate strength (by 20%) and increase the barrier properties of nanocomposites [11], increase the elastic modulus of PE (by 75%) [15]. The authors of [16–19] noted that the use of a layered filler as a flame retardant significantly reduces (by 30–50%) the rate of heat release during the combustion of nanocomposites, relative to an unfilled polyolefin.

The structure of the Na + -MMT modifier used by us differs in its structure from previously used surfactants as clay modifiers.

N, N-diallylacrilamide (DAA) is a monomer, a derivative of acrilamide, into the molecule of which introduces a carbonyl group. The IR spectrum of Na + MMT contains characteristic absorption bands of 3635 and 3400 cm-1 of the free and associated forms of Si-OH, 1635 cm-1 (deformation vibrations of adsorbed H2O molecules), a wide absorption band in the range of 970–1100 cm-1 related to Si-OH groups and Si-O-Si fragments. There are practically no absorption bands in the region of 1374–1453 cm-1, which are characteristic for the carbonate groups.

In the IR spectrum of the modified organoclay, Na + -MMT- DAA, a certain shift of the characteristic absorption bands is observed for the free and associated forms of Si-OH -OH, a wide absorption band belonging to Si-OH -OH groups and Si–O–Si fragments. An intense absorption band appears in the region of 1679 cm – 1, which is characteristic for stretching vibrations of the NH and –CO- carbonil groups. The presence of the polar –CH2–CO– terminal group in DAA facilitates the penetration of layered silicate into the interlayer space, and it is possible that the carboil anion changed its to an inorganic cation from Na+ – MMT. This is also indicated by the almost complete disappearance of the 2630–2640 cm – 1 band from the ammonium cation . The presence of residual moisture, the OH groups of Na + - MMT cause a sufficiently intense absorption band in the region of 3450 cm – 1 and OH groups of silicate correspond to the absorption band at 3626 cm – 1. By the ratio of the intensities of these bands and bands in the absorption region, it can be assumed that the charge on the nitrogen atom of DAA is partially extinguished due to an increase in the concentration of a negatively charged hydroxyl OH - group, which is contained in Na + - MMT, due to which organoclay acquires hydrophobic properties. Intense absorption bands appear in the 1329-1577 cm-1 region caused by bending vibrations of methylene groups characteristic of the long-alkyl substituent on the IR spectrum of the modified organoclay Na + -MMT- DAA , which were confirmed by XRD. Fig. 1. X-ray diffraction patterns of a composite with 5 wt.% organoclay, Na + - MMT containing N, N-diallylacrilamide and pure PP are presented.

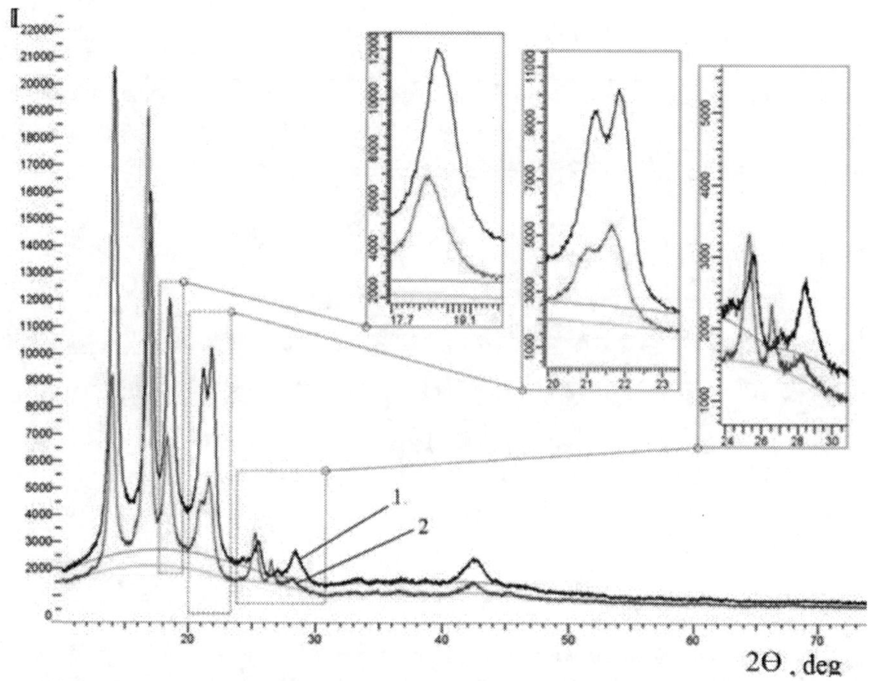

Fig. 1. X-ray diffraction patterns obtained for pure PP (1) composite with 5 wt.% Organoclay Na^+-MMT containing DAA (2)

It can be seen from Fig. 2.2 that for a PP composite with 5 wt% organoclay, the formation of an intercalated structure is 2Θ = 18.530 (d001 = 4.63), 2Θ = 25.310 (d001 = 3.51), 2Θ = 26.650 (d001 = 3, 34).

SAR results are confirmed by SEM images (Fig. 2)

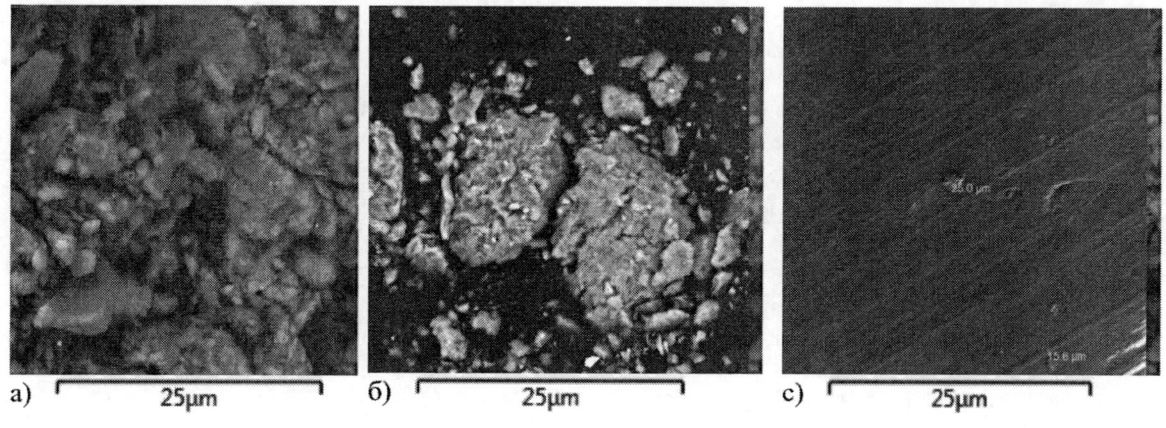

Fig. 2. SEM images: a - Na^+ –MMT; b –Na^+-MMT + DAA; c - composite PP with 5 wt.% organoclay

Comparing the presented images, one can notice that clay particles are homogeneously distributed in the polymer matrix in the form of small tactoids (Fig. 3b). The distance between the clay layers in the nanocomposite increased by 2.3A0 compared to montmorillonite. These data additionally indicate the formation of an intercalated structure of the composite.

The efficiency of composite modification was evaluated by changing the physical and mechanical properties (Tab. 1).

Table 1. Mechanical properties of composites

Composition	MFI g/10 min	Shore hardness number, D 1/15c	A, kJ / m^2 Izod (notched)	Modulus of flexure mPa	Extensional modulus mPas, (1/10)mm/min	Creep limit mPa	Elongation, ε,%	Rupture strength, mPa
PP indust.	6,12	69/63	4000	800	900/1050	28,5	724	28,5
PP(ext)	5,9	69/65	46500 /2953	950	1100	29,5	550	29,5
PP (ext twice)	5,73	61/56	42412/2685	1014	1058/1131	29,3	414	29,3
PP +3% organoclay DAA	5,16	61/59	34929/5602	1176	1163/1345	30	454,6	30
PP(ext)	5,9	69/65	46500 /2953	950	1100	29,5	550	29,5
PP +3% organoclay DAA	5,16	61/59	34929/5602	1176	1163/1345	30	454,6	30
PP + 5% organiclay DAA	5,26	61/56	34912/4766	1241	1174/1364	30,2	254,6	30,2
PP +7% DAA	4,86	61/56	37938/3773	1192	1218/1306	29,3	202,8	29,3

When filling with PPP 3 wt.% organoclay increases: impact strength by 14% (without notch) and by 10% (with a notch of 5 mm); modulus of elasticity in bending by 11.5%; tensile modulus by 11%. In the case of filling of the teaching staff with 5 wt. % organoclay, the toughness increases by 33.5% (without a notch) and by 10,6% (with a notch of 5 mm); the modulus of elasticity in bending and tension is stored as in the previous case. A further increase in the content of organoclay to 7 wt.% leads to a slight decrease in mechanical properties. The intercalated structure formed in the composite suggests a change in the conformation of the molecular chains of polypropylene, which leads to a denser packing of interfacial regions compared to a bulk polymer matrix, therefore, in all cases, the elongation decreases.

In conclusion, it should be noted that the resulting composites are useable as structural materials. The most optimal is the addition of 3-5% by weight of modified organoclay in polypropylene.

The combustibility of the composites was evaluated by the parameters of the flame propagation velocity υ and the burning time of the samples (tab. 2).

Table 2. Thermal characteristics of composites

Composition	Ci	υ burning mm / s (mm/min)
PP indust.	17,3	0,567/34,13
PP + 3% organoclay	19,0	0,483/28,92
PP + 5% organoclay	20,2	0,518/27,29
PP + 7% organoclay	20,4	0,539/28,14

As can be seen from Table 2, with the introduction of 3-5 wt.% Of the modified Na + -MMT + DAA into the PP, a decrease in the flame propagation rate by 74-79% is observed. The slowing down of the burning rate of the composite is associated with a limitation of the diffusion of atmospheric oxygen into the material due to the presence of a barrier layer created by plane MMT particles. The presented data are consistent with the results previously obtained in [19-25].

4. Conclusion

Thus, the assessments of the mechanical properties and fire resistance of the composites allow us to conclude that N, N-diallylacrilamide, whose structure differs from the previously used surfactants, works as a modifier. The type of modifier stacking in Na + - MMT promotes the formation of an intercalated structure of composites and provides an increase in the elastic modulus, impact strength, and lower combustibility.

The reported study was funded by RFBR according to the research project № 18-29-18063.

References

[1] Borodulin A, Kalinnikov A, Tereshkov A, Kharaev A New polymeric binders for the production of composites. *Materials Today: Proceedings* 2019. 139-143.

[2] Delozier D.M., Orwol R.A., Cahoon J.F., Johnston N.J., Smith J.G., Connel J.W. 2002. *J. Polymer*, P. 813-822.

[3] Gerasin V A, Bakhov F A, Merekalova N D, Korolev Yu M, Fischer H R and Antipov E M 2005. *Vysokomolek. Soedien*. Series A V. **47-9** 1635-1651.

[4] Khashirova S Yu, Musaev Yu I, Mikitaev A K, Malkanduyev Yu A, Ligidov M Kh 2009.. *Vysokomolek. Soedien*. Series B. **51-9** 1723-1728.

[5] KojimaY, Usuki A, Kawasumi M, Okada A, KurauchiU and Kamigaito O. 1993. *J. Polym. Sci., Polym. Chem*. **31** 983.

[6] Galimov E R, Tukbaev E E, Fedyaev V L, Ya Galimova N, Samoylov V M, Orlov M A, Borodulin A S, Mavleev I R. Production technology of heat-resistant high-strength syntactic carbon foams for operation in extreme conditions. *IOP Conference Series: Materials Science and Engineering* 2018. 012018.

[7] Kharaev A M, Bazheva R Ch, Begieva M B, Nelyub V A, Borodulin A S Polyethersulfones with improved thermophysical properties *Polymer Science. Series D*. 2019. **12**(1). pp. 24-28.

[8] Davis C H, Mathias L J, Gilman J W, Schiraldi D A, Shields J R and Trulove P, Delong 2002. *J. Polym. Sci., Polym. Phys*. **40.** p. 2661.

[9] Borodulin A S, Kalinnikov A N, Bazheva R Ch, Kharaev A M, Beshtoev B Z Receipt and investigation of performance characteristics of super constructions polyesters *International Journal of Mechanical Engineering and Technology*. 2018. **9(13)**. 1117-1127.

[10] Hasegawa N, Kawasumi M, Kato M, Usuki A and Okada A. 1998. *J. Appl. Polym. Sci.* **67**. p 87

[11] Wang K H, Choi M H, Koo C M, Choi Y S and Chung I J. 2001. *J. Polym*. **42** p. 9819

[12] Begieva M B, Kharaev A M and Malkanduyev Yu A. 2019 Poly-N,N-diallylacrilamide. Patent (RF)

[13] Clareym J, Edwards J, Tzipursre S J, Beal G W and Eisenhour D.D. 2001.Pat. 6050509 USA.

[14] Ashurov N R, Sadykov Sh G and Dolgov V V *Vysokomolek. Soedien*. Series A. 2012. **47-9** 1403-1408.

[15] Deev I S, Sokolov I I, Panina N N, Kim M A, Nelyub V A, Buyanov I A, Chudnov I V, Borodulin A S, Mironov Y M. Properties and structural features of nanodisperse powder fillers *Polymer Science. Series D*. 2013. **6(4).** pp. 285-290.

[16] Borodulin A.S. Polyester resins for production of goods from polymer composite materials by pressing methods *Polymer Science. Series D*. 2013. **6(3).** pp. 269-270.

[17] Borisov V A, Bedanokov A Yu, Karmokov A M, Mikitaev A K, Mikitaev M A and Turaev E R 2007. *Plastic. Masses* **5** 30-33.

[18] Nelyub V A, Borodulin A S Properties of epoxy materials used for production of glass-reinforced plastics by the winding method *Polymer Science. Series D*. 2018 **11(2).** pp. 147-153.

[19] Khashirova S Yu, Beslaneeva Z L, Musaev Yu I, Musaeva E B, Pakhomov S I, Ligidov M Kh and Mikitaev A K. 2012. *News of higher educational institutions. Series: Chemistry and Chemical Technology*. **24-5** 86-88.

[20] Baker S S, Khashirova S Yu, Mikitaev A K 2011. *High technology*. **12-10** 79-81.

[21] Murzakanova M A, Borukaev T A, Khashirova S Yu and Mikitaev A K 2013. *Modern problems of science and education.* **3** 48

[22] Begieva M B, Bazheva R C, Khasanov V V, Malkanduev Y A and Kharaev A M 2018. *Fibre Chemistry* **50-1** 45-48

[23] Bazheva R C, Kharaev A M, Begieva M B, Beslaneeva Z L and Shogenov V N. 2018. *Fibre*

Chemistry. **49-6** 420-424

[24] Nelyub V A, Borodulin A S, Kobets L P, Malysheva G V. Thixotropy hysteresis and structure formation in elastomeric suspensions *Inorganic Materials: Applied Research*. 2018. **9(4).** pp. 603-608.

[25] Begieva M B, Nyrova F M, Tsoroeva Z I, Gudova Yu Kh, Begieva M Kh, Blyashev A V and Amshokova D B. 2018. *Materials Science Forum*., **935** 66-69

IOP Conf. Series: Materials Science and Engineering **934** (2020) 012005 doi:10.1088/1757-899X/934/1/012005

Analysis of the non-isothermal resin impregnation process of carbon fibre reinforced composite wing spar structure

Pyi Phyo Maung[1], G V Malysheva[2], Tun Lin Htet[3] and G V Drobyshev[4]

[1,2,3]Bauman Moscow State Technical University, 2nd Baumanskaya street, Moscow, Russia

[4]School №1387, Moscow state budget educational institution, Moscow, Russia

E-mail: [1]pyaephyo@mail.ru, [4]a@inginirium.ru

Abstracts. This paper presents non-isothermal resin impregnation process of high-performance light weight composite wing spar structure for trainer aircrafts. The manufacturing process of the spar is simulated in PAM-RTM software. In this research developed a reliable model of resin impregnation process simulation to determine the exact infusion time and optimize the resin inlet ports for large scale aircraft structure without using the expensive trial and error experimental approches. The work reported the use of different heating temperature processes to manufacture wing spar from carbon fibre-reinforced epoxy resin matrix composite laminates by using VARI. The influence of heating temperature during the infusion stage on the resin viscosity and impregnation time was analyzed. These results indicate that a proper heating temperature during the infusion process is necessary for quality control and manufacturing time reduction. The error between theoretical simulation and experimental studies is no more than 10%.

1. Introduction

Advanced composite materials have been increasingly used in aviation, aerospace, automobile, shipbuilding due to their complex characteristic of specific strength and modulus, ultralight weight, good fatigue properties and corrosion resistance [1,2]. With the rising demand, manufacturing cost and time are most important factors in the production of composite materials. Generally, composite aircraft structures have been produced by using the prepreg and autoclave forming. Despite the prepregs can offer high quality of product, their applications have been limited due to high costs and service life limitation of prepregs and expensive tooling such as autoclave [3,4]. Vacuum assisted resin infusion (VARI) technologies are widely used to produce the low cost and high-performance aerospace quality large scale complex integral structures such as wing spar, fuselage and wing skin with minimal initial capital costs. In the resin infusion stage, the application of the VARI process depends on resin viscosity so as to achieve uniform and complete impregnation process of fabric [5-6]. Additionally, a proper heating temperature during the infusion process is useful to reduce resin viscosity and consequently reduce the impregnation time, enhance the mechanical properties and increase the adhesion characteristics. However, an improper heating temperature might result in poor impregnation of the preform and produce voids in final cured composite [7,8].

Shojaei and colleagues examined the role of initiators with different reactivities on the process cycle of RTM using numerical simulation to minimize the impregnation time [9]. The results presented that a temperature distribution was created along the resin flow direction for the preheating mould. In order to increase the composite properties, Kedari studied the effect of temperature and inlet pressure of impregnation process on fiber volume content and void inside the composite laminate manufactured by vacuum assisted resin infusion process (VARI) [10]. The results showed that a high processing temperature and low inlet pressure can produce final parts with high fibre volume fraction and low void content. It is detrimental to the uniformity of curing degree and mechanical properties. In summary, heating process during impregnation is an important factor to change processing efficiency for VARI process, so it should be analyzed carefully to obtain both short cycle time and high-quality products.

Additionally, heating process during the resin impregnation might result in obvious crosslinking reactions during the resin infusion stage, and affect the impregnation of the fibre and the temperature distribution due to the curing characteristics of the resin [11-13]. The aim of this study was to investigate the influence of the heating temperature during the resin infusion stage on the resin viscosity and impregnation time of composite wing spar structure. For this purpose, four kinds of processing cycles with different heating temperature were adopted. This is especially important when using fabrics or tapes with metal coatings [14-16]. The temperature influence on the process and accuracy of simulation were investigated, the authors [17-20] used new types of binders. Simulation of the vacuum infusion process is a required analysis tool to optimize the technological regime and to minimize th cost of trial and error experiment.

2. Flow in a porous medium

In the VARI process, the resin impregnates through a reinforcing material, which can be considered as a porous medium. The flow of resin is described by Darcy's law, which states that the flow rate of resin per unit area is directly proportional to the pressure gradient and inversely proportional to the viscosity of the resin. The constant of proportionality is called the permeability of the porous materials. It depends on the fabric geometry.

Darcy's law can be written by

$$\vec{V} = -\frac{K}{\mu}\vec{\nabla}P \tag{1}$$

Where K – permeability tensor, μ – resin viscosity, V – Darcy's velocity, P – pressure.

In order to preserve the balance of resin mass, the velocity field must satisfy the divergence condition:

$$\nabla.\vec{V} = 0 \tag{2}$$

By combining these two equations,

$$\nabla.\left(\frac{K}{\mu}\nabla P\right) = 0 \tag{3}$$

In vacuum assisted resin infusion process flexible plastic vacuum bag were used to reduce the void contents and accelerate the resin flow. In the case of deformable media, considering the resin and the fibre material as incompressible, the equation is decried as

$$div\,(\emptyset.\vec{V_r}) = -div(\vec{V_s}) \tag{4}$$

Where ϕ – porosity, Vr , Vs – velocity of the resin and solid velocity.

Darcy's law is written as

$$div\left(\frac{\lfloor K \rfloor}{\mu}\langle \nabla P \rangle\right) = div(\vec{V_s}) \tag{5}$$

The resin viscosity depends on the temperature during the impregnation process. Temperature dependence of the resin can be written as follow:

$$\mu(T,\alpha) = B \cdot exp\left(\frac{T_b}{T}\right) \cdot \left(\frac{\alpha_{gel}}{\alpha_{gel}-\alpha}\right)^{c_1+c_2 \cdot \alpha} \tag{6}$$

Where B, T_b, α_{gel}, c_1 and c_2 – constant of the resin characteristic

These characteristic constants of resin system were determined by using the differential scanning calorimetry.

IOP Conf. Series: Materials Science and Engineering **934** (2020) 012005 doi:10.1088/1757-899X/934/1/012005

3. Design and materials

3.1 Wing spar structure
Composite wing spar is one of the most critical parts in airframe. The vacuum assisted resin infusion process was used to produce wing spars to ensure the quality of the products and reduce the voids content inside the structure. The length of the spar is 11500mm. Spar wall high is varied from 300mm to 80mm and thickness is 2mm. Spar cap width is 80mm and thickness varied from 10mm at the middle section and to 3mm at the end of the wing tips. The 3D model of wing spar structure was shown in figure 1.

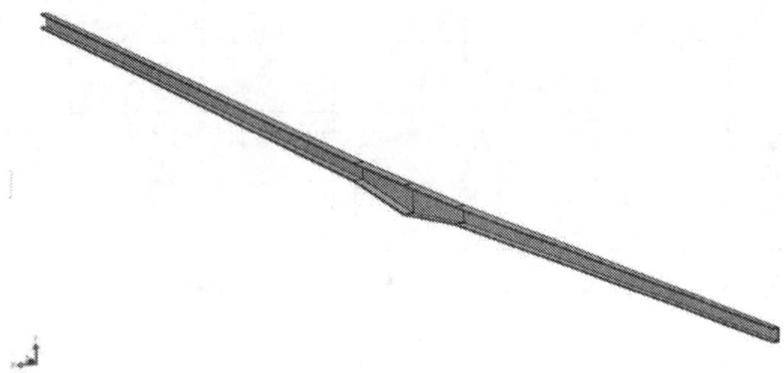

Figure 1. Rear spar structure

3.2 Materials
The epoxy resin used in this study was RIM 935 (supplied by Hexion). The Cure agents RIM RIM H 936 are a modification of aliphatic and cycloaliphatic amines. Resin properties are shown in table 1. The temperature dependence of rein mixture was determined by using viscometer Brookfield cap 2000 (Fig.2). The reinforcing materials are KDU 1034 unidirectional carbon fibre fabric and INTGL 98141 carbon fibre fabric. The properties of reinforcing materials are presented in table 2.

Table 1. Resin System properties

Properties	
Density [g/cmÑ]	1.14 – 1.2
Viscosity [mPas]	400 - 800
Maximum Tg [C]	150 °C
Parts by weight ratio RIM 935/cure agent RIM 936	100:29±2
Gel-time (100 grams at 30◦C)	120 min
Thermal conductivity [W/mK]	0.11
Specific heat [J/kg.K]	1205

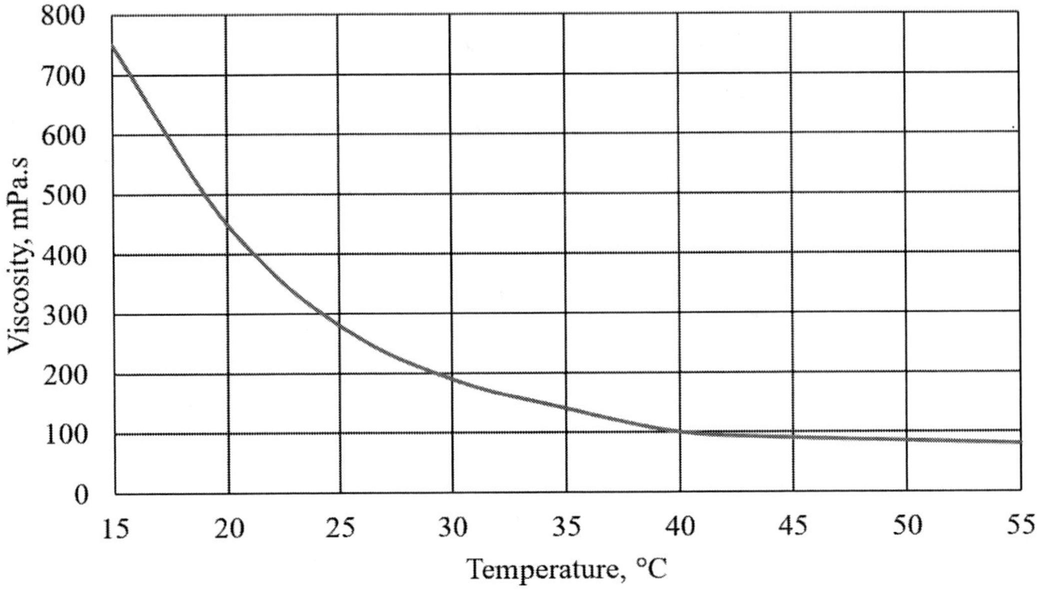

Figure 2. Temperature dependence of resin mixture

Table 2. Reinforcing materials properties

Reinforcing materials	Weaving type	Nominal areal weight, g/m^2	Resin content in g/m^2 in laminate	Thickness, mm	Fibre volume content, %
Interglas 98141	Twill 2/2	204	140	0,3	39
KDU 1034	UD	370	260	0,4	53

4. Modeling

Modeling of the vacuum infusion process is a necessary tool to optimize the process parameters, location of the resin inlet and outlet lines, and to minimize costly and time-consuming trial and error approches for large scale structure. A three-dimensional mathematical model has been developed for flow simulation and implemented in the PAM-RTM finite element software to optimize the flow parameters. and to predict the resin flow front during the infusion process. The process of impregnation has been taken place inside the heating oven with different values of the heating temperature (Fig. 2). Boundary conditions are shown in figure 3.

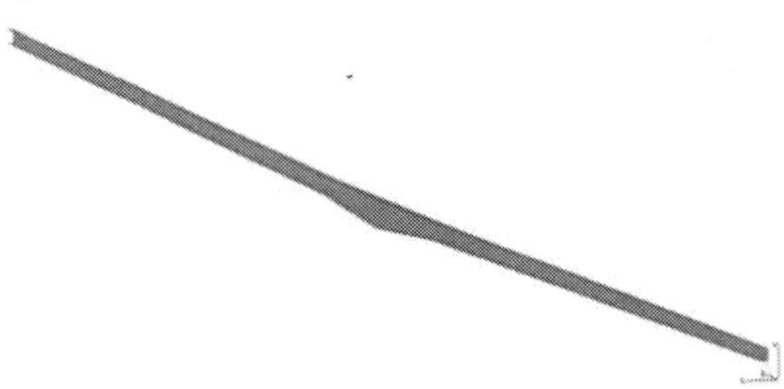

Figure 3. Boundary conditions of resin impregnation process

The resin inlet flow channel is location at the middle line of the spar wall mold. The resin viscosity function relates temperature depended viscosity and cure degree. Injection pressure is 1 Pa and vent pressure 0 Pa. Initial temperature of heating room is 25°C. Modelling temperature are 35°C, 50°C and 55°C. Filling pressure along the wing spar structure is also shown in figure 5.

Figure 4. Filling time of rear spar structure with heating temperature: 50°C

Figure 5. Filling pressure of rear spar structure with heating temperature: 50°C

The simulation reuslts showed that the heating tempeature of the oven during the impregnation process accelerate the impregnation process. More higher temperature allows to start the curing process. The optimal heating temperature must be chosen. The comparison results of rear, front and middle inner spar impregnation processes are shown in table 3.

Table 3. Simulation results of impregnation process

Wing spars	Heating temperature, °C			
	25	35	50	60
	Duration, [s]			
Rear spar	4600	3980	2330	2210
Front spar	4340	3740	2020	1970
Middle inner spar	3450	2430	1090	1040

5. Experimental

The composite laminates were prepared inside the wing spar mold by the VARIM method, as illustrated in figure 6. Three layers of INTGL 98141 carbon fibre fabrics were stacked overall inside the mold. And 31 layers unidirectional carbon fibre fabrics KDU 1034 were stacked on the vertical wall of the mold (Spar cap). The Peel ply, permeable medium and distribution mesh were laid over the surface of the fabric successively. The distribution mesh was terminated at a gap 40 mm from the end of the fabric laminate preform. In order to monitor the temperature variation inside the preform during the curing process, 5 thermocouples were inserted into the fabric layers, as shown in figure 6. The experimental results are shown in table 4.

Figure 6. Experimental VARI process of wing spar structure

Table 4. Experimental results of impregnation process

Wing spars	Heating temperature inside the oven, °C			
	25	35	50	60
	Duration, [s]			
Rear spar	4940	4230	2540	2480
Front spar	4750	4010	2435	2215
Middle inner spar	3880	2650	1320	1250

The manufacturing process of the spar is simulated in PAM-RTM software, which uses the vacuum infusion technique, unlike the traditional wet lay-up method. In this research developed a reliable model of resin impregnation process simulation to determine the exact manufacturing time and optimize the resin inlet ports for large scale aircraft structure without using the expensive trial and error experimental approches.

6. Conclusion

The results of theoretical and experimental studies of the kinetics of the impregnation process depending on the heating temperature were carried out. The manufacturing process of the spar is simulated in PAM-RTM software. In this research developed a reliable model of resin impregnation process simulation to determine the exact infusion time and optimize the resin inlet ports for large scale aircraft structure without using the expensive trial and error experimental approches. The work reported the use of different heating temperature processes to manufacture wing spar from carbon fibre-reinforced epoxy resin matrix composite laminates by using VARI. The relationships among heating temperature, resin viscosity and impregnation time were analyzed. It was found that, in infusion stage a high heating temperature was useful to reduce the total impregnation time. In the case of 60 °C heating temperature, the impregnation time is shortest. However, too high heating temperature effects pre-crosslinking process of resin and to the mechanical properties and nonuniform properties along the resin flow direction. When a higher heating temperature was used, poor resin and fibre impregnation at the resin outlet were responsible for the lower mechanical properties. These results indicate that a proper heating temperature during the infusion process is necessary for quality control and production time reduction. The error between theoretical simulation and experimental studies is no more than 10%.

Acknowledgement
The research results were partially obtained within the framework of the Russian Foundation for Basic Research (RFBR) grant No. 18-29-19034 / 18 "Development of theoretical and technological foundations for the manufacture of high-strength carbon plastics with a set of new functional properties based on metallized fibers"

References

[1] Nelyub V A 2018 Adhesive-strength evaluation via the pull-out method in a binder-elementary-filament system at various treatments of filaments. *Polymer Science – Series D*. vol. **11(3).** pp. 263-266.
[2] Nelyub V A 2018 An Impact Assessment of Metallization of a Carbon Tape Surface on the Mechanical Characteristics of the Filaments. *Polymer Science – Series D*. vol. **11(4).** pp. 423-425
[3] Baurova N I, Zorin V A and Prikhoko V M 2015 Description of scenarios of transition of material from an operable to inperable state using an operable to inperable stage using an equation of fold catastrophe theory Polymer Science-Series D vol. **8(10)** pp. 1–5.
[4] Reznik S V, Prosuntsov P V and Azarov A V 2015 Substantiation of the structural-layout scheme of the mirror-space-antenna reflector with a high shape stability and a low density per unit length *Eng. Phys. Thermophy*. vol. **88(3)** pp. 699-705.
[5] Baurova N I, Zorin V A, Prikhodko V M 2016 Taking into account the factors of technological heredity by technological mechanics *Polymer Science-Series D* **9(4)** pp 402–406.
[6] Sergeev A Y, Turusov R A and Baurova N I 2016 Strength of the Joint of an Anisotropic Composite and a Cylindrical Element of the Exhaust System of Road Vehicles *Mechanics of Composite Materials* **52(1)** pp 89–98
[7] Liang L J and Young W B 1999 The effect of preheater on the resin transfer molding. *J. Reinf. Plast. Compos.* **18** pp954–965.
[8] Young W B and Tseng C W 1994 Study on the pre-heated temperatures and injection pressures of the RTM process *J. Reinf. Plast. Compos.* **13** pp 467–482.
[9] Shojaei A, Farrahinia H and Pishvaie S M R 2010 Effect of system of initiators on the process cycle of non-isothermal resin transfer molding–numerical investigation *Compos. Pt. A.* **41** pp 138–145.
[10] Kedari V R, Farah B I and Hsiao K T 2011 Effects of vacuum pressure, inlet pressure, and mold temperature on the void content, volume fraction of polyester/e-glass fiber composites manufactured with VARTM process *J. Compos. Mater.* **45** (2011) pp 2727–2742.
[11] Sergeyev A Y, Turusov R A, Baurova N I and Kuperman A M 2015 Stresses arising during cure of the composite wound on the cylindrical surface of an element of exhaust system *Mechanics of Composite Materials* **51(3)** pp 321–332.
[12] Agirregomezkorta A, Martínez A B and Sánchez-Soto M 2012 Impact behaviour of carbon fibre reinforced epoxy and non-isothermal cyclic butylene terephthalate composites manufactured by vacuum infusion. *Compos. Pt. B.* **43** pp 2249–2256.
[13] Zhang K M, Gu Y Z, Li M, et al. 2014 Effect of rapid curing process on the properties of carbon fiber/epoxy composite fabricated using vacuum assisted resin infusion molding. *Mater. Design* **54** pp 624–631.
[14] Nelyub V A 2018 Technologies of merallization of carbon fabric and the properties of the related carbon fiber reinforced plastics. *Russian Metallurgy (Metally)*. Vol. **13**. pp. 1199-1201.
[15] Nelyub V A and Kovalev S P 2016 A mesh-free approach to multiscale multiphysical calculations of polymeric composite structures. *Polymer Science – Series D*. vol. **9(1).** pp. 101-105.
[16] Nelyub V A 2016 Quantitative assessment of adhesive contact between carbon fiber and epoxy binding. *Russian Aeronautics*. Vol.**59(2)** pp. 259-262.

[17] Bessonov I, Polezhaev A, Kusnetsova M, Nelyub V and &. 2013 Rheological and thermal analysis of low-viscosity epoxy-furan composites. *Polymer Science-Series D*. vol. **6(4)** pp.308-311.

[18] Kobets L, Malysheva G and Borodulin A 2016 Rheological properties of system "elastomer binder-discrete carbon fiber". Inorganic Materials: Applied Research. vol. 7(1) pp. 15-19.

[19] Nelyub V, Borodulin A, Kobets L and Malysheva G 2017 Viscous hysteresis in filled siloxane binders. Polymer Science – Series D. vol. 10(1) pp. 19-22.

[20] Kharaev A M, Bazheva R C, Begieva M B, Nelyub V A and Borodulin A S 2019 Polyethrsulfones with improved thermophysical properties. Polymer Sience – Series D. vol. 12(1) pp. 24-28.

IOP Conf. Series: Materials Science and Engineering **934** (2020) 012006 doi:10.1088/1757-899X/934/1/012006

Composites Based on Carbon Tapes with a Copper coating Based on Organic and Inorganic Matrices

V A Nelyub

Bauman Moscow State Technical University, 2nd Baumanskaya street, Moscow, Russia

E-mail:mail@emtc.ru

Abstracts. The results of experimental investigations into the impregnating kinetics when using the carbon tape with copper coating according to the magnetron sputtering technology are presented. The epoxy and aluminum-boron phosphate materials are used as binders. The composites based on these binders are evaluated by characteristics of porosity, matrix content, and thermal stability. It is established that metallization insignificantly affects the thermal stability but does not make it possible to increase the impregnation rate and leads to lowering the composite porosity.

1. Introduction

The application fields of polymer composite materials, in the first place carbon-filament composites, are broadened annually. This is associated with their unique mechanical and thermal characteristics, which provides their effective use as structural materials [1-3]. Main advantages of constructions made of carbon-filament composites are low weight, high strength, stiffness and wear resistance, and a whole complex of functional properties (thermal conductivity, electrical conductivity, etc.) [4-6].

The necessity of improving the operational characteristics of occurring constructions made of carbon-filament composites originates the problematic of extending for a complex of their properties. Various technologies such as oxidation, dressing, galvanic treatment, etc. are used to modify the properties of carbon tissues. One method of controlling the properties of carbon fillers is the metallization technology. The main goal of metallization is changing the microstructure, topology, and physicochemical properties of the surface of carbon tissues, which makes it possible to control their reaction ability, thermal and electrical properties, biocompatibility, and many others.

The development of carbon-filament composites with a complex of functional properties will make it possible to increase substantially the competitiveness of products. Such materials will provide an increase in the service life, a decrease in the prime cost, and an increase in efficiency of using the carbon-filament composites. The development of theoretical methods of the investigation [7-9], including the use of the apparatus of the theory of catastrophes, made it possible to predict the service life of constructions made of composites formed based on organic matrices.

The organic matrices, in the first place epoxy matrices, are widely used in composite materials, while inorganic matrices have the substantially higher thermal stability. However, there is no economically effective technologies to form of them the products based on carbon tapes, which is also an urgent problem of the modern materials science.

For a long time, when producing the parts from carbon-filament composites, prepregs [3] and autoclave solidification technologies were used, which provided the specified quality but substantially increased the cost. The high cost of producing the carbon-filament composite parts is in many aspects associated with a very high power intensity of autoclaves. One effective method of decreasing the

prime cost of fabricating the carbon-filament composite parts is the use of direct formation technologies, of which the vacuum infusion technology becomes most widespread for forming the carbon-filament composite parts [10]. Largely due to the vacuum infusion technology, the regions of applying the carbon-filament composites are broadened annually, and there is no the modern branch of industry, where they would not be used. These are aircraft construction, production of space-rocket hardware, automobile industry, mechanical engineering, building, etc. [11-15].
The goal of this work is the experimental investigation into the thermal stability and rheological properties of composites based on carbon tapes with a copper coating fabricated according to the vacuum infusion technology when using organic and inorganic binders.

2. Objects and Methods

We used the import carbon tapes of the FibArmTape-230/300 brand 0.21 mm in thickness (the surface density is 230 g/m and the elasticity modulus is 230 GPa) as the carbon filamentary fillers.
To deposit the copper coating on a carbon tape, the industrial magnetron sputtering installation (Fig. 1) with a vacuum chamber volume of 15 m3 was used. A carbon tape roll was placed onto a special facility and air was pumped from a chamber using vacuum pumps to a pressure of 3 □ 10 mmHg. Then plasma-forming gas, argon in our case, was puffed into the chamber. The glow discharge was ignited by the potential supply to the magnetron cathode from the dc power supply unit. The discharge current, voltage, pressure, and flow rate of the plasma-forming gas were controlled during the deposition of metallic coatings.

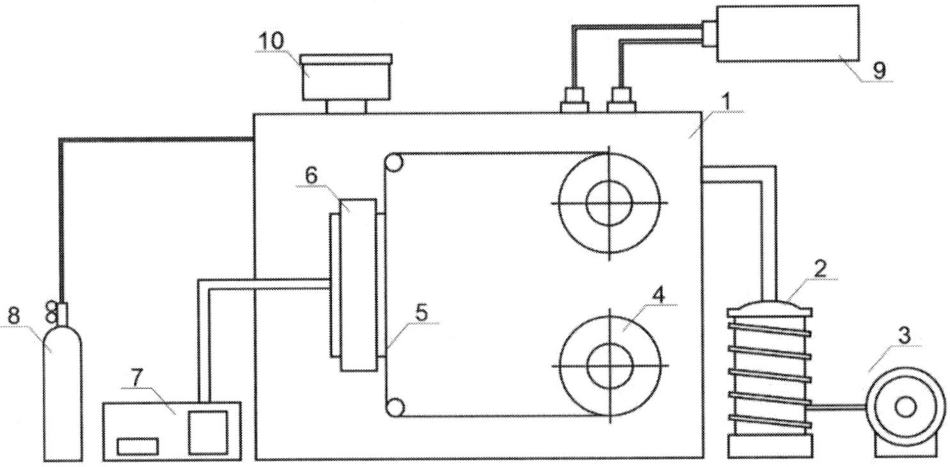

Figure 1. Layout of the MMR-1800M magnetron sputtering installation: 1 – vacuum chamber, 2 – diffusion pump, 3 – forevacuum pump, 4 – facility for tape rewinding, 5 – treated material, 6 – magnetron, 7 – magnetron supply unit, 8 – argon supply system, 9 – gas supply system, and 10 – supply unit

Copper of the M1 brand was used as a target material. The thickness of a copper coating on a carbon tape was 100 nm, which is the optimal value [4].
Two types of binders such as organic (epoxy) and inorganic (aluminum-boron phosphate) were used in the work.
The composition based on epoxy resin of the ED-20 brand was used as an organic binder, and isomethyl tetrahydrophthalic anhydride was used as a hardener; diethyleneglycol was used to decrease the viscosity.
Ruskon-355K was used as the hardener of the inorganic binder (aluminum-boron phosphate).

Composites based on a metallized carbon tape were fabricated following the uniform vacuum infusion technology but using two sets of auxiliary materials. The samples of one type were impregnated using a conductive mesh (this technology was called no. 1 in our work). The samples of the second type were impregnated without using the mesh (this technology was called no. 2 in our work). A mesh made of polyether of the CCVM-FM-KNITW-100-1.25 brand was used as a conductive mesh. This mesh had an increased heat stability, which allowed us to use it during hardening the epoxy binder at a temperature of 200°C.
When fabricating the products according to the vacuum infusion technology, the conductive mesh is always used (technology no. 1) because it provides the spread of a binder over the surface of the first layer of the reinforcing filler with a maximal speed [1]. However, it is very difficult to evaluate visually the influence of the metallic layer that coats the carbon tape on the impregnating process kinetics, which is associated with a very high moving speed of a binder front. The use of technology no. 2 allowed us to decrease substantially (practically by an order of magnitude) the impregnating process rate, which allowed us to evaluate the influence of the type of used binders and influence of metallization of the carbon tape.
The composite porosity was evaluated by hydrostatic weighing.
We evaluated the values of limiting wetting angles experimentally. They were determined under the conditions maximally approached to equilibrium. To increase the determination accuracy of limiting wetting angles, not binders but their separate components were used, for example, the ED-20 resin and a liquid component of aluminum-boron phosphate binder (Russian abbreviation ABFS). To determine the limiting wetting angles, a silicon plate was used as a substrate. The copper coating was deposited on it under the same process modes as for the carbon tape.
To determine the thermal stability, a TG 209 F1 Libra thermogravimetric analyzer was used.

3. Results and Discussion

The limiting wetting angles for different binder components and coating types are presented in Table 1.

Table 1
Limiting wetting angles

Metallic coating type	Limiting wetting angle, °, when using	
	ED-20	ABFS
No coating	56	74
Copper	48	65

Thus, the presence of a metallic coating on the substrate surface makes it possible to decrease the limiting wetting angles. The wetting angle decreased by 14% for epoxy resin and by 12% for the inorganic material.
The impregnation kinetics for the epoxy and inorganic binders was evaluated for the samples with the same geometric shape (100 □ 50 mm). The number of the carbon tape layers is 5, and the reinforcing angle is 0. The impregnation was performed at room temperature without the conductive mesh (technology no. 2). The experimental results are presented in Fig. 2.

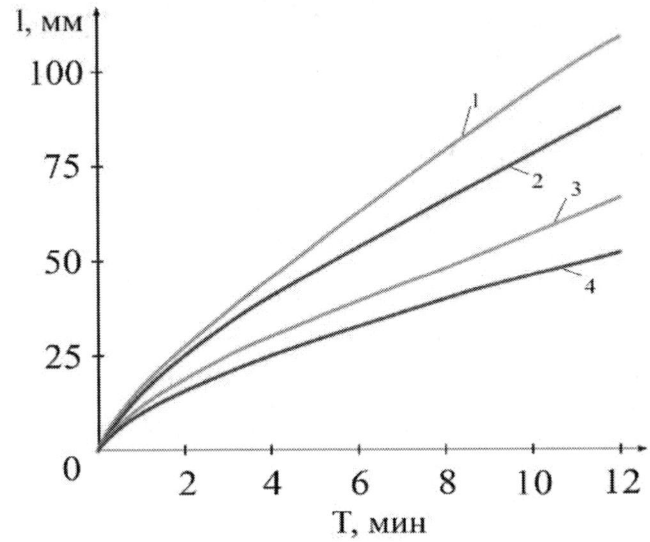

Figure 2. Time dependence of the length of the impregnated carbon tape segment for epoxy (1, 2) and ABFS (3, 4) binders as well as for carbon tapes with no metallic coating (2, 4) and with the copper coating (1, 3).

The analysis of the experimental results shows that the use of a carbon tape with a metallic coating allowed us to increase the impregnation rate for the organic (epoxy) and inorganic (aluminum-boron phosphate) binders. However, the impregnation rate is higher when using the organic binder. For example, the carbon tape sample without the metallic coating, when using the epoxy binder, was impregnated to a length of 76 mm for 10 min, while the coated sample was impregnated to a length of 92 mm. The difference in the impregnation duration was 17.4%. The impregnation rate for the inorganic binder is lower due to its higher viscosity. The difference in the impregnation duration for a length of 100 mm was 26%.

Thus, the metallization of a carbon tape makes it possible to increase largely the impregnation process when using the inorganic binder.

Characteristics of the composites fabricated according to the vacuum infusion technology when using the carbon tape with no coating and with a copper coating are presented in Table 2. The previously used organic and inorganic materials were used as binders.

Table 2

Properties of composites fabricated based on organic and inorganic binders

Characteristics	Carbon tape	
	With no metallic coating	Copper coating
Epoxy binder		
Matrix content, %	33,6	31,8
Porosity, %	4,9	3,9
Weight loss at 200°C, %	4,5	4
Weight loss at 800°C, %	66,4	68,2
Aluminum-boron phosphate binder		
Matrix content, %	48,4	49,8
Porosity, %	24	21
Weight loss at 200°C, %	8	8
Weight loss at 800°C, %	8	9
Weight loss at 1000°C, %	12	11

It is established based on our investigations that the matrix content in composites based on inorganic binders is higher than that one for epoxy binders. The substantially higher porosity is characteristic of inorganic materials, the magnitude of which exceeds 20%, while this quantity for the composites based on the epoxy matrix is lower than 5%. The presence of a copper coating on a carbon tape leads to a decrease in porosity for two types of used binders, notably, it decreases by 26% for epoxy binders and by 14% for aluminum-boron phosphate binders. It seems likely that so high porosity of the composites based on an inorganic matrix are characteristic of only the infusion formation technology and this quantity can be decreased when using other production processes, for example, pressing.
The thermal stability of composites based on an inorganic matrix is substantially higher and does not exceed 9% at 800°C, while similar weight losses for composites based on an organic matrix at this temperature are higher than 66%. The presence of a copper coating on the carbon tape insignificantly affects the thermal stability.

4. Conclusions

It is established based on our investigations that it is possible to fabricate composites based on organic and inorganic binders according to the vacuum infusion technology when using carbon fibrous fillers covered with a metallic coating.
When evaluating the kinetics of impregnation, it was established that the presence of the metallic coating on a carbon tape leads to an increase in the impregnation rate by 17--26% for the organic and inorganic binders. The presence of a copper coating on the substrate surface makes it possible to decrease the limiting wetting angles by 12-14%.
The presence of metallic coatings of a carbon tape makes it possible to decrease the composite porosity for two used types of matrices; however, the porosity for the inorganic matrix is higher than that one for the organic matrix by a factor of several times.
The presence of the metallic coating on a carbon tape most weakly affects the thermal stability of composites; this quantity varies insignificantly. The matrix type affects the thermal stability much more strongly. The thermal stability of composites with the inorganic matrix is higher than 1000°C because the weight loss at this temperature does not exceed 12%.

Acknowledgement
The research results were partially obtained within the framework of the Russian Foundation for Basic Research (RFBR) grant No. 18-29-19034 / 18 "Development of theoretical and technological foundations for the manufacture of high-strength carbon plastics with a set of new functional properties based on metallized fibers"

References

[1] Baurova, N.I. The use of polymer composite materials in the manufacture and repair of machines: textbook. allowance / N.I. Baurova, V.A. Zorin. - M .: MADI, 2016 .-- 264 p.
[2] Petrova, A.P. Glues, adhesive binders and adhesive prepregs: a training manual / A.P. Petrova, G.V. Malysheva / under the total. ed. E.N. Kablova. - M .: VIAM. 2017 .-- 472 p.
[3] Polymer composite materials / S.L. Bazhenov [et al.] - Dolgoprudny: Intellect Publishing House, 2010. - 352 p.
[4] Nelyub, V.A. Electrical properties of carbon plastics and the method of their regulation / V.A. Nelyub, A.A. Berlin // *Chemical industry today.* 2019. No. 1. pp. 38-43
[5] Nelyub, V.A. An Impact Assessment of Metallization of a Carbon Tape Surface on the Mechanical Characteristics of the Filaments // *Polymer Science – Series D.* – 2018. – Vol. 11. – № 4. – pp. 423-425.
[6] Nelyub, V.A. Adhesive-strength evaluation via the pull-out method in a binder-elementary-filament system at various treatments of filaments // *Polymer Science – Series D.* – 2018. – Vol. 11. – №3. – pp. 263-266.
[7] Baurova N.I., Zorin V.A., Prikhodko V.M. Taking into account the factors of technological hereduty by techniligical mechanics // *Polymer Science-Series D.* – 2016. – 9(4). – pp. 402-406.

[8] Baurova N.I. Microstructural investigations of surfaces of destruction of carbon plastic // *Polymer Science-Series D*. – 2013. – 6(3). – pp. 246-249.
[9] Baurova N.I. Durability models of adhesive material using catastrophe theory // *Polymer Science-Series D*. – 2009. – 2(2). – pp. 130-132.
[10] Terekhina S.M., Malysheva G.V., Bulanov I.M., Tarasova T.V. Investigation of tribological properties of polymer composite materials based on bismaleimide binder // *Polymer Science-Series D*. – 2011. – 4(2). – pp. 136-137.
[11] Malysheva G.V., Shimina Y.Y. Influence of preparation technology on carbon fiber composite component composition // *Fibre Chemistry*. – 2014. – 46(4). – pp. 237-240.
[12] Nelyub V.A., Borodulin A.S., Kobers L.P., Malysheva G.V. Viscous hysteresis in filled siloxane binders // *Polymer Science – Series D*. – 2017. – 10(1). – pp. 19-22.
[13] Borodulin A.S. Plasticizers for epoxy adhesives and binders // *Polymer Science – Series D*. – 2013. – 6(1). – pp. 59-62.
[14] Borodulin A.S. Polyester resin for production of goods from polymer composite materials by pressing methods // *Polymer Science - Series D*. 2013. – 6(3). – pp.269-270.
[15] Kobets L.P., Malysheva G.V., Borodulin A.S. Rheological properties of system "elastomer binder-discrete carbon fiber" // *Inorganic Materials: Applied research*. 2016. – 7(1). – pp. 15-19.

IOP Conf. Series: Materials Science and Engineering **934** (2020) 012007 doi:10.1088/1757-899X/934/1/012007

Selection of the best available technology based on the Analytic Hierarchy Process

A Brom, M Stoyanova and I Sidelnikov
Bauman Moscow State Technical University, Moscow, Russia Federation

E-mail: allabrom@bmstu.ru, mstoyanova@bmstu.ru, sid@bmstu.ru

Abstract. The article investigates the problem of choosing the best available technology that must match the concept of sustainable production. When choosing a technology, the difficult task of comparing alternative technological options should be solved. It is necessary to assess the indicators of environmental, economic and social criteria, but which often contradict each other. The article presents an approach to select the best technology for the production of metal grid, composite grid and basalt grid with the use of Analytical Hierarchic Process.

1. Introduction

Modern society is developing in the context of the growing crisis of the relationship between man and the environment. In recent years, this has been demonstrated by a noticeable increase in the number of man-made and socially caused disasters, various crisis situations. The global deterioration of the ecological situation poses the problem of transition to waste-free or low-waste production technology for humanity [3].

All enterprises must comply with the concept of sustainable development, that is, use environmentally friendly technologies and minimize waste [1]. EU states have signed a number of international conventions and agreements, according to which they are obliged to reduce both the existing and the potential negative impact of economic activity on the environment, which can be achieved by introducing the best available technologies (BAT). The best available technology is the technology of production of products (goods), performance of work, rendering of services, determined on the basis of modern achievements of science and technology and the best combination of criteria for achieving the goals of environmental protection subject to the availability of the technical possibility of its application. In this case, all BAT must meet the concept of resource-saving sustainable production [4]. Significant contribution to informing all Russian stakeholders about BAT, EU reference books on BAT and methodological approaches to determining BAT was made by experts of the EU Harmonization of Environmental Standards II, Russian Federation project (EuropeAid / 123157 / C / SER / RU identification number) implemented in Of the Russian Federation in 2007-2009 [2].

When choosing BAT, the difficult task of comparing alternative technological options should be solved, in which the criteria of sustainable production - environmental, economic and social - should be taken into account [7]. The problem of solving such a problem is that criteria often contradict each other - for example, an economic criterion, and in particular, minimizing production costs in most cases conflicts with environmental requirements. Therefore, the question arises of the selection of key criteria when choosing a technology for a particular production while observing the requirements of Sustainability.

2. Research Method

The article presents an approach to the selection of BAT in the production of nets on the example of three technologies: the production of metal grid, the production of composite grid and the production of basalt grid.

Comparing alternative technologies is a complex task, which should take into account numerous environmental, technological, economic and social criteria. When choosing, consideration should be given to the effects on various components of the environment; the choice can be made between emissions or discharges into different environments. The probable effects of various pollutants on the seven priority environmental problems (human toxicity, global warming, toxicity to water bodies, acidification, eutrophication, depletion of the ozone layer, the potential of formation of tropospheric ozone) should be taken into account. Each option is evaluated in terms of affordability from an economic point of view. However, in order to determine the best technology in specific practical conditions, criteria are needed, depending on subjective decisions. For example, when choosing between a car, train and bicycle (as a vehicle), time, convenience, cost and, most importantly, for the scope of this standard, minimizing damage to the environment can be criteria. If the only criterion is to prevent / minimize damage to the environment, the bike will be the best choice. Also, the bike is beneficial in terms of economic costs. However, if the travel time is limited, the best option would be a train or car. If take into account comfort when traveling, the first place will be a car - for those who do not like to travel in a large company, or the train - for those who like to read or sleep on the road. This simple example shows that different technologies will be the best for different users.

In almost all cases, at all stages of the BAT selection process, complex, poorly formalized tasks, described by ambiguous characteristics, arise; in order to prepare a conclusion, it is necessary to resort to expert judgment. The lack of scientifically based environmental decision-making methods makes this process more difficult. Officially accepted methodologies today successfully work only in the simplest situations and cannot give an unequivocal choice of alternative solutions in complex environmental and economic problems [9]. Due to the complexity, multi-criteria and multidimensionality of the problem, the development of ecological-economic-mathematical models and methods for evaluating alternative options and identifying the best technology is an important direction in this subject area [10].

In determining the best available technology, the following requirements are taken as a basis:

1. Application of low-waste technology;
2. Elimination of the use of toxic raw materials;
3. Stimulating the involvement in the economic turnover of discharges, emissions and waste generated in the course of economic activity;
4. Availability of comparable technological processes, production equipment or methods of operation, successfully tested at the industrial level;
5. Consideration of the time required to implement the best available technology;
6. The volume of consumption and efficiency of use of raw materials used in the process;
7. Prevent accidents and minimize their environmental consequences.

To select the basic requirements, we apply the AHP (Analytical Hierarchic Process - Hierarchical Analysis Process) method, which is necessary for decision-making by prioritization in the presence of multiple criteria and subcriteria [6, 8, 12]. This method allows the user to structure complex problems in a hierarchical order or to divide them into integrated levels. In any case, the hierarchy should consist of at least three levels: goal, criteria and alternatives. The priority of the technology will depend on the priority of the criterion in each specific case [5].

3. Results

The criteria were compared in pairs and a degree of importance was set for each.

Table 1. Scale of relative importance

Degree of importance	Numerical value
Identical	1
average	3
High	5
Extremely high	7
Emergency	9

Table 2. Criteria indices

Index	Criterion
D_1	Low waste technology
D_2	Elimination of the use of toxic raw materials
D_3	Stimulate waste involvement
D_4	Availability of comparable technologies
D_5	Implementation time
D_6	Consumption and efficiency of raw materials
D_7	Minimizing the environmental impact of accidents

Table 3. Matrix of pairwise comparisons of criteria with numerical values

Index	D_1	D_2	D_3	D_4	D_5	D_6	D_7	Amount
D_1	1,00	0,20	0,14	9,00	0,14	1,00	0,20	11,69
D_2	5,00	1,00	1,00	9,00	3,00	7,00	1,00	27,00
D_3	7,00	1,00	1,00	9,00	0,33	5,00	0,33	23,67
D_4	0,11	0,11	0,11	1,00	0,11	0,11	0,11	1,67
D_5	7,00	0,33	3,00	9,00	1,00	3,00	3,00	26,33
D_6	1,00	0,14	0,20	9,00	0,33	1,00	0,20	11,88
D_7	5,00	1,00	3,00	9,00	0,33	5,00	1,00	24,33

Table 4. Normalized matrix of pairwise comparisons of criteria with approximate priorities and weights

Index	D_1	D_2	D_3	D_4	D_5	D_6	D_7
D_1	0,09	0,02	0,01	0,77	0,01	0,09	0,02
D_2	0,19	0,04	0,04	0,33	0,11	0,26	0,04
D_3	0,30	0,04	0,04	0,38	0,01	0,21	0,01
D_4	0,07	0,07	0,07	0,60	0,07	0,07	0,07
D_5	0,27	0,01	0,11	0,34	0,04	0,11	0,11
D_6	0,08	0,01	0,02	0,76	0,03	0,08	0,02
D_7	0,21	0,04	0,12	0,37	0,01	0,21	0,04
Amount	0,17	0,03	0,06	0,51	0,04	0,15	0,04

Using the hierarchy method, the following criteria were highlighted:

1) Application of low-waste technology;

Waste - residues of substances and objects resulting from household or industrial activity of a person and not used on site.

Production wastes - remnants of raw materials, materials, semi-finished products, formed during the production of goods or performance of work and have lost all or part of their consumer properties, newly formed in the production process passing substances that are not used. Production waste includes enclosing and overburden rocks formed during the extraction of minerals, by-products and by-products, agricultural waste.

2) Availability of comparable technological processes, production equipment or methods of operation, successfully tested at the industrial level;

3) Consideration of the time required to implement the best available technology, which may include the following items:

- Conclusion of a supply contract with the manufacturer;
- Advance payment;
- Equipment manufacturing supplier;
- Payment as soon as the equipment is ready;
- Shipping and customs clearance, VAT payment;
- Installation and commissioning;
- Payment on the fact of the NDP;
- Conduct training in Germany;
- Conduct training for other employees in Russia;
- Selection of suppliers and conclusion of a supply contract;
- Production plan; Technology 1
- Determination of needs for raw materials and components in accordance with the PP;
- Purchase of raw materials and components/

4) The volume of consumption and efficiency of use of raw materials used in the process (Minimum consumed resources and Minimum amount of raw materials per m2 of finished product)

After a detailed review of each technology, the analyzed data is presented in the table:

Table 5. Technology comparison

BAT selection criteria	**Metallic**	**Composite**	**Basalt**
Waste	Dross, slag, trim, carbon dioxide	Carbon dioxide is 40 times less than metal	Carbon dioxide is released less than the metal and composite
Implementation time	7.5 months	7 months	6 months
Consumed resources	6 kW per 1 kg	2 kW per 1 kg	2 kW per 1 kg
The volume of raw materials per m2 output of finished products in monetary terms	≈0.85 dollar	≈0.69 dollar	≈0.62 dollar

Thus, the criteria for selecting BAT production of the construction grid will be: waste, implementation time, resources consumed, start-up costs, and also the volume of raw materials per 1 m2 of finished products. The alternatives are 3 technologies - the production of metal mesh, the production of composite mesh Rockmesh and the production of basalt mesh.

The first step is to build a hierarchical structure (refer with: Fig.1).

Thus, the criteria for selecting BAT production of the construction grid will be: waste, implementation time, resources consumed, start-up costs, and also the volume of raw materials per 1 m2 of finished products. The alternatives are 3 technologies - the production of metal mesh, the production of composite mesh Rockmesh and the production of basalt mesh.

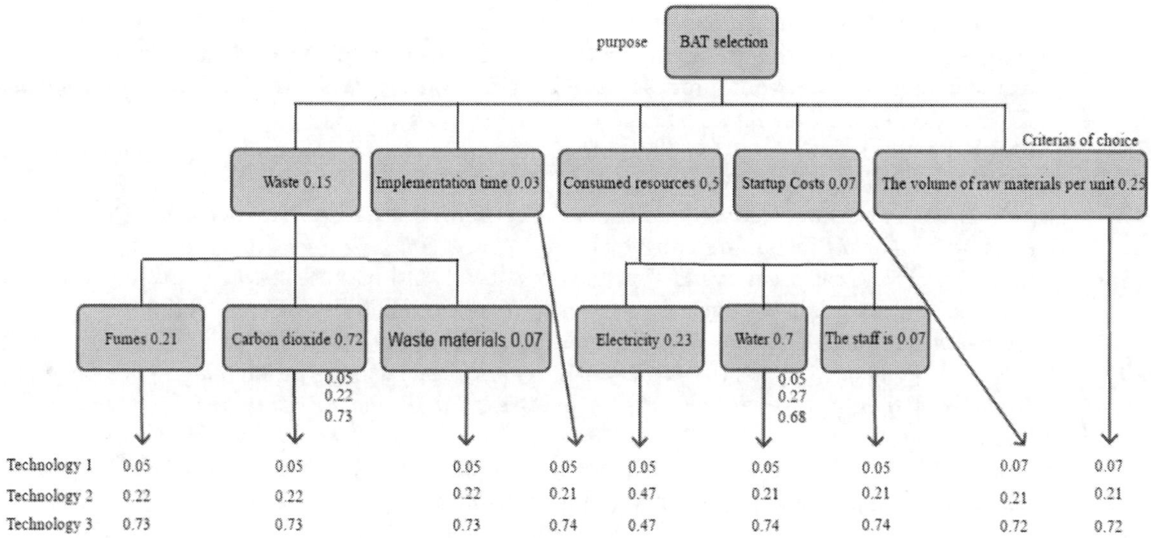

Figure 1. Hierarchical structure

Calculation of weight metal grid:

$$Technology\ 1. = (0{,}15 \cdot 0{,}05) + (0{,}03 \cdot 0{,}05) + (0{,}05 \cdot 0{,}5) + (0{,}07 \cdot 0{,}07) + (0{,}25 \cdot 0{,}07) = 0{,}056.$$

Calculation of weight composite grid:

$$Technology\ 2. = (0{,}15 \cdot 0{,}22) + (0{,}03 \cdot 0{,}21) + (0{,}5 \cdot 0{,}27) + (0{,}07 \cdot 0{,}21) + (0{,}25 \cdot 0{,}21) = 0{,}241.$$

Calculation of weight basalt grid:

$$Technology\ 3. = (0{,}15 \cdot 0{,}73) + (0{,}03 \cdot 0{,}74) + (0{,}5 \cdot 0{,}68) + (0{,}07 \cdot 0{,}72) + (0{,}25 \cdot 0{,}72) = 0{,}701.$$

During the calculations, the following results were obtained: metal mesh - 6%, composite mesh - 24%, basalt mesh - 70%.

Thus, it is clear that the basaltic net greatly surpasses its competitors and, when choosing BAT, the basalt net shows the best performance.

4. Conclusion

We also see that using the method of analyzing hierarchies you can get sufficiently clear information to select the best available technology.

The chosen technology for the production of building grids is the best according to the criteria used in the work, it is waste, time of implementation, consumed resources, start-up costs and the volume of raw materials to produce 1 kg of finished products.

References

[1] Borovskii V G and Ryzhikova T N 2015 Upgrading plants in the processing industry in Russia: Approaches to design *Studies on Russian Economic Development* **26-5** 470-475

[2] Ivlev V Y, Barkova E V, Ivleva M I and Buzskaya O M 2016 Environmental approach to the study of the modern stage of information society development: Research prospects *International Journal of Environmental and Science Education* **11-16** 9113-9124

[3] Krokhina J A, Vinogradova T, Grishnova E Y, Popova O V and Kutlin N G 2018 Environmental risk management system projecting of industrial enterprises *Ekoloji* **27-106** e106181 735-744

[4] Raharjo H, Hendry, Min Xie and Aarnout C. Brombacher 2009 On modeling dynamic priorities in the analytic hierarchy process using compositional data analysis *European Journal of Operational Research* **194-3** 834-846

IOP Conf. Series: Materials Science and Engineering **934** (2020) 012007 doi:10.1088/1757-899X/934/1/012007

[5] Saaty T 2008 Relative measurement and its generalization in decision making why pairwise comparisons are central in mathematics for the measurement of intangible factors the analytic hierarchy/network process *RACSAM-Revista de la Real Academia de Ciencias Exactas Fisicas y Naturales Serie A Matematicas* **102-2** 251-318
[6] Golińska P and Kawa A 2015 *Technology Management for Sustainable Production and Logistics* Berlin Springer-Verlag 65-93
[7] Saaty T L 2012 *Decision Making for Leaders: The Analytic Hierarchy Process for Decisions in a Complex World* Pittsburgh Pennsylvania
[8] Pozdnyakova E D, Maslennikova E V, Komshin A S and Orlova S R 2019 Measuring control of construction materials parameters in order to increase reliability of engineering objects *IOP Conference Series: Materials Science and Engineering* **489-1** 012008
[9] Vasiliev A S and Goncharov A A 2018 Some aspects of problematics in designing technological complexes *IOP Conference Series: Earth and Environmental Science* **194-6** 062033
[10] Directive 2008/1/EC of the European Parliament and of the Council of 15 January 2008 concerning integrated pollution prevention and control
[11] European Commission. Integrated Pollution Prevention and Control. Reference Document on Economics and Cross-Media Effects. July 2006
[12] Nelyb V A, Borodulin A S, Kobets L P and Malysheva G V 2016 Research of the binding structure formation mechanism depending on the surface micro-relief of the carbon fiber *Adhesives Sealants Technologies BMSTU* **2** 19-23

IOP Conf. Series: Materials Science and Engineering **934** (2020) 012008 doi:10.1088/1757-899X/934/1/012008

The influence of the use of composite materials on the rationing of material resources in mechanical engineering

I Sidelnikov, A Brom and I Omelchenko

Bauman Moscow State Technical University, Moscow, Russia Federation

E-mail: sid@bmstu.ru, allabrom@bmstu.ru, logistic@ibm.bmstu.ru

Abstract. The majority of machine-building enterprises, when planning material and technical supply (MTS), use various statistical data collected and systematized for previous periods that underlie the calculation of standards for resource consumption. Depending on the level of technical development of production, used technologies and changes in the design of products, factors are identified that determine the need for adjustments to consumption rates in a minimum time. The article analyzes structural and technological factors, substantiates their influence on the determination of various types of consumption rates during the formation of MTS.

1. Introduction

The active use and introduction of new technologies and materials in mechanical engineering in the last decade, on the one hand, confirms the effectiveness of their use in industry, but, on the other hand, leads to serious changes in the design of products and existing specifications for standardizing material resources in accordance with technological maps processes [1-6].

Due to significant design and technological changes, it is very difficult at factories to adjust the directories and normative documents on the consumption of material resources every year. As a result, situations arise that have a negative impact on the efficiency of an enterprise's production activity — excess stock or a deficit of required types of resources, which is unacceptable in machine-building production, since simple equipment often arises.

2. Research Method

Based on the analysis of statistical information, the enterprise management makes the following decisions regarding various aspects of production activities:

production aspect:

- formation of requirements for the quality of products based on the analysis of similar types of products from competitors;
- determination of the nomenclature of production;

technological aspect:

- introduction of new technologies and industrial equipment;
- implementation of effective production and MTS system management, adjustment of material resource expenditure standards and development of new approaches to assessing consumption norms and standards, optimizing the supply plan by taking into account the overlapping needs of services and departments, differentiating the supply

plan by the delivery dates of certain nomenclature groups of material resources (organizational character).

Due to the tightening of environmental requirements, requirements for material consumption, energy and resource consumption for technical systems, enterprises are forced to resort to changes in design, constantly refining certain blocks and assemblies [8].

Moreover, when planning the MTS, it is necessary to take into account the return logistics flows associated with the return of the defect and with the processing of parts accepted for recycling [7].

Changing the design of the product in order to reduce material consumption lead to a change in the rate of consumption of material on the part.

Often, engineers are faced with the task of upgrading the product in order to endow it with new physical properties. The most effective way to solve this problem is to replace the workpiece material from which the product is manufactured. These operations lead to a change in individual consumption rates for two positions.

Technical products are also characterized by the development of modifications for various purposes based on the base product.

The refinement of the design in order to increase reliability by narrowing the functionality of the product is characteristic of special-purpose products. In this situation, the need for materials and components is reduced. An extension of functionality, on the contrary, leads to an increase in resource requirements. To avoid both of these situations, as well as to minimize cases of excess stock of resources, it is necessary to carry out the adjustment of material consumption in regulatory documents.

3. Results

Technological changes include [9]:

- replacement of the production site;

In connection with changes in the assembly sequence, the assembly principle (for example, replacing a hydraulic circuit with an electric one), energy consumption, and a number of other characteristics, a production workshop is being replaced.

- change in production technology.

In connection with the advent of new resource-saving technologies, this technological change becomes the most relevant. As a result of the introduction of new technologies, individual consumption rates and consolidated material sizes are significantly reduced [10; 11].

For the correct use of regulatory costs, it is necessary to consider the elements of the standardization process in the areas of calculation: work in progress and finished products, physical parameters of standardization, and additional work.

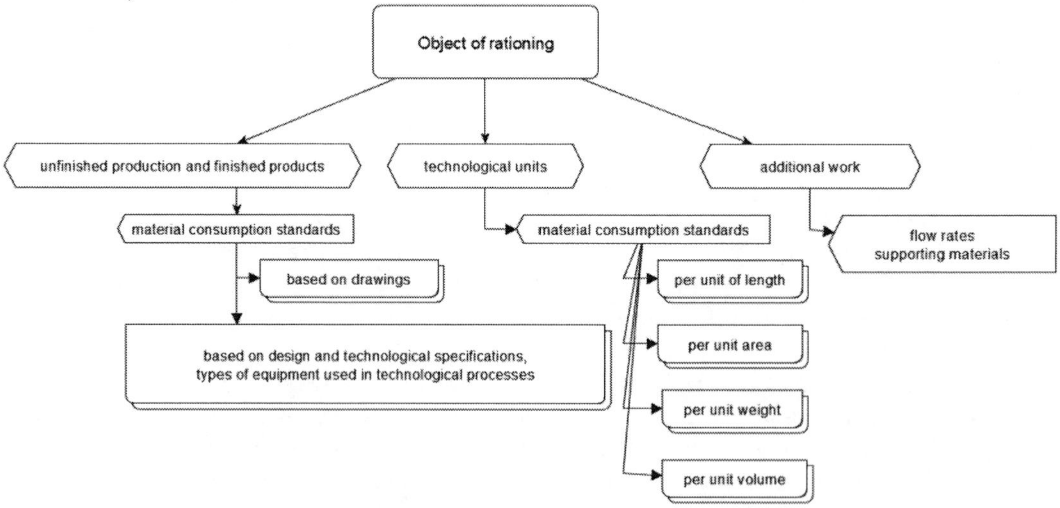

Figure 1. Rationing process

Moreover, material resources, the consumption of which is normalized, have their own classification, which is based on the types of resources used in figure 2.

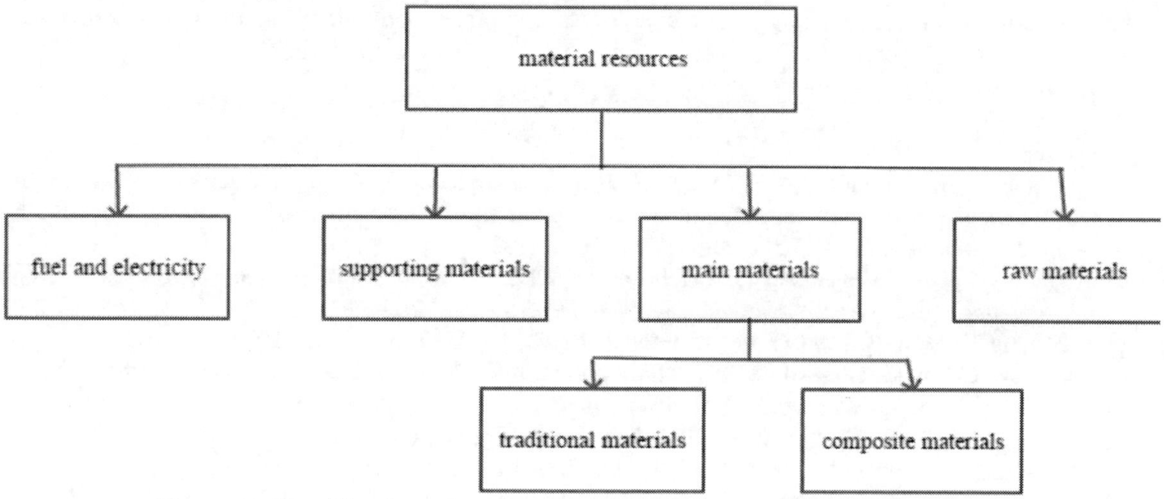

Figure 2. Classification of consumption rates by type of material resources

Rational rationing is a big problem in mechanical engineering, since there are four levels of detailing the norms of resource consumption, presented in Figure 3.

Individual consumption rates are at the base of rationing and planning of material and technical supply. At this planning level at each enterprise, depending on production technologies and available equipment, the unit of regulation is such units as the part, unit, product in units of mass, volume, etc. The generated and systematized data at this level form the basis for the planning of material and technical supply.

The specified consumption rate includes planning not only material resources for each type of output (in terms of unit), but also basic and auxiliary materials, fuel and energy for technological processes for each type of product (parts, products, etc.)

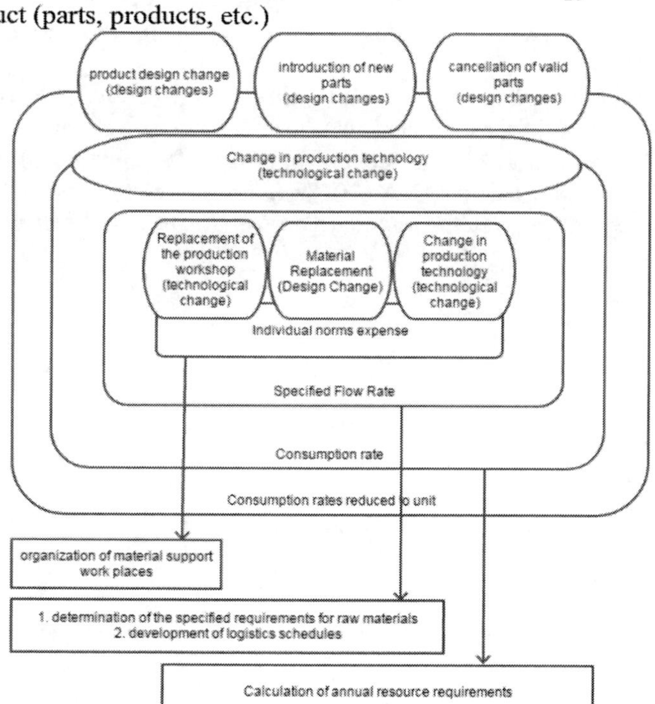

Figure 3. Levels of detail for flow rates

4. Conclusion

Given the above, it is worth noting that today it is impossible to organize an effective MTO system without taking into account design and technological factors. To optimize the supply of resources and ensure the smooth operation of production capacities, it is necessary to timely update the regulatory documents governing consumption rates.

References

[1] Nelyb V A, Borodulin A S, Kobets L P and Malysheva G V 2016 Research of the binding structure formation mechanism depending on the surface micro-relief of the carbon fiber *Adhesives Sealants Technologies BMSTU* **2** 19-23

[2] Baurova N I, Zorin V A and Prikhodko V M 2016 Manifestation of a synergistic effect in technological heredity *Polymer Science Series D* **9-2** 209-211.

[3] Maung P P, Malysheva G V and Gusev S A 2016 A study of the effect of network angle of fabrics on kinetics of impregnation upon molding of articles made from carbon plastics *Polymer Science* Series D **9-4** 407-410.

[4] Yangyang C and Malysheva G 2019 Method for Determining the Rational Regimes of Curing Products from Polymer Composite Materials *Materials Today: Proceedings* **11** 128-133.

[5] Reznik S V, Timoshenko V P, Prosuntsov P V and Mial' L V 2014 Theoretical Principles of Determining the Longitudinal Thermal Conductivity of Thin-Walled Structural Elements from Composite Materials *Eng. Phys. Thermophy* **87-4** 864–870.

[6] Borodulin A S, Marycheva A N and Malysheva G V 2015 Simulation of impregnation kinetics of fabric fillers in the production of fiberglass articles *Glass Physics and Chemistry* **41-6** 660-664.

[7] Borovskii V G and Ryzhikova T N 2015 Upgrading plants in the processing industry in Russia: Approaches to design *Studies on Russian Economic Development* **26-5** 470-475

[8] Krokhina J A, Vinogradova T, Grishnova E Y, Popova O V and Kutlin N G 2018 Environmental risk management system projecting of industrial enterprises *Ekoloji* **27-106** e106181 735-744

[9] Golińska P and Kawa A 2015 *Technology Management for Sustainable Production and Logistics* Berlin Springer-Verlag 65-93

[10] Pozdnyakova E D, Maslennikova E V, Komshin A S and Orlova S R 2019 Measuring control of construction materials parameters in order to increase reliability of engineering objects *IOP Conference Series: Materials Science and Engineering* **489-1** 012008

[11] Vasiliev A S and Goncharov A A 2018 Some aspects of problematics in designing technological complexes *IOP Conference Series: Earth and Environmental Science* **194-6** 062033

Aromatic polyethersulphones with improved performance characteristics

R Ch Bazheva [1], A.M. Kharaev[1], A S Borodulin and A N Kalinnikov[2]

[1]Kabardino-Balkaria State University named after Kh.M. Berbekov, Nalchik, Chernyshevsky street, 173, Russian

[2]Moscow State Technical University named after N.E. Bauman, Moscow, 2nd Bauman, 5/3, Rusian

Abstract. Synthesized modified aromatic polyethersulfones containing dichloroethylene and arylate groups in the main chain. The kinetics of the synthesis of polyethersulfones by the method of acceptor-catalytic polycondensation was studied. It is shown that the property of the obtained polymers depends on the ratio of the starting monomers. It is shown that the property of the obtained polymers depends on the ratio of the starting monomers. The presence of two chlorine atoms in the initial monomer in combination with an unsaturated bond provides the polyethersulfones based on it with high fire resistance and good physical and chemical properties, including frost resistance.

Keywords: oligosulfone, polyethersulfone, bisphenol A, 1,1-dichloro-2,2-di (4-hydroxyphenyl) ethylene, dichloroethylene group, acceptor-catalytic polycondensation.

1. Introduction

The rapid development of technology requires the creation of new polymers with a predetermined set of valuable properties and working under the influence of various external conditions. One of these polymers are aromatic polysulfones and polyethersulfones, which have a range of valuable performance properties. At the same time, the cost and energy costs of their synthesis are relatively low [1-9].

However, polysulfones are not without drawbacks, one of which is a partial or complete loss of performance at low temperatures, that is, limited frost resistance.

Frost resistance of polymers is the ability of materials to maintain their performance properties at low (negative) temperatures.

The problem of frost resistance is relevant for almost all areas of application of polymeric materials: household, construction and utilities, in the transport sectors (automotive, shipbuilding, space and aviation industries). The chemical structure, as well as the physical state of the polymer material, can significantly affect the frost resistance of materials. The possibility of using polymer materials well proven in a temperate climate in the North is a topical issue in modern materials science.

The main ways to increase the frost resistance of amorphous polymers is plasticization by low molecular weight and polymer plasticizers, and crystalline ones by copolymerization and modification, that is, by changing the properties of polymers by chemically reacting them with low molecular weight modifiers [8-15].

The purpose of this work is to produce polyethersulfones (PES) with a long-term operation temperature of 2000C and higher, with high fire resistance and frost resistance, and are resistant to UV radiation.

Numerous studies have shown that polymers containing dichloroethylene groups, in addition to fire resistance, have high mechanical and thermal properties [16-21].

2. Experimental

X-ray diffraction studies were carried out on powder samples using a DRON-6 diffractometer in copper K-radiation with a graphite monochromator on a diffracted beam.

The IR spectra of the materials were measured on a SPECTRUM TWO FTIR spectrometer with a computer control system at a frequency domain of 400-5000 cm^{-1}.

Mechanical properties were investigated under uniaxial tension conditions on a GT-2000 testing machine at a temperature of 20°C, a relative stretching rate of 0,017 s^{-1} and various strain rates.

Phase transitions were studied on a Perkin Elmer DSK 4000 differential scanning calorimeter at a heating rate of 5 deg / min in air.

The thermal stability of the samples was studied by the method of thermogravimetric analysis using a Perkin Elmer TGA 4000 derivatograph in an atmosphere of air at a heating rate of 5 degrees min^{-1} from 30 to 700^0 ° C.

Ubbelohde viscometer was used to determine the intrinsic viscosity. Solutions with relative viscosity = 1.1 were diluted for subsequent measurements of viscosity in such a way that the intervals of the obtained concentrations were approximately the same, and the η of the most diluted solution should not be less than 1.1. These limits of measurement specific viscosity η = 1.1-1.5 correspond to concentrations of polymer solutions for which the dependence of [η] on the velocity gradient of fluid outflow from the capillary is not observed and with dilution there is no deviation from the linear dependence of ηsp / C or ln ηrel. / C from S.

3. Results and discussions

Modified polyethersulfones synthesized according to the following scheme:

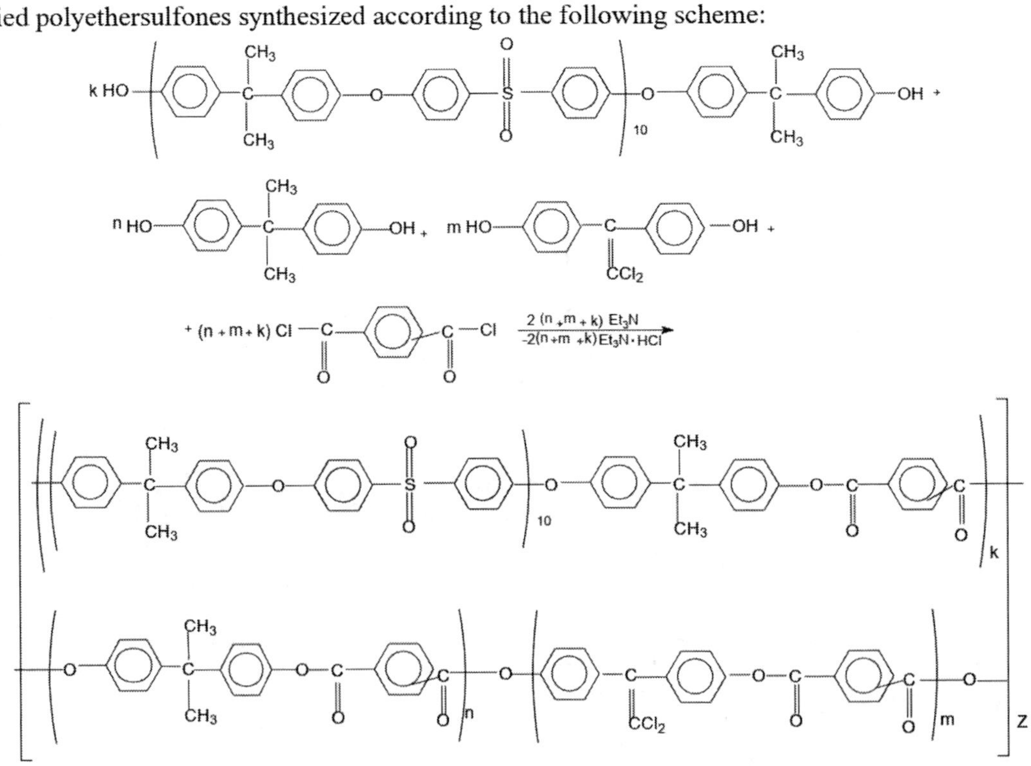

Oligosulfone was synthesized by high temperature polycondensation in dimethyl sulfoxide [8, 14]. Polymers were synthesized by acceptor-catalytic polycondensation in 1,2-dichloroethane. The ratio of bisphenols varied from 0 to 100 molar percent. The amount of oligosulfone with a degree of polycondensation of n = 10 (OASO-10D) was taken constant and amounted to 10 wt.% Of the total number of bisphenols.

The structure of the obtained copolyesters was confirmed by IR spectroscopy, an X-ray phase analysis was carried out, and polydispersity was studied.

Polyesters were obtained with high viscosity indices (0.90-1.25 dl/g) and quantitative yield. The structure of the obtained copolyesters is confirmed by IR spectroscopy and elemental analysis. The presence on the spectra of the corresponding absorption bands and the absence of bands for hydroxyl groups confirms the preparation of polymers of this structure. The IR spectra of the polymers contain absorption bands corresponding to: ester bonds (1200-1250, 1735, 1740 cm^{-1}); the isopropylidene group in the remainder of bisphenol A (1290-1365, 1385, 1415-1465, 2930-2980 cm^{-1}), the group> C = CCl_2 (980 cm^{-1}), sulfonyl groups of oligosulfones (560-570, 1150-1170, 1290, 1300, 1320 cm^{-1}), which indicates the formation of polymers of a given structure. X-ray diffraction analysis of film samples of polymers obtained by irrigation from solution showed that they have an amorphous structure. The studies were carried out using a DRON-6 diffractometer in the step scan mode with a wavelength of 1.54051 Å.

The study of the polydispersity of the copolymers showed that the coagulation threshold of polyesters based on bisphenol 1,1-dichloro-2,2-di(4-hydroxyphenyl)ethylene and bisphenol A increases with an increase in the content of the chlorine-containing component, due to the polarity of > C = CCl_2-groups, and, therefore, the better solubility of these polymers.

The presence on the differential curves of turbidimetric titration of copolymers of one maximum confirms the formation of copolymers, rather than a mixture of homopolymers (Fig. 1). The polydispersity of the copolymers was determined by turbidimetric titration on a FEC 56M photocalorimeter. As solvent, 1,2-dichloroethane was used, and precipitant was isopropyl alcohol.

On differential titration curves constructed from integral ones by graphic differentiation, there is only one maximum.

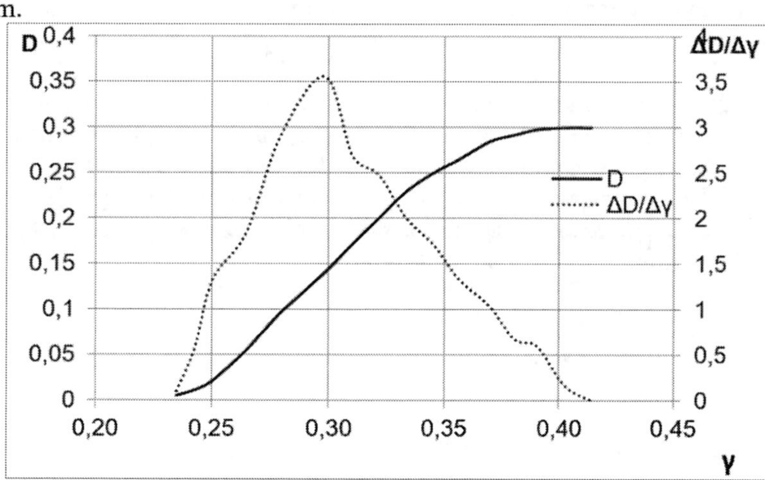

Fig.1. Integral and differential curves of turbidimetric titration of polyethersulfones based on equimolar amounts of starting monomers

This method is based on measuring the optical density of a medium using a photoelectrocolorimeter and is a very convenient and fast method for distinguishing homopolymers and their mixtures from copolymers of various structures (regular and irregular structures, block and familiar copolymers). In addition, the solubility of the polymers can be judged by the deposition curves.

All obtained copolymers are readily soluble in a number of organic solvents: dichloroethane, tetrachloroethane, tetrahydrofurane, chloroforme, methylene chloride, etc. The molecular weight of the copolyesters is in the range of 40,000-70000.

The results of thermomechanical tests of the synthesized copolymers based on bisphenol A bisphenol 1,1-dichloro-2,2-di(4-hydroxyphenyl) ethylene showed that the latter have relatively low thermomechanical properties. Depending on the ratio of the starting bisphenols in the BSP series, the glass transition temperature changes only by 100 ° C. With an increase in the proportion of bisphenol 1,1-dichloro-2,2-di(4-hydroxyphenyl) ethylene, the pour point changes from 250^0 ° C to 350^0 ° C (table 1).

In polymers where the content of sulfonic O = S = O groups prevails over ester arylate –C (O) –O– groups, the heat resistance is higher. The increase in heat resistance in this case is due to the fact that polycondensation aromatic polymers are characterized by the occurrence of branching and crosslinking processes in the region of relatively low degradation temperatures (250-400 ° C), which ultimately leads to the formation of a three-dimensional structure and a change in the operational characteristics of the polymers.

Table 1. Some properties of polyethersulfones

2,2 bis-(4,4′-oxyphenyl)propane, %	1,1-dichloro-2,2-di(4-hydroxyphenyl) ethylene, %	η,. dl / g	T_g, ^{0}C	σ_t, MPa	ε_t, %	A, kJ / m^2 Izod (notched)
100	0	1,20	207	93	14	8,0
75	25	1,25	210	97	20	8,5
50	50	1,06	212	124	25	8,5
25	75	1,02	215	113	20	9,0
0	100	0,95	217	87	15	9,5

4. Conclusion

The possibility of obtaining high molecular weight polyethersulfones of block structure using various polycondensation methods: high temperature and acceptor-catalytic was studied. By the methods of IR spectroscopy, elemental, X-ray phase analysis, the formation of polymers of a given structure is proved. Among copolymers, the highest values of the deformation-strength characteristics are exhibited by a copolymer based on equimolar amounts of the starting bisphenols. A further increase in the content of this does not lead to an increase in tensile strength (σ) and elongation (ε), which is most likely associated with a decrease in the reduced viscosity, and, consequently, the molecular weight of these polymers. Studies have shown that synthesized after exposure for 48 hours at a temperature of -50^0 ° C, retain their basic physical and mechanical characteristics. The complex of physico-chemical characteristics allows us to recommend these polyethersulfones as fire-resistant, heat- and frost-resistant structural and film materials.

Acknowledgement
The reported study was funded by RFBR according to the research project № 18-29-18063.

References

[1] Borodulin A S, Kalinnikov A N , Bazheva R C, Kharaev A M and Beshtoev B Z 2018. Receipt and investigation of performance characteristics of super constructions polyesters. *International Journal of Mechanical Engineering and Technology (IJMET)* **9-13** 1117–1127,Article ID: IJMET_09_13_117.
[2] Borodulin A S, Kalinnikov A N , Bazheva R C and Beshtoev B Z 2018. Synthesis and properties of Aromatic Polyethersulfones. *International Journal of Mechanical Engineering and Technology (IJMET)* **13** 1109–1116, Article ID: IJMET_09_13_116.
[3] Kharaev A M , Mikitaev A K, Bazheva R Ch, Khasbulatova Z S, Kharaeva R A, Chumachenko T N and Istepanov M I 2008 Modified aromatic copolyesters. *International Polymer Science and Technology*, **12** 17–20. ISSN: 0544-2901.

[4] Mikitaev A K, Rusanov A L, Koumykov R M 2013.The novel polynaphthylimides of improved Solubility on the basis of derivatives from chloral and dichlordiphenyltrichlorethane //*Characterization and Development of Novel Materials Research Compendium*. 186-192.

[4] Ozden S, Kharaev A M and Bazheva R C 2009 Synthesis and Modification of Aromatic Polyesters With Chloroacetyl 3,5-dibromo-p-hydroxybenzoic Acid. *J. of Applied Polymer Science*, **111-4** 1755–1762. DOI: https://doi.org/10.1002/app.29147.

[5] Baranov A B, Peksimov O Ye, Prudskova T N, Andreyeva T I, Yemel'yanov I D and Shembel N L 2016, Study of the technological characteristics of materials based on polysulfone. *Thin chemical technologies*, **11-5** 87–90

[6] Zenitova L A and Shteynberg Ye M 2012, Polysulfone as a functional polymer material and its production. *International Research Journal*, **6-6** 23–28. DOI: 10.18454/IRJ.2227-6017.

[7] Kumykov R M and Vologirov A K 2018. New aromatic dinitro-derivatives of chloral as monomers for the synthesis of polyesters and polyheteroarylenes // *News of higher educational institutions. Series: Chemistry and Chemical Technology*. **61-20** 4-14.

[8] Patent № 2382756 (RF). Kharaev A M, Bazheva R Ch, Kharaeva R A, Khasbulatova Z I and Istepanov M.I. Chlorine-containing aromatic oligoethers. Publ. 02.27.2010. Bul 6.

[9] Shteynberg Ye M, Sergeyeva Ye A, Zenitova L A and Abdullin I Sh 2012, The use and production of polysulfone. Overview. *Bulletin of Kazan National Research Technological University,* **15-20** 168–171.

[10] Kharaev A M, Shaov A Kh, Bazheva R Ch 2013.The synthesis and stabilization of polymers. Monograph /Saarbrücken, – 300 p.

[11] Kharayev A M, Bazheva R C and Chayka A A 2006 Aromatic block-copolyethers as prospective heat resistant constructive materials: *Polymers, Polymer Blends, Polymer Composites and Filled Polymers: Synthesis, Properties and Applications* 115-120.

[12] Barokova E B, Bazheva R Ch and Haraev A M 2010 Oligosulphones on the basis of 1,1-Dichlor-2,2-di(4-oxyphenyl)-Ethylene and 4,4'-dichlordiphenyl-sulphone obtained by high-temperature polycondensation. //*J. of the Tribological Association*. **16-2** 284-287.

[13] Kharaev A M, Bazheva R C and Barokova E B 2008 Fireproof aromatic block copolymer resin on the basis of 1,1-dichlor-2,2 di(n-oxyphenyl) ethylene. *Modern Tendencies in Organic and Bioorganic Chemistry: Today and Tomorrow (New York)*. 219-222.

[14] Barokova E B, Haraev A M, Bazheva R C and Haraeva R A 2010 Oligosulfons on the basis of 1,1-dichlor-2,2- di(4-oxyphenyl)ethylene and 4,4'-dichlorodiphenylsulfone obtained by high temperature polycondensation. *Handbook of Condensed Phase Chemistry*. 233-236.

[15] Nelyub V A 2016 A study of the microstructure of dressed glass fibers, *Polymer Science - Series D* **9** 96-100

[16] Nelyub V A 2015 Determination of adhesion interaction between carbon fiber and epoxy binder,*Polymer Science - Series D* **8-1** 6-8.

[17] Nelyub V A, Borodulin A S, Kobets L P and Malysheva G V 2016 Capillary hydrodynamic of oligomer binder, *Polymer Science – Series D* **9-3** 322-325

[18] Nelyub V A, Borodulin A S, Kobets L P and Malysheva G V 2016 A study of structure formation in a binder depending on the surface microrelief of carbon fiber, *Polymer Science – Series D* **9-3** 286-289.

[19] Borodulin A S and Kalinnikov A N, Bazheva R C 2018 Receipt and investigation of performance characteristics of super constructions polyesters, *International Journal of Mechanical Engineering and Technology* **9-13** 1117-1127.

[20] Borodulin A S, Kalinnikov A N and Bazheva R C 2018 Synthesis and properties of aromatic polyethersulfones, IJMET **9** 1109–1116.

[21] Borodulin A S, Kalinnikov A N and Kharaev A M 2018 New Polymeric Binders for the Production of Composit, *Materials today:proceedings* **11** 1107–1111.

IOP Conf. Series: Materials Science and Engineering **934** (2020) 012010 doi:10.1088/1757-899X/934/1/012010

Polyetherimides for creation of heat-resistant polymeric composite materials with high physical and mechanical properties

A S Borodulin[1], A N Kalinnikov[1], A G Tereshkov[1] and S S Muzyka[1]

[1] Intersectoral engineering center "Composites of Russia"MSTU. N. E. Bauman, Moscow state technical University named after N. E. Bauman (national research University). Russia, 105005, Moscow, 2nd Bauman, house 5.

E-mail: atereshkov@emtc.ru

Abstract. Thermoplastic and soluble in organic solvents polyetherimides are of considerable interest for the creation of heat-and heat-resistant composite materials with high physical and mechanical properties. In recent decades, the main scientific approaches to the synthesis of polyimides with the ability to process from melt and solutions in organic solvents have been developed. This methodology is to reduce the glass transition temperature of rigid-chain polyimides by increasing the flexibility of the main polymer chain and / or by reducing intermolecular interaction. Evolution of fine organic synthesis is one of the priority directions of chemical technology development. Chemistry of high-molecular compounds is the subject of research by various scientists, both in the domestic scientific field and abroad. This interest is due to the prospects for the use of materials for industrial purposes. The polyetherimides are of this group of materials contain in their chain of various aromatic, heterocyclic and cycloaliphatic groups, and systems. This article describes the main approaches to the creation of such polymers with improved, compared with polyimides, technological properties and prospects for their application for the production of filled composite materials based on them. The paper presents the properties of the materials under consideration as a justification for the creation of heat-resistant composite materials.

Key words: polyimides, and polyetherimides, glass transition temperature.

1. Introduction

The Russian industry at present has no production capabilities for superstructural polyether materials based on polyetherketones, polyethersulfones, polyaryletherketones and polyetherimides. Critically important for aerospace and electronics products, these materials have to be sourced abroad at prices tens of times higher than the cost of commonly used structural materials [1, 2].

Work on Russian technology for the production of superstructural polymers has gained up pace recently [3, 4, 5, 6, 7, 8, 9, 10, 11, 12, 13, 14, 15]. However we believe that materials such as polyetherimides hardly get the attention they deserve. Thus it would be appropriate to assess the state of the global technology for their synthesis as well as relationships between the structure/composition of the polymer molecule and particular physical/mechanical properties of final materials.

IOP Conf. Series: Materials Science and Engineering **934** (2020) 012010 doi:10.1088/1757-899X/934/1/012010

2. Methodology

We study polyimide-based composites including LaRC-TPI and Aurum, Kapton and Upilex-RN polymers, Ultem series polyetherimides, Aurum polyetherimide, LaRC-TPI polyetherketone (NASA), Upilex-RN polyetherimide, Kapton polyetherimide, Upilex-S polyimide (Ube Industries).

3. Main Discussion

Polyimides are a well-known class of engineering plastics featuring a highly successful combination of several application-relevant properties such as high heat and thermal stability, tensile strength and elastic modulus, good dielectric properties, high chemical and radiation resistance [16,17,18]. Yet, with these advantages of polyimides comes a setback of poor processability: their highly rigid molecular chains and strong intermolecular interaction virtually preclude processing from molten or dissolved form. For that reason the choice of manufacturing processes for polyimide products is limited to a low-performance hot-stamping technology or a two-stage process whereby an acidic polyimide solution is first produced and the polymer is then cyclized into polyimide as solvent evaporates.

Recent decades have seen the development of basic scientific approaches to polyimide synthesis with a capability for processing from melted form or solutions based on organic solvents. This methodology involves lowering of the glass-transition temperature (TGT) of hard-domain polyimides by increasing the flexibility of the main polymer chain and/or reducing intermolecular interaction. The following synthesis methods are used to that end:

1. Introducing chemical groups with larger rotational degree of freedom into the main chain, such as –O–, –S–, –SO2–, –CO–, –C(CH3)2–,–C(CF3)2
2. Introducing asymmetrical structures into the main chain using meta- or ortho- bonds in aromatic groups
3. Introducing bulky groups such as –C6H5, -CH3 and –CF3 into main aromatic rings to reduce cross-molecular packing.

Input substances must be synthesized first in order for approaches stated above to become feasible. With that in mind, synthesizing monomers containing –O– ether groups between aromatic groupings [19] proved to be the most convenient from practical standpoint. Table 1 contrasts glass transition and melting temperatures of polyimides and polyetherimides synthesized using the method described above.

Table 1: Glass transition and melting temperatures for polyimides (PI) and polyetherimides (PEI)

No.	PI/PEI brand name	TGT (°C)	Tmelt (°C)
1	Upilex-S [14] a)	>500 g)	–
2	Kapton [14-16] b)	360–410	–
3	Upilex-RN [17, 14, 18] c)	285	–
4	LaRC-TPI [19] d)	250–260	330–350
5	Aurum [20, 21] e)	250	388
6	Ultem [11, 12, 21] f)	217	–

Where: a, b, c, d, e – see the text for polymer descriptions

g) – glass transition temperature exceeds polymer decomposition temperature.

The Upilex-S polymer (manufactured by Ube Industries) features the following repeat unit structure [20]:

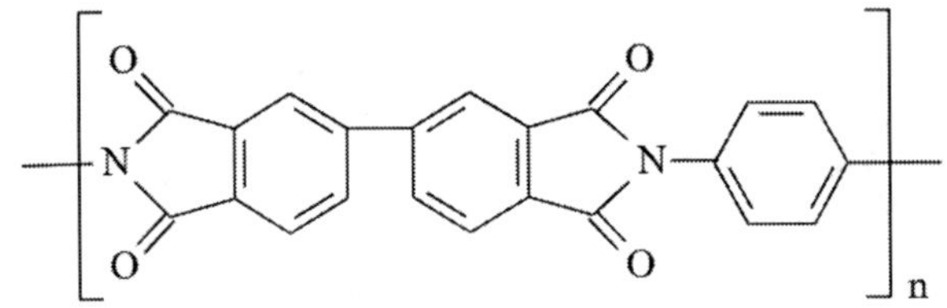

Fig. 1: Repeat unit structure for the Upilex-S polymer

Kapton, a polyetherimide, is produced from pyromellitic acid dianhydride and diaminodiphenyl oxide [21, 22]. This polymer turns out to have a rigid structure as its repeat unit contains a single ether bridge bond making the polymer insoluble in organic solvents and impossible to process from molten form. It is only processed at the polyamide acid formation stage by impregnating fabrics or casting films from a reaction solution in N-methylpirrolidone or diacetylamide, followed by cyclization with incremental heating from 200°C to 270°C. The reaction equation is shown below:

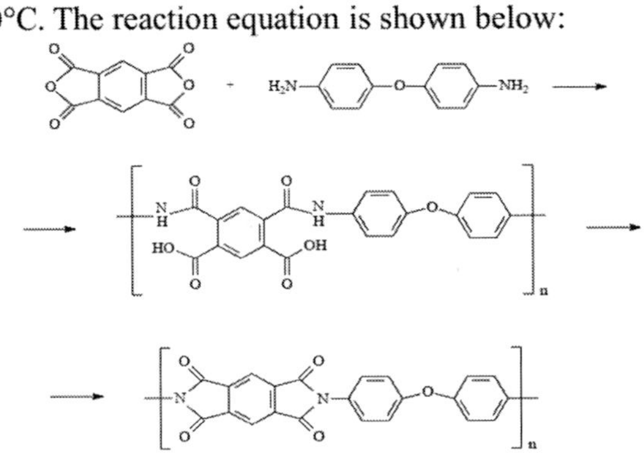

Fig. 2: Kapton (Du Pont) Upilex-RN polyetherimide with the repeat unit structure depicted below [23, 24]:

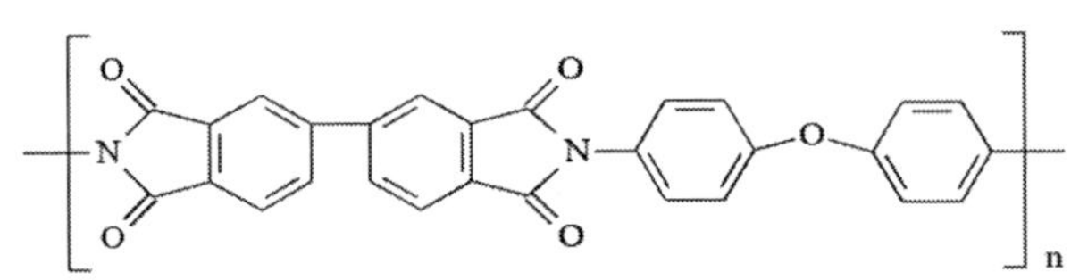

Fig. 3: Upilex-RN (Ube Industries)

Similar to Kapton, this polymer is disadvantaged by a rather inflexible chain and has to be processed via polyamide acid solutions.
LaRC-TPI, a polyimide ketone developed by NASA, features the following repeat unit structure [25]:

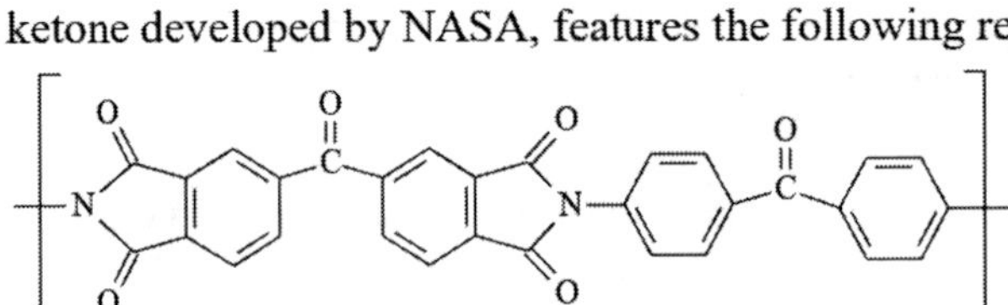

Fig. 4: LaRC-TPI, a polyimide ketone by NASA

Thanks to two pivot groups (C=O) in its repeat unit, this polymer can be softened and processed in molten form under high pressure. It is recommended for application in the form of molding powder, composite-matrix resin, high-temperature film and fiber.

Aurum, a brand of thermoplastic and organically soluble polyetherimide, is composed of pyromellitic acid and diamine with a diphenyl group and two ether groups [26]:

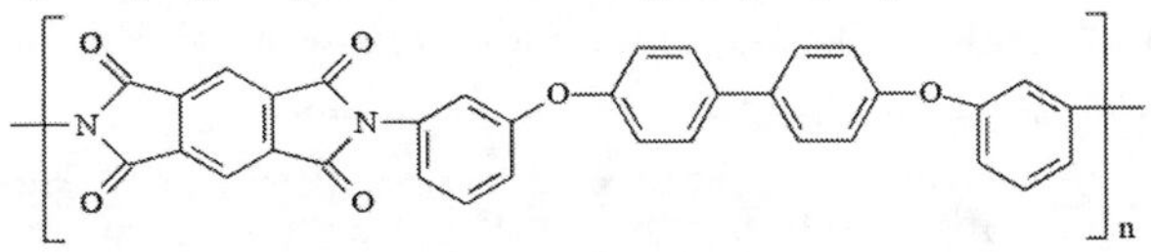

Fig. 5: Aurum (Mitsui Chemicals)

Aurum is a polyetherimide originally developed by DuPont, manufactured currently by Mitsui Chemicals. This polymer has a glass-transition temperature of 250°C and a crystalline phase melting point of 388°C. Products made of this polymer can be used at temperatures as high as 240°C. This polymer can be processed by usual pressure-casting and extrusion methods owing to its low crystallization rate. This polymer stands out with high thermal stability, low friction coefficient, high plasma stability, radiation resistance, good dielectric properties, excellent dimensional stability (better than PEEK and aluminum) and good retention of elastic modulus at elevated temperatures together with low creepage.

Aurum PL450C is used for heat-resistant insulation of electric wires and high-voltage cables, pipeline insulation and fiber insulation. Cast products made of the polymer or its glass-filled brands can serve as substitutes for parts made of metal, ceramic and other, more expensive plastics. Example applications include heat-resistant gears, seals, hubs, mounting parts, reels, electrical insulators, circuit-board cores for semiconductor applications, jet engine parts, check valve spheres, slotted couplings, blades, wear-resistant strips, valve seats, carriers for aluminum hard disk plates and silicon plates, sleeve bearings, bearing retainers etc. [27].

Ultem series polyetherimides were developed by General Electric's Plastics Division. After SABIC had purchased GE's Plastics Division in 2007, Ultem ended up being an asset of Saudi Arabia's largest company.

Ultem 1000 has the following structural formula:

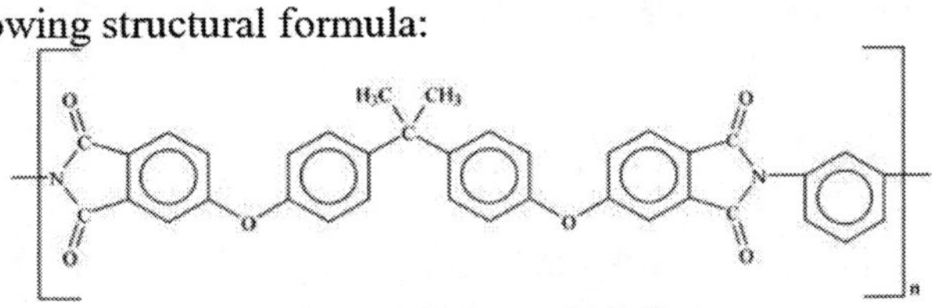

Fig. 6: Ultem 1000

As its repeat unit features three pivot groups, this polymer can be easily processed from molten or organically dissolved form. This is a high-strength thermoplastic with exceptional thermal and heat stability, good mechanical and dielectric properties together with fire resistance. The material comprises a more affordable alternative to polyetherketones at the cost of reduced impact toughness and lower operating temperature. Ultem 9085 has established a good track record in 3D printing thanks to its highly prized combination of properties.

Ultem 1000 can be used for extended time at +170°C. It is fire-resistant without the addition of antipyrenes and releases little smoke when exposed to an open flame, enabling its application in airliners. Key properties of Ultem-1000: superior strength, high stiffness and hardness, high deformational stability, superb creep resistance, excellent thermal resistance of molten material, low thermal expansion coefficient, high chemical stability, excellent resistance to weather, resistance to gamma-ray and hydrolysis, very high torque strength, compatibility with foods. A major drawback is poor crack resistance as the material is prone to internal stresses.

Ultem 1000 articles for mechanical engineering and machinery: carriage travel hubs, camshaft gears, rims for RCh-32 and RCh-37 worm-gear gearboxes, clasp nuts of cross-feed carriages, movable guides, transmission box gears, drive unit gearbox worm gears, hubs, inserts in friction assemblies (skids, mandrels, slide blocks etc.), spindle liners, skids, hubs, gear wheels, sleeve bearings, 7/12/14-ray sprockets.

Ultem 1000 is also used in products manufactured for the chemical industry such as pumps (blades, impellers, gears, sleeve bearings, rolling bearings, packing glands, rotor housing); filters (wheels and hubs); mesh tables; dispensers, separators (friction assembly liners, housings).

Ultem 1000 can be processed with all standard methods depending on subsequent end uses. Most often the material is used for casting of thin-walled products as it belongs to the class of superstructural materials.

As Table 1 shows, LaRC-TPI and Aurum polymers have orderly melting temperature ranges – setting them apart from other polyimides. Kapton and Upilex-RN feature less orderly aggregate structures. While distinct from crystalline phase, their structures are inadequate for making the materials thermoplastic due to rather stiff chains and strong intermolecular interactions [27]. In contrast the Ultem polyetherimide is completely amorphous and has good workability from molten or dissolved form.

Considering the foregoing, we have developed copolyetherimide formulations based on benzophenonetetracarboxylic acid dianhydride, pyromellitic acid anhydride, 4,4'-diaminodiphenyl oxide and a mixture of phenylene diamine isomers. The resulting copolyimide products have been characterized using IR spectroscopy data, elemental analysis, gel permeation chromatography, differential scanning calorimetry and dynamic mechanical analysis. Our findings show that cyclized copolyetherimides of optimal composition have glass transition temperatures in the range of 200°C to 240°C, softening above this temperature, and can be processed from molten form. Moreover, all of these substances are soluble in dipolar aprotonic solvents while some are also soluble in chlorohydrocarbon solutions.

4. Summary of Findings

To summarize, we have developed a laboratory process for the production of long molecular chain copolyetherimides based on benzophenone tetracarboxylic acid dianhydride, pyromellitic acid dianhydride, 4,4'-diaminodiphenyl oxide and mixtures of isomers of phenylene diamine, with glass transition temperatures between 200°C and 240°C suitable for processing from molten form and solutions in commercial organic solvents. Copolyetherimides synthesized by us are promising source materials for the development of high-modulus cast and extrusion composites with disperse glass or carbon fiber reinforcement, as well as carbon fabric-based highly heat-resistant thermoplastic preparations. The latter, having been the subject of intense recent development, are of special application relevance as they make feasible the manufacturing of various structural components with highly affordable vacuum/pressure thermal molding technology.

Acknowledgments

This study has been sponsored financially by the Russian Foundation for Basic Research as a part of scientific project No. 18-29-18035 (Contract No. 18-29-18035/18).

References

[1] Nelyub V A 2019 CFRPs with functional properties based on metallized carbon tapes [*Ugleplastiki s funkcional'nymi svojstvami na osnove metallizirovannyh uglerodnyh lent]*. Abstracts of reports of VII scientific youth school-conference. M 304 p. (Rus)

[2] Nelyub V A and Berlin A A 2019 Electrical properties of carbon plastics and methods for their regulation [Jelektricheskie svojstva ugleplastikov i sposoby ih regulirovanija] *Chemical industry today*. **1** 38-43. (Rus)

[3] RF patent No. 2653058. Nelyub V A, Borodulin A S and Kalinnikov A N and other 2017 *Flame retardant aromatic polyethersulfone* (Rus)

[4] RF Patent No. 2683268. Borodulin A S and Kalinnikov A N and other 2019 *Aromatic polyesters*. (Rus)

[5] RF Patent No. 2683270. Borodulin A S and Kalinnikov A N and other 2019 *Fire-resistant*

aromatic polyester (Rus)

[6] Borodulin A S, Kalinnikov A N and Bazheva R C at all. 2018 Receipt and investigation of performance characteristics of super constructions polyesters *International Journal of Mechanical Engineering and Technology*. **9-13** 1117-1127.

[7] Borodulin A S, Kalinnikov A N and Bazheva R C at all. 2018 Synthesis and properties of aromatic polyethersulfones *International Journal of Mechanical Engineering and Technology (IJMET)*, **9-13** 1109–1116.

[8] Borodulin A S, Kalinnikov AN and Kharaev A M at all. 2019 New Polymeric Binders for the Production of Composit *International Conference on Modern Trends in Man-ufacturing Technologies and Equipment 2018. International Journal of Materials today: proceedings. –, V. 11. Issue P1*, 3107–3111.

[9] Nelyub V A and Borodulin A S et al. 2018 Polyethersulfones with improved thermophysical properties. [Polijefirsul'fony s uluchshennymi teplofizichesimi svojstvami] *M.: Publishing House of Science and Technology LLC, Journal of Adhesives, Sealants, Technologies*, **7** 15-20.

[10] Kharaev A M, Bazheva R Ch, Begieva M B, Nelyub V A and Borodulin A S 2019 Polyethersulfones with Improved Thermophysical Properties *Polymer Science, Series D, 2019, Volume 12, No. 1*, pp. 24–28

[11] Borodulin A S and Kalinnikov A N 2020 Super engineering polyesters: synthesis and performance characteristics *IOP Conf. Series: Materials Science and Engineering, Volume 709, Issue 2.*

[12] Borodulin A S, Kharaev A M, Kalinnikov A N, Bazheva R Ch, Kvashin V A and Beshtoev B Z 2019 Synthesis and Performance Characteristics of Superstructure Polyethers *Key Engineering Materials Volume 816,* pp. 307-311.

[13] Borodulin A, Kalinnikov A, Kharaev A and Shcherbin S 2019 Aromatic polysulfone to create polymer materials with high resistance to frost *IOP Conf. Series: Earth and Environmental Science Volume 302.*

[14] Kalinnikov A N, Borodulin A S, Kharaev A M, Bazheva R Ch, Balkarova S B and Kharaeva R A 2019 Polyether-Ketones Based on 1,1-Dichloro-2,2-Di(3,5-Dibromo-4-Hydroxyphenyl) Ethylene *Key Engineering Materials Volume 816,* pp. 302-306.

[15] Borodulin A S and Kalinnikov A N 2020 Super engineering polyesters: Synthesis and performance characteristics *IOP Conf. Series: Materials Science and Engineering, Volume 709, Issue 2,* 022038.

[16] Bessonov M I, Koton M M, Kudryavtsev V V and Laius L A 1983 Polyimides - a class of heat-resistant polymers. [Poliimidy - klass termostojkih polimerov] - L .: Nauka,328 p.

[17] Johnson R O and Burlhis H S 1983 Polyetherimide: a new high-performance thermoplastic resin *J. Polym. Sci.: Polym Symp.*, **70** 129–143.

[18] Parker D, Vussink J, Grampel H T, Wheatley G W, Dorf E U, Ostlinning E, Reinking K, Schubert F, Junger O 2012 Polymers, high-temperature. In: Ullmann's encyclopedia of industrial chemistry.-Wiley//Weinheim.

[19] Saudi Basic Industries Corporation. URL://www.sabic-ip.com/gep/Plastics/en/ProductsAndServices/ProductLine/ultem.html,http://www.sabic-ip.com/gepapp/eng/weather/weatherhtml?sltRegionList=1002002000%26sltPrd=1002003018%26sltGrd=1002011252%26sltUnit=0%26sltModule=DATASHEETS%26sltVersion=Internet%26sltType=Online

[20] Salamone J.C. (1996) Polymeric materials encyclopedia, vol 8. CRC Press, Boca Raton, 2018

[21] DuPont//URL:http://www2.dupont.com/Kapton/en_US/URL assets/downloads/pdf/HN_datasheet.pdf

[22] Takekoshi T. Chapter 2: Synthesis of polyimides. In: Ghosh M.K., Mittal K.L. (eds) Polyimides: fundamentals and applications, plastic engineering//Marcel Dekker, Inc., New York, 1996 7–48.

IOP Conf. Series: Materials Science and Engineering **934** (2020) 012010 doi:10.1088/1757-899X/934/1/012010

[23] Polyimide products which has developed by Ube Industries, Ltd.//Ube Industries//URL://www.upilex.jp/e_index.html

[24] Encyclopedia of Polymeric Nanomaterials//URL: Ube Industries. http://www.ube.es/archivos/pdfs/UpilexRN-566.pdf

[25] Hou T.H., Bai J.M. (1990) LaRC-TPI 1500 series polymers. NASA contractor report 181899//MitsuiChemicals//URL:www.mitsuichem.com/service/functional_polymeric/compound/aurum/DuPont.http://www2.dupont.com/Vespel/en_US/assets/downloads/aurum/pl450c.pdf

[26] McKeen L.W. Ch. 4.1.8. Polyetherimide. In: Fluorinated Coatings and Finishes Handbook//2016, 112 p.

[27] Isoda S, Shimada H, Kochi M and Kambe H 1981 Molecular aggregation of solid aromatic polymers. I. Small-angle X-ray scattering from aromatic polyimide film. *J. Polym. Sci. Polym. Phys. Ed.* **19** 1293–1312.

Mandatory form: please fill out

Type of paper	Science
Scientific area associated with the paper*	Technical sciences
Group of disciplines associated with the paper*	04.03.02: Chemistry, physics and material mechanics
Discipline associated with the paper*	04.05.01: Pure and applied chemistry

IOP Conf. Series: Materials Science and Engineering **934** (2020) 012011 doi:10.1088/1757-899X/934/1/012011

Numerical simulation of the stability of three-dimensional elastic composite structures based on the finite element method

Yu I Dimitrienko and I O Bogdanov

Department of Computational Mathematics and Mathematical Physics, Bauman Moscow State Technical University, 5 Baumanskaya 2-ya, Moscow, 105005, Russia

E-mail: dimit.bmstu@gmail.com, ibogdanov@bmstu.ru

Abstract. A numerical method for solving the problem of the stability theory of linearly elastic bodies with small deformations in a general three-dimensional formulation is considered. The problems of this class are poorly studied in contrast to the two-dimensional problems of stability theory. At the same time, classical approaches do not allow one to take into account the effect on the structural stability of various three-dimensional effects: areas of compounds, zones of defects, etc. The study formulates a variation formulation of the problem of the three-dimensional stability theory. Based on the finite element method, a numerical statement is obtained in the form of a generalized eigenvalue problem with symmetric global stiffness matrix. The application of the proposed method is demonstrated by the example of calculating the stability of a composite plate under longitudinal compression. The simulation was carried out using the SMCM software package developed at the Department of Calculus Mathematics and Mathematical Physics of Bauman Moscow State Technical University.

1. Introduction

When operating products made of composite materials, of particular importance is the assessment of the influence of various three-dimensional effects (defects, joint areas, fastenings) on the stability of structures [1-5]. Using the classical approaches of one-dimensional and two-dimensional stability theories [6-17] does not allow to provide a sufficient level of accuracy in solving problems of this class. In this regard, the problem of studying the structures stability within the framework of the general three-dimensional theory has recently become relevant [18].

This paper describes a method for solving the three-dimensional problem of the stability theory of linearly elastic bodies with small deformations in a general three-dimensional formulation [19-27].

2. The mathematical model of the three-dimensional stability theory

We introduce three body configurations: reference $\overset{\circ}{K}$, actual K, and varied $\hat{K}$, which differs from the true actual configuration by small displacement. The indicated configurations are shown in Figure 1.

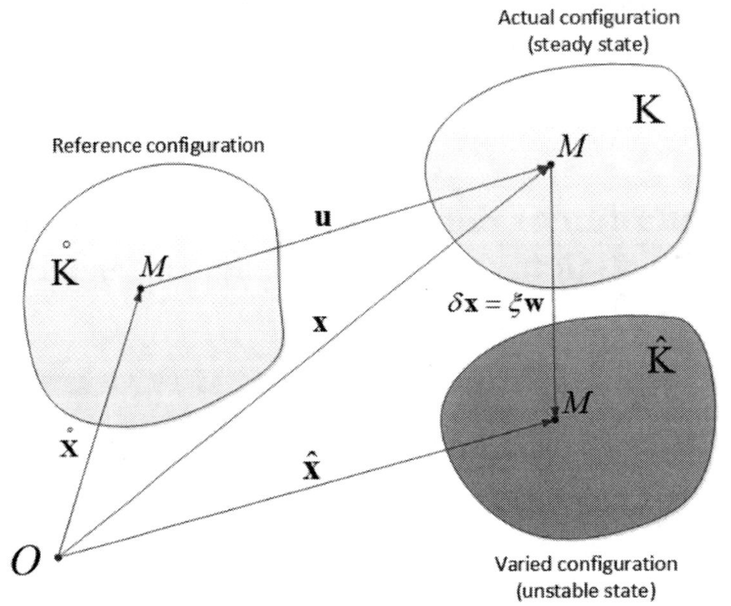

Figure 1. Reference, actual and varied body configurations

The stability problem consists in finding a varied configuration $\hat{K}$ (i.e., a vector $\mathbf{w}$), and using it to search for a possible non-unique solution that determines the unstable state of the body. As shown in [18], in this case, the mathematical model of the stability theory of a linearly elastic body with small deformations consists in the formulation of two main problems solved together. The first of them is the *equilibrium problem for the ground (stable) state* has the form:

$$\begin{gathered} \nabla \cdot \boldsymbol{\sigma}^0 = 0, \\ \boldsymbol{\sigma}^0 = {}^4\mathbf{C} \cdot\cdot \boldsymbol{\varepsilon}^0,\ \boldsymbol{\varepsilon}^0 = \frac{1}{2}\left(\nabla \otimes \mathbf{u}^0 + \nabla \otimes \mathbf{u}^{0T}\right), \\ \mathbf{n} \cdot \boldsymbol{\sigma}^0\big|_{\Sigma_\sigma} = \lambda \mathbf{S}_e,\ \mathbf{u}^0\big|_{\Sigma_u} = \lambda \mathbf{u}_e, \end{gathered} \tag{1}$$

where $\boldsymbol{\sigma}^0$ is stress tensor; $\boldsymbol{\varepsilon}^0$ is small strain tensor; $\mathbf{u}^0$ is displacement vector; ${}^4\mathbf{C}$ is 4th rank elastic modules tensor; $\mathbf{S}_e$ and $\mathbf{u}_e$ are vectors of external surface forces and displacements, respectively; λ is coefficient for vectors of external surface forces and displacements. The second task is actually *the problem of stability theory*:

$$\begin{gathered} \nabla \cdot \boldsymbol{\sigma} - \boldsymbol{\sigma}^0 \cdot\cdot (\mathbf{B} \cdot \mathbf{e}) = 0, \\ \boldsymbol{\sigma} = {}^4\mathbf{C} \cdot\cdot \boldsymbol{\varepsilon}(\mathbf{w}),\ \boldsymbol{\varepsilon}(\mathbf{w}) = \frac{1}{2}\left(\nabla \otimes \mathbf{w} + \nabla \otimes \mathbf{w}^T\right), \\ \mathbf{B} = \nabla \otimes \boldsymbol{\omega},\ \boldsymbol{\omega} = \frac{1}{2}\mathbf{e} \cdot\cdot \Omega(\mathbf{w}),\ \Omega(\mathbf{w}) = \frac{1}{2}\left(\nabla \otimes \mathbf{w} - \nabla \otimes \mathbf{w}^T\right) \\ \mathbf{n} \cdot \left(\boldsymbol{\sigma} - \boldsymbol{\sigma}^0 \cdot \mathbf{e} \cdot \boldsymbol{\omega}\right)\big|_{\Sigma_\sigma} = 0,\ \mathbf{w}\big|_{\Sigma_u} = 0, \end{gathered} \tag{2}$$

where $\boldsymbol{\sigma}$ is stress tensor; $\boldsymbol{\varepsilon}$ is small strain tensor; $\mathbf{w}$ is displacement vector in a varied configuration; $\mathbf{e}$ is Levi-Civita tensor.

The solution of problems (1) – (2) is carried out in accordance with the following algorithm:

1. Solve the problem (1) for the ground state with the parameter value $\lambda = 1$.
2. Calculate the $\sigma^0(1)$ stress tensor field. Due to the linearity of the problem, the field of the stress tensor $\sigma^0(\lambda) = \lambda\sigma^0(1)$ corresponds to any other value of the parameter λ.

3. Substituting the field $\sigma^0(\lambda)$ in (2), we obtain the stability theory problem (eigenvalue problem).
4. Find the system of eigenvalues λ and eigenfunctions $\mathbf{w}$.

3. Variation formulation of the problem

Consider the equilibrium problem for the ground (stable) state [18, 19]. We introduce the kinematically admissible field $\mathbf{\Psi}^0 = \delta\mathbf{u}^0$, where $\delta\mathbf{u}^0$ is the variation of the displacement vector $\mathbf{u}^0$, understood as the difference of two kinematically admissible fields. This field must satisfy the zero boundary condition on part of the surface Σ_u of the region V. Multiplying the equilibrium equation from the system (1) by $\mathbf{\Psi}^0$ and integrating the expression over the region V, taking into account the Gauss-Ostrogradsky theorem, we obtain the variation equation for the equilibrium problem in the ground state:

$$\int_V {}^4\mathbf{C}\cdot\cdot\boldsymbol{\varepsilon}^0\left(\mathbf{u}^0\right)\cdot\cdot\delta\boldsymbol{\varepsilon}^0\left(\mathbf{u}^0\right)dV - \int_{\Sigma_\sigma} \mathbf{S}_e\cdot\delta\mathbf{u}^0 d\Sigma = 0. \tag{3}$$

Applying a similar approach to the stability problem (2), we arrive at the variation equation:

$$\int_V \left({}^4\mathbf{C}\cdot\cdot\boldsymbol{\varepsilon}(\mathbf{w}) + \boldsymbol{\sigma}^0\cdot\Omega(\mathbf{w})\right)\cdot\cdot\delta\nabla\otimes\mathbf{w}^T dV. \tag{4}$$

The relation (4) is an eigenvalue problem in which it is required to find the eigenvalues λ and the corresponding eigenfunctions $\mathbf{w}$. The least eigenvalue $\lambda_{\min}$ is of most practical interest since it corresponds to the critical load $\mathbf{S}^e_{cr}$, which leads to the first form of stability loss. All other λ values will correspond to other forms of structural stability loss.

4. The finite element method for solving the problem of stability theory

Consider the numerical formulation of the problem of stability theory. To formulate it in this paper, we used the finite element method [20, 21]. We assume that for the triangulation of the computational domain, a tetrahedral simplex element with three degrees of freedom at each node is used.

Introducing the matrix analogues of the quantities included in the variation formulation (3), and then applying the classical procedure of the finite element method, we obtain a numerical formulation of the equilibrium problem in the ground state, which can be written in the form:

$$\underset{12\times12}{[K]}\underset{12}{\left\{u^0\right\}} = \underset{12}{\{F\}}, \tag{5}$$

where the notations are introduced:

$$\underset{12\times12}{[K]} = \int_V \underset{12\times6}{[B_1]^T}\underset{6\times6}{[C]}\underset{6\times12}{[B_1]}dV,\ \underset{12}{\{F\}} = \int_{\Sigma_\sigma}\underset{12\times3}{[N]^T}\underset{3}{\{S_e\}}dV,\ \underset{6\times12}{[B_1]} = \underset{6\times3}{[L_1]}\underset{3\times12}{[N]}.$$

Here $[C]$ is the matrix of the elastic modulus tensor ${}^4\mathbf{C}$ components, $[N]$ is the matrix of the shape functions, $[L_1]$ is the differentiation operator, $\{S_e\}$ is the vector of external surface forces.

Let us consider the numerical formulation of the problem for a varied state. To do this, we introduce the following notation:

$$\underset{6}{\{\sigma\}^T} = \left(\sigma_{11}\ \ \sigma_{22}\ \ \sigma_{33}\ \ \sigma_{23}\ \ \sigma_{13}\ \ \sigma_{12}\right) \tag{6}$$

– vector of stress tensor components;

$$\underset{6}{\{\varepsilon\}^T} = \left(\varepsilon_{11}\ \ \varepsilon_{22}\ \ \varepsilon_{33}\ \ 2\varepsilon_{23}\ \ 2\varepsilon_{13}\ \ 2\varepsilon_{12}\right) \tag{7}$$

– vector of the small strain tensor components;

$$\underset{3}{\{W\}^T} = \left(W_1\ \ W_2\ \ W_3\right) \tag{8}$$

– finite element (FE) displacement vector in a varied configuration;

$$\underset{12}{\{w\}}^T = \begin{pmatrix} w_{11} & w_{12} & w_{13} & w_{21} & w_{22} & w_{23} & w_{31} & w_{32} & w_{33} & w_{41} & w_{42} & w_{43} \end{pmatrix} \tag{9}$$

– the displacement vector in the nodes of the finite element in a varied configuration;

$$\underset{9\times9}{\left[\sigma^0\right]} = \begin{pmatrix} \sigma_{11}^0 & 0 & 0 & \sigma_{12}^0 & 0 & 0 & \sigma_{13}^0 & 0 & 0 \\ 0 & \sigma_{11}^0 & 0 & 0 & \sigma_{12}^0 & 0 & 0 & \sigma_{13}^0 & 0 \\ 0 & 0 & \sigma_{11}^0 & 0 & 0 & \sigma_{12}^0 & 0 & 0 & \sigma_{13}^0 \\ \sigma_{21}^0 & 0 & 0 & \sigma_{22}^0 & 0 & 0 & \sigma_{23}^0 & 0 & 0 \\ 0 & \sigma_{21}^0 & 0 & 0 & \sigma_{22}^0 & 0 & 0 & \sigma_{23}^0 & 0 \\ 0 & 0 & \sigma_{21}^0 & 0 & 0 & \sigma_{22}^0 & 0 & 0 & \sigma_{23}^0 \\ \sigma_{31}^0 & 0 & 0 & \sigma_{32}^0 & 0 & 0 & \sigma_{33}^0 & 0 & 0 \\ 0 & \sigma_{31}^0 & 0 & 0 & \sigma_{32}^0 & 0 & 0 & \sigma_{33}^0 & 0 \\ 0 & 0 & \sigma_{31}^0 & 0 & 0 & \sigma_{32}^0 & 0 & 0 & \sigma_{33}^0 \end{pmatrix} \tag{10}$$

– matrix of stress tensor components in the ground state;

$$\underset{9}{\{R\}} = \begin{pmatrix} R_{11} & R_{12} & R_{13} & R_{21} & R_{22} & R_{23} & R_{31} & R_{32} & R_{33} \end{pmatrix} \tag{11}$$

– a vector whose components are derivatives of the form $R_{ij} = \dfrac{\partial w_j}{\partial x_i}$;

$$\underset{9\times9}{[T]} = \begin{pmatrix} 1 & 0 & 0 & 0 & 0 & 0 & 0 & 0 & 0 \\ 0 & 0 & 0 & 1 & 0 & 0 & 0 & 0 & 0 \\ 0 & 0 & 0 & 0 & 0 & 0 & 1 & 0 & 0 \\ 0 & 1 & 0 & 0 & 0 & 0 & 0 & 0 & 0 \\ 0 & 0 & 0 & 0 & 1 & 0 & 0 & 0 & 0 \\ 0 & 0 & 0 & 0 & 0 & 0 & 0 & 1 & 0 \\ 0 & 0 & 1 & 0 & 0 & 0 & 0 & 0 & 0 \\ 0 & 0 & 0 & 0 & 0 & 1 & 0 & 0 & 0 \\ 0 & 0 & 0 & 0 & 0 & 0 & 0 & 0 & 1 \end{pmatrix} \tag{12}$$

– transpose matrix.

Given the notation introduced, the generalized Hooke's law (the second of the equations in (2)) can be written in the following equivalent form:

$$\underset{6}{\{\sigma\}} = \underset{6\times6}{[C]}\underset{6}{\{\varepsilon\}}, \tag{13}$$

and the Cauchy relations (the third of the equations in (2)) take the form:

$$\underset{6}{\{\varepsilon\}} = \underset{6\times3}{[L_1]}\underset{3}{\{W\}}, \tag{14}$$

where the displacement vector of a finite element (FE) in a varied configuration is related to the displacement vector of its nodes as:

$$\underset{3}{\{W\}} = \underset{3\times12}{[N]}\underset{12}{\{w\}}. \tag{15}$$

Variations of the vector of the small strain tensor $\underset{6}{\{\varepsilon\}}$ components and vector $\underset{9}{\{R\}}$ can be expressed as follows, respectively:

$$\underset{6}{\{\delta\varepsilon\}} = \underset{6\times3}{[L_1]}\underset{3}{\{\delta W\}}, \quad \underset{9}{\{\delta R\}} = \underset{9\times3}{[L_2]}\underset{3}{\{\delta W\}}, \quad \underset{3}{\{\delta W\}} = \underset{3\times12}{[N]}\underset{12}{\{\delta w\}}. \tag{16}$$

Performing the transformation of the integrand from (4), taking into account the introduced matrix quantities, we obtain a numerical statement of the problem for a varied state:

$$\underset{12\times12}{[K]}\underset{12}{\{w\}}=\lambda\underset{12\times12}{[S]}\underset{12}{\{w\}}, \tag{17}$$

where the following notations are introduced:

$$\underset{12\times12}{[K]}=\int_V \underset{12\times6}{[B_1]^T}\underset{6\times6}{[C]}\underset{6\times12}{[B_1]}dV,\quad \underset{12\times12}{[S]}=\frac{1}{2}\int_V \underset{12\times9}{[B_2]^T}\underset{9\times9}{\left[\tilde{\Sigma}^0\right]}\underset{9\times12}{[B_2]}dV,$$

$$\underset{9\times9}{\left[\tilde{\Sigma}^0\right]}=\frac{1}{2}\left(\underset{9\times9}{\left[\Sigma^0\right]}+\underset{9\times9}{\left[\Sigma^0\right]^T}\right),\ \underset{9\times9}{\left[\Sigma^0\right]}=\underset{9\times9}{\left[\sigma^0\right]}\left(\underset{9\times9}{[T]}-\underset{9\times9}{[E]}\right),\ \underset{9\times12}{[B_2]}=\underset{9\times3}{[L_2]}\underset{3\times12}{[N]}.$$

Here $[L_2]$ is the differentiation operator, $[B_1]$ and $[B_2]$ are the matrices of derivative functions of the form.

5. Stability of the composite plate

Let us demonstrate the application of the considered model by the example of a multilayer composite plate with a hole.

The test plate consists of two layers of ST-12026 twill weaving carbon fabric with reinforcement structure $(0^\circ/90^\circ)$ and one layer of biaxial diagonal fibreglass SM-42020 with reinforcement structure $(+45^\circ/-45^\circ)$ (see Figure 2). The total thickness is $0{,}01\ \mathrm{m}$. The effective elastic constants of this material are shown in Table 1 and were calculated separately [22].

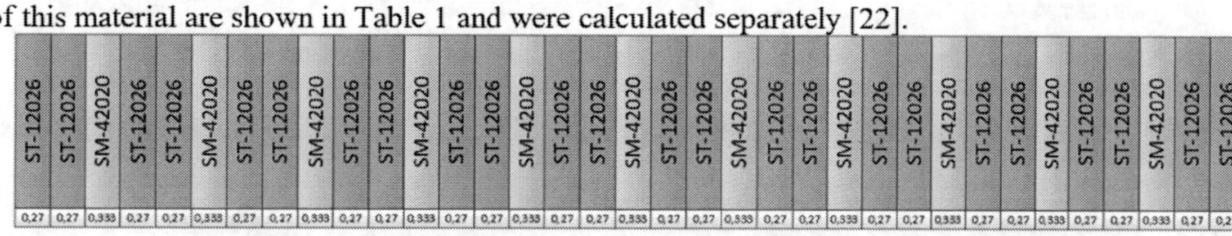

Figure 2. Reinforcement scheme of a multilayer composite plate

Table 1. Values of effective elastic constants

Effective elastic constants	Designation	Value
Young's modulus along the x axis, MPa	E_1	47752.5
Young's modulus along the y axis, MPa	E_2	9039.68
Young's modulus along the z axis, MPa	E_3	47752.5
The shear modulus in the xy plane, MPa	G_{12}	3831.02
The shear modulus in the xz plane, MPa	G_{13}	10430.6
The shear modulus in the yz plane, MPa	G_{23}	3831.02
Poisson's ratio in the xy plane	ν_{12}	0.315505
Poisson's ratio in the xz plane	ν_{13}	0.158262
Poisson's ratio in the yz plane	ν_{23}	0.0597261

The paper considers two calculations for stability under longitudinal compression. In each of the calculations, the plate had a size of $0{,}5\ \mathrm{m}\times1\ \mathrm{m}$. The initial load, set at the short end of the plate, was $0{,}2\ \mathrm{GPa}$, the opposite end was rigidly fixed (zero components of the displacement vector were specified). In the first case, a hole with a diameter of $0{,}05\ \mathrm{m}$ was located in the centre in the upper part of the plate, in the second, in the lower right.

Table 2 shows the calculated critical loads. Figure 3 shows examples of the results of solving the problem for the ground state with symmetrical location of the hole, Figure 4 shows examples of the results of solving the problem for the varied state with symmetrical location of the hole, Figures 5-6 show examples of results, respectively, for the ground and varied states with an asymmetric location of the hole.

a) b) c)

Figure 3. Examples of fields obtained by solving the problem for the ground state with the symmetrical location of the hole: a) component u_1^0 of the displacement vector; b) component u_3^0 of the displacement vector; c) component σ_{33}^0 of the stress tensor

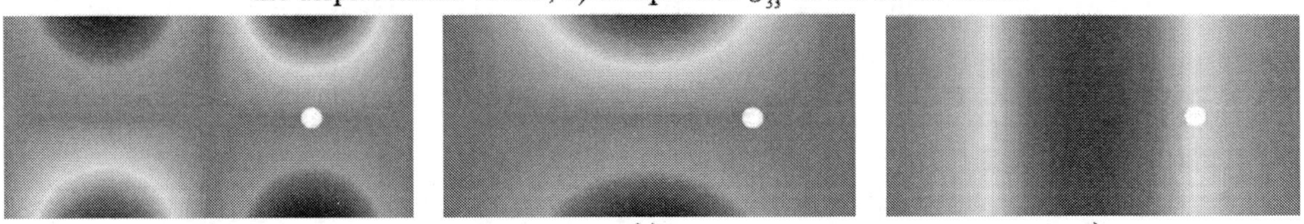

a) b) c)

Figure 4. Examples of fields obtained by solving the problem for a varied state with the symmetrical location of the hole: a) component w_1 of the displacement vector; b) component w_2 of the displacement vector; c) component w_3 of the displacement vector

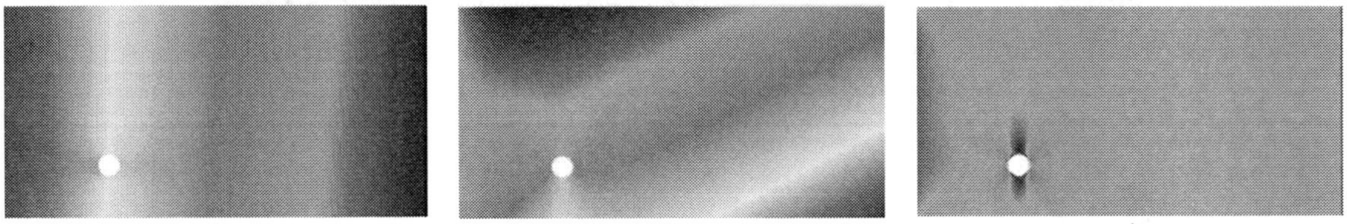

a) b) c)

Figure 5. Examples of fields obtained by solving the problem for the ground state with an asymmetric location of the hole: a) component u_1^0 of the displacement vector; b) component u_3^0 of the displacement vector; c) component σ_{33}^0 of the stress tensor

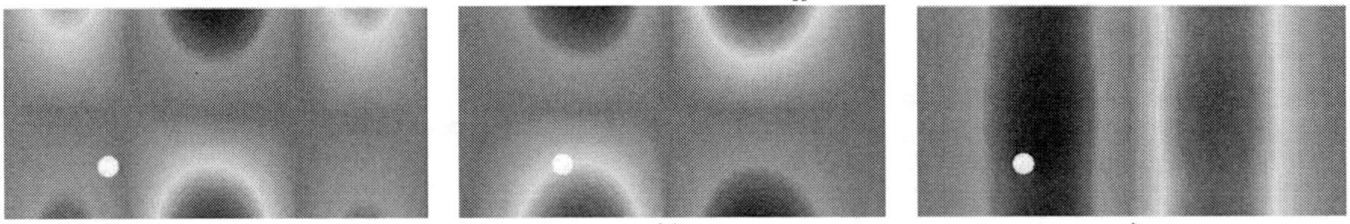

a) b) c)

Figure 6. Examples of fields obtained by solving the problem for a varied state with an asymmetric location of the hole: a) component w_1 of the displacement vector; b) component w_2 of the displacement vector; c) component w_3 of the displacement vector

From the results shown in Table 2, it is seen that the critical loads at different positions of the holes are quite close.

Table 2. Values of calculated critical loads

Hole position	**The value of the critical load, MPa**
Top center	17,449

IOP Conf. Series: Materials Science and Engineering **934** (2020) 012011 doi:10.1088/1757-899X/934/1/012011

Bottom right	17,457

6. Conclusions

The work considers a method for solving the stability theory of linearly elastic bodies with small deformations in a general three-dimensional formulation. The main advantage of this approach is the ability to take into account the influence of three-dimensional effects on the loss of structural stability. In the paper, a variation formulation of the problem is formulated and reduced to a generalized eigenvalue problem based on the use of the finite element method. The results of applying the considered model are presented on the example of studying the stability of a multilayer composite plate with a hole.

References

[1] Belov P A, Borodulin A S, Kobets L P and Malysheva G V 2016 Kinetics of fiber impregnation by a binder. Gradient generalization of Navier–Stokes–Darcy equations *Polymer Science - Series D* **9** 205-208

[2] Golovatov D, Mikhaylov M and Bosov A 2016 Computational studies of connections of spatial composite structures of space-rocket techniques *Indian Journal of Science and Technology* **9** 107495

[3] Magnitsky I V, Odinabekov F R and Sergeeva E S 2018 The development of multi-directional spatially reinforced material structural theory *Solid State Phenomena* **284 SSP** 146-151

[4] Belyaeva A A and Evseev K B 2020 Analysis viscoelastic properties of the composite leaf spring. *IOP Conference Series: Materials Science and Engineering* **709** 033011

[5] Barinov D Y and Prosuntsov P V 2019 Modelling of destruction of carbon-ceramic composite materials with variable density *AIP Conference Proceedings* **2171** 170011

[6] Timoshenko S P and Gere J M 1961 *Theory of elastic stability. 2nd ed.* (New York/Toronto/London: McGraw-Hill) p 356

[7] Timoshenko S P 1971 *Stability of cores, plates and covers* (Moscow: Nauka Publ.) p 808

[8] Simitses G J 1976 *An introduction to the elastic stability of structures* (Prentice Hall, NJ) p 256

[9] Donnell L G 1982 *Beams, plates and covers* (Moscow: Nauka Publ.) p 568

[10] Iyengar N.G.R. 1986 *Structural stability of columns and plates* (New Delhi: Affiliated East-West Press) p 284

[11] Bazant Z P and Cedolin L 1990 *Stability of structures* (Oxford: Oxford University Press) p 316

[12] Bazant Z P 2000 Stability of Elastic, Anelastic, and Disintegrating Structures: A Conspectus of Main Results *ZAMM, Z Angew.Math.Mech.* **80(11-12)** pp 709-32.

[13] Pikul V V 2007 To the theory of stability of covers *Proceedings of the Academy of Sciences*, **416**(3) pp 341-43

[14] Zhukov A E and Volkov V M 2007 Experimental and theoretical research of stability of plates with cracks *Problems of durability and plasticity: Interuniversity collection.* **69**, pp. 13-17

[15] Pikul V V 2008 Current state of the theory of stability of covers *Bulletin of the Far Eastern Branch of the Russian Academy of Sciences* **3** pp. 3-9

[16] Vanko V I 2014 *Sketches according to the theory of stability of elements of designs* (Moscow: BMGTU Publ.) p 220

[17] Solomonov Yu S Georgievskij V P Nedbaj A Ya and Andryushin V A 2014 *Applied problems of mechanics of composite cylindrical covers* (Moscow: Fizmatlit Publ.) p 408

[18] Kohanenko Yu V 2009 Three-dimensional stability of the cylinder at a non-uniform initial state *Reports of the National Academy of Sciences of Ukraine* **1**, pp. 60-62

[19] Dimitrienko Yu I and Bogdanov I O 2016 Finite-element method for three-dimensional problems of elastic structures buckling theory *Herald of the Bauman Moscow State Technical University, Series Natural Sciences* **6** pp 73-92

[20] Dimitrienko Yu I 2013 Generalized three-dimensional theory of elastic bodies stability. Part 1: finite deformations *Herald of the Bauman Moscow State Tech. Univ., Nat. Sci.* **4 (51)** pp. 79-95

[21] Dimitrienko Yu I 2014 Generalized three-dimensional theory of elastic bodies stability. Part 2:

small deformations *Herald of the Bauman Moscow State Tech. Univ., Nat. Sci.* **1** pp. 17-26

[22] Dimitrienko Yu I 2014 Generalized three-dimensional theory of elastic bodies stability. Part 3. theory of shell stability *Herald of the Bauman Moscow State Tech. Univ., Nat. Sci.* **2** pp 77-89

[23] Dimitrienko Yu I 2013 *Continuum mechanics* vol. 4. *Fundamentals of solid mechanics* (Moscow: BMSTU Publ.) p 624

[24] Dimitrienko Yu I 2011 *Continuum mechanics.* vol. 1. *Tensor analysis* (Moscow: BMSTU Publ.) p 367

[25] Dimitrienko Yu I and Yurin Yu V 2015 Finite element modelling of the intense deformed condition of rocks taking into account creep *Mathematical modelling and Computational methods* **3** pp 101-118

[26] Dimitrienko Yu I and Kashkarov A I 2002 Calculation of effective characteristics of composites with periodic structure by final element method *Herald of the Bauman Moscow State Tech. Univ., Nat. Sci* **2**, pp. 95-108

[27] Dimitrienko Yu I and Dimitrienko I D 2019 Computations of stresses and energy dissipation in composite thin laminates with the asymptotic vibration theory *Computers & Mathematics with Applications*. **78(8)** pp.2541-2559

IOP Conf. Series: Materials Science and Engineering **934** (2020) 012012 doi:10.1088/1757-899X/934/1/012012

Universal models of the constitutive relations for transversely isotropic compressible composites with finite strains

Yu I Dimitrienko, E A Gubareva, S B Karimov and D Yu Kolzhanova

Bauman Moscow State Technical University, Moscow, Russia

E-mail: dimit.bmstu@gmail.com

Abstract. A model of a transversely isotropic compressible elastic medium with finite strains is proposed. The model belongs to the class of the so-called universal models, which are formulated in terms of energy (conjugated) pairs of stress and strain tensors - simultaneously for several types of pairs. An algorithm is proposed for calculating the constants included in the constitutive relations of this model, based on a comparative analysis of the results of calculating deformation diagrams under uniaxial loading according to this model and using the asymptotic averaging method. An example of numerical modeling for a layered composite is given.

1. Introduction

To calculate the effective characteristics of composites with small deformations, there are a large number of different methods, we only note some work in this field [1-10].Asymptotic homogenization method (AHM, asymptotic averaging method) is the most promising method for solving the problems of calculating the effective characteristics of composites. AHM was proposed by N.S. Bakhvalov, G.P. Panasenko, Bensoussan A., Lions JL, Papanicalaou G [2] Sanchez-Palencia E [3]. The method of asymptotic averaging is quite well developed at the present time and has been successfully numerically implemented for various problems of mechanics, but mainly for linear problems [4,5]. In [11-14] this AHM method was applied to calculate the nonlinear elastic properties of layered composites with finite strains. An algorithm was proposed for constructing strain diagrams of layered composites with finite strains based on the asymptotic theory of averaging of nonlinear-elastic composites with a periodic structure, for the case of finite strains, and also taking into account the incompressibility of the layer materials. In [15,16], a method was developed for constructing defining relations for layered incompressible composites with finite strains. In this method, first, the effective defining relations for the composite are selected as for a homogeneous anisotropic medium, and then the constants included in these relations are calculated from the condition of the best fit to the calculated strain diagrams obtained using the asymptotic homogenization method. To construct a model of an effective anisotropic medium with finite strains, we use the theory of universal models proposed in [15,17], in which several energy (conjugated) pairs of stress-strain tensors are applied at once. The aim of this work is to apply this method to construct defining relations for layered compressible composites with finite strains.

2. Model of transversely isotropic compressible media with finite deformations.

Let us consider a nonlinear elastic layered composite material (NLCM), which in a initial configuration $\overset{0}{K}$ characterized by a Cartesian basis $\mathbf{e}_\alpha$, is a system of periodically repeating layers orthogonal to the axis $O\mathbf{e}_3$. We introduce the Cartesian coordinates $\overset{0}{x}{}^k$ as the Lagrangian coordinates X^i in the basis, i.e. $X^i = \overset{0}{x}{}^i$, and define the law of motion of the composite in the form $x^k = x^k(X^i)$, where x^k are the Cartesian coordinates of the material points in the current configuration K. We introduce $\mathbf{r}_\alpha = Q^k{}_\alpha \mathbf{e}_k$ - local basis vectors in K, where $Q^k{}_\alpha = \partial x^k / \partial X^\alpha$ is the Jacobian matrix, and also $\mathbf{F} = \mathbf{r}_k \otimes \mathbf{e}^k$ is the deformation gradient, $\otimes$ is the sign of the tensor product. For the deformation gradient $\mathbf{F}$, a polar expansion $\mathbf{F} = \mathbf{U} \cdot \mathbf{O}$ takes place [15], where $\mathbf{U}$ is the left distortion tensor and $\mathbf{O}$ is the rotation tensor. We introduce energy measures of deformation [15]

$$\overset{(n)}{\mathbf{G}} = \frac{1}{n - III} \mathbf{U}^{n-III}, \quad n = I, II, IV, V \tag{1}$$

Energy stress tensors $\overset{(n)}{\mathbf{T}} = {}^4\overset{(n)}{\mathbf{E}}{}^{-1} \cdot\cdot \mathbf{T}$ correspond to these measures, where ${}^4\overset{(n)}{\mathbf{E}}$ are the energy equivalence tensors [16], which depend only on $\mathbf{F}$, and $\mathbf{T}$ is the Cauchy true stress tensor.

Since the properties of the NLCM in the reference configuration do not change during rotations in the plane orthogonal to $O\mathbf{e}_3$, the NLCM can be considered as transversally isotropic medium with the symmetry group T3 [17].

For constructing analytic constitutive relations of NLCM, we apply universal models of compressible transversely isotropic media, the general theory of which is formulated in [15]. The most general form of writing these universal models in tensor form for the so-called models of the class An has the following form [15]:

$$\overset{(n)}{\mathbf{T}} = \rho \frac{\partial \psi}{\partial \overset{(n)}{\mathbf{C}}}, \quad n = I, II, IV, V \tag{2}$$

where n is the model number in the class An, $\overset{\circ}{\rho}$ is the averaged density of the composite, $\overset{(n)}{\mathbf{T}}$ is the energy stress tensor, $\overset{(n)}{\mathbf{C}}$ is the deformation energy tensor (rank 2 tensor), ψ is the Helmholtz free energy, which is a function of the tensor invariants $I_k(\overset{(n)}{\mathbf{C}})$ with respect to the symmetry group transversally -isotropic medium T3

$$\psi = \psi(I_1(\overset{(n)}{\mathbf{C}}), \ldots, I_5(\overset{(n)}{\mathbf{C}})) \tag{3}$$

Then, for a compressible transversely isotropic medium, the following representation of the constitutive relations holds (2)

$$\overset{(n)}{\mathbf{T}} = \sum_{\gamma=1}^{5} \varphi_\gamma \mathbf{I}_{\gamma C}, \qquad \varphi_\gamma = \rho \frac{\partial \psi}{\partial I_\gamma}, \qquad I_{\gamma C} = \frac{\partial I_\gamma(\overset{(n)}{\mathbf{C}})}{\partial \overset{(n)}{\mathbf{C}}}. \tag{4}$$

For a transversely isotropic medium, symmetric tensors of the second rank have 5 invariants. Let us choose the following invariants as a functional basis for invariants of a transversely isotropic medium [15-17]

$$I_1 = \left(\mathbf{E} - \mathbf{c}_3^2\right) \cdot\cdot \overset{(n)}{\mathbf{C}}, \quad I_2 = \mathbf{c}_3^2 \cdot\cdot \overset{(n)}{\mathbf{C}}, \quad I_5 = \det(\overset{(n)}{\mathbf{C}}),$$

$$I_3 = \left(\left(\mathbf{E} - \mathbf{c}_3^2\right) \cdot \overset{(n)}{\mathbf{C}}\right) \cdot\cdot \left(\mathbf{c}_3^2 \cdot \overset{(n)}{\mathbf{C}}\right), \quad I_4 = \overset{(n)}{\mathbf{C}}{}^2 \cdot\cdot \mathbf{E} - I_2^2 + 2I_3, \tag{5}$$

where $\mathbf{c}_3^2 = \mathbf{c}_3 \otimes \mathbf{c}_3$ - is the directing tensor of the 2nd rank, and $\mathbf{c}_i$ - is the orthonormal basis of the principal axes of transversal isotropy.
Defining relation (3) for the Helmholtz free energy we will set in an analytical form in the following form

$$\overset{\circ}{\rho}\psi = \overset{\circ}{\rho}\psi_0 + \frac{1}{2}\left(l_{11}I_1^{2n_1} + 2l_{12}I_1^{n_1}I_2^{n_2} + l_{22}I_2^{2n_2}\right) + l_{33}I_3^{n_3} + l_{44}I_4^{2n_4}, \quad (6)$$

where ψ_0 is the initial value (constant) of the potential ψ, and n_1, n_2, n_3, n_4 and $l_{11}\, l_{12}\, l_{22}\, l_{33}\, l_{44}$ are elastic constants.

We calculate the derivatives $\varphi_\gamma = \rho\partial\psi / \partial I_\gamma$:

$$\varphi_1 = \rho\partial\psi / \partial I_1 = J(n_1 l_{11} I_1^{2n_1 - 1} + n_1 l_{12} I_1^{n_1 - 1} I_2^{n_2})$$

$$\varphi_2 = \rho\partial\psi / \partial I_2 = J(n_2 l_{22} I_2^{2n_2 - 1} + n_2 l_{12} I_1^{n_1} I_2^{n_2 - 1}) \quad (7)$$

$$\varphi_3 = \rho\partial\psi / \partial I_3 = J n_3 l_{33} I_3^{n_3 - 1}, \quad \varphi_4 = \rho\partial\psi / \partial I_3 = J n_4 l_{44} I_4^{n_4 - 1},$$

$$J = \rho / \overset{o}{\rho}$$

The derivatives $I_{\gamma\mathbf{C}}$ are calculated using the rules of differentiation of the tensor invariants [17]. Then, in the component record in the Cartesian coordinate system, the constitutive relations (4) will take the form:

$$\overset{(n)}{T}_{ij} = \varphi_1\left(\delta_{ij} - \delta_{i3}\delta_{j3}\right) + \varphi_2\delta_{i3}\delta_{j3} + \quad (8)$$

$$+\left(\varphi_3 - 2\varphi_4\right)\left(\overset{(n)}{C}_{13}\left(\delta_{i1}\delta_{j3} + \delta_{i3}\delta_{j1}\right) + \overset{(n)}{C}_{23}\left(\delta_{i2}\delta_{j3} + \delta_{i3}\delta_{j2}\right)\right) + 2\varphi_4\left(\overset{(n)}{C}_{ij} - \overset{(n)}{C}_{33}\delta_{i3}\delta_{j3}\right)$$

Where $\overset{(n)}{T}_{ij}$, $\overset{(n)}{C}_{ij}$ are the components of the tensors $\overset{(n)}{\mathbf{T}}$ and $\overset{(n)}{\mathbf{C}}$ in the basis $\mathbf{c}_i$. After finding the energy stress tensor, the Cauchy $\mathbf{T}$ and Piola-Kirchhoff stress tensors $\mathbf{P} = J\mathbf{F}^{-1} \cdot \mathbf{T}$ are determined.

3. Method for finding the model parameters of a transversely isotropic medium

The constants $l_{11}\, l_{12}\, l_{22}\, l_{33}\, l_{44}$ and $n_1\, n_2\, n_3\, n_4$ entering into relations (6), (7) are found from the condition of the best approximation using the model (8) of the strain curves $\mathbf{P} = \mathcal{F}(\mathbf{F})$ obtained on the basis of direct numerical solution of problems in the nuclear field that arise in the method of asymptotic averaging (AHM) [12,13]. Comparison of deformation diagrams using the effective anisotropic medium model (EAM model) and the AHM method is carried out for some standard macroscopic deformation problems in which a uniform stress-strain state is realized with $\mathbf{P}$ and $\mathbf{F}$- independent of coordinates. A numerical method for solving the local problem over periodicity cell (PC) was proposed in [12,13]. This method was used in this work.

As standard problems of macroscopic deformation, we consider the class of problems of 3 axis stretching — compression of a plate in the form of a parallel-facet which is parallel to the coordinate axes. The law of movement of the plate with such movements is given in the form $x^\alpha = k_\alpha X^\alpha$, where

$k_\alpha(t)$ are unknown functions of time. The strain gradient for this motion has the form $\mathbf{F}=\sum_{\alpha=1}^{3}k_\alpha \mathbf{e}_\alpha^2=\mathbf{U}$. The energy strain tensors have the form

$$\overset{(n)}{\mathbf{C}}=\frac{1}{n-III}(\mathbf{U}^{n-III}-\mathbf{E})=\frac{1}{n-III}\sum_{\alpha=1}^{3}(k_\alpha{}^{n-III}-1)\mathbf{e}_\alpha{}^2 \qquad (9)$$

Substituting (9) into (5), we find the expressions for the invariants

$$I_1=\overset{(n)}{C}_{11}+\overset{(n)}{C}_{22}=\frac{1}{n-III}\left(k_1{}^{n-III}+k_2{}^{n-III}-2\right), \qquad I_2=\overset{(n)}{C}_{33}=\frac{1}{n-III}(k_3{}^{n-III}-1)$$

$$I_3=0, \qquad (10) \quad I_4=\overset{(n)}{C}_{11}{}^2+\overset{(n)}{C}_{22}{}^2=\frac{1}{n-III}\left((k_1{}^{n-III}-1)^2+(k_2{}^{n-III}-1)^2\right).$$

Substituting (9) and (10) into (8), we find that the energy stress tensors, as well as the Cauchy stress tensor, will also have a diagonal form:

$$\mathbf{T}=\sum_{\alpha=1}^{3}T_{\alpha\alpha}\mathbf{e}_\alpha{}^2,\ \mathbf{P}=\sum_{\alpha=1}^{3}Jk_\alpha T_{\alpha\alpha}\mathbf{e}_\alpha{}^2,\ T_{\alpha\alpha}=k_\alpha{}^{n-III}\overset{(n)}{T}_{\alpha\alpha}, \qquad (11)$$

$$T_{\alpha\alpha}=k_\alpha{}^{n-III}\left(\left(\varphi_1+\frac{\varphi_4(k_1^{n-III}-1)}{n-III}\right)\delta_{1\alpha}+\left(\varphi_1+\frac{\varphi_4(k_2^{n-III}-1)}{n-III}\right)\delta_{2\alpha}+\varphi_2\delta_{3\alpha}\right).$$

4. Uniaxial tension of a composite

Let us consider the case of stretching a plate in the plane of transversal isotropy in the direction of the axis $O\mathbf{e}_1$. In this case, from the system of equations (15) we obtain the relationship between stress P_{11} and k_1, i.e., the strain diagram under uniaxial tension (compression)

$$P_{11}=Jn_4 l_{44}\frac{k_1{}^{n-III+1}}{n-III}\left(k_1{}^{n-III}-f(k_1{}^{n-III}-1)-1\right)I_4{}^{n_4-1}$$

where

$$I_4=\frac{1}{n-III}\left((k_1{}^{n-III}-1)^2+(f(k_1{}^{n-III}-1))^2\right)$$

and the function $y=f(x)$ is a solution to the equation

$$(l_{11}-\frac{l_{12}^2}{l_{22}})\left(\frac{x+y}{n-III}\right)^{2n_1-1}=-\frac{l_{44}}{n-III}\frac{n_4}{n_1}y\left(\frac{x^2+y^2}{n-III}\right)^{n_4-1}$$

From the solution of this problem, we also find the dependence $k_2=k_2(k_1)$ of transverse compression on longitudinal tension

Similarly, we find the strain diagram $P_{33}=P_{33}(k_3)$ and the diagram $k_3=k_3(k_1)$ of the transverse compression of the composite under tension in the direction of the axis $O\mathbf{e}_3$.

5. Calculation of model parameters for NLCM

Calculations were carried out according to the developed technique for a layered composite, the PC of which consisted of 2 materials: duotan and rubber SKN. Figures 1 and 2 show graphs of deformation diagrams $P_{11} = P^{(э)}{}_{11}(k_1)$ and $k_2 = k^{(э)}{}_2(k_1)$ obtained using a numerical "experiment" (based on the direct solution of local problems in nuclear materials) for the AI and AV models and using approximations based on the AHM model and using the EAM model.

The approximation error for the considered "experimental" data using the EAM model was 10-12%.

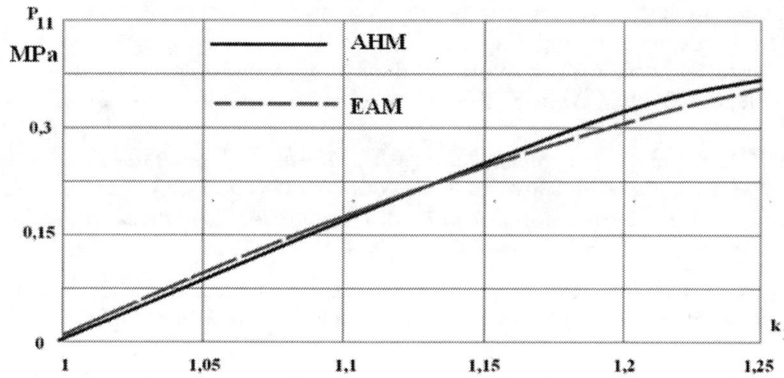

Figure 1. Deformation diagrams $P_{11} = P^{(э)}{}_{11}(k_1)$ for uniaxial deformation of a plate using the AV model with the help of AHM and EAM methods for a 2-layer SKN duotan rubber composite

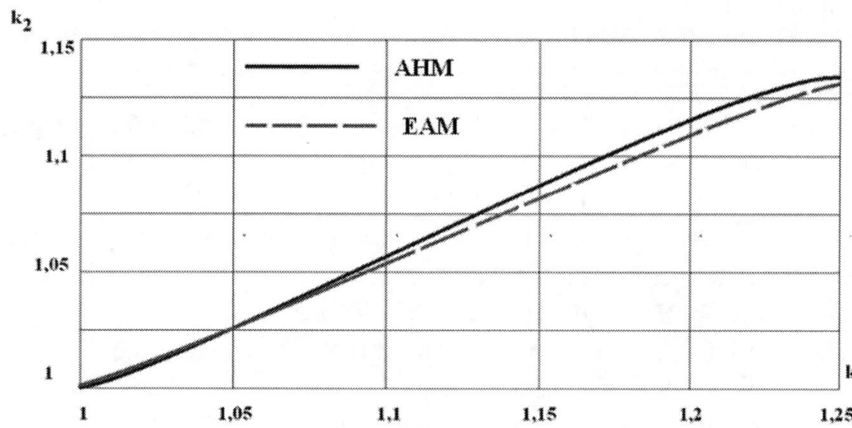

Figure 2. Deformation diagrams $k_2 = k^{(э)}{}_2(k_1)$ for uniaxial plate tension using the AV model according to AHM and EAM methods for a 2-layer SKN duotan rubber composite

6. Conclusions

A model of an effective transversely isotropic nonlinear elastic compressible medium with finite strains is proposed, which belongs to the class of universal models formulated in terms of simultaneously 4 classes of energy pairs of stress – strain tensors. The model is used to model the deformation diagrams of layered compressible elastic composites with finite deformations and a periodic structure, using the universal representation of the defining relations for composite layers.

A method is proposed for finding the elastic effective constants of a composite model by solving the problem of minimizing the deviation of the composite strain diagrams obtained for a set of standard problems using direct numerical solution of problems on a periodicity cell using the AHM method and using an approximation diagram using the EAM method.

The numerical simulation of the strain diagrams of compressible layered composites with finite strains was carried out using the AHM and EAM methods. It is shown that the EAM method allows one to obtain defragmentation diagrams with acceptable engineering accuracy.

References

[1] Bakhvalov N S and Panasenko G P 1984 *Osrednenie protsessov v periodicheskikh sredakh* [Averaging processes in periodic media]. Moscow, Nauka Publ.,352 p.
[2] Bensoussan A, Lions J L and Papanicalaou G 1978 *Asymptotic analysis for periodic structures.*Amsterdam; New York: North-Holland Pub. Co. 396 p.
[3] Sanchez-Palencia E. 1984 *Non-Homogeneous Media and Vibration Theory*. N.Y. etc.: Springer, 1980
[4] Pobedrya B E *Mekhanika kompozitsionnykh materialov* [Mechanics of composite materials]. Moscow, Lomonosov Moscow State University Publ. 324 p.
[5] Dimitrienko Yu I 2015 *Thermomechanics of Composites Structures under High Temperatures*. Springer. 367 p.
[6] Golovatov D, Mikhaylov M and Bosov A 2016 Computational studies of connections of spatial composite structures of space-rocket techniques *Indian Journal of Science and Technology* 9 107495
[7] Nelyub V A and Kovalev S P 2016 A mesh-free approach to multiscale multiphysical calculations of polymeric composite structures *Polymer Science - Series D* 9 **101-105**
[8] Zarubin V S and Sergeeva E S 2019 Mathematical modeling of the structure-sensitive composite elastic properties AIP Conference Proceedings 2116 380010
[9] Savelyeva I Y, Cherednichenko A V and Shukhtin A P 2019 Stress-strains state calculation of a rod at uniaxial tension with non-local effects *IOP Conference Series: Materials Science and Engineering* 489 012025
[10] Bagherinejad Zarandi S, Wang Y-C and Novozhilova O V 2016 Plastic behavior of circular discs with temperature-dependent properties containing an elastic inclusion *Structural Engineering and Mechanics* 58 731-743
[11] Dimitrienko Yu I, Gubareva E A and Kolzhanova D Yu 2015 Modeling of layered composites with finite strains by the method of asymptotic homogenization *Engineering Journal: Science and Innovation* 5. DOI 10.18698/2308-6033-2015-5-1405
[12] Dimitrienko Yu I, Gubareva E A, Karimov S B and Kolzhanova D Yu 2018 Modeling the effective characteristics of transversely isotropic incompressible composites with finite strains. *Mathematical Modeling and Computational Methods*.. **4**. pp.72-92.
[13] Dimitrienko Yu I, Karimov S B and Kolzhanova D Yu 2019. Modeling of the effective universal constitutive relations for elastic laminated composites with finite strains *IOP Conference Series: Material Science and Engeneering*, **683(012006)** doi:10.1088/1757-899X/683/1/012006 pp.1-6.
[14] Dimitrienko Yu I, Gubareva E A, Kolzhanova D Yu and Karimov S B 2017 Modeling of the Incompressible layered composites with finite deformations on the basis of the asymptotic averaging method. *Mathematical modeling and Computational Methods*, **1(13),** pp. 32–54. 2017, DOI: https://doi.org/10.18698/2309-3684-2017-1-3254
[15] Dimitrienko Yu I 2010 *Nonlinear Continuum Mechanics and Large Inelastic Deformations*. Springer.. 722 p.
[16] Dimitrienko Yu I 2018. Universal models for effective constitutive relations of laminated composites with finite *strains IOP Journal of Physics: Conference Series,* **1141** 012098 doi:10.1088/1742-6596/1141/1/012098
[17] Dimitrienko Yu I 2002 *Tensor analysis and Nonlinear Tensor Functions*. Kluwer Academic Publishers.- Dordrecht/Boston/London.. 680 p.

Computational conjugate modeling of aerodynamical flow and thermal stresses in ablative composite structures

Yu I Dimitrienko, M N Koryzkov, Yu I Yurin, A A Zakharov and S V Sborschikov

Bauman Moscow State Technical University, Moscow, 105005, Russia

E-mail: dimit@bmstu.ru

Abstract. The conjugate problem of aerodynamics and thermomechanics for a composite structures made of ablating composite material such as carbon/phenolic or glass/phenolic is considered. The problem includes a three-dimensional Navier-Stokes system of equations - for a high-speed flow around the structure, as well as a system of internal heat and mass transfer in an ablating composite and a three-dimensional theory of elasticity of anisotropic material with variable properties. To solve the conjugate system of equations, a numerical algorithm is developed based on the finite-volume method for solving the Navier-Stokes equations and the finite-element method for solving the equations of heat and mass transfer and thermomechanics. Examples of computations for modeling reentry vehicles are presented. The features of the distribution of heat and mass transfer fields and thermal stresses in the structures caused by the effects of thermal decomposition of the composite during non uniform heating are shown.

1.Introduction

Composite polymer materials, mainly ablative carbon/phenolic and glass/phenolic composites, are still the important types of materials used in heat-shielding systems of re-entry vehicles (RV) moving in dense layers of the Earth's atmosphere, as well as other planets (Mars, Venus). In [1-3], a mathematical model was developed for the processes of internal heat and mass transfer in ablative polymeric composite materials during non-stationary heating, taking into account the processes of thermal decomposition, gas generation and gas filtration in the pores of the material. Since at different times of the descent of the RV in the atmosphere, varying power and thermal loads act on its ablative structure, in order to set the boundary conditions for the problem of thermomechanics, it is necessary to first solve the problem of aerodynamics.
The general formulation of the conjugate problem of aerodynamics and thermomechanics of a ablative heat-shielding structure consists of three systems of equations:

- Navier-Stokes equations for an external high-speed flow;
- equations of internal heat and mass transfer in ablative composite structure;
- equations of thermoelasticity of the ablative heat-shielding structure.

2. The mathematical formulation of the conjugate problem.

Consider the system of equations for a viscous heat-conducting gas (Navier-Stokes equations) [4,5]:

$$\begin{aligned}
&\frac{\partial \rho}{\partial t}+\nabla\cdot\rho\mathbf{v}=0,\\
&\frac{\partial \rho\mathbf{v}}{\partial t}+\nabla\cdot\left(\rho\mathbf{v}\otimes\mathbf{v}+p\mathbf{E}-\mathbf{T}_v\right)=0, \qquad (1)\\
&\frac{\partial \rho\varepsilon}{\partial t}+\nabla\cdot\left(\left(\rho\varepsilon+p\right)\mathbf{v}-\mathbf{T}_v\cdot\mathbf{v}+\mathbf{q}\right)=0,
\end{aligned}$$

where ρ – is the gas density, t – is time, **v** –is velocity vector, p – is pressure, **E** – is the unit tensor, ε is the density of the total energy of the gas, $\mathbf{T}_v$ - is the viscous stress tensor and **q** - is the heat flux vector, ∇ - nabla-operator. We add to this system the determining ratios of the perfect gas:

$$p=\rho\frac{R}{M}\theta,\quad e=c_v\theta,\quad \varepsilon=e+|\mathbf{v}|^2/2,$$
$$\mathbf{T}_v=\mu_1(\nabla\cdot\mathbf{v})\mathbf{E}+\mu_2(\nabla\otimes\mathbf{v}+\nabla\otimes\mathbf{v}^T),\qquad (2)$$
$$\mathbf{q}=-\lambda\nabla\theta,$$

where R - is the universal gas constant, M - is the molecular mass, θ - is temperature, c_v - is the heat capacity at constant volume, μ_1 and μ_2 - are viscosity coefficients, λ - is the gas thermal conductivity. Coefficients of viscosity and thermal conductivity are functions of temperature.

The boundary conditions for the system (1), (2) on the surface of a solid with a gas flow are as follows:

$$\mathbf{v}=\mathbf{0},\quad \theta=\theta_s,\qquad (3)$$

where θ_s - is the surface temperature of the structure.

Next, we consider a heat-shielding structure made of a ablative polymer composite material. Under the influence of high temperatures, usually above 300 ° C, internal physicochemical transformations — thermal decomposition processes — occur in polymer composite materials. As a result, the composite during heating is a multiphase system; in [1-3], a 4-phase system model was proposed in which the 1st phase is a filler (fibers, reinforcing particles), the 2nd phase is a polymer matrix; The 3rd phase is the solid pyrolytic phase, which is formed from the polymer at high temperatures, the 4th phase is the gas, also formed due to thermal decomposition of the initial phase 2 and located in the pores.

The mathematical model of internal heat and mass transfer in a thermal decomposition composite consists of a system of equations for changing the mass of the polymer phase, the equation for filtering gaseous products in the pores of the material, and the heat conduction equation

$$\rho_b\frac{\partial\varphi_b}{\partial t}=-J;$$
$$\frac{\partial\rho_g\varphi_g}{\partial t}+\nabla\cdot\rho_g\varphi_g\mathbf{v}_g=J\Gamma;\qquad (4)$$
$$\rho c\frac{\partial\theta}{\partial t}=-\nabla\cdot\mathbf{q}-c_g\nabla\theta\cdot\rho_g\varphi_g\mathbf{v}_g-J\Delta e^0;$$

The notation is introduced here: φ_b, φ_g — volumetric concentrations of the polymer and gas phases; ρ_b is the density of the initial polymer matrix; ρ_g is the density of the gas phase; c_g is the heat capacity of the gas at a constant volume, ρ and c - are the density and heat capacity of the composite, **q** - is the heat flux vector, θ - is the temperature of the composite; $\mathbf{v}_g$ - is the gas phase velocity; Δe^0 - is the specific heat of thermal decomposition of the matrix; J - is the mass rate of thermal decomposition of the matrix; G is the coefficient of gasification of the matrix.

The heat and mass transfer equations are closed by the defining relations that relate the vector functions $\mathbf{q}$, $\mathbf{v}_g$ to the temperature and pressure gradients $\nabla\theta$, ∇p, using the Fourier and Darcy laws, as well as the Arrhenius relation for the mass rate of thermal decomposition:

$$\mathbf{q}=-\mathbf{\Lambda}\cdot\nabla\theta,\quad \rho_g\varphi_g\mathbf{v}_g=-\mathbf{K}\cdot\nabla p_g,\qquad (5)$$
$$J=J_0\varphi_b\exp\left(-\frac{E_A}{R\theta}\right),\quad p_g=\frac{\rho_g R\theta}{M_g},$$

where J_0 - is the preexponential factor, E_A - is the activation energy of thermal decomposition processes, M_g is the molecular mass of the gas, **Λ** - is the thermal conductivity tensor, and **K** - is the composite gas permeability tensor.

The boundary conditions for equations (1) - (3) on the heated surface are as follows:

$$p_g = p_e,$$
$$-\lambda\nabla\theta\cdot\mathbf{n} = -\nabla\theta_s\cdot\mathbf{\Lambda}\cdot\mathbf{n} + \varepsilon_g\sigma_{SB}\theta^4_{max} - \varepsilon_s\sigma_{SB}\theta^4_w, \tag{6}$$

where p_e, θ_e is the pressure and temperature of the gas flow on the surface of the body, θ_{max} is the temperature of the gas flow on the surface, ε_g and ε_s - are the emission coefficients of the gas and solid, σ_{SB} is the Stefan-Boltzmann constant.

On the inner side of the heat-shielding material, the conditions of tightness and thermal insulation are set:

$$\mathbf{n}\cdot\nabla p_g = 0, \quad \mathbf{n}\cdot\mathbf{\Lambda}\cdot\nabla\theta = 0. \tag{7}$$

To calculate the thermal stresses in the heat-shielding structure, we use the statement of the mechanics of elastic composite media with the presence of thermal decomposition of the matrix [1-3]. These equations consist of equilibrium equations, defining relations, Cauchy relations

$$\nabla\cdot\boldsymbol{\sigma} - \nabla(\varphi_g p) = 0,$$

$$\boldsymbol{\sigma} = \mathbf{C}\cdot\cdot(\boldsymbol{\varepsilon} - \overset{0}{\boldsymbol{\varepsilon}}), \qquad \overset{0}{\varepsilon} = \boldsymbol{\alpha}(\theta - \theta_0) - \boldsymbol{\beta} \tag{8}$$

$$\boldsymbol{\varepsilon} = \frac{1}{2}(\nabla\otimes\mathbf{u} + \nabla\otimes\mathbf{u}^T)$$

Here are designated: $\boldsymbol{\sigma}$ - stress tensor, $\boldsymbol{\varepsilon}$ - small strain tensor, $\mathbf{u}$ - displacement vector, $\mathbf{C}$ - elastic modulus tensor, depending on temperature and the local anisotropy basis, $\overset{0}{\boldsymbol{\varepsilon}}$ - thermal strain tensor, $\boldsymbol{\alpha}$ - thermal expansion tensor, $\boldsymbol{\beta}$ - chemical shrink tensor. At high temperatures, the elastic modulus tensor substantially depends on temperature and duration of heating. Various modules for taking into account the effect of temperature on the properties of polymer composites were proposed in 7-11 and others. In this article, we used a model developed in [1-3].

Boundary conditions on the heated surface of the structure

$$\mathbf{n}\cdot\boldsymbol{\sigma} = -(p_e - \varphi_g p_g)\mathbf{n} \tag{9}$$

Boundary conditions on the remaining parts of the surface - free surface or hard termination

$$\mathbf{n}\cdot\boldsymbol{\sigma} - \mathbf{n}(\varphi_g p) = 0 \quad \text{or} \quad \mathbf{u} = 0 \tag{10}$$

3.Numerical methods for solving the conjugate problem.

To solve the aerodynamics problem (1) - (3), a finite-volume second-order approximation method with TVD reconstruction is used [6].

To solve the problem of internal heat and mass transfer (4) - (7), we write the variational statement of the problem

$$\begin{cases} \int\limits_V \rho_b\delta\varphi_b\frac{\partial\varphi_b}{\partial t}dV = -\int\limits_V \delta\varphi_b J dV, \\ \int\limits_V \delta p\frac{M_g}{R\theta}\frac{\partial p\varphi_b}{\partial t}dV - \int\limits_V \delta p\frac{pM_g}{R\theta^2}\frac{\partial\theta\varphi_b}{\partial t}dV = -\int\limits_V \nabla p\cdot\mathbf{K}\cdot\nabla\delta p dV + \int\limits_V \delta p\cdot J\Gamma dV, \\ \int\limits_V \rho c\delta\theta\frac{\partial\theta}{\partial t}dV = \int\limits_S \delta\theta q_e dS - \int\limits_V \nabla\delta\theta\cdot(\mathbf{\Lambda}\cdot\nabla\theta)dV + \int\limits_V \delta\theta\varphi_g c_g\nabla\theta\cdot(\mathbf{K}\cdot\nabla p)dV - \int\limits_V \delta\theta\cdot J\Delta e^0 dV \end{cases} \tag{11}$$

where the variations of unknown functions are indicated through: $\delta\varphi_b$, δp, $\delta\theta$ - the concentration of the polymer phase, pressure and temperature. According to the finite element method, we divide the computational domain into tetrahedra, introduce barycentric coordinates in each element (which coincide with the shape functions in the case of a linear approximation) and represent unknown functions in the following form:

$$\varphi_b = \{\Phi\}^T\{u\}, \quad p = \{\Phi\}^T\{y\}, \quad \theta = \{\Phi\}^T\{\vartheta\}, \tag{12}$$

where u, y, ϑ - the values of the corresponding functions at the nodes of the finite element, and $\{\Phi\}^T = (L_1, L_2, L_3, L_4)$ - is the string of form functions. Then from the variational formulation, we can obtain a local system of equations for each tetrahedron in the form:

$$\begin{cases} [M]\dfrac{\partial\{u\}}{\partial t} + [R]\{u\} = 0, \\ [M_p]\dfrac{\partial\{y\}}{\partial t} - [S]\dfrac{\partial\{\vartheta\}}{\partial t} + [K_p]\{y\} - [R_p]\{u\} = 0, \\ [M_\theta]\dfrac{\partial\{\vartheta\}}{\partial t} + [K_\theta]\{\vartheta\} - [R_\theta] = 0, \end{cases} \qquad (13)$$

designations are introduced here for the following matrices:

$$[M_p] = \int_V \{\Phi\}\otimes\{\Phi\}^T \frac{M_g}{R\theta} dV, \quad [M_\theta] = \int_V \{\Phi\}\otimes\{\Phi\}^T \rho c dV, \quad [M] = \rho_b \int_V \{\Phi\}\otimes\{\Phi\}^T dV,$$

$$[S] = \int_V \{\Phi\}\otimes\{\Phi\}^T \frac{pM_g}{R\theta^2} dV, \quad [K_p] = \int_V [B]^T \cdot \mathbf{K} \cdot [B] dV, \quad [K_\theta] = \int_V [B]^T \cdot \mathbf{\Lambda} \cdot [B] dV,$$

$$[R] = J_0 \int_V \exp\left(-\frac{E}{R\theta}\right)\{\Phi\}\otimes\{\Phi\}^T dV, \quad [R_p] = \Gamma J_0 \int_V \exp\left(-\frac{E}{R\theta}\right)\{\Phi\}\otimes\{\Phi\}^T dV, \qquad (14)$$

$$[R_\theta] = \int_S q_e \{\Phi\} dS - \int_V \{\Phi\}\psi dV, \quad [B] = \{L\}\otimes\{\Phi\}^T, \quad \{L\} = \left(\frac{\partial}{\partial x}, \frac{\partial}{\partial y}, \frac{\partial}{\partial z}\right),$$

$$\psi = c_g \nabla\theta \cdot \mathbf{K} \cdot \nabla p - J\Delta e^0 = c_g \{\vartheta\} \cdot [B]^T \cdot [K] \cdot [B] \cdot \{y\} - J\Delta e^0.$$

Collecting system (13) for all finite elements and applying the Runge-Kutta method to solve it, we obtain a linear system of equations with a sparse, asymmetric matrix. To solve it, a numerical bi-conjugate gradient method is used.

To solve the problem of the theory of elasticity (8) - (9), the finite element method is also used.

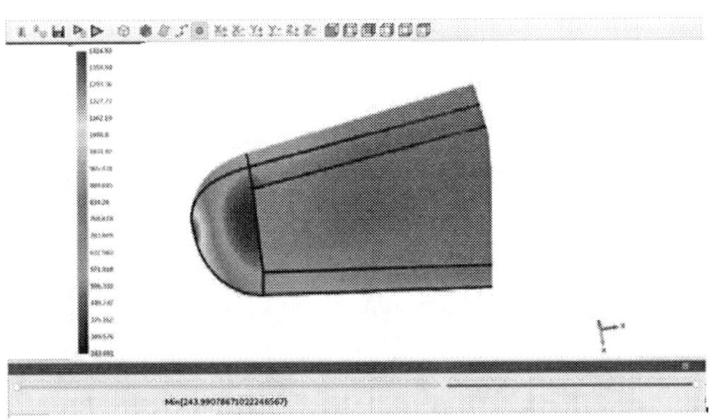

Figure 1. Temperature distribution in the nose cone of modeling RV, K

4.Results of computational simulation.

A numerical solution of the problem was carried out for a model RV moving at a speed of M = 6. The design of the model RV consisted of two materials: phenolic fiberglass (critical blunting area), and epoxy fiberglass (shell). In fig. 1 shows the temperature distribution over the thickness of the structure of RV. The most heat-loaded region is critical blunting, where the temperature reached 1400 K. Under the influence of high temperatures, the phenolic matrix of the composite began to decompose, resulting in the formation of a large amount of gas in the pores of the composite. In fig. 2. shows the distribution of pore pressure. The maximum values were achieved in the material with an epoxy matrix, despite the fact that in this region the temperature is lower than in the region of blunting. This effect is explained by the presence of higher primary porosity in the phenolic matrix, compared with epoxy. In fig. 3. The transverse stresses in the RV structure are presented. Their maximum value is achieved in the region of blunting at a certain depth from the heated surface.

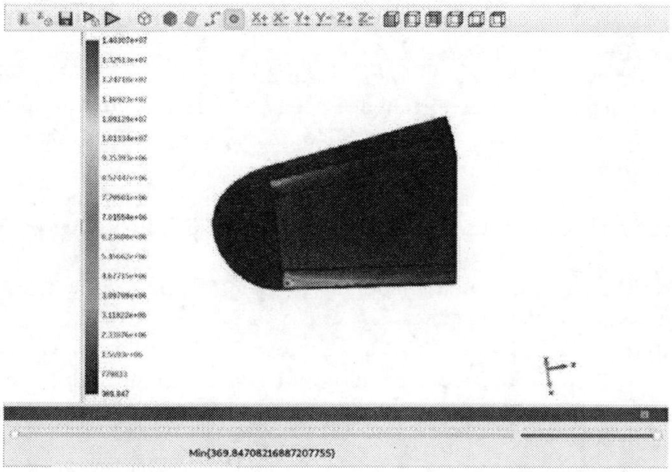

Figure 2. Pore pressure distribution in the nose cone of modeling RV, Pa.

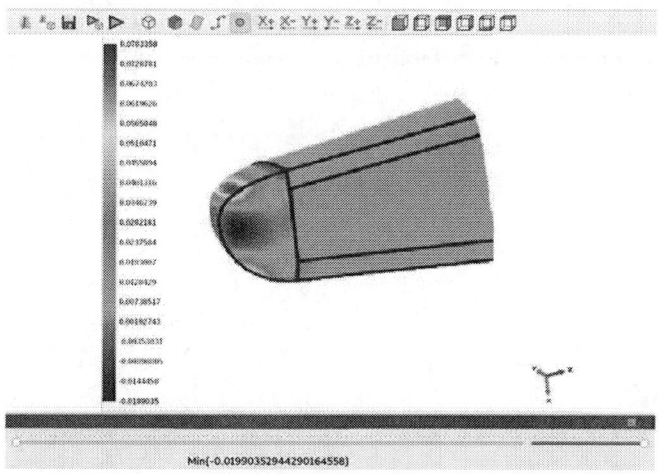

Figure 3. Transverse normal stress distribution in the nose cone of modeling RV, GPa.

5.Conclusions

The formulation of the conjugate problem of aerodynamics and thermomechanics of a ablative structure made of thermally destructive composites is considered. Numerical finite-volume finite-element methods for solving this problem are developed. An example of computational simulation for the modeling RV is given. Some effects of the stress state of structures due to thermal decomposition of the composite at high temperatures have been established.

References

[1] Dimitrienko Yu.I. *Thermomechanics of Composites Structures under High Temperatures.* Springer.2015. 367 p.
[2] Dimitrienko Yu.I. A structural thermomechanical model of textile composite materials at high temperatures. *Composite science and technologies.* 1999. Vol.59. P. 1041-1053.
[3] Dimitrienko Yu.I. Thermomechanical Behaviour of Composite Materials and Structures under High Temperatures. Part 2. Structures. *Composites. Part A: Applied Science and Manufacturing.* 1997. Vol.28A. P. 463-471.

[4] J. D. Anderson, Hypersonic and High-Temperature Gas Dynamics. 2nd ed. American Institute of Aeronautics and Astronautics, Reston, Virginia, 2006.
[5] V. V. Kuzenov et al Calculating processes of laminar and turbulent heat transfer around the elements of the aircraft, J. Phys.: Conf. Ser. 980 012023, 2018
[6] Dimitrienko Yu I, Koryakov M N, Zakharov A A Application of Finite Difference TVD Methods in Hypersonic Aerodynamics. Finite Difference Methods,Theory and Applications. *Lecture Notes in Computer Science. Springer.* Volume 9045, 2015, pp 161-168 Date: 17 Jun 2015 (DOI: 10.1007/978-3-319-20239-6_15
[7] Yanyan C, Nelyub V A, Malysheva G V 2019 An Investigation of the Kinetics of the Heating Process for Parts Made of Carbon Fiber in the Process of Curing *Polymer Science - Series D* **12** 296-299
[8] Valishin A A 2017 Management and modeling of the influence of cracks on the temperature distribution in polymers and composites *Espacios* **38** 38
[9] Galimov E R, Tukbaev E E, Fedyaev V L, Ya Galimova N, Samoylov V M, Orlov M A, Borodulin A S, Mavleev I R 2018 Production technology of heat-resistant high-strength syntactic carbon foams for operation in extreme conditions *IOP Conference Series: Materials Science and Engineering* 412 012018
[10] Reznik S V, Kolesnikov A F, Prosuntsov P V, Gordeev A N, Mikhailovskii K V 2019 Development of Elements of a Reusable Heat Shield from a Carbon–Ceramic Composite Material. 2. Thermal Tests of Specimens of the Material *Journal of Engineering Physics and Thermophysics* **92** 306-313
[11] Mikhaylovskiy K, Prosuntsov P 2018 Modelling of thermal and stress-strain state of transformable space structures from hybrid composite materials *MATEC Web of Conferences* **194** 01039

Modeling the nonlinear permeability of porous composite structures with non-Newtonian fluids

Yu I Dimitrienko and Li Shuguang

Department of Computational Mathematics and Mathematical Physics, Bauman Moscow State Technical University, 5 Baumanskaya 2-ya, Moscow, 105005, Russia

Abstract. This work is related to the modelling filtration flow of an incompressible non-Newtonian viscous fluid in porous composite structures. A physical mathematical model of an incompressible non-Newtonian fluid flowing in a porous composite structure has been proposed. The pore–scale description of flow in a reinforced composite is obtained using asymptotic homogenization method. Then, the nonlinear filtration law is investigated theoretically using the anisotropic tensor function representation of the tensor independent variable. The finite element method was used to calculate the local problem, and the distribution of single hole velocity, pressure and non-Newtonian viscosity was obtained. Based on the numerical results verified by the famous Darcy's law, the nonlinear filtration law of Carreau viscosity fluid was explored and the effective permeability under different parameters was obtained.

1. Introduction

In the production process of polymer composites, techniques based on impregnation of reinforcing cages with liquid binders such as epoxy resins are commonly applied and produced. A more representative of these process technologies is the RTM method [1,2]. The RTM method allows the production of high-quality polymeric products, but at the same time, there is a need for expensive injection equipment and the manufacture of rigid equipment. And the quality of composite structures obtained using RTM methods largely depends on the parameters of the manufacturing technology: geometric parameters of the structure, permeability and porosity of the dry frame, impregnation pressure, and also the viscosity of the liquid binder.

The liquid binder is usually a macromolecular polymer solution containing various additives (coagulation promoters, modifiers, nanoparticles, etc., for increasing the mechanical properties of the composite) [1,3]. Therefore, the viscosity characteristics of liquid binder are generally non-Newtonian. Therefore, it is extremely important to study the filtration flow of non-Newtonian liquid binders in a porous composite structure having a spatial geometry of complex pores.

In phenomenological theories of filtration based on Darcy's law and its modification, rather rough estimates of real processes occurring inside pores with complex geometry inherent in composite materials are obtained, which leads to large deviations in determining permeability. In this regard, an important part of the filtration study is the analysis of fluid flow local processes in a single pore and the derivation of averaged filtration equations based on continuum mechanics, and not based on phenomenological approaches [4-8]. In this work, this multi-scale modeling method based on the asymptotic averaging method was used to model the filtration flow of non-Newtonian fluids in a porous composite structure. This method can obtain a mathematically reasonable averaging equation for a homogeneous medium based on an asymptotic analysis of the exact initial equation of continuum mechanics ("according to the first principle"). The asymptotic homogenization method has been well applied to the local transport process modeling of Newtonian fluids in porous systems [9,10], determining the strength characteristics of reinforced composites [11-15].

2. General expression of the incompressible non-Newtonian fluid flow in a porous composite structurel

2.1. Geometric model

In this paper, a three-dimensional orthogonal composite structure is considered [10] (as shown in Figure 1). It is assumed that the porous composite structure is a uniform medium having no dead holes and having a periodicity. And periodic cells have geometric and physical symmetry about the local coordinate system plane, which allows for the consideration of 1/8 of periodic cells(as shown in Figure 2)..

Fig. 1. V - periodic porous structure;.

Fig. 2. V_ξ - *1/8* periodicity cells.

2.2. Non-Newtonian fluid assumption

For the properties of non-Newtonian fluids, the following assumptions are made: (1) fluids are isotropic non-Newtonian viscous incompressible media; (2) the porous skeleton is assumed to be non-deformable, i.e. his movement is not considered; (3) fluid movement is isothermal; (4) mass forces are absent.

We introduce the following notation: l_0-characteristic size of periodicity cell (PC) V_ξ; x_0-characteristic size of the global entire porous medium V; $\kappa = l_0/x_0 << 1$-small parameter; $\vec{\xi} = \vec{x}/\kappa$-dimensionless local coordinates varying within a periodicity cell (PC) V_ξ. .

3. Mathematical model of non-Newtonian fluid motion

Under the framework of the hypothesis, the motion of the non-Newtonian fluid in the porous composite structure is described by the continuity equation and the momentum equation of the incompressible fluid, which together with the attachment conditions on the solid surface have the following form [9]:

$$\nabla \cdot \vec{v} = 0 \tag{1}$$

$$\rho\left(\frac{\partial \vec{v}}{\partial t}+\vec{v}\cdot\nabla\otimes\vec{v}\right)=-\nabla p+\nabla\cdot\mu\left(\nabla\otimes\vec{v}+\nabla\otimes\vec{v}^T\right) \tag{2}$$

$$\vec{v}\big|_\Sigma=0 \tag{3}$$

$$p\big|_{t=t_0}=p_0 \tag{4}$$

where ρ -density; $\vec{v}$ -velocity of fluid; p -pressure; μ -non-newtonian viscosity and $\otimes$ -tensor product.
The strain rate tensor:

$$D=\frac{1}{2}\left(\nabla\otimes\vec{v}+\nabla\otimes\vec{v}^T\right)$$

The A_I and A_V models of Voigt isotropic viscous media [13] are considered, and non-Newtonian viscosity satisfies the following defining relation [1]:

$$\frac{\mu-\mu_\infty}{\mu_0-\mu_\infty}=\left(1+\lambda^2 I_2^2\right)^{\frac{n-1}{2}} \tag{5}$$

which depends on the second invariant of the strain rate tensor [14]:

$$I_2(D)=\sqrt{2D\cdot\cdot D}$$

Then we used a dimensionless form, where:

$$\bar{\rho}=\frac{\rho}{\rho_0},\ \bar{v}=\frac{\vec{v}}{v_0},\ \bar{p}=\frac{p}{p_0},\ \bar{t}=\frac{t}{t_0},\ \bar{x}=\frac{\vec{x}}{x_0},\ t_0=\frac{x_0}{v_0}.$$

Here $\bar{p}$, $\bar{\rho}$, $\bar{v}$ are the dimensionless pressure, density and velocity of fluid, respectively; p_0, ρ_0, v_0 are their typical magnitudes. t_0 is the dimensionless time.

The equations for the flow of an incompressible fluid can be written in a dimensionless form as follows (the symbol is omitted below):

$$\nabla\cdot\vec{v}=0 \tag{6}$$

$$\frac{\partial \vec{v}}{\partial t}+\vec{v}\cdot\nabla\otimes\vec{v}=-Eu\nabla p+\frac{1}{\mathrm{Re}}\nabla\cdot\mu\left(\nabla\otimes\vec{v}+\nabla\otimes\vec{v}^T\right) \tag{7}$$

$$\mu=s+(1-s)\left(1+Cu^2 I_2^2\right)^{\frac{n-1}{2}} \tag{8}$$

$$\vec{v}\big|_\Sigma=0,\ p\big|_{t=t_0}=p_0 \tag{9}$$

where $Eu = \frac{p_0}{\rho_0 v_0^2}$ - Euler number, $\mathrm{Re} = \frac{\rho_0 v_0 x_0}{\mu_0}$ - Reynolds number, $Cu = \frac{\lambda v_0}{x_0}$ - Carreau number.

4. Local problems in non-Newtonian fluid filtration process

In the framework of the method of asymptotic averaging [15], the application of the asymptotic homogenization method to the initial systems of equations for the incompressible non-Newtonian fluid leads to local filtration problems on the periodicity cell. It is noteworthy that the local problems of incompressible non-Newtonian viscous fluid:

$$\nabla_{\xi} \cdot \vec{v}^{(0)} = 0 \tag{10}$$

$$-\nabla_{\xi} p^{(1)} + \eta_0 \nabla_{\xi} \cdot \left(2\mu^{(0)} D^{(0)}\right) = \nabla_x p^{(0)} \tag{11}$$

$$D^{(0)} = \frac{1}{2}\left(\nabla_{\xi} \otimes \vec{v}^{(0)} + \nabla_{\xi} \otimes \vec{v}^{(0)T}\right) \tag{12}$$

$$\mu^{(0)} = s + (1-s)\left(1 + Cu^{0\,2} Y^{(0)}\right)^{\frac{n-1}{2}}, \quad Y^{(0)} = 2D^{(0)} \cdot\cdot D^{(0)} \tag{13}$$

$$\vec{v}^{(0)}\Big|_{\Sigma_{\xi g}} = 0, \quad \left\langle p^{(1)} \right\rangle = 0, \quad [\![\vec{v}^{(0)}]\!] = 0, \quad [\![p^{(1)}]\!] = 0 \tag{14}$$

where unknown $\vec{v}^{(0)}$ and $p^{(1)}$, the symbol $[\![\cdot]\!]$ denotes the periodicity conditions. $\nabla_x p^{(0)}$ is considered as "input data". And $\mathrm{Re} = \kappa^0 \mathrm{Re}_0$, $Eu = \kappa^2 Eu_0$ and $Cu = \kappa^0 Cu_0$. Then $\eta_0 = \frac{1}{\mathrm{Re}^0 \, Eu^0}$.

5. Nonlinear filtering law

From system (10-14) it can be seen that the effective filtration law associated with the average velocity and pressure gradient can be written in the following form of function:

$$\left\langle \mathbf{v}^{(0)} \right\rangle = F\left(\nabla_x p^{(0)}\right)$$

where $\left\langle \mathbf{v}^{(0)} \right\rangle$ - flow rate (filtration), F - functions depending on $\nabla_x p^{(0)}$.

Function F is indicated if there is information about the geometric symmetry type of the periodicity cell of the composite and the anisotropy type of the fibers and matrix [14]. Since the basic assumption of symmetry is accepted, the solution of all local problems will have the indicated type of symmetry. This means that the function F must have this symmetry. But the transformations listed above and the identical transformations form an orthotropy group, therefore the function F will be a tensor function that is indifferent with respect to the orthotropy group. But then for such a function, we can use its representation in the tensor basis of the orthotropy group, which has the form:

$$\left\langle \mathbf{v}^{(0)} \right\rangle = \sum_{\alpha=1}^{3} \varphi^{(\alpha)}\left(I_1^p, I_2^p, I_3^p\right) \mathbf{e}_{\alpha}$$

Or

$$\left\langle v_i^{(0)} \right\rangle = \sum_{\alpha=1}^{3} \varphi^{(\alpha)}\left(I_1^p, I_2^p, I_3^p\right) \delta_{i\alpha}$$

where $I_i^p = \dfrac{\partial p^{(0)}}{\partial x_i}$ -scalar functions of pressure gradient invariants.

These $\varphi^{(\alpha)}$ functions in fact represent the desired effective permeability of porous media. For orthogonal fields, we write the formulas $\varphi^{(\alpha)}$ explicitly:

$$\varphi^{(\alpha)}\left(I_1^p, I_2^p, I_3^p\right) = \widetilde{\varphi}^{(\alpha)}\left(\left|I_1^{p(\alpha)}\right|, \left|I_2^{p(\alpha)}\right|, \left|I_3^{p(\alpha)}\right|\right) I_\alpha^p$$

$$= \widetilde{\varphi}^{(\alpha)} p_{,\alpha}^{(0)}$$

Therefore:

$$\left\langle \mathbf{v}^{(0)} \right\rangle = -\mathbf{K} \cdot \nabla_x p^{(0)}$$

where $\mathbf{K} = \sum_{\alpha=1}^{3} \widetilde{\varphi}^{\alpha} \mathbf{e}_\alpha^2$ - permeability tensor. $\mathbf{e}_\alpha^2 = \mathbf{e}_\alpha \otimes \mathbf{e}_\alpha$ - dyadic basis. .

6. Results

Next, the finite element method will be used to solve the local problems (16)-(19). We only consider finite elements that meet LBB conditions [16]. Here, the chemical industry raw material benzene is taken as an example for numerical study, so $\eta_0 = 0.0652$.

6.1. The adequacy of the mathematical model

The validity of the model is first verified using Newtonian fluid. It can be seen from the results that the permeability of the orthogonal structure is isotropic and linearly depends on the pressure gradient. This is consistent with existing results.

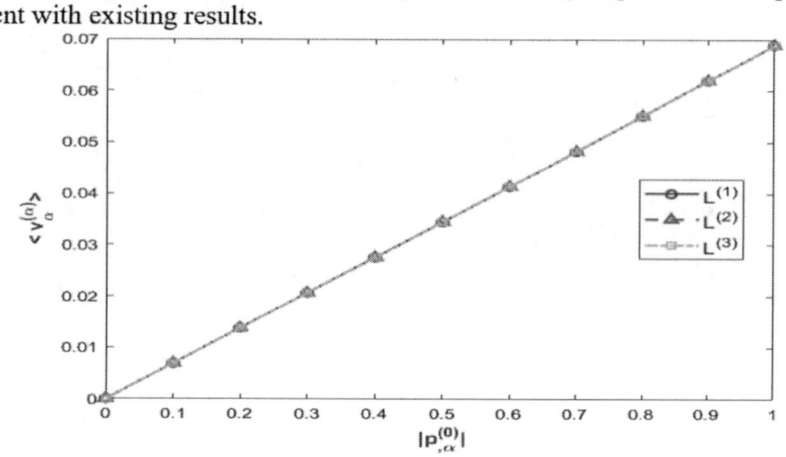

Figure 3. The Darcy filtration law.

6.2. Numerical simulation of local problems

Next, in the orthogonal structure, the results of calculating the local problem $L^{(1)}$ for pseudoplastic fluids at $\nabla_x p^{(0)} = (1,0,0)^T$, showing the distributions of the velocity $\tilde{v}_1^{(0)}$ and viscosity $\tilde{\mu}^{(0)}$ components in 1/8 of the periodic cells, respectively, follow.

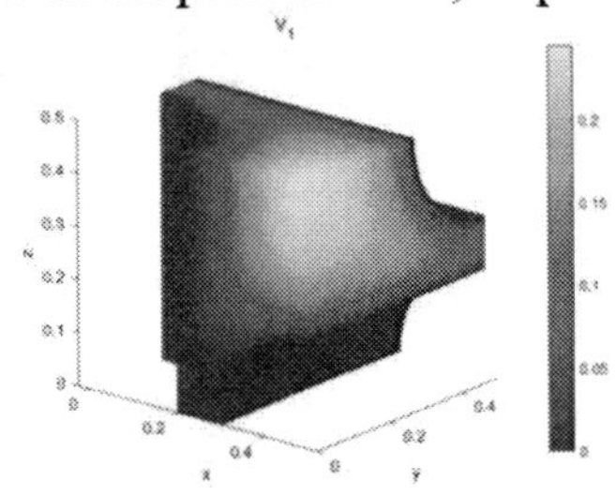

Figure 4. Distribution of the velocity $\tilde{v}_1^{(0)}$.

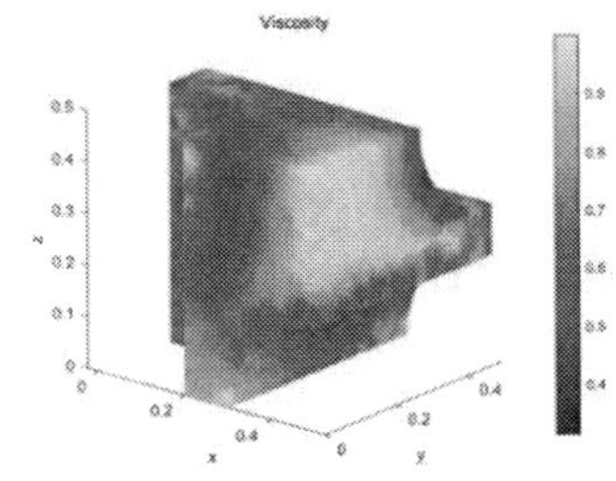

Figure 5. Distribution of the viscosity $\tilde{\mu}^{(0)}$.

6.3. The nonlinear filtration law

Based on the multiscale model, we then analyze the non-linear law of filtration of non-Newtonian fluids. Figure 6 shows the relationship between the effective filtration rate and the pressure gradient for non-Newtonian fluids at various power indices in the orthogonal structure.

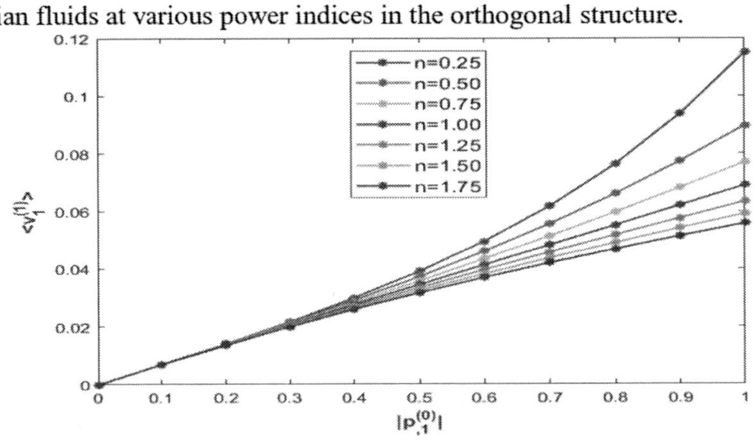

Figure 6. Distribution of permeability $\langle v^{(0)} \rangle$ and macro pressure gradient $\nabla_x p^{(0)}$ and power law index n .

7. Conclusions

Development and application of the asymptotic homogenization method for filtering non-Newtonian fluid in porous media. A local problem of the non-Newtonian fluids flow on a periodicity cell of porous media is obtained. The local problem of the non-Newtonian fluids flow is solved by the finite element method. Formulation of the nonlinear filtration law of the non-Newtonian fluids flow in porous media. The filtration law of non-Newtonian fluids has a nonlinear relationship with the modulus of the macro pressure gradienta.

References

[1] Bird R B, Armstrong R C and Hassager O 1987 *Dynamics of polymeric liquids. Vol. 1: Fluid*

mechanics. John Wiley, 649p.

[2] Idris Z, Orgéas L, Geindreau C, et al. 2004Microstructural effects on the flow law of power-law fluids through fibrous media. *Modelling and Simulation in Materials Science and Engineering*, **12(5)** pp. 995.

[3] Dimitrienko Yu.I and Li Shuguang 2018 Mathematical simulation of non-isothermal steady flow of non-Newtonian fluid by finite element method. *Mathematical Modeling and Computational Methods*, **2** pp. 70-95.

[4] Taraskin N Y and Prosuntsov P V 2019 Impact analysis of the pore shapes on radiation heat transfer in porous carbon-ceramic composite materials *AIP Conference Proceedings* **2171** 030011

[5] Barinov D Y and Prosuntsov P V 2019 Modelling of destruction of carbon-ceramic composite materials with variable density *AIP Conference Proceedings* **2171** 170011

[6] Zarubin V S and Sergeeva E S 2017 Porosity influence of power generating equipment structural materials on its thermoelastic characteristics and thermal *conductivity Journal of Physics: Conference Series* **891** 012321

[7] Nelyub V A, Borodulin A S, Kobets L P and Malysheva G V 2018 Thixotropy Hysteresis and Structure Formation in Elastomeric Suspensions *Inorganic Materials: Applied Research* **9** 603-608

[8] Komarov K A, Yarkov A V and Yurchenko S O 2019 Diagrammatic method for tunable interactions in colloidal suspensions in rotating electric or magnetic fields *Journal of Chemical Physics* **151** 244103

[9] Dimitrienko Yu I and Dimitrienko I D 2013 Simulation of local transfer in periodic porous media. *European Journal of Mechanics/B-Fluids*, **37** pp. 174-179.

[10] Dimitrienko Yu I and Bogdanov I O 2017 Multiscale modeling of filtration liquid binding processes in composite designs at RTM production method. *Mathematical Modeling and Computational Methods*, **2** pp. 3–27.

[11] Dimitrienko Yu I and Dimitrienko I D 2017 Modeling of the thin composite laminates with general anisotropy under harmonic vibrations by the asymptotic homogenization method *Journal for Multiscale Computational Engineering* **15(3)** P. 219-237

[12] Dimitrienko Yu I and Dimitrienko I D 2019 Computations of stresses and energy dissipation in composite thin laminates with the asymptotic vibration theory *Computers & Mathematics with Applications*. **78(8)** pp.2541-2559

[13] Dimitrienko Yu I 2010 *Nonlinear continuum mechanics and large inelastic deformations*. Springer 747p.

[14] Dimitrienko Yu I 2002 *Tensor analysis and nonlinear tensor functions*. Springer 662p.

[15] Dimitrienko Yu I 2015 *Thermomechanics of composites under high temperatures*. Springer, 434p.

[16] Zienkiewicz O C, Taylor R L and Zhu J Z 2005 *The Finite Element Method: Its Basis and Fundamentals: Its Basis and Fundamentals*. Elsevier, , 733p.

IOP Conf. Series: Materials Science and Engineering **934** (2020) 012015 doi:10.1088/1757-899X/934/1/012015

Modeling of stresses in inorganic composite plates under non uniform high temperature heating

Yu I Dimitrienko, E S Egoleva, D O Yakovlev and S V Sborschikov

Bauman Moscow State Technical University, Moscow, 105005, Russia

E-mail: dimit@bmstu.ru

Abstract. An asymptotic theory of thin composite plates with multi-stage high-temperature phase transformations is proposed. The theory is based on an asymptotic analysis of 3-dimensional equations of the mechanics of composite materials, taking into account phase transformations. Phase transformations are described by a system of kinetic equations, which is solved in conjunction with the equations of internal heat and mass transfer. The so-called local problems for plate theory are obtained and the averaged equations of plate theory with phase transformations are derived. An example of a composite based on an aluminum-chromo-phosphate matrix under uneven heating is considered. A numerical-analytical solution showed that although the heating is uneven due to the peculiarities of phase transformations, the stress-strain state changes significantly in time and is variable across the plate thickness. The developed theory allows us to calculate the stress distribution in the plate with high accuracy.

1. Introduction

Composite materials based on inorganic matrices, for example, aluminum phosphate, aluminum chromium phosphate, magnesium phosphate, and ceramic fibers based on SiO2, Al2O3, ZrO2, and others, are promising materials for the electrical, engineering, and aerospace industries [1-3]. These materials are able to withstand high temperatures, for a long time up to 500 ° C, and for a short time, more than 1000 ° C, while maintaining a sufficiently high mechanical and electromagnetic properties. In contrast to polymeric materials, in which mechanical properties are degraded upon heating, inorganic matrix composites (IMC) under certain conditions, temperature hardening is possible, and only then degradation. The reason for such unusual properties of IMC is the complex chain of phase transformations that occur in the matrix and fibers when heated to high temperatures. There are currently very few mathematical models of the mechanical behavior of IMC taking into account phase transformations. Apparently, the first such model was proposed in [3,4]. The aim of this work is to use this model of thermomechanical behavior of IMC to calculate stresses in thin plates under non-uniform high-temperature heating.

To calculate the stresses in thin plates, an asymptotic analysis of 3-dimensional equations of the theory of elasticity [5-15] is used, which provides high accuracy in calculating the fields of stresses, strains and displacements, and is not internally contradictory, unlike many traditional plate theories.

2. Statement of the three-dimensional problem of the linear theory of thermoelasticity for a thin plate with phase transformations

Consider a thin plate of tissue IMC, in which multi-stage high-temperature phase transformations (MHPT) occur in the matrix and fibers upon heating. We introduce a small parameter $h / L = \chi << 1$, where h is the thickness and L is the plate length. We denote the dimensionless coordinates: $x_k = \hat{x}_k / L$, $k = 1,2,3$, where $\hat{x}_k$ are the ordinary rectangular Cartesian coordinates, such that the axis $O\hat{x}_3$ is normal to the upper and lower planes of the plate, and the axis $O\hat{x}_1$, $O\hat{x}_2$ belong to the median plane of the plate. We write in a dimensionless form the non-stationary problem of heat conduction, the three-dimensional problem of the linear theory of thermoelasticity for a plate with MHPT, and also the system of kinematic equations for phase concentrations in the matrix and fibers

$$\rho(\varphi)C(\varphi)\frac{\partial\theta}{\partial t}=-\chi^2\frac{\partial q_i}{\partial x_i},\quad q_i=-\lambda_{ij}(\varphi)\frac{\partial\theta}{\partial x_j}, \tag{1}$$

$$\frac{\partial\sigma_{ij}}{\partial x_j}=0,\quad \varepsilon_{ij}=\frac{1}{2}\left(\frac{\partial u_i}{\partial x_j}+\frac{\partial u_j}{\partial x_i}\right),\quad \sigma_{ij}=A_{ijkl}(\varphi)\left(\varepsilon_{kl}-\varepsilon_{kl}^T\right),\quad \varepsilon_{kl}^T=\alpha_{kl}(\varphi)(\theta-\theta_0), \tag{2}$$

$$\Sigma_{3\pm}:\sigma_{i3}=-\chi^3 p_{\pm}\delta_{i3},\quad \theta=\theta_{e\pm}, \tag{3}$$

$$\Sigma_T: q_I n_I=0;\quad u_i=u_{ei} \tag{4}$$

$$\rho_{sw}\frac{\partial\varphi_{sw}}{\partial t}=\sum_{v=1}^{N-1}J_{svw}\varphi_{vw},\quad w=\{a,b\},\quad s=\overline{1,N},\quad \rho_h\frac{\partial\varphi_h}{\partial t}=J_h{}^0\varphi_h f_h(\theta) \tag{5}$$

$$\sum_{s=1}^{N}\varphi_{sw}=\varphi_w,\quad \varphi_h+\varphi_l=\varphi_f,\quad \varphi_a+\varphi_b+\varphi_f+\varphi_g=1, \tag{6}$$

$$t=0:\quad \varphi_{sw}=\varphi_{sw}{}^0,\quad \varphi_h=\varphi_f,\quad \theta=\theta^0 \tag{7}$$

Dimensionless quantities are indicated here: t - time, u_j - components of the displacement vector, θ - temperature, ρ - density, c - heat capacity of the composite, $\sigma_{ij},\varepsilon_{ij}$ - components of the tensors of stresses and deformations, A_{ijkl} - components of the tensor of elastic moduli, - components of the tensor of thermal conductivity, q_i - components of the heat flux vector, u_{ei} - given components displacement vectors at the ends Σ_T of the plate, $p_{\pm}$, $\theta_{e\pm}$ - specified pressures and temperature on the outer surfaces $\Sigma_{3\pm}$ of the plate, φ_{sw} - volumetric phase concentrations in the matrix, w - index of the initial component in the matrix, s - number of the phase of each component of the matrix ($s=1,\dots,N$), φ_h -is the concentration of the initial (amorphous) phase of the fiber, φ_l -is the crystalline phase in the fibers, φ_g -is the porosity, ρ_{sw}, ρ_h - is the phase density. The matrix J_{svw} of multistage phase transformations rates has a size $N\times(N-1)$ and is characterized by a ribbon structure (2 elements in each row, one with "+", the other with "-"), for example, for the case $N=4$

$$J_{svw}=\begin{pmatrix} -J_{1w}{}^0 f_{1w}(\theta) & 0 & 0 \\ J_{1w}{}^0 f_{1w}(\theta)(1-\Gamma_1) & -J_{2w}{}^0 f_{2w}(\theta) & 0 \\ 0 & J_{2w}{}^0 f_{2w}(\theta) & -J_{3w}{}^0 f_{3w}(\theta) \\ 0 & 0 & J_{3w}{}^0 f_{3w}(\theta)(1-\Gamma_4) \end{pmatrix} \tag{8}$$

The functions $f_{sw}(\theta)$ and $f_h(\theta)$ obey the law of Arrhenius: $f_{sw}(\theta)=\exp(-E_{sw}/\theta)$, $f_h(\theta)=\exp(-E_h/\theta)$, where E_{sw} and E_h are the dimensionless activation energies of phase transformations. All plate characteristics: $A_{ijkl}(\varphi)$, $\lambda_{ij}(\varphi)$, $\alpha_{kl}(\varphi)$, $\rho(\varphi)C(\varphi)$ - depend on phase concentrations $\varphi=(\varphi_{1w},\dots,\varphi_{sw},\varphi_h,\varphi_l)$, $w=\{a,b\}$.

3. Asymptotic expansions

We introduce the local dimensionless coordinate $\xi = x_3 / \chi$, $-0.5 < \xi < 0.5$. The solution to problem (1) - (7) will be sought in the form of asymptotic expansions in terms of a small parameter χ

$$\begin{gathered}\varphi = \varphi^{(0)}(x_I,\xi) + \chi\varphi^{(1)}(x_I,\xi) + \chi^2\varphi^{(2)}(x_I,\xi) + \chi^3 \ldots, \\ \theta = \theta^{(0)}(x_I,\xi) + \chi\theta^{(1)}(x_I,\xi) + \chi^2\theta^{(2)}(x_I,\xi) + \chi^3 \ldots \\ u_k = u_k^{(0)}(x_I) + \chi u_k^{(1)}(x_I,\xi) + \chi^2 u_k^{(2)}(x_I,\xi) + \chi^3 u_k^{(3)}(x_I,\xi) + \ldots \\ \sigma_{ij} = \sigma_{ij}^{(0)} + \chi\sigma_{ij}^{(1)} + \chi^2\sigma_{ij}^{(2)} + \ldots\end{gathered} \tag{9}$$

Substituting expansions (9) into problem (1) - (7) and grouping the terms by degrees χ, we obtain a recurrent sequence of local problems of thermoelasticity with allowance for multistage phase transformations. The problem in the zeroth approximation has the form:

$$\begin{gathered}\rho_{sw}\frac{\partial\varphi_{sw}^{(0)}}{\partial t} = \sum_{v=1}^{N-1} J_{svw}{}^{(0)}\varphi_{vw}{}^{(0)}, \quad w=\{a,b\}, \; s=\overline{1,4}, \quad \rho_h\frac{\partial\varphi_h^{(0)}}{\partial t} = J_h{}^{(0)}\varphi_h^{(0)} f_h\left(\theta^{(0)}\right), \\ \rho C\left(\varphi^{(0)}\right)\frac{\partial\theta^{(0)}}{\partial t} = \left(\lambda_{33}^{(0)}\theta_{/3}^{(0)}\right)_{/3}, \\ \sigma_{i3/3}^{(0)} = 0, \; \sigma_{I3}^{(0)} = 2A_{I3K3}\left(\varphi^{(0)}\right)\varepsilon_{K3}^{(0)}, \; \sigma_{33}^{(0)} = A_{33IJ}(\varphi)(\varepsilon_{IJ}^{(0)} - \varepsilon_{IJ}^{(0)T}) + A_{3333}\left(\varphi^{(0)}\right)(\varepsilon_{33}^{(0)} - \varepsilon_{33}^{(0)T}), \\ \varepsilon_{IJ}^{(0)} = \frac{1}{2}\left(u_{I,J}^{(0)} + u_{J,I}^{(0)}\right), \; \varepsilon_{I3}^{(0)} = \frac{1}{2}\left(u_{3,I}^{(0)} + u_{I/3}^{(1)}\right), \; \varepsilon_{33}^{(0)} = u_{3/3}^{(1)}, \\ \Sigma_{3\pm}: \theta^{(0)} = \theta_{e\pm}, \quad \sigma_{i3}^{(0)} = 0; \quad \int_{-0.5}^{0.5} u_i^{(1)} d\xi = 0; \\ t=0: \quad \varphi_{sw}^{(0)} = \varphi_{sw}{}^{0}, \; \varphi_h^{(0)} = \varphi_f, \quad \theta^{(0)} = \theta^{0}.\end{gathered} \tag{10}$$

Hereinafter, the notation for the derivatives is introduced: $\dfrac{\partial f}{\partial\xi} = f_{/3}$, $\dfrac{\partial f}{\partial x_I} = f_{,I}$, and also the averaging operations

$$< f >= \int_{-0.5}^{0.5} f(\xi)d\xi; \qquad < f >_{\xi} = \int_{-0.5}^{\xi} f d\tilde{\xi} - < \int_{-0.5}^{\xi} f d\tilde{\xi} >, \; \{f\}_{\xi} = \int_{-0.5}^{\xi}\left(f(\tilde{\xi}) - \langle f\rangle\right)d\tilde{\xi} \tag{11}$$

The problems for higher approximations at $m = 1, 2, \ldots$ are as follows

$$\begin{gathered}\rho_{sw}\frac{\partial\varphi_{sw}^{(m)}}{\partial t} = \sum_{v=1}^{N-1}\sum_{m'=0}^{m} J_{svw}{}^{(m-m')}\varphi_{vw}{}^{(m')}, \quad w=\{a,b\} \; s=\overline{1,N}, \quad \rho_h\frac{\partial\varphi_h^{(m)}}{\partial t} = \sum_{m'=0}^{m} J_h{}^{0}\varphi_h^{(m')} f_h^{(m-m')}, \\ \sum_{m'=0}^{m}\left(\rho C^{(m-m')}\frac{\partial\theta^{(m')}}{\partial t} - \left(\lambda_{33}^{(m-m')}\theta_{/3}^{(m')}\right)_{/3}\right) - \sum_{m'=0}^{m-2}\left(\lambda_{IJ}^{(m-m')}\theta_{,J}^{(m')}\right)_{,I} = 0, \\ \sigma_{i3/3}^{(m)} + \sigma_{iJ,J}^{(m-1)} = h_i^{(m)}, \quad \sigma_{I3}^{(m)} = 2\sum_{m'=0}^{m} A_{I3K3}^{(m-m')}\varepsilon_{K3}^{(m')}, \\ \sigma_{33}^{(m)} = \sum_{m'=0}^{m} A_{33IJ}^{(m-m')}(\varepsilon_{IJ}^{(m')} - \varepsilon_{IJ}^{(m')T}) + A_{3333}^{(m-m')}(\varepsilon_{33}^{(m')} - \varepsilon_{33}^{(m')T}), \\ \varepsilon_{IJ}^{(m)} = \frac{1}{2}\left(u_{I,J}^{(m)} + u_{J,I}^{(m)}\right), \; \varepsilon_{I3}^{(m)} = \frac{1}{2}\left(u_{3,I}^{(m)} + u_{I/3}^{(m+1)}\right), \; \varepsilon_{33}^{(m)} = u_{3/3}^{(m+1)}, \\ \Sigma_{3\pm}: \theta^{(m)} = 0, \quad \sigma_{i3}^{(m)} = -p_{\pm}\delta_{i3}\delta_{m3}; \quad < u_i^{(m)} >= 0, \\ t=0: \quad \varphi_{sw}^{(m)} = 0, \; \varphi_h^{(0)} = 0, \quad \theta^{(0)} = 0.\end{gathered} \tag{12}$$

After the introduction of the functions, the equilibrium equations (2) take the form:

$$h_i^{(0)} + \chi h_i^{(1)} + \chi^2 h_i^{(2)} + \ldots = 0 \tag{13}$$

4. The solution of local problems

The problem of non-stationary thermal conductivity and the task of calculating the concentrations in system (10) are solved separately from the problem of mechanical equilibrium, and they are solved numerically. As a result, we find the distribution of temperatures $\theta^{(0)}(\xi, x_J, t)$ and phase concentrations $\varphi_{sw}^{(0)}$, $\varphi_h^{(0)}$. The solution of the equilibrium equations in system (10) has the form

$$\sigma_{i3}^{(0)} = 0, \quad u_I^{(1)} = -\xi u_{3,I}^{(0)}, \quad u_3^{(1)} = -\varepsilon_{KL}^{(0)} U_{3KL} + U_3^T, \tag{14}$$
$$U_{iKL}(\xi) = < Z_{iKL} >_\xi, \quad U_i^T(\xi) = < Z_{ikl} \varepsilon_{kl}^{T(0)} >_\xi,$$

where $Z_{ikl} = A_{i3j3}^{(0)-1} A_{j3kl}^{(0)}$.

Solving local problems of the theory of elasticity in system (12), we find recurrence formulas for stresses of higher approximations

$$\sigma_{i3}^{(m)} = -\{\sigma_{iJ,J}^{(m-1)}\}_\xi, \quad \sigma_{i3}^{(m+1)} = -\left\{\sigma_{iJ,J}^{(m)}\right\}_\xi - \left(p_- + \Delta p(\xi + 0.5)\right)\delta_{i3}\delta_{m3} \tag{15}$$

as well as expressions for constants $h_i^{(m)} = < \sigma_{iJ,J}^{(m)} > - \Delta p \delta_{i3} \delta_{m2}$, где $\Delta p = p_+ - p_-$.

For m = 1,2,3 stress and strain, we calculate explicitly through the derivatives of displacements $u_3^{(0)}$ and $u_1^{(0)}$

$$\sigma_{IJ}^{(1)} = \xi A_{IJKL}^{(0)} \eta_{KL} - A_{IJkl}^{(0)} \varepsilon_{kl}^{T(1)}, \quad \varepsilon_{KL}^{(1)} = \xi \eta_{KL}, \quad \eta_{KL} = -u_{3,KL}^{(0)} \tag{16}$$
$$\sigma_{I3}^{(1)} = -\left\{A_{IJKL}^{(0)}\right\}_\xi \varepsilon_{KL,J}^{(0)} + \left\{A_{IJkl}^{(0)}\right\}_\xi \varepsilon_{kl,J}^{T(0)}, \quad \sigma_{33}^{(1)} = 0.$$

5. Averaged equations of plates with multi-stage phase transformations

We denote the forces N_{IJ}, transvers forces Q_I and moments M_{IJ} in the plate

$$N_{IJ} = < \sigma_{IJ}^{(0)} > + \chi < \sigma_{IJ}^{(1)} > + \ldots, \quad M_{IJ} = \chi < \xi \sigma_{IJ}^{(0)} > + \chi^2 < \xi \sigma_{IJ}^{(1)} > + \ldots$$
$$Q_I = \chi < \sigma_{I3}^{(1)} > + \chi^2 < \sigma_{I3}^{(2)} > + \ldots, \tag{17}$$

Averaging the equilibrium equations in system (12) with allowance for (17), we obtain the moment equations and equilibrium equations for thin plates [15]

$$N_{IJ,J} = 0, \quad Q_{J,J} = \chi^2 \Delta p \quad M_{IJ,J} - Q_I = 0. \tag{18}$$

The averaged constitutive relations for plates with phase transformations have the following form

$$N_{IJ} = \bar{A}_{IJKL}\left(\varphi^{(0)}\right)\varepsilon_{KL}^{(0)} + B_{IJKL}\left(\varphi^{(0)}\right)\eta_{KL} - N_{IJ}^T,$$
$$M_{IJ} = B_{IJKL}\left(\varphi^{(0)}\right)\varepsilon_{KL}^{(0)} + \bar{D}_{IJKL}\left(\varphi^{(0)}\right)\eta_{KL} - M_{IJ}^T, \tag{19}$$

where membrane, mixed, flexural and also shear modules of the plate stiffnesses are indicated

$$\bar{A}_{IJKL}\left(\varphi^{(0)}\right) = < A_{IJKL}^{(0)} >, \quad A_{IJKL}^{(0)} = A_{IJKL} - A_{IJk3}\tilde{A}_{k3i3}^{-1}A_{i3KL}$$
$$B_{IJKL}\left(\varphi^{(0)}\right) = \chi < \xi A_{IJKL}^{(0)}\left(\varphi^{(0)}\right) >, \quad \bar{D}_{IJKL}\left(\varphi^{(0)}\right) = \chi^2 < \xi^2 A_{IJKL}^{(0)}\left(\varphi^{(0)}\right) >, \tag{20}$$
$$N_{IJ}^T = < A_{IJkl}^{(0)} \varepsilon_{kl}^{T(0)} >, \quad M_{IJ}^T = \chi < A_{I3M3} U_{3,M}^T > + \chi < A_{I3kl} \varepsilon_{kl}^{T(1)} >.$$

6. Numerical calculation results for a thin plate with uneven heating and bending

Figures 1 - 3 show the results of numerical modeling of the solution to the problem of deflection of a multilayer plate when exposed to a uniformly diştributed pressure p+ = 0.1 MPa and a uniform temperature field $\theta_{e\pm}$ on the upper surface (ξ = 0.5). The left end of the plate is pinched, the right one is freely supported, the lower surface (ξ = –0.5) is thermally insulated, the plate thickness is 0.025 m, length 1 m.

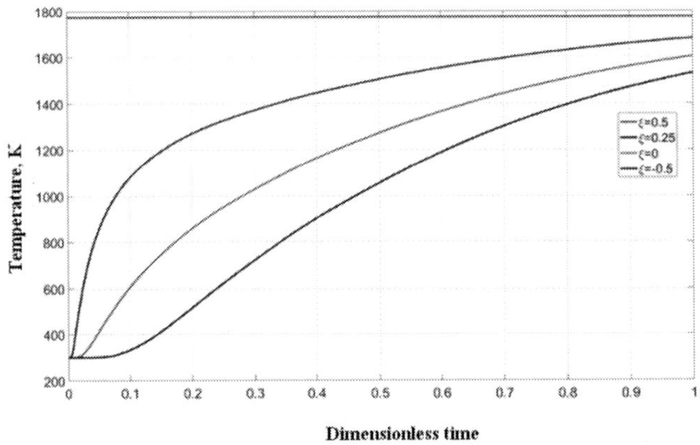

Figure 1. Temperature change from warm-up time at ξ = 0,5; 0,25; 0; –0,5.

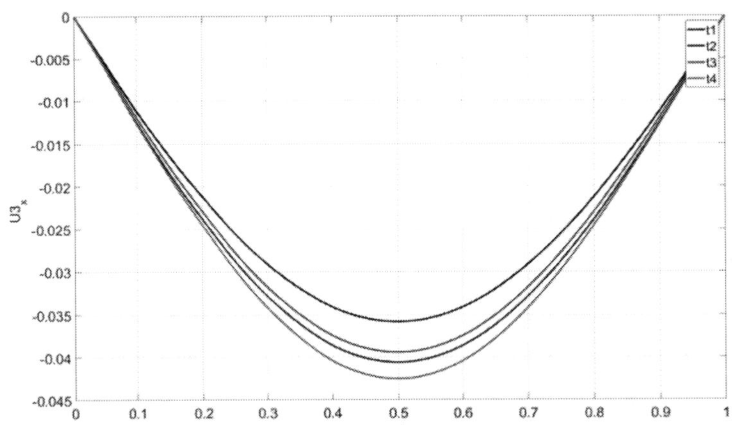

Figure 2. Distribution of the deflection along the length of the plate at different time moments.

Figure 1 shows the temperature change over time at different distances from the heating surface. As shown in Figure 1, a constant temperature is set on the upper surface, and the plate layer located at a distance of 0.25 from the upper surface (ξ = 0.25) warms up faster than the layer in the middle of the plate (ξ = 0) and the lower surface (ξ = - 0.5). At the beginning of heating, only the first phase was present in the composite material matrix; phases 4 and 5 remained at the end of heating. With an increase in the heating time, the concentration of the "amorphous" phase of the fiber decreases and the concentration of the "crystalline" phase of the fiber increases at a distance of 0.25 from the upper surface of the plate.

The maximum temperature of 1800 K is on the upper surface of the plate, and in time t1 the middle surface of the plate has time to warm up to a temperature of about 700 K, while the lower surface warms up only to 400 K. By the time t1, phase transformations in the binder and fiber only started in the part of the plate above the median surface, as evidenced by a constant modulus of elasticity in thickness at the bottom of the plate. At time t1 at the point ξ = 0.3, the elastic modulus of the binder is much higher than on the upper surface of the plate, this shows that there is a phase ratio of the binder in a certain temperature range at which the elastic properties will be greater than at the maximum temperature. This is a feature of the proposed model of phase transformations of the binder, which is characterized by an increase in the elastic properties of the binder in the temperature range 800–1100 K [6].

Figures 2 and 3 show the changes in shear stresses and deflection over the plate thickness in the cross section x1 = 0.75 at time instants t1< t2 < t3 < t4. Plate deflection increases nonlinearly over time.

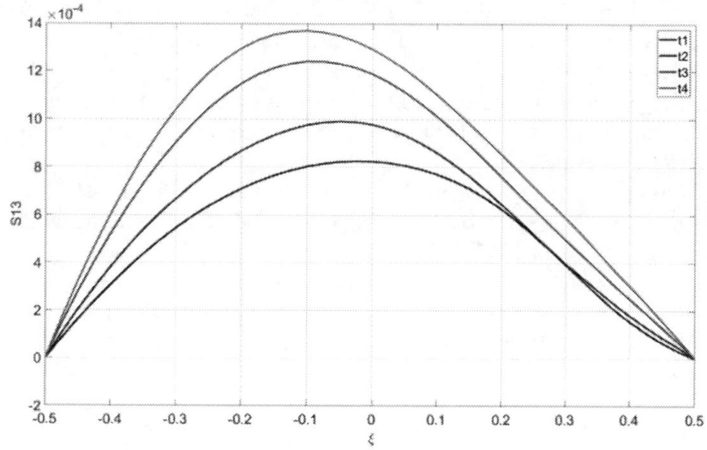

Figure 3. Distribution of the in shear stress σ_{13} along the length of the plate at different time moments in the cross section x1 = 0.75.

7. Conclusions

A modified version of the asymptotic theory of the calculation of thin elastic multilayer plates from high-temperature composites with multi-stage phase transformations is developed, taking into account the finite values of the shear characteristics of the composites.

A numerical-analytical solution to the problem of the stress-strain state of a thin-walled plate made of a fabric composite material with high-temperature phase transformations under non-uniform non-stationary heating and uniform pressure is obtained. The calculations performed to establish the effect of phase transformations on the kinetics of changes in the stress-strain state of the plate for different time instants, in particular, the appearance of internal zones of hardening of the material after preliminary heating to the beginning of the intermediate stages of phase transformations, and, as a result, an increase in the level of bending stresses in these zones.

References

[1] Abzgil'din F Yu and Tesvyatskiy S G 1980 *Asbo-phosphate materials.* Kiev. Naukova dumka 99 p.

[2] Kopeykin V A ed. 1974 *Technology and properties of phosphate materials*. Moscow, Stroyizdat,. 224 p.

[3] Dimitrienko Yu I, Sborschikov S V, Egoleva E S and Matveeva A A 2013 Modeling of thermo-elastic properties of composites with aluminochromic phosphate *matrices Science and Education. Electronic scientific and technical Journal* **11** pp.497-518

[4] Dimitrienko Yu I, Sborshikov S V, Gubareva E A, Egoleva E S and Matveeva A A 2015 Multiscale Modeling of High Temperature elastic-strength properties of inorganic matrix Composites *Mathematical Modeling*. **27 (11)** pp. 3-20.

[5] Dimitrienko Yu I and Yakovlev D O 2015 The Asymptotic Theory of Thermoelasticity of Multilayer Composite Plates. *Composites: Mechanics,Applications. An International Journa.* **6 (1)** pp.13-51.

[6] Dimitrienko Yu I 2015 *Thermomechanics of Composites Structures under High Temperatures* Springer. 367 p.

[7] Dimitrienko Yu I, Gubareva E A and Sborshikov S V 2015 Numerical simulation of composite material thermal expansion by homogenization method *Engineering Journal: Science and Innovation*, **12 (48)**

[8] Dimitrienko Yu I, Sborschikov S V, Belenovskaya Yu V, Aniskovich V A and Perevislov S N 2013 Modeling of Microdestruction and Strength of ceramic Composites on the base of reaction bonded SiC *Science and Education. Electronic scientific and technical Journal.* **11**, pp.475-496.

[9] Dimitrienko Yu I and Yurin Yu V 2017 Finite Element Modeling of Damage and Durability of Composite Structures with Local Delaminations *Mathematical Modeling and Computational Methods*. **3**, pp 49–70.

[10] Dimitrienko Yu I and Dimitrienko I D 2019 Computations of stresses and energy dissipation in composite thin laminates with the asymptotic vibration theory *Computers & Mathematics with Applications.* **78(8)** pp.2541-2559

[11] Gorodetskii M A, Nelyub V A, Malysheva G V, Shaulov A Y and Berlin A A 2018 Technology of Forming and the Properties of Reinforced Composites Based on an Inorganic Binder *Russian Metallurgy (Metally)* **13** 1195-1198

[12] Burchenkova T, Slavkina V and Nelyub V 2019 Modern Technologies for the Production of Composites Based on Inorganic Binders *Materials Today: Proceedings* **11** 112-117

[13] Mikhaylovskiy K and Prosuntsov P 2018 Modelling of thermal and stress-strain state of transformable space structures from hybrid composite materials *MATEC Web of Conferences* **194** 01039

[14] Buyanov I A, Rostovtsev M Y and Shelofast V V 2016 Validation of software for computer-aided engineering of composite constructions in relation to calculating the natural frequencies of asymmetrical laminated plates *Russian Aeronautics* **59** 606-612

[15] Bokhoeva L A, Rogov V E, Chermoshentseva A S and Lobanov D V 2016 Stability and Process of Destruction of Compressed Plate of Layered Composite Materials with Defects *IOP Conference Series: Materials Science and Engineering* **142** 012077

IOP Conf. Series: Materials Science and Engineering **934** (2020) 012016 doi:10.1088/1757-899X/934/1/012016

Modeling of the electromagnetic and elastic properties of composite materials

Yu I Dimitrienko, K M Zubarev, Yu V Yurin and V I Yakovlev

Bauman Moscow State Technical University, Moscow, 105005, Russia

E-mail: dimit.bmstu@gmail.com, zubarev.bmstu@mail.ru.

Abstract. In this paper, we consider the solution of the problem of the theory of piezoelectricity. The effective elastic, piezoelectric, and dielectric characteristics are calculated for a two-layer composite with a volume concentration of layers of 50%. The corresponding characteristics are calculated for a two-layer material with an orthogonal-reinforced structure. For the numerical solution of the problem, the conjugate gradient method for solving systems of algebraic equations was used. The results were visualized using a software package developed at Bauman Moscow state technical University . The paper presents the distribution of the stress field σ_{ij} and the vector $\bar{d}_i$ in problems L_r and L_{pq}.

1. Introduction

The Electromagnetically elastic properties and tools for their study are described in [1-24].

Many areas of production and industry have special requirements for the properties and characteristics of the materials used, such as engineering, shipbuilding, aviation, space industries, construction and others.

The interest in creating new modeling methods is growing, as more and more new, modern computer technology provides more and more opportunities to obtain faster and more accurate results [9].

Significant opportunities for creating such solution techniques are offered by the method of asymptotic averaging (MAO), or the method of homogenization. In this paper, we propose the use of this method in the simulation of composite materials with piezoelastic properties [8].

The solution to this problem is based on such phenomena as the piezoelectric effect (direct and inverse). When mechanical stress is applied to some crystals, an electric moment arises [7]. This phenomenon is the essence of the direct piezoelectric effect . In [10], such quantities as the piezoelectric moduli are determined similarly, and it is also shown that these characteristics are components of the third-rank tensor [12]. Most of the theoretical assumptions of this work are based on the definitions and conclusions described and obtained in [1].

2. Mathematical statement of the problem

We need to simulate a composite material with more effective piezelasticity characteristics . Based on this, in the course of this work, a software package was implemented that made it possible to calculate the effective characteristics of the piezelastic tensor of various composite materials, depending on the class of their symmetry.

We shall denote the cell periodicity for the task piezoelectricity - V_ξ. Suppose PL consists of N phases V_α, $\alpha = 1,..,N$. We introduce the parameter $k = \frac{l}{L}$ — a small parameter (under the assumption, $k << 1$), where l is the characteristic size of the periodicity cell, and L is the characteristic linear size of the composite material as a whole [2].

We pose the problem of piezoelasticity , with the introduction of Cartesian coordinates (x^k), as well as global ($\overline{x}^k = \frac{x^k}{L}$) and local ($\xi^k = \frac{x^k}{k}$) coordinates, and also assume that contact on the interface $\Sigma_{\alpha\beta}$ between the components of the material is ideal :

$$\sigma_{ij,j} = 0; \tag{1}$$

$$-d_{ij} + \rho_e = 0; \tag{2}$$

$$\sigma_{ij} = C_{ijkl}\varepsilon_{kl} + v_{kij}e_k; \tag{3}$$

$$\varepsilon_{ij} = \frac{1}{2}(u_{i,j} + u_{j,i}); \tag{4}$$

$$d_i = v_{ikl}\varepsilon_{kl} - э_{ik}e_k; \tag{5}$$

$$e_k = \varphi_{,k}; \tag{6}$$

$$[u_i] = 0,\ [\sigma_{ij}]n_j = 0 \text{ on } \Sigma_{ij}; \tag{7}$$

$$[\varphi] = 0,\ [d_i]n_i = 0 \text{ on } \Sigma_{ij}; \tag{8}$$

$$\sigma_{ij}|_{\Sigma_\sigma} n_j = S_i; \tag{9}$$

$$d_i n_i|_{\Sigma_d} = d_{ne}. \tag{10}$$

In the system of equations (1) - (10): (1) is the equilibrium equation, (2) is the Gauss equation, (3) is the elasticity relation taking into account the effect of piezoelasticity , (4) is the Cauchy relation, (5) is the electrical relation with taking into account the inverse piezoelectric effect , (6) is the expression for the electric potential, (7) is the condition of the ideal mechanical effect at the interface of the components, (8) is the ratio of the ideal electrical contact at the interface of the components of the material, (9) is the kinematic and static conditions on the external surfaces, (10) - electrical conditions at the boundaries;

3. Problem Solving method

We write the system of equations 1-10 in the form of two systems : a system of local equilibrium equations and a system of averaged equilibrium equations [13].

$$\begin{cases} \frac{\partial}{\partial \overline{x}_j}\left\langle \sigma_{ij}^{(0)}(\overline{x},\xi)\right\rangle = 0 \\ \frac{\partial}{\partial \overline{x}_j}\left\langle \sigma_{ij}^{(1)}(\overline{x},\xi)\right\rangle = 0 \\ \ldots \\ \frac{\partial}{\partial \overline{x}_j}\left\langle \sigma_{ij}^{(n)}(\overline{x},\xi)\right\rangle = 0 \end{cases} \tag{12}$$

$$\begin{cases} \frac{\partial}{\partial \overline{x}_j}\langle d_i^{(0)}(\overline{x},\xi)\rangle = 0 \\ \frac{\partial}{\partial \overline{x}_j}\langle d_i^{(1)}(\overline{x},\xi)\rangle = 0 \\ \dots \\ \frac{\partial}{\partial \overline{x}_j}\langle d_i^{(n)}(\overline{x},\xi)\rangle = 0 \end{cases} \tag{13}$$

We will solve this problem on a periodicity cell. B Udem assume that the periodicity cell is divided into N parts, and each part is assigned its number α. The task can be divided into 9 separate tasks [12, 13], 6 tasks of type " L_{pq} " and 3 tasks of type " L_r ":

$$«L_{pq}»: \begin{cases} \sigma^{(0)}_{ij(pq)/j} = 0; \\ \sigma^{(0)}_{ij(pq)} = C_{ijkl}\varepsilon^{(0)}_{kl(pq)} + v_{kij}e^{(0)}_{k(pq)}; \\ \varepsilon^{(0)}_{kl(pq)} = \frac{1}{2}\left(U_{k/l(pq)} + U_{l/k(pq)}\right). \end{cases}$$

$$\begin{cases} -d^{(0)}_{i(pq)/i} = 0; \\ d^{(0)}_{i(pq)} = v_{ikl}\varepsilon^{(0)}_{kl(pq)} - Э_{ik}\, e^{(0)}_{k(pq)}. \end{cases}$$

$$«L_r»: \begin{cases} \sigma^{(0)}_{ij(r)/j} = 0; \\ \sigma^{(0)}_{ij(r)} = C_{ijkl}\varepsilon^{(0)}_{kl(r)} + v_{kij}e^{(0)}_{k(r)}; \\ \varepsilon^{(0)}_{kl(r)} = \frac{1}{2}\left(U_{k(r)/l} + U_{l/k(r)}\right). \end{cases}$$

$$\begin{cases} -d^{(0)}_{i(r)/i} = 0; \\ d^{(0)}_{i(r)} = v_{ikl}\varepsilon^{(0)}_{kl(r)} - Э_{ik}\, e^{(0)}_{k(r)}. \end{cases}$$

The variational formulation for problems L_{pq} has the form:

$$\int_{V_a} \delta\overline{e}^T\overline{\sigma}dV = \int_{\Sigma_a} \delta\overline{U}^T\overline{S}d\Sigma, \quad \int_{V_a} \delta\overline{e}^T\overline{d}dV = 0 \tag{14}$$

The variational formulation of the problem is formulated in the same form L_r.

We will approximate the displacements $\vec{U}$ and electric potential in finite elements by linear coordinate functions:

$$\overline{U} = \overline{\Phi}\overline{q}, \quad \varphi = \overline{\phi}^T\overline{y}, \tag{15}$$

In formulas (15) $\overline{q}$ and $\overline{y}$ are the coordinate columns of displacements and electric potentials at the nodes of finite elements. $\overline{\Phi}$ and $\overline{\phi}^T$ are the matrix and column of form functions. Then from (14) we obtain the final resolving system of linear algebraic equations [19]:

$$\tilde{K}\overline{q} = \tilde{f}$$

where

$$\tilde{K} = \int_V \tilde{B}^T\tilde{C}\tilde{B}dV,$$

$$\tilde{C}=\begin{pmatrix}\bar{C} & \bar{V}\\ \bar{V}^T & \bar{Э}\end{pmatrix},\quad \tilde{B}=\begin{pmatrix}\bar{B} & 0\\ 0 & \bar{B}_e\end{pmatrix},\ \bar{B}=\bar{D}\bar{\Phi},\quad \bar{B}_e=\bar{D}_e\bar{\phi},$$

$$\tilde{q}=\begin{pmatrix}\bar{q}\\ \bar{y}\end{pmatrix},\ \tilde{f}=\begin{pmatrix}\bar{f}\\ \bar{f}_e\end{pmatrix}.$$

4. Results of a numerical solution

For the test calculation, tourmaline was used, which represents the B- rhombohedral symmetry class according to the classification [20].

We show for this problem also the distribution of the stress tensor field for some problems L_{pq}. Figure 1 shows the distribution σ_{11} in the problem L_{11} for $\bar{e}_1=0$.

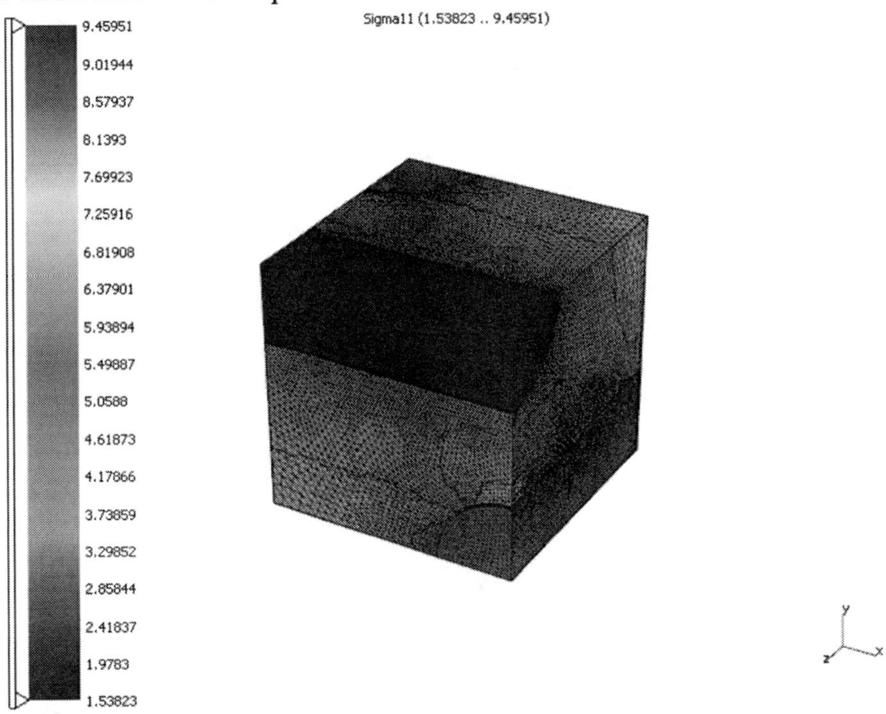

Figure 1. Distribution σ_{11}, GPa , in the task problem L_{11}

The distribution $\bar{d}_1$ in the task L_1 is presented in Figure 2.

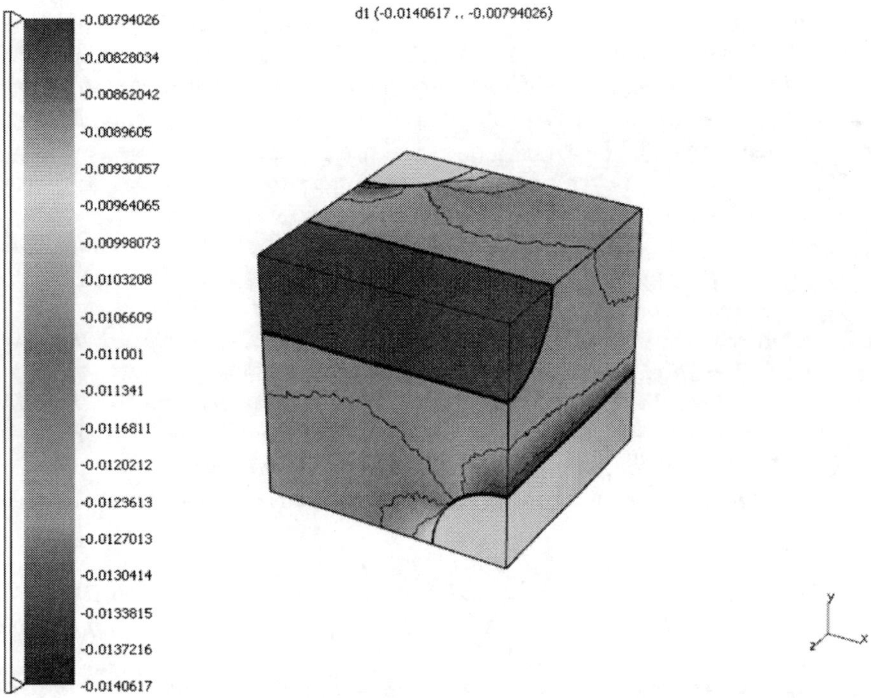

Figure 2. Distribution $\overline{d}_1$, C/m², in the task problem L_1

Similarly, the values of the desired effective characteristics for composite materials whose components belong to other symmetry groups can be obtained. To do this, you only need to specify the matrix form of their constants - elastic, piezoelectric and dielectric [21].

5. Conclusions

In this work, a method for finding the effective characteristics of a composite material was presented and implemented.

The theoretical basis of the method is given. A software package has been developed that provides the calculation of the desired characteristics.

For the numerical solution of the problem, the conjugate gradient method for solving systems of algebraic equations was used.

The test problem of the theory of piezoelasticity for a single-layer material is solved, on the basis of the results of which it is concluded that the presented method is working.

The effective elastic, piezoelectric, and dielectric characteristics are calculated for a two-layer composite with a volume concentration of layers of 50%. The results of this calculation well satisfy the analytical results, taking into account the computational error.

References

[1] Sulimov V D, Shkapov P M and Sulimov A V 2016 *Optimization of singular values of parameter dependent matrices using hybrid algorithms*, Herald of the Bauman Moscow State Technical University, Series Natural Sciences (5), p. 46-66

[2] Zarubin V S and Sergeeva E S 2018 *Application of Mathematical Modeling to Determine the Thermoelastic Characteristics of Nano-Reinforced Composites* Mathematical Models and Computer Simulations 10(3), p. 288-298

[3] Kozlov O S and Skvortsov L M 2016 *MVTU software package in scientific research and applied developments* Mathematical Models and Computer Simulations 8(4), p. 358-368

[4] Fetisov D A 2016 *On the construction of solutions of terminal problems for multidimensional*

affine systems in quasicanonical form Differential Equations 52(12), p. 1638-1649

[5] Tikhomirov, V G, Gudkov A, Petrov V, Yankevich V and Evseenkov A 2017 *Simulation of electric field distribution in GaN HEMTs for the onset of structure degradation* Proceedings of the 2017 11th International Workshop on the Electromagnetic Compatibility of Integrated Circuits, EMCCompo 2017 7998094, p. 115-118

[6] Agasiev T and Karpenko A 2017 *The Program System for Automated Parameter Tuning of Optimization Algorithms* Procedia Computer Science 103, p. 347-354

[7] Dimitrienko Y I, Zubarev K M and Krasnov I K 2018 *Mathematical model operation of an electric field in the nonuniform mediums at intubation by a direct current* // Journal of Physics: Conference Series, 2018 **1141** , Issue 1 .- Art.no 012110

[8] Mikrin E A, Zubov N E, Efanov D E and Ryabchenko V N 2018 *Superfast Iterative Solvers for Linear Matrix Equations* Doklady Mathematics **98(2),** p. 444-447

[9] Karpenko A P and Leshchev I A 2019 *Nature-Inspired Algorithms for Global Optimization in Group Robotics Problems* Studies in Systems, Decision and Contro l174, p. 91-106

[10] Tao-tao Zhang 2016 *Dynamic analysis of 2–2 cement based piezoelectric composite actuator with considering ambient temperatures.* - Piezoelectricity, Acoustic Waves, and Device Applications (SPAWDA),.

[11] Kalamakarov A 2003 *Modeling of anisotroping magneto-piezoelastic materials.* – Direct and Inverse Problems of Electromagnetic and Acoustic Wave Theory, 2016.1990. – 232 p.

[12] Yang Gao, *The refined theory of piezoelectric thick plates.* - Piezoelectricity, Acoustic Waves, and Device Applications (SPAWDA), 2009..

[13] Bian Z, Chen W and Lu C 2008 *Wave propagation in two-layered infinite composite piezoelectric hollow cylinder with imperfect interfaces.* - Piezoelectricity, Acoustic Waves, and Device Applications,.

[14] Burger M and Muhlhuber W 2002. *Numerical approximation of an SQP-type method for parameter identification* / / SIAM J . Numer. Anal. **40(5)** p. 1775-1797.

[15] Forsgren A, G i l l P.E and Wright M H 2002 *Interior methods for nonlinear optimization* / / SIAM Review. **44(4)** p. 525-597.

[16] Zubarev K M and Ivanova T L 2016 *Forward and inverse problems of magnetotellurics.* Polytechnical Student Journal. **3**

[17] Hoppe R H 2002 *Shape and topology optimization in materials sciences* / / Proceeding of ICCM-2002. — Novosibirsk: NCC Publ.,. — p. 17-31.K.M.

[18] Zubarev T L and Ivanova 2016 "Forward and inverse problems of magnetotellurics." Polytechnical Student Journal.**3**

[19] P M van den Berg, A L van Broekhoven, Abubakar A. 1999 *Extended contrast source inversion, Inverse Problems,* , т.15, p.1325–1344.

[20] Ivanova T L and Zubarev K M 2016 *Multilayer media electrophysical parametrs recovery on completion direct current sounding.* Polytechnical Student Journal.**4**

[21] Avdeev A V, Lavrentiev M M and Prijmenko V I 1999 *Inverse problems and some applications.*— Novosibirsk: ICM&MG Publ.,.

[22] Gao B 2009 *The refined theory of torsional deformation of transversely isotropic piezoelectric circular shaft.* - Piezoelectricity, Acoustic Waves, and Device Applications (SPAWDA),.

[23] Purukava T 1985 *Piezoelectricity and pyroelectricity in ferroelectric polymers.* - Electrets (ISE 5),. 5th International Symposium.

[24] Nelyub V 2020 Methods for developing carbon composites with functional properties *IOP Conference Series: Materials Science and Engineering* **709** 022037

IOP Conf. Series: Materials Science and Engineering **934** (2020) 012017 doi:10.1088/1757-899X/934/1/012017

Asymptotic stress analysis of multilayer composite thin cylindrical shells

Yu I Dimitrienko, E A Gubareva and A E Pichugina

Bauman Moscow State Technical University, Moscow, 105005, Russia

E-mail: dimit@bmstu.ru, eagubareva@mail.ru, a.e.pichugina@gmail.com

Abstract. Using the asymptotic theory, explicit analytical expressions were obtained that allow one to calculate the stress-strain state of multilayer thin cylindrical anisotropic elastic composite shells for small deformations. Forenamed expressions allow us to obtain the distribution of all components of the stress-strain state without any assumptions about their nature of the distribution over the thickness. An algorithm was proposed that makes it possible to obtain the total stress tensor in the shell, and all components of the stress tensor are numerically calculated for the case of a cylindrical composite shell with axisymmetric bending by pressure. The character of the distribution of characteristics of the stress-strain state of the shell was analyzed and the corresponding graphs were constructed.

1. Introduction

Thin shells are widely used in various technical industries, while the use of cylindrical thin shells is most widespread, for example, such a shell may be a rocket igniter charge body, aircraft fuselage, satellite power frame, etc. For calculating the bearing capacity of structures made of composite materials it is necessary to sufficiently accurately calculate the distribution of microstresses in the components of a composite material, for example, for layered structures, the stresses in each individual layer, especially interlaminar and trough-thickness stresses, due to low strength of composites in these directions. [1-8]

The classical theories of calculating thin composite shells of the Kirchhoff–Love type, Timoshenko and their numerous modifications are based on certain assumptions regarding the nature of the distribution of stresses and displacements over the thickness of the structure, the adequacy of which must be confirmed for each class of tasks. [9-17] For example, the classical assumption of the linear nature of the distribution of longitudinal displacements over the thickness and the independence of the transverse displacement from the thickness leads to a linear dependence of the interlayer shear stresses on the transverse coordinate. [12,13] At the same time, these same interlayer stresses can be calculated using three-dimensional equations of the theory of elasticity, and a quadratic dependence of these stresses on thickness arises, which indicates the inconsistency of the assumptions made. In [14,15], a method was developed for the asymptotic averaging of equations of the three-dimensional theory of elasticity for thin multilayer shells [16,17], which, without any hypotheses regarding the distribution of displacements or stresses over the thickness, allows one to obtain an explicit, mathematically founded expression for all 6 components of the stress tensor. The aim of this work is the application of the developed general asymptotic theory of shells for the case of cylindrical thin multilayer shells and the analysis of obtained results.

2. The equations of three-dimensional theory of elasticity in curvilinear coordinates

Let's consider a thin cylindrical shell for which the following relation holds: $æ = h / L << 1$: where h is the shell thickness, L is the shell length, $æ$ is a small parameter. [12-13] We introduce $q_k = \{z, s, r\}$ – the cylindrical coordinates (z is directed along the axis of the cylinder, s is the length of the arc in the circumferential direction, r is directed along the radius of the shell), measured from the middle surface of the shell and $\bar{q}_k = q_k / L$ – the dimensionless coordinates, and the local coordinate $\xi = \bar{q}_3 / æ$ along the thickness of the shell. The Lame parameters H_α for a cylindrical thin shell in cylindrical coordinates and their derivatives have the following form [13]: $H_3 = 1$, $\frac{\partial H_\alpha}{\partial \bar{q}_\beta} = A_{\alpha,\beta} + æ(A_\alpha \bar{k}_\alpha)_{,\beta} \approx A_{\alpha,\beta}$, $\frac{\partial H_\alpha}{\partial \xi} = æ H_{\alpha 3}$.

Let's consider the quasistatic problem of the linear theory of elasticity in a general form [18]:

$$\nabla \cdot \sigma = 0, \tag{1}$$

$$\sigma = C \cdot\cdot \varepsilon, \tag{2}$$

$$\varepsilon = \frac{1}{2}(\nabla \otimes \mathrm{u} + \nabla \otimes \mathrm{u}^{\mathrm{T}}) \tag{3}$$

$$\Sigma_{3\pm} : \sigma_{i3} = -\tilde{p}_\pm \delta_{i3} \qquad \Sigma_T : u_i = u_{ei} \qquad \Sigma_S : [\sigma_{i3}] = 0 \qquad [u_i] = 0 \tag{4}$$

where (1) is the equilibrium equations, (2) is the generalized Hooke's law, (3) is the Cauchy relations, (4) are the boundary conditions, $\Sigma_{3\pm}$ - the outer and inner surfaces of the shell ($\xi = \pm 0.5$), Σ_T - the end surface on which the displacement u_{ei} is specified, Σ_S - the interface between the layers of the shell, on which the conditions for the ideal contact of the layers are specified, $[u_i]$ - the jump in the functions, u_i - the dimensionless displacements, $\tilde{p}_\pm$ are dimensionless pressures related to the characteristic value of the modulus of elasticity E_0.

In curvilinear coordinates the system takes form:

$$\begin{aligned}
&\varepsilon_{\alpha\alpha} = O_\alpha u_{\alpha,\alpha} + O_1 O_2 H_{\alpha,\beta} u_\beta + O_\alpha H_{\alpha,3} u_3 \\
&2\varepsilon_{12} = H_1 O_2 (u_1 O_1)_{,2} + H_2 O_1 (u_2 O_2)_{,1} \\
&\varepsilon_{33} = \frac{1}{æ} u_{3/3} \\
&2\varepsilon_{\alpha 3} = \frac{1}{æ} u_{\alpha/3} + O_\alpha (u_{3,\alpha} - H_{\alpha 3} u_\alpha) \\
&\sigma_{IJ} = C_{IJKL}\varepsilon_{KL} + C_{IJ33}\varepsilon_{33}. \\
&\sigma_{I3} = 2C_{I3K3}\varepsilon_{K3}; \\
&\sigma_{33} = C_{33KL}\varepsilon_{KL} + C_{3333}\varepsilon_{33};
\end{aligned} \tag{5}$$

Here $C_{ijkl}(\xi)$ are the dimensionless elastic moduli, ε_{ij} are small deformations. The shell layers are assumed to be elastic, with a cylindrical monoclinic symmetry of elastic properties [18].

3. Asymptotic solution

Let us consider the case when the pressure $\tilde{p}_{\pm}$ on the outer and inner surfaces of the shell is of the order of smallness $O(æ^3)$, $\tilde{p}_{\pm} = æ^3 p_{\pm}$, where $p_{\pm}$ is a finite value. This order of pressure, as a rule, corresponds to the real conditions of loading of thin shells. The solution of the problem (6) is sought in the form of asymptotic expansions over small parameter $æ$:

$$u_k = u_k^{(0)}(\overline{q}_I) + \sum_{n=1}^{\infty} æ^n u_{ij}^{(n)} = u_k^{(0)}(\overline{q}_I) + æ u_k^{(1)}(\overline{q}_I, \xi) + æ^2 u_k^{(2)}(\overline{q}_I, \xi) + \ldots$$

$$\varepsilon_{ij} = \sum_{n=0}^{\infty} æ^n \varepsilon_{ij}^{(n)} = \varepsilon_{ij}^{(0)} + æ\varepsilon_{ij}^{(1)} + æ^2 \varepsilon_{ij}^{(2)} + \ldots$$

$$\sigma_{ij} = \sum_{n=0}^{\infty} æ^n \sigma_{ij}^{(n)} = \sigma_{ij}^{(0)} + æ\sigma_{ij}^{(1)} + æ^2 \sigma_{ij}^{(2)} + \ldots$$

Using these asymptotic expansions, explicit expressions were obtained for the longitudinal σ_{IJ}, shear σ_{I3}, and transverse σ_{33} components of the stress tensor through deformations $\varepsilon_{KL}^{(0)}$ and curvatures η_{KL} of the shell:

$$\begin{gathered}\sigma_{IJ} = \hat{C}_{IJKL}^{(1)} \varepsilon_{KL}^{(0)} + \hat{C}_{IJKL}^{(0)} \xi \eta_{KL};\\ -\sigma_{\alpha 3} = \hat{C}_{\alpha 3KL}^{(2)} \varepsilon_{KL}^{(0)} + \hat{R}_{\alpha 3KLJ}^{(2)} \varepsilon_{KL,J}^{(0)} + \hat{N}_{\alpha 3KL} \eta_{KL} + \hat{V}_{\alpha 3KLJ} \eta_{KL,J};\\ \sigma_{33} = \hat{C}_{33KL}^{(3)} \varepsilon_{KL}^{(0)} + \hat{R}_{33KLJ}^{(3)} \varepsilon_{KL,J}^{(0)} + \hat{E}_{33KLJM}^{(3)} \varepsilon_{KL,JM}^{(0)} + \hat{N}_{33KL}^{(3)} \eta_{KL} + \hat{V}_{33KLM}^{(3)} \eta_{KL,M} +\\ + \hat{U}_{33KLMN}^{(3)} \eta_{KL,MN} + \hat{W}_{KL}^{(3)} \eta_{KL}^{(2)} - æ^3 (p_{\pm} + \Delta p(\xi + 0.5))\end{gathered} \qquad (6)$$

The tensors included in these expressions $\hat{C}_{IJKL}^{(1)}$, $\hat{C}_{IJKL}^{(0)}$, $\hat{C}_{\alpha 3KL}^{(2)}$, $\hat{R}_{\alpha 3KLJ}^{(2)}$, $\hat{N}_{\alpha 3KL}$, $\hat{V}_{\alpha 3KLJ}$ at other depend only on the elastic moduli C_{IJKL} of the composite and local coordinate ξ.

4. Example of calculation of stresses in a composite cylindrical shell

Consider the case of axisymmetric external pressure loading of a 4-layer composite cylindrical shell, cantilever fixed at the ends. For this type of loading the following system was obtained for longitudinal σ_{IJ}, shear σ_{I3}, and transverse σ_{33} components of the stress tensor through displacements $u_3^{(0)}$ of the shell:

$$\begin{gathered}\sigma_{IJ} = \hat{C}_{IJ22}^{(1)} \frac{1}{R} u_3^{(0)}(X^1) - \hat{C}_{IJ11}^{(0)} \xi u_{3,11}^{(0)},\\ -\sigma_{\alpha 3} = \hat{C}_{\alpha 322}^{(2)} \frac{1}{R} u_3^{(0)}(X^1) + \hat{R}_{\alpha 3221}^{(2)} \frac{1}{R} u_{3,1}^{(0)}(X^1) - \hat{N}_{\alpha 311} u_{3,11}^{(0)} - \hat{V}_{\alpha 3111} u_{3,111}^{(0)},\\ \sigma_{33} = \hat{C}_{3322}^{(3)} \frac{1}{R} u_3^{(0)}(X^1) + \hat{R}_{33221}^{(3)} \frac{1}{R} u_{3,1}^{(0)}(X^1) + \hat{E}_{332211}^{(3)} \frac{1}{R} u_{3,11}^{(0)}(X^1) -\\ \hat{N}_{3311}^{(3)} u_{3,11}^{(0)} - \hat{V}_{33111}^{(3)} u_{3,111}^{(0)} - (p_{\pm} + \Delta p(\xi + 0.5))\end{gathered} \qquad (7)$$

Each layer of the shell was a unidirectional fiberglass, the system of corners α_i of the reinforcement of the composite layers changed during the program experiment.

Following figures show the distribution of the flexural stresses σ_{11}, σ_{22} along the thickness of the composite sheath with the system of corners $\alpha_1 = 30°$, $\alpha_2 = 45°$, $\alpha_3 = 45°$, $\alpha_4 = 30°$:

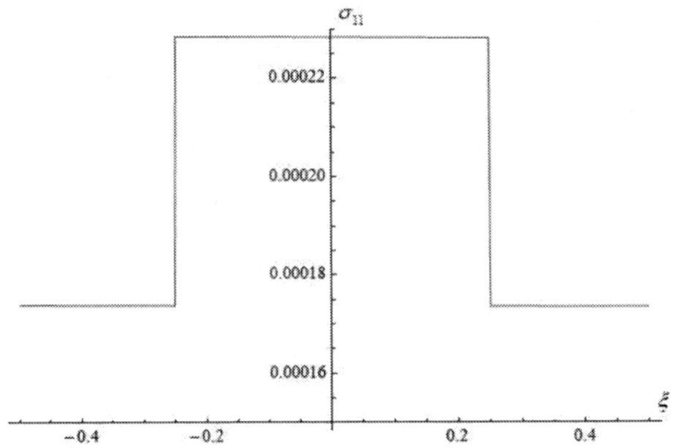

Figure 1 - Distribution of stress σ_{11}, GPa across the thickness of the composite shell under axially symmetric bending

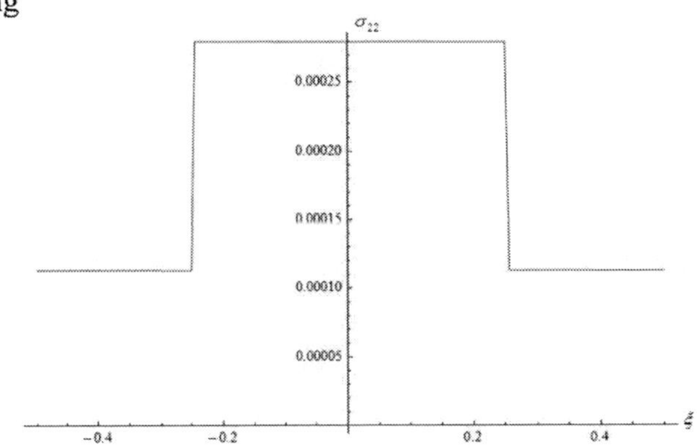

Figure 2 - Distribution of stress σ_{22}, GPa across the thickness of the composite shell under axially symmetric bending

It can be seen that the graphs of the components σ_{11}, σ_{22} of stress tensor have a piecewise linear appearance, which is due to the absence of conditions on the nature of the graph in all directions except the transverse, since the normal is directed along coordinate 3 (see boundary conditions (4)).

Following figures show the distribution of the stresses σ_{13}, σ_{33} along the thickness of the composite sheath with the system of corners $\alpha_1 = 30^\circ$, $\alpha_2 = 45^\circ$, $\alpha_3 = 45^\circ$, $\alpha_4 = 30^\circ$.

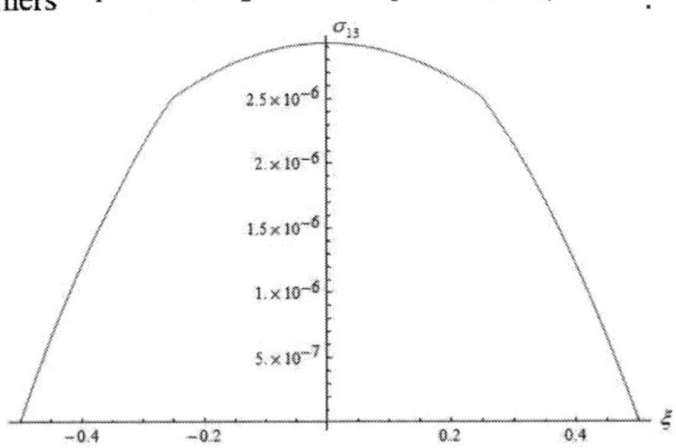

Figure 3 - Distribution of stress σ_{13}, GPa across the thickness of the composite shell under axially symmetric bending

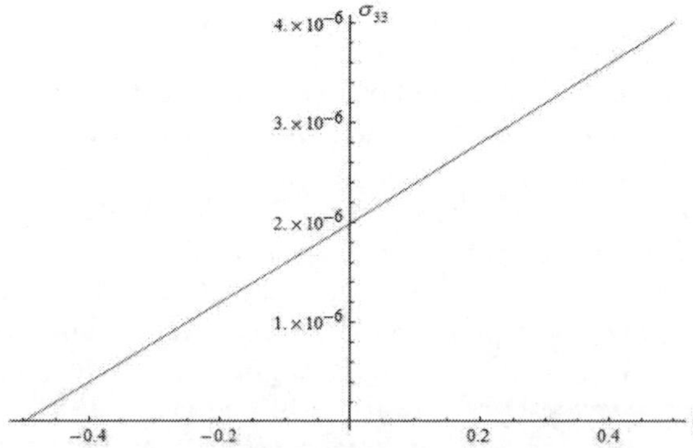

Figure 4 - Distribution of stress σ_{33}, GPa across the thickness of the composite shell under axially symmetric bending

It can be seen that the absolute values for the graphs of the components σ_{13}, σ_{33} of the stress tensor are significantly less than normal stresses, which correlates with the Timoshenko theory, but at the same time, the Timoshenko theory does not allow obtaining expressions for these stresses.

5. Conclusions

A theory of thin cylindrical composite shells is developed, based on the asymptotic analysis of three-dimensional equations of the theory of elasticity without additional assumptions about the distribution of stresses and displacements along the thickness of the shell. Explicit, mathematically accurate analytical formulas are obtained for calculating the distribution of the components of all 6 components of the stress tensor over the thickness of a cylindrical shell. An example is given of calculating stresses in a cylindrical shell under axisymmetric bending of pressures

Thus, local problems of the theory of thin shells were formulated and solved; explicit expressions were obtained for all six components of the stress tensor, including transverse normal stresses and interlayer shear stresses in the shell; a special case of the problem of a loaded axisymmetric cylindrical shell was considered; explicit expressions were obtained for all six components of the stress tensor, including transverse normal stresses and interlayer shear stresses in the shell for the case of a loaded axisymmetric cylindrical shell; a software package was implemented for calculating displacements and stresses using separate procedures that implement all the mathematical apparatus necessary to solve the problem; displacements and all 6 components of the stress tensor were calculated and visualized for the case of axisymmetric pressure loading of a cantilevered cylindrical shell.

References

[1] Vasilyev V V 1988 *Mechanics of constructions from composite materials* (Moscow: Mechanical engineering) 272 p

[2] Van'ko V I 2016 On the load-carrying capacity of structural members *Journal of Applied Mechanics and Technical Physics* **57(5)** 784-8

[3] Bakulin V N and Vinogradov Y I 2017 Analytical and asymptotic solution of boundary value problems in the mechanics of deformed shells under concentrated loading *Russian Aeronautics* **60(1)** 13-20

[4] Marycheva A N, Guzeva T A, P'e P M, Tun L K and Malysheva G V 2019 Reinforcing Fillers for Polymer Composites Based on Organic Unwoven Materials *Polymer Science - Series D* **12** 170-173

[5] Mikhaylovskiy K and Prosuntsov P 2018 Modelling of thermal and stress-strain state of transformable space structures from hybrid composite materials *MATEC Web of Conferences* **194** 01039

[6] Prosuntsov P V, Shulyakovskii A V and Taraskin N Y 2017 Numerical Simulation of a

Thermal-Protection Element of a Promising Reusable Capsule-Type Lander *Journal of Engineering Physics and Thermophysics* **90** 110-116

[7] Bazheva R C, Borodulin A S, Kalinnikov A N and Charaev A M 2019 Polymer composites based on halogen-containing oligoethers *IOP Conference Series: Materials Science and Engineering* **683** 012032
Egorov A V 2019 Estimated delamination of an elastic-plastic liner in a high-pressure metal composite vessel *Journal of Physics: Conference Series* **1399** 044057

[8] Ruslantsev A N, Portnova Ya M, Tairova L P and Dumansky A M 2016 Analysis of mechanical properties anisotropy of nanomodified carbon fibre-reinforced woven composites *IOP Conference Series: Materials Science and Engineering* **153(1)** 012003

[9] Dimitrienko Yu I, Yakovlev D O, Erasov V S, Fedonyuk N N, Sborschikov, Gubareva E A, Krylov V D, Grigoriev M M and Prozorovsky A A 2014 Development of a multilayer polymer composite with a discrete constructive-orthotropic filler *Composites and nanostructures* **6(1)** 32-48

[10] Grigolyuk E I and Kulikov G M 1988 Generalized model of mechanics of thin-walled constructions from composite materials *Mechanics of Composite Materials* **4** 698-704

[11] Levinski T, Telega J J 2000 *Plates, laminates and shells.Asymptotic analysis and homogenization.* (Singapore; London, World Sci. Publ.) 739 p

[12] Dimitrienko Y I, Gubareva E A and Yurin Yu V 2016 Explicit formulas for the calculation of the complete tensor of the stresses in the monoclinic thin composite shells based on the asymptotic homogenization method *Inzhenernyy zhurnal: nauka i innovatsii — Engineering Journal:Science and Innovation* 12 (60) http://dx.doi.org/10.18698/2308-6033-2016-12-1557

[13] Dimitrienko Yu I, Gubareva E A and Pichugina A E 2019 Theory of composite cylindrical shells under quasistatic vibrations, based on an asymptotic analysis of the general viscoelasticity theory equations *IOP Conference Series: Material Science and Engeneering*, volume 683 № 012013

[14] Dimitrienko Yu I, Dimitrienko I D and Sborschikov S V 2015 Multiscale Hierarchical Modeling of Fiber Reinforced Composites by Asymptotic Homogenization Method. *Applied Mathematical Sciences,* **9** 145 7211-7220

[15] Dimitrienko Yu I, Gubareva E A, Markevich M N and Sborschikov S V 2016 Mathematical modelling of dielectric properties of nanostructural composites using asymptotic homogenizing method *Herald of the Bauman Moscow State Technical University Series Natural Sciences*. **1** 76-89.

[16] Dimitrienko Yu I, Gubareva E A and Sborschikov S V 2014 Asymptotic theory of constructively orthotropic plates with a two-period structure *Matematicheskoe modelirovanie i chislennye metody – Mathematical Modeling and Computational Method* **1** 36-57

[17] Dimitrienko Yu I and Dimitrienko I D 2019 Computations of stresses and energy dissipation in composite thin laminates with the asymptotic vibration theory Computers & Mathematics with Applications. **78(8)** pp.2541-2559

[18] Dimitrienko Yu I 2010 *Nonlinear Continuum Mechanics and Large Inelastic Deformations.* Springer. 722 p.

IOP Conf. Series: Materials Science and Engineering **934** (2020) 012018 doi:10.1088/1757-899X/934/1/012018

Asymptotic theory of multilayer plates with imperfect contact between layers

Yu I Dimitrienko and E A Gubareva

Bauman Moscow State Technical University, Moscow, 105005, Russia

E-mail: dimit@bmstu.ru, eagubareva@mail.ru

Abstract. An asymptotic theory of the deformation of thin multilayer elastic plates with non-ideal contact between the layers is proposed. The linearized slip conditions of the contacting layers with different slip coefficients for each pair of layers are considered. An asymptotic theory is an expansion over a small geometric parameter of the original 3-dimensional elasticity problem with sliding conditions of the layers. A recurrent sequence of local problems of elasticity theory is obtained. An analytical recurrent solution to these problems is found, which allows one to determine all the components of the stress tensor in the plate. An averaged system of equations of the theory of thin two-layer plates with slipping layers is derived, which has an increased order of partial derivatives compared with the theory of plates with ideal contact. An analytical solution is found to this problem of bending a 2-layer composite plate under the influence of uniform pressure. Specific effects of plate deformation due to the effect of layer slippage are established

1. Introduction

The classical theories of multilayer plates are based on the assumption of the ideal contact of the layers — the continuity of the displacement and stress vectors on the layer interface [1-4]. However, in a number of engineering problems, a situation arises with imperfect contact, for example, when layers slip freely relative to each other [5-9]. In this case, the main assumption of the classical theory of plates, including Kirchhoff-Love plates and Timoshenko-Mindlin plates, on the linearity of the distribution of longitudinal displacements over the thickness, no longer holds. Currently, asymptotic plate theories proposed relatively recently [10–16] are very promising, in which no assumptions are made regarding the nature of the distributions of displacements and stresses over the thickness, and the equations of the plate theory are derived from general 3-dimensional equations using their asymptotic analysis. The goal of this work is the further development of the asymptotic theory of plates for the case of slippage of layers during deformation.

2. Formulation of a 3-dimensional problem of the theory of elasticity for a multilayer plate with slipping layers.

In the framework of the three-dimensional theory of elasticity [17], we consider the problem of deformation of a thin multilayer elastic anisotropic plate with the presence of slippage of the layers relative to each other under the influence of transverse and shear loads on the side surfaces, with the conditions of the plate edges fixing:

$$\begin{gathered}
\nabla_j \sigma_{ij}^{\{m\}} = 0; \\
\varepsilon_{ij}^{\{m\}} = \frac{1}{2}\left(\nabla_j u_i^{\{m\}} + \nabla_i u_j^{\{m\}}\right); \\
\sigma_{ij}^{\{m\}} = C_{ijkl}^{\{m\}} \varepsilon_{kl}^{\{m\}}; \qquad m = 1,2,...,N, \\
\Sigma_m : \sigma_{i3}^{\{m\}} = \sigma_{i3}^{\{m+1\}}, \ u_3^{\{m\}} = u_3^{\{m+1\}}, \ u_I^{\{m+1\}} - u_I^{\{m\}} = \theta^{\{m\}} \sigma_{I3}^{\{m\}}; \qquad m = 1,2,...,N-1, \\
\Sigma_{3+} : \sigma_{i3}^{\{N\}} = S_{i+}, \ \Sigma_{3-} : \sigma_{i3}^{\{1\}} = S_{i-}; \\
\Sigma_T : u_i^{\{m\}} = u_{ei}, \ m = 1,2,...,N
\end{gathered} \tag{1}$$

Here $\sigma_{ij}^{\{m\}}$ - are the components of the stress tensor; $\varepsilon_{kl}^{\{m\}}$ — components of the tensor of small deformations; $u_i^{\{m\}}$ — components of the displacement vector; $\nabla_j = \partial / \partial x_j$ — differentiation operator with respect to Cartesian coordinates $\tilde{x}_j$ (the axis Ox_3 is directed normal to the plate surfaces); $C_{ijkl}^{\{m\}}$ - components of the tensor of elastic moduli of the layers; {m} - layer index; $m = 1,2,...,N$; $\Sigma_{3\pm} = \{x_3 = \pm h/2\}$ - external and internal surfaces of the plate, h - total thickness of the multilayer plate); Σ_T - end surface of the plate; $\Sigma_m = \{\tilde{x}_3 = \tilde{x}_3^{\{m\}}\}$ - contact surfaces of the layers; $\theta^{\{m\}}$ - slip coefficients of the layers relative to each other, $\theta^{\{m\}} \geq 0$.
The system of equations (1) includes equilibrium equations, Cauchy relations, the generalized Hooke law, conditions on the contact surfaces of the layers Σ_m (imperfect contact with the presence of slip), on the external surfaces of the plate $\Sigma_{3\pm}$ (force vectors $S_{i\pm}$ are given on them) and butt surface Σ_T (displacement vectors u_{ei} are specified). We assume that all layers of the plate are monoclinic materials; therefore, tensors $C_{ijkl}^{\{m\}}$ contain no more than 13 nonzero constants [17]).

3. Asymptotic expansions for a slip plate.

We introduce a small parameter $\kappa = h / L << 1$ as the ratio of the total plate thickness h to the plate length L, and also introduce global x_k and local ξ coordinates:

$$x_k = \tilde{x}_k / L, \ \xi = x_3 / \kappa, \ k = 1,2,3. \tag{2}$$

The coordinate ξ along the plate thickness varies in the range $-0.5 \leq \xi \leq 0.5$, and the contact surfaces of the layers have coordinates $\xi = a_m = \tilde{x}_3^{\{m\}} / h$, $m = 1,2,...,N-1$, $a_0 = -0.5$, $a_N = 0.5$.

Let us suppose, that a pressure $\tilde{p}_\pm$ on the outer surfaces of the plate have the order of smallness is $O(\kappa^3)$, then the force vectors $S_{i\pm}$ have the following form $S_{i\pm} = -\kappa^3 p_\pm(x_I)\delta_{i3}$.
The solution to problem (1) will be sought in the form of asymptotic expansions in the parameter κ in the form of functions depending on global and local coordinates

$$u_k^{\{m\}} = u_k^{(0)}(x_I) + \kappa u_k^{\{m\}(1)}(x_I, \xi) + \kappa^2 u_k^{\{m\}(2)}(x_I, \xi) + ..., \tag{3}$$

where $I = 1,2$; $m = 1,2,...,N$; $k = 1,2,3$.
Substituting expansions (3) into the Cauchy relations in system (1), taking into account the rule of differentiation of functions of local coordinates [7], we obtain asymptotic expansions for deformations

$$\varepsilon_{ij}^{\{m\}} = \varepsilon_{ij}^{\{m\}(0)} + \kappa\varepsilon_{ij}^{\{m\}(1)} + \kappa^2\varepsilon_{ij}^{\{m\}(2)} + ..., \tag{4}$$

where

$$\varepsilon_{IJ}^{\{m\}(0)} = \frac{1}{2}\left(u_{I,J}^{(0)} + u_{J,I}^{(0)}\right), \varepsilon_{I3}^{\{m\}(0)} = \frac{1}{2}\left(u_{3,J}^{(0)} + u_{I/3}^{\{m\}(1)}\right), \varepsilon_{33}^{\{m\}(0)} = u_{3/3}^{\{m\}(1)},$$
$$\varepsilon_{IJ}^{\{m\}(n)} = \frac{1}{2}\left(u_{I,J}^{\{m\}(n)} + u_{J,I}^{\{m\}(n)}\right), \varepsilon_{I3}^{\{m\}(n)} = \frac{1}{2}\left(u_{3,I}^{\{m\}(n)} + u_{I/3}^{\{m\}(n+1)}\right), \varepsilon_{33}^{\{m\}(n)} = u_{3/3}^{\{m\}(n+1)}, \quad n = 1,2... \tag{5}$$

Hereinafter, the following notation is used: $u_{I/3}^{\{m\}(1)} = \dfrac{\partial u_i^{\{m\}(1)}}{\partial \xi}$; $u_{i,j}^{\{m\}(1)} = \dfrac{\partial u_i^{\{m\}(1)}}{\partial x_j}$, and also introduces operations of averaging over layer thicknesses

$$\left\langle u_i^{\{m\}(n)} \right\rangle_m = \int_{a_{m-1}}^{a_m} u_i^{\{m\}(n)} d\xi, \quad \left\langle u_i^{(n)} \right\rangle = \int_{-0,5}^{0,5} u_i^{(n)} d\xi = \sum_{m=1}^{N} \left\langle u_i^{\{m\}(n)} \right\rangle_m \tag{6}$$

$$\left\{f^{\{m\}(n)}\right\}_{1\xi} = \int_{-0.5}^{\xi} \left(f^{\{m\}(n)}(\tilde{\xi}) - \left\langle f^{\{m\}(n)} \right\rangle\right) d\tilde{\xi} \quad \left\{f^{\{N\}(n)}\right\}_{2\xi} = \int_{\xi}^{0.5} \left(f^{\{N\}(n)} - \left\langle f^{\{m\}(n)} \right\rangle\right) d\tilde{\xi}$$

Substituting expressions (4) into Hooke's law in system (1), taking into account the monoclinicity of the layer materials, we obtain asymptotic expressions for the stresses:

$$\sigma_{ij}^{\{m\}} = \sigma_{ij}^{\{m\}(0)} + \kappa\sigma_{ij}^{\{m\}(1)} + \kappa^2\sigma_{ij}^{\{m\}(2)} + ..., \tag{7}$$

where

$$\sigma_{IJ}^{\{m\}(n)} = C_{IJKL}^{\{m\}}\varepsilon_{KL}^{\{m\}(n)} + C_{IJ33}^{\{m\}}\varepsilon_{33}^{\{m\}(n)}; \; \sigma_{I3}^{\{m\}(n)} = 2C_{I3K3}^{\{m\}}\varepsilon_{K3}^{\{m\}(n)};$$
$$\sigma_{33}^{\{m\}(n)} = C_{33KL}^{\{m\}}\varepsilon_{KL}^{\{m\}(n)} + C_{3333}^{\{m\}}\varepsilon_{33}^{\{m\}(n)}, \quad n = 1,2,3... \tag{8}$$

4. Local contact problem.

Substituting the expansions (3), (4) and (7) into the equilibrium equations and the boundary conditions of system (1), and equating the terms at κ^{-1} to zero in the equations of equilibrium, and for other powers of κ, to some quantities $h_i^{(0)}, h_i^{(1)}, h_i^{(2)}$ independent of ξ , we obtain a recurrence sequence local contact problems of the theory of elasticity:

The problem for the zeroth approximation has the form

$$\sigma_{i3/3}^{\{m\}(0)} = 0, \; \sigma_{I3}^{\{m\}(0)} = 2C_{I3k3}^{\{m\}}\varepsilon_{k3}^{\{m\}(0)};$$
$$\sigma_{IJ}^{\{m\}(0)} = C_{IJKL}^{\{m\}}\varepsilon_{KL}^{(0)} + C_{IJ33}^{\{m\}}\varepsilon_{33}^{\{m\}(0)}; \; \sigma_{33}^{\{m\}(0)} = C_{33KL}^{\{m\}}\varepsilon_{KL}^{(0)} + C_{3333}^{\{m\}}\varepsilon_{33}^{\{m\}(0)};$$
$$\varepsilon_{IJ}^{(0)} = \frac{1}{2}\left(u_{I,J}^{(0)} + u_{J,I}^{(0)}\right); \; \varepsilon_{I3}^{\{m\}(0)} = \frac{1}{2}\left(u_{3,I}^{(0)} + u_{I/3}^{\{m\}(1)}\right), \; \varepsilon_{33}^{\{m\}(0)} = u_{3/3}^{\{m\}(1)}; m = 1,2,...,N; \tag{9}$$
$$\Sigma_{3+} : \sigma_{i3}^{\{N\}(0)} = 0; \; \Sigma_{3-} : \sigma_{i3}^{\{1\}(0)} = 0;$$
$$\Sigma_m : \sigma_{i3}^{\{m\}(0)} = \sigma_{i3}^{\{m+1\}(0)}, u_3^{\{m\}(1)} = u_3^{\{m+1\}(1)}, \; u_I^{\{m+1\}(1)} - u_I^{\{m\}(1)} = \theta^{\{m\}}\sigma_{I3}^{\{m\}(1)}, \quad m = 1,2,...,N-1$$

Problems for higher approximations are as follow:

$$\sigma_{i3/3}^{\{m\}(n)} + \sigma_{iJ,J}^{\{m\}(n-1)} = h_i^{(n-1)}; \; \sigma_{I3}^{\{m\}(n)} = 2C_{I3k3}^{\{m\}}\varepsilon_{k3}^{\{m\}(n)};$$
$$\sigma_{IJ}^{\{m\}(n)} = C_{IJKL}^{\{m\}}\varepsilon_{KL}^{\{m\}(n)} + C_{IJ33}^{\{m\}}\varepsilon_{33}^{\{m\}(n)};$$
$$\sigma_{33}^{\{m\}(n)} = C_{33KL}^{\{m\}}\varepsilon_{KL}^{\{m\}(n)} + C_{3333}^{\{m\}}\varepsilon_{33}^{\{m\}(n)};$$

$$\varepsilon_{IJ}^{\{m\}(n)} = \frac{1}{2}\left(u_{I,J}^{\{m\}(n)} + u_{J,I}^{\{m\}(n)}\right); \; \varepsilon_{I3}^{\{m\}(n)} = \frac{1}{2}\left(u_{3,I}^{\{m\}(n)} + u_{I/3}^{\{m\}(n+1)}\right);$$
$$\varepsilon_{33}^{\{m\}(n)} = u_{3/3}^{\{m\}(n+1)}; \qquad m = 1,2,...,N;$$
$$\Sigma_{3+} : \sigma_{i3}^{\{N\}(n)} = S_{i+}^{(n)}; \; \Sigma_{3-} : \sigma_{i3}^{\{1\}(n)} = S_{i-}^{(n)}; \tag{10}$$
$$\Sigma_m : \sigma_{i3}^{\{m\}(n)} = \sigma_{i3}^{\{m+1\}(n)}, \; u_3^{\{m\}(n+1)} = u_3^{\{m+1\}(n+1)}, \; u_I^{\{m+1\}(n+1)} - u_I^{\{m\}(n+1)} = \theta^{\{m\}}\sigma_{I3}^{\{m\}(n+1)},$$
$$m = 1,2,...,N-1.$$

where $S_{i\pm}^{(n)} = S_{i\pm}\delta_{n3}$, $n = 1,2,3...$

The equilibrium equations in system (1) after the introduction of functions $h_i^{(0)}$, $h_i^{(1)}$, $h_i^{(2)}$ take the form:

$$h_i^{(0)} + \kappa h_i^{(1)} + \kappa^2 h_i^{(2)} + ... = 0. \tag{11}$$

The solution to the local zero-approximation problem (9) is a function $u_j^{\{m\}(1)}$, $\varepsilon_{kl}^{\{m\}(0)}$, $\sigma_{ij}^{\{m\}(0)}$, that depends on the local coordinate ξ and input data of this problem – displacements $u_j^{\{m\}(0)}(x_J)$. The solution to problem (11) are functions $u_j^{\{m\}(n+1)}$, $\varepsilon_{kl}^{\{m\}(n)}$, $\sigma_{ij}^{\{m\}(n)}$, and the functions $u_j^{\{m\}(n)}$, $\varepsilon_{kl}^{\{m\}(n-1)}$, $\sigma_{ij}^{\{m\}(n-1)}$ in this problem are input data.

5. Solution of the zero-approximation problem.

The solution of equilibrium equations with boundary conditions in the local zero-approximation problem (9) has the form:

$$\sigma_{i3}^{\{m\}(0)} = 0, \qquad \sigma_{IJ}^{\{m\}(0)} = C_{IJKL}^{\{m\}(0)}\varepsilon_{KL}^{(0)}; \tag{12}$$

$$u_I^{\{m\}(1)} = -\xi u_{3,I}^{(0)} - \sum_{k=1}^{N-1} H_{km}\theta^{\{k\}}\sigma_{I3}^{\{k\}(1)}(a_k), \qquad u_3^{\{m\}(1)} = -\varepsilon_{KL}^{(0)}U_{3KL}^{\{m\}},$$

where indicated

$$C_{IJKL}^{\{m\}(0)} = C_{IJKL}^{\{m\}} - C_{IJ33}^{\{m\}}Z_{3KL}^{\{m\}}, U_{3KL}^{\{m\}} = \xi Z_{3KL}^{\{m\}} + B_{KL}^{\{m\}}, \; Z_{3KL}^{\{m\}} = C_{3333}^{\{m\}-1}C_{33KL}^{\{m\}}$$
$$B_{KL}^{\{m\}} = \sum_{k=1}^{N-1} H_{km}a_m(Z_{3KL}^{\{m+1\}} - Z_{3KL}^{\{m\}}) + \langle \xi Z_{3KL}^{\{m\}} \rangle, \; H_{km} = \begin{cases} H_k - 1, k \le m-1; \\ H_k, \quad k > m \end{cases}, \; H_k = \sum_{m=k+1}^{N} h_m. \tag{13}$$

where $h_m = a_m - a_{m-1}$ are the thicknesses of the layers, $m = 1,2,...,N$.

6. Solving local problems for higher approximations.

The solution of the equilibrium equations in (11) together with the boundary conditions $\sigma_{i3}^{\{N\}(n)} = -p_+\delta_{i3}\delta_{n2}$ and $\sigma_{i3}^{\{1\}(n)} = -p_-\delta_{i3}\delta_{n2}$ on the surfaces $\xi = \pm 0.5$, as well as the conditions $\sigma_{i3}^{\{m\}(n)} = \sigma_{i3}^{\{m+1\}(n)}$ on the contact surfaces $\xi = a_m$, $m = 1,2,...,N-2$ has the form:

$$\sigma_{i3}^{\{m\}(n+1)} = -\left\{\sigma_{iJ,J}^{\{m\}(n)}\right\}_{1\xi} - \left(p_- + \Delta p(\xi+0.5)\right)\delta_{i3}\delta_{n2} \qquad n=0,1,2,...$$

$$m=1,2,...,N-1 \tag{14}$$

$$\sigma_{i3}^{\{N\}(n+1)} = \left\{\sigma_{iJ,J}^{\{N\}(n)}\right\}_{2\xi} - \left(p_+ + \Delta p(\xi-0.5)\right)\delta_{i3}\delta_{n2} \qquad n=0,1,2,...$$

From the boundary condition on the contact surface, we find the expression for the functions

$$h_i^{(n)} = -\Delta p\delta_{i3}\delta_{n2} + <\sigma_{iJ,J}^{\{m\}(n)}>, \tag{15}$$

where $\Delta p = p_+ - p_-$.

Formulas (14) are recurrent, substituting successively these formulas into each other for various n, we calculate the stresses $\sigma_{i3}^{\{m\}(n+1)}$ up to the 3rd approximation, through the stresses $\sigma_{IJ}^{\{m\}(0)}$ and $\sigma_{IJ}^{\{m\}(1)}$. Substituting formula (12) for $\sigma_{IJ}^{\{m\}(0)}$ into expressions (14) for shear stresses, we find

$$\sigma_{I3}^{\{m\}(1)} = -A_{IJKL}^{\{m\}}\varepsilon_{KL,J}^{(0)}, \qquad m=1,2,...,N$$

$$A_{IJKL}^{\{m\}} = \left\{C_{IJKL}^{\{m\}(0)}\right\}_{1\xi}, \quad m=1,2,...,N-1, \quad A_{IJKL}^{\{N\}} = -\left\{C_{IJKL}^{\{N\}(0)}\right\}_{2\xi}. \tag{16}$$

For stresses $\sigma_{IJ}^{\{m\}(1)}$ we obtain the following expressions

$$\sigma_{IJ}^{\{m\}(1)} = -\xi C_{IJKL}^{\{m\}(0)}u_{3,KL}^{(0)} - W_{IJMKLN}^{\{m\}}\varepsilon_{KL,MN}^{(0)}, \tag{17}$$

where

$$W_{IJMKLN}^{\{m\}} = C_{IJSP}^{\{m\}(0)}\tilde{\Phi}_{SPMKLN}^{\{m\}} \tag{18}$$

$$\tilde{\Phi}_{IJMKLN}^{\{m\}} = \frac{1}{2}\sum_{k=1}^{N-1} H_{km}\theta^{\{k\}}(A_{IMKL}^{\{k\}}(a_k)\delta_{JN} + A_{JMKL}^{\{k\}}(a_k)\delta_{IN}).$$

For deformations of the 1st approximation, we find

$$\varepsilon_{IJ}^{\{m\}(1)} = -\xi u_{3,IJ}^{(0)} - \tilde{\Phi}_{IJMKLN}^{\{m\}}\varepsilon_{KL,MN}^{(0)}, \qquad m=1,2,...,N \tag{19}$$

7. Averaged equilibrium equations for a two-layer plate with imperfect contact of the layers.

Substituting expressions (15) into the asymptotic expansion (11) of the equilibrium equations of the plate, we obtain

$$<\sigma_{iJ,J}^{\{m\}(0)}> + \kappa<\sigma_{iJ,J}^{\{m\}(1)}> + \kappa^2(<\sigma_{iJ,J}^{\{m\}(2)}> - \Delta p\delta_{i3}) + ... = 0. \tag{20}$$

Multiplying the asymptotic expansion of the equilibrium equations of the plate (1) by $\xi\kappa$ and integrating them over the thickness, we obtain the following auxiliary equation

$$\kappa(<\xi\sigma_{IJ,J}^{\{m\}(0)}> - <\sigma_{I3}^{\{m\}(1)}>) + \kappa^2(<\xi\sigma_{IJ,J}^{\{m\}(1)}> - <\sigma_{I3}^{\{m\}(2)}>) + ... = 0, \tag{21}$$

It is taken into account that $<\xi\sigma_{I3/3}^{\{m\}(1)}> = -<\sigma_{I3}^{\{m\}(1)}>$, $<\xi\sigma_{I3/3}^{\{m\}(2)}> = -<\sigma_{I3}^{\{m\}(2)}>$ due to boundary conditions on $\Sigma_{3-} : \sigma_{I3}^{\{1\}(n)} = 0$, and $\Sigma_{3+} : \sigma_{I3}^{\{N\}(n)} = 0$.

We introduce in a standard way [7,8] the forces T_{IJ}, moments M_{IJ}, and transvers forces Q_I in the plate

$$T_{IJ} = <\sigma_{IJ}^{\{m\}(0)}> + \kappa<\sigma_{IJ}^{\{m\}(1)}> + \kappa^2...,$$

$$Q_I = \kappa<\sigma_{I3}^{\{m\}(1)}> + \kappa^2<\sigma_{I3}^{\{m\}(2)}> + ..., \tag{22}$$

$$M_{IJ} = \kappa < \xi\sigma_{IJ}^{\{m\}(0)} > + \kappa^2 < \xi\sigma_{IJ}^{\{m\}(1)} > + \kappa^3 \dots .$$

Then from (20) and (21) we obtain the equations of equilibrium and moments for a two-layer plate, which coincide with the corresponding equations for plates without friction

$$T_{IJ,J} = 0, \qquad Q_{J,J} = \Delta\bar{p}, \qquad M_{IJ,J} - Q_I = 0, \tag{23}$$

indicated here $\Delta\bar{p} = \kappa^2 \Delta p$.

We substitute in the formulas (22) for T_{IJ} and M_{IJ} the expressions for the stresses $\sigma_{IJ}^{\{m\}(0)}$ and $\sigma_{IJ}^{\{m\}(1)}$, calculated by formulas (13) and (17), then we obtain the defining relations for the plate with slipping layers

$$T_{IJ} = \bar{C}_{IJKL}\varepsilon_{KL}^{(0)} + B_{IJKL}\eta_{KL} - H_{IJKLMN}\varepsilon_{KL,MN}^{(0)}, \tag{24}$$

$$M_{IJ} = B_{IJKL}\varepsilon_{KL}^{(0)} + D_{IJKL}\eta_{KL} - F_{IJKLMN}\varepsilon_{KL,MN}^{(0)}$$

Here are designated $\bar{C}_{IJKL}$ - membrane, B_{IJKL} - mixed, D_{IJKL} - flexural stiffnesses of the plate, which coincide with the classic stiffnesses of plates without friction, as well as the stiffness of the plate H_{IJKLMN} and F_{IJKLMN} due to slippage

$$\bar{C}_{IJKL} = < C_{IJKL}^{\{m\}(0)} >, \; B_{IJKL} = \kappa < \xi C_{IJKL}^{\{m\}(0)} >, \; D_{IJKL} = \kappa^2 < \xi^2 C_{IJKL}^{\{m\}(0)} >, \tag{25}$$

$$H_{IJKLMN} = \kappa < W_{IJMKLN}^{\{m\}} >, \; F_{IJKLMN} = \kappa^2 < \xi W_{IJMKLN}^{\{m\}} >.$$

In (24), notation was also introduced for the components of the plate curvature tensor η_{KL}, which, together with relations for deformations $\varepsilon_{IJ}^{(0)}$ close the averaged system of equations (23) and (24)

$$\eta_{KL} = -u_{3,KL}^{(0)}, \quad \varepsilon_{IJ}^{(0)} = \frac{1}{2}(u_{I,J}^{(0)} + u_{J,I}^{(0)}) \tag{26}$$

After substituting (26) in (24), and (24) in (23), we obtain a system of 3 equations with respect to 3 unknown functions $u_3^{(0)}$ - deflection and longitudinal displacements $u_I^{(0)}$, as in the classical theory of Kirchhoff-Love plates (KLP) with perfect contact layers. However, unlike the KLP, the system of equations of the asymptotic theory of plates with layer slippage due to layer slippage has a higher (5th instead of 3rd) order of derivatives with respect to longitudinal displacements $u_I^{(0)}$ of functions.

8. The solution to the problem of bending a multilayer plate taking into account slippage of the layers.

Consider the problem of bending a two-layer rectangular plate under the action of uniformly distributed constant pressure $\Delta\bar{p}$. We will consider both layers of the plate orthotropic. The boundary conditions are chosen corresponding to the conditions of hinging the edges of the plate. Numerical calculations were performed for a 4-layer composite plate, each layer of which was a unidirectional carbon / epoxy, the filaments of which were oriented at angles ϕ_m. Layers 1 and 2, as well as 3 and 4, had reinforcement angles, which differed only in sign ($\phi_2 = -\phi_1, \phi_4 = -\phi_3$). The following values were taken: $\phi_1 = 45°$, ϕ_3=60°. The interfaces between layers 1, 2, and also 3 and 4 were ideal, and slippage conditions were fulfilled on the interface between layers 2 and 3.

Numerical calculations were carried out with the parameter value κ = 0.02. The pressure values were chosen as follows: p_+= 10-4 GPa, p_-= 0. The relative thickness of the first layer $h_1 = 0{,}5 + a_1$

ranged from 0.5 to 0.8. The value of the dimensionless slip parameter θ of the layers also varied in the range from 0 to 103. The calculation option was also considered with ideal contact of the layers, when $\theta = 0$.

The results of calculations of displacements $u_3^{(0)}$ depending on the longitudinal coordinate are shown in Figure 1. The distribution of the deflection of the plate is shown for various values of the slip parameter θ, as well as for the case of perfect contact of the layers, for the case. Calculations allow us to conclude that, at the same pressure, a plate with an ideal contact of the layers always has minimal deflection compared with the case of slippage of the layers. The effect of layer slippage leads to a decrease in the total stiffness of the layer system and, therefore, the plate deflection increases. With an increase in the slip parameter from 0.1 to 103, the deflection increases by more than 3 times.

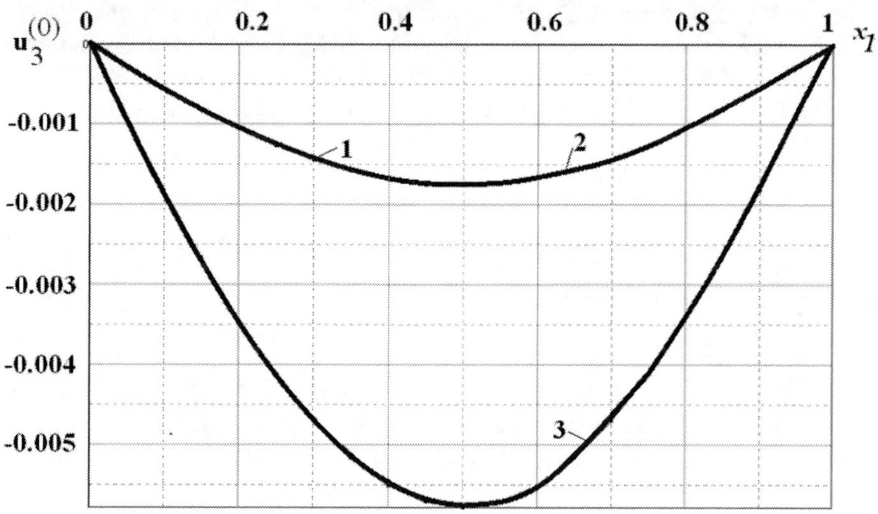

Figure 1. Dependence of the deflection of a two-layer composite plate with slippage on the longitudinal coordinate x1 for different values of the coefficient θ: 1) $\theta = 0$ - ideal contact between the layers, 2) $\theta = 0.1$, 3) $\theta = 1000$.

9. Conclusions.

An asymptotic theory of the deformation of thin multilayer plates with the effect of slipping layers is proposed, for the study of which the linearized slip conditions at the boundary of the contacting layers are considered.

A recurrent sequence of local problems of elasticity theory is obtained for a plate with slippage of layers. An analytical recurrent solution to these problems is found, which allows one to determine all the components of the stress tensor in the plate.

An averaged system of equations of the theory of thin two-layer plates with slipping layers is derived, which has an increased order of partial derivatives compared with the theory of plates with perfect contact.

The problem of bending a two-layer plate with slipping layers under uniform pressure is considered. An analytical solution to this problem is found and a numerical analysis of the solution is carried out using the example of a 2-layer composite plate. Specific effects of plate deformation due to the effect of layer slippage are established.

References

[1] Love A E H 1927 *A treatise on the mathematical theory of elasticity*. Cambridge. University

Press, 674 p.

[2] Timoshenko S P and Voinovsky-Krieger S 1966 *Plastinki i obolochki* [Plates and shells]. Moscow, Nauka Publ., , 635 p.

[3] Vasilyev V V 1988 *Mechanics of constructions from composite materials* Moscow: Mechanical engineering. 272 p.

[4] Grigolyuk E I and Kulikov G M 1988 Generalized model of mechanics of thin-walled constructions from composite materials *Mechanics of Composite Materials* **4** 698-704

[5] Mikhaylovskiy K, Prosuntsov P 2018 Modelling of thermal and stress-strain state of transformable space structures from hybrid composite materials *MATEC Web of Conferences* **194** 01039

[6] Burchenkova T, Slavkina V and Nelyub V 2019 Modern Technologies for the Production of Composites Based on Inorganic Binders *Materials Today: Proceedings* **11** 112-117

[7] Buyanov I A, Rostovtsev M Y and Shelofast V V 2016 Validation of software for computer-aided engineering of composite constructions in relation to calculating the natural frequencies of asymmetrical laminated plates *Russian Aeronautics* **59** 606-612

[8] Bokhoeva L A, Rogov V E, Chermoshentseva A S and Lobanov D V 2016 Stability and Process of Destruction of Compressed Plate of Layered Composite Materials with Defects *IOP Conference Series: Materials Science and Engineering* **142** 012077

[9] Shelofast V V, Kulish G G, Smerdov A A and Rostovtcev M Y 2017 Experimental identification of dynamic parameters of carbon-fiber-reinforced plastic (CFRP) and their use in a national experimental complex for finite element analysis *International Journal of Applied Engineering Research* **12** 728-737

[10] Kohn R V and Vogelius M. 1984 A new model of thin plates with rapidly varying thickness. *Int. J. Solids and Struct*, pp. 333–350.

[11] Levinski T and Telega J J 2000 *Plates, laminates and shells. Asymptotic analysis and homogenization.* Singapore; London, World Sci. Publ. 739 p

[12] Dimitrienko Yu I 2012 Asymptotic theory of multilayer thin plates. *Herald of the Bauman Moscow State Technical University Series Natural Sciences* **№3** pp. 86-100.

[13] Dimitrienko Yu I and Yakovlev D O 2015 The Asymptotic Theory of Thermoelasticity of Multilayer Composite Plates. *Composites: Mechanics,Applications. An International Journal.* V.**6.** № 1 pp. 13-51.

[14] Dimitrienko Yu I Dimitrienko I D and Sborschikov S V 2015 Multiscale Hierarchical Modeling of Fiber Reinforced Composites by Asymptotic Homogenization Method. *Applied Mathematical Sciences,* **9(145)** pp.7211-7220

[15] Dimitrienko Yu I, Gubareva E A, Markevich M N and Sborschikov S V 2016 Mathematical modelling of dielectric properties of nanostructural composites using asymptotic homogenizing method *Herald of the Bauman Moscow State Technical University Series Natural Sciences.* **1** 76-89.

[16] Dimitrienko Yu I, Gubareva E A and Sborschikov S V 2014 Asymptotic theory of constructively orthotropic plates with a two-period structure *Matematicheskoe modelirovanie i chislennye metody – Mathematical Modeling and Computational Method* 1 pp. 36-57

[17] Dimitrienko Yu I 2010 *Nonlinear Continuum Mechanics and Large Inelastic Deformations.* Springer. 722 p.

IOP Conf. Series: Materials Science and Engineering **934** (2020) 012019 doi:10.1088/1757-899X/934/1/012019

Development of a High Strength Polymeric Composite Material Using 3D-Printing and Vacuum Impregnation Technology

Yu A Lopatina[1,2] and V E Slavkina[1]

[1]Federal State Budgetary Scientific Institution "Federal Scientific Agricultural Engineering Center VIM", 1st Institutsky proezd 5, Moscow, Russia
[2]Bauman Moscow State Technical University, 2nd Baumanskaya st. 5, Moscow, Russia

E-mail: lopatina.julia@yandex.ru

Abstract. This paper offers a method for increasing the strength of FDM 3D printed parts by impregnation in a vacuum chamber in an environment of an epoxy compound with a long lifetime. We investigate the impact of the duration of impregnation and the internal geometry of the part being impregnated on the impregnation degree, expressed in the absence of cavities inside the material. The experimental samples are cylinders made of PLA plastic, the impregnation is carried out in an epoxy resin-based compound with PEPA hardener. The use of this strengthening technology may be useful for expanding the scope of FDM 3D printing from prototyping, being currently the most common application, to manufacturing parts for actually functional structures, which, unlike prototypes, carry load while in use, and therefore need increased strength.

1. Introduction

There has been an increasing use of parts made of polymeric materials in the design of modern machinery. Compared to metals, polymers have a number of advantages, such as high chemical resistance, low density combined with strength that is sufficiently high for many applications, ease of processing into products and availability [1]. In many cases, use of polymers is the only feasible solution, taking into account the operating environment or design features. The widespread use of polymeric parts in the design of machinery is complicated by the problem of their quick replacement in the case of failure. Since most of agricultural machinery now in use in Russia is imported from abroad, spare parts for these machines must be ordered through dealers, who send requests for supply of the parts directly to foreign manufacturers. As a result, the downtime between machine failure and repair can be considerably protracted. The machine will remain out of operation for this period of time, causing significant losses for the enterprise. Agricultural enterprises may be particularly affected during the period of seasonal agrotechnical works, as the possible time frames for these works are relatively narrow and even a few days of delay may lead to sizeable financial consequences.
In recent years, technologies such as additive fabrication or 3D printing, have been increasingly developed [2-8]. These technologies make it possible to manufacture parts of any complexity in terms of geometric features and a fairly wide range of materials. One of the main distinguishing features of additive fabrication technology is the possibility of end-to-end use of computer technologies, computer-aided design and 3D modelling systems. If it is necessary to manufacture parts in the conditions of single-part production, these technologies provide for considerable saving of time and

money, compared to the use of conventional technology for the same purpose. Therefore, the use of additive fabrication technologies for the manufacture of spare parts for repair of equipment is very relevant, as it enables quick restoration of machinery to operation in those industries where it is especially relevant, in particular, in agriculture [9-14].

As of today, the most common is the 3D printing technology called FDM (Fused Deposition Modeling) [15-20]. This technology is a method of part fabrication by layer-by-layer fusing of polymeric material, following a pre-arranged control sequence. The raw material is a plastic rod of 1-3mm diameter supplied in coils. The rod is fed to the printing head with guiding rolls or screws and trasnferred via special nozzle (Figure 1). The head heats the material up and feeds it to the product [21-25]. Electronic system adjusts the material feeding and controls the head and deck movement. Yet, the main field utilizing such technology is currently production of non-support structures that are not used in real operation, such as various models and prototypes. This is due to poor strength of the produced parts because of FDM 3D printing: layering of plastic during the 3D-printing process results in poor inter-layer adhesion and expressed anisotropy of the part properties when maximum strength is distributed along the plastic layer, but it is much weaker in transverse direction of laying. Also, this method often doesn't allow to use plastics with high mechanical properties widely used in production of polymer parts because the printing material is selected mainly based on its process properties and recycling ability with this method. Thus, FEM 3D-printed parts strengthening methods must be developed. This paper proposes one of such methods.

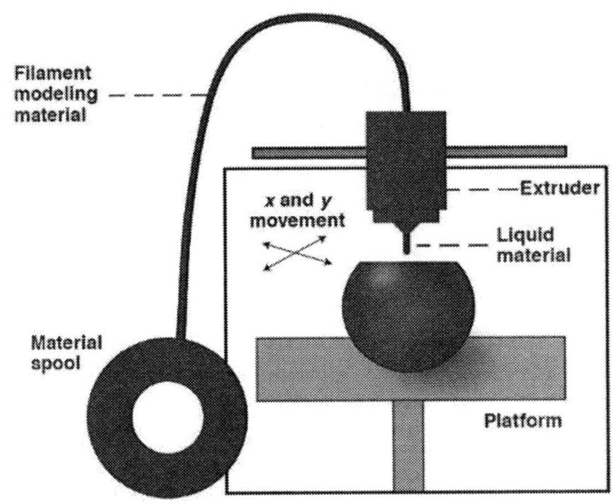

Figure 1. FDM 3D printing scheme.

2. Materials and methods

This paper proposes to impregnate the 3D-printed part with polymer compound in vacuum chamber (Figure 2), thus forming within the part volume composition material of two types of plastic with controlled inner structure and output properties. Impregnating polymer compound will bind plastic layers thus decreasing anisotropy of the part strength properties, and also, being a stronger material, will increase mechanical properties of the part compared to complete plastic 3D printing. Production of such composite part enables to widely vary inner geometry of the part by using various printing modes and wide range of combinations of the materials used.

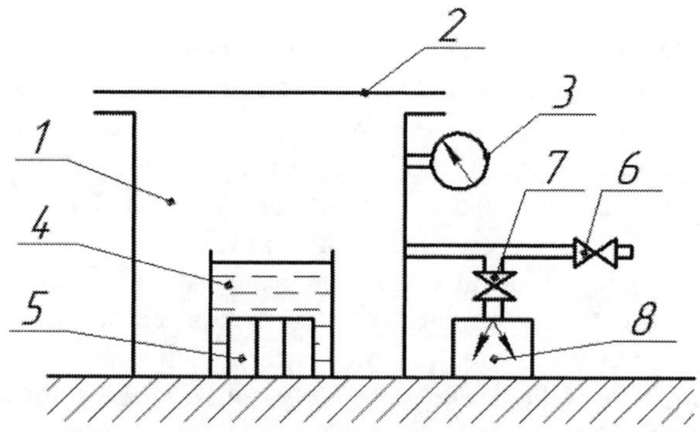

Figure 2. Impregnation of 3D printed parts with polymer compound: 1 – chamber; 2 – transparent lid; 3 – pressure gauge; 4 – polymer compound; 5 – impregnated part; 6 – release valve; 7 – shut-off valve; 8 – vacuum pump.

In this survey, polylactide (PLA plastic) is used as 3D printing material for production of samples. Samples FDM 3D printing mode is presented in Table 1. Compound on basis of ED20 epoxy resin with polyethylenepolyamine solidifier is used as impregnation material. Samples were cylindric-shaped with 15mm diameter and 40mm height with various filling ratio by PLA plastic: 20, 300 and 50% (the parameter is set during 3D printing process, figure 3). 9 samples were made per each filling option. 3 experiments were conducted, each measuring sample compound impregnation time: holding duration was 10, 20 and 30 minutes respectively. Selection of 9 samples was used in each experiment, 3 samples per each filling option. Impregnated samples were cut in several planes in order to examine the impregnation degree.

Table 1. 3D-printing parameters for experimental samples.

Parameter	Value
Nozzle temperature, °C	215
Blow-off ration, %	100
Plastic feeding ratio, %	98
Layer height, mm	0.2
Line width, mm	0.34
Perimeter/filling printing speed, mm/s	60/30

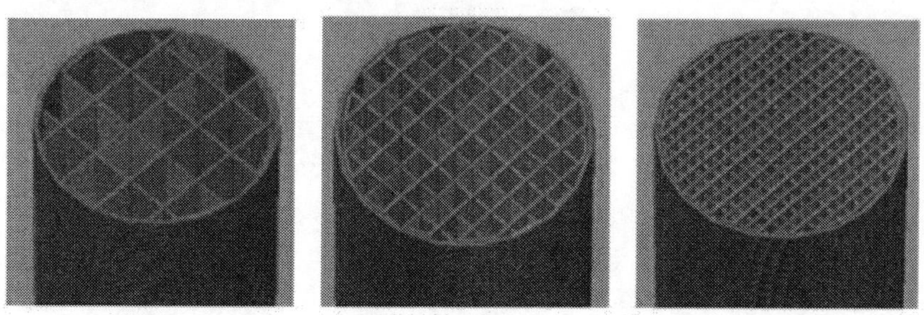

20 % 33 % 50 %

Figure 3. Inner geometry of the examined samples with various filling ratio.

3. Results and discussion

Compound impregnation degree of polymer frames is affected by two groups of factors: process factors related to the selected impregnation mode (holding time, pressure, temperature), and geometric factors related to the impregnated frame geometry. It was suggested that the most important process factor affecting the impregnation degree is impregnation duration, and the main geometric factor is PLA plastic and epoxy resin ratio expressed as ratio of part filling with extracted plastic in the printing process. These two groups of factors were varied in this experiment.

The samples were cut after impregnation in several planes at cut-off machine. Each section plane was photographed and uploaded to PC. The images were further processed in Compas-3D software: total section area was determined, as well as area of pores within such section, the obtained data were used as basis to determine voids ratio in the sample. Impregnability of the samples with 50% of pores and more was deemed as "insufficient"; in case of 50-70% – as "satisfactory"; in case of 70-95% – as "good", and in case of below 95% – as "excellent". The samples held in the compound for 10 minutes had insufficient impregnation degree (Figure 4, a); in case of 20 minutes – good degree (Figure 4, b, c); in case of 30 minutes – excellent degree.

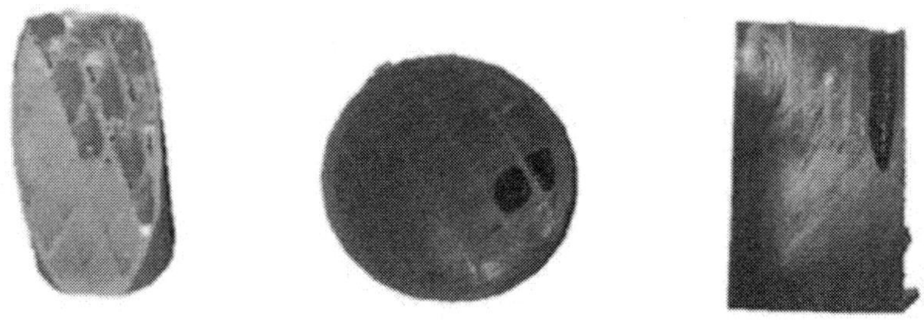

a b c

Figure 4. Samples sections after impregnation: a – 10 minutes holding; b, c – 20 minutes holding.

4. Conclusions

Based on the results of the tests, it was found that the best degree of impregnation corresponding to the minimum number of pores, and therefore the best mechanical properties, corresponds to the test articles with the impregnation time of 30 minutes. There was no impact of the internal geometry on the impregnation process.

Acknowledgements

The study was supported by the Russian Foundation for Basic Research (RFBR), Contract No. 19-38-90037\19 "Development of Scientific Basis for Fabricating Polymeric Composite Materials Based on 3D Printed Frames Reinforced with Polymeric Compounds".

References

[1] Bobovich B B 2014 *Polymeric Construction Materials* Moscow: FORUM: INFRA-M p 400 (in Russian)

[2] Maltseva O V 2018 *Rossiyskiy Vneshneeconomicheskiy Vestnik* **9** pp 88-97 (in Russian)

[3] Zlenko M A, Nagaycev M V and Dovbysh V M 2015 *Posobie Dlya Inzhenerov* Moscow: GNC RF FGUP NAMI p 220 (In Russian)

[4] Igonina E V and Drujinina O V 2017 *Sovremennye Informatsionnye Tekhnologii i IT-Obrazovanie* **13 (2)** 185-193. (in Russian)

[5] Katkar R, Taft R and Grant G 2018 3D volume rendering and 3D printing *Dental Clinics of North America* **62** pp 393-402

[6] Ngo T D, Kashani A, Imbalzano G, Nguyen K T and Hui D 2018 Additive manufacturing (3D printing): A review of materials, methods, applications and challenges *Composites Part B: Engineering* **143** pp 172-196

[7] Mohsen A 2017 The rise of 3-D printing: The advantages of additive manufacturing over traditional manufacturing *Business Horizons* **1399** pp 3-16

[8] Sergeeva O YU 2018 Additive technology and 3D-printing *Nanotekhnologii v Stroitelstve: Nauchnyj Internet-Zhurnal* **4(10)** pp 142-158 (in Russian)

[9] Lopatina Yu A 2019 The use of 3D printing by the FDM method in the repair of machinery and equipment *Tekhnicheskij Servis Mashin* **3(136)** pp 40-45 (in Russian)

[10] Sviridov A S, Tuzhilin S P and Lopatina Yu A 2019 The use of 3D digital farm in the repair production of agricultural machine *Tekhnicheskij Servis Mashin* **1(134)** pp 93-99 (in Russian)

[11] Fedorenko V F and Golubev I G 2018 *Perspective for the use of Additive Technologies in the Production and Technical Service of Agricultural Machine: Scientific Analysis* Moscow: FGBNU Rosinformagrotekh p 140 (in Russian)

[12] Nefyolov I S 2018 Recovery of worn machine parts with the help of additional repair parts made using additive *technology Remont. Vosstanovlenie. Modernizatsiya* **11** pp 15-17 (in Russian)

[13] Lopatina Yu A, Sviridov A S and Plohih A I 2019 Evaluation of the possibility of using 3D printing filament from recycled materials for the manufacture of parts of agricultural machinery and equipment *Tekhnicheskij Servis Mashin* **10** pp 58-64 (in Russian)

[14] Sviridov A S, Lopatina Yu A and Plohih A I 2019 The choice of materials and explanation of the technological parameters of the manufacture of sliding supports for the repair of garden equipment by using additive technologies *Selskohozyaystvennaya Tehnika: Obsluzhivanie i Remont* **11** pp 54-65 (In Russian)

[15] Petrova G N, Platonov M M, Bolshakov V A and Ponomarenko S A 2016 Research of the complex of characteristics of base materials for FDM of technology of the additive synthesis Physico-mechanical and heat-physical properties *Plasticheskie Massy* **5** pp 53-58 (In Russian)

[16] Jason C and others 2017 *Proceedings of the 2016 Annual Conf. on Experimental and Applied Mechanics* **3** pp 89-105

[17] Sofiane G, Sofiane B and Hedi N 2019 Printability and tensile performance of 3D printed polyethylene terephthalate glycol using fused deposition modelling Polymers **11(7)** p 1220

[18] Muammel M, Robert M and Laszlo Z 2019 Anisotropy evaluation of different raster directions, spatial orientations and fill percentage of 3D printed PETG tensile test specimens Key Engineering Materials **821** pp 167-173

[19] Dizon, J, Espera A, Chen Q and Advincula R 2018 Mechanical characterization of 3d-printed polymers Additive Manufacture 20 pp 44–67

[20] Popescu D, Zapciu A, Amza C, Baciu F and Marinescu R 2018 FDM process parameters influence over the mechanical properties of polymer specimens: A review Polymer Test **69** pp 157–166

[21] Bocharov A, Vigivskiy V, Nelyub V 2019 Metal coating of carbon fabric and properties of the carbon-fiber-reinforced plastic *Materials Today: Proceedings* pp 107-111

[22] Nelyub V.A. 2018 Technologies of merallization of carbon fabric and the properties of the related carbon fiber reinforced plastics. *Russian Metallurgy (Metally)*. Vol. **13**. pp. 1199-1201.

[23] Nelyub, V.A., Kovalev, S.P. 2016 A mesh-free approach to multiscale multiphysical calculations of polymeric composite structures. *Polymer Science – Series D*. vol. **9(1).** pp. 101-105.

[24] Nelyub V.A. 2016 Quantitative assessment of adhesive contact between carbon fiber and epoxy

binding. *Russian Aeronautics.* vol.**59(2)** pp. 259-262.
[25] S V Reznik, P V Prosuntsov , Azarov A.V. 2015 Substantiation of the structural-layout scheme of the mirror-space-antenna reflector with a high shape stability and a low density per unit length *Eng. Phys. Thermophy.* vol. **88(3)** pp. 699-705.

IOP Conf. Series: Materials Science and Engineering **934** (2020) 012020 doi:10.1088/1757-899X/934/1/012020

Selection and justification of polymer composite wing structural arrangement using parametrical modeling

S V Baranovski and K V Mikhailovskiy

Space–rocket composite design department, Bauman Moscow State Technical University (BMSTU), 105005, Moscow, 2nd Baumanskaya St., Bldg. 5, Block 1, Russian Federation

E–mail: serg1750@mail.ru

Abstract. Currently, due to the increase in air traffic, the need for modern reliable airliners is increasing, the requirements for which are constantly growing. One of the options for meeting these requirements is the use of polymer composite materials (PCM) in the structure, in particular in the wing. This article is devoted to the urgent task of designing a polymer composite wing power circuit within the framework methods of designing developing. The modeling, calculation and analysis of ninety variants of the wing structural and power schemes. The rational geometric parameters of the primary structural member arrangement are established, including the step and installation direction of ribs, stringer, spars, providing the greatest margin of safety factor. The skin power element is considered in detail. The stringer shape and material based on the calculations made are justified. The work is part of the methodology for designing a polymer composite wing.

1. Introduction

The wing is one of the most critical airliner designs, which create aerodynamic lift and ensure dihedral stability of the airliner. The wings of modern airliners are mainly made of polymer composite materials (PCM), which can provide mass efficiency without compromising performance, as well as achieve geometric characteristics inaccessible to metal structures. However, their design has always been a difficult task, including: developing and analyzing a large number of circuit options, conducting expensive experiments to determine various parameters, which leads to significant time and financial costs. Therefore, the development of a universal methods of designing a polymer composite wing, which allows one to obtain the primary wing geometry in a relatively short time, is a relevant objective.

Worldwide aircraft manufacturing specialists carry out the aircraft structures designing. Attention is being increasingly focused on multidisciplinary and the transition from narrowly focused product design methods to an integrated approach to their creation [1]. Most of the works consider certain objects, wings types and are aimed at solving specific problems. A study is being carried out in the field of forward–swept wing aerodynamics with a surface consideration of the structural arrangement (SA) to assess the overall performance of the wing under load [2]. Optimization of wing stiffness under the airload of both the whole structure [3 – 9] and specific SA [10, 11], including composite materials [12, 13]. Particular attention is paid to the calculation and optimization of the wing structure weight [14], including the use of curved force elements [15]. In addition, PCM panels [16 – 20] with integrated stringers [21, 22] are considered as an optimization object, and the results of optimizing the nose diaphragm are presented in [23]. Research is being conducted in the field of load–bearing element topology optimization taking into account only strength limitations [24, 25]. The design of the entire wing configuration is carried out at aviation enterprises using their own software products that are not available for commercial use [26, 27].

2. Selection of wing structural arrangement

2.1. Source data

The passenger airliner wing with a wingspan of 35 m, with a direct sweep, with a straightened section and an asymmetric airfoil was chosen as the object of study [28].

The calculation was carried out for three design cases — angle of attack are 0°, +11°, –7°. The corresponding load values were obtained when simulating the external aerodynamics. Figure 1 shows the aerodynamic pressure distribution over the wing surface in the mid section for a +11° angle of attack. In the root section, the wing is limited in movement and rotation relative to the axes [29, 30].

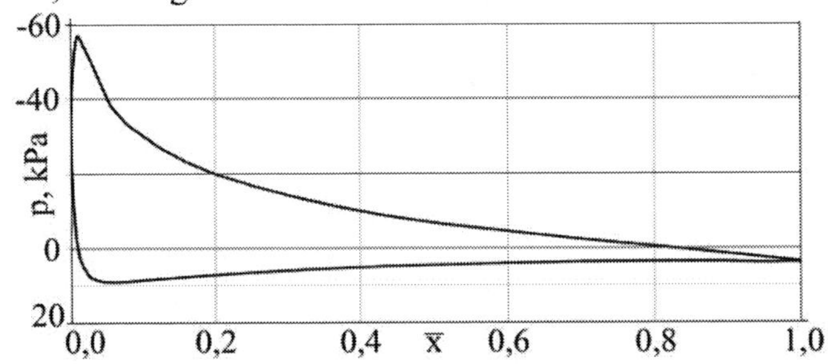

Figure 1. Pressure distribution over the airfoil.

In the calculation, finite Shell type elements were used, the load–bearing element thicknesses were specified in groups: upper skin, lower skin, front spar, rear spar, ribs, stringers. It was simulated 90 variants SA. Table 1 shows the variable parameters.

Table 1. Wing SA variable parameters.

Spar pitch,% of chord (front / rear)	20/70	30/60	25/65	20/60	30/70
Rib pitch,% MAC [a]	20	30	40		
Ribs mounting orientation	I[b]	II[c]			
Stringer pitch, m	0,12	0,16	0,2		

[a] Mean Aerodynamic Chord
[b] Perpendicular to the rear spar
[c] Looking forward

Carbon fiber is the main structural material. Table 2 shows the carbon fabric properties.

Table 2. Carbon fabric stress–related characteristics.

Density, kg/m^3	1580
Layer thickness, mm	0,285
Tensile modulus, along an axis 1[a] / 2[b], GPa	174,3 / 174,3
In–Plane shear modulus, GPa	2,9
Poison's ratio	0,32
Tensile (Compressive) strength, along an axis 1 / 2, GPa	2,7 (2.9) / 2,7 (2.9)
In–Plane shear strength, GPa	0,1

[a] Warp
[b] Weft

Taking into account the varied parameters, 210 design cases of SA are considered in the work.

2.2. Results

As a result of the calculation, stress–strain behavior were determined and the values of structure deformation under the airload were obtained.

IOP Conf. Series: Materials Science and Engineering **934** (2020) 012020 doi:10.1088/1757-899X/934/1/012020

Figures 2 and 3 show the values of mass and maximum deformation in the structure for all considered variants. Hereinafter, the values are presented for the design case with maximum load — flight at an +11° angle of attack.

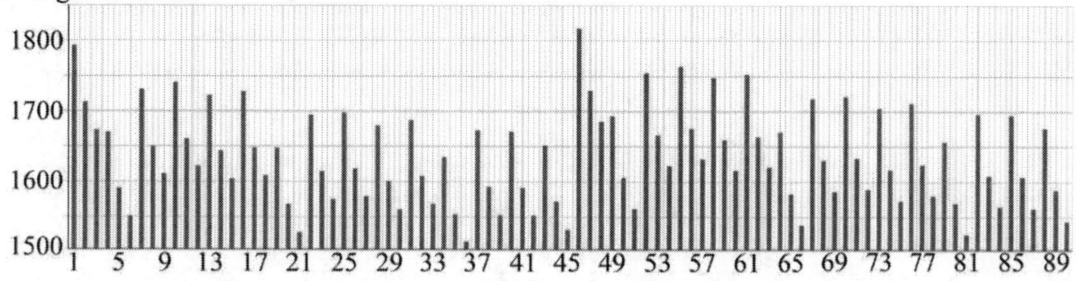

Figure 2. Construction weight, kg.

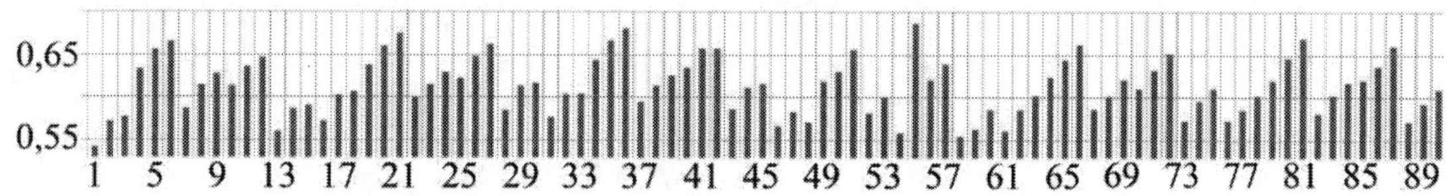

Figure 3. Maximum structure deformation, m.

Figure 4 shows the values of normal stresses along the local Z axis for three groups of elements — sheathing, spars and stringers (longitudinal load–bearing elements), ribs (transverse load–bearing element). The tensile strength along the Z axis for fabric–based carbon fiber reinforced plastic is much smaller than along the other two axes due to the anisotropy of physical and mechanical properties and does not exceed critical values. In the direction of the X and Y axes, the value of the tensile strength is also not exceeded.

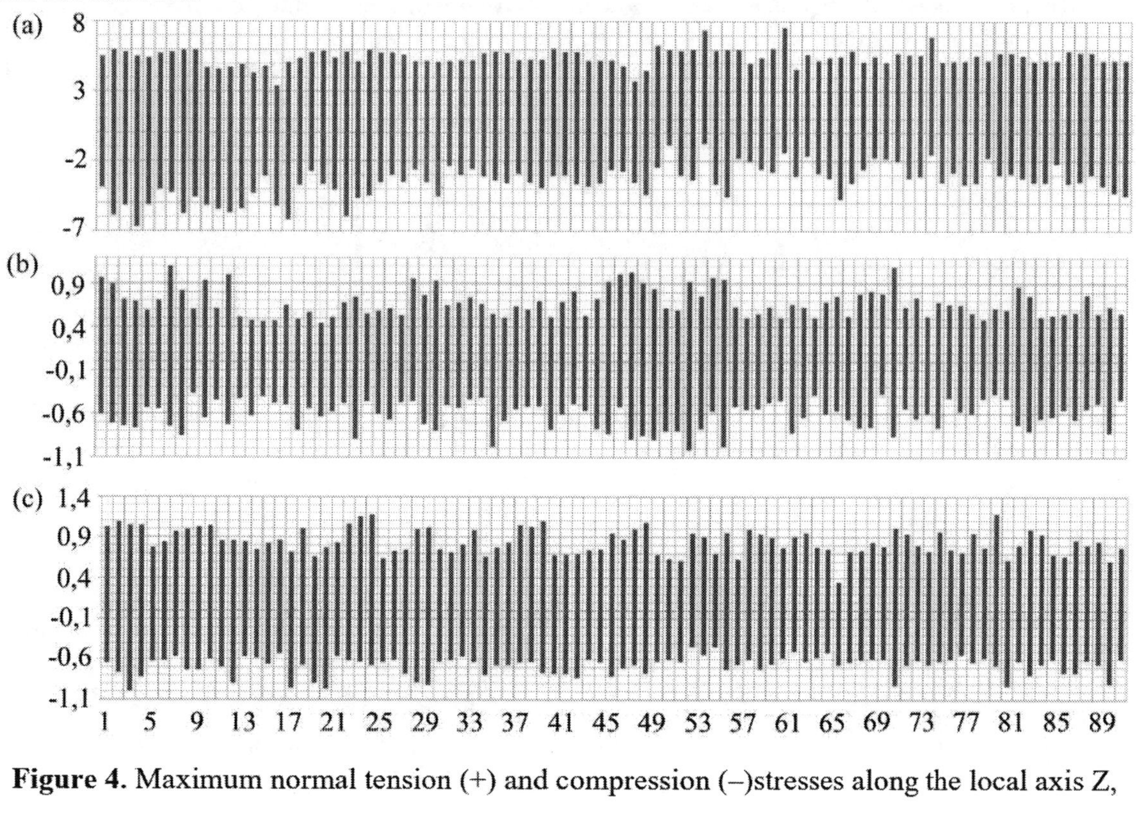

Figure 4. Maximum normal tension (+) and compression (–)stresses along the local axis Z,

MPa: (a) — skins, (b) — longitudinal load–bearing elements, (c) — transverse load–bearing element.

Figure 5 shows the results of choosing the SA of the polymer composite wing, where the mass and the deformation corresponding to the maximum structure load–bearing ability are selected as criteria.

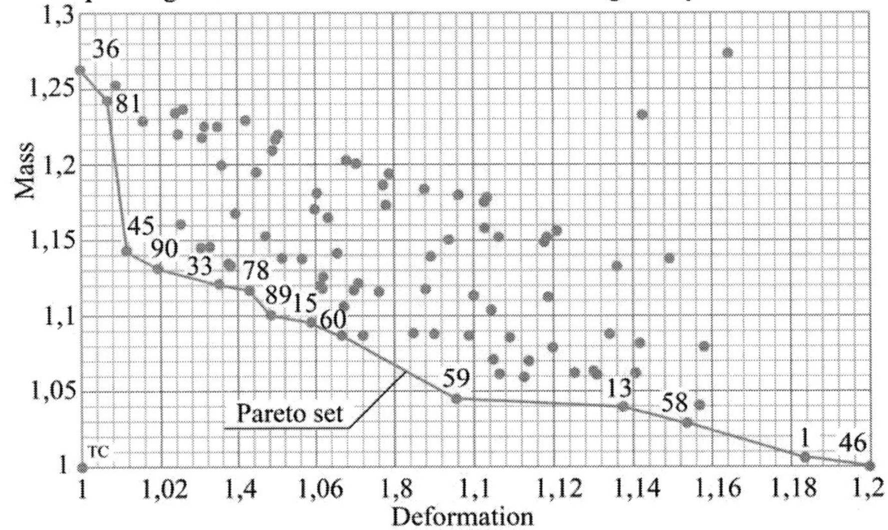

Figure 5. The distribution of values in relative units.

The results of SA calculations show that none of the variants exceeds the values of tensile strengths, the safety factor is not less than 1,8, therefore, the initial approximation of the load–bearing element wall thickness should be reduced by subsequent optimization (provided that operability is maintained as a result of adding additional loads to the structure — weight of fuel, engines, etc.).

A lot of non–dominant alternatives were identified according to two criteria (Pareto set) — deflection and weight of the structure (14 variants). The optimal SA, from the obtained ones, can be determined both by introducing an additional parameter, for example, cost, and by choosing the shortest distance to the theoretical center (TC) according to the equation (1):

$$K = \left[\frac{\left(m_{TC} - m_i\right)^2}{m_{AM}^2} + \frac{\left(d_{TC} - d_i\right)^2}{d_{AM}^2} \right]^{1/2} \tag{1}$$

where m_{TC}, d_{TC} — TC mass and deformation values, m_i, d_i — variant mass and deformation values, m_{AM}, d_{AM} — arithmetic mean mass and deformation values.

It was found that the smallest distance to the IC for variant 33 (Table 3). Figure 6 shows a front view of the deformed and undeformed wings at different angles of attack.

Table 3. The values of the selected wing SA geometric parameters.

Front spar pitch, % of chord	20
Rear spar pitch, % of chord	70
Rib pitch,% MAC [a]	40
Ribs mounting orientation	I[b]
Stringer pitch, m	0,2

[a] Mean Aerodynamic Chord
[b] Perpendicular to the rear spar

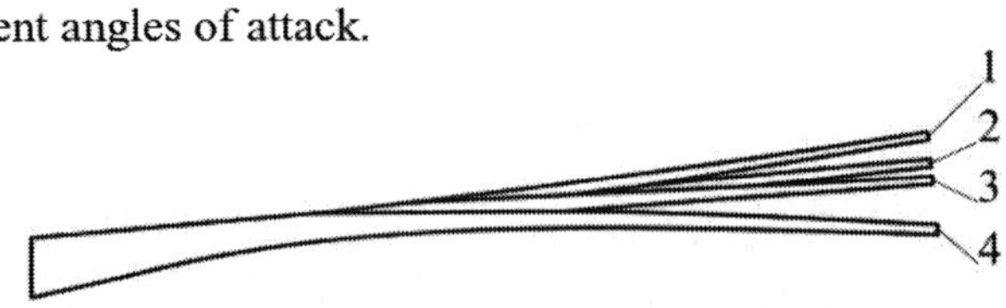

Figure 6. Wing deformations: 1 — angle of attack + 11°, 2 — undeformed wing, 3 — angle of attack 7°, 1 — angle of attack 7°.

It is determined that the maximum deformation from airload occur at the tapering end of the wing. In turn, at root section (where the wing is as wide as possible), deformations are insignificant or absent. It follows that the engines (if placed on the wing) are rationally installed in the area of the straightened section or in the transition zone to the swept part. It has been established that longitudinal and transverse load–bearing elements (ribs, spars and stringers) unload the power schemes, and the stresses along the X axis in the global coordinate system (CS) are informative about the operation of the PCM structure. To assess the overall structure reliability qualification, it is necessary to consider each load–bearing elements (group of elements) in the local CS associated with the layers angles in the part, which is associated with a pronounced PCM properties anisotropy.

3. Selection of stringer

3.1. Source data

As an example of optimization of the load–bearing element, a calculation was made for the choice of the stringers shape and material. The following types of materials were used for the calculations: carbon fiber, fiberglass, aluminum alloy, and lightweight foam filler. Aluminum alloy is given as a comparison of PCM and metal.

Figure 7 shows the considered variants of the stringers cross–sectional shapes.

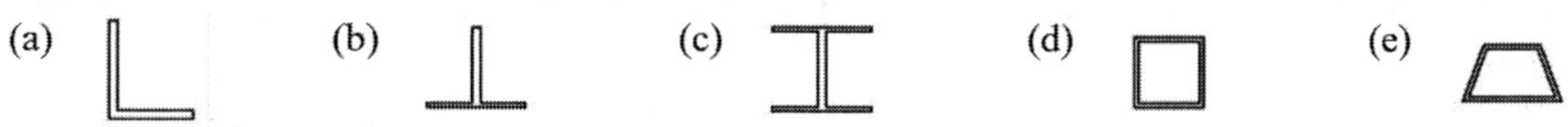

Figure 7. Section shapes: (a) — angle–shape; (b) — T–shape; (c) — I–shape; (d) — square–shape with light foam aggregate; (e) — trapezoid–shape with light foam aggregate.

The calculation was carried out using finite element methods. To construct the finite element mesh, the geometry of stringers was simplified, in particular, fillets were not taken into account. In parametric modeling, 15 design cases were considered.

3.2. Results

As a result of the calculation, the stringers deformation and mass values were obtained. Figure 8 shows the results of stringer shape parametric optimization, where the mass and deformation corresponding to the maximum load–bearing ability are selected as criteria.

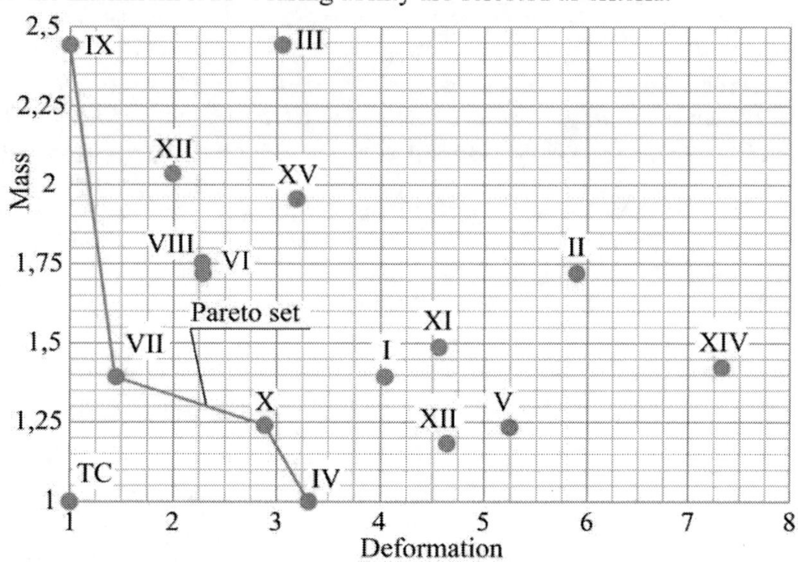

Figure 8. The distribution of values in relative units.

A certain Pareto set consists of four variants. Variant IX is excluded because corresponds to a metal stringer. The manufacture of stringers (panels) of metal implies a mechanical way of attaching the

casing to the stringers, for example, rivets, which requires equipment that is more complex. Variants VII (I–shape) and X (square–shape) are complex from a technological point of view, since for the manufacture of a stringer in the I–shape form, complex equipment is necessary, as in the case of a square–shape profile.
Variant IV is easy to manufacture by laying out using a simple equipment. With equal thicknesses of the shelf and wall, the T–shape is less material–intensive than the I–shape. Based on the foregoing, a carbon–fiber stringer of T–shape was selected (variant IV).

4. Conclusion

In this work, based on the performed parametric modeling of 90 geometric models of the SA of the polymer composite wing at different angles of attack (0°, + 11°, –7°) established:

- rational geometric parameters of the SA wing — the pitches of ribs, stringers, spars;
- shape and material of the panel element — stringer.

From the analysis of the simulation results, a SA variant was chosen in which pitch are: front spar is 20% of the chord, the rear spar is 70% of the chord, the rib is 40% of the MAC, the stringer is 0.2 m and the mounting orientation of the ribs is perpendicular to the rear spar.
Based on the solution of the parametric optimization problem, a T–shape stringer made of carbon fiber is selected.

References

[1] Schuhmacher G, Daoud F, Petersson O and Wagner M 2012 *28th Congress of the International Council of the Aeronautical Sciences(Brisbane)* ICAS 2012–0.4.
[2] Kruse M, Wunderlich T and L Heinrich 2012 *30th AIAA Applied Aerodynamics Conference (New Orleans)* AIAA 2012–3208 — p 1–27.
[3] Gorodetsky M A, Mikhaylovsky K V, Reznik S V 2019 *IOP Conf. Series: Materials Science and Engineering* **683 012075**. 8 p
[4] Dillinger K S, Abdalla M M, Klimmek T and Gurdal Z 2013 *10th World Congress on Structural and Multidisciplinary Optimization (Orlando)* p 19–24.
[5] T. Bach, L Heinrich, S Dähne and C Huehne 2014 *14th AIAA Aviation Technology, Integration, and Operations Conference* (Atlanta) AIAA 2014–2295.
[6] Daochun Li and Shijun Guo 2013 *54th AIAA/ASME/ASCE/AHS/ASC Structures, Structural Dynamics, and Materials Conference* (Boston) AIAA 2013–1934.
[7] Reznik S V, Prosuntsov P V, Mikhailovsky K V, Shafikova I R 2016 *IOP Conference Series: Materials Science and Engineering* **153(1)** 10 p. DOI: 10.1088/1757-899X/153/1/012001.
[8] Reznik S V, Prosuntsov P V, Mikhailovskii K V 2015 *J. Eng. Phys. Thermophy* **88(3)** p. 594-601.
[9] Rhea P Liem, Gaetan K W Kenway and J R R A Martins *AIAA Journal* **53(1)** p 104–22.
[10] Soloshenko V N and Popov Yu I 2013 *Aerospace MAI Journal* **20** p 16–30.
[11] Stanford B, Dunning P 2015 *Journal of Aircraft* **52** pp 1298–1311.
[12] Yu Wang, Xing Ouyang, Hailian Yin and Xiongqing Yu 2016 *Journal of Aircraft* **53** pp 351–59.
[13] Khong Fong Nguen and Biryuk V I 2014 *Proceedings of MIPT* **6** c 133–41.
[14] Chen Bai, Luo Mingqiang, Shen Zhong, Wu Zhe, Man Yiming and Fang Lei 2014 *IJASS* **15** pp 383–95.
[15] Qiang Liu, M Jrad, S B Mulani and R K Kapania 2015 *56th AIAA/ASCE/AHS/ASC Structures, Structural Dynamics, and Materials Conference (Kissimmee)* AIAA 2015-0137.
[16] Q Liu, S B Mulani and R K Kapania 2014 *55th AIAA/ASME/ASCE/AHS/ASC Structures, Structural Dynamics & Materials Conference (National Harbor)* AIAA 2014-0471.
[17] Reznik S V, Prosuntsov P V, Azarov A V 2015 *J. Eng. Phys. Thermophy* **88(3)** p. 699-705
[18] Locatelli D, Mulani S and Kapania R 2011 *Journal of Aircraft* **48** p 1671–84.
[19] Khani A, IJsselmuiden S, Abdalla M and Gürdal Z 2011 *Composites Part B: Engineering* **42** pp

546–52.

[20] Prosuntsov P V, Reznik S V, Mikhaylovskii K V and Belenkov E S 2018 *Journal of Physics. Conference Series.* **1134** 012048 7 p

[21] Stanford B, Beran P and Bhatia M 2014 *Journal of Aircraft* **51** pp 938–44.

[22] Mitrofanov O V and Ognyanova T S 2013 *Natural and technical sciences* **6** c 261–65.

[23] Sandeep D and Nageswara Rao A 2014 *IJMRD* **1** pp 34–41.

[24] Sysoeva V V and Chedrik V V 2011 *TsAGI Science Journal* **42** c 91–102.

[25] Baranovski S V and Mikhailovskiy K V 2019 TsAGI Science Journal **50(3)** p 325-339.

[26] Morgulets S V, Chernetsov A A, Afanas'ev A V and Kosarev V A 2012 *Aviation Industry* **1** c 37–41.

[27] Chedrik V V *Vestnik of Lobachevsky University of Nizhni Novgorod* **4** c. 1847–49.

[28] Mikhailovskiy K V and Baranovski S V 2016 *Proceedings of Higher Educational Institutions. Machine Building* **11(680)** p 86-98.

[29] Mikhaylovskiy K V and Baranovski S.V. 2018 *Vestn. Mosk. Gos. Tekh. Univ. im. N.E. Baumana, Mashinostr* **5** p 15–28.

[30] Mikhailovskiy K V and Baranovski S V 2019 *Proceedings of Higher Educational Institutions. Machine Building* **3** p 61–70

IOP Conf. Series: Materials Science and Engineering **934** (2020) 012021 doi:10.1088/1757-899X/934/1/012021

Selection and justification of polymer composite wing load–bearing elements design parameters with material anisotropy and airload

S V Baranovski and K V Mikhailovskiy

Space–rocket composite design department, Bauman Moscow State Technical University (BMSTU), 105005, Moscow, 2nd Baumanskaya St., Bldg. 5, Block 1, Russian Federation

E–mail: serg1750@mail.ru

Abstract. The rapidly developing aviation industry is looking for new design solutions in the field of designing modern passenger airliners. To ensure reliability and increase the service life of developed products can help polymer composite materials, which are already successfully used in the construction of a number of aircraft. This work is devoted to the urgent task of determining and optimizing the structural parameters of the carbon fiber wing load–bearing elements, taking into account rational reinforcement schemes. The wall thicknesses, layup patterns of unidirectional layers, the safety margins of the wing structural elements, taking into account the current operational loads for several angles of attack corresponding to different flight modes. The work is part of the methodology for designing a wing of polymer composite materials.

1. Introduction

The wing development and design has always been a complex task requiring the consideration of many different factors, which only becomes more complicated with the use of new materials and technologies. In turn, promising developments actively introduced by world manufacturers, including in the field of polymer composite materials (PCM), do not find such widespread use in the Russian aircraft industry, in which metal alloys continue to remain the main structural materials and the design methods developed for them are applied. However, there is a need to improve and change approaches in the design of aircraft structures, including those from composite materials (CM), which can not only reduce the weight of the product being developed units, but also provide fundamentally new and effective layout schemes that could not be achieved with traditional design approach [1]. These include, for example, non–standard structural arrangement (SA), consisting of curved elements, the design of which implies general wing optimization (spars, skin ribs) followed by a particular one (for example, skin parameters in contact zones with other elements) [2]. In other words, the wing design parameters are selected at the level of the entire structure depending on the deformation, while the CSS is optimized as a subsystem from the conditions for satisfying these restrictions [3].

In order to simplify and speed up the calculations, work is carried out on comparing various variants for determining the strength characteristics, for example, the use of various finite element models [4]. Many complex multidisciplinary tasks are solved, such as optimization of a polymer composite and metal wing and a comparison of different types of stringers and their influence on the mass and deformation of the airliner wing [5]. The work performed using the commercial MSC Nastran software aims to optimize for several design parameters, such as weight, deformation, strength, resistance to damage and aeroelasticity with preliminary grouping of layers in the package to reduce design variables [6] and variable stiffness and thickness matrix elements [7–9].

The buckling of a polymer composite wing are solved, for example, for a light sports airplane with vertical take–off and landing [10, 11]. The design work and the calculation of the parameters of thin–walled elements under the influence of airload [12], with the choice of SA and variants for the wing box junction with the center wing [13], panels [14], as well as determining the optimal layup in skins, do not go unnoticed. [15]. Today, designs use typical quasi–isotropic stackings with several different (usually no more than four) layer orientations. Moreover, the choice of the layers direction is one of the key features of the design of products from composite material. Therefore, for this purpose, work is underway to study the properties of PCM stacks other than quasi–isotropic [16, 17]. The effect of the layers orientation on aeroelasticity, for example, a triangular wing, is studied experimentally and numerically, but only angles with a step of 15 ° are considered [18]. When using hybrid PCMs in the wing structure of a suborbital reusable spacecraft, only typical quasi–isotropic stacking are also considered [19]. In this area, the developing technology of directional layup of fibers in layers, which can be carried out using automated fiber placement machines (AFP machines), is very interesting. It is assumed that the studied and developed composite materials with directional fibers in the layer (Tow–steered Composite) will improve the characteristics of wing aeroelasticity, but there is no obvious or intuitive structure of the fiber stacking that would give such a result. To this end, wings from the proposed composite material with directional fibers, conventional unidirectional PCM and metals in terms of fuel and weight efficiency of the structures obtained [20 – 23], as well as from a hypothetical composite material based on carbon nanotubes using optimization in strength and stability are compared [24, 25].

Based on the foregoing, the developed methodology for designing a polymer composite wing will simplify the design at the stage of design calculations. According to the general scheme of the technique, after determining the wing loads and designing the KSS, it is necessary to calculate the parameters of the structural elements from PCM.

The paper discusses the process of determining the structural parameters of the main elements from the PCM for the obtained wing geometric model under the action of operational loads.

2. Source data

We considered a swept wing with a span of 35 m, with a straightened section, an asymmetric airfoil of a mid–range passenger airliner consisting of upper and lower finned panels (which can be divided into skins and reinforcing each eleven stringers), two spars and fourteen ribs (Figure 1) [26].

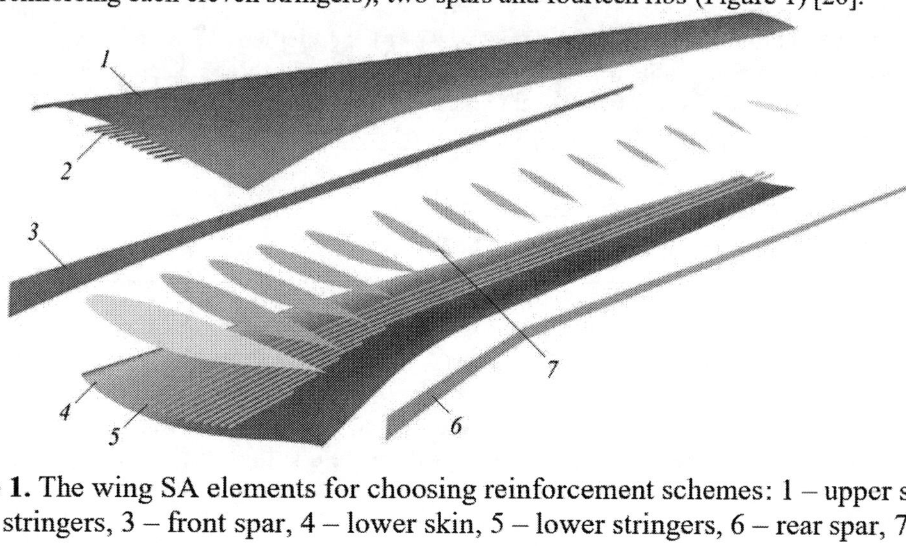

Figure 1. The wing SA elements for choosing reinforcement schemes: 1 – upper skin, 2 – upper stringers, 3 – front spar, 4 – lower skin, 5 – lower stringers, 6 – rear spar, 7 – ribs.

The simultaneous action of the following concentrated and distributed loads was considered [27, 28]:

- dead load;
- engine weight (2400 kg);
- fuel weight of 20,000 kg (it is assumed that the tank is full);
- air–pressure corresponding to one of the considered angles of attack (cruising flight at a speed of 242 m / s, at an altitude of 11500 m, angles of attack: 0°, + 11°, –7°).

When choosing the optimal thicknesses and layup angles, carbon fiber was adopted as the material (Table 1).

Table 1. Carbon fiber stress–related characteristics.

Tensile modulus, along an axis X / Y, GPa	48,57/ 48,57
In–Plane shear modulus, GPa	28,9
Poison's ratio	0,44
Tensile (Compressive) strength, along an axis 1 / 2, MPa	483 (507)/483 (507)
In–Plane shear strength, MPa	262

The thicknesses choice of the wing elements was carried out for two design models:

Model 1 — upper and lower skins, front and rear spars, 14 ribs.

Model 2 — upper and lower skins, front and rear spars, 8 ribs, 6 stringers (3 on each skin).

The step of the ribs and stringers was chosen to ensure uniform distribution over the wing box. Table 2 presents the initial values of the thickness of the elements in a first approximation. The maximum and minimum ranges of thickness variation from 0,002 to 0,015 m, except for stringers for which the maximum value is 0,009 m.

Table 2. Initial thicknesses of elements, m.

Skins	Front spar	Rear spar	Ribs	Stringers
0,007	0,008	0,01	0,007	0,007

When choosing a layup, we considered both each element individually (casing, spars) and groups of elements (ribs, stringers).

The calculation was carried out iteratively for each element under consideration, starting from the upper casing until the installation from the previous step matches the values at the current one, and the error does not exceed 1,5%.

The layup angle ranged from –90° to 90°, and the initial value for all layers of each element was 0°.

The number of cases considered for each individual element or group without taking into account variation in the angle of attack and iteration exceeded 11 thousand.

3. Results

Because of the calculation by choosing the thicknesses of the elements, the deformation and mass of the structure are determined. Table 3 shows the parameters of the three selected variants (taking into account the angle of attack), from which the thickness values of the wing elements were selected.

Table 3. Calculation results for choosing the thickness of the elements.

	Thickness, m						Mass, kg	Angle of attack		
	Skin		Spar		Ribs	String.		0°	+11°	–7°
	Upper	Lower	Front	Rear				Deformation, m		
I	0,002	0,0085	0,0065	0,0046	0,0025	0,0022	1041,1	0,517	0,827	2,030
II	0,008	0,01	0,007	0,008	0,004	0,003	1717,2	0,277	0,397	1,035
III	0,003	0,005	0,006	0,004	0,002	0,0025	823,89	0,510	0,835	2,027

It can be seen that the deflection and mass values of variants I and III are similar and are two times different from II. In this case, in none of the cases considered, the stresses in the structure reach the ultimate strength of the material, and the maximum values are due to stress concentrators and the error of the finite element mesh. Based on the foregoing, option III was chosen with the minimum calculated thicknesses, which can be increased with further calculation of elements taking into account various factors (for example, amplification for mechanization elements).

For the obtained thicknesses, the number of layers was determined based on the layer with rounding upwards Table 4.

Table 4. Number of layers in elements.

Skin		Spar		Ribs	Stringers
Upper	Lower	Front	Rear		
11	18	20	14	7	9

The optimal stackings calculated for different angles of attack were compared, stacking from the initial approximation (orientation of all layers is 0 °), two quasi–isotropic stackings, the ratio of the layers in which is presented in Table 5:

- A — optimal layup corresponding to an angle of attack of 0°;
- B — optimal layup corresponding to an angle of attack of +11°;
- C — optimal layup corresponding to an angle of attack of –7°;
- D — quasi–isotropic layup 1;
- E — quasi–isotropic layup 2;
- F — first approach layup.

Table 5. The ratio of layers with layup angles 0°/±45° /90° in quasi–isotropic layup.

	Layup 1		Layup 2	
	Value	Value, %	Value	Value, %
Upper skin	2/8/1	18/73/9	3/6/2	27/55/18
Lower skin	4/12/2	22/67/11	6/8/4	33/44/22
Front spar	6/12/2	30/60/10	6/10/4	30/50/20
Rear spar	4/8/2	29/57/14	4/6/4	29/43/29
Ribs	2/4/1	29/57/14	2/4/1	29/57/14
Stringers	1/6/2	11/67/22	3/4/2	33/44/22

The evaluation criteria were the values of the entire structure deformation and the safety factor (for stresses, strains and the Tsai – Wu criterion). Each structural element was considered separately. Table 6 shows the results of the optimal layup for each element depending on the angle of attack.

Table 6. Optimum layup for wing elements.

AA[a]	Skin		Spar		Ribs, No.			Stringers	
	Upper	Lower	Front	Rear	1, 14	2–5	6–13	Upper	Lower
0°	C	C	C	A	C	A	C	A	C
–7°	A	B	A	A	F	A	A	F	F
+11°	A	A	A	A	A	A	A	A	A

[a] Angle of attack

It can be seen that at an angle of attack of 0 ° for the structural elements, the optimal variants are C (66.67%) and A (33.33%). For the angle of attack –7° — A (55.5%), F (33.3%) and B (11.2%), and for + 11°, for all structural elements, layup A (100%). Because for all angles of attack, the largest number (63%) of optimal variants correspond to A; a comparison was made with B, C, F for the following elements: upper and lower skin, front spar, ribs No. 1, 6–14, upper and lower stringers. As a result, it was found that the deflection of the wing during layup A increases by 0,004 m (2%), the values of the safety factor are reduced by no more than 17% (where 100% is the destruction of the structural element) compared to option C. However, such a large decrease in the safety factor can be neglected, because the zone of the indicated voltages is incomparably small in comparison with the area of the element in which the supply is high, and is due to the stress concentrator and a coarse grid.

In turn, layup variant B and F show the best values of deflection and safety factor — the gain in deflection is 0,177 m (17%) and 5% in the safety factor for layup B, 0.004 m (2%) in deflection and 16% by coefficient — F, but only for several elements (skin, ribs, stringers) at one angle of attack. Therefore, they can also be neglected.

There is a gain in the mechanical properties of the structure, in the elements of which optimized styling is used compared to quasi–isotropic. For some elements, the difference reaches 40%.
As a result, variant A was selected (Table 7).

Table 7. Layup of wing elements.

Upper skin	28°/–10°/–48°/56°/–2°/–1°/–23°/–3°/–7°/–10°/14°
Lower skin	–90°/0°/–30°/–54°/–64°/–74°/–76°/–79°/–80°/–82°/–84°/–84°/–85°/–86°/–86°/ –86°/–87°/–87°
Front spar	0°/–12°/17°/–17°/44°/27°/34°/13°/–11°/66°/–30°/–52°/14°/–45°/33°/33°/–4°/ 2°/–48°/–1°
Rear spar	–40°/29°/13°/–16°/30°/–52°/–26°/–27°/21°/–78°/11°/80°/–1°/47°
Ribs No. 1,14	–30°/4°/–48°/28°/48°/–85°/10°
Ribs No. 2–5	41°/13°/7°/49°/–1°/33°/5°
Ribs No. 6–13	63°/–37°/0°/–84°/–4°/–40°/–20°
Stringers	14°/2°/–22°/–12°/8°/4°/–15°/90°/–18°

4. Conclusion

Because of polymer composite wing elements, calculations at different angles of attack (0°, + 11°, –7°) determine:

- rational minimum possible number of layers;
- rational orientation layers angles;
- thickness;
- structure deformation values for the selected optimal layup;
- safety factors of structural elements.

The advantage (up to 40%) of layup optimized for operational loads over quasi–isotropic is shown.
The maximum structure deformation is achieved at an angle of attack of –7 °, is 1,22 m and is due to the interference of existing loads. The safety factor does not exceed 0.35, with the exception of several zones in which it reaches 0,47, which is due to the stress concentrator and the error of the finite element mesh.
The obtained values of the characteristics correspond to the level of the outline design and will be optimized upon a detailed examination of each element.
The results of this work are part of the compiled polymer composite wing design methodology and will be taken into account and used in its compilation.

References

[1] Zichenkov M Ch, Kondakov I I and Shanygin A N 2016 *Civil Aviation High Technologies* **19** p 127–36.
[2] Liu Q, Jrad M, Mulani S B and Kapania R K 2015 *56th AIAA/ASCE/AHS/ASC Structures, Structural Dynamics, and Materials Conference (Kissimmee)* AIAA 2015–0137.
[3] Qun Z, Yunliang D and Haibo J 2011 *Chinese Journal of Aeronautics* **24** pp 425–33.
[4] Kaplan E M, Yesilkaya K K and Yaman K 2016 *7th International Conference on Mechatronics and Manufacturing (Singapore)* **45** pp 1–5.
[5] Bach T, Dähne S, Heinrich L and Hühne C 2014 *14th AIAA Aviation Technology, Integration, and Operations Conference (Atlanta)* AIAA 2014–2295.
[6] Li D and Xiang J 2013 *54th AIAA/ASME/ASCE/AHS/ASC Structures, Structural Dynamics, and Materials Conference (Boston)* AIAA 2013–1934.
[7] Dillinger J K S, Abdalla M M and Klimmek T 2012 *12th AIAA Aviation Technology, Integration, and Operations (ATIO) Conference and 14th AIAA/ISSM (Indianapolis)* AIAA 2012–5401.
[8] Silva G H C, A Pereira do Prado, P H Cabral, Roeland De Breuker and Dillinger J K S 2019 Journal of Aircraft **56 (3)** p 990–1004.

[9] Pusch M, Ossmann D, Dillinger J, Kier T M, Tang M and Lübker J 2019 *AIAA Scitech 2019 Forum (San Diego)* AIAA 2019-0131.
[10] Shingel L P 2012 *Izvestia of Samara Scientific Center of the Russian Academy of Sciences* **14** p 514–16.
[11] Reznik S V, Prosuntsov P V and Mikhailovskii K V 2015 *J. Eng. Phys. Thermophy* **88(3)** p. 594-601.
[12] Kasumov E V 2015 *TsAGI Science Journal* **46** p 63–79.
[13] Soloshenko V N and Popov Yu I 2013 *Aerospace MAI Journal* **20** p 16–30.
[14] Khania A, Abdallaa M M, Gürdala Z, Sinkea J, Buitenhuisb A and Van Toorenb M J L 2017 *Composite Structures* **180** p 821–830.
[15] Mitrofanov O V 2017 Current problems of modern science **96** p 49–53.
[16] Peeters D and Abdallay M 2017 *Journal of Aircraft* **54** pp 1454–64.
[17] Reznik S V, Prosuntsov P V and Azarov A V 2015 *J. Eng. Phys. Thermophy* **88(3)** p. 699-705
[18] Sapkal K S and Attar P J 2011 *52nd AIAA/ASME/ASCE/AHS/ASC Structures, Structural Dynamics and Materials Conference, Structures, Structural Dynamics, and Materials and Co–located Conferences (Denver)* AIAA 2011–1743.
[19] Reznik S V, Prosuntsov P V and Ageeva T G 2013 *Vestnik NPO im. S.A. Lavochkina* **17** p 38–43.
[20] Brooks T R, Martins J R R A and Kennedy G J 2017 *58th AIAA/ASCE/AHS/ASC Structures, Structural Dynamics, and Materials Conference (San Diego)* AIAA 2017–1350.
[21] Brooks T R, Martins J R R A and Kennedy G 2019 *Journal of Fluids and Structures* 88 p 122–47.
[22] Gorodetsky M A, Mikhaylovsky K V and Reznik S V 2019 *IOP Conf. Series: Materials Science and Engineering* **683 012075**. 8 p
[23] Baranovski S V and Mikhailovskiy K V 2019 TsAGI Science Journal **50(3)** p 325-339.
[24] Kennedy G J, Kenway G W and Martins J R R A 2014 *52nd Aerospace Sciences Meeting (National Harbor)* AIAA 2014–0596.
[25] Reznik S V, Prosuntsov P V, Mikhailovsky K V and Shafikova I R 2016 *IOP Conference Series: Materials Science and Engineering* **153(1)** 10 p. DOI: 10.1088/1757-899X/153/1/012001.
[26] Mikhailovskiy K V and Baranovski S V 2016 *Proceedings of Higher Educational Institutions. Machine Building* **11(680)** p 86-98.
[27] Mikhaylovskiy K V and Baranovski S.V. 2018 *Vestn. Mosk. Gos. Tekh. Univ. im. N.E. Baumana, Mashinostr* **5** p 15–28.
[28] Mikhailovskiy K V and Baranovski S V 2019 *Proceedings of Higher Educational Institutions. Machine Building* **3** p 61–70

IOP Conf. Series: Materials Science and Engineering **934** (2020) 012022 doi:10.1088/1757-899X/934/1/012022

Mechanics of orthotropic plates with honeycomb filler

A V Berezin [1], M A[2] Lyubchenko and I V Gadolina [1]
[1]IMASH RAS, Maly Kharitonievsky 4, Moscow, Russia
[2]MGTU named after Bauman, Moscow, Russia, ma.ljubchenko@yandex.ru

Abstract. Materials with a honeycomb structure have a cellular geometry, which reduces the amount of material used to achieve mass perfection of the finished structure. Due to the widespread application of honeycomp composite materials (structures), the need for developing the design methods exists. It is especially important to have design methods to guarantee the longevity and reliability of the elements of aviation structures during the exploitation. The peculiarities of operation such complicated elements as honey comb require consideration of existing of tough and soft layers, which work differently. The equations of motion of the three-layer plate with honeycomp filling are given. Some important issues of investigation based on this equation are described. In the paper, the questions of destructive and non-destructive diagnostics of composite constructions are discussed.

1. Introduction

The advantages of cellular structures made of composite materials (CM) today are well known and undeniable. These are high strength characteristics at low density, the combination of high rigidity with low specific gravity, resistance to crack development, the ability to orient the properties in the directions of the acting loads [1,2].

Materials with a honeycomb structure have a cellular geometry, which reduces the amount of material used to achieve mass perfection of the finished structure. In addition, the geometric dimensions of honeycomb structures can vary widely. This circumstance led to the widespread use of products made of CM with honeycomb filler in the rocket and space and aviation industry (Tu-204, Tu-334, Il-96, Il-114, RRJ-95, etc.) [3-5].

Despite the fact that the theory and practice of creation and application of cellular structures with cladding of CM today is quite seriously worked out and continues to develop, the problem of forecasting the durability of structures of such panels under the action of dynamic loads remains an urgent and important practical task [1, 2].

In [6] the model for estimation of elastic and strength characteristic was developed. The model is based on the use of asymptotic homogenization method of hierarchical periodic structures. In [7] three-layer (sandwich) structures consisting of external power sheathing and lightweight filler are described.

To guarantee the stable and reliable exploitation of complicated construction made of composite, the question of diagnostic is very important. In [8] the nowadays approaches to health monitoring including diagnostic of aviation construction are analyzed. As one of the practical solutions, to the problem of ensuring the weight efficiency of modern technical systems with increased reliability requirements can be the use of composite mesh (anisogrid) structures as power elements [9]. For sandwich structure composed of two laminated skins and a honeycomb core was proposed a numerical optimization procedure in order to obtain a true global optimal solution for the considered problem of strength [10].

To study the influence of structural and technological factors, engineering methods of identification for mechanical characteristics of unidirectional composites were developed in [11]. They consider the physical nonlinearity which is observed in complicated composite structures, which may be a consequence of shear stresses influence. Descripting the mechanical properties of a laminate for a given structure was usually carried out based on the regularities of its constituent layers. The analysis of the viscoelastic properties of composite materials a physical nonlinearity was observed, which may be a consequence of shear stresses.

Under production conditions, the shell is modeled from hard (carbon fiber, fiberglass, etc.) and soft layers (honeycomb filler). Soft layers work only on "shear" and "stretch-compression", which are distributed evenly over the thickness of the layer. The arrangement of these layers is modeled in such a way as to fill the space between the panels with the maximum preservation of the mass perfection of the design. This method guarantees the condition of optimal operation of the structure.

2. Considering the dynamic loads in the prediction of durability

The prediction of the durability of cellular structures under dynamic loads should be understood as the definition of the forms and frequencies of natural oscillations. Knowledge of these parameters will allow developing methods for durability estimation.

The equations of motion of a three-layer plate with honeycomb filler have the form:

$$\frac{\partial Q_x}{\partial x}+\frac{\partial Q_y}{\partial y}=q+\rho\frac{\partial^2\omega}{\partial t^2}+2\xi_0\frac{\partial\omega}{\partial t^2},$$

$$\begin{gathered}\left[\frac{1}{S_x}\left(D_x\frac{\partial^2}{\partial x^2}+D_y\frac{\partial^2}{\partial y^2}\right)-1\right]Q_x+\left[\frac{1}{S_y}\left(\frac{D_{xy}}{2}+v_xD_y\frac{\partial^2}{\partial x\partial y}\right)\right]Q_y\\-\left[D_x+\left(D_{xy}+v_xD_y\right)\frac{\partial^3}{\partial x\partial y^2}\right]\omega=0,\\ \\ \left[\frac{1}{S_x}\left(\frac{D_{xy}}{2}+v_yD_x\right)\frac{\partial^2}{\partial x\partial y}\right]Q_x+\left[\frac{1}{S_y}\left(D_y\frac{\partial^2}{\partial y^2}+\frac{D_{xy}}{2}\frac{\partial^2}{\partial x^2}\right)-1\right]Q_y\\-\left[D_y\frac{\partial^2}{\partial y^3}+\left(D_{xy}+v_yD_x\right)\frac{\partial^3}{\partial x^2\partial y}\right]\omega=0.\end{gathered} \qquad (1)$$

Where Dx, Dy, Dxy, – bending and rotating stiffness, Sx, Sy, – shear stiffness, vx, vy,– Poisson's coefficients for orthotropic three-layer plate, ρ- mass of the unit area of the plate, ξ0-damping coefficient.

From the system of equations (1) it is possible to obtain differential equations for transverse displacement for an orthotropic three-layer plate, which were obtained in [12], where the solution of the problem of the orthotropic plate oscillation was also obtained by the asymptotic method of V. V. Bolotin [13]. As a result of these studies, the oscillation frequencies of the panel with all fixed edges were determined for the cases of a honeycomb panel made of carbon fiber with a reinforcement scheme [0/±45°] with a thickness of 0.45 mm and a panel skin made of aluminum alloy AMG-2H with a thickness of 0.5 mm. The filler in both cases was taken by a honeycomb, the honeycombs of which are made of alloy AMG-2H with a cell size of 2.5 mm and a wall thickness of 0.03 mm. The obtained theoretical values of natural frequencies were compared with experimental data for panels with facings of aluminium alloy. The discrepancy between the calculations and the experiment was 8%, which may be due to the idealization of the sealing conditions adopted in the calculation of frequencies.

The assumption that the sheathing works as a membrane is true when considering thin enough compared to the thickness of the honeycomb filler sheathing. It is possible to obtain equations for determining stresses in plate sheathing away from the edges. Using this assumption, it is possible to estimate shear stresses in the honeycomb filler provided that transverse shear forces are perceived only by the honeycomb.

The Puppo-eversen destruction criterion composite materials was adopted as a criterion for damage accumulation [14]. Knowing the voltages, it is simple enough to construct estimates for the mean squares of the voltages, the equivalent voltage, and the durability [15,16].
In the process of cyclic loading of composite materials, their stiffness and damping properties change. The change in the stiffness properties leads to the fact that the coefficients of the system of equations (1) become dependent on the acting stresses and time, or on the number of loading cycles. Changes in the stiffness characteristics and leads to a change in the mode of normal vibrations and self-resonant frequencies of the structure. However, in the case of loads with a wide and continuous spectrum, the most significant contribution to the values of dynamic stresses is made by changing the damping properties of the composite honeycomb structure.
In addition, studies have been conducted on the change in the logarithmic decrement of oscillations depending on the change in the stiffness of the sample in the process of bending fatigue tests with torsion. Torsional bending was realized by means of the asymmetrical load arrangement at the end of the excited cantilever-mounted specimen.
Fig.1 shows the experimental dependence of the change in the logarithmic decrement of oscillations when changing the stiffness of the sample in the process of torsional bending tests. If such tests damage occurs due to honeycomb filler, so in this case, the change of the logarithmic decrement of the oscillations is associated only with the destruction of the honeycomb. The curve I in fig. 1 corresponds to a change in the logarithmic decrement of attenuation with a decrease in bending stiffness in the first form of bending oscillations. Curve II corresponds to a change in the decrement of attenuation with a decrease in torsional stiffness in the first form of torsional oscillations. From the presented graphs it can be seen that the decrements of attenuation, depending on the change in the stiffness of the sample, increase when the honeycomb filler is damaged.

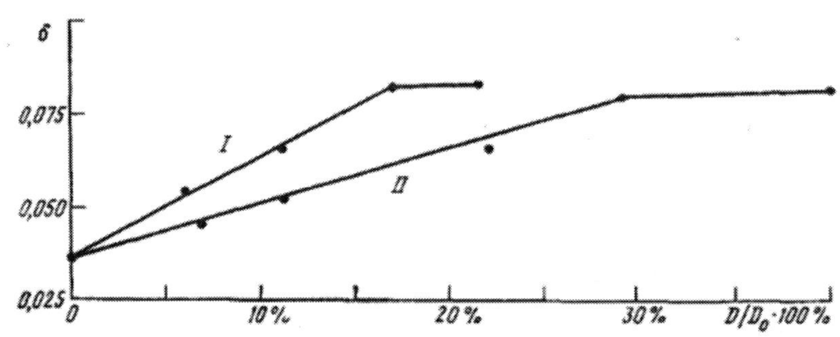

Fig. 1. Dependence of change of logarithmic decrement of attenuation at the change of rigidity of a sample

It can be assumed that the stiffness and damping characteristics do not change much during cyclic loading and take into account only the change in the logarithmic decrement of attenuation over time. This assumption follows from the experimental curves shown in figure 1. Since the change in stiffness is small compared to the change in the decrement of attenuation, it can be concluded that in the pre-collapse state the dynamic stresses at the same load intensity are about three times less than in the undamaged panel.

3. Technological imperfections and their influence on deformation and strength characteristics of structures from CM. Defects diagnostic.

Characteristic defects of composite panels with honeycomb filler and structures made of these materials are delaminations, non-adhesives, voids, cracking of the inner layers, holes and so on. The use of nondestructive testing methods for the diagnosis of defects is limited by the characteristics of the material. For the control of composite panels, specially developed methods and conventional acoustic methods [17-19] are used to diagnose damage. The thermographic methods [18] and acoustic

IOP Conf. Series: Materials Science and Engineering **934** (2020) 012022 doi:10.1088/1757-899X/934/1/012022

emission method [17,18] have been applied. For the most part, these methods give a qualitative character of the geometry and the number of defects.

It should be noted that the influence of the considered defects on the thermophysical characteristics of composite materials is little studied. The effectiveness of thermal methods in the diagnosis of composite materials is reduced due to the significant thermal inertia inherent in most types of composite materials.

The amplitude-frequency characteristics of honeycomb panels with carbon fiber and fiberglass sheathing were studied and their logarithmic attenuation decrement was determined. The frequency dependence of the amplitude of the span of the end of the cantilever-mounted honeycomb panel with carbon fiber sheathing is shown in Fig. 2 at different levels of exciting load. Cellular panels made of carbon fiber have a symmetric dependence of the amplitude of oscillations on the frequency. The logarithmic decrement of attenuation of a honeycomb panel with carbon fiber sheathing was determined, which amounted to $\delta 1 = 0{,}037$.

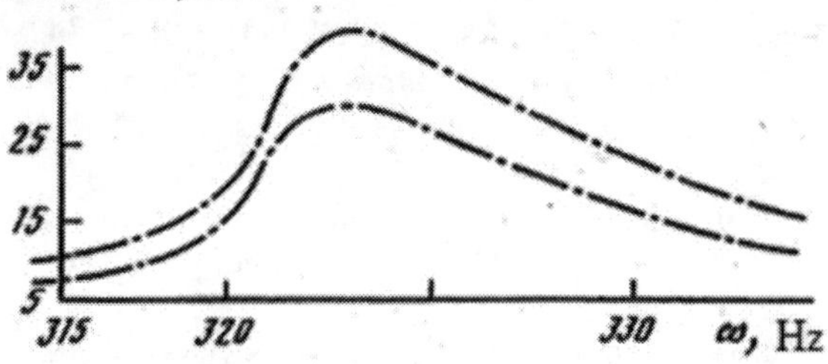

Fig.2. Dependence of amplitude span of the honeycomb beam end on frequency near resonance

It is mentioned though, that when using acoustic methods, surface defects such as delamination can be omitted if the force pressing the finder to the controlled product is greater than necessary and thus assess the product as defect-free. The method of acoustic emission allows the identification of only developing defects. Additionally, the source of acoustic emission cannot be recognized by analyzing the frequency parameters of individual pulses. The main drawback when using non-destructive methods of flaw detection is the complexity of decoding signals and adequate modeling of defects occurring in the material.

As one of the other means to investigate the situation with strength resistance, the electron microscopy of the broken elements might be employed [20]. In case of honeycomp structures in might be applied to the broken elements of stiff components made of carbon plastic. In Fig. 3 the fragments of the electron microscopy of the broken unidirectional sample of carbon fiber are given. Investigating such type of images could help to understand the main types of breaking modes. In Fig.3 the main types of breaking modes of unidirectional carbon fiber are written next to the images.

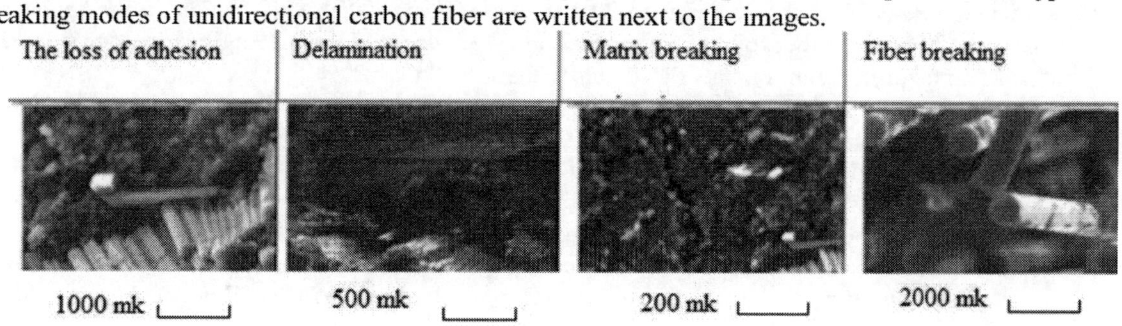

Fig.3. The example of the main types of breaking modes of unidirectional carbon fiber

4. Conclusions

The importance of investigation of dynamic loading in honeycomb panels is characterized and motion equations are presented.

Some important topics concerning defect diagnostic (acoustic emission, studying vibrations and electron microscopy) are discussed.
The use of honeycomb structures, like other uses of composite materials [21,22], is at the forefront of technological development and requires close attention from scientists involved in the formulation of computational and diagnostic methods.

Acknowledgement
The authors are obliged to our coworkers PhDs M.M. Khrushchov and O.S. Antonova (IMASH RAS) who performed the electro-microscope study.

References

[1] Dumansky A M, Alimov M A and Terekhin A V 2019 *Experiment- and computation-based identification of mechanical properties of fiber reinforced polymer composites.* // Journal of Physics: Conference Series 1158, , Issue 1. DOI: 10.1088/1742-6596/1158/2/022037.
[2] Dumansky A.M and Liu Hao 2019 *Deformation regularities of carbon fiber reinforced plastic under time variable loading* // Journal of Physics: Conference Series 1158, , Issue 1. DOI: 10.1088/1742-6596/1158/2/022038.
[3] Berezin A V etc. 2016 *Application of carbon fiber composite materials in airframe components of civil airframe components of civil aircrafts PSP «TUPOLEV»* // Engineering and Automation Problems,**3** pp. 4-8.
[4] Libove C and Batdorf S B 1948 A general small-deflection theory for flat sandwich plates. – Report N 899, National Advisory Committee for Aeronautics,.
[5] Berezin A V 1999 Behavior of three-layer plates made of composite materials under static and vibration influences – Problemy mashinostroenia i nadezhnosti mashin,**6** pp. 93-101.
[6] Dimitrienko Yu I at al. 2014 *Modeling and development of three-layer sandwich composite material with honeycomb core* // Herald of the Bauman Moscow State Technical University, Series Natural Sciences. **5** pp. 66-81.
[7] Fleck N A and Deshpande V S 2004 The resistance of clamped sandwich beams to shock loading // J. Appl Mech. **71** P. 386–401.
[8] Svirskiy Yu A, Bautin A A, Papic L and Gadolina I V 2017 Methods of modern aircraft structural health monitoring and diagnostics of technical state. In book: Diagnostic Techniques in Industrial Engineering. Zwitzerland, P. 1-28.
[9] Azikov N S at al. 2018 *Structural and power scheme of the wing with spatial mesh composite ribs.* //Deformation and Failure of Composite Materials and Structures (DFCMS-2018). Proc. of the Conference. Oct. 23-25. Moscow. P. 5-7.
[10] Catapano A and Marco Montemurro 2014 *Optimal design of sandwich plates with honeycomb core//* Proc. of the Joint. Conference on Mechanical, Design Engineering & Advanced Manufacturing, Jun, France. P.1-7.hal-01063625
[11] Dumansky A M and Hao Liu 2019 *Deformation regularities of carbon fiber reinforced plastic under time variable loading//* J. Phys.: Conf. Ser. 1158 022038
[12] Berezin A V and Sulimenkov V V 1986 *Dynamics of orthotropic plates with honeycomb filler.* – Uchenyye zapiski TSAGI, **XVII** pp. 53-63.
[13] Bolotin V V 1960 Edge effect on vibrations of elastic shells. – PMM, **24(5)**
[14] Roulands R 1978 *Composite Materials floating and loss of bearing capacity in case of biaxial stress conditions: preparing the calculations and experiment data.* / Under. ed. K. Gerakovich. – M.: Mir,.
[15] Bolotin V V 1965 Statistical methods in structural mechanics. – M.: Stroyizdat,.
[16] Il'ichev V D 1965 *Spectral summing of fatigue damage application in case of complex-stressed constructure conditions.* – Uchenyye zapiski TSAGI, x. **X(2).**
[17] Vaynshtok I S and others 1977 *Instruments for material and product non-destructive control.* / – M.: Mashinostroyeniye,. 190 p.

[18] Potapov A I and Tsekker F P 1977 Non-destructive control for composite materials constructions. M.: Mashinostroyeniye,190 p.
[19] Kollikot R D 1990 Damage diagnostics. M.: Nauka,. 135 p.
[20] Gadolina I V at al. 2018 *Using the data of electron microscopy while planning the fatigue testing of carboplastics*//Deformation and Failure of Composite Materials and Structures (DFCMS-2018). Proc. of the Conference. Oct. 23-25. Moscow. P. 22-24.
[21] Zakharov M N, Lyubchenko M A and Magnitskii I V 2019 Threaded Sample for High-Temperature Mechanical Tests of Carbon–Carbon Composites. Russian Engineering Research.. **39** Issue 6. - pp. 447 – 452.
[22] Zakharov M N, Lyubchenko M A and Magnitskii I V 2019 Failure of Threaded Joints between Composite Components. Russian Engineering Research.. **39** Issue 3. - pp. 195 – 198.

Analysis of 3D-reinforced composite annular plate subjected to concentrated forces

B S Sarbayev and D T Bregvadze

Bauman Moscow State Technical University, 2-ya Baumanskaya ul. 5, str. 1, Moscow, 105005, Russia.

E-mail: Znatok-david-bregv@yandex.ru

Abstract. This paper is dedicated to the solution of the problem about 3D-reinforced composite annular plate compressed by concentrated forces. 3D-reinforced composite materials, including high-temperature composites, are widely used in modern space industry, while their properties are insufficiently studied. Analytical solution of considered problem about annular plate compression will be a foundation for a method of determination of anisotropic composite materials stress-strain properties needed when conducting structural analysis: elastic and shear moduli, and Poisson's ratio. Transverse shear deformations and changes in annular plate metrics during deformation are taken into account in the process of finding analytical solution of the problem. The analytical solution of the problem about annular plate compression was obtained by the authors. This solution is used to find relations between required properties of annular plate and radial displacements of annular plate characteristic points.

Keywords: composite materials, annular plate, 3D-reinforcement, stress-strain properties, anisotropic materials, analytical solution

1. Introduction

3D-reinforced composite materials, including high-temperature composite materials engineered to work under temperatures higher than 1000 °C are widely used in modern space industry. They are present in numerous applications in high-heat elements of rocket engines, such as combustion chambers and nozzle heads, as well as in body parts of rockets and hypersonic spacecrafts, that are subject to significant aerodynamical and thermal loads under their operation, such as payload fairings and leading edges. This topic is more fully studied in papers [1], [2] and [3].

At the same time one should notice that physical and mechanical properties of high-temperature anisotropic composites, such as carbon-carbon composite materials (CCCM) and carbon-ceramic materials, are not yet fully studied. A major part of existing data on properties of such materials is obtained solely by experiments, like in extensive studies of high-temperature ceramics [4] and [5]. The same applies to articles on CCCM tensile and shear strength [6] and [7], as well as to papers [8] [9] [10] [11] [12] [13] [14]. In all of those cases obtained data is relevant only for a specific material or composition, and can't be generalized for a corresponding class of materials, that is apparently a substantial drawback. Therefore the problem of development of methods for physical and mechanical properties determination of high-temperature composites (such as elastic and shear moduli, Poisson's ratio) is highly relevant. Radial compression experiments are often carried out to solve this problem, which is studied in [15] and [16]. Apart from that, blade-shaped specimens are often used for tensile tests [17] [18].

Method of physical and mechanical properties determination for high-temperature anisotropic materials, offered by authors of this paper, is based on the problem solution about an annular plate subjected to compressing forces with consideration of transverse shear deformations. One should notice that analytical solution of a similar problem about diametrally compressed thin annular disk is present in paper [19]. At the same time it's highly complicated to obtain the analytical solution for a problem studied by authors for anisotropic, in particular, orthotropic, materials because of equation system difficulty. Analytical solution of the problem about an annular plate subjected to compression is obtained by the authors of this paper. This solution can be used to find relations between required elasticity characteristics and radial displacements of annular plate characteristic points.

2. Statement of the problem

Orthogonally reinforced annular plate with inner radius r_1, outer radius r_2 and thickness h is studied in this paper. Annular plate is subjected to compressing forces F, as shown on figure 1.

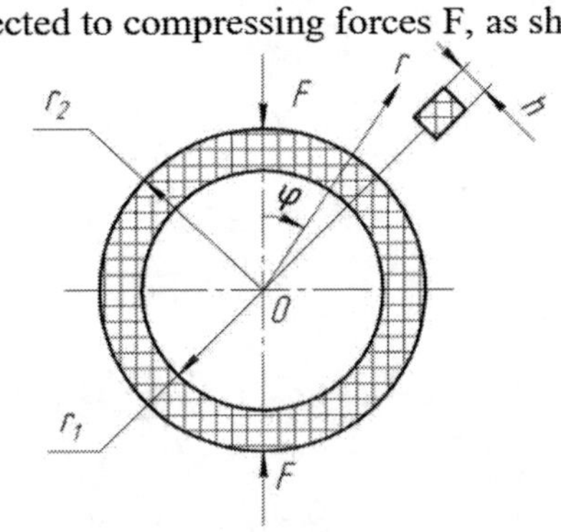

Figure 1. Annular plate subjected to external load.

Annular plate behavior under applied load is governed by following equations:

A) Differential equations of equilibrium.

Let's highlight an annular plate elementary sector of width $d\varphi$ for an arbitrary section φ and examine it's equilibrium (figure 2). The following loads (per unit of width) act in annular plate section: axial force N, transverse force Q and bending moment M.

Having written the equilibrium equations $\Sigma F_{y'} = 0; \Sigma F_{z'} = 0; \Sigma M_A = 0$ in local coordinate system yOz, we obtain the following:

$$\frac{dN}{d\varphi} = -Q; \frac{dQ}{d\varphi} = N; \frac{dM}{d\varphi} = QR \tag{1}$$

B) Kinematic (deformation) equations (the Cauchy equations).

Determine relations between generalized strains ε_{22}, κ_2, Θ_{23}, and displacements v, w, θ_{23} (circumferential, radial and angular respectively) taking into account compatibility relations [20]:

$$\varepsilon_{22} = \frac{1}{R}\left(\frac{dv}{d\varphi} + w\right); \kappa_2 = \frac{d\theta_{23}}{Rd\varphi}; \Theta_{23} = \frac{1}{R}\left(R\theta_{23} + \frac{dw}{d\varphi} - v\right) \tag{2}$$

C) Physical relations (Hooke's law).

Determine relations between strains and stresses in an arbitrary section of annular plate. Let's write them in matrix representation in global and local coordinate systems:

$$\{\varepsilon'\} = [S']\{\sigma'\}; \{\varepsilon\} = [S]\{\sigma\} \tag{3}$$

Here the elastic compliances matrix in global coordinate system:

$$[S']=\begin{bmatrix} s'_{22} & s'_{23} & s'_{26} \\ s'_{32} & s'_{33} & s'_{36} \\ s'_{62} & s'_{63} & s'_{66} \end{bmatrix}=\begin{bmatrix} \frac{1}{E_2} & -\frac{\nu_{23}}{E_2} & 0 \\ -\frac{\nu_{23}}{E_2} & \frac{1}{E_3} & 0 \\ 0 & 0 & \frac{1}{G_{23}} \end{bmatrix} \quad (4)$$

Using the following relations of annular plate stress-strain state parameters in global and local coordinate systems:

$$\{\sigma\}=[T_1]\{\sigma'\};\{\varepsilon\}=[T_2]\{\varepsilon'\} \quad (5)$$

Where matrices of rotation:

$$[T_1]=[T_1](\varphi)=\begin{bmatrix} \cos^2\varphi & \sin^2\varphi & \sin 2\varphi \\ \sin^2\varphi & \cos^2\varphi & -\sin 2\varphi \\ -\frac{\sin 2\varphi}{2} & \frac{\sin 2\varphi}{2} & \cos 2\varphi \end{bmatrix};[T_2]=[T_2](\varphi)=\begin{bmatrix} \cos^2\varphi & \sin^2\varphi & \frac{\sin 2\varphi}{2} \\ \sin^2\varphi & \cos^2\varphi & -\frac{\sin 2\varphi}{2} \\ -\sin 2\varphi & \sin 2\varphi & \cos 2\varphi \end{bmatrix} \quad (6)$$

Through mathematical manipulations we obtain relations for the elastic compliances matrix components in local coordinate system:

$$\begin{gathered} s_{22}=\mathrm{I}_1+\mathrm{I}_2\cos 2\varphi+\mathrm{I}_3\cos 4\varphi;s_{33}=\mathrm{I}_1-\mathrm{I}_2\cos 2\varphi+\mathrm{I}_3\cos 4\varphi;s_{23}=s_{32}=\mathrm{I}_1-\frac{\mathrm{I}_4}{2}-\mathrm{I}_3\cos 4\varphi \\ s_{26}=s_{62}=-(2\mathrm{I}_3\sin 4\varphi+\mathrm{I}_2\sin 2\varphi);s_{36}=s_{63}=2\mathrm{I}_3\sin 4\varphi-\mathrm{I}_2\sin 2\varphi;s_{66}=\mathrm{I}_4-4\mathrm{I}_3\cos 4\varphi \end{gathered} \quad (7)$$

Where:

$$\mathrm{I}_1=\frac{3s_{22}+2s_{23}+3s_{33}+s_{66}}{8};\mathrm{I}_2=\frac{s_{22}-s_{33}}{2};\mathrm{I}_3=\frac{s_{22}-2s_{23}+s_{33}-s_{66}}{8};\mathrm{I}_4=\frac{s_{22}-2s_{23}+s_{33}+s_{66}}{2} \quad (8)$$

D) Boundary conditions

Let's write down boundary conditions for annular plate neutral axis, taking symmetry into account:

$$v(0)=0;\theta_{23}(0)=0;Q(0)=0;v\left(\frac{\pi}{2}\right)=0;\theta_{23}\left(\frac{\pi}{2}\right)=0;Q\left(\frac{\pi}{2}\right)=\frac{F}{2h} \quad (9)$$

3. Solution of the problem

Because of symmetrical geometry and load it's convenient to analyze only 1/4 of annular plate in solving. The problem is solved in polar coordinate system $rO\varphi$ (figure 1). Dependence of elastic properties of annular plate section on a corresponding section angle φ is a specialty of featured problem.

Let's assume $\sigma_{33}=0$ due to small value of interlaminar stresses. Then, having written (3) in expanded form, we obtain through manipulations:

$$\sigma_{22}=\overline{d_{22}}\varepsilon_{22}+\overline{d_{26}}\gamma_{23};\sigma_{23}=\overline{d_{26}}\varepsilon_{22}+\overline{d_{66}}\gamma_{23} \quad (10)$$

Where:

$$\overline{E_2}=\overline{d_{22}}=\frac{1}{\mathrm{I}_1+\mathrm{I}_2\cos 2\varphi+\mathrm{I}_3\cos 4\varphi};\overline{G_{23}}=\overline{d_{66}}=\frac{1}{\mathrm{I}_4-4\mathrm{I}_3\cos 4\varphi};\overline{\eta_{26}}=\frac{\overline{d_{26}}}{\overline{d_{66}}}=\frac{\mathrm{I}_2\sin 2\varphi+2\mathrm{I}_3\sin 4\varphi}{\mathrm{I}_4-4\mathrm{I}_3\cos 4\varphi} \quad (11)$$

Because we take into account changes in annular plate metrics (its curvature radius change), we have before and after deformation (figure 2):

$$CD = ds \rightarrow AB = ds\left(1+\frac{z}{R}\right); C'D' = ds\left(1+\varepsilon_{22}\right) \rightarrow A'B' = ds\left(1+\varepsilon_{22}\right)\left(1+\frac{z}{R^*}\right) \qquad (12)$$

Curvature of annular plate neutral axis before and after deformation, considering (12):

$$\frac{1}{R}=\frac{d\theta_{23}^{*}}{ds};\ \frac{1}{R^{*}}=\frac{d\theta_{23}^{**}}{ds\left(1+\varepsilon_{22}\right)}=\frac{d\left(\theta_{23}^{*}+\psi_{23}\right)}{ds\left(1+\varepsilon_{22}\right)}=\frac{1}{1+\varepsilon_{22}}\left(\frac{1}{R}+\kappa_{2}\right) \qquad (13)$$

Where $\theta_{23}^{*},\theta_{23}^{**}$ – angles describing position of annular plate section before and after deformation respectively, ψ_{23} – angle of rotation of annular plate section normal line, $\kappa_2=\frac{d\psi_{23}}{ds}$ – annular plate curvature change, $R=r_1+b_0$. As a result circumferential strain ε_{22}^{z} of annular plate layer, separated by z from its neutral axis, considering (12) and (13):

$$\varepsilon_{22}^{z}=\frac{A'B'-AB}{AB}=\frac{ds\left(1+\varepsilon_{22}\right)\left[1+\frac{z}{1+\varepsilon_{22}}\left(\frac{1}{R}+\kappa_{2}\right)\right]-ds\left(1+\frac{z}{R}\right)}{ds\left(1+\frac{z}{R}\right)}=\frac{1+\varepsilon_{22}+\frac{z}{R}+z\kappa_{2}-\left(1+\frac{z}{R}\right)}{1+\frac{z}{R}}=\frac{\varepsilon_{22}+z\kappa_{2}}{1+\frac{z}{R}} \qquad (14)$$

Let's assume that annular plate neutral axis matches with its middle surface, i.e. $b_0=\frac{b}{2}$, then $R=r_1+\frac{b}{2}$ (figure 3).

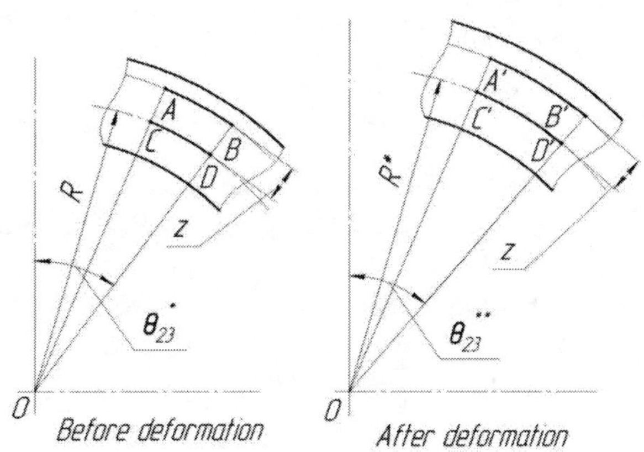

Figure 2. To obtaining of annular plate equidistant layer strain

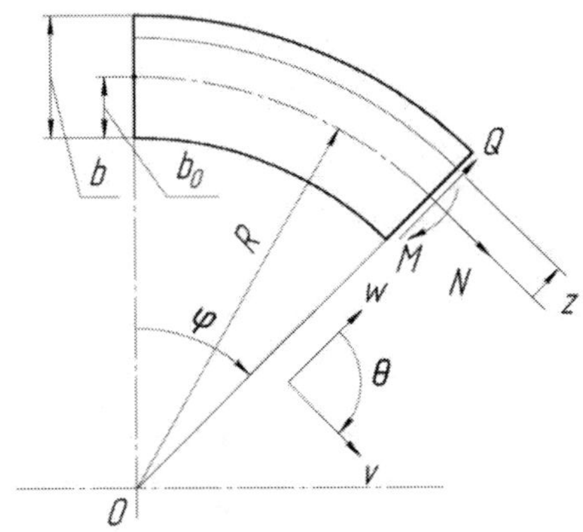

Figure 3. To obtaining of physical relations for annular plate

From the second equation of system (10) we obtain:

$$\gamma_{23}^{z} = \frac{\sigma_{23}}{d_{66}} - \frac{\overline{d_{26}}}{d_{66}}\varepsilon_{22}^{z} \tag{15}$$

Where ε_{22}, γ_{23} – circumferential and shear strains of annular plate neutral axis. Then, considering (14) and (15), we obtain transverse shear deformation averaged to section height [21]:

$$\Theta_{23} = \frac{1}{b}\int_{-\frac{b}{2}}^{\frac{b}{2}} \gamma_{23}^{z}\left(1+\frac{z}{R}\right)dz = \frac{1}{b}\int_{-\frac{b}{2}}^{\frac{b}{2}}\left(\frac{\sigma_{23}}{d_{66}} - \frac{\overline{d_{26}}}{d_{66}}\left(\frac{\varepsilon_{22}+z\kappa_2}{1+\frac{z}{R}}\right)\right)\left(1+\frac{z}{R}\right)dz = \frac{Q}{d_{66}b} - \frac{\overline{d_{26}}}{d_{66}}\varepsilon_{22} \tag{16}$$

Where transverse force per unit of width $Q = Q_{23} = \sigma_{23}b$, then from (16):

$$Q = bd_{26}\varepsilon_{22} + d_{66}\Theta_{23} = C_{23}\varepsilon_{22} + K_2\Theta_{23} \tag{17}$$

Having inserted (10), (15) and (17) into integral relations for loads per unit of width [22], we obtain:

$$N = \int_{-\frac{b}{2}}^{\frac{b}{2}} \sigma_{22}^{z}dz = B_{22}\varepsilon_{22} + C_{22}\kappa_2 + C_{23}\Theta_{23}; M = \int_{-\frac{b}{2}}^{\frac{b}{2}} \sigma_{22}^{z}zdz = C_{22}\varepsilon_{22} + D_{22}\kappa_2 N \tag{18}$$

Where:

$$B_{22} = \left(r_1 + \frac{b}{2}\right)\ln\left(\frac{b}{r_1}+1\right)\left[\overline{d_{22}} - \frac{\left(\overline{d_{26}}\right)^2}{d_{66}}\right] + b\frac{\left(\overline{d_{26}}\right)^2}{d_{66}} \text{ – tensile stiffness.} \tag{19}$$

$$C_{22} = -\left(r_1 + \frac{b}{2}\right)\left[\left(r_1 + \frac{b}{2}\right)\ln\left(\frac{b}{r_1}+1\right) - b\right]\left[\overline{d_{22}} - \frac{\left(\overline{d_{26}}\right)^2}{d_{66}}\right] \text{ – combined stiffness.} \tag{20}$$

$$D_{22} = \left(r_1 + \frac{b}{2}\right)^2\left[\left(r_1 + \frac{b}{2}\right)\ln\left(\frac{b}{r_1}+1\right) - b\right]\left[\overline{d_{22}} - \frac{\left(\overline{d_{26}}\right)^2}{d_{66}}\right] \text{ – bending stiffness.} \tag{21}$$

Having written (17) and (18) combined in matrix representation, we obtain:

$$\{F\}=[B]\{\varepsilon\} \tag{22}$$

Where:

$$\{F\}=\begin{Bmatrix} N \\ M \\ Q \end{Bmatrix};[B]=\begin{bmatrix} B_{22} & C_{22} & C_{23} \\ C_{22} & D_{22} & 0 \\ C_{23} & 0 & K_2 \end{bmatrix};\{\varepsilon\}=\begin{Bmatrix} \varepsilon_{22} \\ \kappa_2 \\ \Theta_{23} \end{Bmatrix} \tag{23}$$

(23) combined with (1) and (2) form a resulting equations system of our problem, which is resolved related to unknown forces and displacements considering boundary conditions (9). Having inserted (2) in (23), we obtain a system of 6 equations with 6 unknowns (N,M,Q,v,w,θ_{23}). Analytical solution of the problem was obtained by the authors of this paper through mathematical manipulations.

4. Results of calculations and their discussion

Analytical solution of the problem is the following:

$$\begin{gathered}
N=-\frac{F}{2h}\sin\varphi;M=\frac{FR}{h}\left[\frac{\sin\varphi}{2}-\frac{\theta_0}{\pi I_1(k_1+1)}\right];Q=\frac{F}{2h}\cos\varphi \\
v=\frac{FR}{2bh}\left[\theta_0\left(1-2\frac{\varphi}{\pi}\right)+v_s(\pi-2\varphi)\sin\varphi+v_{s2}\sin 2\varphi+v_{s4}\sin 4\varphi+v_c\cos\varphi+v_{c3}\cos 3\varphi+v_{c5}\cos 5\varphi\right] \\
w=-\frac{FR}{2bh}\left[w_0+w_s\sin\varphi+w_{s3}\sin 3\varphi+w_{s5}\sin 5\varphi+v_s(\pi-2\varphi)\cos\varphi+w_{c2}\cos 2\varphi+w_{c4}\cos 4\varphi\right] \\
\theta_{23}=\frac{-F}{2bh}\left[-\theta_0\left(1-2\frac{\varphi}{\pi}\right)+\theta_{s2}\sin 2\varphi+\theta_{s4}\sin 4\varphi+\theta_c\cos\varphi+\theta_{c3}\cos 3\varphi+\theta_{c5}\cos 5\varphi\right]
\end{gathered} \tag{24}$$

Where:

$$\begin{gathered}
k_1=\frac{b}{R\ln\left(\frac{2R+b}{2R-b}\right)-b};\theta_0=k_1\left[I_1-\frac{I_2}{3}-\frac{I_3}{15}\right]-\frac{2I_2}{3}-\frac{8I_3}{15};w_0=-\frac{2\theta_0 k_1}{\pi(k_1+1)};w_{c2}=\frac{2\theta_0 I_2}{3\pi I_1}\left(\frac{k_1+2}{k_1+1}\right) \\
v_s=\frac{k_1(2I_1-I_2)+2I_4}{8};v_{s2}=\frac{\theta_0 I_2}{3\pi I_1}\left(\frac{k_1-1}{k_1+1}\right);v_{s4}=\frac{\theta_0 I_3}{30\pi I_1}\left(\frac{k_1-7}{k_1+1}\right);w_{c4}=\frac{2\theta_0 I_3}{15\pi I_1}\left(\frac{k_1+8}{k_1+1}\right) \\
v_c=k_1\frac{-48I_1+15I_2+4I_3}{48}+\frac{3I_2+2I_3}{6};v_{c3}=\frac{k_1[I_2-I_3]+8I_2+4I_3}{48};v_{c5}=\frac{I_3(k_1+28)}{240} \\
w_s=k_1\frac{24I_1-3I_2-4I_3}{48}-\frac{2I_3+3I_4}{6};w_{s3}=-\frac{k_1[I_2-I_3]-12I_3}{16};w_{s5}=-\frac{I_3(k_1-20)}{48} \\
\theta_{s2}=\frac{\theta_0 I_2}{\pi I_1};\theta_{s4}=\frac{\theta_0 I_3}{2\pi I_1};\theta_c=\frac{(k_1[2I_1-I_2]-I_2)}{2};\theta_{c3}=\frac{(k_1[I_2-I_3]-I_2-2I_3)}{6};\theta_{c5}=\frac{I_3(k_1-2)}{10}
\end{gathered} \tag{25}$$

Numerical solution of the problem was also obtained by authors using MathCAD 15.0 software. Apart from that, the problem was solved through finite element analysis (FEA) in MSC Patran + MSC Nastran 2014 on a mesh of 45*14 (radial*circumferential directions respectively) QUAD4 elements. All solutions mentioned were obtained and compared with the analytical solution (24) for the following initial values (corresponding to figure 1):

$$\begin{gathered}
r_1=40\,\text{mm}=0.04\text{ m};r_2=60\text{ mm}=0.06\text{ m};h=10\text{ mm}=0.01\text{ m};E_2=15\text{ GPa}=15*10^9\text{ Pa}; \\
E_3=10\text{ GPa}=10*10^9\text{ Pa};G_{23}=5\text{ GPa}=5*10^9\text{ Pa};\nu_{23}=0.1;F=2\text{kN}=2*10^3\text{ N}
\end{gathered} \tag{26}$$

Figures 4–6 illustrate how displacements of annular plate change with the angle φ [deg]. Here the the analytical solution, corresponding to relations (24) and matching the numerical solution obtained in MathCAD 15.0 software, correspond to markers ○. The FEA results correspond to markers ●.

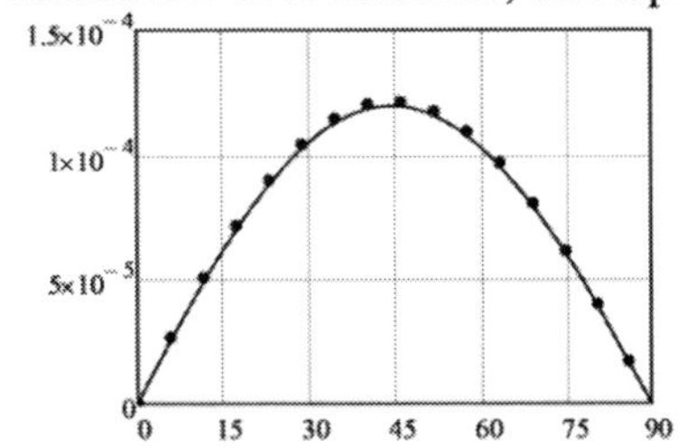

Figure 4. The distribution of linear displacement $v(\varphi)$ [m].

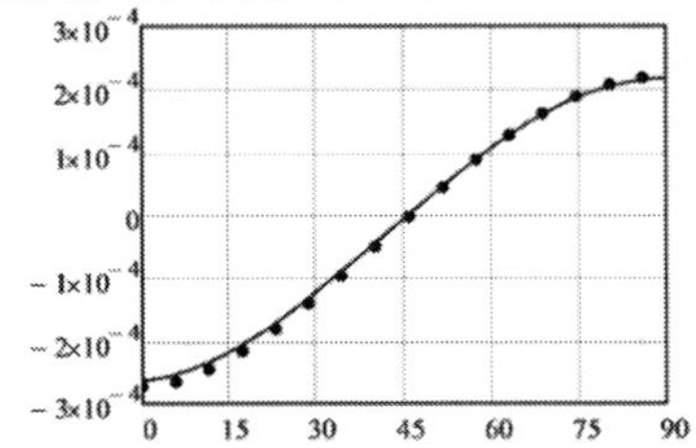

Figure 5. The distribution of linear displacement $w(\varphi)$ [m].

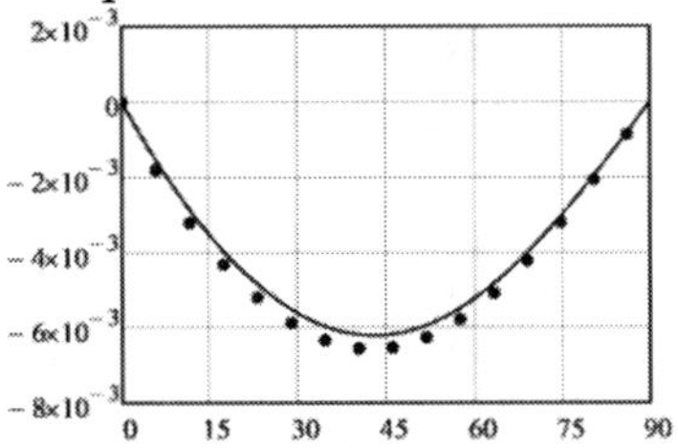

Figure 6. The distribution of angular displacement $\theta_{23}(\varphi)$ [rad].

Comparison of the obtained results has shown that FEA solution are almost identical to the analytical solution. Maximum error of FEA solution for angular displacements doesn't exceed 5 %, while maximum error for linear displacements doesn't exceed 1 %. This fact, along with absolute agreement of analytical solution and numerical solution obtained in MathCAD 15.0, confirms the accuracy of relations (24). The results of comparison described above show that relations (24) allow for determination of unknown elasticity characteristics of annular plate from 3D-reinforced composite material. This might be done by using the relations between required elasticity characteristics and displacements of annular plate characteristic points.

5. Conclusions

The analytical solution of the problem about 3D-reinforced composite annular plate compressed by concentrated forces was obtained. Transverse shear deformations and changes in annular plate metrics during deformation are taken into account in the obtained solution.

The problem was also solved numerically using 2 techniques: the FEA in MSC Patran + MSC Nastran software, and analysis in MathCAD 15.0 software. The comparison of obtained results demonstrates the offered method applicability for unknown characteristics determination of 3D-reinforced composite materials.

References

[1] Kostikov V I and Varenkov A N 2003 *Ultrahigh-temperature composite materials* (Moscow: Intermet Engineering) pp 20–23
[2] Baker A A, Dutton S and Kelly D 2004 *Composite materials for aircraft structures* (Michigan: American Institute of Aeronautics and Astronautics) pp 449–462
[3] Guniaev G M and Gofin M Ya 2013 *Aviat. Mater. Technol.* **S1** pp 62–90
[4] Neuman E W, Hilmas G E and Fahrenholtz W G 2016 *J. Am. Ceram. Soc.* **99(2)** pp 597–603
[5] Neuman E W, Fahrenholtz W G and Hilmas G E 2019 *Int. J. Appl. Ceram. Technol.* **16(5)** pp 1715–22
[6] Hatta H, Aoi T, Kawahara I, Kogo Y and Shiota I 2004 *J. Compos. Mater.* **38(19)** pp 1667–84
[7] Goto K, Ohkita H, Hatta H, Iseki H and Kogo Y 2007 *Proc. Int. Conf. on Composite Materials (Kyoto)* pp 1–6
[8] Guo S 2019 *J. Am. Ceram. Soc.* **102** pp 3843–48
[9] Hua G, Zhong J, Qi Y and Cheng X 2018 *J. Am. Ceram. Soc.* **101** pp 5717–31
[10] Nelyub V 2020 *IOP Conf. Ser.: Mater. Sci. Eng.* **709(2)** 022037
[11] Bocharov A, Vigovskiy V and Nelyub V 2019 *Materials Today: Proc* pp 107–111
[12] Nelyub V A, Borodulin A S, Kobets L P and Malysheva G V 2016 *Polym. Sci. Ser. D* **9(3)** pp 286–289
[13] Malysheva G V, Kobets L P and Borodulin A S 2016 *Fibre Chemistry* **47(6)** pp 482–485

IOP Conf. Series: Materials Science and Engineering **934** (2020) 012023 doi:10.1088/1757-899X/934/1/012023

[14] Yanyan C, Nelyub V A and Malysheva G V 2019 *Polym. Sci. Ser. D* **12(3)** pp 296–299
[15] Bakulin A A, Magnitskii I V, Ponomarev K A and Tashchilov S V 2015 *Const. Comp. Mater.* **138** pp 46–51
[16] Ponomarev K A 2015 *Const. Comp. Mater.* **146** pp 62–67
[17] Kobayashi H, Goto K, Hatta H, Koyama M and Fukuda H 2009 *Proc. Int. Conf. on Composite Materials (Edinburgh)* pp 1–8
[18] Smerdov A A Tairova L P Mironikhin A N Tashchilov S V and Magnitskii I V 2011 *Const. Comp. Mater.* **146** pp 35–50
[19] Tokovyy Y V, Hung K M and Ma C C 2010 *J. Mathem. Sci.* **165(3)** pp 342–354
[20] Vasiliev V V 1988 *Mechanics Of Composite Structures* (Moscow: Mashinostroenie) p 215
[21] Vasiliev V V 1988 *Mechanics Of Composite Structures* (Moscow: Mashinostroenie) p 67
[22] Vasiliev V V 1988 *Mechanics Of Composite Structures* (Moscow: Mashinostroenie) p 69

IOP Conf. Series: Materials Science and Engineering **934** (2020) 012024 doi:10.1088/1757-899X/934/1/012024

Determination of the permeability coefficient in the Darcy formula for liquid molding of thermal insulation products from quartz fibers

M A Komkov [1,4] , Yu V Badanina[2,5] and P O Yanko[3,6]

[1] D. Sc, Prof, SM-12, Bauman Moscow State Technical University,105005, Russia

[2] D. Sc, SM-12, Bauman Moscow State Technical University, 105005, Russia

[3] Student SM-12, Bauman Moscow State Technical University, Moscow, 105005, Russia

E-mail: :[4]cm12@cm.bmstu.ru; m_komkov@list.ru, [5]julia555-90@yandex.ru, [6]panteleymon.yanko@mail.ru

Abstract. The results of water pouring of flat samples from chopped quartz fibers with a diameter of 12 microns and an elongation of 560...1680 are presented. Experimental values of water permeability coefficient through highly porous flat samples were obtained. The constant in the Kozeni-Karman permeability formula is determined depending on the porosity of the fiber layer.

Keywords: Short quartz fibers, material porosity, filtration method, water permeability coefficient

1. Introduction

Difficulties in the development of a process for producing heat-insulating or heat-shielding parts from short chopped or ground elementary glass, quartz or basalt fibers are associated with their small size [1-3] and specific methods of forming products from them [4-7]. The most developed technological process is the formation of a layer of fibers of the required thickness and density according to the shape of the designed part by liquid filtration [8-10]. Liquid technology is widely used in paper and construction industries, in the production of non-woven materials, asbestos cement products [11], as well as in the aerospace industry when creating heat-resistant tiles of the Buran type VKS [12, 13]. However, the permeability of the medium [14-18], which depends on the structure of the fibrous material, its porosity [19-21] and the geometric characteristics of pores, has a great influence on the duration of the filtration deposition of fibers from a sparse pulp.

The aim of the work is to determine the permeability coefficient in the Darcy equation for the transverse filtration of water through a flat sample of chopped quartz fibers of the same diameter and different lengths.

2. Statement of the problem, experiments studied

The permeability of a liquid through a layer of short fibers depends, first of all, on the uniformity of the structure and porosity of the material, the diameter and length of the fibers, as well as the presence of non-fibrous inclusions and components of the bundle (Fig. 1, a). A great influence on the permeability is exerted by the direction of fluid flow around the fibers — along the fiber axis or across the fibers (Fig. 1, b). Transverse flow around a fiber with a viscous liquid can be accompanied by a separation of the jet from its surface with the formation of stagnant regions [11].

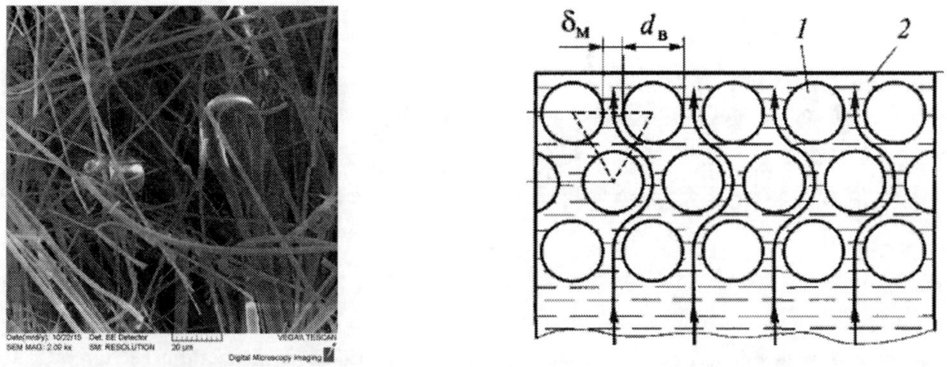

a   b

Figure 1. Microscopic view of the crushed superthin basalt fiber under the microscope (a) and the scheme of fluid flow through the porous fiber layer (b): 1 - fiber; 2 - interfiber pore space

The rate of filtration or fluid flow through a porous layer of already deposited fibers (Fig. 2a) is determined by Darcy's law [8, 10]:

$$U_{fl}(x) = \frac{K_D(P_1 - P_2)}{\eta_{lq} L} \tag{1}$$

Where, K_D is the coefficient of permeability of the fluid through the porous medium, m^2; $\Delta P = (P_1 - P_2)$ is the fluid pressure drop across the fiber layer, Pa; η_{lq} is the dynamic viscosity of the filtered fluid, Pa·s; L is the thickness of the permeable layer of fiber sediment, m. To determine the permeability coefficient, technical water spills were carried out according to the scheme of Figure 2, and through the sediment layers of different thicknesses (Figure 2, b) from chopped quartz fibers with a diameter of 12 μm, a length of 20.1; 13.4 and 6.7 mm with fiber elongation of 1680 - 1120 - 560.

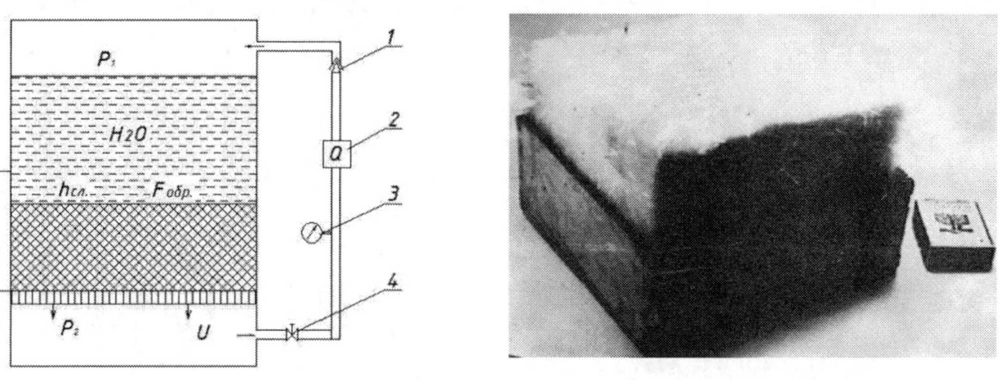

a   b

Figure 2. Scheme of pouring plant (a): 1 - pump; 2 - counter; 3 - vacuum gauge; 4 - valve and (b) - a layer of quartz fibers

When samples were spilled with water, the pressure drop across the layer was measured $\Delta P = P_2 - P_1$ by vacuum gauge, fiber layer thickness h_{lr} and fluid flow rate $Q = F_{smpl.} \cdot U$, cm3/s - by flow meter, where $F_{smpl.}$ = 220 cm2 is the area of the sample. The porosity of the fibrous sample was determined from the equation:

$$n = 1 - M_{smpl} / \left(F_{smpl} h_{lr}\, \rho_f\right),$$

Where, M_{smpl} is the mass of the dry sample, h_{lr} is the thickness of the fiber layer, ρ_f = 2.19 g/cm3 is the density of quartz fibers. The permeability coefficient in the Darcy equation was determined by the formula (2), and the calculation results in the form of experimental points are presented in Figure 3,a.

$$K_D = Q\,L\,\eta_{lq}/\left(F_{smpl}\Delta P\right) \quad (2)$$

3. Results and discussion

The permeability coefficient is inherently a structural characteristic, determined mainly by the geometry of the pore space and according to I. Czerny's hypothesis [10] depends on the specific volume of open pores $n = V_p/V_{lr}$ in a cube divided by the square of the specific surface of fibers Sy in a porous material slowly flowing around with a liquid:

$$K_D = \frac{c_k\, n^3}{S_S^2} = \frac{c_k\, n^3}{(S_0\, \vartheta_f)^2} = \frac{0{,}5\; d_f^2\; n^3}{16\;(1-n)^2}, \quad (3)$$

Where, $S_0 = S_f/\,f_f = 4/d_f$, 1/м – specific surface of one fiber; d_f - average (efficiently streamlined) fiber diameter, m; ϑ_f - volumetric fiber content in the sediment layer, $\vartheta_f = 1 - n$.

The coefficient c_k in the Cozeny-Karman formula (3), determined from the solution of the hydrodynamic equation for the flow of viscous liquid around round fibers along their axis, is $c_k = 0.5$. However, when the fluid flows across the fibers (Fig. 1, b), this coefficient obtained from experiments is many times smaller than the theoretical one and depends, first of all, on the surface of the flow around round fibers with the formation of stagnant regions and on the porosity of the fiber layer. Processing of experimental data made it possible to determine the value of the coefficient c_k, which linearly depends on the porosity n (Fig. 3, b)

$$c_k = 0{,}76(1-n) \quad (4)$$

In fig. Figure 3a shows the dependences of the permeability coefficient K_D on the porosity of the fibrous material (curve 1), calculated by formula (3) for $c_k = 0.5$. Curve 2 is an approximation of the experimental values of the coefficient K_D on porosity, constructed according to formula (3) taking into account dependence (4):

$$K_D = \frac{0{,}76(1-n)\; d_f^2\; n^3}{16\;(1-n)^2}, \quad (5)$$

From the analysis of Fig. 3a, it follows that dependence (5) agrees relatively well with the results of spilling of fibrous samples from short quartz fibers with a diameter of 12 μm.

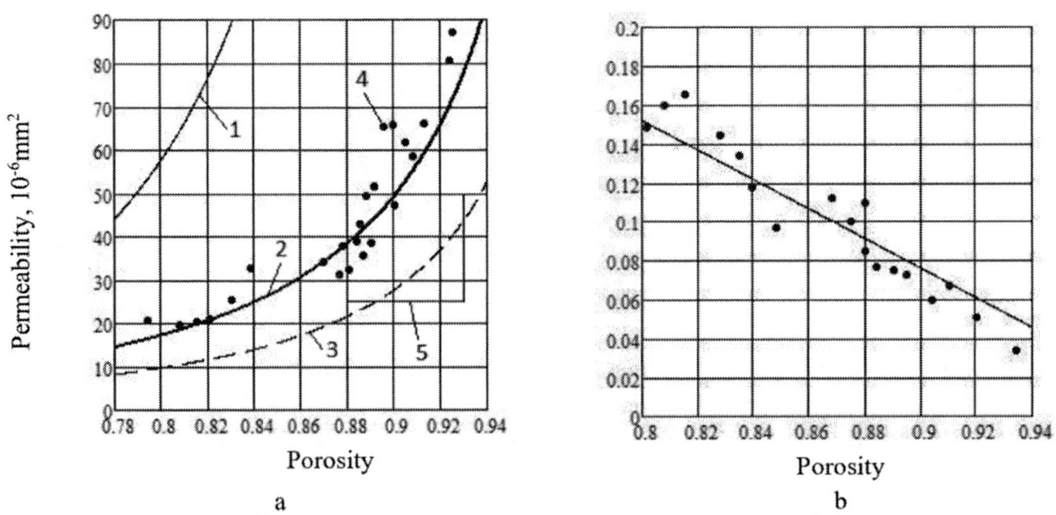

Figure 3. The dependences of the permeability coefficient K_D on the porosity of the fiber layer: 1 - curve constructed by the formula (3); 2 - by the formula (5); 3, 5 - the region of experimental values of the coefficient of permeability from porosity obtained on samples of fiberglass [10, p. 26]; 4 - experimental values of the spilling of samples of quartz fibers

IOP Conf. Series: Materials Science and Engineering **934** (2020) 012024 doi:10.1088/1757-899X/934/1/012024

Conclusion

When filtering liquid through a layer of fibers oriented in the plane of the sample, the use of the coefficient $c_k = 0.5$ in the Cozeny-Karman formula is not correct, since a transversely flowing water from the back of the fibers produces a shaded region, an uneven and slowed down fluid flow rate.

The linear dependence of the coefficient $c_k = 0.76$ (1-n) on porosity obtained from processing the results of experimental spills allows the Darcy equation to be used in the calculation of technological parameters for the manufacture of heat-shielding, heat-insulating, and filtration products from short ceramic fibers by liquid filtration.

References

[1] Komkov M A, Badanina Yu V, Tarasov V A and Filimonov A S 2017 Analysis of structural and thermophysical characteristics of highly porous basalt thermal insulation of tubing (*Engineering Journal: science and innovation*) No 1 (61)

[2] Badanina Yu V 2017 *Technological design of highly porous insulating structures made of short basalt fibers based on the liquid filtration method* (Abstract. dis. Cand. tech. sciences. Moscow, BMSTU) pp 16

[3] Rheological properties of binders filled with discrete carbon fibres Malysheva, G.V., Kobets, L.P., Borodulin, A.S. 2016 Fibre Chemistry 47(6), c. 482-485

[4] Timofeev M P 2007 *Development and research of filtration technology for manufacturing products from fibrous inorganic materials* (Abstract. dis. Cand. tech. sciences. Moscow, BMSTU) pp16

[5] Komkov M A, Bandanna Yu V, Tarasov V A, Timofeev M P and Moiseev A V 2015 *Modeling and experimental determination of technological parameters of liquid molding of basalt thermal Isulation of tubing* (Science and Education. BMSTU. Electron. Journal) No 04 pp 13–28

[6] Denisov P and Badanina Yu V 2015 Research and preparation of hydromass for molding heat insulation coating of tubing from short basalt fibers (*Youth Scientific and Technical Bulletin BMSTU. Electron. Journal*) No 04

[7] Filimonov A S, Tarasov V A, Komkov M A, Moiseev M A, Timofeev M P and Gerasimov N V 2012 An experimental analysis of the properties of promising heat-insulating materials of mechanical engineering obtained by filtration deposition (*Engineering Journal: Science and Innovation*) No 9

[8] Leontiev N E 2009 Fundamentals of the theory of filtration (*M.: Publishing House of the CPI at theFaculty of Mechanics and Mathematics of Moscow State University*) pp 88

[9] Musket M 2006 *The flow of homogeneous liquids in a porous medium* (M. - Izhevsk: Institute f Computer Research) pp 640

[10] Collins R 1964 *The flow of fluid through porous materials* (Per. from English; Ed. G.I. Barenblata. M .: Mir) pp 351

[11] Berney I I 1969 *Basics of molding asbestos cement products* (M.: Stroyizdat) pp 336

[12] Lozinsky G E and Bratukhin A G 1997 *Aerospace systems* (ed. Moscow, MAI Publishing House) pp 416

[13] Kablov E N 2013 *VIAM materials and technologies for the Energia*-Buran (ISS. M .: VIAM) pp 144

[14] Impregnation kinetics of fibers with liquids: Simulation within the generalization of Navier-Stokes equationsBelov, P.A., Kobets, L.P., Borodulin, A.S. 2014 Inorganic Materials: Applied Research 5(4), c. 403-406

[15] Kinetics of fiber impregnation by a binder. Gradient generalization of Navier–Stokes–Darcy equations Belov, P.A., Borodulin, A.S., Kobets, L.P., Malysheva, G.V. 2016 Polymer Science - Series D 9(2), c. 205-208

[16] Tarasov V A and Nelyub V A 2018 Rheological model of viscous incompressible fluid withdisperse fillers (*edited by International Journal of Mechanical Engineering and Technology, Moscow*) pp 488-497

[17] Badanina Yu V, Komkov M A, Tarasov V A and Timofeev M P 2018 *Study of the effect of acidityon the process of filtration and deposition of aluminum hydroxide on short-fiber basalt thermalinsulation* (All materials. Encyclopedic reference book) No 8 pp 23-28

[18] Analysis of the Effect of Hydro mass Acidity on the Process of Filtration and Precipitation of Aluminum Oxide in Short-Fiber Basalt Heat Insulation. / Yu.V. Badanina, M.A. Komkov, V.A. Tarasov, M.P. Timofeev // Polymer Science. Series D 2019. - Vol. 12, Issue 1. - C. 72-76. DOI: 10.1134/S1995421219010027

[19] Basalt superthin fiber "MINOL" // BSTV MINOL Plant 2019: company website. Access mode:http://uteplitel-minol.ru/holst/

[20] Komkov M A, Badanina Yu V, Tarasov V A and Filimonov A S. 2017 Analysis of structural and thermophysical characteristics of highly porous basalt thermal insulation of tubing (Engineering Journal: Science and Innovation) No 1 (61)

[21] Komkov M A, Badanina Y V, Garashchenko A H. Reduction of heat loss of pump-compressor pipes for oil production due to optimal thermal insulation based on basalt and glass fibers // IOP Conference Series: Materials Science and Engineering 683 (2019) 012026 IOP Publishing doi:10.1088/1757-899X/683/1/012026

IOP Conf. Series: Materials Science and Engineering **934** (2020) 012025 doi:10.1088/1757-899X/934/1/012025

Thermal Regulation of Space Platform Panels by Using Electric Heaters

M A Gorodetsky, K V Mikhaylovsky and S V Reznik

BMSTU, Moscow, Russia

E-mail: mikgord@gmail.com

Abstract. In orbital flight, the temperature state of spacecraft and their constituent space platforms complexly changes in time and space. The inhomogeneous temperature field of the space platform can cause shape distortion, which negatively affects the performance of targets. To equalize the temperature field of the space platform, it is proposed to use electric heaters with differentiated power distributions. The formulation of the boundary inverse heat conduction problem is presented and the specific power of the heaters necessary for uniform distribution of the temperature field of a panel of composite material is determined.
Key words: space platform, composite materials, inverse heat conduction problem

1. Introduction

In rocket and space technology, dimensionally stable structures are widely used, in particular, spacecraft platforms, spacecraft reflectors and other objects, the main influencing factor of which are external heat fluxes [1-4]. The influence of the external thermal environment on the spacecraft is rapidly changing due to objective factors such as the orbital motion of the spacecraft, a change in the orientation of the satellite relative to the sun and the earth [5]. The influence of these factors has a negative impact on the dimensional stability of the spacecraft supporting structure, including on the panels that make up the spacecraft platform. The stress-strain state of the spacecraft surfaces depends on the correct choice of methods for monitoring and regulating temperature fields.

There are several ways to control the temperature field, one of which is the use of electric heaters. Differentiated electric heaters are able to change the heating power depending on the temperature requirements for the heated surface, combining the distribution of power. Thus, with the help of differentiated electric heaters, it is possible to create an adjustable heating system and regular the temperature of the spacecraft surfaces, based on the readings of temperature sensors.

The purpose of this work is to improve the thermal stability of spacecraft platforms through the use of differentiated electric heaters, as well as to reduce the time spent on designing, as part of the methodology for determining the specific power of heaters, necessary to equalize the temperature field on the outer side of the spacecraft panel.

2. The spacecraft platform temperature field simulation

An integral part of design studies is the modeling of the orbital motion of the spacecraft, corresponding to their intended purpose. To assess the change in the temperature field, conditions corresponding to the motion along the low Earth orbit with the following parameters were set: orbital altitude 514 km, orbital inclination 97.4 °, local transit time of the ascending node 6 h. The type of orbit is near-circular (with an eccentricity of 0.00135), solar-synchronous, multiple (the multiplicity period is 14 days).

The investigated fragment of the spacecraft platform was in the form of a parallelepiped of a three-layer panel with a total thickness of 20 mm, including 2 ligaments of carbon fiber 4 mm thick and a filler made of polyimide foam. The temperature state was simulated for a representative structural element with dimensions of 500x500x20 mm in the NX / Nastran software package using the NX / Space Systems Thermal module with the Monte Carlo method. The initial data included orbital parameters, spacecraft orientation in space, position of the Sun relative to the orbit. To take into account the redistribution of radiation from structural elements, the Radiation module was used. Thermal loading conditions were defined in the Orbital Heating module. A three-dimensional type of mesh was used, with an element size of 5 mm (Fig. 1). The simulation was carried out on one orbit in 45 s increments.

Fig. 1. The finite element model of the platform representative element.

The temperature field of the platform was described by the heat equation:

$$c\rho(T)\frac{\partial T}{\partial \tau} = \frac{\partial}{\partial x}\lambda_x(T)\frac{\partial T}{\partial x} + \frac{\partial}{\partial y}\lambda_y(T)\frac{\partial T}{\partial y} + \frac{\partial}{\partial z}\lambda_z(T)\frac{\partial T}{\partial z} + q_e, \quad (1)$$

where: qe – specific power of electric heaters, W/m3; λx, λy, λz – thermal conductivity coefficients in the direction of the main axes, W/(m·K); T – temperature of the outer side of the panel, K; C – specific heat, J/(kg·K); ρ – density of the material, kg/m3.

With initial and boundary conditions:

$$\tau = 0;\ T = T_0; \quad (2)$$

$$x = 0;\ y = 0;\ \frac{\partial T}{\partial x} = 0;\ \frac{\partial T}{\partial y} = 0; \quad (3)$$

$$x = \frac{l_2}{2};\ -\lambda_x(T)\frac{\partial T}{\partial x} = q_{ex} - \varepsilon\sigma_0(T_w^4 - T_f^4); \quad (4)$$

$$y = \frac{l_1}{2};\ -\lambda_y(T)\frac{\partial T}{\partial y} = q_{ey} - \varepsilon\sigma_0\left(T_w^4 - T_f^4\right); \quad (5)$$

$$z = 0;\ -\lambda_z(T)\frac{\partial T}{\partial z} = A_s q_s - \varepsilon\sigma_0\left(T_w^4 - T_f^4\right); \quad (6)$$

$$z = \delta;\ \lambda_z(T)\frac{\partial T}{\partial z} = \varepsilon\sigma_0\left(T_w^4 - T_f^4\right), \quad (7)$$

where: l1 – length of the face side, located along the X axis, m; l2 – length of the face side, located along the Y axis, m; δ – length of the face side, located along the Z axis, m; Tw – temperature at the wall boundary, K; Tf – flow temperature, K; qs – solar heat flux, W/m2; As – absorption coefficient of the solar flux by the surface material; ε – degree of blackness; σ0 – Stefan-Boltzmann constant, W/(m2·K4). Equations (1)-(7) are the formulation of the direct heat conduction problem.

When creating a heating system and equalizing the temperature of the spacecraft surfaces with the help of electric heaters, it is necessary to know in advance what specific power is required to inform each heating element. To do this, you can use the methodology for assessing the specific power of heaters, based on the solution of the boundary inverse heat conduction problem in the extreme setting.

A meaningful formulation of the inverse heat conduction problem is to determine such a distribution function of the heat flux on the boundary surface q(x, y, z), which delivers a minimum of discrepancies in the calculated and set temperature values $\left(T^*(q,x,y,z,\tau),T(\tau)\right)$ at time τ.

Choosing the standard deviation as the measure of deviation, we obtain the optimal solution:

$$\bar{q}_e(x,y,z)=\arg\min\int_0^{\tau}\sum_1^{N}\left(T^*(q,x,y,z,\tau)-T(\tau)\right)^2 d\tau\le\delta^2 \quad (8)$$

where: δ – total error, including the error in the temperature information δT and the approximation error of the boundary heat conduction problem δA:

$$\delta^2=\delta_T^2+\delta_A^2. \quad (9)$$

Equations (8)-(9) are the formulation of the boundary inverse heat conduction problem.

The sensitivity of the model to external thermal influences reproducing the operation of differentiated heaters is analyzed. Two cases of heat transfer are considered: 1. When exposed to external heat fluxes without heaters. 2. When exposed to external heat fluxes with side differentiated heaters.

From the simulation results it was determined that the temperature values on the platform surface vary from plus 29 °C to plus 39 °C when exposed to external heat fluxes without heaters (Fig. 2).

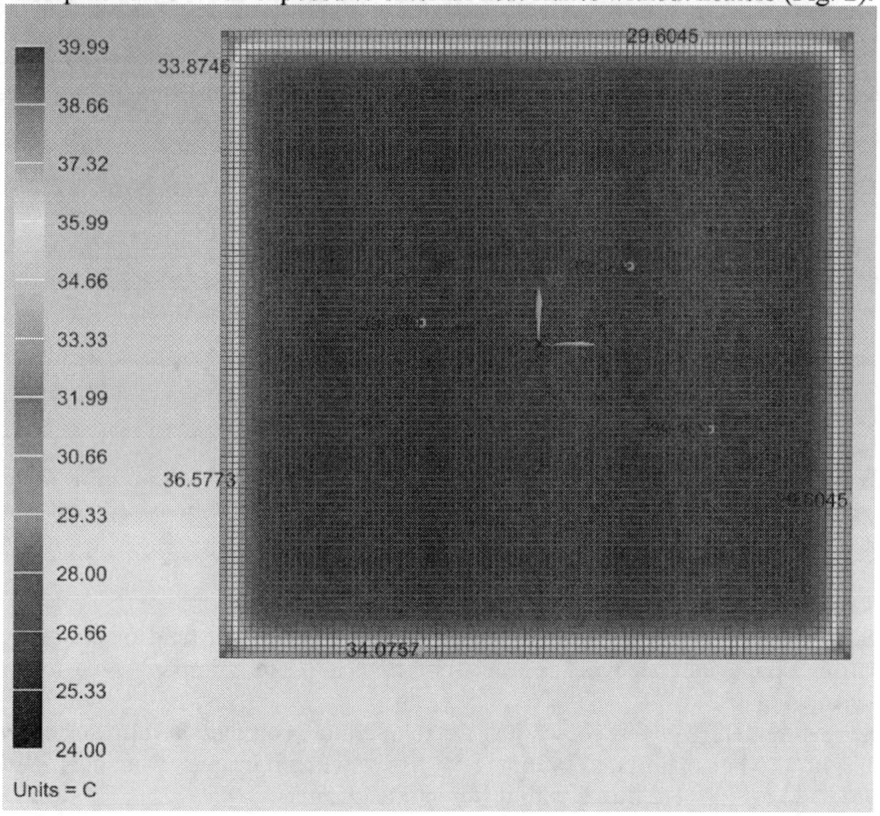

Fig. 2. Distribution of temperature fields on a platform panel located in orbit under the influence of external heat fluxes.

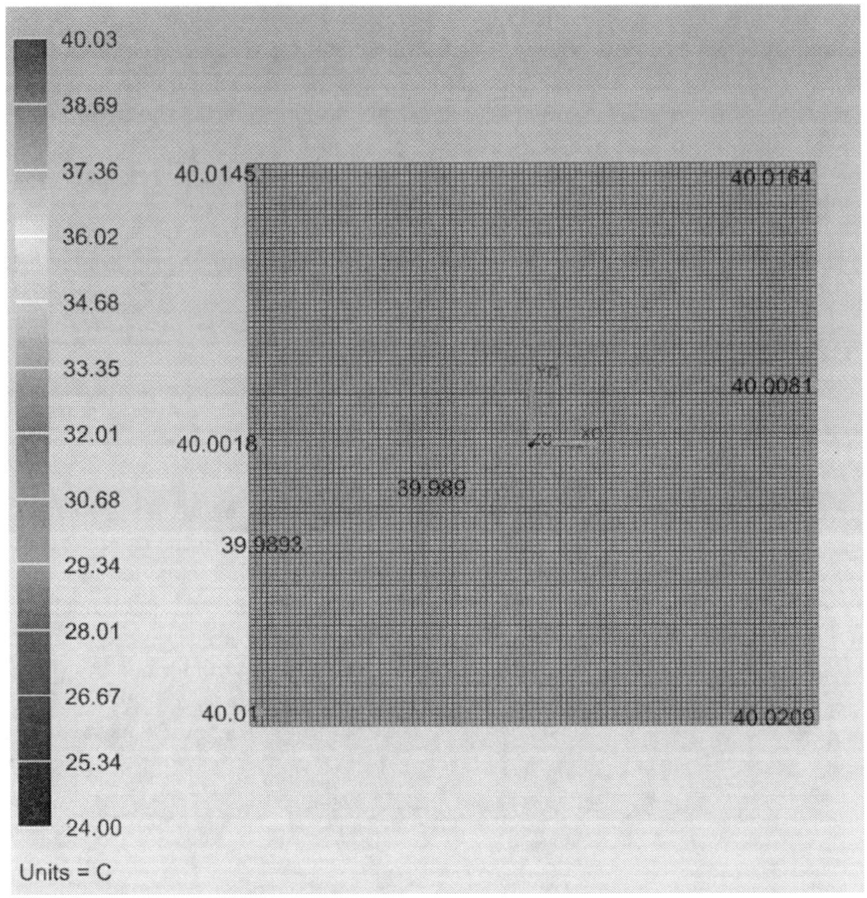

Fig. 3. Distribution of temperature fields on a platform panel located in orbit under the influence of external heat fluxes and differentiated electric heaters.

The temperature values on the platform surface vary from plus 39 °C to plus 40 °C when exposed to external heat fluxes with differentiated heaters (Fig. 3).

With the help of mathematical models implemented as algorithms in the Matlab software package, the ERS SC working orbit parameters were determined: orbit height h=514 km, orbit inclination i=97,4°, local time at ascending node m=6 h. The orbit type is near-circular– (with eccentricity e=0,00135), sun-synchronous, multiple (the multiplicity period is 14 days).

3. Conclusion

A mathematical model is proposed for determining the temperature field of the spacecraft platform surface in orbit, taking into account the effects of electric heaters, implemented in the Siemenx NX software package.

Using numerical modeling, it was shown that the temperatures on the platform surface vary: from plus 29 °C to plus 40 °C when exposed to external heat fluxes without heaters; from plus 39 °C to plus 40 °C when exposed to external heat fluxes with differentiated heaters.

As a result of the simulation, it was found that using differentiated heaters it is possible to achieve a uniform temperature distribution on the spacecraft surface.

The developed methodological techniques can be useful at the stage of technical proposals for the materials selection for the ERS SC constructions with stable form and dimensions.

The developed methodological techniques can be useful at the stage of technical proposals for the selection of electric heaters operating modes for dimensionally stable spacecraft structures.

References

[1] Reznik S V 2013 Actual problems of design, production and testing of rocket-space composite constructions *Engineering Journal: Science and Innovation*. 3 p 16

[2] Reznik S V, Mikhailovsky K V and Prosuntsov P V 2012 Development of high thermal polymer composite materials for space constructions *Bulletin of BMSTU. Ser. Engineering. Special release Progressive materials, designs and technologies of rocket and space mechanical engineering* 98–106

[3] Reznik S V, Prosuntsov P V and Azarov A V 2015 Substantiation of the structural-layout scheme of the mirror-space-antenna reflector with a high shape stability and a low density per unit length J. *Eng. Phys. Thermophy* **88-3** 699–705. DOI:10.1007/s10891-015-1239-x

[4] Reznik S V and Novikov A D 2016 Comparative analysis of the honeycomb and thin-shell space antenna reflectors. MATEC *Web of Conferences: Thermophysical Basis of Energy Technologies* 01012 5 p. DOI: 10.1051/matecconf/2017920101292

[5] Mikhailovsky K V and Gorodetsky M A 2017 Development of methods for determining and correcting the parameters of the working orbit of the Earth remote sensing spacecraft *Bulletin of RUDN. Engineering research* **18-3** 361–372

[6] Schaub H and Junkins J L 2009 *Analytical mechanics of space systems. 2-nd edition* (Reston: AIAA), 819 p.

[7] Reznik S V, Prosuntsov P V, Mikhailovsky K V and Shafikova I R 2016 Material science problems of building space antennas with a transformable reflector 100 m in diameter IOP *Conference Series: Materials Science and Engineering* **153-1**. 012012 10 p. DOI: 10.1088/1757-899X/153/1/012001

[8] Bolton W 2007 *Construction materials: metals, alloys, polymers, ceramics, composites. Pocket Guide* translation from English (Moscow: Metallurgy) 336 p

[9] Latyev L N, Petrov V Ya, Chekhovskaya E N and Shestakov E N 1974 *Radiative properties of solid materials* ed. A. E. Sheindlin. (Moscow: Energy) 472 p

[10] Sayapin S N and Shkapov P M 2016 Kinematics of deployment of petal-type large space antenna reflectors with axisymmetric petal packaging J. *of Machinery Manufacture and Reliability* **45-5** 387-397

[11] Reznik S. V., Prosuntsov P. V. and Azarov A.V. 2015 Modeling of the temperature and stressed-strained states of the reflector of a mirror space antenna *J. Eng. Phys. Thermophy* **88-4** 978–983. DOI:10.1007/s10891-015-1273-8

[12] Golovatov D, Mikhaylov M and Bosov A 2016 Optimization of technological parameters of impregnation of load-bearing rod elements of reflector made of polymer composite materials by transfer molding method *Indian J. of Science and Technology* **9-46** No 107492 DOI: 10.17485/ijst/2016/v9i46/107492

[13] Prosuntsov P V, Reznik S V, Mikhailovsky K V, Novikov A D and Zaw Ye Aung 2016 Study variants of hard CFRP reflector for intersatellite communication *IOP Conf. Series: Materials Science and Engineering* **153-1.** 012012 7 p. DOI: 10.1088/1757-899X/153/1/012012

[14] Reznik S V and Novikov A D 2016 Comparative analysis of the honeycomb and thin-shell space antenna reflectors MATEC Web of Conferences 01012 (2017). *Thermophysical Basis ofEnergy Technologies* 5 p. DOI: 10.1051/matecconf/2017920101292

[15] Website: https://www.energia.ru/rus/news/news-2006/press_release-07-26.html

[16] Montenbruck O and Gill E 2000 *Satellite orbits: models, methods, applications* (Berlin: Heidelberg Springer – Verlag) 369 p

[17] Hastings D and Garrett H 2004 *Spacecraft-Environment interactions* (Cambridge: Cambridge University press) 320 p.

IOP Conf. Series: Materials Science and Engineering **934** (2020) 012026 doi:10.1088/1757-899X/934/1/012026

Carbon nanofillers in polypropylene: in situ synthesis and properties

O M Palaznik[1], P M Nedorezova[1] and V G Shevchenco[2]

[1]Semenov Institute of Chemical Physics, Russian Academy of Sciences, Kosygina 4, Moscow, 119991 Russia
[2]Institute of Synthetic Polymeric Materials, Russian Academy of Sciences, Profsoyuznaua 70, 117393 Moscow, Russia

E-mail: siloxia@rambler.ru

Abstract. Composites of polypropylene and different carbon nanofillers (graphene nanoplatelets and carbon nanotubes) were prepared by in situ polymerization using a homogeneous isospecific metallocene catalyst rac-Me2Si(2-Me-4-PhInd)2ZrCl2 activated by methylaluminoxane. The electrophysical, mechanical, thermophysical properties of the composites have been investigated. Even small amounts of carbon nanofillers significantly improve the mechanical and thermophysical characteristics of composites. The use of carbon nanofillers makes it possible to prepare composites with controlled electrophysical characteristics.

1. Introduction

Various types of carbon nanoparticles are often used as fillers for polymer composite materials. They have high electrical conductivity, high specific surface area and chemical stability. Their introduction into the polymer matrix significantly improves functional properties of the material: increase electrophysical, mechanical characteristics, thermal conductivity, and heat resistance [1-3]. Highly anisotropic fillers are particularly interesting, such as graphene nanoplatelets and carbon nanotubes.
The objective of this work was to develop methods for producing PP composites with various types of carbon nanoparticles and controlled electrophysical characteristics by in situ polymerization and to investigate the properties of the composites.

2. Synthesis of composites

In situ polymerization (polymerization filling) was used to create composite materials based on PP and various nanocarbon fillers [4]. This method allows to obtain composites with a uniform distribution of the filler in matrix volume over a wide range of concentrations. Different types of fillers were used to create composites: graphene nanoplates, prepared by thermal reduction of graphite oxide (TRGO, 3-5 graphene layers, Ssp = 620 m^2/g) [5], multi-walled carbon nanotubes (MWCNT, Shenzhen Nanotech Port Co., LTD (China); d <10 nm, l ~ 10 μm; Ssp = 160 m^2/g) and single-walled carbon nanotubes (SWCNT, Tuball™, OCSiAl, d = 1.4 nm, l > 5 μm, Ssp = 395 m^2/g).
Composites were synthesized using a metallocene catalytic system (*ansa*-zirconocene *rac*-Me_2Si(2-Me-4-PhInd)$_2ZrCl_2$ activated by methylaluminoxane (MAO)), which is a universal system for producing composites with isotactic PP (iPP). This system has high activity and provides PP with high stereoregularity and molecular weight.

IOP Conf. Series: Materials Science and Engineering **934** (2020) 012026 doi:10.1088/1757-899X/934/1/012026

The synthesis was carried out in the presence of fillers at 60° C and pressure of ca. 2.5 MPa in a steel reactor 200 cm^3 in volume and equipped with a high-speed stirrer. The carbon fillers used in this work were subjected to preliminary ultrasonic treatment with the addition of MAO [6]. Data on the conditions and results of polymerization are shown in Table 1. It can be noted that the introduction of carbon nanofillers into the reaction medium leads to a decrease in the activity of the catalytic system, but it remains sufficient to obtain composites of given composition. Table 1 also presents data on the stereoregularity of PP in composites according to IR spectroscopy [7]. The macrotacticity index, characterizing the proportion of propylene in long isotactic sequences, varies little with composition. The degree of macrotacticity is 87-91%, that is, the filler does not affect the stereoregularity of the polymer chain.

Table 1. Polymerization conditions and PP stereoregularity in composites.

Filler	Activity, kgPP/(mmolZr·hr)	Filler content in composite, wt.%	PP stereoregularity (macrotacticity) D998/D973
-	125	-	0.9
TRGO	118	0.2	0.91
	69	3.3	0.92
	45	10	-
MWCNT	107	0.2	0.9
	84	0.6	-
	104	1.4	0.9
	95	2.1	-
SWCNT	13	1.2	0.91
	20	2.56	0.9
	14	8	-
	0.7	8.43	-

Figure 1 presents TEM micrographs of the composites, showing a fairly uniform distribution of filler particles.

3. Properties of composites

The electrophysical properties of composites were studied in the low-frequency and in the microwave regions. The measurements in the frequency range $10^{-2} - 10^{6}$ Hz were carried out using Novocontrol Alpha-A impedance analyzer and a ZGS Alpha Active Sample Cell dielectric cell with gold-plated disc electrodes with a diameter of 20 and 30 mm. Figure 2 shows the frequency dependence of conductivity for the iPP/TRGO, iPP/SWCNT, and iPP/MWCNT composites with different filler concentrations. The frequency independence of conductivity indicates that charge transfer occurs over an infinite cluster of contacting filler particles, i.e. at this concentration, the percolation threshold is exceeded. A linear frequency dependence indicates hopping conductivity (charge transfer between isolated filler particles), i.e. the percolation threshold has not been reached at this concentration.

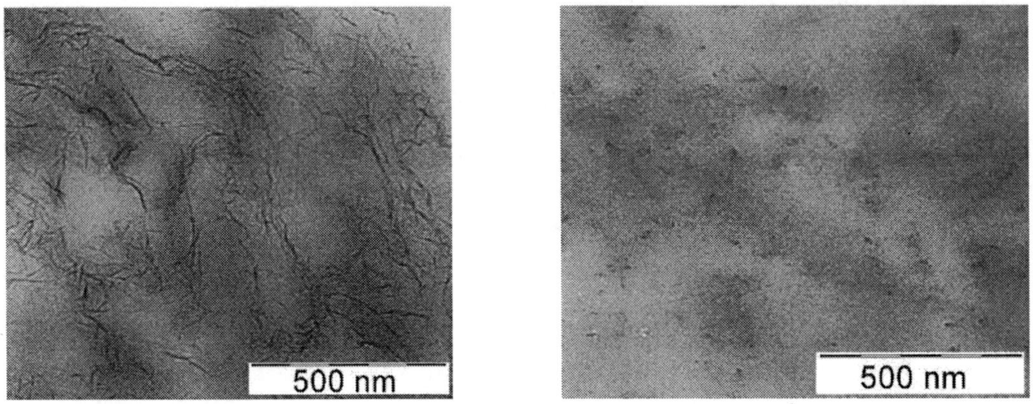

a b

Figure 1. TEM micrographs of isotactic PP composites: 3.3 wt.% TRGO (a) and 1.2 wt.% SWCNT (b).

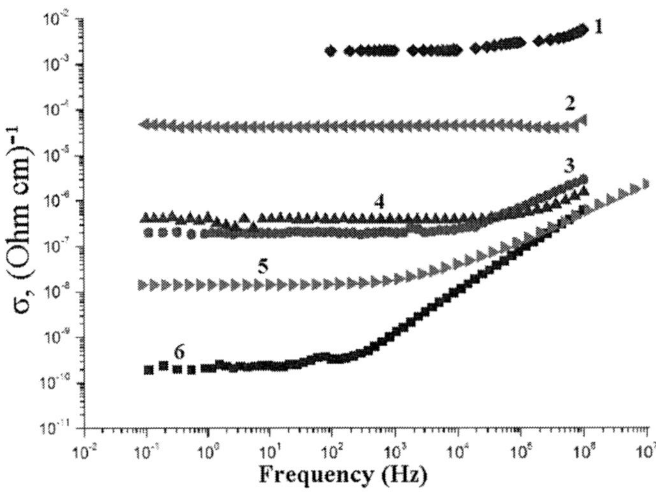

Figure 2. Frequency dependence of conductivity of the composites: iPP/TRGO (2 — 10 wt.%; 4 — 3.3 wt.%), iPP/SWCNT (1 — 8 wt.%; 3 — 2.56 wt.%; 6 — 1.2 wt.%), iPP/MWCNT (5 — 3.5 wt.%).

The results indicate that compositions with SWCNT have lower conductivity thresholds. The percolation threshold for composites with SWCNT is ca. 2 wt.%, with TRGO ca. 3 wt.%, and MWNT ca. 3.5 wt.%. At similar filler content in the composites (8-10 wt.%), compositions with SWCNT are characterized by higher conductivity values.
One of the prospective properties of conductive materials is their ability to reflect or absorb electromagnetic radiation. For iPP composites with TRGO, SWCNT and MWCNT, the reflection coefficients of microwave electromagnetic waves at different frequencies were determined for film samples placed on a metal substrate (the thickness of the composite films was 200-300 μm). The data obtained (Table 2) show that all composites have good electrodynamic properties. It can be seen that composites with SWCNT have better characteristics compared to composites with TRGO and MWCNT.

IOP Conf. Series: Materials Science and Engineering **934** (2020) 012026 doi:10.1088/1757-899X/934/1/012026

Efficient absorbers of dielectric type are usually thick plates. In our case, thin films have characteristics comparable to thick absorbers. The resulting materials are promising for the creation of absorbing screens.

Table 2. Reflection coefficient of composite films in microwave range.

Sample	Reflection coefficient, %	
	30 GHz	35 GHz
iPP	-	95
iPP/TRGO (3.3 wt.%)	91	79
iPP/MWCNT (0.6 wt.%)	-	79
iPP/MWCNT (2.1 wt.%)	-	66
iPP/SWCNT (1.2 wt.%)	69	57
iPP/SWCNT (2.56 wt.%)	48	46

The tensile characteristics of nanocomposites were studied in the regime of quasi-static tension using an Instron-1122 tensile testing machine. Figure 3 shows how the relative modulus of elasticity varies with filler content in the composite, indicating that all carbon nanofillers increase the elastic modulus by 10-30%.

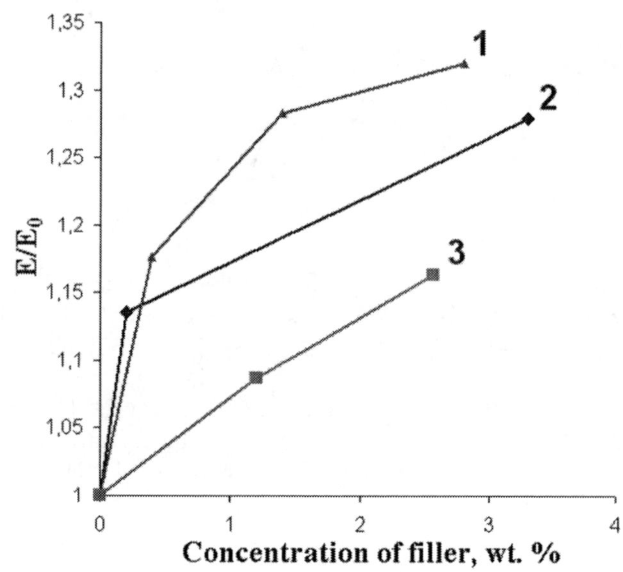

Figure 3. The dependence of the relative elastic modulus on filler content in the composite: 1 – iPP/MWNT, 2 – iPP/TRGO, 3 – iPP/SWCNT.

Figure 4 shows the mass loss according to TGA data obtained in air for iPP and the composites. Even at a low content of filler, a marked increase in the temperature of the maximum mass loss rate is observed, indicating that the thermal stability of composite materials is higher compared to iPP.
It is clear that carbon fillers increase the thermal stability of the materials (Tmax increases by 30-60° C). The increase in thermal stability is probably associated with a slowdown in the thermo-oxidative

decomposition of the polymer due to a decrease in the rate of oxygen diffusion in the presence of the filler and with the termination of radicals on filler particles surface.
It is known that polymers are subject to aging due to exposure to temperature and oxygen. Often, antioxidants are added during the synthesis to prevent the loss of properties of polymer products. It is known that various forms of carbon can slow down the oxidation of polymers [8]. Using the kinetic method for studying thermo-oxidative degradation by oxygen absorption at 130° C, the resistance to thermal oxidation of the initial isotactic PP and all the composites was studied. Kinetic oxidation curves are shown in Figure 5. It is obvious that all the fillers act as inhibitors of thermal oxidative decomposition of PP — an increase in the induction period of thermal oxidative degradation and a decrease in the rate of thermal oxidation are observed.

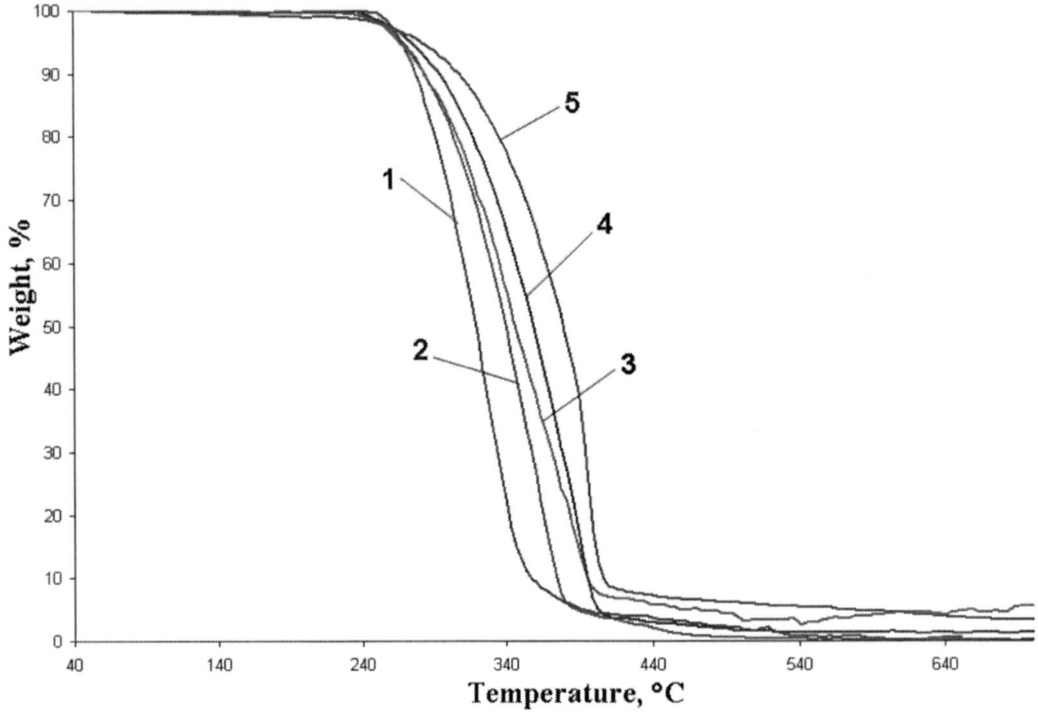

Figure 4. TGA curves in air for isotactic PP (1) and composites: 2 – iPP/TRGO (0.2 wt.%), 3 – iPP/SWCNT (2.56 wt.%), 4 – iPP/SWCNT (1.2 wt.%), 5 – iPP/TRGO (3.3 wt.%).

The DSC method was used to study the effect of filler content on the thermophysical properties of composites. For all types of carbon nanofillers, an increase in the crystallization temperature by 8-15° C is observed, which indicates their nucleating effect. Increase of the crystallization temperature facilitates the processing of materials.

4. Conclusion

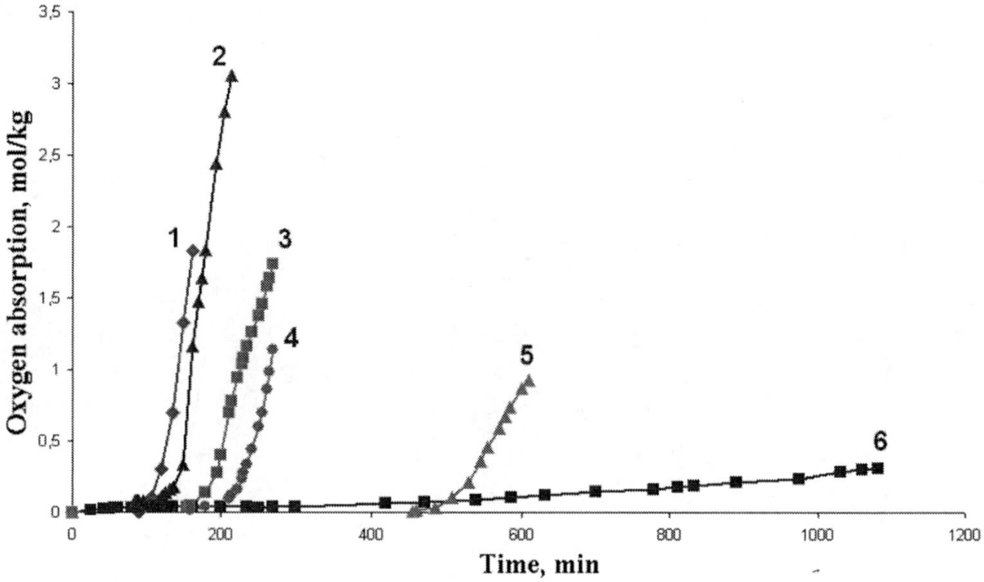

Figure 5. Kinetic oxidation curves of isotactic PP (1) and composites: 1 – iPP, 2 – iPP/TRGO (3.3 wt.%), 6 – iPP/TRGO (10 wt.%), 3 – iPP/SWCNT (1.2 wt.%), 4 – iPP/SWCNT (2.56 wt.%), 5 – iPP/SWCNT (8 wt.%).

The composites based on iPP and nanocarbon fillers in a wide range of compositions with a fairly uniform distribution of the filler were obtained by in situ polymerization. All composites have a low percolation threshold and good electrodynamic properties. They can be used as materials for absorbing screens and filters of electromagnetic radiation in the microwave range. The introduction of carbon nanofillers can increase the thermal stability and resistance of materials to thermal oxidative decomposition.

This work was financially supported by the Russian Foundation for Basic Research in the framework of scientific project 18-33-00573 mol_a.

References

[1] Zhong Y, Shu Z and Zhen H 2017 Graphene: Fundamental research and potential applications *Flat Chem* **4** 20-32

[2] J Phiri, P Gane and T C Maloney 2017 General overview of graphene: Production, properties and application in polymer composites *Mater. Sci. Eng., B* **215** 9-28

[3] Muradyan V E, Arbuzov A A, Sokolov E A, Babenko S D and Bondarenko G V 2013 The effect of addition of functionalized graphene oxide on the dielectric properties of epoxy composite *Tech. Phys. Lett.* **39** 798–800

[4] Nedorezova P M, Tsvetkova V I, Shevchenko V G, Shchegolikhin A N and Korolev Yu M 2004 Polymerizationally filled conducting polypropylene-graphite composites prepared with highly efficient metallocene catalysts *Polym. Sci. - Ser. A* **46** 242-49

[5] Arbuzov A A, Muradyan V E, Tarasov B P, Sokolov E A and Babenko S D 2016 Epoxide composites with thermally reduced graphite oxide and their properties *Russ. J. Phys. Chem.* **90** 907

[6] Kovalchuk A A, Nedorezova P M, Klyamkina A N, Aladyshev A M, Shevchenko V G and Shchegolikhin A N 2008 Isotactic and syndiotactic polypropylene/multi-wall carbon nanotube composites: synthesis and properties *Journal of Materials Science* **43** 7132-40

[7] Kissin Yu V 1985 *Isospecific Polymerization of Olefins* (New York; Berlin; Heidelberg; Tokyo: Springer-Verlag)

[8] Monakhova T V, Nedorezova P M, Bogaevskaya T A, Tsvetkova V I, Shlyapnikov Yu A 1988 Thermooxidative destruction of polypropylene-graphite composites *Polym. Sci. - Ser. A* **30** 2589-94

IOP Conf. Series: Materials Science and Engineering **934** (2020) 012027 doi:10.1088/1757-899X/934/1/012027

Matrix curing in the workpieces made of polymer composites using microwave radiation

P V Prosuntsov[1]*, S V Reznik[1], A R Gareev[1,2] and P V Polsky[1]

[1]Bauman Moscow State Technical University, 2nd Baumanskaya str., b. 5, Moscow, 105005, Russia
[2]Joint-stock company "State Research Institute of Constructional Materials Graphite "NIIgraphite", Electrodnaya str., b. 2, Moscow, 111524, Russia

E-mail: pavel.prosuntsov@mail.ru

Abstract. The physical and mathematical models of curing of polymer matrix composite under microwave radiation are presented. The models describe the processes of the electromagnetic radiation propagation in the volume of the operating area, volumetric heat generation due to the electromagnetic energy absorption and the exothermic curing of the matrix as well as conductive heat transfer in the workpiece and radiation heat transfer between non-uniformly heated surfaces of the workpiece and the walls of the operating area. The modeling of the curing in thick-walled and thin-walled cylindrical workpieces is performed. The challenges of processing of thin-walled parts are presented.

1. Introduction

The growing applications of polymer matrix composites (PMC) in state-of-the-art technology are being limited by high cost of manufacturing. Matrix curing is a time- and energy-consuming stage of the manufacturing. The application of the electromagnetic radiation in microwave wavelength range has a potential for heating and curing of the matrix [1-8]. This method can extremely reduce energy consumption of the PMC products manufacturing as well as enhance their quality. This technology is expected will be the high-demand in the manufacturing of force rods of the flexible space reflectors, trusses and solar panels [9–15].

2. Object of study

The cylindrical PMC workpieces (reinforcement – glass fiber, matrix – epoxy resin, volume fraction of glass fiber was 0.5) were the object of current study. The length of the cylindrical workpiece was 250 mm with an outer diameter of 37 mm. Two types of the workpiece were considered: thick-walled with a wall thickness of 12 mm and thin-walled with a wall thickness of 1 mm. The workpiece curing was performed in the operating area of the microwave heating equipment. The dimensions of the equipment were: height – 100 mm, width and length – 350 mm. Three 1200 W magnetrons that operated at frequency of 2.45 GHz were a heating source. In order to ensure uniformity of heating during the curing, the rotation of the workpiece around its transverse vertical axis at a speed of 1 rpm was performed (figure 1).

It was assumed that the workpiece material was homogeneous and isotropic and had the following properties: density 2100 kg/m3, thermal conductivity 0.41 W/(m·K), heat capacity 950 J/(kg·K), complex permittivity 4.75·(1-0.023·j). The electrical resistance of the workpiece material was equal to zero while the magnetic permeability – to 1.

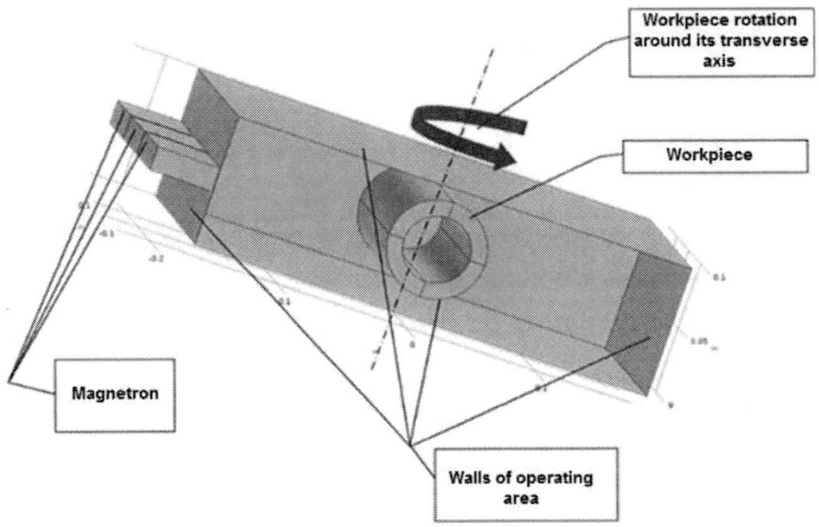

Fig. 1. The geometric model for the positioning of PMC workpiece in the operating area of the microwave heating equipment

3. The mathematical model of the curing

There are several heterogeneous interconnected physical processes proceeded simultaneously during matrix curing in the operating area of the equipment: propagation of electromagnetic radiation in volume of the operating area, its absorption in the workpiece with volumetric heat generation, liquid-solid transition of the matrix in the workpiece, volumetric heat generation in the workpiece due to the proceeding of exothermic process of matrix curing, conductive heat transfer in the workpiece, the radiation heat transfer between non-uniformly heated workpiece surfaces and the walls of the operating area.

The electromagnetic wave propagation in the operating area of the microwave radiation equipment was described by the Helmholtz equations:

$$\nabla^2\mathbf{H} - \varepsilon\varepsilon_0\mu\mu_0\frac{\partial^2\mathbf{H}}{\partial\tau^2} = 0 \quad (1)$$

$$\nabla^2\mathbf{E} - \varepsilon\varepsilon_0\mu\mu_0\frac{\partial^2\mathbf{E}}{\partial\tau^2} = 0, \quad (2)$$

where E – electric field vector, V/m; H – magnetic field vector, A/m; ε_0 – permittivity, F/m; μ_0 – permeability, H/m; ε – relative permittivity; μ – relative permeability; τ – time, s.

The following boundary conditions that ensured the continuity of the tangent components of the electric and magnetic fields vectors were set at the boundary surface between the internal volume of the operating area (air) and the PMC dielectric workpiece:

$$H_{\tau1} = H_{\tau2} \quad (3)$$

$$E_{\tau1} = E_{\tau2} \quad (4)$$

$$D_{\tau1} = \frac{\varepsilon_1}{\varepsilon_2}D_{\tau2} \quad (5)$$

$$B_{\tau1} = \frac{\mu_1}{\mu_2}B_{\tau2}, \quad (6)$$

where $H_{\tau 1}$ and $H_{\tau 2}$ – tangent vectors of magnetic field strength in the chamber volume and workpiece; $E_{\tau 1}$ and $E_{\tau 2}$ – tangent vectors of electric field strength in the chamber volume and workpiece; $D_{\tau 1}$ and $D_{\tau 2}$ – tangent vectors of electric induction in the chamber volume and workpiece; ε_1 and ε_2 – dielectric constants in the chamber volume and workpiece; $B_{\tau 1}$ and $B_{\tau 2}$ – tangent vectors of magnetic induction constants in the chamber volume and workpiece; μ_1 and μ_2 – magnetic constants in the chamber volume and workpiece.

During the energy absorption of the electromagnetic radiation, volumetric heat generation in the workpiece took place. The intensity of heat generation for dielectrics depends on the frequency and power of the radiation:

$$q_{mw}(T) = \varepsilon_0 \varepsilon \mathrm{tg}\delta f E^2, \tag{7}$$

where T – temperature, K; f – linear frequency of microwave radiation, Hz; tgδ – dielectric dissipation factor.

The walls of the operating area were considered as metal. Also the condition of perfect electric conductor was set:

$$E_{\tau} = 0. \tag{8}$$

The wave-propagating system with three guides was used to inject the microwave radiation into the volume of the operating area. The blind facet of each waveguide was a rectangular port.

The process of the electromagnetic energy transfer in the operating area of the equipment was analyzed for 8 workpiece positions. These positions corresponded to rotation angles of the workpiece around its transverse vertical axis – 0°, 45°, 90°, 135°, 180°, 225°, 270° and 315°. The 0°corresponded to the direction at which the workpiece axis was parallel to the wall of the operating area on which the magnetron waveguides were located.

It was considered that the matrix curing in the PMC could be described by the one-stage Arrhenius equation:

$$\frac{\partial h}{\partial \tau} = -h^a \cdot (1-h)^n \cdot k \cdot \exp(-\frac{E}{RT}), \tag{9}$$

where h – is degree of matrix cure; a, n – reaction order; k – pre-exponential factor, 1/s; E – activation energy of the curing reaction, J/mol; R – gas constant, J/(K·mol). The initial degree of cure of the PMC workpiece was equal to zero.

During matrix curing in the workpiece the volumetric heat generation took place with a power:

$$q_{exo}(T) = \upsilon_m \cdot \rho \cdot h^a \cdot (1-h)^n \cdot k \cdot \exp(-\frac{E}{RT}), \tag{10}$$

where q_{exo} – specific power of the exothermal effect of matrix curing, W/m3; υ_m – volume ratio of the matrix in PMC; ρ – PMC density, kg/m3.

A three-dimensional transient non-linear process of heat transfer in the PMC workpiece that was exposed by microwave radiation was simulated. The matrix curing with heat generation in the PMC workpiece was occurred. Convective and radiation heat release to ambient medium (walls of the operating area of the equipment) with a constant temperature T_f on each workpiece surface was considered. Mutual radiation of the non-uniformly heated surfaces in the internal cavity of the cylindrical workpiece was also taken into account. It was taken that volumetric heat generation in the volume of the workpiece caused by the absorption of microwave radiation and the exothermic curing reaction was proceeded. To control the magnetrons power, the PID controller principle was used.

In this case the mathematical model of heat transfer could be described by the three-dimensional thermal conductivity equation:

$$C(T,h)\cdot\rho\frac{\partial T(r,z,\phi,\tau)}{\partial\tau}=\frac{\partial}{\partial z}(\lambda_z(T,h)\frac{\partial T(r,z,\phi,\tau)}{\partial z})+\frac{1}{r}\frac{\partial}{\partial r}(r\lambda_r(T,h)\frac{\partial T(r,z,\phi,\tau)}{\partial r})+ +\frac{1}{r^2}\frac{\partial}{\partial\phi}(\lambda_\phi(T,h)\frac{\partial T(r,z,\phi,\tau)}{\partial\phi})+q_{mw}+q_{exo} \quad (11)$$

with an initial condition:

$$T=T_0 \quad (12)$$

and a boundary condition on the workpiece surface:

$$-\vec{n}\cdot\lambda(T,h)\frac{\partial T(R,z,\phi,\tau)}{\partial n}=\alpha_f(T\cdot(R,z,\phi,\tau)-T_f)+\varepsilon_w\sigma_0(T^4(R,z,\phi,\tau)-T_f^{\ 4}), \quad (13)$$

where ε_w – workpiece emissivity factor; σ_0 – Stefan-Boltzmann constant, W/(m2·K4); T_f – ambient temperature (wall of the operating area), K; T_0 – workpiece initial temperature, K; α_f – heat-transfer coefficient, W/(m2·K); λ – workpiece thermal conductivity, W/(m·K); C – workpiece heat capacity, J/(kg·K); $\vec{n}$ – direction of the inward normal to a workpiece surface; R,z,ϕ – cylindrical coordinates.

4. Modeling results

The modeling of the heating and curing of the thick-walled and thin-walled workpieces have been performed. Figures 2 and 3 illustrate the distribution of the electric field intensity in the workpieces for different angles of rotation. It is apparent that during the workpiece rotation there was a significant redistribution of the electric field intensity and the positions of standing waves "peaks" were shifted. Moreover, it is shown that non-uniformity of electric field intensity for the thin-walled workpiece is significantly higher. For each angle of rotation there are areas with the extremely high and extremely low values of the electric field intensity. Such non-uniform distribution of the electric field intensity leads to the non-uniform heat generation in the workpiece volume that is partially smoothed by the displacement of the maxima of the electric field intensity during the workpiece rotation.

Figures 4-7 depict the distribution of the cure degree and temperature for thick-walled and thin-walled workpieces at different time moments. Figures 8 and 9 demonstrate the dynamics of the degree of matrix cure in the workpiece and dynamics of the temperature at four test points of the sample. The test point № 1 is located in the center of the upper part of the workpiece on the its axis of rotation, the point № 2 is also located in the upper part of the workpiece at a distance of 120 mm from the axis of rotation (almost on the edge of the workpiece), points № 3 and 4 are located similarly to the points 1 and 2 in the horizontal diametrical plane of the workpiece.

It is apparent that to the moment of 3600 s the thick-walled workpiece has been almost completely cured (degree of cure is not less than 95%) and to the moment of 9000 s the degree of cure has reached the level of 99%, that should have a positive impact on the products quality. Figure 8 clearly shows the temperature peaks related to the exothermic curing process. It is shown that peaks reach the level of 30-40°C, therefore it should be taken into account by providing the magnetrons power control in order to avoid "overburning" of the matrix.

At the same time to the moment of 9000 s the thin-walled workpiece has not been completely cured (the degree of cure in the edge area is not more than 30%) (figure 8), and the maximum temperature change in the test points has reached 100°C (figure 9). Moreover, there have been significant temperature fluctuations caused by the workpiece rotation (up to 5°C). These fluctuations result from the fact that parts of the thin-walled shell are in or out the areas with high electric field intensity. High heterogeneity of temperature field in thin-walled workpiece is connected with increased role of the radiation and convective heat release on cold walls of the operating area and to the ambience. Heat supply by thermal conductivity from adjacent areas with a higher temperature is constrained by the small workpiece thickness.

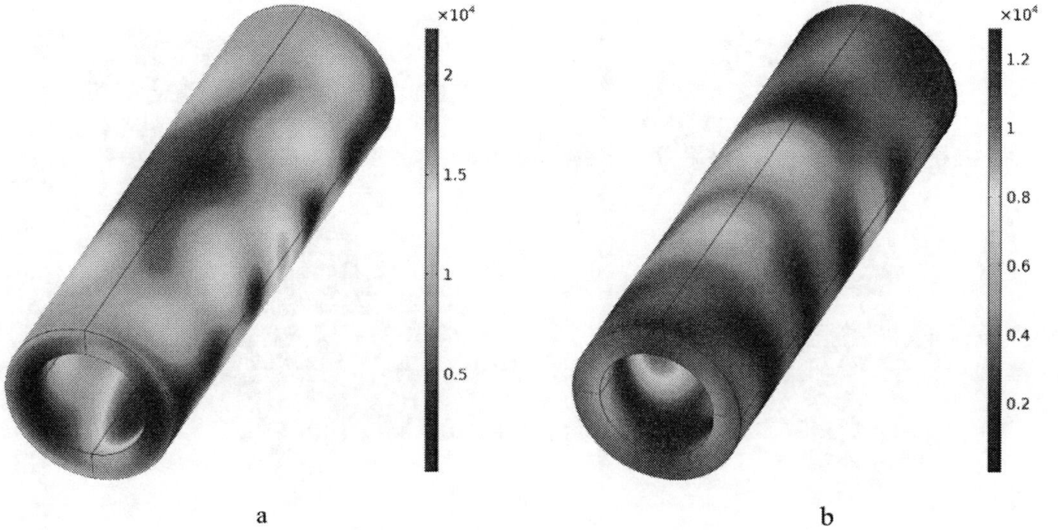

Fig. 2. Distribution of the electric field intensity in the thick-walled workpiece at the rotation angles of 0° (a) and 90° (b), V/m

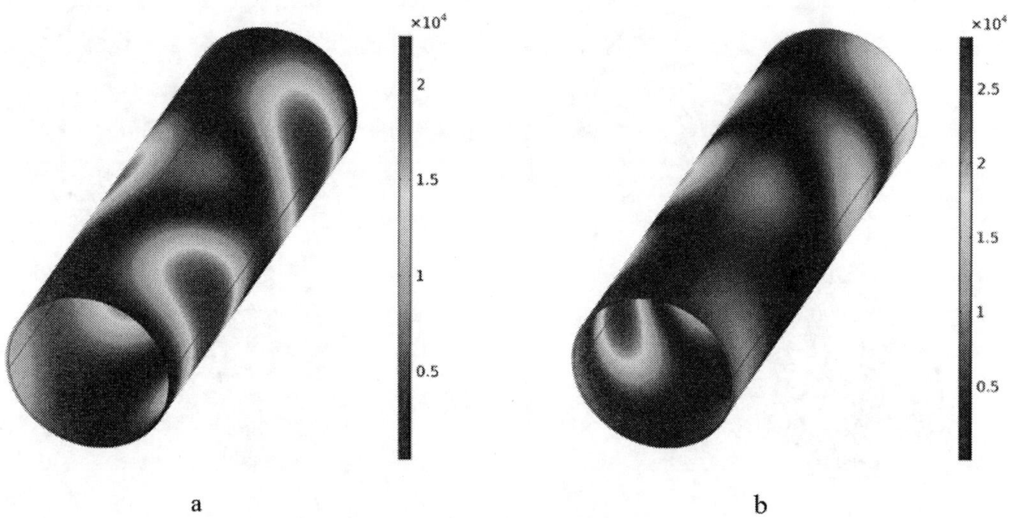

Fig. 3. Distribution of the electric field intensity in the thin-walled workpiece at the rotation angles of 0° (a) and 90° (b), V/m

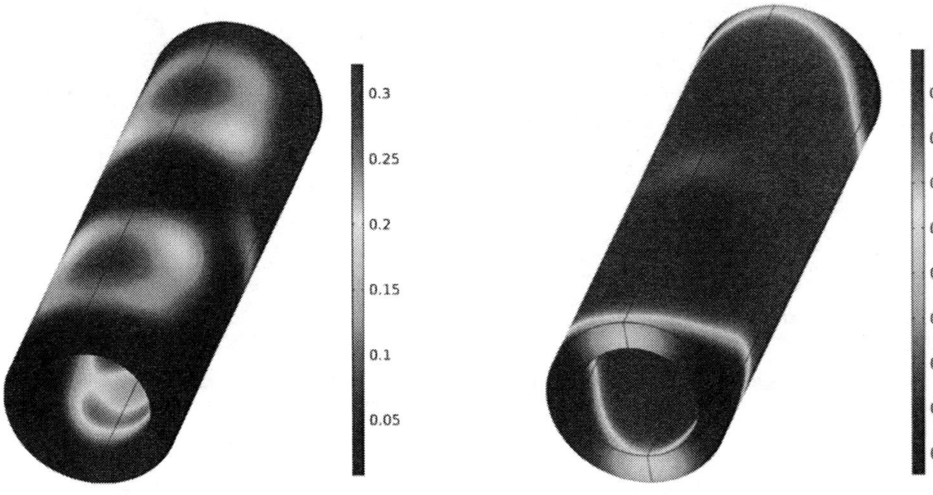

a b

Fig. 4. Degree of matrix cure in the thick-walled workpiece at the time of 600 (a) and 1200 (b) s

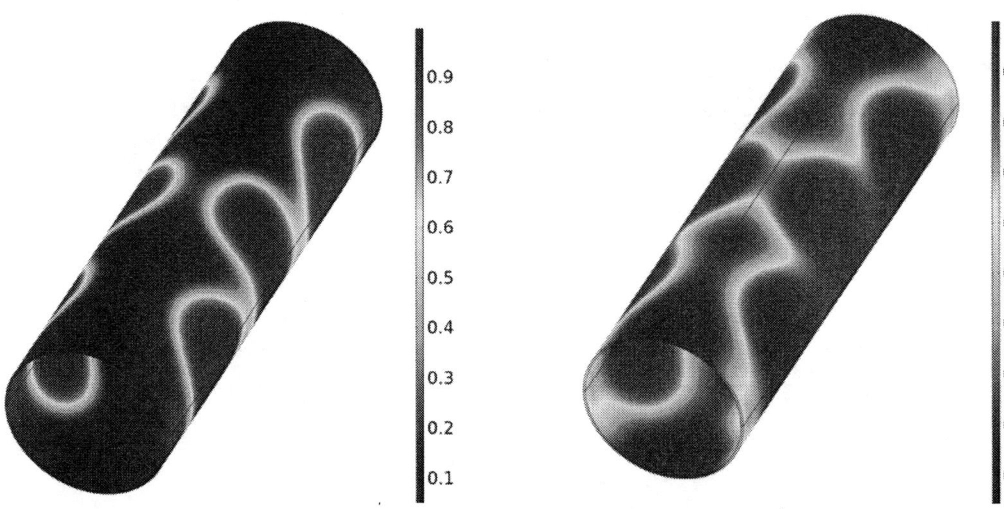

a b

Fig. 5. Degree of matrix cure in the thin-walled workpiece at the time of 600 (a) and 1200 (b) s

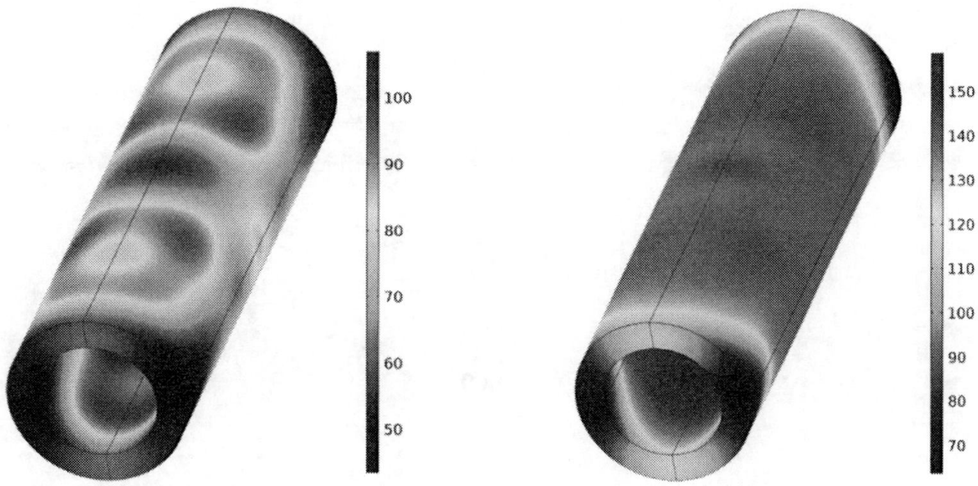

a b

Fig. 6. Temperature of the thick-walled workpiece at the time of 600 (a) and 1200 (b) s, °C

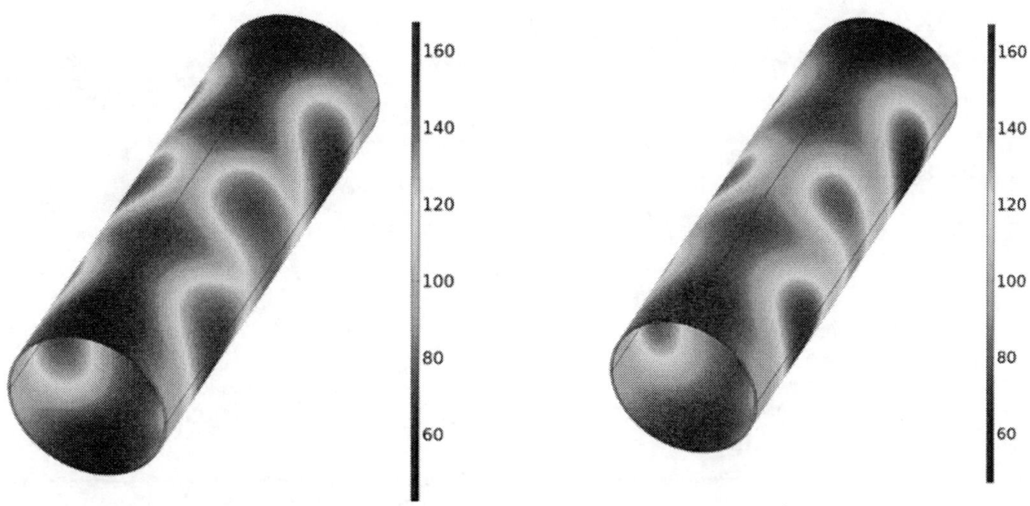

a b

Fig. 7. Temperature of the thin-walled workpiece at the time of 600 (a) and 1200 (b) s, °C

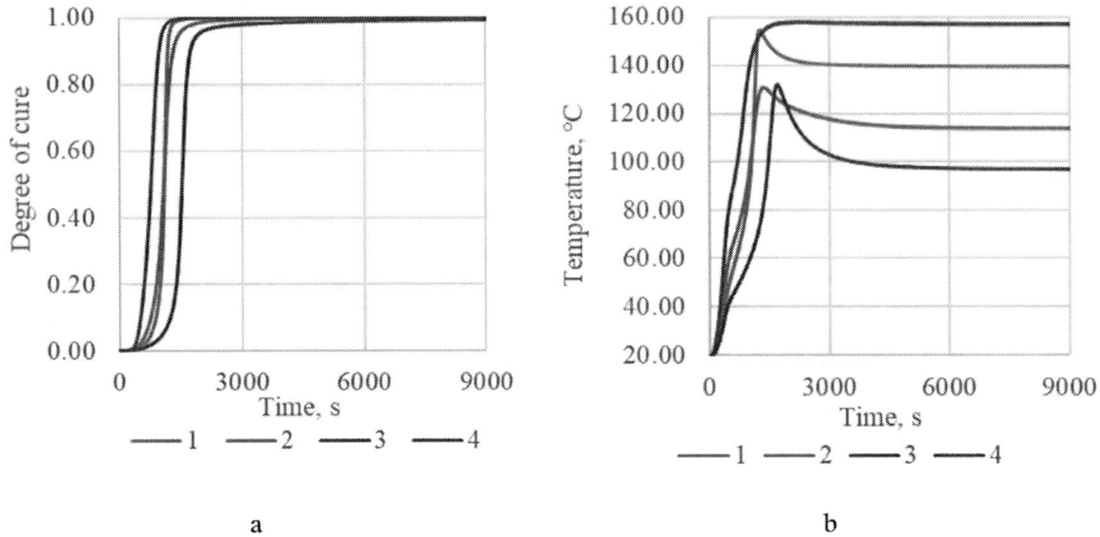

Fig. 8. Change in the degree of cure (a) and temperature (b) at the test points of the thick-walled workpiece (the curve number corresponds to the test point number)

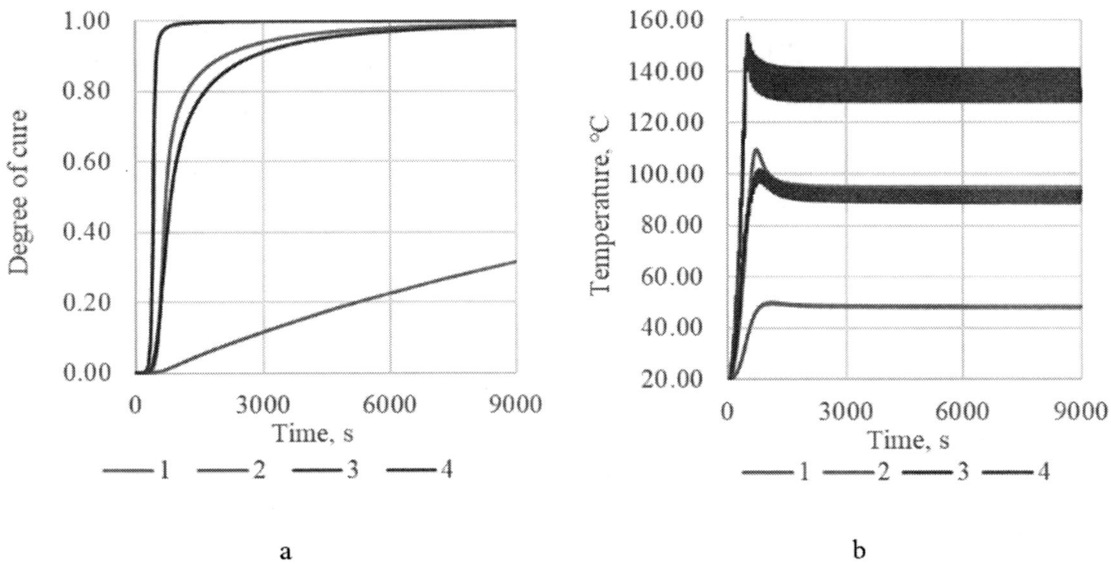

Fig. 9. Change in the degree of cure (a) and temperature (b) at the test points of the thin-walled workpiece (the curve number corresponds to the test point number)

5. Conclusion

The physical and mathematical models of curing of the polymer composite matrix composite under microwave radiation are developed. The models describe the heterogeneous physical processes occurring in the operating area of the equipment: propagation of electromagnetic radiation in volume of the operating area, its absorption in the workpiece with volumetric heat generation, liquid-solid transition of the matrix in the workpiece, volumetric heat generation in the workpiece due to the proceeding of exothermic process of matrix curing, conductive heat transfer in the workpiece, the radiation heat transfer between non-uniformly heated workpiece surfaces and the walls of the operating area. The modeling of the cure of thick-walled and thin-walled cylindrical workpieces has been performed. The results have suggested that the thick-walled part can be cured quite quickly and successfully using microwave radiation, whereas the curing of the thin-walled part can occur the difficulties caused by the increasing influence of the convective and radiative heat release from the workpiece surface.

References

[1] Museok K 2016 *Microwave curing of carbon-epoxy composites: Process development and material evaluation* (London: Imperial College London, Department of Aeronautics) p 150

[2] Day R J and Samoladas E 1998 Micromechanics of model carbon epoxy composites cured using microwave heating *Science and Engineering of Composite Materials*. **7** (1-2) 23–26

[3] Navarrete A, Mato R B and Cocero M J 2012 A predictive approach in modeling and simulation of heat and mass transfer during microwave heating. Application to SFME of essential oil of Lavandin Super Chemical Engineering Science. **68** 192–201

[4] Morozov G A et al 2011 Microwave treatment thermoreactive and thermoplastic polymers Physics of Waves Processes and Radio Engineering Systems (Fizika Volnovikh Processov i Radiotekhnicheskie Sistemi) 3 114–121

[5] Mikhailovskii K V and Reznik S V 2014 A prediction of the temperature regimes binder curing of details from polymer composite materials by microwave radiation *Thermal Processes in Engineering* (Teplovie Processi v Technike) **6-8** 363–368

[6] Reznik S V and Rumyantsev S A 2014 Mathematical modelling of temperature state of cylinder details from polymer composite materials under microwave heating *Science and Education (Nauka i Obrazovanie)* **1** 6–21

[7] Guzeva T A 2013 *Upgrade of the curing organic plastics technological regimes under microwave radiation* Thesis on Cand. of Sci. (Moscow: Bauman MSTU) p 163

[8] Prosuntsov P V, Reznik S V, Mikhailovskii K V and Belenkov E S 2018 Multiscale Modeling Of the Binder Polymer Composite Materials Heating Using Microwave Radiation *Journal of Physics: Conference Series. VII International Conference: Functional Nanomaterials and High Purity Substances* **1134** 012047 DOI: 10.1088/1742-6596/1134/1/012047

[9] Zimin V N, Koloskov I M, Meshkovsky V E and Usyukin V I 2001 Investigation of natural oscillations for self-deployable truss space antennae *Trans. Mod. and Sim*. **30**, 497–504

[10] Gottero M, Sacchi E, Scialino G L, Reznik S V and Kalinin D Y 2005 The large deployable antenna (LDA) a review of thermal aspects SAE Technical Papers 2005. *Proc. 35th Int. Conf. on Environmental Systems*, ICES 2005 (11-14 July 2005, Rome, Italy) DOI: 10.4271/2005-01-2985

[11] Kotik A, Usyukin V, Vinogradov I and Arkhipov M 2017 Simulation of reflecting surface deviations of centimeter-band parabolic space radiotelescope (SRT) with the large-size mirror *Proc. of SPIE - The Int. Soc. for Optical Engineering*, **10567**, 105671A 6 p DOI: 10.1117/12.2308118

[12] Meguro A, Harada S and Watanabe M 2003 Key technologies for high-accuracy large mesh antenna reflectors *Acta Astronautica* **53-11** pp. 899-908 DOI: 10.1016/S0094-5765(02)00211-4

[13] Reznik S V, Timoshenko V P, Shulyakovsky A V and Denisov O V 2013 Thermal-vacuum tests of hollow composite rods intended for structures in space Polymer Science D 6 3 pp. 242–245

[14] Reznik S V, Prosuntsov P V, Mikhailovsky K V and Shafikova I R 2016 Material science problems of building space antennas with a transformable reflector 100 m in diameter *IOP Conference Series: Materials Science and Engineering* **153-1**. 012012 10 p. DOI: 10.1088/1757-899X/153/1/012001

[15] Golovatov D, Mikhaylov M and Bosov A 2016 Optimization of technological parameters of impregnation of load-bearing rod elements of reflector made of polymer composite materials by transfer molding method Indian *J. of Science and Technology 9*, **46-107492** DOI:

10.17485/ijst/2016/v9i46/107492

Dynamic characteristics of modified bitumen systems

L V Fedorova[1], Z O Tretyakova[2], M V Voronina[2]

[1]Bauman Moscow State Technical University, 5/1, ul. Baumanskaya 2-ya, Moscow, 105005, Russia.

[2]Saint-Petersburg Mining University, 2, 21st Line, St Petersburg, 199106, Russia

E-mail: momd@yandex.ru

Abstract. Thiokol-modified bitumens are investigated. Based on the measurements of time-dependent viscosity parameters (melting point, penetration, and spin-spin relaxation time T2), it is established that at a content of thiokol more than 25%, formation of a continuous phase is possible, which shows up as a sudden change in the bitumen properties and is associated with the phase inversion. The method of bitumen modification suggested can be recommended for obtaining roofing materials.

Keywords: bitumen, thiokol, modification, rheology

1. Introduction

Bitumen is produced by distillation of crude oil and is a complex mixture of high molecular weight petroleum hydrocarbons and their heter derivatives containing oxygen, sulfur, nitrogen and metals [1]. The properties of petroleum bitumen depend mainly on their chemical composition. Oil bitumen due to a valuable set of technical properties are widely used in building materials [2]. Bitumen is a thermoviscoelastic material in which the temperature and speed of application of the load have a great influence on their behavior [3-5].

It is well known that bitumen in the conditions of their storage and operation are subject to aging.

The durability of bituminous materials is limited mainly by their unsatisfactory deformation stability-stiffness and brittleness at low and fluidity at elevated temperatures. The leading direction in the modification of bitumen in order to eliminate these shortcomings is their compatibility with polymers [6, 7].

There is a steady trend of increasing consumption of polymer composite materials used in atmospheric conditions, primarily in construction. The most common among them as sealing, waterproofing and roofing materials that can long-term and effectively resist the effects of aggressive factors such as ultraviolet (UV), ozone, radiation, water, in a wide temperature range from -60 to +100°C, are bitumen compositions based on elastomers [8]. A special place among them is occupied by compositions based on polysulfide oligomers, the main representative of which is liquid thiocol. This is due to their properties-high resistance to climatic influences (ultraviolet, ozone, water), in the case of liquid thiols - unique oil and gas resistance and the ability to provide reliable sealing and operation under the influence of these aggressive factors in conditions of significant alternating deformations in a wide temperature range: from-60 to +100°C [9].

The analysis of modern publications suggests that not enough attention is paid to the problem. Most of the works (from a huge number of publications) devoted to the modification of bitumen polymers, is empirical in nature. The theory of compatibility of polymers with bitumen has not yet been created. Meanwhile, knowledge of the structure of bitumen and polymers in combination with the accumulated empirical data, allows you to get a General idea of the processes occurring during their combination, and explain in accordance with these ideas, the observed changes in the properties of composites [10].

The rheological properties of bitumen essentially depend on their composition and primarily on the content of the main structure-forming components, which include the most high-molecular components, resinous-asphaltene substances, as well as paraffins at temperatures below their crystallization temperature, which can significantly influence on the properties of bitumen [11–15].

Viscosity increases in proportion to the volume of particles of the dispersed phase and the degree of their aggregation. Aggregation of particles leads to an increase in apparent viscosity, since aggregates reduce the effective cross-section of the flow channel and create additional inhibition of the displacement of the fluid layers. Destruction of the structure or deformation of solvate shells causes an anomaly of viscosity [16].

One of the indispensable conditions for improving the properties of bitumen when modified by polymers is their compatibility, estimated by such an indicator as the solubility parameter.

In connection with the above, the modification of bitumen thiocols, characterized by high heat resistance and frost resistance.

2. Methods and materials

We used oil bitumen of trademark BN-70/30 with the following characteristics: softening point 69C, penetration 13x0.1 mm at 25^{0}C and 8 x 0.1 mm at 0^{0}C, ductility 3 cm at 25^{0}C and 0 cm at 0^{0}C, brittleness temperature -80^{0}C, and flash point 220^{0}C. As a modifier, we used thiokol of trademark II: content of end SH-groups 2,15%, viscosity of 38 Pa·s, and glass transition temperature -59^{0}C. The thiokol-based compounds are resistant to the action of corrosive media, atmospheric action, and ultraviolet (UV) radiation; they are moisture- and gas-impermeable, and ozone and frost resistant.

Thiokol was introduced into bitumen in the form of a sealing paste containing, besides the thiokol, 30 weight parts of P-803 technical carbon. The thiokol was cured in the bitumen medium by two curing agents of oxidizing type, namely by zinc oxide or manganese dioxide in the form of a curing paste (10 wt.% of the sealing paste). The thiokol content in bitumen varied from 5 to 50%. The properties of modified bitumens were estimated after holding the compounds at 23±2^{0}C for one day to one year.

The dynamic viscosity and other rheological characteristics of bitumens and modified compounds were determined on a Polymer RPE-1m.2 rotational viscometer with a measuring cell of «cylinder—cylinder» type at 95^{0}C.

3. Results and discussion

The modification of bitumen by thiocols, characterized by resistance to swelling in various solvents and oils, as well as a sufficiently high frost resistance, is investigated.

However, studies have shown that thiocols alone in bitumen systems do not give the desired result. Therefore, we proposed another method of such modification, namely the introduction of thiocol into bitumen together with a curing paste and a vulcanization accelerator (diphenylguanidine), which allowed to improve the complex properties of bitumen systems. However, as in other cases, the elastication of bitumen, it is possible to significantly increase the viscosity of bitumen systems [17, 18].

As can be seen from figure 1, the softening temperature of bitumen increases both with increasing thiocol content in the composition and with aging of bitumen-thiocol systems in time (1 day, 20 days, 40 days). The obtained result testifies to the dynamic nature of the process of modification of bitumen thiocols together with curing paste, since the process of vulcanization of thiocols in time in the medium of bitumen at a temperature of 20±20C proceeds slowly and does not end, probably, even for 40 days.

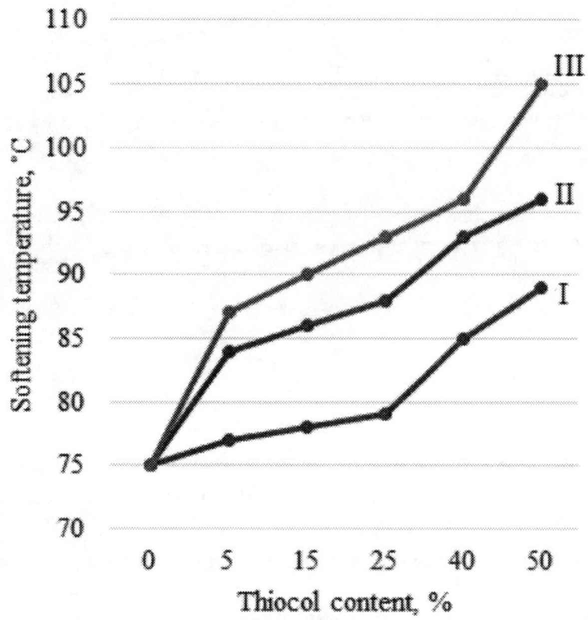

Figure 1. Concentration dependences of bitumen softening temperature on thiocol content during its vulcanization during different number of days: I-1 day, II-20 days, III-40 days

Flow curves of the initial bitumen grade bn-70/30 and its modifications by thiocols depending on the exposure time of the systems are presented in figure 2 for the most optimal composition with 40% modifier content. At low values of shear rates (γ) for initial bitumen it is possible to allocate area γ in which the current close to Newtonian is realized, and area of sharp decrease in viscosity at the expense of destruction of coagulation structure of bitumen (non-Newtonian current) [19, 20].

Flow curves of the initial bitumen grade bn-70/30 and its modifications by thiocols depending on the exposure time of the systems are presented in figure 2 for the most optimal composition with 40% modifier content. At low values of shear rates (γ) for initial bitumen it is possible to allocate area γ in which the current close to Newtonian is realized, and area of sharp decrease in viscosity at the expense of destruction of coagulation structure of bitumen (non-Newtonian current).

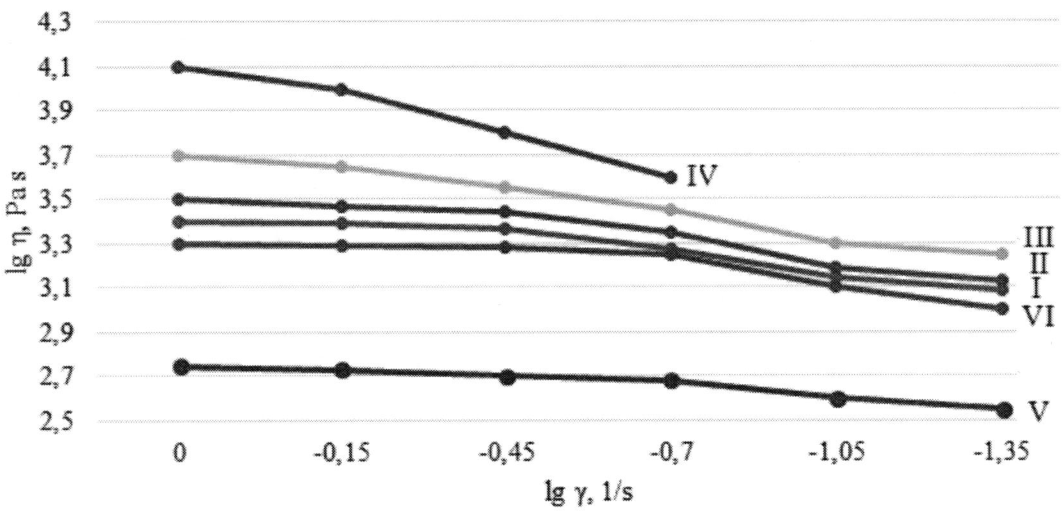

Figure 2. Flow curves of bitumen-thiocol systems with 40% content of vulcanizing thiocol depending on the time of vulcanization (I-IV) and 40% content of non-vulcanized thiocol (V); bitumen with 4% content of vulcanizing paste (VI).
I-initial bitumen, II-exposure 6 days, III-exposure 9 days, IV-exposure 32 days

When modified by thiocol, the character of the bitumen flow in the comparable region γ changes to non-Newtonian, while the values of σ are higher than that of the original bitumen. With increasing exposure time systems increases, and the degree of deviation from the Newtonian flow increases, indicating the destruction of structural formations of bitumen-thiocol systems with increasing γ.

The introduction of unvulcanized thiocol (40%) leads to a significant decrease in the content of bitumen, while the nature of the flow is practically unchanged. Similarly, the action of hardening of the paste (4%). Therefore, in pure form, both thiocol and its modifier have a plasticizing effect on bitumen, more precisely, its Malten fraction.

The change in the content of vulcanizing thiocol in bitumen from 5 to 50% (figure 3) leads to both an increase in σ and a greater dependence of γ on γ in the studied range of shear rates. Even at small values of γ, the flow of bitumen-thiocol systems is carried out due to the destruction of structural formations in this system. Thus, bitumen-thiocol systems are characterized by significant kinetic instability, and to the greater extent, the higher the content of vulcanizing thiocol and the longer the holding time of these systems.

4. Conclusions

Thus, the data indicate an increase in the hardness of the systems, from which it can be concluded that thiocol vulcanization occurs in the systems and the greater the concentration of thiocol in bitumen.

We also studied the change in the main performance indicators of bitumen-thiocol systems from the concentration of thiocol in them and from the time of their formation. However, it should be borne in mind that in the summer in natural conditions, when the roof surface is heated to 80-900C, the vulcanization time of thiocol in bitumen can be significantly reduced.

As a result of the conducted researches it is established that at modification of bitumen by thiocol, the continuous phase of thiocol sealant defining, eventually, the basic properties of the modified bitumen is formed.

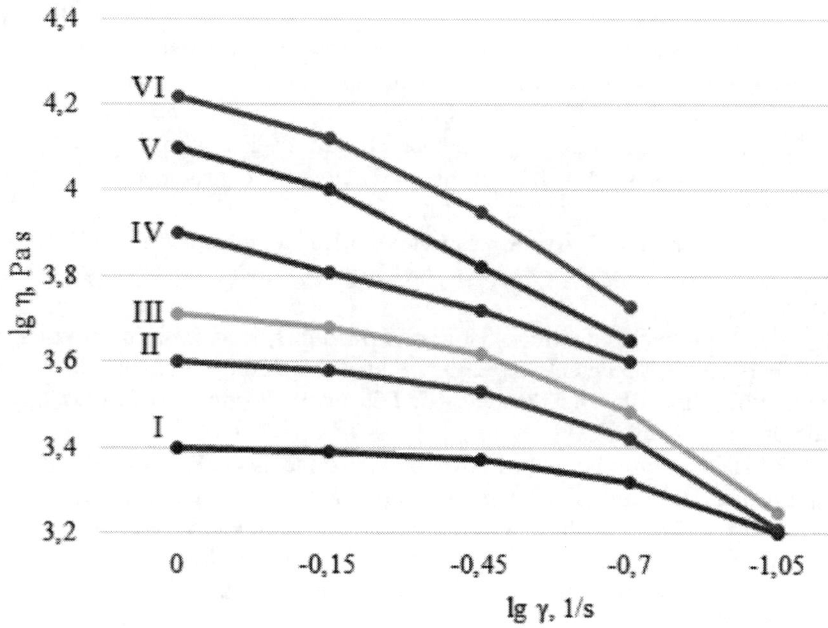

Figure 3. Liquid curves of bitumen modified by different contents of vulcanizing thiocol after 30 days of vulcanization at a temperature of 200C.
I-initial bitumen, II-bitumen+5% thiocol, III-bitumen+15% thiocol, IV-bitumen+25% thiocol, V-bitumen+40 thiocol, VI-bitumen+50% thiocol

References

[1] Gun R B 1973 Petroleum Bitumen (Moscow: Chemistry) 432

[2] The bitumen industry - a global perspective. Production, chemistry, use, specification and occupational exposure. Third Edition 2015 (USA: Asphalt Institute Inc. and European Bitumen Association – Eurobitume) **230 (is-230)** 60 ISBN 978-1-934154-73-1 https://www.eurobitume.eu/public_downloads/General/The%20Bitumen%20Industry%203rd%20edition.pdf

[3] Arafat Yero S and Hainin M R 2012 Viscosity Characteristics of Modified Bitumen *ARPN Journal of Science and Technology* **2** 5

[4] Kristjansdottir O 2006 Warm Mix Asphalt for Cold Weather Paving *Masters of Science in Civil Engineering thesis* (University of Washington) 127 https://www.wsdot.wa.gov/research/reports/fullreports/650.1.pdf

[5] L V Fedorova, Z O Tretyakova and M V Voronina Modified Oil Bitumen // Published under licence by IOP Publishing Ltd *IOP Conference Series: Materials Science and Engineering, Volume 683, Advances in Composite Science and Technology 5–8 December 2018, Bauman Moscow State Technical University*

[6] Teltaev B B, Rossi Ch O and Ashimova S Z 2018 Composition and rheological characteristics of bitumen for short-term and long-term aging *Civil Engineering Journal* **5(81)** 93–101

[7] Kemalov A F and Kemalov R A 2012 Study of disperse polymer systems for producing high-quality polymeric-bituminous materials *Chemistry and Technology of Fuels and Oils* **48(5)** 339–343

[8] Sungatova Z O 1999 Modification of petroleum bitumen by elastomers *Candidate of Technical Sciences thesis* (Kazan) 150 https://www.dissercat.com/content/modifikatsiya-neftyanykh-bitumov-elastomerami

[9] Makarov A M 2006 The processes of formation of the structure and properties of composite

highly porous cellular materials *Doctor of Technical Sciences thesis* (Permian) 348 https://www.dissercat.com/content/protsessy-formirovaniya-struktury-i-svoistv-kompozitsionnykh-vysokoporistykh-yacheistykh-mat

[10] Khmurovska Y and Štemberk P 2018 Comparison of Engineered cementitious composites and concrete for strengthening of concrete structural details using RBSM *MATEC Web of Conferences* **146** 02004 1-6 https://www.matec- conferences. Org /articles /matecconf/ pdf/2017/07 /matecconf _bd2017 02004 pdf

[11] Minkin V S, Sukhanov P P, Averko-Antonovich L A and Kirpichnikov P A 1989 Obtaining and curing reactive oligomers based on liquid thiocol and oligoesteracrylates *Radical polymerization* 168

[12] A method of preparing thermoplastic compositions based on a mixture of bitumen and elastomers 1975 (US Patent 3890263 MKI C 08 B 95/00)

[13] Fuks G I 2003 Viscosity and Plasticity of Petroleum Products (Moscow-Izhevsk: Institute for Computer Research) 328

[14] Petrov S M 2009 Polyfunctional modifiers for producing oxidized road bitumen with improved properties *Candidate of Technical Sciences thesis* (Kazan: Kazan State Technological University) 187 https://www.dissercat.com/content/modifikatory-polifunktsionalnogo-deistviya-dlya-polucheniya-okislennykh-dorozhnykh-bitumov-s

[15] Tumanyan B P 2000 Scientific and applied aspects of the theory of oil dispersed systems (Moscow: Tekhnika) 335

[16] Khakimullin Y N, Khozin V G, Murafa A V, Sungatova Z O and Nagumanova E I 2000 Investigation of the structure and characteristics of curing bitumen-thiokol compositions *Mechanics of composite materials* 423-428

[17] Khakimullin Y N, Murafa A V, Sungatova Z O and Khozin V G 1998 Thiocol's modified bitumen compositions *International conférences on Polymer characterization* (Texas) 62

[18] Kemalov A F, Kemalov R A, Abdrafikova I M, Fakhretdinov P S and Valiev D Z 2018 Polyfunctional Modifiers for Bitumen and Bituminous Materials with High Performance *Advances in Materials Science and Engineering* **7913527** 15 https://www.hindawi.com/journals/amse/2018/7913527/

[19] Rheological and plastoelastic properties of rubbers and rubber compounds 2018 https://msd.com.ua/metody-issledovaniya-struktury-i-svojstv-polimerov/reologicheskie-i-plastoelasticheskie-svojstva-kauchukov-i-rezinovyx-smesej/

[20] Bosykh M S 2004 Rheological properties of polymers under periodic mechanothermal exposure *Candidate of Technical Sciences thesis* (Voronezh) 149 https://www.dissercat.com/content/reologicheskie-svoistva-polimerov-pri-periodicheskom-mekhanotermicheskom-vozdeistvii

IOP Conf. Series: Materials Science and Engineering **934** (2020) 012029 doi:10.1088/1757-899X/934/1/012029

Parametric and topology optimization of load bearing elements of aircraft fuselage structure

Tun Lin Htet and P V Prosuntsov

Bauman Moscow State Technical University (BMSTU), 105005, Moscow, Russia

E-mail: tunlinhtet64509@gmail.com

Abstract. This paper presents the weight reduction of load bearing elements of aircraft DA-62 fuselage structure using parametric and topology optimizations methods. The structural analysis of rear part of aircraft fuselage in various load calculation conditions is performed. In the first step using parametric optimization of structure, allow to reduce its mass to 15.84 %. In addition using topology optimization of optimized structure by parametric modeling added mass reduction to 32.87% of total structural mass.

1. Introduction

Reducing the weight of the structure is one of the priorities of the development of modern aviation technology, as it allows increasing the economic efficiency of aircraft. To achieve this goal in the design of aircraft are increasingly introduced composites, superior in many parameters, especially specific strength and stiffness, traditional metal materials. For example, the Boeing 787 fuselage is built in five main sections consisting of composite material that account for 50% of the aircraft's total structural weight.

A perspective approach to reducing the weight of the power structure is the use of optimization methods. Optimization is a process of selecting or converging onto a final solution amongst a number of possible options, such that a certain requirement or a set of requirements is best satisfied i.e. a design in which some quantifiable property is minimized or maximized (e.g., strength, weight, strength-to-weight ratio). In recent years, traditional approaches based on parametric optimization have been added by topological methods [1–3]. Topology optimization is a mathematical method that optimizes material layout within a given design space, for a given set of loads, boundary conditions and constraints with the goal of maximizing the performance of the system.

The purpose of topological optimization is to determine the optimal material distribution in the design of structure under specified loads, and meet the optimization criteria, in other words, finding the optimal placement of material in design, so that the target function have a maximum or minimum value of boundary conditions. In topology optimization, the desired result in a material distribution function in the design area [4]. In this study, the ANSYS Topology Optimization [5, 6] module based on SIMP method is used.

2. Objects and materials

The object of our work is rear part of DA-62 fuselage structure that consists of fuselage skin and load bearing elements: four oval shaped ribs (figure 1). The aircraft is available in two weight versions. The European version has five seats and maximum takeoff weight of 1,999 kg, the US version has seven seats and a 2,300 kg [7]. The later version is used as a model object of our work. The general characteristics of DA 62 are shown in table 1. The fuselage skin and its load bearing elements ribs are

IOP Conf. Series: Materials Science and Engineering **934** (2020) 012029 doi:10.1088/1757-899X/934/1/012029

made of carbon fiber-reinforced plastic (CRFP) based on carbon fabric WL-Blatt 8.3520.80 that made it possible to simultaneously provide strength and light weight structure of aircraft.

Table 1. Performance characteristics of DA 62 [7]

Performance	Index (SI)
Maximum speed (14,000 ft.)	353 km/hr
Cruise speed at 75% (12,000 ft.)	317 km/hr
Maximum stall speed (V_s)	191 km/hr
Maximum rate of climb (MSL)	5.2 m/s
Empty Weight	1590 kg
Useful Load	710 kg
Maximum take-off mass	2300 kg

Physical-mechanical characteristics of CFRP are shown in the table 2.

Table 2. Physical-mechanical properties of CFRP

Parameter	Unit	Values
Density	kg/m^3	1480
Young's Modulus, E_1	GPa	91.82
Young's Modulus, E_2	GPa	91.82
Young's Modulus, E_3	GPa	9.00
Shear Modulus, G_{12}	GPa	19.500
Tensile strength	MPa	829
Compressive strength	MPa	439
Shear Strength	MPa	120

It is necessary to determine the loads acting on the rear part of fuselage structure for the design optimization process. Firstly, 3D model (figure 1) of the structure is modelled in NX software. After that simulation calculation is performed in ANSYS module Fluent to determine the aerodynamic loads acting on the aircraft in flight condition.

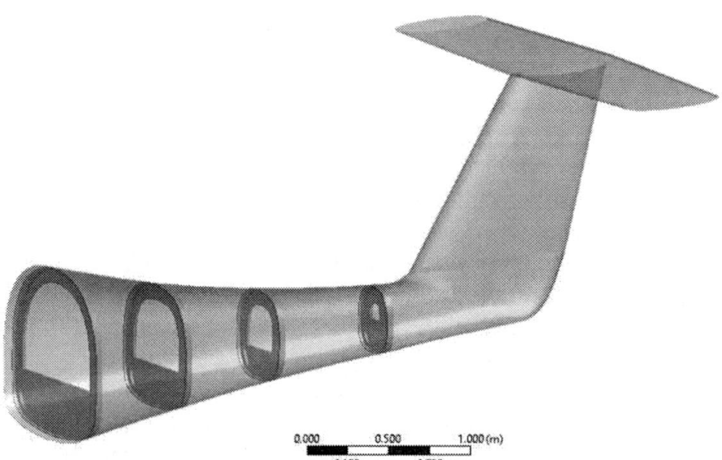

Figure 1. 3D model of rear part of fuselage structure

3. Optimization of load bearing elements of fuselage structure

3.1 Calculation load conditions

In accordance with Aviation Rules of Russian standards [8], the calculation of the static strength of aircraft is considered upon the external loads acting on the aircraft and its individual units in various calculation cases. In the description of those calculation conditions show the aircraft position, characteristic of flight, failure load factor and safety factor, direction and distribution of loads. Basic calculation conditions according to Aviation Rules are classified as A, A', B, C, D and D' (figure 2).

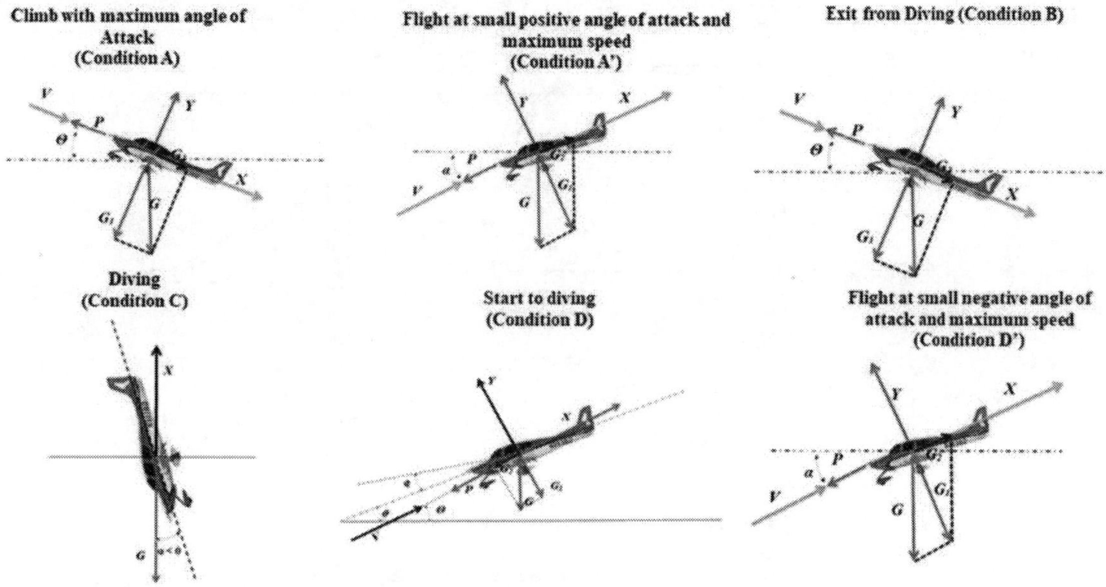

Figure 2. Loads acting on the aircraft in various load cases

3.2 Finding aerodynamic loads acting on the aircraft

For the simulation calculation processes, the 3D geometry of aircraft is modelled using FEM software SolidWorks [9] and it is imported to ANSYS Fluent. For the airflow simulation process the mesh model of more than 8 million elements and about 2 million nodes was created. The mesh should also include a boundary layer to solve lamina flow equations. For the model of our analysis the amount of layers in the boundary is set to 5 and the total thickness is 50 mm. Since the aerodynamic load acting on the aircraft will be the maximum when performing maneuvers at sea level, the ambient pressure was assumed to be equal to p = 101325 Pa and the air density ρ= 1,225 kg/m3 [10–15].

As a result of the simulation, the pressure distributions on the surface of the aircraft were determined for all six calculated cases (table 3).

Table 3. Maximum total pressure acting on the aircraft in various load conditions

Conditions	A	B	C	D	A'	D'
Pressure, Pa	103219	107602	10700	105625	107747	107244

3.3. Parametric optimization of fuselage structure

Simulation of the stress-strain analysis of the fuselage structure was carried out in the ANSYS Static Structural. It was considered that both aerodynamic and mass-inertial loads are subjected on the structure. At the same time, the change in the direction and magnitude of the load factor (f) was also considered in calculation process. As a results, the stresses and displacements in ribs of the rear part of fuselage structure were determined.

At the first stage of the study, the question of choosing rational values of geometric dimensions of the profile of frames was solved based on the parametric optimization method (figure 3). Firstly, three variants of the width (b) of ribs: 50, 100 and 150 mm were compared and the variant in which the stresses in ribs set a minimum value was chosen for further analysis. The obtained values of stresses (σ, MPa), displacements (δ, mm) and the total mass of frames (m, kg) are given in the table 4.

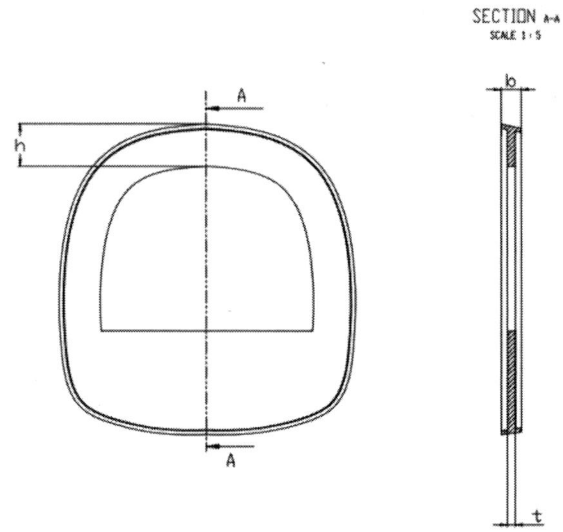

Figure 3. Geometric parameters of oval-shaped rib

Table 4. Structural modeling results of load-bearing elements of fuselage structure with different width

I b – 50 mm			II b - 100 mm (Base model)			III b - 150 mm		
m, kg	δ_{max}, mm	σ_{max}, MPa	m, kg	δ_{max}, mm	σ_{max}, MPa	m, kg	δ_{max}, mm	σ_{max}, MPa
34.80	2.09	62.38	41.35	1.083	89.15	47.87	1.813	68.787

Next, the problem of optimizing the thickness (t) of the rib was solved, and variants with a thickness of 10 mm to 25 mm were analyzed (table 5).

Table 5. Structural modeling results of load-bearing elements of fuselage structure with different thickness

I t - 10 mm			II t - 15 mm			III t - 20 mm			IV t - 25 mm		
m, kg	δ_{max}, mm	σ_{max}, MPa	m, kg	δ_{max}, mm	σ_{max}, MPa	m, kg	δ_{max}, mm	σ_{max}, MPa	m, kg	δ_{max}, mm	σ_{max}, MPa
20.7	3.455	79.9	2.77	27.7	64.5	34.8	2.09	62.38	41.9	1.57	76.6

The second step analysis results show that the variant with the thickness of 20 mm has the minimum stress value and it is selected for the third step. In the third step, we varied the height of ribs (h): 60 mm, 80 mm, 10 mm and 120 mm respectively (table 6). As a result, it was found that the rational set of geometric parameters of frames is: width (b) -50 mm, thickness (t) -20 mm and height (h) – 100 mm.

In total, the parametric optimization of the size of the ribs of rear part of fuselage was able to reduce its mass by 15.84%, from 41.35 kg to 34.8 kg

Table 6. Structural modeling results of load-bearing elements of fuselage structure with different thickness

I h – 60 mm			II h - 80 mm			III h - 100 mm			IV h - 120 mm		
m, kg	δ_{max}, mm	σ_{max}, MPa	m, kg	δ_{max}, mm	σ_{max}, MPa	m, kg	δ_{max}, mm	σ_{max}, MPa	m, kg	δ_{max}, mm	σ_{max}, MPa

29.9	2.12	68.42	32.5	2.12	68.07	34.8	2.09	62.38	37.1	2.05	68.59

3.4 Topology optimization in ANSYS

At the second stage, the topology optimization of the shape of the ribs was carried out, and the result of parametric optimization was used as the starting point. In this paper, the total weight of ribs of the fuselage structure is chosen as the target function, and the maximum allowable stress in the structure is set as a limit. For the topology optimization algorithm, determined the design area, the topology of which can be changed and the area for which such changes are not allowed [6]. Here, the entire volume of the ribs was used as a design area, except for the outer surfaces of the ribs, which are needed to join with the skin. As a result of topology optimization, a new variant of load bearing elements of rear part of the aircraft was obtained (figure 4).

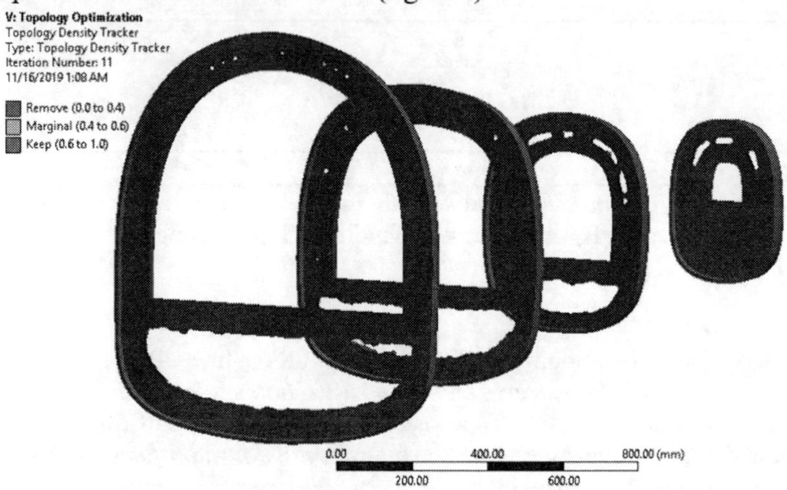

Figure 4. Geometrical model of load bearing element of fuselage structure, obtained by using topology optimization

After topology optimization process we had to perform a validation step, in which we rerun our static simulation with the loads and constraints, which are mentioned above. So, this output model is imported to Siemens NX in order to modify to get smooth faces and optimized shapes and then transfer it to structural analysis (figure 5).

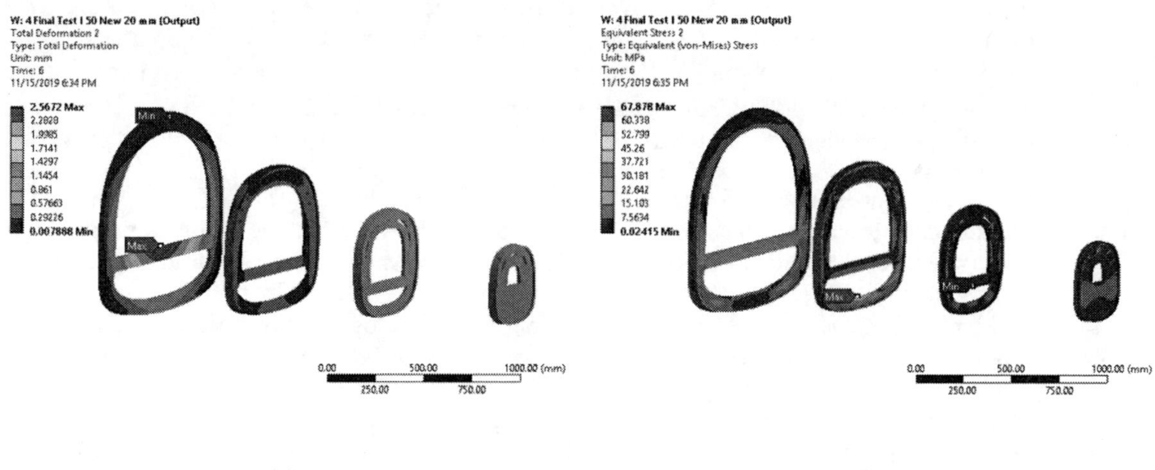

(a) (b)

Figure 5. Results of structural analysis of validation step: a) deformation, mm, b) equivalent stress, MPa

According to the results of validation step, the values of maximum deformation and von-Mises stress obtained by the structural strength modeling of rear part fuselage structures with new design ribs are not far different from the results of initial design model modeling results.

4. Results and Conclusion

Thus, a new version of the design of the load bearing elements of fuselage structure is proposed and its total mass is 32.87 % less than the original version for the same loading conditions (table 7).

Table 7. Comparison of mass of initial and new design of structure

Parts	Mass of initial design model kg	Mass of new design model, kg	Mass reducing, %
1st rib	17.98	11.483	36.13
2nd rib	11.75	7.7039	34.13
3rd rib	6.307	4.8584	29.90
4th rib	4.876	3.7124	20.80
Total	41.3483	27.7577	32.87

As a result of a complex of computational and theoretical studies, a new original design of load bearing elements of DA62 aircraft was proposed, which has higher weight characteristics compared to the original version.

Reference

[1] Reznik S V 2017 Development of Satellite Carbon-fiber Reinforced Plastics Antennas *Innovation in materials for extreme environments (Huddersfiedl, UK)* p 23

[2] Vinoth Raj A and Raghu T 2015 Topology design optimization of ribs, spars, wing attachmetn bulkhead and nose landings gear bracket *Internatinal Journal of research in aeronautical and mechanical engineering* pp 16–35

[3] Abramova E N and Reznik S V 2019 Small Spacecraft's Inflatable Aerodynamic Decelerator Design Issues Analysis *AIP Conference Proceedings* **2171, 04002** DOI: https://doi.org/10.1063/1.5133188

[4] Bashin K A, Torsunov R A and Semenov S V 2017 Topology optimization methods in aerospace industry PNRPU *Aerospace Engineering Bulletin* No.51 pp 39-51

[5] ansys.com [Internet]. Topology Optimization, [updated 2018]. Available from: https://www.ansys.com/products/structures/topology-optimization

[6] Baranovski S V 2019 Topological Optimization of the Basic Elements of the Wing from Polymeric Composite Materials *XLV Gagarin Science Conference 2019* MAI pp 28

[7] diamondaircraft.com [Internet] Diamond DA-62 [updated 2018] Available from: https://www.diamondaircraft.com/en/private-pilots/aircraft/da62/overview/

[8] Airworthiness standards of civil light aircraft 2014 *Interstate Aviation Committee (IAC)* p 207 (in Russian)

[9] Introducing SolidWorks 2015 (Dassault Systemes SE Company USA) p128

[10] Ashikhmina E R and Prosuntsov P V 2020 Coupled CFD-based Shape Optimization of Wing of Reusable Space Vehicle of Tourist Class *IOP Conference Series: Materials and Science and Engineering* ***709 (2020)*** DOI: 10.1088/1757-899X/709/2/022108

[11] Ashikhmina E R and Prosuntsov P V 2019 Wing Box Topology Optimization of Tourist Class Reusable Space Vehicle *AIP Conference Proceedings* ***2171, 030014*** DOI: https://doi.org/10.1063/1.5133180

[12] Ashikhmina E R and Prosuntsov P V 2019 Wing Airfoil Selection and Optimization for the Tourist Class Reusable Space Vehicle *AIP Conference Proceedings* ***2135, 020005*** DOI: https://doi.org/10.1063/1.5120642

[13] Abramova E N and Reznik S V 2019 Small Spacecraft's Inflatable Aerodynamic Decelerator Design Issues Analysis *AIP Conference Proceedings* ***2171, 04002*** DOI:

https://doi.org/10.1063/1.5133188
[14] Tun Lin Htet 2020 Structurl Analysis and Topology Design Optimization of Load Bearing Elements of Aricraft Fuselage Structure *IOP Conference Series: Material and Science and Engineering* ***709 (2020)*** DOI: 10.1088/1757-899X/709/4/044113
[15] Mikhailovskiy K V and Baranovski S V 2018 Determining Aerodynamic Loads Affecting and Aircraft Wing During Parametric Modelling Taking the Main Airliner Components into Account *Hearld of the Bauman Moscow State Technical University* ***(5) 2018*** *No(5)* pp 15 – 28 DOI: 10.18698/0236-3941-2018-5-15-28 (in Russian)

Development and implementation of a technology for introducing reinforcement into aluminum matrix composite

Yijin Chen, Yu A Kurganova and S P Scherbakov

Department of Engineering Technology, Bauman Moscow State Technical University, 5/1, ul. Baumanskaya 2-ya, Moscow, 105005, The Russian Federation

E-mail: chenyijin@yandex.ru

Abstract. In this study, a new method for introducing reinforcement was developed and used in stir casting to fabricate aluminum matrix composites. Different reinforcements (microparticle SiC and nanofiber Al2O3) were incorporated into the technical aluminum AD0 (1050 ISO) alunimum alloy by a mechanical stirrer and then cylindrical specimens were cast. The microstructural characterization of the composite samples showed grain refinement of aluminum matrix with the introduction of reinforcement. The effects of reinforcement on the mechanical properties and impact behavior of the composites were investigated. Based on experiments, it was revealed that the presence of Al2O3 nanofiber reinforcement led to improvement in hardness. This was due to introduction of reinforcement and grain refinement of aluminum matrix. Studies on the fracture surfaces revealed that AMC with reinforcement is brittle, but in the case of nanofibers fillers, tough component was also present, which eliminated the risk of brittle fracture.

1. Introduction

Aluminum matrix composites (AMCs) offer high specific strength, high specific stiffness and hardness, good wear resistance, low thermal expansion coefficient. AMCs are successfully used in the aerospace, automotive, thermal management and bio-medical industries [1-7]. Normally the size of reinforcements generally ranges from a few nanometers to several hundred micrometers. Recently it is of interest to use nano-sized reinforcements to improve the properties of AMCs.

Various of methods for producing AMCs reinforced with nano-sized Al_2O_3 have been developed, including powder metallurgy [8, 9], spark plasma sintering [10], in-situ routes [11], stir casting [20]. For mechanical alloying, it normally involves mechanical mixing of Al powders and nano-Al_2O_3 for fabrication of bulk. Although the powder metallurgy techniques are more effective than the casting methods in terms of reinforcement distribution, uniform mixing of nano-sized reinforcements is lengthy, expensive, and energy consuming [12]. Furthermore, solid techniques seriously suffer from limitations in size and complexity of the components.

Melt processing which involves the stirring of nano-Al_2O_3 into Al melts, has some important advantages such as better matrix–reinforcement bonding, easier control of matrix structure, simplicity, low cost of processing, and nearer net shape and the wide selection of materials [12,13]. However, it is extremely difficult for the mechanical stirring method to distribute and disperse nano-sized reinforcements uniformly in Al melts due to high viscosity, low wettability in Al melts and large surface-to-volume ratio of nano-sized reinforcements. Furthermore, when the nano-Al_2O_3 is wetted in the aluminum melt, it will tend to sink to the melt due to the density difference between Al_2O_3 and aluminum. So nano-Al_2O_3 is hard to introduce into aluminum melt and easily agglomerated and clustered.

Regarding the mentioned challenges, introducing nano-Al_2O_3 and achieving to uniform dispersion is a principal purpose in production process of these composites.

The method combined with ball milling and stir casting is developed to fabricating Al-nano Al_2O_3 metal matrix composites. Different metallic particles are mixed with nano-Al_2O_3 as the reinforcements. Mazahery A. et al. [12] used the mixture of nano-Al_2O_3 particles (50 nm) and aluminium particles (16μm) as the reinforcements. Akbari et al. separately milled nano-Al_2O_3 particles with aluminum and copper powders and incorporated the mixture into A356 alloy using vortex method [13-16]. Copper acts as strengthening factor and, also affected particle dispersion [17].

In this research, different fillers were introduced into the Al melt, i.e., untreated SiC_p, milled particles Cu, and milled nanofiber Al_2O_3 - Cu composite powder.

2. Experimental procedure

2.1. Materials

In the current research, technical aluminum AD0 (1050 ISO) was chosen as the matrix alloy. The chemical composition of AD0 according to GOST 4784-97 is shown in table 1. Table 2 shows the properties of technical aluminum AD0.

Table 1. Chemical composition of technical aluminum AD0 (wt.%).

Composition	Al	Fe	Si	Zn	Cu	Mg	Ti	Mn
wt.%	>99.5	<0.4	<0.25	<0.07	<0.05	<0.05	<0.05	<0.02

Table 2. Properties of cast technical aluminum AD0.

Material	Density (g/cm³)	Brinell Hardness (MPa)	Young's modulus (GPa)	Tensile strength (MPa)
AD0	2.71	200	71	60

Discontinuous nanosized alumina fibers with an average diameter of 10-20 nm (ANF Technology Ltd, Estonia, table 3) and micro copper powders with average size of 20 μm were used as components of reinforcement powders. The powders were mixed in the mass ratio of $Cu/Al_2O_3 = 1$. Mixed powders were milled in a mortar for 20 min.

Table 3. Properties of Al_2O_3 nanofibers.

Material	Density (g/cm³)	Thermal resistance (°C)	Young's modulus (GPa)	Tensile strength (GPa)
Al_2O_3-nanofibers	3.98	1200	400	12

For comparative analysis, unreinforced alunium АД0 and composite reinforced with micro powders Cu, composite reinforced with microparticle SiC (17μm) were fabricated (table 4).

Table 4. Composition of the experimental samples.

Specimens	Composition of the samples (wt.%)	
	Matrix	Reinforcement
1	AD0	-
2	AD0	1%SiC (17μm)
3	AD0	1%Cu (20μm)
4	AD0	1%Cu (20μm)+1%Al_2O_3 nanofibers

2.2. Fabrication of composites

The composites were produced by stir casting method. Technical aluminum AD0 was charged into the ceramic crucible and heated up to 800°C (above the alloy liquidus temperature) for melting. The graphite stirrer fixed on the mandrel of the drilling machine was introduced into the melt. Reinforcements were introduced into the melt at a temperature of 800°C. Pure argon was used as the carrier gas for injection of the reinforcements. After completion of the injection, the slurry was stirred at a rate of 600 rpm for 10 min. The composite slurry was poured into cast iron mold and shaped in the form of cylinder of 20 mm outer diameter and height of 80 mm. Figure 1 shows the schematic of the experimental set-up used in the production of the composites and the as-cast specimens.

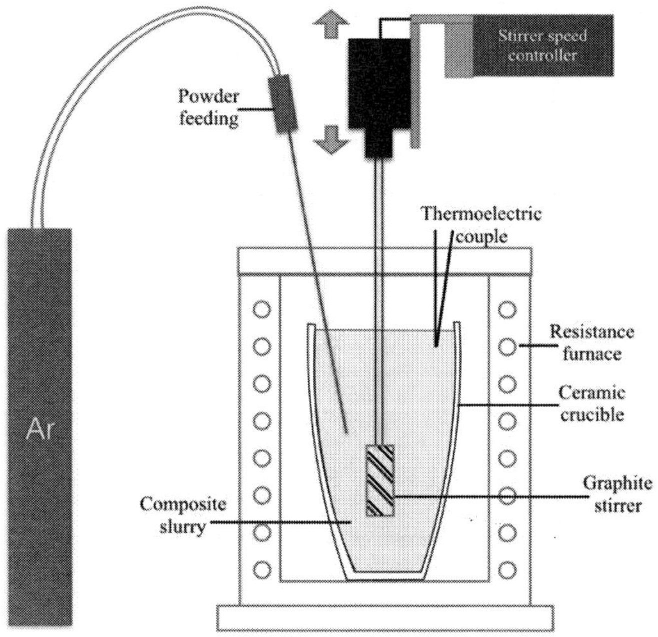

a b

Figure 1. (a) Schematic of the experimental set-up used in production of the composites. (b) As-cast specimens.

2.3. Composite characterizations

The morphology of the Cu powder, nanofiber Al_2O_3 and the Cu-nanofiber composite powders were observed using a optical microscope (Olympus GX51).

To study the microstructure, samples were cut and prepared by grinding through 800 and 1200 grit papers followed by polishing with 9,3,1μm diamond paste, and etched with Keller's reagent (2 ml HF (48%), 3 ml HCl (conc.), 5 ml HNO_3 (conc.) and 190 ml water). Microscopic examinations of the composites and matrix alloy were characterized by optical microscope (Olympus GX51).

To study the microhardness, the Vickers microhardness values of the samples were measured on the polished samples using an DURASCAN 70 microhardness tester with a load of 100 g. For each material, 100 points on an area of 100 mm^2 were tested in order to study the frequency distribution of microhardness values.

Charpy impact test specimens pre-crack notches were machined from each casting according to GOST 9454-78. Fractographic examinations were also carried out on the fracture surfaces of the broken specimens by means of Tescan Vega II LMH scanning electron microscope (SEM).

3. Results and discussion

3.1. Powder characterization

Figure. 2 shows the morphology of as-received Cu powders and nanofibers Al_2O_3, milled Cu powders and milled nanofiber Al_2O_3–Cu composite powders. After milling, most of the Cu powders become smaller. Because of the Van der Waals force nanofibers Al_2O_3 are easily agglomerated (figure. 2b). In addition, these aggregates may have entrapped air, which could lead to difficulty in their incorporation

IOP Conf. Series: Materials Science and Engineering **934** (2020) 012030 doi:10.1088/1757-899X/934/1/012030

into liquid Al melt, as well as result in the incorporation of gas bubbles within the microstructure of the composite as the aggregates are drawn beneath the surface of the melt [10]. Also, these nanofibers Al_2O_3 have poor wettability with melt [18,21]. Figure. 2d reveals that nanofibers Al_2O_3 have covered the surface of Cu powders. In addition, due to the milling effect, the nanofibers are not agglomerated and the distribution is more uniform. Furthermore, milling remove air on alumina nanofibers, free oxygen on the surface of alumina nanofibers reacts with Cu, leading to the formation of CuO/Cu_2O. When composite powders are incorporated into the molten alloy, CuO/Cu_2O will react with Al melt, leading to the formation of new Al_2O_3.

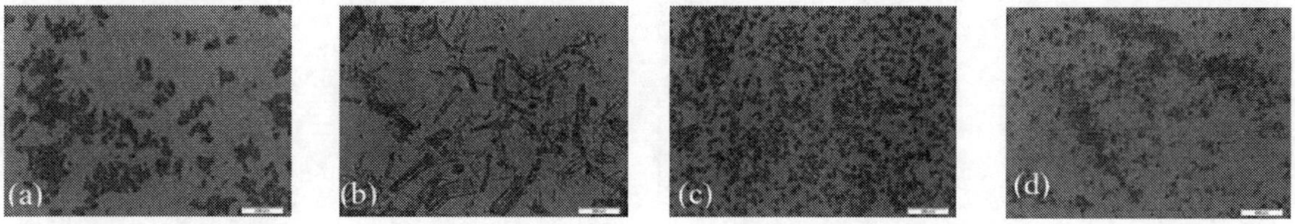

Figure 2. Image of as-received Cu powders (a), nanofibers Al_2O_3(b), milled Cu powders (c) and milled nanofiber Al_2O_3–Cu composite powders (d).100x

3.2. Comparative analysis of the microstructure

The microstructural analysis was performed on the technical aluminum AD0, AD0+1%SiC, AD0+1%Cu, AD0+1%Cu+1%Al_2O_3. As shown in figure 3, after the introduction of SiC microparticles or Cu, the structure is refined. When Al_2O_3 nanofibers are added, decreasing is more efficient.

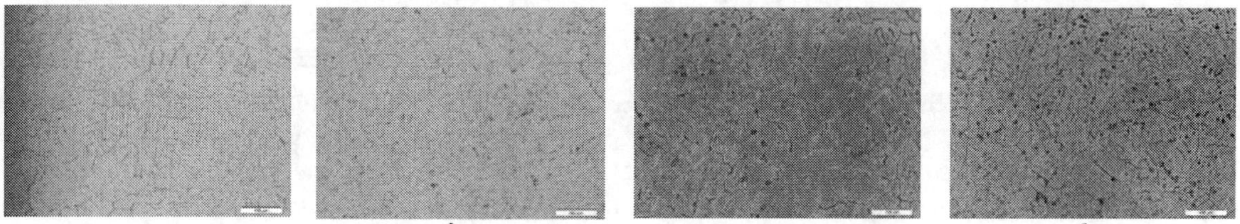

a b c d

Figure 3. Microstructure of experimental samples:
(a)AD0; (b)AD0+1%SiC; (c)AD0+1%Cu; (d)AD0+1%Cu+1%Al_2O_3 (200x)

3.3. Microhardness

Figure 4 displays the frequency distribution of microhardness values of the specimens. In figure 4 (a) the frequency distribution of microhardness values of technical aluminum AD0 is approximately uniform and symmetric. After introducing 1% micro-sized SiC particles (Figure 4 (b)), the hardness is increased but the frequency distribution symmetry is destroyed, which is because composite materials contain a soft matrix and hard reinforcement phase. When introducing 1% Cu, the hardness distribution is wider (Figure 4 (c)). After introducing 1% Cu (20μm) + 1% Al_2O_3 nanofibers into the Al melt (Figure 4 (d)), the hardness is increased 18% . The frequency distribution obviously consists of two parts: soft matrix part and reinforced part. The dispersion of Al_2O_3 nanofibers enhances the hardness, as Al_2O_3 is harder than aluminum matrix. And the higher hardness of the composites could be attributed to the fact the Al_2O_3 nanofibers act as obstacles to the motion of dislocation. The hardness increment can also be attributed to reduced grain size.

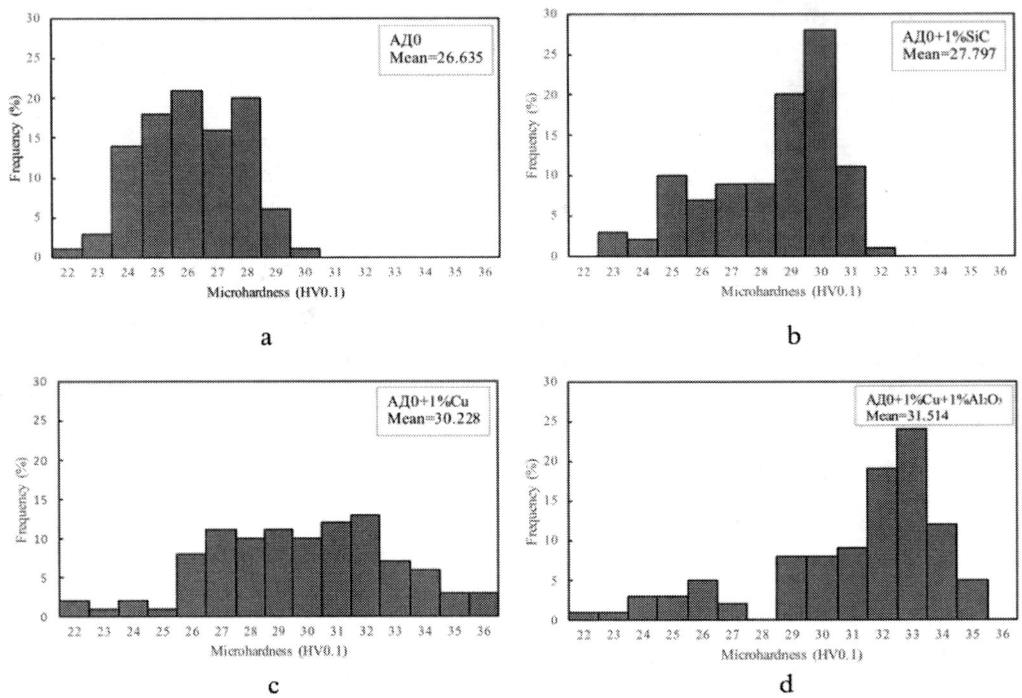

Figure 4. Histogram of distribution of Vickers microhardness values:(a)AD0; (b)AD0+1%SiC; (c)AD0+1%Cu; (d)AD0+1%Cu+1%Al_2O_3

3.4. Impact behavior

As depicted in table 5, the absorbed impact energy of aluminum matrix composites with 1% SiC, 1% Cu and 1%Cu+1%Al_2O_3 is found to be lower compared to technical aluminum AD0. The reduction in impact energy can be explained by the fact that a stress concentration may exist near the reinforcements. In addition, the agglomeration of the reinforcements is easily separated under impact loading. Similar findings have been observed in other studies of aluminum matrix composite [19]. Compared with the sample of AD0+1%Cu, the absorbed impact energy of aluminum matrix composites with Al_2O_3 is increased. The introduction of Al2O3 fibers has a positive effect on impact behavior.

The fracture surface morphologies of the samples were shown in figure 5. Elongated dimples were visible on the fracture surface of the sample of matrix composition AD0. A rough surface indicates that the matrix alloy exhibits serious plastic deformation before fracture. The presence of hard reinforcements SiC_{17} and Al_2O_3 embrittle the material, which is expressed by the nature of the development of the crack. The AD0 + 1% Cu sample has a relatively smooth fracture surface, and there are a large number of tiny dimples on the surface. Compared to the sample, after the introduction of Al_2O_3, the nature of the crack propagation changes, but the size of a single dimple increases, which indicates the presence of a toughness component of the fracture. The results of fractographic studies are consistent with the data of mechanical properties under impact.

Table 5. Absorbed impact energy of the experimental samples.

Materials	Absorbed impact energy (J)
AD0	67.4±5.9
AD0+1%SiC	34.9±9
AD0+1%Cu	22.5±2.2
AD0+1%Cu+1%Al_2O_3	31.3±1.1

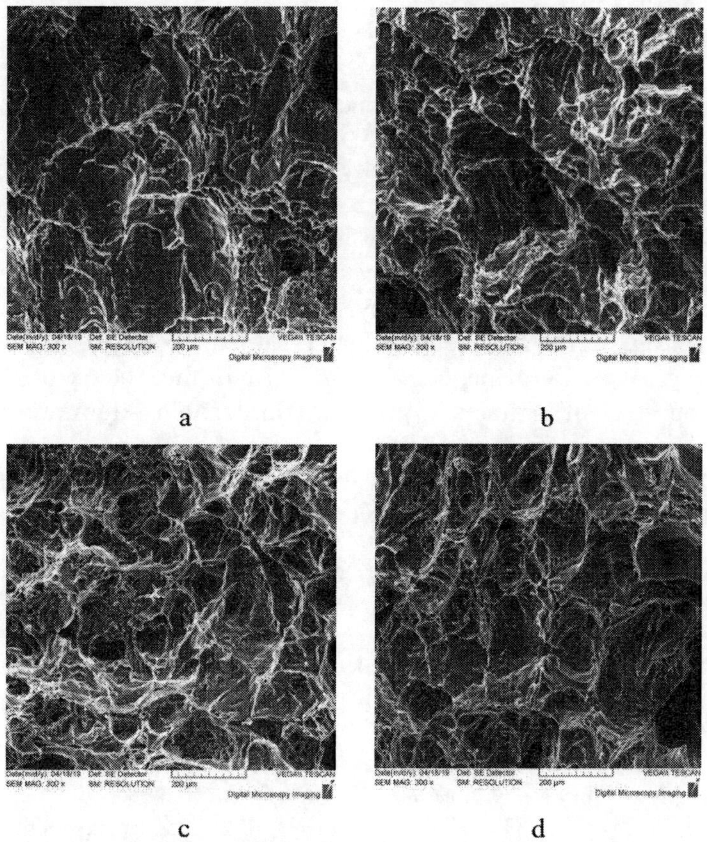

Figure 5. Fracture surface morphologies of (a)AD0, (b)AD0+1%SiC, (c)AD0+1%Cu, (d)AD0+1%Cu+1%Al_2O_3 (300x)

4. Conclusion

In general, Al_2O_3 nanofibers are incorporated into aluminum matrix with the help of micro-Cu particles mixing and under the processing conditions of stir casting used in this study.

Based on the present study, the following conclusions can be drawn:

(1) As a result of the work, samples of a composite material with alumina fibers with a diameter of 10-20 nm are obtained.

(2) In the course of the study, the effect of introduction of reinforcements on structure refinement is established.

(3) The general patterns of changes in hardness with a change in component composition are established.

(4) Analysis of impact behavior show that AMC with hard reinforcement is brittle, but in the case of nanofibers fillers, tough component is also present, which eliminates the risk of brittle fracture.

Acknowledgments

The authors would like to thank Mr. S.D. Karpukhin for the use of scanning electron microscope and useful discussion. The lead author Yijin Chen gratefully appreciates the financial support of China Scholarship Council (CSC).

References

[1] Miracle D B 2005 Metal matrix composites–from science to technological significance. Composites science and technology, *Compos. Sci. Technol.* **65** 2526-2540

[2] Kurganova Y A, Chernyshova T A, Kobeleva L I, and Kurganov S V 2011 Service properties of aluminum□-matrix precipitation-hardened composite materials and the prospects of their use on the modern structural material market *Russian Metallurgy (Metally)* **7** 663-666

[3] Chen Itszin 2017 Metal-matrix composite materials. Prospect of effective application and production (review) *Metal technology* **10** 25-30

[4] Tjong S C 2013 Recent progress in the development and properties of novel metal matrix nanocomposites reinforced with carbon nanotubes and graphene nanosheets *Mater Sci. Eng. R. Rep* **74** 281–350

[5] Artyukh, V. A., Yusupov, V. S., Zelensky, V. A., Kholin, M. S., & Fakhurtdinov, R. S. (2017). Structural characteristics of a mechanically alloyed Al–Fe powder composite. Inorganic Materials: Applied Research, 8(3), 459-463.

[6] Puchkov, Y. A., Yanlun, W., Polyanskii, V. M., Yanwei, W., Zhiqian, W., & Eliseev, Z. A. (2010). Study of alloy V91 of the system Al–Zn–Mg–Cu system supercooled solid solution decomposition. Metal Science and Heat Treatment, 52(7-8), 362-367.

[7] Puchkov, Y. A., & Berezina, S. L. (2014). Use of the Theory of Hardening Factor for Predicting the Properties of Articles from Heat-Hardenable Aluminum Alloy V91. Metal Science and Heat Treatment, 56(3-4), 131-136.

[8] Prabhu B, Suryanarayana C, An L and Vaidyanathan R 2006 Synthesis and characterization of high volume fraction Al–Al_2O_3 nanocomposite powders by high-energy milling *Mater. Sci. Eng. A Struct. Mater.* **425** 192–200

[9] Mobasherpour I, Tofigh A A, Ebrahimi M 2013 Effect of nano-size Al_2O_3 reinforcement on the mechanical behavior of synthesis 7075 aluminum alloy composites by mechanical alloying *Mater. Chem. Phys.* **138** 535-541

[10] Babu N K, Kallip K, Leparoux M, Alogab K A, Maeder X and Dasilva Y A R 2016 Influence of microstructure and strengthening mechanism of AlMg5–Al_2O_3 nanocomposites prepared via spark plasma sintering *Mater. Des.* **95** 534-544

[11] Sreekumar V M, Hari Babu N and Eskin D G 2017 Prospects of in-situ α-Al_2O_3 as an inoculant in aluminum: a feasibility study *J. Mater. Eng. Perform.* **26** 4166-76

[12] Mazaherya A, Abdizadeha H and Baharvandi H R 2009 Development of high-performance A356/nano-Al_2O_3 composites *Mater. Sci. Eng. A Struct. Mater.* **518** 61–64

[13] Akbari M K, Baharvandi H R and Mirzaee O 2013 Nano-sized aluminum oxide reinforced commercial casting A356 alloy matrix: Evaluation of hardness, wear resistance and compressive strength focusing on particle distribution in aluminum matrix *Compos. B Eng.* **52** 262–268

[14] Akbari M K, Baharvandi H R and Mirzaee O 2013 Fabrication of nano-sized Al_2O_3 reinforced casting aluminum composite focusing on preparation process of reinforcement powders and evaluation of its properties *Compos. B Eng.* **52** 426–432

[15] Akbari M K, Mirzaee O and Baharvandi H R 2013 Fabrication and study on mechanical properties and fracture behavior of nanometric Al_2O_3 particle-reinforced A356 composites focusing on the parameters of vortex method *Mater. Des.* **46** 199–205

[16] Tahamtan S, Halvaee A, Emamy M and Zabihi M S 2013 Fabrication of Al/A206–Al_2O_3 nano/micro composite by combining ball milling and stir casting technology *Mater. Des.* **49** 347–359

[17] Chen Y, Qin H, Kurganova Yu A and Gaaze V K 2018 A method for introduction of Al_2O_3 nanofiber into aluminum alloy *IOP Conf. Ser. Mater. Sci. Eng.* **347** 012050

[18] Schultz B F, Ferguson J B and Rohatgi P K 2011 Microstructure and hardness of Al_2O_3 nanoparticle reinforced Al–Mg composites fabricated by reactive wetting and stir mixing *Mater. Sci. Eng. A Struct. Mater.* **530** 87-97

[19] Ozden S, Ekici R and Nair F 2007 Investigation of impact behaviour of aluminium based SiC particle reinforced metal–matrix composites. *Compos Part A Appl Sci Manuf.* **38** 484–494

[20] Kurganova Y A and Scherbakov S P 2017 Influence of a discrete additive of aluminum oxide on structure and properties of aluminium alloy *J. Min. Inst.* **228** 717–721

[21] Leon C A, Mendoza-Suarez G and Drew R A L 2006 Wettability and spreading kinetics of molten aluminum on copper-coated ceramics *J. Mater. Sci.* **41** 5081–87

A study of physical and mechanical characteristics of polymer composite materials by ultrasonic technique

D I Chulkov[1], A B Terekhin[1] and A M Dumansky[2, 3]

[1]ORPE Technologiya named after A.G. Romashin", Obninsk, Russia
[2]Mechanical Engineering Research Institute of the RAS, Moscow, Russia
[3]BMSTU named after N.E. Bauman, Moscow, Russia

E-mail: DIChulkov@gmail.com

Abstract. This work is concerned with the determining of physical and mechanical characteristics (dynamic elastic modulus along the direction of reinforcement, porosity, density, percentage of binder, strength) of fiberglass plastics by ultrasonic technique. The techniques, which are currently used for evaluation of the above-mentioned parameters of polymeric composite materials, are presented. The ultrasonic technique of physical and mechanical characteristics determining of glass fiber reinforced plastics for radio engineering purpose on the base of quarzitic, glass and siliceous fillers is proposed.

1. Introduction

Polymer composite materials (PCM) are of great commercial and technological interest due to their special properties: high specific strength, stiffness and elastic modulus; good damping ability and corrosion resistance; high fatigue life and low coefficient of thermal expansion [1-4].
Of the entire variety of PCMs, fiberglass plastics — PCMs based on inorganic oriented fabric quartz, glass, and silica fillers — should be distinguished. Fiberglass is one of the most common PCMs, combining high strength, low density, good dielectric properties and reasonable price [5-7].
To ensure the high quality of fiberglass products, it is necessary to use new solutions in the development and creation of methods and means of non-destructive testing (NDT). This will eliminate the flaw at the early stages of manufacturing products, control the correctness of the technology parameters, and evaluate the reliability, manufacturability [8, 9].
An important area of NDT quality of materials, parts, products and structures made of PCM is the determination of physical and mechanical characteristics, which allows us to switch from spot defining these properties on specially manufactured samples for their control on final products without their destruction or damage. This increases the reliability of product quality assessment and reduces material costs.

2. Techniques of non-destructive testing of physical and mechanical characteristics of PCM

The determination of the physical and mechanical characteristics of PCMs using NDT techniques is a difficult task, which is due to the wide variety of PCMs, specific design features of them and manufacturing technology, a wide range of physical and mechanical characteristics, and a variety of types of defects that arise during the manufacturing process [10].
The main techniques of non-destructive testing physical and mechanical characteristics of PCM [11]: dielectric constant; on thermal effects; electrical resistance; acoustic methods.

Acoustic techniques have been widely used to determine the following physical and mechanical properties of PCMs: elastic properties (elastic modulus, shear modulus, Poisson's ratio), strength (tensile, compression, bending, torsion, shear, etc.), technological (density, humidity, the content of individual components, etc.), structural (anisotropy of the material), the content of inclusions and other properties [12].

In standard [13], non-destructive testing of PCMs was carried out using ultrasonic waves by acousto-ultrasonic (AU) technique according to the correlation dependence of the SWF criterion (stress wave factor) on the diagnostic parameter. The following main variants of the SWF criterion can be selected: amplitude, when the criterion is defined as the maximum of the amplitude signal; countable criterion based on the number of the signal exceeds the threshold value; energy criterion determined by the energy of the signal (the amplitude integral over time or the integral of the signal spectrum in the frequency range of interest). The disadvantage of this method is that it does not allow us to determine the strength characteristics, but only makes it possible to identify structural heterogeneities.

In [12, 14], to determine the strength characteristics of PCM, it was proposed to use the integral of the optimal signal spectrum as the SWF criterion:

$$\sigma=\varphi(\int_{fmin}^{fmax}[S(f)]\frac{1}{n}\sum_{i=n}^{n}[S_{i1}(f)-S_{i2}(f)]\,df), \qquad (1)$$

where σ, φ are strength and the sign of correlation consequently; $S(f), S_{i1}(f)$ and $S_{i2}(f)$ are spectra of pulses of ultrasonic vibrations transmitted in the PCM control object, witnesses specimens after manufacture and subjected to damaging influences; n – number of witness specimens; i is serial number of measurements; fmin and fmax are the range of spectral boundary frequency. The technique allows determining only the strength characteristics of carbon plastics.

In [15–18], techniques for determining the physical and mechanical characteristics of PCMs from the correlation dependence of the velocity, amplitude, or attenuation of ultrasonic waves on the diagnostic parameter with a through or echo-pulse monitoring scheme are presented. The disadvantage of this technique is the low accuracy of determining of PCMs characteristics in the direction of reinforcement. Determination of physical and mechanical characteristics with the aid of acoustic resonance and the method of acoustic emission described in [19-22]. The disadvantages of these techniques are the need of loading the structure, the complexity of the equipment and the low reliability of the control.

Each of the above listed techniques has its own limitations, making impossible to use it for NDT determining of physical and mechanical characteristics of different-thickness parts of ogival form made from fiberglass with woven angle-ply lay-up and high porosity.

To solve this problem, it was proposed to use a comprehensive control technique, which consists in applying two techniques for measuring the parameters of ultrasonic waves: along the direction of reinforcement and perpendicular to the PCM layers. Measurement of the parameters of ultrasonic wave's perpendicular to the layers makes it possible to increase the control sensitivity due to the possibility of using a higher signal frequency and allows local control (region with a diameter of 5 mm) of the material structural characteristics, which seems impossible to control by the standard method [13]. It is also possible to increase the accuracy of determining the parameters in the direction of reinforcement, by means of increasing the information content of the SWF criterion, which can be used to calculate not only the parameters of the ultrasonic signal obtained by the AU method, but also the parameters of ultrasonic waves propagating perpendicular to the PCM layers at higher frequencies. Figure 1 shows the schemes of the control techniques proposed for use.

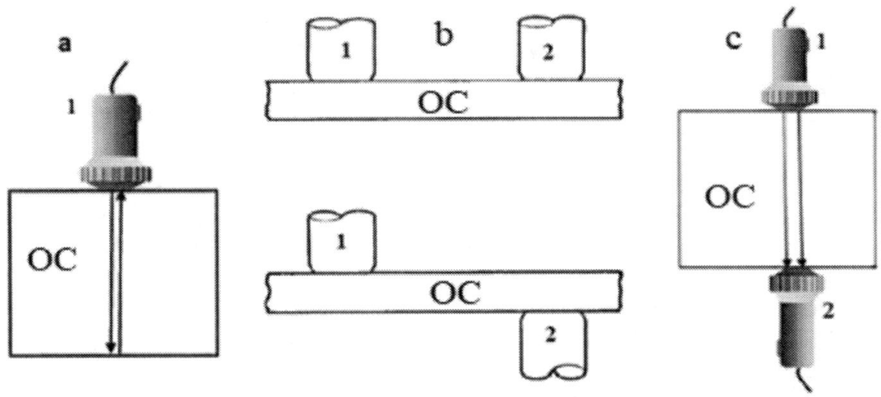

Figure 1. Measurement schemes of acoustic signal parameters; a – echo method, b –AU method, c — passage method; 1, 2 - source and receiver of ultrasonic waves.

As the receiver and source of ultrasonic vibrations, direct contact piezoelectric transducers are used. Using the echo technique and the transmission technique, the speed of ultrasonic wave's propagation (C) and the signal amplitude (A) are determined perpendicular to the PCM layers. Moreover, control is carried out in the frequency range from 2 to 10 MHz. At higher frequencies, the sensitivity of the technique increases, but at the same time, the attenuation of the ultrasonic signal in the PCM increases. Therefore, the exact value of the working frequency is determined experimentally depending on the wall thickness, product shape, ultrasonic wave attenuation, acoustic contact and other factors.

There is a correlation between the physical and mechanical characteristics of PCM and the parameters of ultrasonic vibrations propagating in it, in particular, ultrasonic waves interact with pores: they scatter, bend around them, etc. With an increase in the number of pores, these interactions occur more and more, which leads to more attenuation and decrease of the speed of ultrasonic waves. Thanks to such relationships, the physical and mechanical characteristics of PCM can be determined from the previously constructed correlation dependencies of the diagnostic parameters on the parameters of the ultrasonic waves.

When controlled with the help of acoustic-ultrasonic technique, the velocity and amplitude of the ultrasonic waves are determined along the direction of reinforcement. The control is determined at frequencies from 0.2 to 1.5 MHz, the operating frequency is obtained experimentally for a specific fiberglass. Using the AU method, it is possible to determine the direction of layer reinforcement, the elastic and strength properties of PCM from the previously constructed correlation dependencies of the diagnostic parameter on the calculated SWF criterion. To obtain a multi-parameter correlation, one can calculate the SWF criterion, which contains information about the parameters of acoustic signals measured by various techniques. A promising direction is the construction of correlation dependencies of the PCM characteristics on the SWF criterion calculated from the spectral components of acoustic signals.

The proposed comprehensive ultrasonic technique for determining the physical and mechanical characteristics of the fiberglass can be outlined by the following steps:

1. Control of discontinuity (non-glued, delamination, cracks, etc.) by NDT techniques (impedance, thermovision, ultrasonic shadowing, computer radiation tomography, etc.);
2. Measurement of the parameters of ultrasonic waves in the direction perpendicular to the layers by the passage technique or echo one using direct probes at a frequency of 5-10 MHz;
3. Determination of the structural characteristics of PCMs according to the correlation dependencies on the SWF criterion;
4. Measurement of the ultrasonic waves parameters along the direction of reinforcement by the acousto-ultrasonics technique (if necessary, on a different sounding base) using a probe at a frequency of 0.2-1.5 MHz;

5. Determination of physical and mechanical characteristics of PCMs in the direction of reinforcement from preliminary constructed correlation dependencies of the control parameter of the SWF criterion. For each specific type of PCM, it is necessary to determine the most informative SWF criterion, taking into account the structural features of the material.

The proposed techniques are implemented using commercially available equipment and can be used for evaluation and determination of the elastic, strength and structural parameters of PCM parts.

3. The results of experimental work

Using the AU examination technique, the direction of reinforcement can be determined. For experimental investigation of the proposed technique, two batches of 6 samples of 120×120×1.5 mm in size were made, which are the PCMs based on a phenol-formaldehyde binder reinforced with six layers of TS 8/3 K-TO quartz fabric. The first batch was unidirectional, the second one [0; +33; -33; +63; -63; 0] lay-up.

Using an ultrasonic flaw detector and broadband ultrasonic probes in the specimens, the velocities (C) and amplitudes (A) of ultrasonic waves were measured at a frequency of 500 kHz on the base of 100 mm at different angles (αo) to the direction of the reinforcement. The results obtained are presented in Figure 2. From the presented data, it is seen, that for all specimens the maximum of ultrasonic waves velocity are observed when αo corresponds to the direction of reinforcement (Figure 2, b, d). This can be explained that maximum of modulus of elasticity in the warp direction.

On the Figure 2 (d) it is seen that despite the change in the average velocity caused by different density and porosity of the specimens, the dependence of the velocity of ultrasonic waves in the direction of reinforcement is clearly visible. According to the data in Figure 2 (a, c), one can conclude that using the signal amplitude as an informative parameter can be applied only for unidirectional layering.

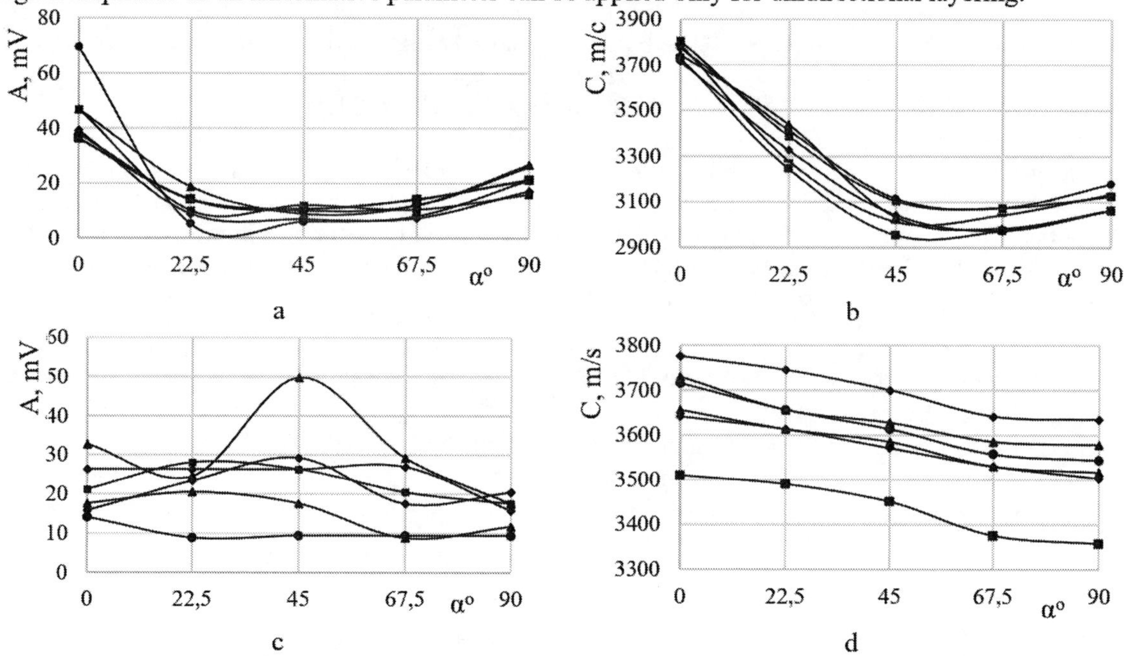

Figure 2. Velocity and amplitude of the ultrasonic signals; a, b - in specimens with unidirectional calculation; c, d - [0; +33; -33; +63; -63; 0] lay-up.

The density (ρ), porosity (ε), and binder content (γ) were determined on specimens of 20×10×1.5 mm size consisting of six layers of TS 8/3 K-TO fabric with [0; +33; -33; +63; -63; 0] lay-up with varying density at the manufacturing stage. The velocity of the ultrasonic waves were measured by the passage technique using a probe at a frequency of 5 MHz. The obtained dependences of the density, porosity

IOP Conf. Series: Materials Science and Engineering **934** (2020) 012031 doi:10.1088/1757-899X/934/1/012031

and content of the binder samples on the measured velocities of ultrasonic waves are presented in figure 3.

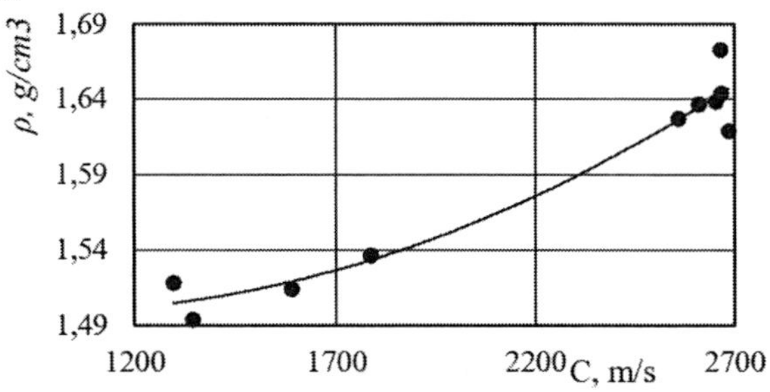

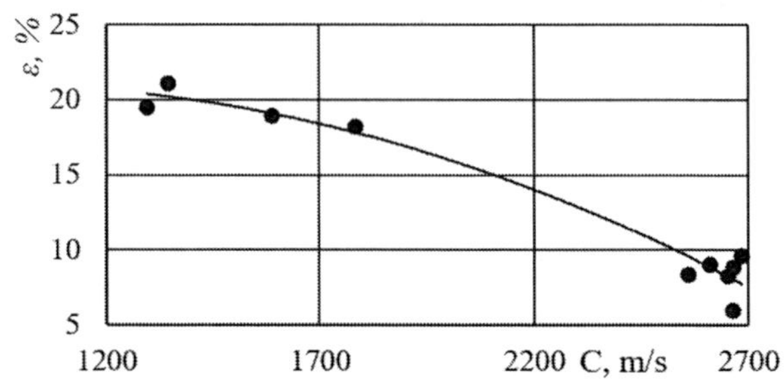

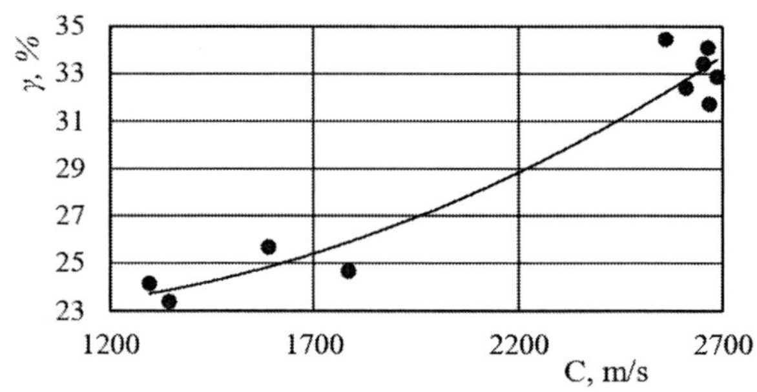

Figure 3. Correlation dependence of the structural characteristics of PCM samples on the speed of ultrasonic waves;

The resulting correlation equations:

$$\rho = 0{,}00000005C^2 - 0{,}00009250C + 1{,}54227852$$
$$R^2 = 0{,}95$$
$$\upsilon = -0{,}000004C^2 + 0{,}007974C + 17{,}229608 \qquad (2)$$
$$R^2 = 0{,}96$$
$$\gamma = 0{,}000003C^2 - 0{,}004693C + 24{,}828420$$
$$R^2 = 0{,}94$$

The determination of the elastic modulus in the direction of reinforcement was carried out on specimens 250×25×1.5 mm size having six layers of fabric TS 8/3 K-TO with a unidirectional lay-up. The ultrasonic wave velocity were measured with the aid of acoustic-ultrasonic technique on 120 mm base and frequency 1 MHz in the direction of the warp direction. The calculation of the dynamic modulus of elasticity was defined according to the relation [20]:

$$E_D = \frac{\rho C^2(1+\upsilon)(1-2\upsilon)}{(1-\upsilon)}, \qquad (3)$$

where υ - Poisson's ratio equal to 0.11 in the warp direction.

After measuring the velocities of ultrasonic waves in the specimens, the tensile modulus ET was determined according to GOST 11262-2017 (ISO 527-2: 2012). The results is presented in Figure 4.

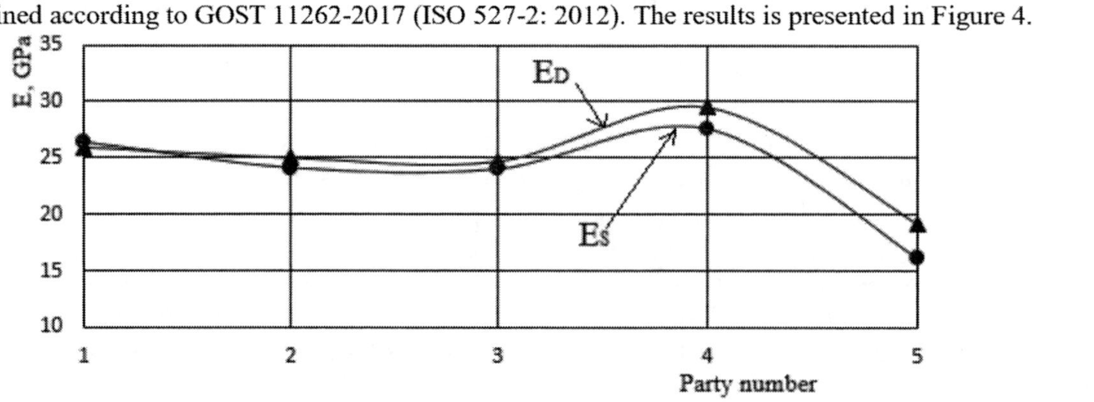

Figure 4. Average values of elastic moduli in lots (lot size - 10 pcs).

The Figure 4 shows that the difference between the elastic moduli in the direction of reinforcement determined in tensile testing and the non-destructive ultrasonic technique does not exceed 15%, which indicates the possibility of using the proposed technique for assessing the elastic characteristics of PCMs structures.

4. Conclusions

To find the direction of PCMs reinforcement acoustic-ultrasonic technique was used. It has been experimentally confirmed the possibility to determine the direction of fiberglass reinforcement with varying degree of impregnation.

A technique for determining the structural characteristics of PCM (density, porosity, binder content) by an ultrasonic passage technique with determination coefficient from 0.94 to 0.96 was proposed and tested.

A technique for determining the elastic modulus of unidirectional PCMs in the direction of reinforcement based on the theory of acoustoelasticity was experimentally confirmed. The error in the elastic modulus determining did not exceed 15%.

To increase the accuracy of determining the elastic and strength characteristics in the direction of reinforcement in angle-ply lay-up fiberglass plastics, it is recommended to use the correlation dependencies of PCMs characteristics on the SWF criterion, which are calculated from the parameters of the acoustic signals that passed through the control object in different directions.

References

[1] Bogolyubov V S et al. 2006 *Technology for the production of products from composite materials, plastics, glass and ceramics* (Engineering: encyclopedia. 40 t. Section III. Technology for the production of machines. T. III-6; edited by V .S. Bogolyubova. M.: Mechanical Engineering) p.576

[2] Kondrat'yev A V and Kovalenko V A 2011 *Review and analysis of world trends and problems of expanding the use of polymer composite materials in rocket and space technology units Design and production of aircraft structures.* **3 (67)** FROM. 7-18

[3] Ashby M and Jones D 2010 *Engineering Materials 1. An Introduction to Properties, Applications and Design* (Elsevier Butterworth-Heinemann) p.472

[4] Gurtovnik IG et al. 2003 *Radiolucent fiberglass products* (edited by V. I. Sokolov. - M.: Mir) p.368

[5] Shayakhmetdinov U Sh 1999 *Composite materials based on silicon nitride and phosphate binders* (M .: SP interment engineering) p.128

[6] Golovkin G S 1997 *Reinforced plastics: Ref. allowance* (M .: MAI) p.404

[7] Lubin J. 1988 *Handbook of CM*: 2 vols. (T. 1 - M.: Mechanical engineering) p.467

[8] Kolganov VI 2001 *Ultrasonic non-contact method and software and hardware for automated non-destructive quality control of products from polymer composite materials* (dissertation for the degree of Doctor of Technical Sciences) p.249

[9] Aleshin N P, Grigoriev M V and Schipakov N A 2015 *Modern equipment and technologies for non-destructive testing of PCM.* Engineering Bulletin **01**, January 2015

[10] Ermolov I N and Lange Yu. 2006 *Non-destructive testing: Handbook of 8 volumes* (under the general editorship of V.V. Klyuyev. T. 3. Ultrasonic testing, 2nd ed., Moscow: Engineering) p.864

[11] Potapov A I and Pekker FP 1977 *Non-destructive testing of structures made of composite materials* (L., "Mechanical Engineering") p.192

[12] Generalov A S 2015 *Determination of the strength properties of carbon fiber reinforced plastics through an ultrasonic reverberation-through method* (thesis for the degree of candidate of technical sciences) p.158

IOP Conf. Series: Materials Science and Engineering **934** (2020) 012031 doi:10.1088/1757-899X/934/1/012031

[13] ASTM E 1495 - 02 Standard Guide for Acousto-Ultrasonic Assessment of Composites, Laminates, and Bonded Joints
[14] Patent RU 2461820 C1 Method for determining the strength characteristics of polymer composite materials
[15] Patent EP 0322446 Bond strength measurement of composite panel products
[16] USSR patent 77708 A method for studying bodies by elastic or electromagnetic waves
[17] USSR patent 808930 a method for determining the physical and mechanical characteristics of a material
[18] Gorbunov A I 1968 *Control of bonding strength by ultrasonic echo method.* Flaw detection **2** p.42-50
[19] Schliekelmann R I 1975 Non-destructive testing of bonded joints Nondestructive testing, April, v. 8, No. 2, pp. 100-103
[20] Berezin A V and Kozinkina A I 1993 Analysis of damage accumulation in layered composites by acoustic emission taking into account the stress-strain state of Sat. XIII All-Russian ScientificTechnical Center: Non-Destructive Physical Methods and Means of Control, St. Petersburg,p.113
[21] Aleshin N P, Belyi V E and Vopilkin A X 1989 *Methods of acoustic control of metals* (M.:Mechanical engineering) p.456
[22] Kholodov S S et al. 2016 Acoustic emission control of fiberglass damage. All materials. Encyclopedic reference book **5** pp. 7 - 12

IOP Conf. Series: Materials Science and Engineering **934** (2020) 012032 doi:10.1088/1757-899X/934/1/012032

Calculation and experiment approach to predicting the properties of C/SiC and SiC/SiC composites

D V Sapronov[1,2], M A Mezentsev[2] and D S Palchikov[2]

[1]Bauman MSTU
[2]Central Institute of Aviation Motors

Abstract. One of the current trends in aviation and space technology is higher flight speed. An increase in flight speed means an increase in the temperature of the aircraft components and tougher requirements for the physical and mechanical characteristics of the materials used. As metals are reaching their limits, heat-resistant non-metallic materials are attracting more and more attention. Of all heat-resistant non-metallic materials, the most appropriate for this purpose are monolithic ceramic materials, carbon-carbon and ceramic-matrix composite materials. This paper proposes an approach to predicting the mechanical characteristics of ceramic composite materials based on experiment and numerical simulation of the material microstructure. Multi-stage impregnation and pyrolysis (PIP) method was used to manufacture C/SiC and SiC/SiC ceramic samples, which were later subjected to tensile testing and 4-point bend testing. The results of numerical modeling and the results of experiments were compared, and the mechanical properties of the materials were determined.

1. Introduction.

Ceramic composite materials (CCM) based on carbon fiber reinforcement and silicon carbide matrices, or C/SiC composites possess a unique combination of high thermal and mechanical properties and low density. C/SiC composite structures can retain their working efficiency at up to 2700°C. However, the lifespan of such structures is limited by the oxidation of carbon fibers at temperatures above 450 °C and the oxidation of SiC matrix at temperatures above 1000-1900 °C, depending on the properties of the environment. C/SiC composites have found wide application in rocket and space technology: thermal protection elements and liquid rocket engines [1, 2].

NASA was planning to use C/SiC composite in the nose cone of the X-38 reusable spacecraft project (Figure 1 [3]). C/SiC composite is also used in the advanced liquid-propellant rocket engines: Vulcain main engine nozzle (Ariane 5 launch vehicle), EAM space platform correction engine nozzle and combustion chamber [4] (Figure 2). SNECMA used C/SiC composite in the nozzle flaps of the M88-2 gas turbine engine for Rafale fighter jet. As part of the SIMPLEX Turbo pump Blisk program, NASA is developing a disk wheel from C/SiC composite for a turbo-pumping unit of a liquid rocket engine [5]. When reinforcing the ceramic matrix with carbon fibers, both the fiber oxidation and the carbide formation along the boundaries of the fibers must be prevented. Therefore, for structures that will operate in the Earth's atmosphere for a prolonged period of time, the CCMs reinforced with coreless SiC fibers appear the most promising. However, the strength of the modern SiC fibers starts declining at temperatures above 1250°C [6-15].

NASA is conducting CCM research under the NASA Fundamental Aeronautics Program [16]. General Electric Company is developing SiC/SiC turbine blades as part of the Adaptive Engine Technology Development program. The combustion chamber demonstrator from SiC/SiC for the CFM56 engine by

Snecma and Herakles has been tested successfully. Pratt & Whitney's thrust vectoring nozzle flaps have successfully completed over 11,000 test cycles /5,000 hours.

Fig. 1. C/SiC composites in the X-38 spacecraft

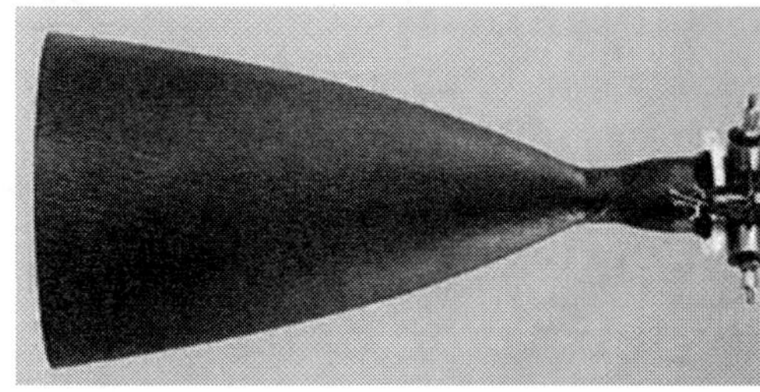

Fig. 2. C/SiC composites in the EAM engine

2. C/SiC and SiC/SiC manufacturing technology.

The main methods for producing silicon carbide matrix for reinforced ceramic composite materials are gas vapor deposition (CVI), silicon melt impregnation (LSI) and polymer solution impregnation followed by multi-stage pyrolysis (PIP).

Each of the methods has a different effect on the properties and cost of the final composite. LSI causes the greatest degradation of the physical mechanical properties of the material, and CVI means the highest labor intensity. PIP allows obtaining silicon carbide matrix without expensive equipment, and there is no significant degradation of properties, which makes this method the most popular. However, it should be noted that in the case of PIP, the porosity of the material is the highest and decreases only with an increase in the impregnation and pyrolysis stages.

The main components used in this research were a polycarbosilane precursor (a carbide structure after pyrolysis), 2D carbon fabric with equal strength, fabric, and 3rd-generation coreless SiC fibers of the Hi-Nicalon Type S (OX) type. Mechanical properties of the components are in Table 1.

Table 1. Mechanical properties of the components of ceramic composites.

Characteristic	Ceramic composite component		
	SiC matrix	C fabric	SiC fiber
Density, g/sm^3	1.9	1.81	2.85
Young's modulus, GPa	7/7*	230	200*
Poison's ratio	0.2	0.2	0.22
Ultimate stress, MPa	75-100	5490	2300
Fiber content, %	–	43	30
Filament diameter, μm	–	5	10
Warp/ weft cross-section, mm	–	2 x 0.15	–

* – characteristic at 1000°C

3. Calculation and experiment investigation of the strength of C/SiC and SiC/SiC samples.

An approach is proposed for determining the mechanical properties of ceramic composites based on the comparison of experimental results and numerical simulations. First, the strain in characteristic zones must be determined experimentally. Then a finite element calculation of the loads must be conducted for samples with mechanical properties obtained by separate modeling of the material microstructure. After comparing the results, it is necessary to adjust the data in the material microstructure model.

3.1 Strength analysis of C/SiC samples. For experimental studies, we used two samples, with the working area of 250 mm x 4 mm x 14 mm. Fiberglass pads were glued to the either side of each sample to ensure grip during testing (Figure 3). The samples were made from polycarbosilane and carbon fibers (2D isotropic fabric) using PIP technology.

The tests were carried out using the Instron 5985 as per the GOST 25.601-80 standard. The sample was clamped into wedge-shaped hydraulic grips and stretched until it fractured. The longitudinal strain in the center of the working area of the sample was measured using the Epsilon extensometer (Figure 4).

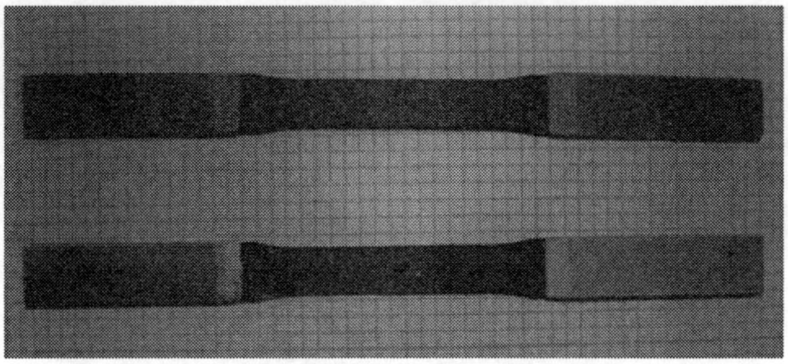

Fig. 3. C/SiC samples

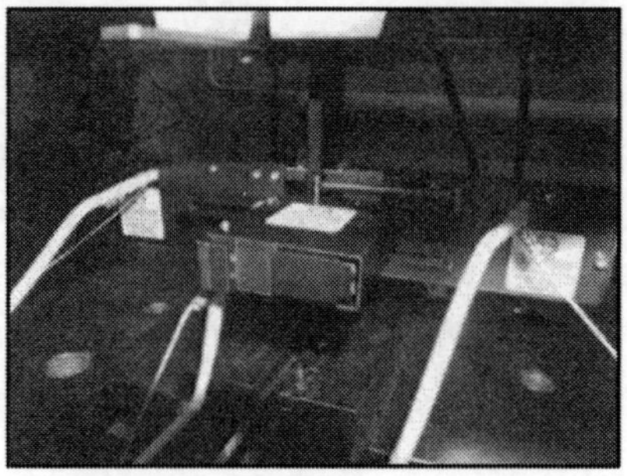

Fig. 4. Longitudinal strain test

The tests demonstrated the linear increase in the load up to the initial failure of the samples. The fracture point for both samples corresponded to the stress level in the section of the working area of ~ 60 MPa and longitudinal strain of 0.11%. The fracture area is the junction of the working area and the grip (Figure 5).

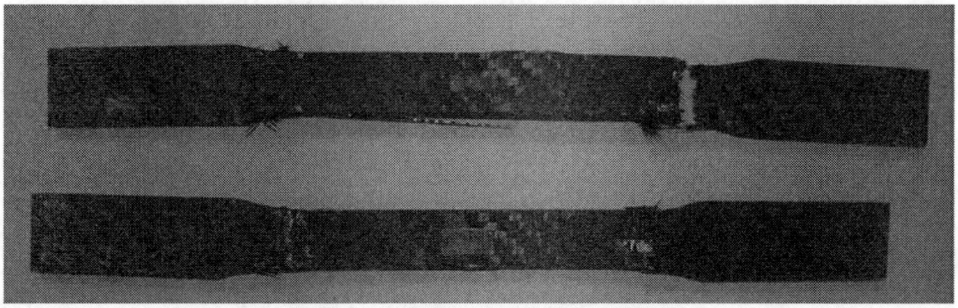

Fig. 5. C/SiC samples after tests

To determine the mechanical properties of the C/SiC composite, a 2D woven structure model was created in the Digimat software package (Figure 6). The input data were the component characteristics (Table 1). The properties of the obtained material are in Table 2. A finite element model of the sample was created in ANSYS and progressive loading was simulated numerically. The boundary conditions in the model are shown in Figure 7: in the lower end part, the sample is fixed in the X and Y directions, a force distributed between the nodes of the top line was applied to the upper end part. Elements simulating a multilayer composite were used.

IOP Conf. Series: Materials Science and Engineering **934** (2020) 012032 doi:10.1088/1757-899X/934/1/012032

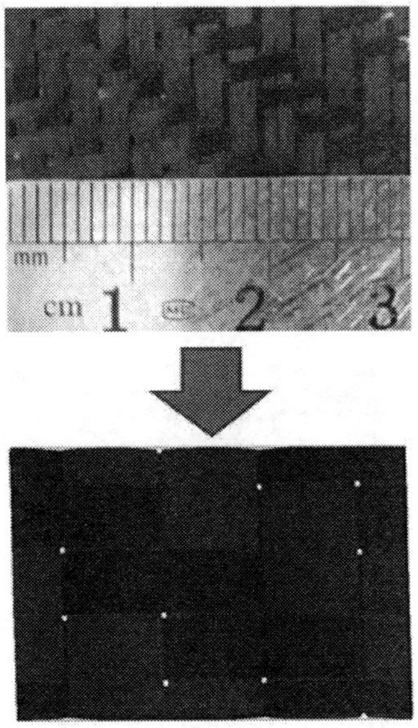

Fig. 6. C/SiC microstructure modelling

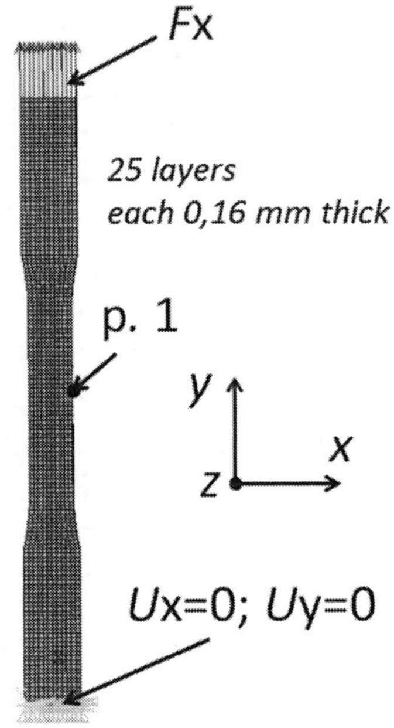

Fig. 7. C/SiC microstructure modelling

The comparison of the numerical and experimental results of longitudinal strain at point 1 (Figure 7) is shown in Figure 8. The following points are marked: A - the moment of initial fracture and B -- the moment of final fracture. Up to A, the deformation is close to elastic. In section A-B, progressive destruction of the sample occurs, but this section is negligible. The stress distribution at the moment of the initial fracture (point A, Figure 8) in one layer is shown in Figure 9.

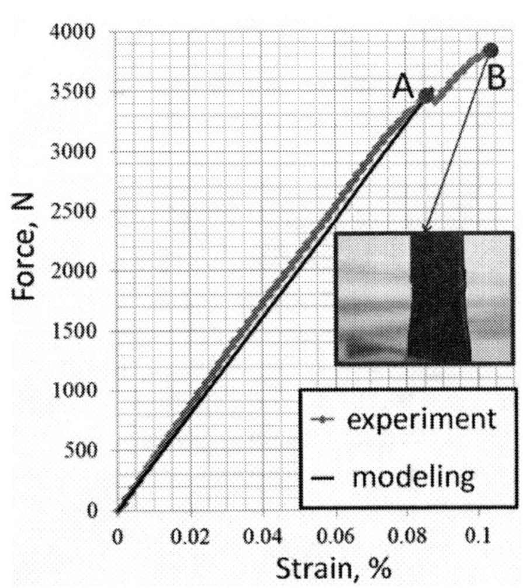

Fig. 8. Comparison of experiment and numerical results for C/SiC

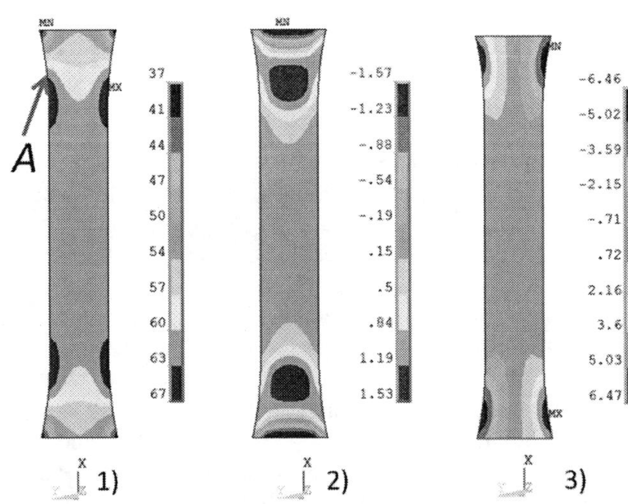

Fig. 9. Stress distribution at the moment of fracture for C/SiC (1 – longitudinal; 2 – lateral; 3 –in plane), MPa

There is good agreement between the results of finite element modeling and experiment up to the initial fracture moment for the C/SiC composite. The maximum stress zone along the X axis (A, Figure 9) qualitatively coincides with the fracture zone of the sample in the experiment.

3.2 Strength analysis of SiC/SiC samples. Three test pieces with dimensions of 4.00 x 2.5 x 48 mm were used for the experiment (Figure 10). These samples were manufactured using PIP technology. The manufacturing process involved four stages of polymer impregnation and pyrolysis.

Fig. 10. A – SIC fiber Hi-Nicalon Type,
B – SiC/SiC samples

Four-point bend test for SiC / SiC samples was conducted at 1000° C on the Instron 5985 testing machine with a EuroTherm high-temperature furnace. The lower support span was L1 = 40 mm, and the upper support span was L2 = 20 mm. Analysis of the loading diagrams showed that after a certain threshold value, a load drop was observed due to the fiber failure, with the load instantly transferred to other fibers. The testing terminated automatically when the load fell by 50% of the maximum. Thus, the failure is quasiplastic. Upon visual inspection of the tested samples, no damage was observed.
The mean value of the ultimate flexural strength from four-point bend test was ~ 157 MPa and the ultimate strain was ~ 0.22%.
To determine the mechanical characteristics of the SiC/SiC samples, a 3D structure model was created in the Digimat software package (Figure 11). The input data were the component characteristics (Table 1). Properties of the resulting material are in Table 2.
A finite element model of the sample was created in ANSYS to simulate progressive loading. The boundary conditions are shown in Figure 12: at the point of contact with the lower supports, the sample was fixed in the Y direction, the central part of the sample in the YZ plane was fixed along the X axis, and in the YX plane, along the Z axis; the force was applied at the point of contact with the upper supports. 3D elements with specified properties in the local coordinate system were used.
The comparison of the numerical and experimental results of longitudinal strain at point 1 (Figure 12) is shown in Figure 13. The following points are marked: A - the moment of initial fracture and B -- the moment of 50% decrease in the load. Up to A, the deformation is close to elastic. In section A-B, progressive fracture is clearly visible.
There is good agreement between the results of finite element modeling and experiment up to the initial fracture moment of the SiC/SiC composite.

Fig. 11. SiC/SiC microstructure modelling

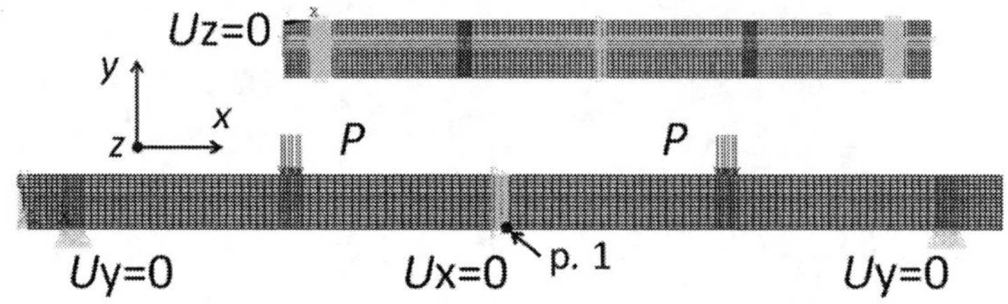

Fig. 12. Numerical modelling of 4-point bend test, SiC/SiC

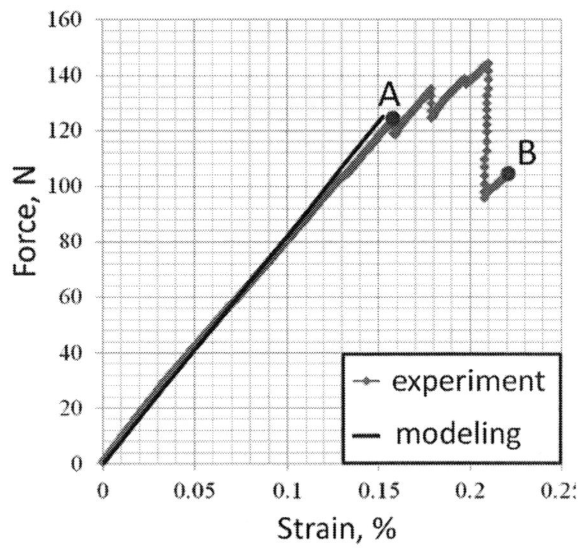

Fig. 13. Comparison of experiment and numerical results for SiC/SiC

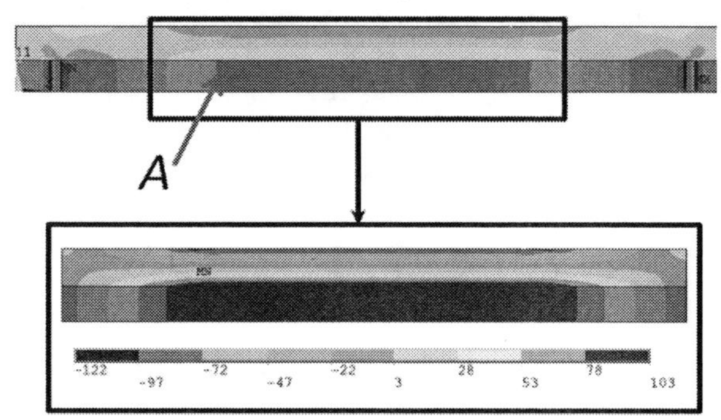

Fig. 14. Stress distribution along longitudinal axis of SiC/SiC sample at the moment of load drop, MPa

Table 2. Obtained properties (1 – C/SiC, 2 – SiC/SiC).

№	E_1, GPa	E_2, GPa	E_3, GPa	ν_{12}	ν_{21}	ν_{13}	ν_{31}	ν_{23}	ν_{32}	G_{12}, GPa	G_{13}, GPa	G_{23}, GPa	ρ, g/sm³
1	71,5	71,5	23,1	0,06	0,06	0,25	0,08	0,25	0,08	10,7	9,8	9,8	1,85
2	64,9	11,5	11,93	0,21	0,02	0,21	0,02	0,25	0,25	5,1	5,1	4,6	2,1

4. Conclusion.

The C/SiC and SiC/SiC composite samples were fabricated using PIP technology. Ultimate tensile strain for C/SiC at room temperature was ~0,11%. Ultimate flexural strain from the four-point bend test for

IOP Conf. Series: Materials Science and Engineering **934** (2020) 012032 doi:10.1088/1757-899X/934/1/012032

SiC/SiC at 1000°C was ~ 0.22%. It was demonstrated that up to the initial failure moment, the C/SiC and SiC/SiC are subjected to linear strain, and the progressive fracture is characteristic only of SiC/ SiC. Numerical modelling of the microstructure and loading compared to the strain gauge readings allowed obtaining the mechanical properties of the ceramic composites. With the proposed models, the sample strain can be predicted with a high degree of accuracy up to the initial failure moment.

References

[1] Narottam P Bansal and Jacques Lamon 2015 *Ceramic Matrix Composites: Materials, Modeling and Technology* (New Jersey: John Wiley&Sons) p 695

[2] Narottam P Bansal *Handbook of Ceramic Composites* 2015 (Boston: Kluwer Academic Publishers) p 554

[3] Hua-Tay Lin and Mrityunjay Singh 2011 *Advanced Ceramics, Materials and Structures* (Westerville: American Ceramic Society) p 853

[4] Walter Krenkel 2008 *Ceramic Matrix Composites: Fiber Reinforced Ceramics and their Applications* (Weinheim: WILEY-VCH Verlag GmbH&Co) p 421

[5] Min J B, Harris D L and Ting J M 2011 Advances in Ceramic Matrix Composite Blade Damping Characteristics for Aerospace Turbomachinery Applications *Proc. Int. Conf. 52nd AIAA/ASME/ASCE/AHS/ASC Structures, Structural Dynamics and Materials (Electronic Materials)* (ASME pres) AIAA 2011-1784

[6] Christopher D Newton, Martin R Bache, Steven P Jordan and Louise Gale 2019 High Temperature Fatigue Assessment of a SiCf/SiC Ceramic Matrix Composite Using Advanced Monitoring Techniques Proc. Int. Conf. ASME Turbo Expo 2019 (Electronic Materials vol 6) (Phoenix, Arizona: ASME pres) GT2019-90355

[7] Rajesh S. Kumar and Matthew M Mordasky 2019 Foreign Object Impact Damage in Ceramic Matrix Composites: Experiments and Computational Predictions *Proc. Int. Conf. ASME Turbo Expo* 2019 (*Electronic Materials*) (Phoenix, Arizona: ASME pres) **6** GT2019-90368

[8] Sapronov D, Mezencev M, Karimbaev T, Reznik S and Prosuntsov P 2019 Complex Design Method of Ceramic Blades and Metal Disk Connection *Proc. Int. Conf. ASME Turbo Expo* 2019 (*Electronic Materials*) (Phoenix, Arizona: ASME pres) GT2019-90875 **6**

[9] Hayao Sato, Yusuke Ueda, Masaharu Andoh and Tatsuhito Honda 2019 Validation of Fracture Prediction Model to Design CMC Turbine Vane *Proc. Int. Conf. ASME Turbo Expo* 2019 (*Electronic Materials*) (Phoenix, Arizona: ASME pres) GT2019-91233 **6**

[11] Robert J Boyle, Ankur H Parikh, Vinod K Nagpal P 2019 Design Considerations for Ceramic Matrix Composite High Pressure Turbine Blades *Proc. Int. Conf. ASME Turbo Expo* 2019 (*Electronic Materials*) (Phoenix, Arizona: ASME pres) GT2019-91787 **6**

[13] Frank Adbi, Cody Godines, Michael J. Presby, Amir Eftekharian, Jalees Ahmad, Sung Choi, Gregory N Morscher, Steve Gonczy, Jun Shi 2019 CMC Mode II Crack Growth Resistance Prediction of ASTM Specimen and Test Validation *Proc. Int. Conf. ASME Turbo Expo* 2019 (*Electronic Materials*) (Phoenix, Arizona: ASME pres) GT2019-90875 **6**

[14] Li Longbiao P 2019 Thermomechanical Fatigue of Ceramic Matrix Composites *Proc. Int. Conf. ASME Turbo Expo* 2019 (*Electronic Materials*) (Phoenix, Arizona: ASME pres) GT2019-90067 **6**

[15] Reznik S, Sapronov D and Prosuntsov P 2018 Computational and Experimantal Investigation of Mechanical and Thermal Contacts of Ceramic Blades with the Metal Disk in Gas Turbine Engines *Journal of Engineering Physics and Thermophysics* **91** (New York: Springer US) 1365-1372

[16] Ferber M, Richerson D and Roode M 2003 *Ceramic Gas Turbine Component Development and Characterization* (New York: ASME Press) p 425

IOP Conf. Series: Materials Science and Engineering **934** (2020) 012033 doi:10.1088/1757-899X/934/1/012033

Development of a method of hydroerosive saturation of liquids with microparticles of target materials using ultra-jet

A L Galinovisky [1], Kyaw Myo Htet [2], S A Erokhin [2] and Jia Zhenyuan[2]

[1] D. Sc, Prof, Head of department of SM-12, Bauman Moscow State Technical University, 105005, Russia

[2] PhD. SM-12, Bauman Moscow State Technical University, Moscow, 105005, Russia

E-mail: galcomputer@mail.ru, htet2066@gamil.com

Abstract. The article deals with the questions of assessing the applicability of the method of ultra-jet diagnostics of carbon plastics. In the work, phenomenological ideas about the process of destruction of carbon plastics under the action of a high-speed-jet of liquid are considered. The experimental part of the work is related to the implementation of the diagnostic effect of the ultra-jet on the surface of the carbon fiber reinforced plastic, the assessment of the geometric parameters of the hydraulic caverns and the qualitative assessment of the state of the fibers and the adhesion in the binder-fiber structure. In this experiment, carbon fiber samples were used both in the initial state and after the inhibitory effect of x-ray radiation. Such a technique made it possible to carry out a comparative analysis of the samples and evaluate the information content of the ultra-jet diagnostics method as applied to the materials under consideration.

1. Introduction

One of the most promising areas in the development of modern technology is the use of composite materials [1]. These materials require a special approach and the formation of new scientific and technical solutions in the development and creation of methods and means of assessing their operational reliability. This is due to the wide variety of types of such materials, the specific features of the structures made of them, manufacturing technology, a random change in the physio-mechanical and strength characteristics, and a large variety of types of defects that arise during the manufacturing process [2].

Composites are characterized by technological stresses arising in the elements at the stages of their manufacture. In addition, these materials in most industries operate under static and dynamic loads [3]. In constructions made using polymer composite materials (PCM) and especially carbon plastics experiencing dynamic loads, in order to obtain complete and reliable information about their condition, it is necessary to take special control and safety measures, for example, in mission-critical products: rocket and space technology , aircraft, shipbuilding, various transport structures, etc [4].

It should be specially noted that ensuring the safe operation of polymer composite material products is an important and urgent task [5].

Despite the fact that composite structures have many advantages, they can sometimes have various defects, such as delamination between composite layers, the waviness of composite layers or bends, in which at least partial twisting of the composite bundle on the upper part occurs, resulting in inner curls are formed in the composite structure. Considering that some of these defects can be detected

IOP Conf. Series: Materials Science and Engineering **934** (2020) 012033 doi:10.1088/1757-899X/934/1/012033

as a result of visual inspection of the composite structure, many defects can be located in the inner part of the composite structure, therefore, they cannot be detected during visual inspection [6].

In this regard, non-destructive methods for monitoring and diagnosing such structures are of great importance. They allow you to objectively determine the actual state of the structure, evaluate the reliability of their operation and give recommendations for its repair or restoration.

Because of this, many verification technologies have been developed, using, for example, X-rays [6], ultrasonic signals [7], and others for a detailed study of the structure of a composite structure [8]. Although these validation technologies can detect many defects, such as layer delamination, other defects that can be caused by disorientation or displacement of structural fibers, it is more difficult to detect defects in the resin of a composite construction.

Attempts to solve this problem with the help of flaw detection using various methods - ultrasonic, radio wave, etc. did not lead to the desired results. This is due to several reasons [9]:

1. Flaw detection methods can detect macrodefects, while a violation of the decrease in strength can be caused by microdefects (microcracks, micropores, etc.) and a number of other factors, for example, a violation of the composition of the material during application of force loads, a violation of manufacturing technology, etc.
2. Microdefects, which cause a decrease in reliability, are mainly formed in the process of loading the controlled structure with some loads (static or dynamic force, internal pressure for cylinders, etc.), and flaw detection methods mainly do not allow non-destructive testing in the process loading structures.
3. When controlling complex spatial structures, or objects that did not occupy the entire field of view of the recording system, along with informative parameters, interference is recorded that significantly reduces the reliability of the control results.

It is necessary to take into account the fact that during the production and operation of carbon fiber products, areas may appear where there is a significant weakening or disappearance of adhesion in the transition zone between adjacent layers of material, which leads to a significant deterioration in strength and performance [10]. In case of violations of the technological process, adhesion in the binder-fiber pair is also possible and a decrease in adhesive strength. Therefore, it is extremely important to enrich the existing arsenal of diagnostic technologies with new methods that would allow us to study the micro and macrostructure of carbon fiber reinforced plastics, indirectly assess the adhesive strength [7].

However, for polymer composite materials, adhesive bonds determine their quality, being its most important indicator. The determination of these bonds and the obtaining of objective information on adhesion in the binder-fiber structure are an important diagnostic task.

At present, methods of uneven tearing [10], in which the load is applied to the edge of the joint, due to which failure occurs non-uniformly, have become most widespread in testing materials for adhesive strength. The advantage of such methods is the ability to study adhesion fluctuations in different areas of the test sample.

The disadvantage of existing methods is the need to manufacture special samples, not high measurement accuracy, complexity of implementation and the availability of special equipment. In addition, these methods are not suitable for solving the problems of assessing the effect of binder destruction on adhesive strength. Given the existing potential of tools and engineering algorithms for the implementation of ultra-jet technologies developed at the department of SM-12 of BMSTU [8-9], of scientific and practical interest is the study of the possibility of their adaptation for the study of polymer composite material quality indicators (see Fig. 1).

Thus, the aim of the study is to develop a technique for ultra-jet diagnostics (UJD) of carbon plastics and assess its applicability for the diagnosis of material damage, including exposure to adverse environmental factors.

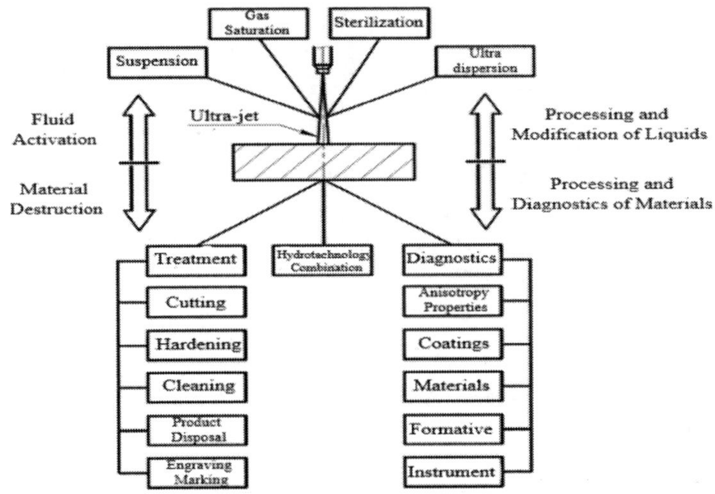

Figure 1 - The main directions of development of ultra-jet technologies of the scientific school of the Department of SM12 in BMSTU.

2. Experimental studies of ultra-jet diagnostics of a polymer composite material

The hypothesis of the study was that the action of an ultra-jet water on the surface of the sample witness will make it possible to obtain a hydrocavity on its surface, the parameters of which will be associated with a number of physical and mechanical characteristics of the material, its reinforcement pattern, the state of the binder (degradation), and the presence of a number of defects (damage), adhesive strength. A study of the fibers at the site of exposure to the ultrajet will allow us to evaluate the adhesive bonds in the binder-fiber structure by determining the ratio of the projection area of the binder particles to a plane parallel to the fiber axis perpendicular to the axis of the impact of the ultrasound to the surface of the sample.

To implement this technique, it is possible to effectively use the technological capabilities of the ultra-jet diagnostics, in particular by increasing the distance from the cut of the focusing tube to the surface of the sample, it is possible to provide a gentle effect on the material by the action of the spray skirt, and not the body of the ultrajet itself. All experiments on the ultra-jet diagnostics on a machine for waterjet cutting brand Flow Mach 3 3131b (USA).

Based on the experience of conducting experiments on the ultra-jet diagnostics of materials and coatings (see Fig. 1), an experimental scheme adapted to the stated task of polymer composite materials diagnostics was proposed, the experimental design shown in Fig. 2. The height h (from the cut of the focusing tube to the surface of the sample) varied from (recommended values for waterjet cutting of materials) 2–4 mm to 70–160 mm. Cutting head feed rate from 1000 to 7500 mm / min.

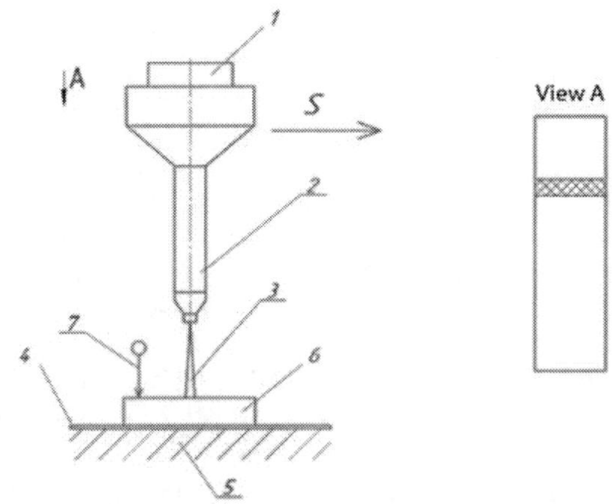

1 - cutting head, 2 - focusing tube, 3 - high pressure water jet, 4 - substrate, 5 - installation table, 6 - composite material sample, 7 - clamp, S - cutting head feed.

Figure 2 - Scheme of the experiment on the ultra-jet diagnostics samples from polymer composite materials

For the experiment, a carbon fiber sample was used, mounted on a table of a water-jet installation by means of a clamp using sheet equipment and clamps. The carbon fiber samples used in the experiments were made on the basis of a LU-P/0.1 carbon tape impregnated with the following characteristics: laying - (0°/45°/135°/90°/90°/135°/45°/0°) n, modulus of tensile elasticity - 59 166 MPa, tensile strength - 263 MPa., thickness 3.6 mm

As a result of test passes, the following diagnostic mode was chosen experimentally: feed of the cutting head 7500 mm / min, water pressure 130 MPa, the distance from the cut of the focusing tube to the surface of the sample 160 mm (hereinafter referred to as mode 1). When choosing these parameters, it was possible to ensure surface destruction of the binder without noticeable destruction of the fiber (Fig. 3a). An increase in pressure, a decrease in the distance h, and feed rate was accompanied by penetrating destruction of the material with rupture and damage to the fibers (Fig. 3b). To form a cavern in accordance with the scheme shown in Fig. 3 b the following diagnostic mode was selected: feed of the cutting head 6000 mm / min, water pressure 150 MPa, the distance from the cut of the focusing tube to the surface of the sample 80 mm (hereinafter referred to as mode 2).

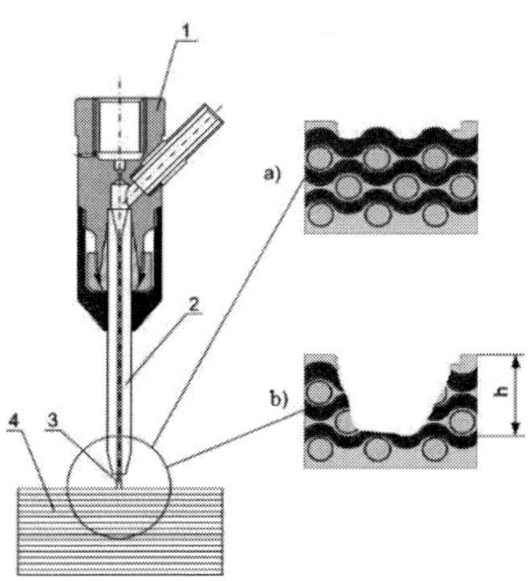

a) Surface destruction, b) penetrating destruction (with fiber damage) 1 – jet head, 2 – focus tube, 3 – high speed jet, 4 – composite sample, h – the formed cavern depth.

Figure 3 – The scheme of the experiment on ultra-jet diagnostics of polymer composite materials

Both of the presented variants of the results of the diagnostic effect of the ultrajet are of practical interest. In the first case (Fig. 3a), a spray of water (ultra-jet) is used to "bind" the binder (matrix) from the fibers, which, as described above, makes it possible to evaluate the adhesive properties of polymer composite material. In the second case (Fig. 3, b), the trunk of the ultra-jet of water carries out hydroerosive effect on the material on the local surface area of the sample. In this case, we can talk about the operation of the entire package. The higher cohesive strength of the binder and the adhesive strength of the binder-fiber structure will obviously prevent the formation of a cavern and, therefore, its geometrical parameters, especially (Fig. 3b) will be deliberately lower than in the case of poor adhesive strength, reduced in the result of a combination of adverse environmental factors leading to the destruction of the binder.

To enable a comparative analysis of the results of ultra-jet diagnostics to assess adhesion in the binder-fiber structure, part of the carbon fiber samples was exposed to x-ray radiation, and part was left in its original state. Samples were cut from a carbon fiber panel 80 mm long and 20 mm wide. X-ray irradiation of the samples was carried out in an industrial system of computed tomography and fluoroscopy XT N 450, manufactured by Nikon Metrology (Japan). The time and parameters of x-ray exposure are presented in table. 1. A large number of experiments and different exposure parameters are explained by the parallel conduct of other studies in which carbon fiber samples were placed in the camera of a computer tomograph.

Table 1. - Parameters of x-ray radiation used as pre-treatment

№	Voltage	Current	Experiment time
1	360	185	109
2	300	100	135
3	360	100	74
4	350	145	105
5	375	185	74
6	325	120	57
7	380	200	26

At the next stage, the ultra-jet diagnostics procedure of both groups of samples was implemented, in accordance with the scheme presented in Fig. 2. To form a cavern, Fig 3b. Mode 2 was used.

The study of samples (hydro-caverns) after conducting an experiment on the ultra-jet diagnostics of polymer composite materials was carried out using a VHX-6000 digital microscope (Japan). In fig. Figures 4-5 show images of (a) hydro-caverns and a sample after the ultra-jet diagnostics (b).

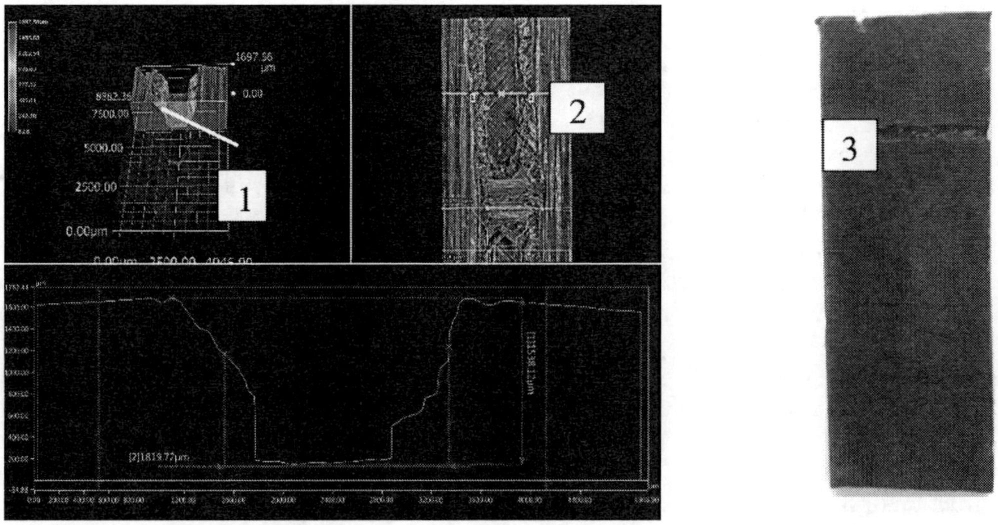

a b

Figure 4 - The results of the ultra-jet diagnostics of polymer composite materials sample after an X-ray exposure using a digital microscope (a - 3D profile of a cavern (1), a section with tracing (2), a profilogram; b - image of a sample with a cavern (3)

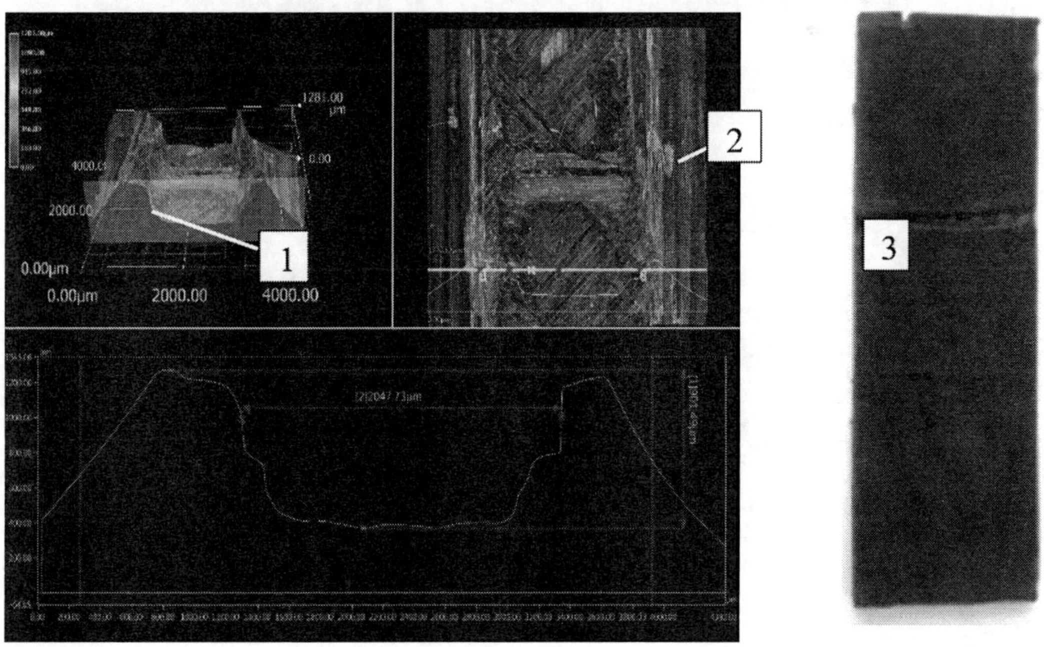

a b

Figure 5 - The results of the ultra-jet diagnostics of polymer composite materials sample in its initial state using a digital microscope (a - 3D profile of a cavern (1), plot with tracing (2), profilogram; b - image of a sample with a cavern (3))

The results of determining the geometric parameters of the cavern after ultra-jet diagnostics of polymer composite materials in the initial state and after x-ray exposure are presented in table. 2.

Table 2. - The results of determining the geometric parameters of the cavern

Sample	The depth of the cavity, microns	The width of the cavity, microns

After exposure	1538,12/1600,34*	1819,72/1910,12*
initial state	1235,26/1165,33*	1447,12/1580.19*

* The average value of 10 measurements of 3 samples

Further, mode 1 was used to form the hydro-cavern in Fig. 3a. As a result of the ultra-jet diagnostics, surfaces with traces of hydroerosion were obtained. For the study, an electron microscope manufactured by Phenom-World (Netherlands) was used.

Images of sections of carbon fiber samples after the ultra-jet diagnostics are shown in Fig. 6. It can be seen from the figure that in the case of the initial sample (Fig. 6a), the surface of the sample (zone 1) practically did not suffer from the action of the peripheral zone of the jet — the spray. This cannot be said about a sample exposed to radiation (Fig. 6b), where there are practically no traces of a binder on the surface. In the fiber images (Figs. 6c and 6d) obtained at 2000x magnification when focusing in the zone of the cavern (zone 2), one can notice a significantly larger amount of binder on the fiber surface for the initial sample (Fig. 6c). Obviously, the causes of this state of the surface of the samples and fibers are associated with the destruction of the binder, its embrittlement and removal of the products of destruction of the ultra-fluid from the zone of the ultra-jet diagnostics.

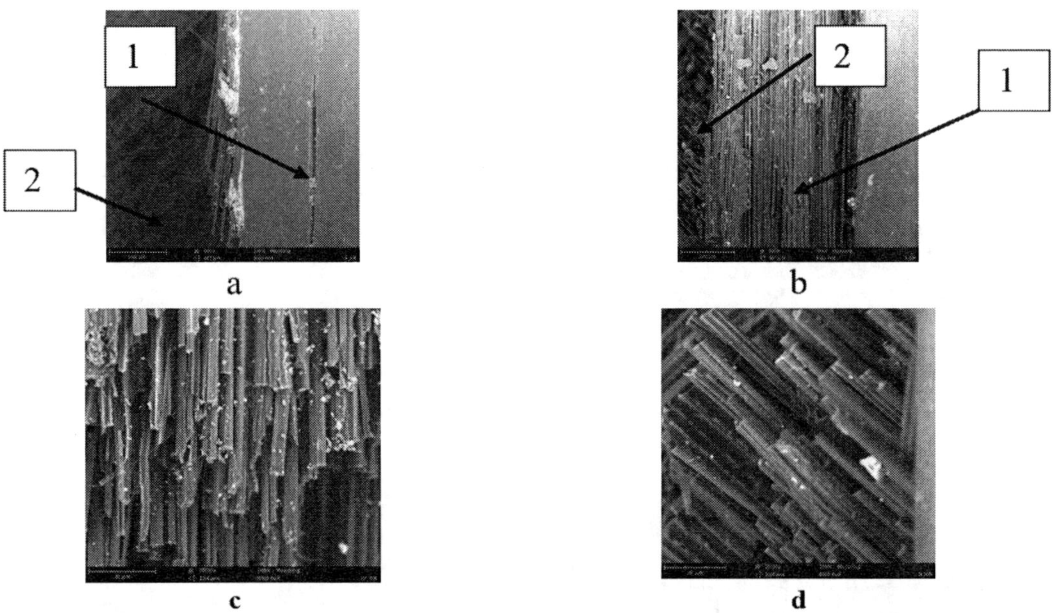

Figure 6. - Sites of surface hydroerosion of samples after the ultra-jet diagnostics (mode 1) at 600x magnification: a - initial sample, b - sample subject to x-ray irradiation; at 2000x magnification: c - initial sample, g - sample subject to x-ray irradiation.

Studies have shown that the ultra-jet diagnostics method demonstrates the presence of various adhesive bonds in polymer composite materials, which may be the basis for its refinement, the creation of an engineering diagnostic mechanism, the formation of hydraulic fracture databases for various composite materials, and the introduction at the stage of technological preparation of production or for selective control of batches details or samples of witnesses. The possibility of varying the technological parameters of the ultra-jet diagnostics will allow a diagnostic effect on the test object to be given at a given depth or number of fiber layers, which expands the potential capabilities of the method, for example, if it is necessary to study veils or tapes embedded in the package and intended to change the specified characteristics and properties of the polymer science materials, in particular, thermal conductivity, electrical conductivity, etc.

Conclusions

In conclusion, it should reflect the capabilities, features and advantages of the developed method of the ultra-jet diagnostics of polymer composite material:

- a simulated local wave and shock-force hydraulic action of an ultrajet of a liquid appears on the material;
- the implementation of the ultra-jet diagnostics method is possible with a combination of external forces and factors, such as preloading, for example, tensile or bending of the sample, temperature effect (complex or local);
- The ultra-jet diagnostics does not provide for the production of special samples for testing, and any products made from polymer composite material can be examined; the possibility of both qualitative and quantitative assessment of the quality of polymer composite material and its adhesive properties is provided, in particular by analyzing the area of the binder remaining after the ultra-jet diagnostics on the fiber surface;
- it seems possible to evaluate both the binder-fiber structure and the polymer composite material components separately, in particular, the cohesive properties of the binder;
- for the implementation of the method does not require the creation of special technological equipment or instruments, research is carried out on standard measuring equipment and installation for waterjet cutting of materials;
- a diagnostic effect can be exerted in several sections of the sample at a given depth and area, taking into account the direction of reinforcement, winding, laying, etc

Thus, there is the possibility of a fairly widespread implementation of the ultra-jet diagnostics of polymer composite material method, first of all, at the stage of testing the technological preparation of production, diagnostics and control of witness samples. The method is of particular importance in assessing damage to polymer composite material, which is exposed to various factors of influence, such as temperature, thermal cycling, climatic, X-ray, ultraviolet, etc.

Acknowledgments
The studies were carried out within the framework of grants of the Russian Foundation for Basic Research 18-29-18081 and 19-38-90228 \ 19, a grant of the President of the Russian Federation for state support of leading scientific schools of the Russian Federation NSh-3778.2018.8 and grants from the Fund for the Promotion of Innovations under the UMNIK-18 program in accordance with the contract No. 14727GU/2019 and No. 14549GU/2019.

References

[1] Bhagwan D A, Lawrence J B and Chandrashekhara K 2017 Analysis and Performance of Fiber Composites. *4-th ed.: John Wiley & Sons* p 576

[2] Leighton H 2018 *Carbon Fibers: Formation, Structure, and Properties*. Boka-Raton: CRC Press p 218

[3] Verma R, Shukla A, Kandasamy J, Kar V R, Rajesh M and Thirumaliniand S 2019 Structural health monitoring of aerospace composites ,Structural HealthMonitoring of Biocomposites, Fibre-Reinforced Composites and Hybrid Composites. *Amsterdam : Elsevier* pp 33-52

[4] Baurova N I and Zorin V A 2016 *The use of polymer composite materials in the manufacture and repair of machines.* textbook. allowance. M: MADI p 264

[5] Abashin M L, Barzov A A and Denchik A L 2018 Design and technological solutions for ultra jet processing of liquid-phase structures. *submitted to Science and Technology of Kazakhstan journal*

[6] Galinovstky A L, Iyukhina A A, and Kolpakov V I 2018 The Features of Hydroabrasive Cutting of Honeycomb Panels of Space Vehicles. *Moscow University Physics Bulletin, Moscow.* pp 441-446

[7] Galinovsky A L, Bethena G F and Petrov I V 2017 Improving the performance of waterjet cutting of materials by selecting rational modes of processing by acoustic emission *.edtied by metal technology,Moscow.* pp 18–27

[8] Galinovsky A L and Ilyukhina A A 2018 Analysis of the interconnectedness of the functional physical capabilities of additive and ultra-jet technologies. *edtied by Fundamental and applied problems of engineering and technology*. pp 73–81
[9] Galinovskiy A L and Mugla D R 2019 Selection of Rational Technological Modes and Parameters of Underwater Waterjet Cutting. *Lecture Notes in Mechanical Engineering*. pp 267-276
[10] Tarasov V A and Nelyub V A 2018 Rheological model of viscous incompressible fluid with disperse fillers. *edited by International Journal of Mechanical Engineering and Technology, Moscow*. pp 488-497
[11] Ilyukhina A A 2018 Analysis of the interconnectedness of the functional physical capabilities of additive and ultra-jet technologies .*edtied by Fundamental and appliedproblems of engineering and technology*. pp 73–81
[12] Iyukhina A A and Kolpakov V I 2018 The Features of Hydroabrasive Cutting of Honeycomb Panels of Space Vehicles. *Moscow University Physics Bulletin, Moscow*. pp 441-446
[13] Puzakov V S 2007 Development and analysis of the functional capabilities of ultra-jet activation of hydrotechnological media for mechanical processing production. *Vetnik BMSTU publication,Moscow*.
[14] Galinovsky A L, Provators A S and Kyaw Myo Htet 2019 On the question of the effectiveness of various methods of dispersing nanosuspension suspensions .*All materials: Encyclopedic reference*. No 11 S 2 – 7
[15] Galinovsky A L, Provators A S and Kyaw Myo Htet 2020 Prospects for the Development of Ultra-Jet Dispersion Technology for Nanocontaining Suspensions .*IOP Conf. Ser.: Mater. Sci. Eng. 709 044092*. volume 709
[16] Galinovsky A L, Provators A S and Barzov A A 2019 Information-physical mechanism of ultrahydrojet diagnostics of the quality of functional coatings .*Hardening Technologies and Coatings*. No 2 S 51 – 57
[17] Mugla D R and Golubev E S 2019 Surface express ultrajet diagnostics of space vehicle materials .IOP Conference Series: Materials Science and Engineering. Vol 683 Issue 1
[18] Galinovskij A L, Abashin M I and Khafizov M V 2014 Rapid Determining of the Optimum Operation Mode for Abrasive Waterjet Cutting Process by Means of Acoustic Emission (*Applied Mechanics and Materials*) Vol 698
[19] Zh u J, Imam A, Crane R, Lozano K, Khabasheku V N and Barrera E V 2017 Processing a glass fiber reinforced vinyl ester composite with nanotube enhancement of interlaminar shear strength. *Compos Sci Technol*. pp 1509–17
[20] Tarasov V A and Baskakov V D 2016 Mathematical modeling of the collision process of flat jets of an ideal fluid .*sumitted to Vestnik BMSTU, Moscow*. p 79–90
[21] Tarasov V A, Stepanishchev N A and Boyarskaya R P 2011 Methods of experimental determination of the characteristic points in time of the technological process of preparing nanosuspensions under conditions of ultrasonic action .*submitted to Vestnik BMSTU*. p 53–65
[22] Nelyub V A and Tarasov V A 2020 Ion-beam processing of metallized carbon fiber-reinforced plastics .*Materials and Manufacturing Processes*. Vol 35 Issue 2 pp 172 – 178
[23] Nelyub V A and Borodulin A S 2019 Polyethersulfones with Improved Thermophysical Properties .*Polymer Science. Series D*. Vol 12 Issue 1 pp 24 – 28
[24] Nelyub V A and Malysheva G V 2019 Properties and technology of applying metal coatings to carbon tape .*Fibre Chemistry*. Vol 50 Issue 6 pp 524 – 527
[25] Nelyub V A and Borodulin A S 2018 Properties of Epoxy Materials Used for Production of Glass Reinforced Plastics by the Winding Method .*Polymer Science. Series D*. Vol 11 Issue 2 pp 147 – 153
[26] Kobets L P , Borodulin A S and Malysheva G V 2016 Study of Microcapillary Impregnation of Carbon Fibres by Epoxy Binders. *Fibre Chemistry*. Vol 48 No 4 pp 311 – 315

[27] Kobets L P , Borodulin A S and Malysheva G V 2016 Rheological properties of system "elastomer binder–discrete carbon fiber" .*Inorganic Materials: Applied Research*. Vol 7 Issue 1 pp 15 – 19

[28] Nelyub V A, Borodulin A S, Kobets L P and Malysheva G V 2016 Capillary hydrodynamics of oligomer binders .*Polymer Science. Series D*. Vol 9 Issue 3 pp 322 – 325

[29] Nelyub V A 2018 Technologies of Metallization of Carbon Fabric and the Properties of the Related Carbon Fiber Reinforced Plastics. *Russian Metallurgy (Metally*). Vol 2018 Issue 13 pp 1199 – 1201

[30] Nelyub V A 2015 Method for Assessing the Chemical Reaction Between Carbon Fibre and Epoxide Binder. *Fibre Chemistry*. Vol 47 Issue 1 pp 40 - 42

IOP Conf. Series: Materials Science and Engineering **934** (2020) 012034 doi:10.1088/1757-899X/934/1/012034

Stability of Shape Memory Alloy rods under reverse phase transformation in case of initial tensile phase-structural strains

S A Dumanskiy

Institute of Applied Mechanics of Russian Academy of Sciences, Moscow, Russian Federation

E-mail: stanislavdym@mail.ru

Abstract. Stability of a round cross section rod of Shape Memory Alloys under reverse phase transition where the initial strain was obtained with the help of reverse loading and subsequent unloading was investigated. It was shown that the additional structural transformation connected with buckling could not take place within the framework of a given model of the constitutive equations for Shape Memory Alloys. Numerical solutions in various problem statements and concepts were obtained.

1. Introduction

In last decades, smart materials become more and more widespread. Their performance may be controlled by means of change in outside conditions. Shape Memory Alloys (SMAs) of which thermomechanical properties depend on temperature and stress level may be considered as an example of such materials. This paper does not consider any magnetic and other specific SMAs. One of the most well-known unique capability of SMAs is the shape memory effect that consists in almost completely recovery of nonlinear (phase-structural) strains accumulated in low temperature (martensitic) state when heating at a certain temperature range. Such heating is accompanied by a change in the type of crystal lattice and is called the reverse phase transition. Together with it, structural transformations associated with change in the degree of orientation of martensitic elements at the micro level and martensitic structures in the form of plates or needles at the meso-level, also occur.

Constructions and devices built on the base of the properties of SMAs are widely applied in automotive, railway, biomedical and aerospace industries [1-5]. Some of these devices work as actuators using the shape memory effect and once or many times transform thermal energy to mechanical one.

One of the very important characteristics of any construction is stability with respect to different influences: stress level changes, magnetic or temperature fields, etc. During the phase transition under constant load [6], rods of SMAs loss stability at stress levels being many times smaller than those computing by using the Euler formula for the minimum value modulus of elasticity corresponding to the martensitic phase state. This prompts the consideration of a wider set of constitutive equations that allow the complex mechanisms of phase-structural deformation to be described and can explain such a large discrepancy between the elastic solution and experimental data. The model of constitutive equations that is used hereinafter was offered in [7-8]. Only rods with constant cross section area F in the martensitic phase state with compressive constant load $P < 0$applied to their upper ends are considered hereinafter. The method creating the initial strain namely its phase-structural component $\varepsilon^{phst}(q = 1, \sigma = \frac{P}{F}) = \varepsilon^*$, where q is the phase composition parameter, plays an important role for solving the problem of stability during the reverse phase transition. The case where the initial strain ε^* is negative and the point (ε^*, σ) defining the state rod before starting the reverse phase transition and belonging to the martensitic non-elasticity diagram (Fig 1. line I.) or located at the left of such diagram (Fig 1. left part area III.) was investigated in [9] for the rod with rectangle cross section. But this does not exhaust all the ways of preparation of specimens. The point (ε^*, σ) can be located in area (III) as a result of reverse loading with further unloading and the strain ε^* generated thereafter can be positive or negative.

An analog of the theory of plasticity flow with combined hardening [10] is used hereinafter for describing structural deformation. The effect of difference resistance of SMAs that observed in experiments [11] also is taken into account.

In this work, the Euler quasi-static method is regarded as a stability loss criterion. It consists in considering the form of equilibrium adjacent to the initial (linear) one. If a nontrivial form of equilibrium exists, this means that the rod buckles or loses stability. Such an approach allows the problem to be solved in the linearization formulation with respect to the little variation of deflections.

Let us consider two conceptions: fixed load (FL) and variated load (VL). In the conception VL, let us assume that load variation and curvature increment associated with rod buckling have one and the same order of smallness. In this case, critical force is defined as the least of all the forces for which at least at one value of the load variation a loss of stability occurs. In the conception FL with the appearance of adjacent form equilibrium, the cross section may be divided into two single-connected areas where stress variations have different signs. In the conception VL, one area occupies the whole cross section and sign of stress variation within the whole cross section is constant.

The problem is investigated in three statements depending on the variables that can be variated. Phase composition parameter in the non-coupled statement is not changed during buckling and its increment is equal to zero. Solution that was obtained in the non-coupled statement coincides with the Euler critical force for elastic rod with martensitic (minimum) modulus of elasticity. As stated below, the elastic solution is the precise upper bound of the set of all critical forces for the considered statements. In the singly-coupled statement, the volume fraction of the martensitic phase depends on stress increment and is changed in case of loss of stability but the temperature remains constant. The twice-coupled statement includes the singly-coupled one and takes into account absorption of latent heat and heat emission related to dissipative terms during buckling.

2. Constitutive equations

The one-dimensional nonlinear model of the phase-structural deformation of SMAs is as follows:

$$d\varepsilon^{phst} = d\varepsilon^{e} + d\varepsilon^{phst},\ d\varepsilon^{phst} = \varepsilon^{0}dq + d\varepsilon^{ph'} + d\varepsilon^{st}$$

$$\varepsilon^{e} = \frac{\sigma}{E(q)}, \frac{q}{E_M} + \frac{1-q}{E_A}, 0 \le q \le 1 \tag{1}$$

$$d\varepsilon^{ph'} = \omega dq, \omega = \frac{\varepsilon^{phst'}}{q} \tag{2}$$

$$d\varepsilon^{st} = \rho_D(\mu_\varepsilon) q_{st} \varphi_2'(|\sigma|) d\sigma,\ \mu_\varepsilon = \frac{4\det(\varepsilon_{ij}')}{\varepsilon_i^3} \tag{3}$$

$$\varphi(S) = \left[\frac{\varepsilon^{phst'}}{\rho_D(\mu_\varepsilon)}\right]_{max},\ S = \sigma' - r,\ d\sigma < 0 \tag{4}$$

$$dr = g d\varepsilon^{st}, g = const \tag{5}$$

$$d\varepsilon^{st} = \frac{q_{st}}{g} d\sigma \tag{6}$$

$$q = \frac{1}{2}(1 - cos(\pi t)) \tag{7}$$

$$t = 1 - \frac{T - A_s^\sigma}{A_f^\sigma - A_s^\sigma}, A_s^\sigma = A_s^0 + \frac{\omega\sigma + Z(\sigma) + \sigma\varepsilon^0}{\Delta S}, A_f^0 - A_s^0 = A_f^\sigma - A_s^\sigma, Z(\sigma) = \frac{E_A - E_M}{2E_A E_M}\sigma^2$$

$$k_q \Delta T = C_\sigma \dot{T} + T\theta\dot{\sigma} - (\Delta U + \varepsilon^0\sigma + \omega\sigma + Z(\sigma))\dot{q} - \sigma\dot{\varepsilon}^{st} \tag{8}$$

$\varepsilon, \varepsilon^e, \varepsilon^{phst}$ are the axial components of total, elastic and phase-structural strains, accent " ' " indicates the deviator part; ε^0 is the linear part of the deformation of the bulk effect of phase transition; σ is the axis component of stress; q is the volume fraction of the martensitic phase or phase composition parameter that is defined by equation (7), q_{st} is the part of volume that undergoes structural transformation; $E(q), E_A, E_M$ are the modulus of elasticity and its values for austenitic and martensitic phase states; $\varphi_2(|\sigma|)$ is the material function approximating the diagram of martensitic non-elasticity; $\rho_D(\mu_\varepsilon)$ is the material function returning the maximum value of structural deformation corresponding to fully detwinned martensite and depending on the deformed state, μ_ε is the parameter defining a type of deformed state, $\rho_{D_2} = \rho_D(-1), \rho_{D_1} = \rho_D(1)$; $\varphi(S)$ is the material function defining a type of loading surface, S is the active stress, r is the translation stress; A_s^0, A_f^0, A_f^σ are the temperatures of start and finish of the reverse phase transition for unloading state and the temperature of finish the reverse transition under a constant stress σ, ΔS is the difference between volume densities of entropy for austenitic and martensitic phases; formula (8) is the equation of energy balance, $k_q, C_\sigma, \theta, \Delta U$ are the heat conductivity coefficient, heat capacity per unit volume under a constant stress, thermal expansion coefficient and volume density of latent heat of phase transition; g is the material constant defining the slope of the line approximating the segment of the reverse loading diagram with only translational hardening.

Equation (5) sets the position of the center of the loading surface, equations (3) and (6) determine the increment of structural deformation during combined and kinematic hardening, if conditions (4) are satisfied.

3. Process of formation of the initial phase-structural strain

At the beginning, unloaded rod was in martensitic phase with zero forming strain, i.e. in the state of twinned martensite. Tensile stress σ_1 is applied to the rod, under the action of which phase-structural strain $\varepsilon_1^{phst'} > 0$ appears. Then there is complete unloading, followed by loading with compressive stresses up $\sigma_2 < 0$. The strain reaches the value $\varepsilon_2^{phst'}$ that is less or equal to $\varepsilon_1^{phst'}$. The equal sign is achieved only if the stress $\sigma_2$is greater than σ_s at which the point $(\varepsilon_1^{phst'}, \sigma_2)$ first appears on the loading surface defined by condition (4). If $\sigma_2 < \sigma_s$, then with a monotonic decrease in stress according to (4), purely translational hardening first occurs, which is subsequently replaced by combined hardening. At the last stage of preparation, the stresses increase from σ_2 to $\sigma < 0$, while the strain $\varepsilon^* = \varepsilon_2^{phst}$ does not change.

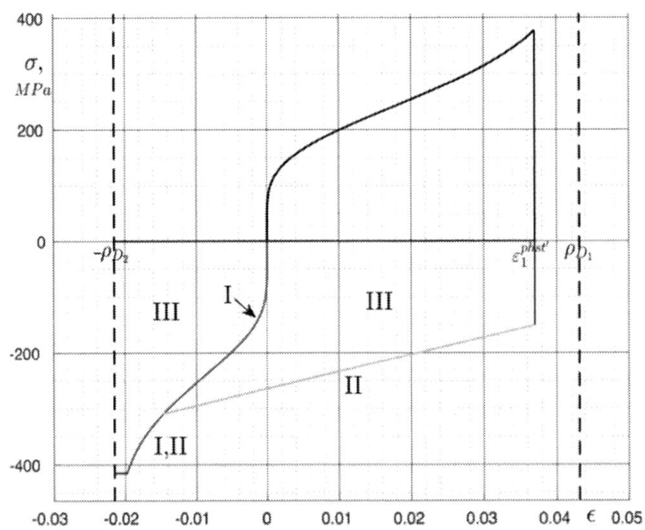

Fig. 1. Loading path

With the described method for formation the initial strain, the natural restrictions appear on the interval in which ε^*can change under a fixed value of stress . Let $(\varepsilon^{phst'}, \sigma_3)$ be the intersection point of the line describing purely translational hardening and the diagram of martensitic non-elasticity. Then, when $\sigma_2 \in [\sigma_3, \sigma_s]$ decreases, kinematic hardening occurs, and when $\sigma_2 \in [\sigma_p, \sigma_3]$ decreases, combined hardening occurs. Here σ_p is the yield stress. Let us assume that if the stress σ_2 reaches or exceeds σ_p, then using unloading, it is possible to return to the region of pre-yield stresses $\sigma < \sigma_p$, while preserving the properties of SMAs. This assumption allow us to limit the initial phase-structural strain ε^* from below to the value $\varepsilon^0 - \rho_{D_2}$.

The initial strain $\varepsilon^* = \varepsilon^0 + \epsilon(\varepsilon_1^{phst'}, \sigma)$, where the deviator component ϵ belongs to the following intervals:

a. $(-\rho_{D_2}, \varepsilon_1^{phst'}]$ if $\sigma_s(\varepsilon_1^{phst'}) < \sigma < 0$

b. $\left(-\rho_{D_2}, \frac{\sigma + S(\varepsilon_1^{phst'})}{g}\right)$ if $\sigma_3(\varepsilon_1^{phst'}) < \sigma \leq \sigma_s(\varepsilon_1^{phst'})$

c. $(-\rho_{D_2}, -\rho_{D_2}\varphi_2(-\sigma))$ if $\sigma \leq \sigma_3(\varepsilon_1^{phst'})$

Radius of the loading surface (4) depends on deformation $\varepsilon_1^{phst'}$ and does not depend on way of loading from 0 to σ_1. This hypothesis is confirmed by experimental data describing the cross hardening phenomena. Therefore, it is possible to expand many ways of preparation the rod by considering various ways of creating the initial tensile strain $\varepsilon_1^{phst'}$.

It is easily to see that because of unloading on the last step of preparation of the rod, the point (ε^*, σ) is always located inside the loading surface on a finite distance to its bound. It is clear that the point (ε^*, σ) may not appear on the loading surface because of infinitely small increment of stresses. This means that additional structural deformation may not occur. It is worth noting that, for $\sigma > \sigma_3$, the point located on the diagram of martensitic non-elasticity does not belong to the loading surface, and there is no structural deformation for it.

4. Phase-structural strain of trivial process

Prior to the loss of stability, the rod is uniformly heated under the action of a constant compressive force, while maintaining a rectilinear shape. During this process, the reverse phase transition not accompanied by structural transformation takes place. The structural deformation does not occur because none of conditions (4) is satisfied. Phase-structural component of strain deviator is defined from differential equation (2) with the initial condition $\varepsilon^{phst'}(q = 1, \sigma) = \epsilon$. After integration and substitution the initial condition, we obtain $\varepsilon^{phst'} = \epsilon q$ or $\varepsilon^{phst} = \varepsilon^* q$.

5. Increment of phase composition parameter and additional phase transition

Hereinafter, we regard only SMAs with wide hysteresis for which $M_s^0 \leq A_s^0$. Therefore, additional phase deformation occurs only if increment of phase composition parameter is less than zero. In order to evaluate variation of q, equation (7) is variated. Incrementation of temperature is determined from equation (8) by replacing time differentiation with variation. It can be assumed that temperature field is uniform; then $\Delta T \equiv 0$. Likewise, we neglect $T\theta\dot{\sigma}$ in comparison with other components. As the additional structural deformation may not occur, $\dot{\varepsilon}^{st} \equiv 0$.

$$\delta q = \frac{\Upsilon(\varepsilon^*,\sigma)}{\Theta(q,\varepsilon^*,\sigma)}\delta\sigma, \frac{1}{\Theta(q,\varepsilon^*,\sigma)} = \frac{\lambda(q)}{\Delta S\left(1+\frac{\lambda(q)}{C\sigma}(\Delta U+\varepsilon^*\sigma+Z(\sigma))\right)} \tag{9}$$

$$\lambda(q) = \frac{\pi\sqrt{q(1-q)}}{A_f^0-A_s^0}, \Upsilon(\varepsilon^*,\sigma) = \varepsilon^* + B\sigma, B = \frac{E_A-E_M}{E_A E_M}$$

Function $\Theta > 0$ for nitinol and many other SMAs regardless of the values of its variables. It is easily shown that $sign(\delta q) = sign(\Upsilon) sign(\delta\sigma)$.

Further, a rod of round cross section with radius h is considered. In every cross section of the rod we can enter Cartesian coordinate system axes of which are lines of symmetry. Let l be significantly greater than h; then we can use the plane section hypothesis for total strain incrementation:

$$\delta\varepsilon = \delta\varepsilon^e + \delta\varepsilon^{ph} + \delta\varepsilon^{st} = \delta\varepsilon_0 + y\delta\kappa \tag{10}$$

where $\delta\varepsilon_0$ is the increment of axis component of strain, $\delta\kappa > 0$ is the increment of curvature of deflection of the rod.

Incrementations of elastic and phase components of strain are calculated from equations (1) and (2):

$$\delta\varepsilon^e = \frac{\delta\sigma}{E(q)} + B\sigma H(-\delta q)\delta q, \quad \delta\varepsilon^{ph} = \varepsilon^* H(-\delta q)\delta q, \quad \delta\varepsilon^{st} \equiv 0 \tag{11}$$

where $H(x)$ is Heaviside unit step function.

Relations for incrementation of stresses can be obtained from equations (9), (10) and (11).

6. Increment of stresses

$$\delta\sigma = \begin{matrix} \frac{\zeta-\zeta_0}{\Psi_1} h\delta\kappa, -1 < \zeta < \zeta_0 \\ \frac{\zeta-\zeta_0}{\Psi_2} h\delta\kappa, \zeta_0 < \zeta < 1 \end{matrix}, \quad \Psi_k = \frac{1}{E(q)} + \frac{\Upsilon^2(\varepsilon^*,\sigma)}{\Theta(q,\varepsilon^*,\sigma)} H((-1)^{k+1}\Upsilon) \tag{12}$$

$\zeta = \frac{y}{h} \in [-1,1]$ is the dimensionless coordinate in cross section of the rod, $\zeta_0 = \frac{y_0}{h} = -\frac{\delta\varepsilon_0}{\delta\kappa}$ is the value of ζ in the conception FL determining the location of neutral axis on which $\delta\sigma = 0$. Subscript k to function Ψ corresponds to two areas of cross section where stress increments from the loss of stability have different signs. If < 0 , then additional loading takes place and $k = 1$ and $k = 2$ when $\delta\sigma > 0$. It is easy to prove that $sign(\delta q) = (-1)^k sign(\Upsilon)$ whence it follows that $H(-\delta q) = H((-1)^{k+1}\Upsilon)$.

7. Location of neutral axis

In the conception FL, if $\Psi_1 \neq \Psi_2$, the coordinate of neutral axis can be obtained from an equilibrium equation for external load $\delta P = \int_F \delta\sigma dS = 0$. After integration, the condition of FL for the round cross section assume the following form:

$$\frac{1}{3}(1+\frac{1}{2}\zeta_0^2)\sqrt{1-\zeta_0^2}+\frac{1}{2}\left(arcsin(\zeta_0)-\frac{\pi}{2}\frac{\Psi_1+\Psi_2}{\Psi_1-\Psi_2}\right)\zeta_0=0 \tag{13}$$

This equation is transcendental therefore its solution can be obtained only numerically. Otherwise $\Psi_1 = \Psi_2$ and we must accept that $\zeta_0 = 0$.

For the conception VL, based on the hypothesis that $0 < \frac{|\delta P|}{\delta\kappa} < \infty$ and from the definition of critical load given at the beginning of this paper, it follows that the smallest critical force is reached if the sign of the load variation is constant in the entire cross section, i.e. $\zeta_0 = \pm 1$

8. Critical length

From the equation for the variation of the bending moment

$$\delta M = h^2 \int_F \sigma\zeta dS = P\delta w$$

we can obtain the differential equation for the variation of the deflection after integration

$$\delta w'' + \eta \delta w = 0$$

where $\eta = 2\sqrt{\pi s \Psi_1 \Psi_2}\left[(\Psi_1-\Psi_2)\left(\frac{1}{3}(2\zeta_0^2-5)\zeta_0\sqrt{1-\zeta_0^2}-arcsin(\zeta_0)+\frac{\pi}{2}\frac{\Psi_1+\Psi_2}{\Psi_1-\Psi_2}\right)\right]^{-\frac{1}{2}}$ in the conception FL and $\eta = \frac{2\sqrt{|\sigma|\Psi}}{h}$ in the conception VL, when $\Psi = \max_k\{\Psi_k\}$. According to the boundary conditions for hinged rod $\delta w(0) = \delta w(l) = \delta w''(0) = \delta w''(l) = 0$ the critical length is calculated by the formula:

$$l = \frac{\pi}{\max_q\{\eta(q,\varepsilon^*,\sigma)\}} \tag{14}$$

If we fix an arbitrary stress value $\sigma = \sigma_0$, then according to (9), the critical length will be a function of the initial strain ε^*.

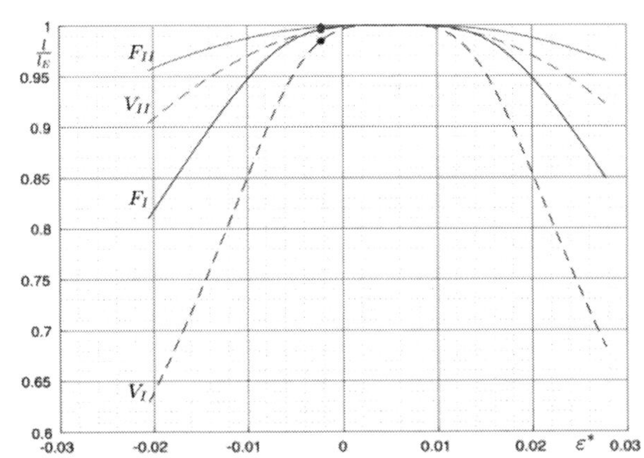

Fig. 2. Critical length

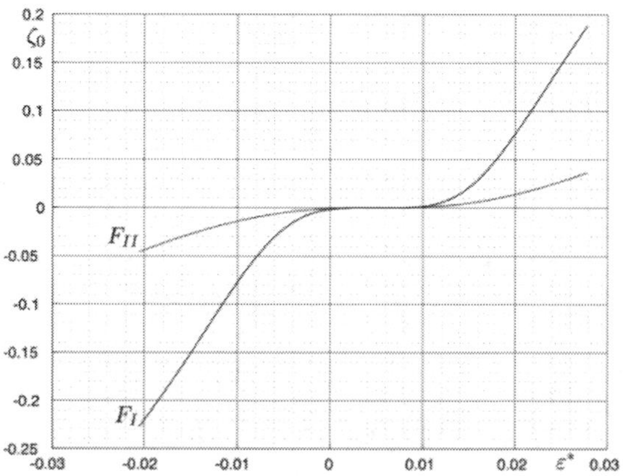

Fig. 3. Neutral line

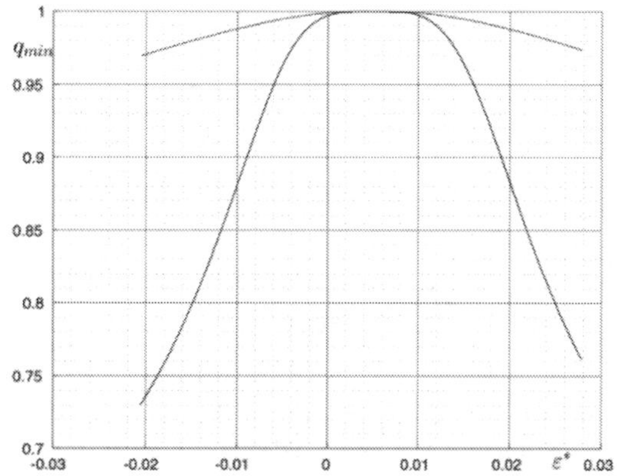

Fig. 4. Minimum value of phase composition parameter

Fig. 2 shows the dependence of the critical length related to the Euler length $l_E = \frac{\pi h\sqrt{E_M}}{\sqrt{3|\sigma|}}$ when $\sigma_0 = 173$ [MPa] and $\varepsilon_1^{phst'} = 0.036$. The Latin letters F and V mean the conceptions of FL and VL of the external load, and the Roman numerals I and II are singly and twice coupled statements. Dots mark the strains corresponding to the diagram of martensitic non-elasticity under compression. It is easy to see that the curve is symmetric with respect to the straight line $\varepsilon^* = B|\sigma|$. This is due to the fact that the functions Ψ_k quadratically depend on Υ. This means that for equal critical lengths, the tensile deformation should be greater than the compressive one.

Fig. 3 shows the relationship between the critical position of the neutral line depending on the initial strain. Fig. 4 shows the relationship between the critical value of the phase composition parameter and the initial strain, upper line is twice-coupled statement and lower line – singly-coupled statement.

As in [12], the Gamma distribution is proposed to use as a function approximating the diagram of martensitic non-elasticity, $\gamma(x, d, \alpha) = \frac{1}{\Gamma(\alpha)}\int_0^{\frac{x}{d}} t^{\alpha-1}e^{-t}dt$ with scale parameter $d = 34.66$ [MPa] and shape parameter $\alpha = 7.9$. Material constants: $\rho_D(1) = \rho_{D_1} = 2\rho_{D_2} = 0.043$, $g = 3066.4$ [MPa], $\rho_{NiTi} = 6440$ [kg/m³],

$A_f^0 - A_s^0 = 20$ [K°], $\Delta U = 30000$ [J/kg], $\Delta S = 100$ [J/(kg•K)], $C_\sigma = 500$ [J/(kg•K)].

It easy to calculate $\sigma_s = 140.9$ [MPa] and $\sigma_3 = 289{,}2$ [MPa]. For the stress under consideration, the initial strain ε^* is bounded above by the value 0.0278, which corresponds to the strain belonging to the diagram describing the purely kinematic hardening for reverse loading at stress $\sigma = \sigma_0$.

9. Conclusion

Numerical solutions of the stability problem are obtained for the method of setting the initial strain described in this paper. It is established that, when passing to the adjacent form of equilibrium, an additional structural transformation, by virtue of condition (4), cannot occur. The critical length in the conception FL is greater than for the conception VL. Solutions in the singly coupled statement (isothermal buckling) lead to smaller critical lengths than in the twice-coupled statement (adiabatic buckling). If the point describing the initial stress-strain state belongs to the martensitic non-elasticity diagram for compression, then at stresses $\sigma < \sigma_3$, additional structural deformation occurs if it is obtained by monotonous loading from the state of chaotic martensite and is absent for reverse loading.

It should be noted that for the strains ε^* identical in absolute value, the case when one of the strains is compressive and another is tensile would correspond to the smallest value of the critical length.

Acknowledgments

The work was performed out within the framework of the state budget topic (Registration No. AAAA-A19-

119012290118-3) with the partial financial support from the Russian Foundation for Basic Research (Project 20-01-00240).

References

[1] Williams E A, Show G and Elahinia M 2010 *Mechatronics* **20** Issue 5 pp 527-534
[2] Kato T 2011 *Applications and Technologies. Woodhead Publishing Series in Metals and Surface Engineering* p 120-124
[3] Melton K R 1999 *Cambridge University Press (Cambridge)* chapter 10 pp 220-239 ed. Otsuka K and Wayman C M
[4] Morgan N B *Materials Science and Engineering: A* vol 378 issues 1-2 pp 16-23
[5] Kumar P and Lagoudas D 2008 *Shape Memory Alloys Modeling and Engineering Applications.*(Springer Science+Business Media) pp 1-51
[6] Movchan A A and Kazarina S A 2002 *Problemy Mashinostraeniya i Nadezhnos'ti Mashin* issue 6 pp 82-89
[7] Movchan A A, Movhan I A and Silchenko L G 2010 *Mechanics of Solids* vol 45 issue 3 pp 406-416
[8] Movchan A A and Kazarina S A 2012 Physical Mesomechanics **15** issues 3-4 pp. 214-223
[9] Dumanskiy S A and Movchan A A 2019 *Mechanics of Solids* issue 4 pp 94-108
[10] Mishusrin I V2018 *Nanoscience and Technology: An International Journal* **9** issue 2 pp 165-181
[11] Movchan A A, Movhan I A and Silchenko A L 2015 *Mekhanika kompozitsionnykh materialov i konstruktsii* **21** issue 1 pp 83-9 Movchan A A, Movhan I A and Silchenko A L 2019 *Russian Metallurgy (Metally)* **2019** issue 4 pp 301-308

IOP Conf. Series: Materials Science and Engineering **934** (2020) 012035 doi:10.1088/1757-899X/934/1/012035

Homogenization of nanocontaining suspensions using ultra-jet methods

Kyaw Myo Htet [1,4] and A L Galinovskiy[2,3]

[1] PhD. SM-12, Bauman Moscow State Technical University, Moscow, 105005, Russia

[2]D. Sc, Prof, Head of department of SM-12, Bauman Moscow State Technical University,105005, Russia

E-mail: [3]galcomputer@mail.ru, [4]htet2066@gamil.com

Abstract. The article discusses the preparation of suspension and dispersion of liquids using ultra-jet technology. The applied method allows for a sufficiently high level of suspension and dispersion.

Keywords: liquid, ultra-jet technology, suspension, target, dispersion

1. Introduction
Dispersion is a system in which distributed particles of one material are dispersed in a continuous phase of another material. The two phases may be in the same or different states of matter. Most dispersion methods are based on mechanical destruction (grinding, dispersion) of the solid-phase material of the suspension filler [1]. Suspensions are a two-phase fairly homogeneous hydrostructure, consisting of a liquid (colloidal solution, gel, etc.) and finely divided solid particles that are under normal conditions in suspension. Most hydrotechnological media, including ordinary water, can be considered as specific ultrafine suspensions with a very low concentration of solid-phase particles [2].

2. Experimental studies for dispersion and suspension for processing and modifying liquids
According to [3] for dispersion, to prepare a mixture for ultrasonic treatment, 3 g of nanopowders (Boehmite and nanodiamonds) mixed with 1 litre of water in a cup and measured with the Microtrac Bluewave S3500, where in the laboratory hall of the SM-12 department of BMSTU in Moscow. The characteristics of boehmite was mass fraction of the main substance 99%, specific surface according to the BET method 400 m2/g, density, 3.06 g/cm3, size of particles, 0.1-0.8 nm and took from Research Institute of Pulse Processes with Experimental Production "OCP NII IP with OP", Minsk, Belarus . The chareacteristics of nanodiamonds were external diameter 10-20 nm, tube length 2 microns, specific surface according to the BET method 300 m2/g, content of impurities not more than 1% and took from Arkema (France).

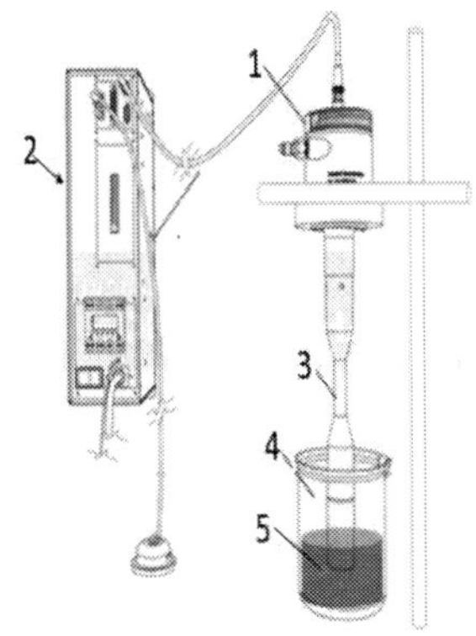

Figure 1. Scheme of ultrasonic treatment: 1- body, 2- ultrasonic generator, 3- ultrasonic waveguide, 4- glass for suspension, 5- nano-containing suspension.

At the next stage of research, the possibilities of the method of ultra-jet treatment of suspensions were studied. Figure 2 presents the experimentally schematic diagram of ultra-jet dispersion of suspensions [4]. The studies were based on the methodological base, which was formed in the framework of the scientific school "Ultrajet processing and diagnostics of materials and liquids" (NSh-3778.2018.8) at the Department of SM-12 of BMSTU [5, 6].

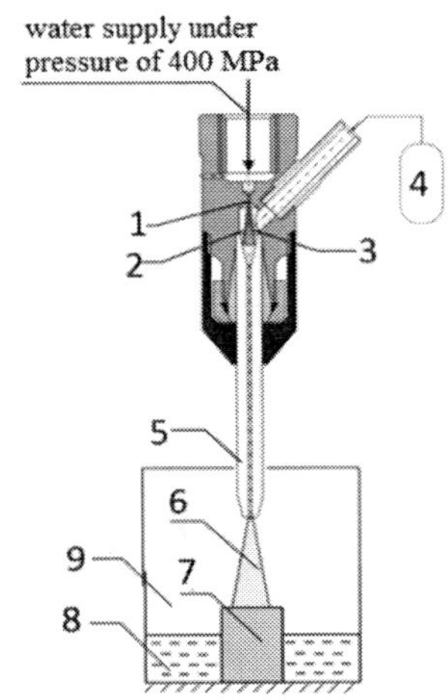

Figure 2. Schematic diagram of ultra-jet dispersion of nanosuspension suspensions: 1 - hydraulic nozzle; 2 - mixing chamber; 3 - high-speed liquid stream; 4 – container for suspension with nanomaterials; 5 - focusing tube; 6 - high-speed jet of suspension; 7 - target made of synthetic diamond; 8 - processed suspension; 9 - container-trap for collecting suspension..

In the process of experimental development of ultra-jet technology, nanosized suspensions were subjected to processing - dispersion according to the scheme of Fig. 2. Suspensions based on distilled water with boehmite, carbon nanotubes were fed into the mixing chamber, ultra-jet installation using a special measuring dispenser presented to the Fig 2. Acceleration of the suspension was 600 m/s, which corresponds to a maximum working pressure in the hydraulic system of 400 MPa. It was previously established that the speed of the ultrajet determines the efficiency of the liquid treatment [7]. The target is a synthetic diamond fixed in a mandrel, which, in turn, using screws is fixed in the nozzle of the container cover (Fig. 3) [7,8].

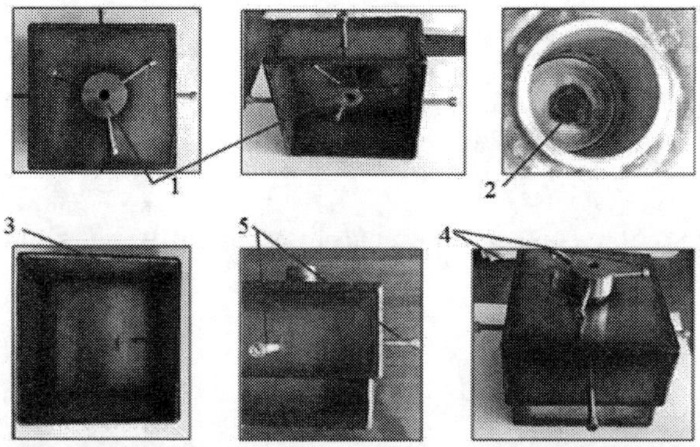

Figure 3. Tooling for dispersing micro and nanosuspensions: 1- a pipe for installing a focusing tube in it; 2 - a cylindrical target with a diamond single crystal fixed in it; 3 - tank for collecting suspension (bottom of the tank); 4 - screws - an element for fixing the upper cover to the bottom of the container for collecting suspension; 5 - screws - target fixation element..

The results of the analysis were the average particle sizes for the quantitative and volume distributions, as well as the minimum recorded particle size in the samples. After ultrasonic treatment and ultra-jet processing, the treated nanosized suspensions were studied using a Microtrac Bluewave laser particle size analyzer (Microtrac S3500), using the Tri-laser technology shown in Fig. 4 [8,9].

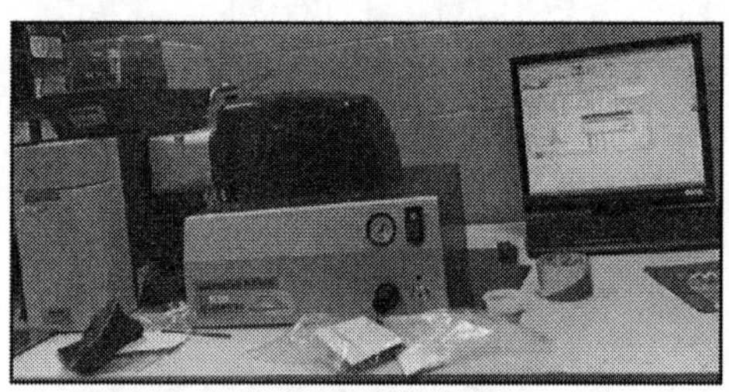

Figure 4. Laser Microtrac S3500 particle analyzer.

The results of a comparative analysis of the particle size in the composition of the suspensions are shown in the following figures.

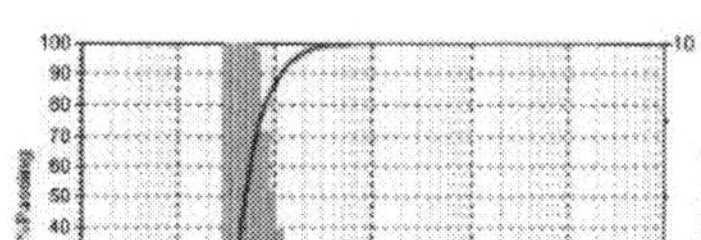

boehmite AL (OOH)

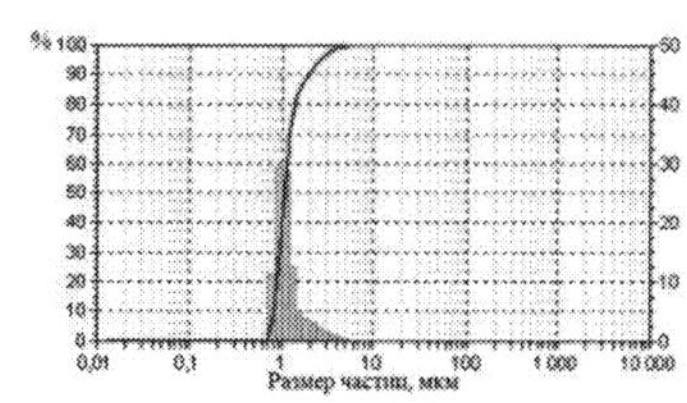

cabon nanotubes

Figure 5. Graphs of values of average particle sizes by quantitative distribution for initial samples

IOP Conf. Series: Materials Science and Engineering **934** (2020) 012035 doi:10.1088/1757-899X/934/1/012035

boehmite AL (OOH) cabon nanotubes

Figure 6. Graphs of values of average particle sizes by quantitative distribution after ultrasonic treatment

boehmite AL (OOH) cabon nanotubes

Figure 7. Graphs of values of average particle sizes by quantitative distribution after ultrajet processing

For suspension, the main technological parameters and informative data on ultra-jet suspension are presented in table 1. The nozzle head was moved along the surface of the sample in a spiral of Archimedes, the path length was 575 mm. This made it possible to ensure uniformity of removal of the surface layer of the target material (sample) in comparison with the point impact on the surface and circular motions of small radius, as can be seen in Fig. 8 [9,10]. The materials used in the experiments on the preparation of suspensions were selected: pure silver (Ag 999.9) and electrolytic copper (Cu-ETP) of high purity (99.95%) [10,11].

(a) (b) (c)

Figure 8. Tooling for receiving and collecting suspension((a) Fixing the sample in technological equipment, (b) The process of combining an ultra-jet nozzle with a hole in the disk of the top cover of the container for collecting the suspension, , (c) 1 - carbide target (VK8); 2 - a cavern)

The experiments were carried out at two pressures in the hydraulic system of 200 and 350 MPa, with a feed rate of 2 mm / s and a distance from the cut of the focusing tube to the sample surface of 2 mm, unit operation time t = 4 min, feed rate S = 2 mm/s, the distance from the nozzle to the target h = 2 mm (see Table 1) [12,13].

Table 1 - The results of experimental studies on the preparation of suspensions

Target Material: Copper (Cu)		
Suspension mode	Pressure P = 200 MPa	Pressure P = 350 MPa
The mass of the target before processing, gr	m=138,67	m=139,04
The mass of the target after processing, gr	m_1=138,63	m_1=138,67

The temperature of the suspension, ° C	54,0	73,0
Diameter of material removed from the target, mm	18	18
The speed of the jet at the nozzle exit, m/s	447	592
The flow rate of the working fluid, l/min	4,7	6,3
power, kWt	45	60
The mass of carried away material, gr	0,040	0,460
Target Material: Silver (Ag)		
The mass of the target before processing, gr	m=69,62	m=72,29
The mass of the target after processing, gr	m_1=69,38	m_1=69,62
The temperature of the suspension, ° C	55,6	78,4
The speed of the jet at the nozzle exit, m/s	447	592
The flow rate of the working fluid, l/min	4,7	6,3
power, kWt	45	60
The mass of carried away material, gr	0,240	2,670

At the next stage of experiments using a particle analyzer, the following were determined: mass concentration of ultra-jet suspensions; shape, size and distribution of solid particles formed during hydrodispersion of targets, the particle size distribution of the materials under consideration was determined in the volume of the suspension under investigation (Table 2).

Table 2 - The results of the analysis of the geometric parameters of the microparticles of the material after ultra-jet hydraulic action

Size		Number of particles, %		Average area in the considered range (μm²)
Over (μm²)	Before (μm²)			
Target Material - Silver				
0	9000	69	89,60	2184
9000	18000	6	7,80	12974
18000	27000	0	-	-
27000	36000	1	1,30	28608
36000	45000	1	1,30	41418
Target Material - Copper				
0	2500	136	85,50	816
2500	5000	17	10,50	3556
5000	7500	4	2,48	5363
7500	10000	0	-	-
10000	12500	1	0,62	11962
12500	15000	3	1,90	13117

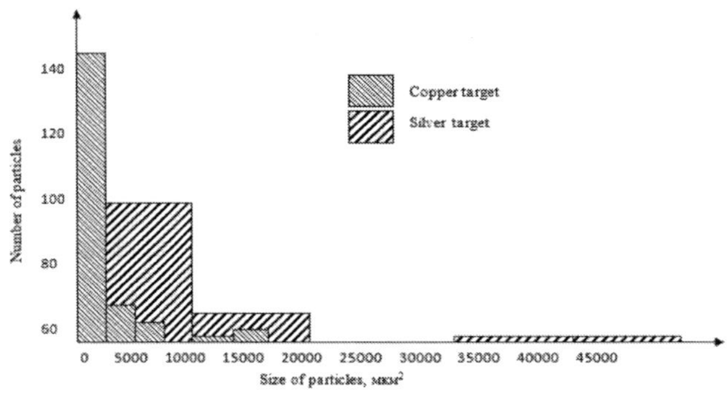

Figure 9. Typical distributions of copper and silver particles of various fractions depending on their area.

According to the results of experiments on ultrasonic suspension, the following main conclusions were made.

1. Changes in technological conditions and ultra-jet suspension modes significantly affect the mass concentration of the final product. In particular, it was found that ultra-jet suspension at high pressure (350-400 MPa) of the hydraulic system increases the concentration of the target materials in the liquid by an order of magnitude, which was also revealed using "soft" targets (Cu, Ag) instead of steel plates St45.
2. The shape, size and development of the surface of the particles significantly depends on the physico-mechanical characteristics of the sample material.
3. The fractional composition of the solid phase of suspensions obtained by ultra-jet technology is very different, which obviously can affect their functional properties.

Taking into account the results of previous studies [12], it can be assumed that the wide possibilities of the ultra-jet suspension technology for varying the input and output processing parameters will allow a very flexible approach to solving a wide variety of specific tasks, such as, for example, increasing the physicochemical activity of a suspension, low-temperature sterilization, changes hydrogen indicator, changes in microbiological and other vitological properties, etc.

(a) (b)

Figure 10. Images of microparticles of target materials after hydroerosion of their surface as a result of ultra-jet impact, (a) image of silver microparticles at 425x magnification, (b) image of copper microparticles at 415x magnification

3. Conclusion

A comparative analysis of the obtained particle size distributions of the studied suspension samples with their initial distributions showed that the ultra-jet treatment method can be used in practice and is quite effective. Its advantages were noted for dispersing suspensions with carbon nanotubes and boehmite, where the minimum particle size was an order of magnitude smaller in comparison with initial state of samples.
A comparative analysis of the obtained particle size distributions of the studied suspension samples with their initial distributions showed that the ultra-jet treatment method can be used in practice and is quite effective. Its advantages were noted for dispersing suspensions with carbon nanotubes and boehmite, where the minimum particle size was an order of magnitude smaller in comparison with initial state of samples.
Conducted experimental studies for suspension and analysis of the results showed that in general we can talk about the possibility of using ultra-jet technologies to obtain suspensions. This technology has a number of advantages and has a number of fundamental differences, in particular, unlike others, it is based on a powerful shock-dynamic effect, which exerts an ultra-strong on a solid-state target being processed. The power density of the ultrajet is comparable with industrial lasers for this indicator. Moreover, it was previously proved that the high-speed action of the ultra-jet affects the properties of liquids, which is close in its results to the technology of ultrasonic processing. Considering the previously performed studies can be said that the energy of the newly formed surfaces of the target material — the surface energy of microparticles of the solid phase — is very significant. As a result, these particles are very reactive in physicochemical respect. This makes it possible to create new hydraulic technologies for using ultra-jet micro- and nanosuspensions and the corresponding gel structures, in particular, based on water-soluble polymers, high molecular weight organic liquids and liquid crystals.

Acknowledgments

The studies were carried out within the framework of grants of the Russian Foundation for Basic Research 18-29-18081 and 19-38-90228\19, a grant of the President of the Russian Federation for state support of leading scientific schools of the Russian Federation NSh-3778.2018.8 and grants from the Fund for the Promotion of Innovations under the UMNIK-18 program in accordance with the contract No. 14727GU/2019 and No. 14549GU/2019.

References

[1] Tarasov V A and Nelyub V A 2018 Rheological model of viscous incompressible fluid with disperse fillers (*edited by International Journal of Mechanical Engineering and Technology, Moscow*) pp 488-497

[2] Galinovsky A L, Bethena G F and Petrov I V 2017 Improving the performance of waterjet cutting of materials by selecting rational modes of processing by acoustic emission (*edtied by metal technology,Moscow*) pp 18–27

[3] Ilyukhina A A 2018 Analysis of the interconnectedness of the functional physical capabilities of additive and ultra-jet technologies (*edtied by Fundamental and applied problems of engineering and technology*) pp 73–81

[4] Mugla D R 2019 Selection of Rational Technological Modes and Parameters of Underwater Waterjet Cutting (*Lecture Notes in Mechanical Engineering*) pp 267-276

[5] Iyukhina A A and Kolpakov V I 2018 The Features of Hydroabrasive Cutting of Honeycomb Panels of Space Vehicles (*Moscow University Physics Bulletin, Moscow*) pp 441-446

[6] Nelyub V A 2015 Determination of adhesion interaction between carbon fiber and epoxy binder (*Polymer Science - Series D*) pp 6-8

[7] Nelyub V A, Borodulin A S, Kobets L P and Malysheva G V 2016 Capillary hydrodynamic of oligomer binder (*Polymer Science – Series D*) pp 322-325

[8] Nelyub V A, Borodulin A S, Kobets L P and Malysheva GV 2016 A study of structure formation in a binder depending on the surface microrelief of carbon fiber (*Polymer Science – Series D*) pp 286-289

[9] Nelyub V A 2016 A study of the microstructure of dressed glass fibers (*Polymer Science – Series D- 9 (1)*) pp 96-100
[10] Nelyub V A 2015 Determination of adhesion interaction between carbon fiber and epoxy binder (Polymer Science - Series D - 8 (1)) pp 6-8
[11] Nelyub V A, Borodulin A S, Kobets L P and Malysheva G V 2016 Capillary hydrodynamic of oligomer binder (Polymer Science – Series D 9(3)) pp 322-325
[12] Puzakov V S 2007 Development and analysis of the functional capabilities of ultra-jet activation of hydrotechnological media for mechanical processing production (*Vetnik BMSTU publication, Moscow)*
[13] Galinovsky A L, Provators A S and Kyaw Myo Htet 2019 On the question of the effectiveness of various methods of dispersing nanosuspension suspensions (*All materials: Encyclopedic reference*) No 11 S 2 – 7
[14] Galinovsky A L, Provators A S and Kyaw Myo Htet 2020 Prospects for the Development of Ultra-Jet Dispersion Technology for Nanocontaining Suspensions (*IOP Conf. Ser.: Mater. Sci. Eng. 709 044092)* volume 709
[15] Galinovsky A L, Provators A S and Barzov A A 2019 Information-physical mechanism of ultrahydrojet diagnostics of the quality of functional coatings (*Hardening Technologies and Coatings*) No 2 S 51 – 57
[16] Mugla D R and Golubev E S 2019 Surface express ultrajet diagnostics of space vehicle materials (IOP Conference Series: Materials Science and Engineering) Vol 683 Issue 1
[17] Galinovskij A L, Abashin M I and Khafizov M V 2014 Rapid Determining of the Optimum Operation Mode for Abrasive Waterjet Cutting Process by Means of Acoustic Emission (*Applied Mechanics and Materials*) Vol 698
[18] Zh u J, Imam A, Crane R, Lozano K, Khabasheku V N and Barrera E V 2017 Processing a glass fiber reinforced vinyl ester composite with nanotube enhancement of interlaminar shear strength (*Compos Sci Technol*) pp 1509–17
[19] Tarasov V A and Baskakov V D 2016 Mathematical modeling of the collision process of flat jets of an ideal fluid (*sumitted to Vestnik BMSTU, Moscow*) p 79–90
[20] Tarasov V A, Stepanishchev N A and Boyarskaya R P 2011 Methods of experimental determination of the characteristic points in time of the technological process of preparing nanosuspensions under conditions of ultrasonic action (*submitted to Vestnik BMSTU*) p 53–65

IOP Conf. Series: Materials Science and Engineering **934** (2020) 012036 doi:10.1088/1757-899X/934/1/012036

Innovative processes of production functional gradient layered compositions with enhanced tribological properties

R S Mikheev

Faculty of Engineering Technologies, Bauman Moscow State Technical University
5, 2-ya Baumanskaya str., Moscow, 105005, Russian Federation

E-mail: mikheev.roman@mail.ru

Abstract. Arc surfacing processes are proposed as innovative methods for the synthesis of functionally gradient layered compositions. Existing composite filler materials based on aluminum and tin have been investigated. The schemes and technological parameters of arc welding allowing to produce the coatings having a composite structure with a given reinforcement ratio, satisfactory reinforcements distribution, and enhanced tribotechnical properties have been shown.

1. Introduction

The friction units used in machines and mechanisms have a wide range of constructions and represent both mono- and bimetallic structures: for example, bushings or blocks of sliding bearings. Moreover, surface layers mainly work in the friction units and their damage due to wear leads to emergency situations [1-3]. Therefore, an extremely important and promising scientific direction meeting modern requirements is the creation of functionally gradient layered compositions with enhanced tribotechnical properties based on steels or alloys with surface layers consisting of particle-reinforced composite materials (CM) based on non-ferrous alloys. The variety of CM forms allows adjusting the basic tribotechnical properties by the composition selection, changes in the ratio of components and the formation processes choice [4-9]. This will significantly reduce manufacturing costs, as well as provide the opportunity to carry out repairs during the operation of products made from similar materials [8-10]. However, the experience of obtaining and using such functionally gradient layered compositions in friction units is rather limited.

2. The choice of processes for creating functionally gradient layered compositions

Existing processes for the formation of CM coatings on the basis of steels or alloys can be divided into three groups [8]:

- processes occurring in the solid phase (for example, rolling packages of matrix material and CM, explosion welding, thermomechanical processing of the CM surface layer);
- processes occurring in the liquid phase (for example, centrifugal casting, arc, laser, plasma surfacing, modifying treatment);
- coating processes from the gas phase.

Among them the methods of creating functionally gradient layered compositions by arc welding processes are the most economical and promising according to the criteria: productivity, universality, implementation level and accessibility in industry [8-11]. The use of such processes will allow the formation of the necessary structure and the specified properties of CM coatings due to layer-by-layer synthesis. An additional features of these methods are the possibility to provide the necessary level of wetting and to limit the chemical interaction between the filler, matrix, and substrate material due to the correct choice of temperature-time characteristics (heating temperature, time at a given temperature), the introduction of alloying elements, or the use of intermediate transition layers [5, 8]. All these will allow to solve the important problem of thermodynamic and kinetic compatibility of CM components.

3. Composite filler material development

According to modern concepts, the CM working capacity is an integral characteristic that depends on the physical and chemical properties of the phases that make up the materials and their ratio. Therefore, the lack of appropriate filler materials until recently limited the wide using of arc processes for composite coatings deposition in the manufacture of functionally gradient layered compositions. To date, there is positive experience in the manufacture and testing of CM based on aluminum or tin, which satisfy the basic requirements for filler materials, including:

- high-quality formation of a deposited coatings with saving and uniform distribution of reinforcing particles;
- lack of reinforcement degradation during the deposition of composite coatings;
- absence of defects in the deposited coatings.

Such CMs are characterized not only by a high level of tribotechnical properties, but also by sufficient manufacturability [8, 10, 12]. This makes it possible to obtain from them filler materials in the form of rods meeting the requirements of GOST 21449 for the arc welding processes in automatic or manual mode with using standard welding equipment, which is especially valuable when carrying out repair operations (Fig. 1).

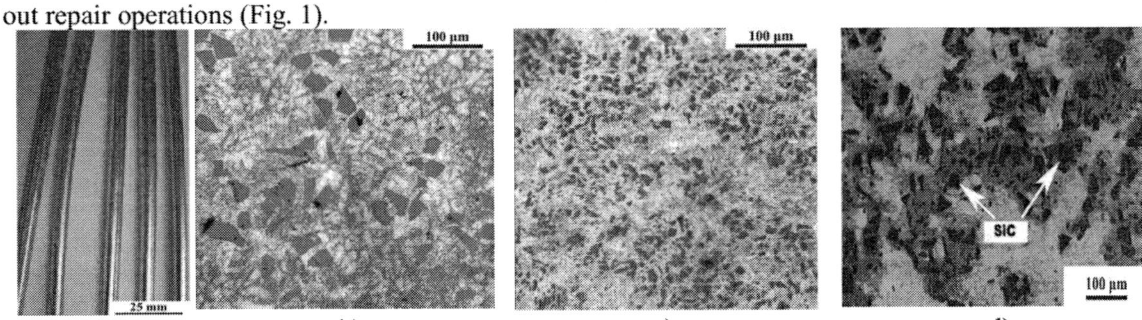

a) b) c) d)

Figure 1. Appearance (a), microstructure (b, c, d) of the filler rods with compositions wt.%: Al-12Si + $10SiC_{(40)}$ (a, b), Al-12Si-2Cu-1Mg-1Ni + $10SiC_{(14)}$ (c), B83 + $5SiC_{(40)}$ (d).

The use of foundry technology for the manufacture of aluminum-based CM filler materials ensures uniform distribution of reinforcing particles, the absence of intense interfacial interaction and contamination at the particle-matrix interface (Fig. 1, b, c). To achieve the required quality of the composite coatings formation, such materials should contain no more than 10 wt.% reinforcement particles (SiC) of the minimum size (14 microns) and (11-13) wt.% silicon. Therefore, widely used in industry casting alloys Al-12Si and Al-12Si-2Cu-1Mg-1Ni are recommended as a matrix for producing of composite filler rods [8].

The technology of B83-based CM filler materials production has been first developed and tested to introduce and ensure uniform distribution in the tin matrix of poorly wettable reinforcements and modifying additives of various types and fraction size. The technology consists of stages: mechanical alloying and cold pressing of the powder composite mixture; compact sintering and hot extrusion [11]. Produced according the technology filler rods from CM based on B83 alloy with a reinforcement mass fraction not more than 5% are characterized by the absence of defects (Fig. 1, d).

4. Deposition of composite coatings

The application of the produced new filler CMs required to solve a number of primary problems. The main of them are related to the choice of schemes and determination of technological parameters of the deposition processes, allowing to form high-quality coatings and ensure uniform reinforcement distribution and the absence of its degradation. In addition, the development of technologies for the formation of such coatings on low-carbon steel substrates is of particular importance. The technology will make it possible to obtain bimetallic compositions widely used in friction units, the main requirement of which is to provide standard values of adhesive strength and high level of tribotechnical properties.

The schemes and technological parameters of the arc welding process that differ in the temperature-time effect on the filler material were tested. It was established that the scheme with introduction of composite filler rods into the head of the weld pool (along the leading edge) was preferred (Fig. 2, a) [10]. The maximum values of the melt heating temperature and the duration of the reinforcement in it lead to an increase in fluidity, which contributes to the formation of coatings with a uniform reinforcement distribution (Fig. 2, c, d).

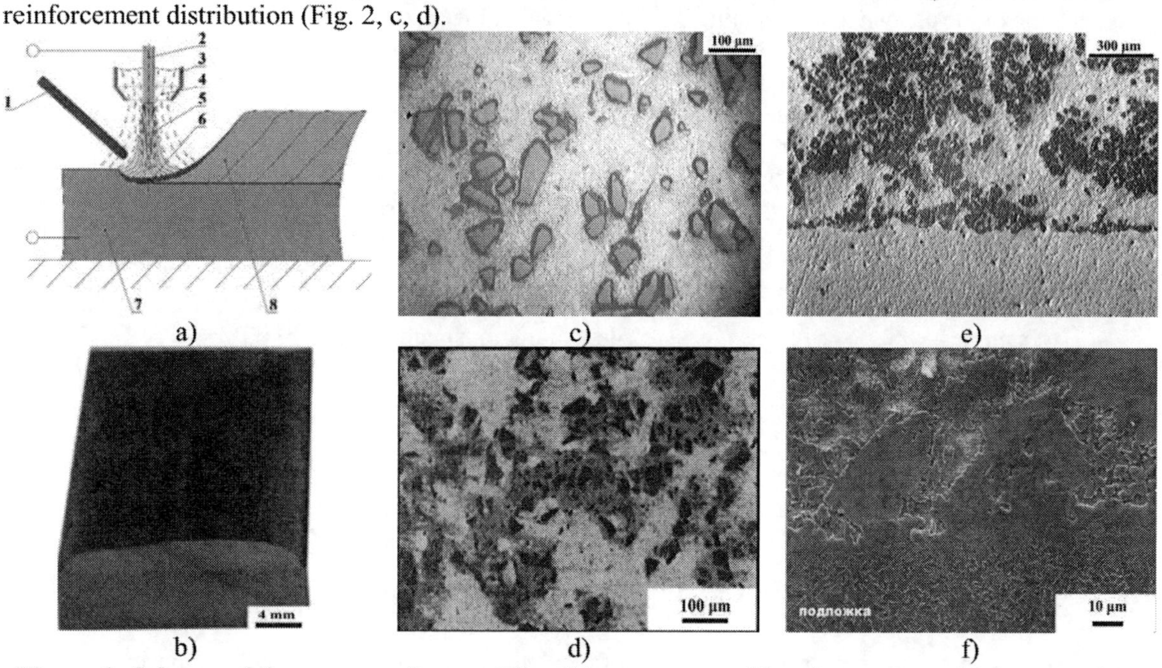

Figure 2. Scheme of the process of arc welding (a), appearance (b), microstructure in the central part (c, d) and along the fusion line (e, f) of the coating compositions, wt.%: b, d,f) B83 + $5SiC_{(40)}$; c) Al-12Si + $10SiC_{(40)}$; e) Al-12Si-2Cu-1Mg-1Ni + $10SiC_{(14)}$. The numbers indicate: 1 — composite filler rod; 2 - non-consumable electrode; 3 - welding torch; 4 - shielding gas; 5 - electric arc; 6 - weld pool; 7 - substrate; 8 - deposited layer.

Technological parameters of the arc welding process were selected according to minimum degree of penetration of the substrate from Al-3Mg alloy (GOST 4784), saving of the reinforcing code and high-quality formation of coatings (welding current - I = 150-160 A, arc voltage - U = 18 -20 V, welding speed - V = 7-12 m/h, shielding gas (argon) consumption - V = 12-15 l/min). The technological parameters allow receiving 2,5-3,0 mm thickness coatings in one pass, which crystallize epitaxially and, as a result, have a high level of adhesive strength (Fig. 2, e) [10]. The established scheme and regimes of the arc welding process can be used to form coatings of similar composition on the low-carbon steel bases. It will make possible to obtain bimetallic compositions widely used in friction units. However, in such cases for preventing the formation of a continuous intermetallic layer, it is necessary to limit the thermal effect of the electric arc on the steel-aluminum interface by providing partial penetration of the intermediate aluminum layer formed by the explosion welding process [13].

Tin-based CM coatings are characterized by a smooth surface with a smooth transition to the metal of the steel substrate. The coatings have 2.0-3.0 mm thickness of each layer after the machining required to level the working surface. Preliminary tinning of the steel (St3sp in accordance with GOST 380) substrate surface provided provided fluidity increase during the formation of composite coatings due to the formation of an intermediate layer of pure tin with a thickness of not more than 0,5 mm. The technological parameters of the arc welding process are determined (welding current - I = 40-60 A, arc voltage - U = 16-18 V, welding speed - V = 7-12 m/h, shielding gas (argon) consumption - V = 12-15 l/min). The technological parameters ensure the stability of the phase composition and structure, the absence of alloying elements burnout and minimal mixing of the base and deposited metal (Fig. 2, f) [14].

5. Tribological properties of deposited coatings

To evaluate of the deposited CM coating working capacity, the results of their wear tests were compared with those for well-known and widely used in industry antifriction materials based on aluminum and tin. Tests were carried out using machines (tribometers): MTU-01 and CERT UMT Multi-Specimen Test System under dry sliding friction conditions according to scheme "bush on disk" at a sliding speed of 0,39 m/s. The formed aluminum-based CM coatings are characterized by an extended range of tribo-loading. The coatings have 2-3 times lower values of wear rate and wear coefficient compared to the anti-friction alloy Al-20Sn-1Cu, which goes into intensive wear at specific pressures of more than 0,46 MPa (Table 1).

Table 1. Wear rate and wear coefficient of antifriction alloy Al-20Sn-1Cu and deposited CM coatings.

Type and composition of samples, wt%	Wear rate Iv, x10^{-3} mm^3/m / Wear coefficient K, x10^{-4}, depending on specific pressure p, MPa				
	0,2	0,33	0,46	0,59	0,7
Al-20Sn-1Cu	10,2 / 1,69	15 / 1,54	18,9 / 1,4	intensive wear	
Al-12Si + 5SiC$_{(40)}$	4,24 / 1,41	6,3 / 1,35	7,72 / 1,19	8,58 / 1,03	10,98 / 1
Al-12Si-2Cu-1Mg-1Ni + 5SiC$_{(14)}$	4,09 / 1,95	6,21 / 1,9	7,66 / 1,68	9,35 / 1,6	11,44 / 1,63
Al-12Si-2Cu-1Mg-1Ni + 10SiC$_{(14)}$	4,28 / 2,09	5,95 / 1,87	6,65 / 1,5	6,92 / 1,22	7,38 / 1,08
AK12 + 10SiC$_{(40)}$	4,69 / 1,7	5,88 / 1,37	6,65 / 1,11	7,2 / 0,94	7,38 / 0,8

A larger number of reinforcing particles in the coatings reduces the matrix fraction in the friction surface. It reduces the risk of intensive wear and helps to reduce the load on each particle individually when they play the role of supporting elements on the contact surface. This increases the wear resistance, which is especially noticeable at specific pressures of more than 0,46 MPa. Also the tribo-loading parameters range (specific pressures, sliding speeds and loading durations) is significantly expanded.

The friction process of tin-based CM coatings obtained by arc welding is characterized by high stability (stability coefficients characterizing the level of vibration and noise achieved by friction, not less than 0,8) in the entire tribo-loading range. It is associated with an increase in hardness due to dispersion of the matrix alloy structure during deposition process [15]. The presence in the composition of submicron-sized particles (e.g. boron or boron carbide) increases the wear resistance of coatings by 35% and 30% (from Im = 2,62x10^{-5} g/m to Im = 1,71x10^{-5} g/m and Im = 1,77x10^{-5} g/m, respectively). Also it reduces the value of the friction coefficient by 40-60% compared with coatings made of antifriction alloy B83 (Fig. 3). The use of developed composite filler rods containing SiC particles provides an increase in wear resistance by 30% and a decrease in the friction coefficient by 30–40% even at the maximum specific pressure tested (0,7 MPa) [15].

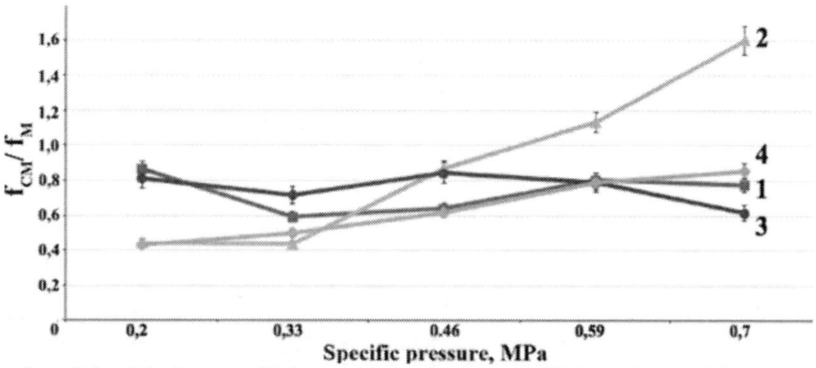

Figure 3. The ratio of the friction coefficients of the deposited CM coatings with composition, wt.%: 1 – B83 + 0,25$B_{(<0,5)}$; 2 – B83 + 0,25$B_4C_{(<1)}$; 3 – B83 + 5$SiC_{(40)}$; 4 - B83 + 5$SiC_{(40)}$ + 0,25$B_4C_{(<1)}$ and coatings from babbit B83 depending on specific pressure.

Comprehensive studies of the relief and composition of the friction surfaces, the morphology of the wear products made it possible to determine that the wear of the deposited composite coatings occurred according to the abrasive and oxidative mechanisms. These mechanisms are characteristics of the initial and steady-state stages of the friction regime. After testing at maximum specific pressures of 0,7 MPa, plastic deformation bands and setting zones are formed on fairly smooth friction surfaces of the samples. These zones are characterized by fragmentation and small area in comparison with samples of antifriction alloys (Fig. 4, a, b, c). The secondary cellular structures of 5-10 µm in size are formed in the plastic deformation bands at the stage of steady wear. The structures are contoured by ceramic particles with an average size of 1 µm, which minimize the destruction of surface layers and screen unacceptable setting processes. Moreover, the friction surfaces of composite coatings containing different reinforcement particles are characterized by the minimally pronounced plastic deformation bands, as well as the largest area occupied by the mechanical mixture from the transition layer composition formed during the friction process. The small (up to 10 µm) particles of equiaxial shape formed as a result of wear by the oxidation mechanism are dominated among the wear products (Fig. 4, d) [10, 15].

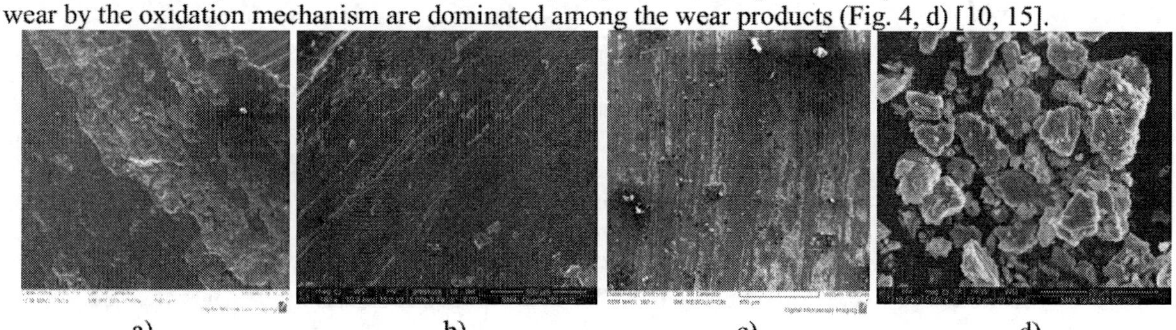

a) b) c) d)

Figure 4. Friction surfaces (a, b. c) и wear products (d) of the coatings after testing at a specific pressure of 0,7 MPa. The composition of the samples, wt.%: a) B83; b, d) Al-12Si-2Cu-1Mg-1Ni + 5$SiC_{(14)}$, c) B83 + 5$SiC_{(40)}$ + 0,25$B_4C_{(<1)}$.

Thus, the proposed processes make it possible to create functional gradient layered compositions with high tribological properties in a wide range of tribo loading: specific pressure from 0,2 to 3,3 MPa, sliding speed from 0,39 to 1,0 m/s. This allows us to recommend the developed materials for use in friction units of machines and mechanisms with high work stresses, which include bearings and bearing shells of steam and gas turbines, marine small and medium-sized diesels.

6. Conclusion

Arc surfacing processes are proposed as innovative methods for the synthesis of functionally gradient layered compositions. Existing composite filler materials based on aluminum and tin were investigated. The schemes and technological parameters of arc welding were shown. They allow to produce the coatings having a composite structure with a given reinforcement ratio, satisfactory reinforcements distribution, and enhanced tribotechnical properties (increased wear resistance up to 3 times, reduced friction coefficient 60% compared with traditional antifriction alloys Al-20Sn-1Cu and B83).

References

[1] Drokonov A M, Osipov A and Birykov A V 2009 *Turbine Bearings: Tutorial* (Bryansk: BGTU)
[2] Pogodaev L I, Kuzmin A A and Lopyrev Y K 2015 *Reliability of Ship Equipment* (Saint-Petersburg: GUMRF)
[3] Yazykov A E, Kazanskiy V N and Buskunov R Sh 2010 *Bearings and Oil Systems: Tutorial* (Chelyabinsk: Cicero)
[4] Chernyshova T A, Mikheev R S, Kalashnikov I E, Akimov I V and Kharlamov E I 2011 *Inorg. Mater. Appl. Res.* **2** 322-9
[5] Mikheev R S and Chernyshova T A 2008 *Zagotovitel'noye Proizvodstvo v Mashinostroyenii* **11** 44
[6] Kalashnikov I E, Bolotova L K and Chernyshova T A 2011 *Nanotechnologies in Russia* **6** 144
[7] Kalashnikov I E, Bolotova L K, Bykov P A, Kobeleva L I, Katin I V, Mikheev R S and Kobernik N V *Russian Metallurgy (Metally)* **7** 669
[8] Mikheev R S and Chernyshova T A 2013 *Aluminum Matrix Composite Materials for Solving Problems of the New Technology* (Moscow: Maska)
[9] Kurganova Yu.A., Chen Y., Shcherbakov S.P. Evaluation of the effectiveness of the introduction of a discrete filler in the aluminum matrix melt // IOP Conference Series: Materials Science and Engineering. 2019. V. 683. 012034.
[10] Mikheev R S, Kobernik N V, Kalashnikov I E, Bolotova L K and Kobeleva L I 2015 *Inorg. Mater. Appl. Res.* **6** 493
[11] Mikheev R S, Kobernik N V and Kovalev V V 2014 *Fizika i Khimiya Obrabotki Materialov* **4** 35
[12] Kalashnikov I E, Bolotova L K, Katin I V, Kobeleva L I, Kolmakov A G, Mikheev R S and Kobernik N V 2017 *Inorg. Mater. Appl. Res.* **8** 335
[13] Mikheev R S, Kobernik N V, Kovalev V V, Kalashnikov I E, Kolmakov A G, Bolotova L K and Kheifets M L 2019 *Svarka i Diagnostika* **2** 48
[14] Mikheev R S, Kobernik N V, Kalashnikov I E, Bolotova L K, Bykov P A, Kobeleva L I and Kolmakov A G 2017 *Inorg. Mater. Appl. Res.* **8** 745
[15] Bolotova L K, Kalashnikov I E, Kobeleva L I, Katin I V, Kolmakov A G, Mikheev R S, Kobernik N V and Podymova N B 2018 *Russian Metallurgy (Metally)* **1** 63

Simulation and Optimization of a CFRP and a GFRP floating pontoon

M N Lotfy[1], E Fathallah[1], Y A Khalifa[1] and A K Dessouki[2]

[1]Department of Civil Engineering, Military Technical College, Cairo, Egypt.

[2]Ain Shams University, Cairo, Egypt.

Abstract. According to the unique mechanical properties of composite materials, the replacement of the conventional materials like steel with composite materials is increasing in all fields. The low weight / capacity characteristics of composite materials make it possible to accommodate much more loads. Because of the great importance of the floating bridges for both military and civilian purposes especially in crises, the evolution of floating bridges is necessary. The traditional transportation steel ferry is composed of floating steel pontoons to accommodate the MLC-70 (Tank load) of weight equals (63.5 t). In this study, the steel floating pontoon is replaced with composite pontoon simulated for both carbon fibres polymers and glass fibres polymers with the laminate configuration$[\alpha_1^{\circ}/\alpha_2^{\circ}/\alpha_3^{\circ}/\ldots/\alpha_n^{\circ}]$. The finite element analysis is performed using ANSYS software. The capacity of the ferry is increased to reach (90 t) instead of (70 t). The total deformation is determined under the applied load. The failure criteria is investigated for both composite models (Tsai-Wu, Tsai-Hill, maximum stress and maximum strain). The nonlinear buckling analysis is also, investigated. The optimization process is designed and performed to get the optimum number of layers and angles orientation of the composite layers as well as possible.

Keywords: Floating; Pontoon; Composite; Optimization; Modelling

1. Introduction

Composites materials have been increasingly used in marine and civil engineering applications instead of conventional construction materials such as steel. [1, 2]. Many Studies conducted on the designing and optimization of composite structures using different types of materials [3-8]. Marín et al.[9], presented an optimization methodology for a geometric design of a composite stiffened panel, with optimum stacking sequence using static analysis and hydrothermal effects. Herath et al.[10], illustrated an optimization method to optimize a composite marine propeller blades using genetic algorithm and finite element approach. Costaetal [11], described an application of a genetic algorithm to a material and sizing optimization problem of a plate. Kim et al. [12], optimized a layup using genetic algorithm for composite wing and maximize the failure index of Tsai–Hill criterion. Akbulut [13], optimized a composite plates for weight minimization under in-plane loading. Hirono et al. [14], determined the vertical displacement of a floating bridge. There are many researchers considered the pontoon as a beam rested on elastic foundation subjected to applied loads [15]. Fu and Cui [16], studied numerically the hydro elastic responses of connected pontoons under dynamic and static cases of loading and introduced a numerical method for analysis of a ribbon floating bridge. Alnahhala and Aref [17], illustrated that the combination of FRP composites with concrete present a promising improvement in civil infrastructure applications additionally, the stiffness of the hybrid bridge was 35% higher than FRP-only bridge. Khalifa [18], investigated the floating bridge under eccentric static loads. Raftoyiannis et al. [19], studied numerically the dynamic response of floating bridges under moving load with constant velocity. Siwowski and Rajchel, [20], studied the structural performance of hybrid FRP composite concrete bridge girder for evaluating the strength, stiffness and the global safety of the girder. Wang et al. [21], investigated the dynamic response of floating bridge subjected to moving load. Panteleev [22], optimized a metallic sandwich structure for decreasing its weight. Abozaid et al. [23], investigated numerically the structural performance of pontoon with 70 ton capacity made of hybrid composite FRP. The results illustrated that the PVC foam has a great impact to reduce the web buckling and the weight. In this study, a numerical simulation and optimization of a CFRP and a GFRP floating pontoon is carried out to investigate the structural performance of composite pontoon to overcome the disadvantages of the traditional transportation steel ferry. Finite element model is constructed using ANSYS Composite Prep Post (ACP). The failure criteria are incorporated in the optimization process for composite models (Tsai-Wu, Tsai-Hill, maximum stress and maximum strain). The optimum number of layers and the best orientation of the angles are determined.

2. Composite Failure Criteria

Based on these failure mechanisms it can be more appropriate to consider the composite as a structure rather than as a material. Fibre failure is one of the simplest failure mechanisms to identify and quantify which occurs, when the loads applied to a composite structure cause fracture in the fibers. Matrix cracks are an intralaminar form of damage and involve cracks or voids between fibers within a single composite layer or lamina. Additionally, buckling is a structural phenomenon that occurs in compression or shear and though not necessarily resulting in failure. The large deformations, bending and loss of structural capacity involved typically promotes other types of damage and leads to structural collapse [24] . A successful design of a structure requires efficient and safe use of materials. Theories need to be developed to compare the state of stresses and strains in a material to failure criteria. So, that one needs to use reliable failure criteria.

2.1. Maximum Stress Failure Theory

In this case, the principal stresses in each ply are compared with their corresponding strength values X_t, X_c, Y_t, Y_c, and S. Where X_t and X_c are the longitudinal tensile and compressive strengths, respectively, Y_t and Y_c are those in the transverse direction, S is the ultimate in-plane shear strength. The failure index is defined as [25, 26]:

$$I_F = \max \begin{cases} \sigma_{11} / X_t \quad \text{if} \quad \sigma_{11} > 0 \quad or \quad -\sigma_{11} / X_c \ \text{if} \quad \sigma_{11} < 0 \\ \sigma_{22} / Y_t \quad \text{if} \quad \sigma_{22} > 0 \quad or \quad -\sigma_{22} / Y_c \quad \text{if} \quad \sigma_{22} < 0 \\ |\tau_{12}| / S \end{cases} \tag{1}$$

2.2. Maximum Strain Failure Criteria

The maximum strain failure criterion can be stated as:

$$I_F = \max \begin{cases} \varepsilon_{11} / \varepsilon_t^x \quad \text{if} \quad \varepsilon_{11} > 0 \quad or \quad -\varepsilon_{11} / \varepsilon_c^x \ \text{if} \quad \varepsilon_{11} < 0 \\ \varepsilon_{22} / \varepsilon_t^y \quad \text{if} \quad \varepsilon_{22} > 0 \quad or \quad -\varepsilon_{22} / \varepsilon_c^y \ \text{if} \quad \varepsilon_{22} < 0 \\ |\gamma_{12}| / \gamma_{xy} \end{cases} \tag{2}$$

Where ε_{11} , ε_{22} and γ_{12} are ply strains in the ply coordinate system (1 parallel to fibers and 2 perpendicular to fibers). Also, ε_t^x , ε_c^x , ε_t^y , ε_c^y and γ_{xy} are allowable strains in the corresponding direction and loading (tensile or compressive).

2.3. *Tsai-Hill Failure Criteria*

Tsai-Hill failure criterion is one of the most practicable criterions for the prediction of damage progress in laminates, which assumed that, there is an important interaction between longitude strength, traverse strength and shear strength in the damage progress. The criterion can be expressed as [6, 27, 28]:

$$\sigma_{11}^2 / X^2 + \sigma_{22}^2 / Y^2 - \sigma_{11}\sigma_{22} / X^2 + \tau_{12}^2 / S^2 = 1 \tag{3}$$

Where X and Y are the longitude strength and traverse strength, respectively whether tension or compression which is depended on the stress status in laminates.

2.4. *Tsai-Wu Failure Criteria*

The Tsai-Wu failure criterion is the most generalized criterion since it distinguishes between the compressive and tensile strength of a lamina and can be written as the following expanded form [29, 30]:

$$I_F = F_{11}\sigma_{11}^2 + F_{22}\sigma_{22}^2 + F_{66}\tau_{12}^2 + F_1\sigma_{11} + F_2\sigma_{22} + 2F_{12}\sigma_{11}\sigma_{22} \tag{4}$$

Where the symbols σ_{11} and σ_{22} are the stresses in the longitudinal and transverse fiber directions, respectively, τ_{12} is the in-plane shear stress

3. Model definition

3.1. *Steel ferry definition*

The steel ferry is designed to accommodate the dynamic load of the tank (MLC-70) with capacity up to (70 t). The floating ferry configuration is composed of 16 repeated floating pontoons. These pontoons dimensions are 5.25 m length, 2.4 m width and 1.5 m height for every pontoon. The pontoons are separated from each other's by a connection system separating every pontoon from the other one by 10 cm. As a result to this configuration the ferry has a total length of 21.3 m and a total width of 9.9 m. The ferry is as shown in **Figure 1**.

Figure 1: The configuration of the ferry composed of 16 pontoon

The floating pontoon is composed of longitudinal, transverse, and vertical stiffeners simulated as beam elements. These stiffeners have different cross sections as shown in Figure 2. The upper, lower and side plates are simulated as shell elements with different thicknesses.

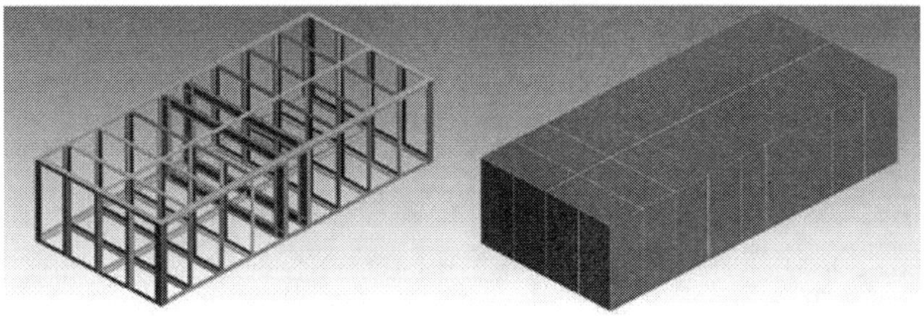

Figure 2: The Isometric sections of one pontoon.

The used material is steel 37 with the properties tabulated in **Table 1**. [18]:

Table 1: The properties of structural steel 37

Density	7850 kg/m^3
Tensile yield strength	2.4 × 10^8 (Pa)
Compressive yield strength	2.4 × 10^8 (Pa)
Tensile ultimate strength	4.6 × 10^8 (Pa)

The steel ferry is supported with elastic supports that represent the supporting water. These elastic supports have stiffness value that depends on the stiffness of both the structure and the water. To calculate the elastic support stiffness for any structure, it is obligatory to draw a non-linear curve between the elastic support stiffness values and their corresponding effect to the draft of the structure under its own weight and then choose the elastic support stiffness that achieve the draft of the floating structure under its own weight from the analytical analysis [18]. In our case the stiffness is calculated as follows:
The ferry follows the floating principles and Archimedes law as it is a floating structure. This law states that the weight of the displaced water by the submerged volume of the floating body is equal to the weight of the floating structure [31].

$$W_W = W_F \tag{5}$$

Where, (W_W) is the weight of the displaced water, and (W_F) is the own weight of the ferry. The own weight of the steel ferry (W_F) can be calculated so, (W_w) must be equal to that weight to follow Archimedes' law and make the ferry float. The weight of the displaced water (W_w) can be calculated as follows:

$$W_w = F_B * F_L * F_{SD} * \gamma_w \tag{6}$$

Where, (F_B) represents breadth of the ferry without connections (9.6 m), (F_L) represents the ferry length without connections (21 m), (F_{SD}) represents the submerged depth of the ferry in water which is still unknown for now and (γ_w) represents the density of water (1000 kg/m^3). By substituting in equation (2), (F_{SD}) is calculated. So, the the elastic support stiffness (k) that makes the deformation of the ferry under its own weight equals (F_{SD}) is computed as follows:

$$W_w = k * F_B * F_L * F_{SD} \quad (7)$$

By substituting in equation (2)

$$k = \gamma_w \quad (8)$$

So, the elastic support stiffness (k) is 9810 (N/m^3) for this simulation model.

4. The composite pontoon

The composite pontoon is simulated with the same exterior dimensions as the steel pontoon (5.25 m length, 2.4 m width and 1.5 m height) but, with different internal configuration. The stiffeners of the composite pontoon are designed to be shell elements. These shell elements are simulated to act as longitudinal and transverse stiffeners as shown in **Figure 3**. The simulation is performed using advanced composite programming (ACP).

Figure 3: The configuration of the composite pontoon

5. Boundary conditions

5.1. Loads and supports

The composite ferry is designed to accommodate a total dynamic load of (90 t). Because of the optimization process and the number of layers, it takes too much time to be simulated as a dynamic load so, the load is applied as a static load. The static load is increased by a factor (1.3) to represent the dynamic load so, the ferry is subjected to a static load of about (115 t). The equivalent static load that affect one pontoon is calculated to be a uniform load (P) of (0.54 t/m^2). The pontoon is supported by elastic supports that represent water.

5.2. Validation

The elastic support stiffness is calculated to be (9810 N/m3) as mentioned before. The verification is performed on the loaded pontoon.

$$W_T = O.W + F_{app.} \quad (9)$$

Where, ($F_{app.}$) is the applied load and ($O.W$) is the own weight of the composite pontoon which is calculated from the software to be equal (0.853 t) for the glass fiber-resin polymer (GFRP) model.

$$F_{app.} = P \times F_B \times F_L \tag{10}$$

By substituting in equation (5), the total weight is calculated so, the draft can be calculated as follows.

$$\Delta = \frac{W_T}{F_B \times F_L \times \gamma_w} \tag{11}$$

Where, (Δ) is the draft of the pontoon. By substitution (Δ) equals (62 cm).
The pontoon total deformations is determined from simulation to be (63.6 cm) which is 97.5% matching with the analytical value noting that this difference is the deflection of the upper shell. The total deformation is illustrated at **Figure 4**

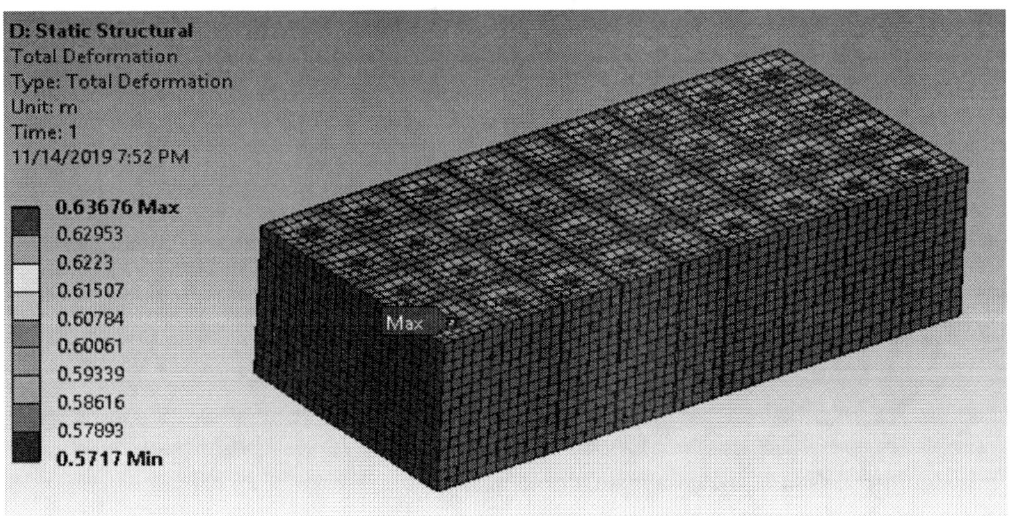

Figure 4: The total deformation of the loaded composite pontoon

6. Optimization

There are many techniques of optimization supported by ANSYS such as the screening method, non-dominated sorting genetic algorithm (NSGA), multi-objective genetic algorithm, non-linear programming by quadratic lagrangian, etc. The type of the technique is determined automatically in ANSYS according to the number of parameters, the objective function and the number of the constraints[32]. In our study the screening optimization is the most proper method to get the optimum weight of the ferry as well as possible. The Screening optimization method depends on using a simple approach based on sorting and sampling. The screening method supports multiple objectives and constraints as well as all types of input parameters. The screening method is the first process used in the new developed optimization strategy called the multiphase optimization strategy (MOST)[33, 34]. Usually, it is used for preliminary design, which may lead the designers to reach to the predicted results in a quick way instead of try and error methods.

6.1. The objective function

The pontoon is divided into 4 categories; the upper deck, the lower deck, the side transverse stiffeners and the longitudinal stiffeners. Each stiffener of these stiffeners is a composite laminate

The laminate is defined as a set of lamina oriented in a specific manner to achieve a desired result and this laminate's response depends on the mechanical properties of each lamina [35]. The objective function is to minimize the buoyancy factor (B F) which is defined as the ration between the total weight of the structure to the weight of the displaced water by the full body of the structure as if it is fully submerged [36] as the following equation illustrates.

$$The\ buoyancy\ factor(B\ F) = \frac{W_T}{V_t \times \gamma_w} \quad (12)$$

Where, (V_t) represents the total volume of the structure and (γ_w) represents the water density. The objective function is defined in the following equation as follows:

$$F_1(X): minimize\ \ B\ F = \frac{W_T}{V_t \times \gamma_w} \quad (13)$$

6.2. The design variables(Parameters)

The number of layers and the angle of orientation are the input parameters for the optimization process. The laminate parameters are defined as:

$$W = [\alpha_1^{\circ} / \alpha_2^{\circ} / \alpha_3^{\circ} / \ldots . / \alpha_n^{\circ}\ ],\ (-90^{\circ}) \leq \alpha \leq (90^{\circ})\ ,\ 1 \leq n \quad (14)$$

Where, (W) represents the laminate configuration, (α) represents the orientation angle of the layer, the subscript numbers beside (α) (1,2,3,….,n) represents the layer number not the number of layers used and (n) is the number of layers for the laminate.

6.3. Constraints

The optimization process is constrained by the failure criteria (FC), the total deformations and the buckling load factor (BLF). Tsai-Wu, Tsai-Hill, Max. Stress and Max. Strain are used to determine the failure of the composite model. The buckling load factor (BLF) is the ration of the critical buckling load (P_{cr}) to the applied load (P).

$$BLF = \frac{P_{cr}}{P} \quad (15)$$

So, the constraints are defined as follows:

$$g_1 = BLF = \frac{P_{cr}}{P} \geq 1 \quad (16)$$

$$g_2 = FC\ \leq 1 \quad (17)$$

$$g_3 = \frac{\Delta}{P_H} \leq 0.8 \quad (18)$$

Where, (g_1), (g_2) and (g_3) represent the constraints and (P_H) represents the pontoon height that equals (1.5 m).

7. Composite models

7.1. The properties of the used composite materials

In this study 3 composite materials are used; Glass /Epoxy, Carbon/Epoxy and with the following properties [32] shown in **Table 2**.

IOP Conf. Series: Materials Science and Engineering **934** (2020) 012037 doi:10.1088/1757-899X/934/1/012037

Table 2: The properties of composite materials

Properties	**Glass/Epoxy (S - UD)**	**Carbon/Epoxy woven wet (230)**
Density (kg/m³)	2000	1420
Young's modulus X direction (MPa)	50000	61340
Young's modulus Y direction (MPa)	8000	61340
Young's modulus Z direction (MPa)	8000	6900
Tensile X direction (MPa)	1700	805
Tensile Y direction (MPa)	35	805
Tensile Z direction (MPa)	35	50
Compressive X direction (MPa)	-1000	-509
Compressive Y direction (MPa)	-120	-509
Compressive Z direction (MPa)	-120	-170
Shear XY (MPa)	80	125
Shear YZ (MPa)	46.154	65
Shear XZ (MPa)	80	65

8. The Glass/Epoxy pontoon

The pontoon is simulated using GFRP in ACP and the optimization process is performed and the results are as follows:
The configuration of the upper deck is the same as the lower deck arranged as $[0_2 / 22.5 / 45_2 / 90]_S$. These laminate is composed of 12 layer as mentioned in the configuration. Every layer of this type of fabric (single layer glass fiber) has a thickness of (0.2 mm) so, the total thickness of the upper deck is (2.4 mm) and the lower deck is (2.4 mm). The configuration of the transverse stiffeners is arranged as $[0_2 / 45 / 90 / 45]_S$. The total thickness of the shell is (2 mm) for the used 10 layers. The configuration of the longitudinal stiffeners is arranged as $[0_2 / 45 / 90 / 45_2 / 90 / 45 / 0_2]_S$. The total thickness of this stiffener is (4 mm) for the used 20 layers. The total weight of the glass pontoon is (804 kg).

8.1. Failure criteria analysis

The upper shell is found to be the critical shell to the stress generated from the applied load so, the failure index is higher than the other shells. The failure index for the first layer of the upper shell is as shown in **Figure 5**.

IOP Conf. Series: Materials Science and Engineering **934** (2020) 012037 doi:10.1088/1757-899X/934/1/012037

Figure 5: The failure criteria for the critical layer of the upper shell

The failure index equals (0.75389) which is safe. If the load is increased till failure occurs, the failure will occur at the coincidence of the longitudinal stiffener with the transverse stiffener. The failure is determined for every layer and found that the failure is not of the same shape for every layer but, all layers are safe. The critical layer in the lower shell is the layer number (12) which is the the upper layer connected to the longitudinal and transverse stiffeners. The failure criteria is safe and has a failure index equals (0.786) and will occur as shown in **Figure 6**.

Figure 6: The failure criteria for the critical layer of the lower shell

The critical layer for the transverse stiffeners is safe with failure index equals (0.257) at the layer number (10). The critical layer for the longitudinal stiffeners is also safe with failure index equals (0.177) at the layer number (17). So, the critical shells that are affected with the stresses of the applied load are the upper and the lower shells.

8.2. Nonlinear buckling analysis

The nonlinear buckling analysis is performed as a constraints of the optimization process to be about (1.5) for military standards [37] noting that the minimum allowed value of (BLF) is (1). The buckling is illustrated as follows in Figure 7.

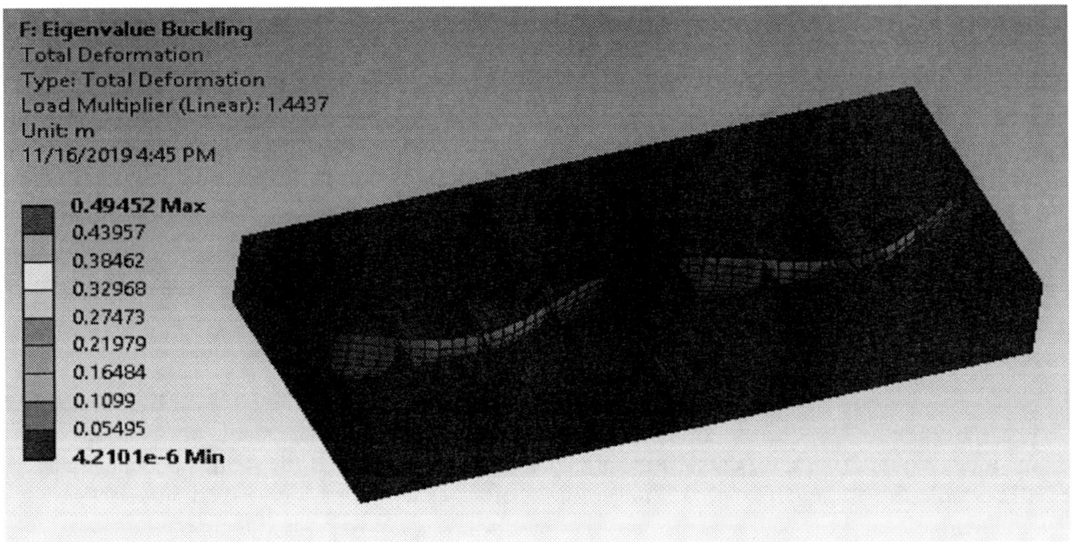

Figure 7: The nonlinear buckling of the longitudinal stiffener

The (BLF) equals (1.44) which is safe for the model. The longitudinal stiffener is the critical laminate for this configuration and the (BLF) is the critical constraints that affects the results more than the failure criteria.

8.3. The total deformation

The deformation constraint is applied in the optimization process and the value is determined. The total deformation equals (63.67 cm) as shown in Figure 4.

9. The Carbon/Epoxy pontoon

The optimization is performed and the optimum number of layers and orientation angles are determined as follows to satisfy the constraints conditions of the failure and the buckling. The shell optimized configuration is $[0_2 / 45_2 / 90]_S$ with total thickness from those (10) layers equals (2 mm) for the upper lower and transverse stiffeners shells. The configuration of the longitudinal shell is $[0 / 90 / 0_{13}]$ with total thickness of (3 mm) from those (15). The total weight of the carbon pontoon is (583 kg).

9.1. Failure criteria analysis

The failure index of the upper shell is (0.233) at the layer number (10) as shown in **Figure 8**.

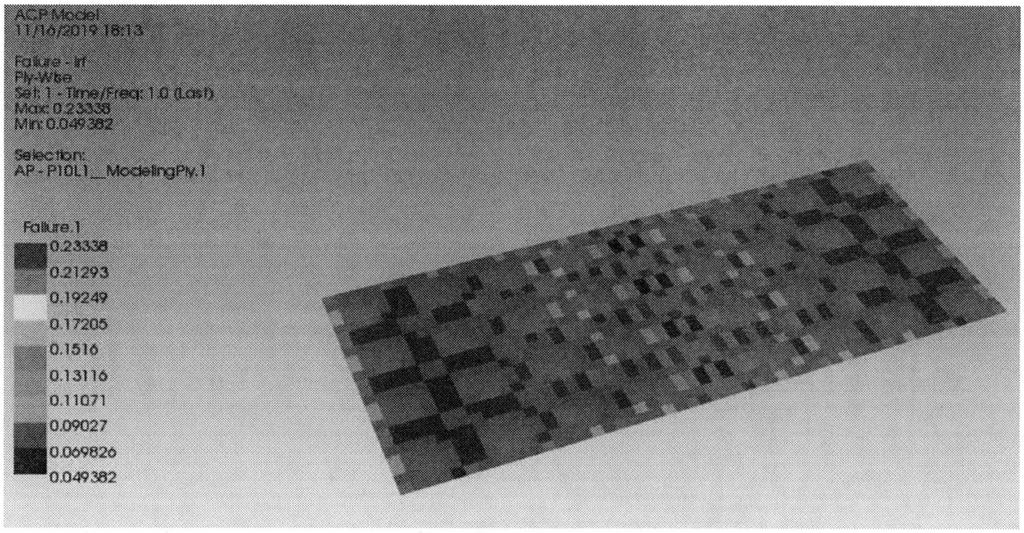

Figure 8: The failure criteria for the critical layer of the upper shell

The failure index for the lower shell is (0.24) and the same as the upper failure in the shape. The failure index of the longitudinal and the transverse stiffeners also are not effective.

9.2. The nonlinear buckling analysis

The buckling of the laminate is the critical constraints that affects the optimization results. The critical shell for the buckling analysis is also the longitudinal stiffener and the buckling load factor is (1.5). The nonlinear buckling is illustrated in **Figure 9**.

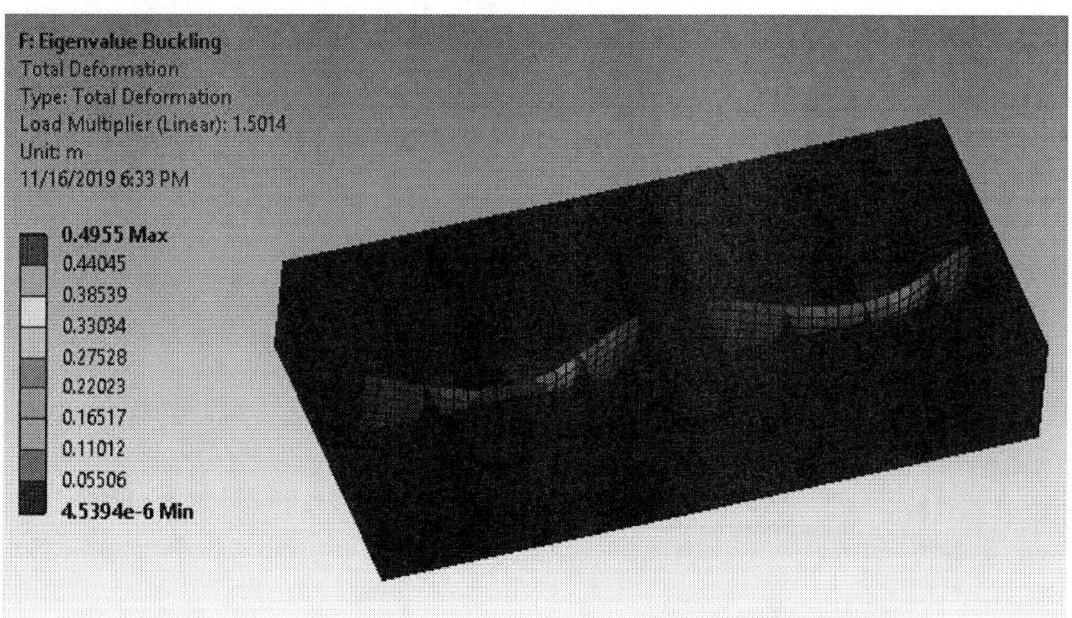

Figure 9: The nonlinear buckling of the longitudinal stiffener

9.3. The total deformation

The total deformation of the carbon pontoon equals (61 cm) as shown in **Figure 10**.

IOP Conf. Series: Materials Science and Engineering **934** (2020) 012037 doi:10.1088/1757-899X/934/1/012037

Figure 10: The total deformation of the carbon pontoon

10. Conclusion

In this paper the simulation of the composite materials is performed. The unique mechanical properties of composites make them candidate to replace the traditional materials. The pontoon is simulated for both GRFP and CRFP instead of structural steel. The capacity of the composite ferry is increased to be (90 t) instead of (70 t) of the steel ferry. The optimum number of layers and the best orientation of the angles are determined. The critical analysis of the composite pontoon is the buckling analysis. The failure criteria is effective for the loaded shells that are subjected to bending. The vertical stiffeners are the critical sections for buckling. The total deformations are determined for the different materials. The weight of the carbon pontoon is reduced by 27.5% from the glass pontoon that results in the reduction of the buoyancy factor. The reduction of buoyancy factor results in more increase in the capacity of the pontoon.

References

[1] Vinson J R 1993 The Behavior of Shells Composed of Isotropic and Composite Materials ed. S.M.a.I. Applications: Springer

[2] Fathallah E 2019 Finite Element Modelling and Multi-Objective Optimization of Composite Submarine Pressure Hull Subjected to Hydrostatic Pressure *Materials Science Forum* **953** p 53-58

[3] Imran M et al 2019 Design optimization of composite submerged cylindrical pressure hull using genetic algorithm and finite element analysis *Ocean Engineering* **190** p 106443

[4] Fathallah E et al.2014 Design Optimization of Composite Elliptical Deep-Submersible Pressure Hull for Minimizing the Buoyancy Factor *Advances in Mechanical Engineering*

[5] Fathallah E et al. 2014 Multi-Objective Optimization of Composite Elliptical Submersible Pressure Hull for Minimize the Buoyancy Factor and Maximize Buckling Load Capacity *Applied Mechanics and Materials* **578** p75-82

[6] Helal M et al. 2019 Numerical Analysis of Sandwich Composite Deep Submarine Pressure Hull Considering Failure Criteria *Journal of Marine Science and Engineering* **7(10)** p 377

[7] Vinson J R and Vinson R L 2008 The Behavior Of Structures Composed Of Composite Materials (Springer)

[8] Kassapoglou C 2013 Design and Analysis of Composite Structures: With Applications to Aerospace Structures (John Wile)

[9] Marín L et al. 2012 Optimization of Composite Stiffened Panels under Mechanical and Hygrothermal Loads using Neural Networks and Genetic Algorithms *Composite Structures* **94(11)** pp 3321-3326

[10] Herath M T et al 2014 Smoothed Finite Element and Genetic Algorithm Based Optimization for Shape Adaptive Composite Marine Propellers *Composite Structures* **109** pp 189-197

[11] Costa L et al. 2004 Multiple and Single Objective Approaches to Laminate Optimization with Genetic Algorithms *Structural and Multidisciplinary Optimization* **27(1-2)** pp 55-65

[12] Kim T U, Shin J W and Hwang I H 2007 Stacking Sequence Design of a Composite Wing Under a Random Gust Using a Genetic Algorithm *Computers and Structures* **85(10)** pp 579-585

[13] Akbulut M and Sonmez F O 2008 Optimum Ddesign of Composite Laminates for Minimum Thickness *Computers and Structures* **86 (21-22)** pp 1974-1982

[14] Hirono Y et al. 2016 Positional displacement measurement of floating units based on aerial images for pontoon bridges. in International Conference on Advanced Engineering Theory and Applications (ed Springer)

[15] Wu J-S and Shih P-Y 1998 Moving-load-induced vibrations of a moored floating bridge *Computers & structures* **66(4)** pp435-461

[16] Fu S and W. Cui W 2012 Dynamic responses of a ribbon floating bridge under moving loads *Marine Structures* **29(1)** pp246-256

[17] Alnahhal W and Aref A 2008 Structural performance of hybrid fiber reinforced polymer–concrete bridge superstructure systems *Composite Structures* **84(4)** pp319-336

[18] Khalifa Y A 2008 Study the structural system effect on the stability of floating metallic bridges

[19] Raftoyiannis I G, Avraam T P and Michaltsos G T 2014 Analytical models of floating bridges under moving loads *Engineering Structures* **68** pp144-154

[20] Siwowski T and Rajchel M 2019 Structural performance of a hybrid FRP composite lightweight concrete bridge girder *Composites Part B: Engineering* **174** p107055

[21] Wang H-H and Jin X-l 2016 Dynamic analysis of maritime gasbag-type floating bridge subjected to moving loads *International Journal of Naval Architecture and Ocean Engineering* **8(2)** pp 137-152

[22] Panteleev A D 1984 Optimal Design of Minimum Weight Sandwich Plates and Shallow Shells *Applied Mechanics* **20(11)** pp 103-107

[23] Abozaid M A, Elbeblawy M S A and Sayed-Ahmed E Y 2016 Structural Performance of Hybrid Composite Pontoon Compared to Steel *The Int Conf on Civil and Architecture Engineering* **11**(11PthP International Conference on Civil and Architecture Engineering) pp1-14

[24] Orifici A C, Herszberg I and Thomson R S 2008 Review of Methodologies for Composite Material Modelling Incorporating Failure *Composite Structures* **86(1-3)** pp 194-210

[25] Barbero E J 2013 Finite Element Analysis of Composite Materials Using Abaqus (CRC PressTaylor & Francis Group)

[26] Fathallah E et al. 2015 Design optimization of lay-up and composite material system to achieve minimum buoyancy factor for composite elliptical submersible pressure hull *Composite Structures* **121** pp 16-26

[27] Jingxuan H et al. 2011 Failure Prediction on Advanced Grid Stiffened Composite Cylinder Under Axial Compression *Composite Structures* **93(7)** pp 1939-1946

[28] Helal M et al. 2019 Numerical Analysis and Dynamic Response of Optimized Composite Cross Elliptical Pressure Hull Subject to Non-Contact Underwater Blast Loading *Applied Sciences* **9(17)** pp 3489

[29] Lopez R H et al. 2014 Advantages of Employing a Full Characterization Method over Form in the Reliability Analysis of Laminated Composite Plates *Composite Structures* **107** pp 635-642

[30] Fathallah E et al. 2015 Numerical investigation of the dynamic response of optimized composite elliptical submersible pressure hull subjected to non-contact underwater explosion *Composite Structures* **121** pp 121-133

[31] King R 1982 Principles of flotation (South African Institute of Mining and Metallurgy Johannesburg)

[32] 2019 ANSYS Inc. Products Release 19.0

[33] Collins L M et al. 2005 A strategy for optimizing and evaluating behavioral interventions. Annals of Behavioral Medicine **30(1)** pp 65-73

[34] Collins L M, Murphy S A and Strecher V 2007 The multiphase optimization strategy (MOST) and the sequential multiple assignment randomized trial (SMART): new methods for more potent eHealth interventions *American journal of preventive medicine* **32(5)** pp S112-S118

[35] Sendeckyj G P 2016 Mechanics of Composite Materials: Composite Materials **2** ed Elsevier

[36] Elsayed F et al. 2014 Optimal Design Analysis of Composite Submersible Pressure Hull. in Applied Mechanics and Materials (Trans Tech Publ.)

[37] Hornbeck B, Kluck J and Connor R 2005 Trilateral design and test code for military bridging and gap-crossing equipment *Tacom research development and engineering center warren mi.*

IOP Conf. Series: Materials Science and Engineering **934** (2020) 012038 doi:10.1088/1757-899X/934/1/012038

Research of change of mechanical properties of the ionic-alloyed samples in the corrosion environment

R R Giniyatullin and N M Yakupov

Institute of Mechanics and Engineering Subdivision of FIC KazanSC of RAS, Kazan, Lobachevsky st., house 2/31

E-mail: yzsrr@kfti.knc.ru

Abstract. Currently, it is becoming urgent to create composites from various materials using their strengths to obtain the desired properties. The method of modification of the surface layer of a structural element using ion alloying is one the most promising. This method is used for obtaining corrosion-resistant composites. In this article we can find the results of an experimental-theoretical study of the mechanical properties of thin-walled structural elements whose surface is alloyed with carbon ions. It is used to reduce the corrosive deterioration.

1. Introduction

Due to the peculiarity of operation there are different requirements for various structural elements in terms of characteristics of materials. Some elements require increased deterioration resistance, while another elements require increased strength or corrosion resistance. The composite material structure of structural elements largely determines their properties [1, 2].

The structure and the property of the surface layer are of particular importance. The morphology, phase composition, and dislocation structure of the surface layers of the metal elements determine the deterioration resistance, endurance limit, heat resistance and other characteristics. In this regard, technological processes allowing adjustment or rebuilding of surface layers have a great future. The ion alloying method allowing introducing a metered amount of ions of almost any elements into the surface layer of a metal element is of great interest.

Ion alloying is the method of introducing ions of any other substances, including gases. It happens with the help of generating a beam and accelerating it in an electrostatic field when the surface is heated.

In the 70 - 80s of the twentieth century, along with traditional methods of volumetric exposure to metal materials, such as thermomechanical treatment, hardening, shock wave loading and other methods, it became possible to conduct surface treatment with concentrated energy flows (from 103 to 108 W / cm2) [3-6] using electron and ion beams, laser radiation and plasma flows.

When processing with concentrated energy flows, radiation, thermal and shock - mechanical effects are simultaneously carried out. The processes of restructuring which are developing during this process make it possible to obtain surface layers with a unique set of physicomechanical properties.

The use of laser radiation for improving the operational properties of materials is well reflected in the following research works [7, 8]. The structures of various types of steel subjected to laser heating are presented in that books.

Compared to laser radiation, processing by electron and ion beams has several advantages: a higher efficiency, lower cost per unit of energy, the ability to process large areas, as well as a high degree of absorption of the input energy by all materials.

IOP Conf. Series: Materials Science and Engineering **934** (2020) 012038 doi:10.1088/1757-899X/934/1/012038

Many works are devoted to the study of changes in the mechanical properties of metal surfaces by ion alloying [9]. Preliminary alloying of iron with doses of 1016, 5x1016 ion / cm2 on the surface of the steel leads to an increase in the mechanical and tribological resistance of the sample material. Ion alloying of copper on the surface of samples made of 30KhGSN2A steel, when tested for low-cycle fatigue, led to an improvement in the results by 5-22% compared with the control sample. Also, deterioration resistance of such samples was 2–2.5 times greater [10]. It was noted in [11] that the maximum hardening of 12Kh18N10T steel is achieved by alloying boron atoms on its surface.
An increase in the microhardness of steel was observed after implantation of nitrogen, argon, boron, and carbon ions [12]. Upon implantation of nitrogen ions, the microhardness of steel increases according to Vickers from 300 to 400 kg [13]. The implantation of these ions prevents the formation of cracks on the metal surface and reduces the corrosion and friction properties of steel. Alloying of nitrogen atoms on the surface of industrial titanium leads to a 2-fold increase in its corrosion resistance in an acidic medium [11].

2. The purpose of the work

A study on the basis of the experimental - theoretical method [16–18] changes in the stiffness of thin-walled metal samples treated by the ion alloying method under the influence of a corrosive medium [14,15].

3. The effect of ion surface alloying on corrosion deterioration

Corrosion tests were carried out for three groups of steel samples from sheet structural steel (08ps) with a thickness of t = 0.5 mm. The surfaces of steel samples were modified by ion alloying with carbon atoms C. The samples were aged in a medium of 10% hydrochloric acid. In each group of samples, there was a control sample with an non alloyed surface, which was nearby under identical conditions. The exposure time of the corrosive medium: for group №1 - 48 hours, for group №2 - 72 hours, for group № 3 - 96 hours.
Figures 1 and 2 show surface images of samples of groups №1 and №2 after the influence of corrosive medium, increased by 40 times. Control samples are on the left and alloyed samples are on the right.

Figure 1. Samples of the first group:
on the left - non alloyed, on the right - alloyed

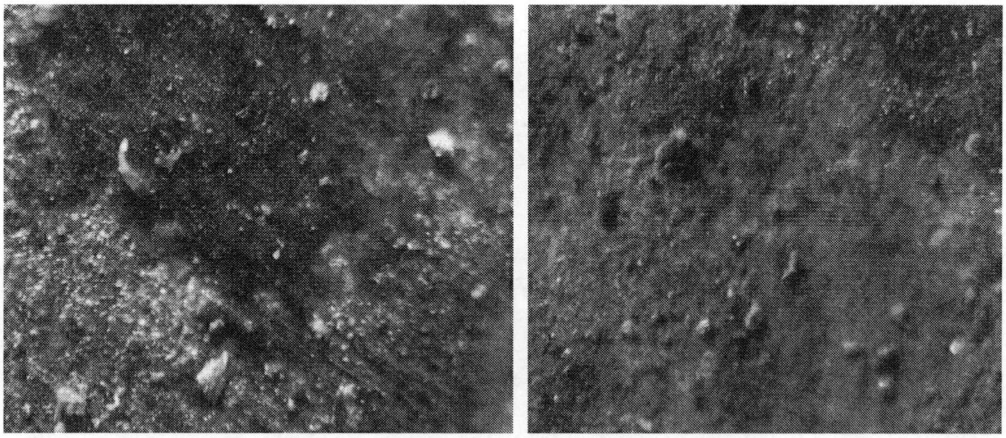

Figure 2. Samples of the second group:
on the left - non alloyed, on the right - alloyed

The results of measurements of the thickness of the samples after aging in an aggressive environment are presented in table 1.

Table 1. Experimental data.

Group Samples	Group №1		Group №2		Group №3	
Corrosion exposure, *hour*	48		72		96	
Surface modification	+	-	+	-	+	-
Thickness after deterioration, *mm*	0.481	0.469	0.478	0.464	0.468	0.452

After analyzing figures №1 and №2, we discovered that the surface of non alloyed samples has deeper cavities with large dimensions in plan. From table 1, you can see the geometric changes in the samples in thickness: surface-modified samples have a larger total thickness compared to unmodified samples. All this indicates that alloyed samples are more corrosion resistant.

For analyzing the mechanical characteristics of the samples we used the experimental - theoretical method, which is described in following research works [16–18].

Based on experimental data, the dependences "rigidity B - pressure P" were constructed by us. They are presented in tables 2-4 and Figures 3-5. The graphs show two curves: for a alloyed and non- alloyed sample.

Table 2. The dependence «rigidity **B** - pressure **P**» for the sample in group №1

***P**, MPa*	0.02	0.04	0.06	0.08
B_+, kg/cm	3.38 x10^5	3.01 x10^5	2.74 x10^5	2.39 x10^5
B_-, kg/cm	1.44 x10^5	1.44 x10^5	1.50 x10^5	1.40 x10^5

Table 3. The dependence «rigidity **B** - pressure **P**» for the sample in group №2

***P**, MPa*	0.02	0.04	0.06	0.08

B_+, kg/cm	2.48 x10^5	2.04 x10^5	1.64 x10^5	1.40 x10^5
B_-, kg/cm	1.51 x10^5	1.35 x10^5	1.11 x10^5	1.02 x10^5

Table 4. The dependence «rigidity **B** - pressure **P**» for the sample in group №3

P, MPa	0.02	0.04	0.06	0.08
B_+, kg/cm	2.25x10^5	2.04 x10^5	1.85 x10^5	1.69 x10^5
B_-, kg/cm	1.51 x10^5	1.49 x10^5	1.5 x10^5	1.44 x10^5

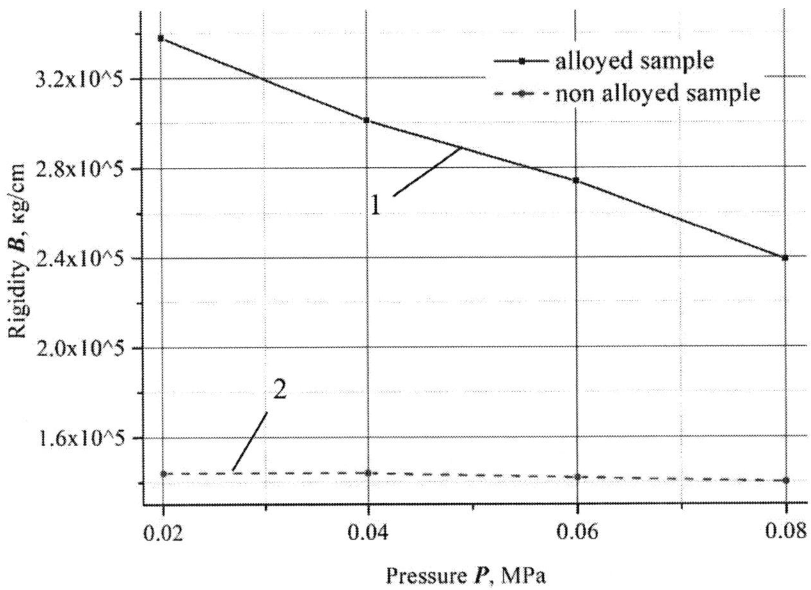

Figure 3. The dependence «rigidity **B** - pressure **P**» for the group of samples №1: curve 1 - alloyed samples; curve 2 – non alloyed samples.

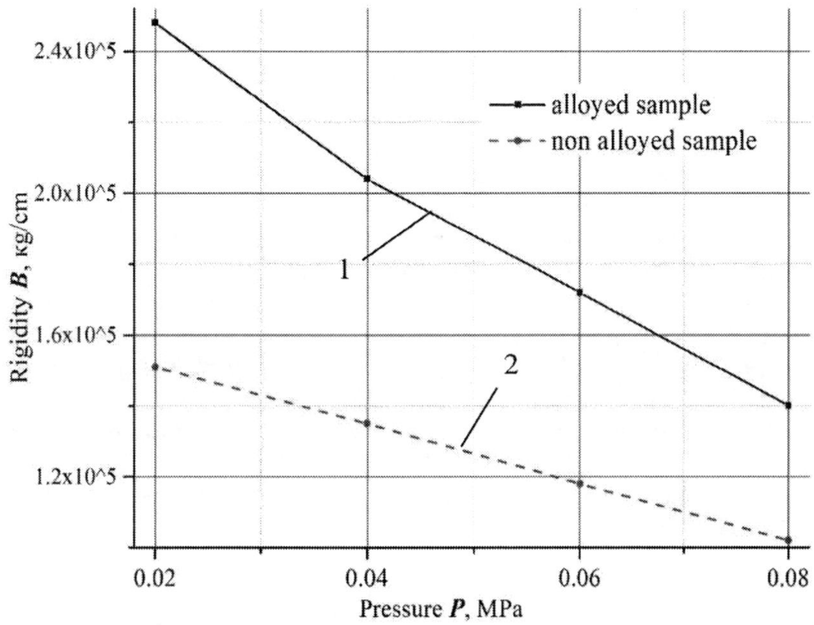

Figure 4. The dependence «rigidity **B** - pressure **P**» for the group of samples №2: curve 1 - alloyed samples; curve 2 – non alloyed samples.

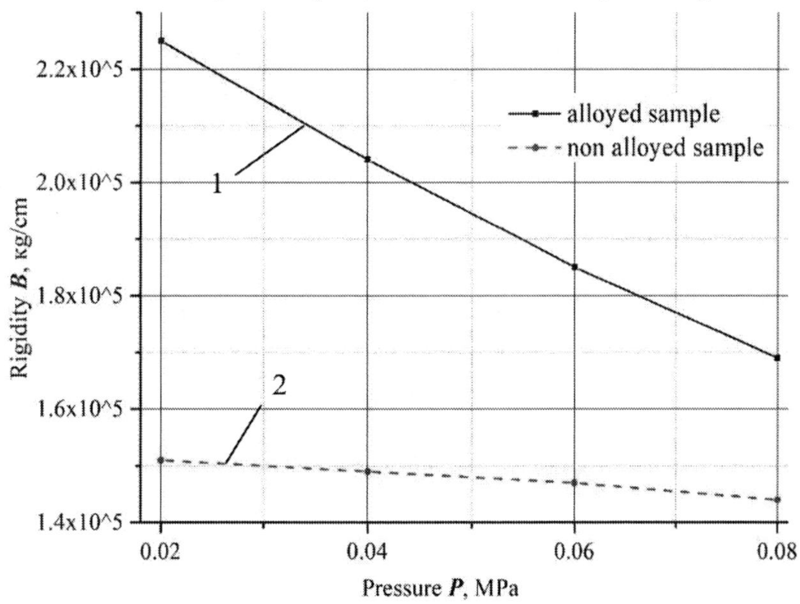

Figure 5. The dependence «rigidity **B** - pressure **P**» for the group of samples №3: curve 1 - alloyed samples; curve 2 – non alloyed samples.

As shown in the figures 3-5, the rigidity of the alloyed samples is greater than that of non alloyed samples. Thus the surface modification of steel samples by alloying with carbon ions increases corrosion resistance.

4. Conclusion

Surface treatment of the sample by ionic alloying helps to reduce corrosion wear, while with increasing aging time in an aggressive environment, the effect decreases due to dissolution of the alloyed layer.

The implanted layer allows you to maintain a passivating protective layer. The implanted layer is an effective way to ensure protection against corrosion.

References

[1] Dumansky A M, Alimov M A and Terekhin A V 2019 Experiment- and computation-based identification of mechanical properties of fiber reinforced polymer composites. // Journal of Physics: Conference Series 1158, Issue 1. DOI: 10.1088/1742-6596/1158/2/022037.

[2] Dumansky A M, Liu Hao and Alimov M A 2019 Strain-rate effect on mechanical behavior of unidirectional carbon fiber reinforced plastic // Materials Today:Proceedings.

[3] Anisimov S I, Imas E A and Romanov G S 1970 *Effect of high-power radiation on metals* (M.: Nauka) p 272

[4] Rykalin N N, Zuev I V and Uglov A A 1978 *Fundamentals of electron beam processing of materials* (M.: Mechanical Engineering) p 249

[5] Misaelides P 1994 *Application of particle and laser beams in materials technology* (Dordrecht-Boston-London: Kluwer Academic Publishes) p 678

[6] Rykalina N N 1985 *The effect of concentrated energy flows on materials* (Moscow Nauka) p 246

[7] Rykalin N N, Uglov A A and Kokora A N 1975 *Laser processing of materials* (Moscow Mechanical Engineering) p 496

[8] Sadovsky V D, Schastlivtsev V M , Tabatchikova T I and Yakovleva I L 1989 *Laser heating and steel structure: Atlas of microstructures* (Sverdlovsk: Ural Branch of the Academy of Sciences of the USSR) p 101

[9] Trillat J J and Haymann I N 1961 *Le bombardement ionique* (Theories et applications. Paris) p 13

[10] Kozlov D A Features of alloying the surface of steel 30KhGSN2A with copper by methods of electrospark alloying and ion implantation // Abstract of the dissertation M. 2009.

[11] Kovalenko I A Investigation of the process of surface hardening of steels and alloys by non-metal ions // Abstract

[12] Pavlov A V, Pavlov P V, Zorin E I and Tetel'baum D I 1974 *All Soviet Meeting on Ion Beam Physics* Kiev p 558

[13] Gabovich M D, Budernaya L D, Poritskii V Y and Protsenko I M 1974 *All Soviet Meeting on Ion Beam Physics* (Kiev) p 559

[14] Yakupov N M and Giniyatullin R R 2011 Corrosion wear of thin-walled elements processed by ion implantation method *Structural mechanics of engineering structures and structures* Vol 3 pp74-79

[15] Yakupov N M and Giniyatullin R R 2011 Corrosion wear of thin-walled elements under the influence of external factors *Actual problems of continuum mechanics* (*To the 20th anniversary of the IME KazSC RAS* Vol 2) pp 203-212

[16] Yakupov N M, Galimov N K and Leontiev A A 2000 Experimental-theoretical method for studying the strength of polymer films *Mechanics of composite materials and structures* Vol 6 No 2 Pp 238-243

[17] Yakupov N M, Galyaviev Sh Sh, Nurgaliev A R and Yakupov S N 2006 The state of the designs of cooling towers and the prevention of their destruction (*News of universities. Problems of Energy)* No8 Pp 36-42

[18] Yakupov N M, Nurgaliev A R and Yakupov S N 2008 Methods of testing films and membranes under conditions of uniform distributed surface pressure (*Factory Laboratory. Diagnostics of materials* Vol 74) No 11 pp 54-56.

IOP Conf. Series: Materials Science and Engineering **934** (2020) 012039 doi:10.1088/1757-899X/934/1/012039

Development of imid-containing appretes for carbon fibers and carbon fabrics

I P Storozhuk [1,2], N G Pavlukovich [2] and A S Egorov [3]

[1] N.E.Bauman Moscow State Technical University (BMSTU), 105005, 2nd Bauman str., 5/1, Moscow, Russia

[2] A.N.Nesmeyanov Institute of Organoelement compounds RAS, 28 Vavilova str., 119991, Moscow, Russia

[3] Research Center "Kurchatov Institute-IREA", 107076, Bogorodsky Val str., 3, Moscow, Russia

E-mail: storozhuk-ip@inbox.ru

Annotation. The results of research on synthesis and approbation of new imide-containing high-molecular and oligomeric compounds as appretiruyuschih compositions for carbon fibers under thermoplastic polyetherimide matrix ("Ultem-1000") are presented. It is shown that both polymeric and oligomeric compositions can be used for carbon fiber appreting, but the greatest effect is manifested when using oligomeric apprets.

1. Introduction

Reinforced plastics based on glass, polymer, carbon and ceramic fibers are widely used in special areas of modern technology. Without their application, it is now impossible to imagine rocket or aviation, they are increasingly used in other areas of the national economy-engineering, shipbuilding, construction, etc. [1-2]. Such widespread polymer composites are due to its high strength, combined with low density, good moisture and chemical resistance, corrosion resistance, resistance to cracking.

In recent years, actively developing research in the development, production and application of composite materials based on super-strong (5000-7000 MPa) carbon fibers with high elongation limits(1,5-2,0%) [3-5].

Of particular interest is the work on the use of binders based on structural thermoplastic materials with increased heat resistance, namely-polyimide materials that allow the operation of products made of them at temperatures above + 250°C (depending on the duration of operation and the environment), at cryogenic temperatures, the impact of penetrating radiation doses of 104 Mrad and more, high mechanical loads and a combination of the above conditions [6-11]. Of no less interest as a binder are superconstructions thermoplastics, e.g. polyether sulfones and polyetherimides [12-19].

The low shear strength of composite materials is a limiting factor in the use of carbon fiber on a thermoplastic polymer matrix. It is known that on the surface of carbon fibers there are small micro-scales containing an increased amount of crystalline graphite, as well as inclusions and microcracks. This structure of the surface of carbon fibers leads to low adhesion of the binder to their surface [4,20, 21].

Thus, it is obvious that the creation of high-quality composite carbon plastics is impossible without a purposeful synthesis of not only heat-resistant polymer matrices, but also materials that ensure their adhesion to carbon fiber, known as textile and operational appretes.

Appretiruyuschie coating increases the adhesive and cohesive properties of the interface between the phases of the carbon fiber-polymer matrix. Composite materials with well-chosen appendages well perceive and transmit loads, have a high resistance to cracking of the material at the interface at high loads.

It should be noted that in the literature there are rarely specific data on the compositions of the appreting compositions and the conditions of their application, since these data are often the subject of know-how.

This paper presents the results of research on the development and testing of new imide-containing high-molecular and low-molecular compounds as appretiruyuschih compositions for carbon fibers under thermoplastic polyimide or polyethyrimide matrix.

As macromolecular coupling agents were synthesized copolyimide (SPI) with the General formula:

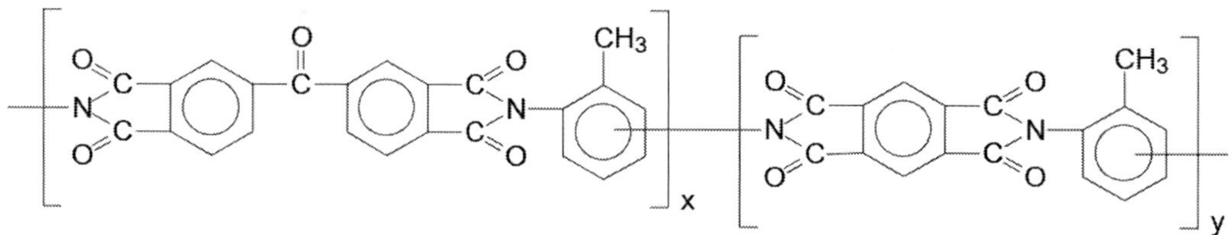

(1)

To assess the effect on the appreting properties of the chemical composition and molecular weight (MM) of copolymers, the ratio of "x:y" links based on benzophenonetetracarboxylic (BPDA) and pyromellitic (PMDA) acids varied in the range of 70:30% by weight. (SPI-1 and SPI-2, MM), 60: 40% by weight. (SPI-3), 50: 50% by weight. (SPI-4 and SPI-5) and 40: 60% by weight. (SPI-6). These copolymers have solubility in dipolar aprotic solvents (dimethylformamide, dimethylacetamide, N-methylpyrrolidone, dimethylsulfoxide) and can be applied to carbon fiber in the form of solutions of different concentrations.

In addition, low molecular weight copolymers based on BPDA of the following structure (SPI-7) were synthesized to compare the appreting properties):

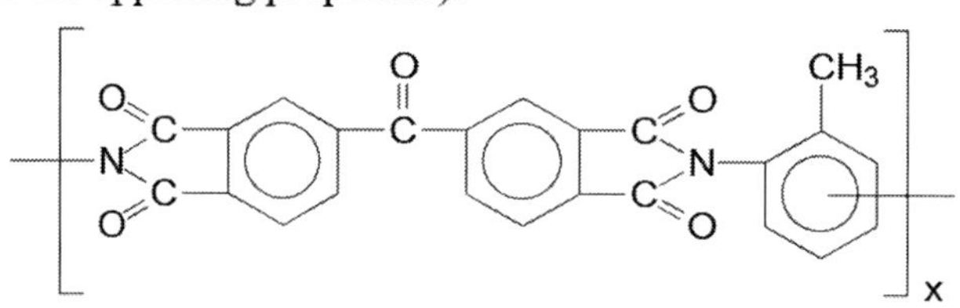

(2)

and based on PMDA (SPI-8):

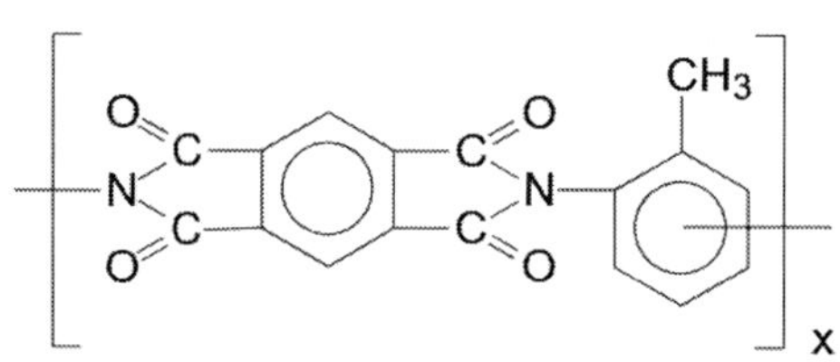

(3)

The BADA-based oligomeric appret SPI-9 with terminal anhydride groups and x=3 was a compound of the following structure:

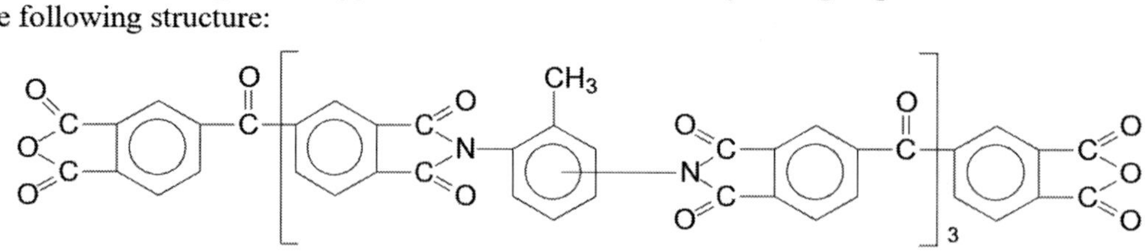

(4)

SPI-10 oligomer based on PMDA with terminal anhydride groups and x=3 had the following structure:

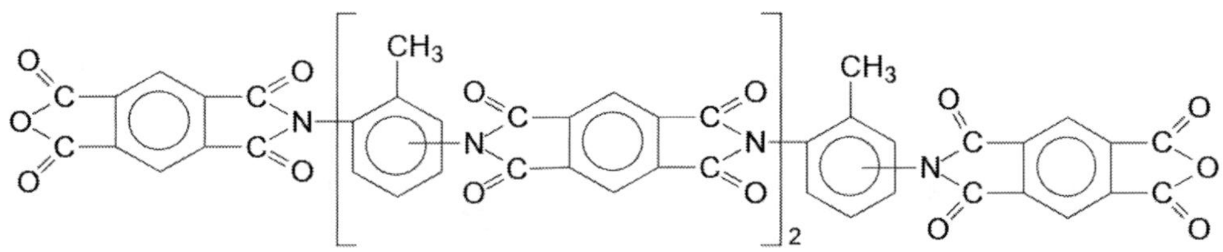
(5)

The synthesized compounds were characterized by the methods of elemental analysis, IR spectroscopy, thermogravimetry, the viscosity values of their 0.5% solutions in N-methylpyrrolidone at 25°C.
The effectiveness of the appreting action of the synthesized compounds was evaluated by SEM-photos of the surface and cut-off samples of the manufactured prepregs. Micrographs were obtained using a scanning electron microscope "Hitachi SU 1510" (Japan) at magnification of 200, 500, 1000 and 2000 times. As a model binder was selected industrial polyethyrimide brand " Ultem-1000 "(firm "SABIC").

2. Receiving samples

For the experiment, we used 1.5 K carbon fiber (made in Taiwan), appretirovannoe under epoxy resin.
Samples of prepregs for the study of coupling efficiency of the synthesized products were made according to the scheme consisting of two stages: 1) cleaning carbon from a source of size and dressing of new compounds and 2) impregnating the treated fibers with a solution of polyetherimide "Ultem-1000".
In the first step, the original carbon fiber was successively passed through a bath of chloroform to remove the factory "textile" appret, and then through a bath with a solution of the new appret in dimethylacetamide. The impregnated thread was passed through a step-heating furnace (first stage +250°C, second +300°C) for 90 seconds for each heating stage. At this stage, it was important to select a suitable concentration of appret in the solvent. The result was evaluated visually: dry appretirovannaya thread with a suitable concentration of appret was soft, bent into a loop without creases and retained solidity (not fluffed) when stretched through the impregnation plant at the second stage of processing. At excessive concentration of an appret the thread appeared too rigid and was wrung at bending; at insufficient concentration of a polymer or oligomer in an appretating solution the thread was torn or stratified at attempt of its broach through impregnating installation. After determining the optimal concentration of the appreting solution, a prototype of the appreted fiber was developed, which was then directed to impregnation with a matrix polymer solution.
The second stage is to obtain a fiber prepreg. The pre-appretirovannoe fiber was successively stretched through two baths filled with 20% solution "Ultem-1000" in chloroform. Between the first and second impregnation bath, the thread was pressed using Teflon rollers to achieve uniform impregnation of the filaments throughout the fiber volume. The geometry of the impregnated fiber was formed by means of a die, then the fiber was stretched through a step-heated furnace (the first stage +350°C, the second +430°C) for 45 seconds for each stage. The hot fiber was wound on a receiving coil with a diameter of 12 cm and removed after complete cooling.

3. Discussion of results

The characteristics of the synthesized compounds are given in tables 1 and 2.

Table 1. Reduced viscosity of synthesized compounds

Шифр полимера	η пр., дл/г	Шифр полимера	η пр., дл/г	Шифр полимера	η пр., дл/г
SPI -1	0,24	SPI -5	0,47	SPI -8	0,14
SPI -2	0,38	SPI -6	0,41	SPI -9	0,10

SPI -3	0,53	SPI -7	0,16	SPI -10	0,12

Table 2. Results of thermogravimetric analysis of thermoplastic polyimides

Шифр полимера	The temperature loss 5% weight of the sample, °C		The temperature loss 10% weight of the sample, °C	
	In argon	In the air	In argon	In the air
SPI-1	416	442	513	522
SPI-2	484	460	540	490
SPI-3	435	454	530	488
SPI-5	486	460	537	486
SPI-6	480	452	526	452
SPI-7	452	438	506	511
SPI-8	458	440	508	515
SPI-9	465	454	521	526
SPI-10	460	448	511	508

Using a scanning electron microscope, the density of the composite material in the cross sections of the samples destroyed during bending was determined. Analysis of the obtained data allows to conclude that the effectiveness of the coupling action is highly dependent on the molecular weight of the synthesized copolyimides. So, when using a macromolecular size of SPI-3 in the cross section of the fiber observed spalls, voids, gaps between the fiber and the matrix, the looseness of the structure of the surface shell fibers (figure 1), in the case of oligomeric size SPI-10 (figure 2) in the section of the fault gouges are practically absent, the fracture surface is smooth, without voids, the contact between the matrix and fibers is complete.

Thus, oligomeric compounds with reduced viscosity of 0.5% solutions in N-methylpyrrolidone from 0.10 to 0.16 DL/g were more effective devices. These oligomers were used to appretirovaniya carbon fibers and tissues before applying melts of thermoplastic polyethyrimide brand "Ultem-1000" and high-molecular copolymides SPI-3 and SPI-5. The resulting thermoplastic carbon composites are currently being tested.

Figure 1. SEM - photo of the section of the prepreg obtained using high molecular weight appret SPI-3.

Figure 2. SEM - photo of the section of the prepreg obtained using low molecular weight appret SPI-10.

Acknowledgments
The study was performed with financial support RFBR under contract No. 18-29-18087/18 grant winner and the research project.

References

[1] Kablov E N 2002 Aviation materials science: results and prospects *Bulletin of the Russian Academy of Sciences* **72** №1 3-12
[2] Komarov G A 2009 State, prospects and problems of PCM application *Polymer materials* №2 5-9
[3] Varshavsky V A 2005 Carbon fiber M.: VINITI, 498
[4] Fitcer E 1988 Carbon fibers and carbon composites M.: Mir, 336
[5] Gunyaev G M, Gofin M Ya 2013 Carbon-carbon composite materials *Aviation materials and technologies* №1 62-90
[6] Trostyanskaya E B 1975 Thermoplastics for structural purposes M.: Chimiya, 240
[7] Richardson M 1980 Industrial polymer composite materials M.: Chimiya, 472
[8] Berlin A A, Volfson S A, Oshmyan V G, Enikolopov N S 1990 Principles of creating composite polymer materials M.: Chimiya, 238
[9] Berlin A A, Pahomova L K 1990 Polymer matrices for high- strength reinforced composites. Review *Polymer Science Series A* **32** №7 1354-70
[10] Mihailin Yu A, Kerber M L, Gorbunova I Yu 2002 Binders for polymer composite materials *Plastics* №2 14-21
[11] Petrova G N 2010 Thermoplastic elastomers are a new class of polymeric materials *Aviation materials and technologies* №4 20-25
[12] Borodulin A S, Kalinnikov A N, Bazheva R C at all. 2018. Synthesis and properties of aromatic polyethersulfones//International Journal of Mechanical Engineering and Technology (IJMET). –v. 9 (13), pp. 1109–1116.
[13] Borodulin A S, Kalinnikov A N, Bazheva R C at all. 2018. Receipt and investigation of performance characteristics of super constructions polyesters. - International Journal of Mechanical Engineering and Technology. - v.9 (13), pp. 1117-1127.
[14] Borodulin A S, Kalinnikov A N, Kharaev A M at all. 2019. New Polymeric Binders for the Production of Composit// International Conference on Modern Trends in Manufacturing Technologies and Equipment 2018. International Journal of Materials today: proceedings. –V. 11. Issue P1, pp. 3107–3111.

[15] Nelub V A, Borodulin A S at all. 2018. Poliefirsul'fony s uluchshennymi teplofizichesimi svoystvami ed. nauka i tekhnologii, *Klei* **7** 15-20.
[16] Borodulin A S, Kalinnikov A N 2020. Super engineering polyesters: synthesis and performance characteristics. – IOP Conf. Series: Materials Science and Engineering - v.709, pp. 1-5.
[17] Borodulin A S, Kalinnikov A N at all. 2019. New Polymeric Binders for the Production of Composit. – Materials Today: Procttdings – v. 11, pp. 139–143.
[18] Borodulin A S, Kharaev A M, Kalinnikov A N at all. 2019. Synthesis and Perfomance Characteristics of Superstructure Polyesters. - Key Engineering Materials- v.**816**, pp. 307-311.
[19] Borodulin A S, Kalinnikov A N, Kharaev A M, at all. 2019. Aromatic polysulfone to create polymer materials with high resistance to frost. - IOP Conf. Series: Earth and Environmental Science – v. **302**, pp. 1-5.
[20] Gunyaev G M 1981 Structure and properties of polymer fiber compositesM.: Chimiya, 232
[21] Brandshtetter F 2005 Trends and prospects of polymer materials development *Polymer materials: products, equipment, technologies* №5 2-4

Comparison of low cost lasers for graphene oxide thin films reduction

I A Komarov [1], N S Struchkov[2], I A Polikarpova[1], V G Peretiyagin[1], A D Buyanov[1], E A Danilova[1], E I Denisenko[1] and E A Onoprienko[1]

[1] Interbranch Engineering Centre "Composites of Russia" Bauman Moscow State Technical University, Moscow, 105005, 2nd Bauman str, bld.5, Russian Federation
[2] Research and Educational Center for Probe Microscopy and Nanotechnology, National Research University of Electronic Technology, 124498 Moscow, Zelenograd, Shokin sq. 1, Russian Federation

E-mail: master_kom@mail.ru, struchkov.nikolaj@gmail.com, ipolikarpova@emtc.ru, struchkov.nikolaj@gmail.com, ipolikarpova@emtc.ru, vlperetiagin@gmail.com, alex.d.buyanov@gmail.com, d.elena.aleksandrovna@yandex.ru, evgeniiadenisenko@gmail.com, evgenia.onoprienko97@yandex.ru

Abstract. In this work we made reduction of graphene oxide films with different low-cost lasers (near IR (1060 nm) and blue (445 nm) to investigate possibility of such films usage for flexible electronic and nanoelectronic applications. We successfully showed possibility of graphene oxide reduction with both types of laser but blue laser showed better uniformity of reduced graphene oxide film parameters including film morphology, resistance and Raman intensity ratios. We showed that reduction with near IR laser spills out into large nonuniformity of resistance with relatively high values. Thus the usability of commercially available laser facility for graphene oxide modification without adjusting control settings (on hardware and software levels) is poor. On the other hand, relatively laboratory device based on low-power blue laser showed much better usability an it's perspective to future market of low-cost modification facilities for thin carbon-based films.

Keywords: graphene oxide, reduced graphene oxide, laser reduction, aerosol deposition, nanoelectronic, flexible electronic.

1. Introduction

New modern research and engineering fields like flexible [1] and organic electronics [2], energy storage [3], sensors [4], functional [5] and classic composites [6] etc. requires new approaches in material processing and modification. One of the other important trends in material processing today is digital manufacturing technologies like 3D printing [7], CNC machines [8] etc. But most of such devices for processing operates with bulk or macrosized materials. And there is a need in processing methodic [9] and devices adapted for functional materials processing. There are some new approaches [10] and methodics [11] that allows to modify composite materials to give them additional functionality [12]. But main lack of these approaches is its nonlocality which is not enough for electronic and sensing applications.

Among possible methods of functional materials processing one of the most perspective is laser modification. Depending on laser parameters it is possible to treat such perspective components of functional materials as nanoparticles [13], carbon nanotubes [14], graphene [15] and its derivatives [16] etc.

For the future functional materials (in particular for usage in wearable electronics) one of the most promising class of components that meet these conditions is graphene and its derivatives like graphene oxide due to its flexibility, conductivity and presence of functional groups [17] in case of graphene oxide or reduced graphene oxide.

In this article we experimentally showed possibility of successful graphene oxide processing. Graphene oxide up to day is better decision for conductive film formation due to much easier transfer process with variety of possible substrates. Pure graphene transfer is not an easy process [18] and it doesn't work well with flexible polymer substrates or such basis of composite materials as glass and carbon fibers. Another great advantage of graphene oxide (and reduced graphene oxide) is abovementioned presence of different functional groups that give possibility to couple with nanoparticles [19], biological molecules [20] or different polymers [21].

There are two key operation in graphene oxide based conductive film formation: graphene oxide deposition (that should be as uniform as possible) and reduction (that should be precisely controllable). Graphene oxide can be deposited on the substrate by many techniques including drop casting [22], rod coating [23], Langmuir-Blodgett technique [24], spin-coating [25] etc. But among them all one most promising is aerosol deposition due to possibility of formation of very thin but uniform and large area film [26] on variety of substrates including glass and carbon fibers. Moreover, this method has an advantage on drop-casting and spin-coating in less solution consumption. Nevertheless, to obtain best theoretical uniformity we tried to use spin - coating method [27].

Development of wearable electronic devices requires formation of the conductive layer and it's preferably to use such method that give possibility to precisely control modification parameters. In case of graphene oxide based wearable electronic devices there is a need to form conductive pathways by reduction of graphene oxide with predetermined pattern.

Graphene oxide reduction could be done by different methods, but two most known methods are chemical [28] and thermal [29] reduction. Conventional reduction methods include graphene oxide processing in reducing agents as hydrazine, ascorbic acid etc., or thermal annealing in inert atmosphere. Nevertheless, these methods don't allow to perform local reduction and, thus, they are not suitable for functional device mass fabrication.

Promising reduction method being developed last decade is graphene oxide laser treatment [30], realizing photochemical [31] or thermal reduction [32] depending on laser wavelength, energy density and pulse width. Femtosecond laser pulses allows to attach or remove various functional groups [33] via photochemical reaction, but such method has significant drawback: high price of laser system. On the other hand, laser induced thermal reduction, that can be performed by much cheaper micro/nanosecond pulse lasers or even continuous wave laser, allows to reduce graphene oxide directly on flexible substrate due to rapid local heating. Concept of laser reduction combines fast processing, controllable reduction degree, possibility of patterning and applicability for a wide range of substrates.

In this article we show possibility of local modification of graphene oxide films by using microsecond pulsed laser with different wavelength (blue and near IR). As we assume it's possible to use both type of this lasers for local modification of thin graphene oxide films with better results obtained by blue laser.

2. Materials and methods

Water GO suspension (4,1 mg/ml) produced by improved Hummers method was purchased at LLC "MIP Graphene" (Yakutsk, Russian Federation) [34]. Suspension was further diluted in deionized water to 0,5 mg/ml concentration for spin coating deposition.

We used poly(ethylene terephthalate) (PET) substrates as a promising material for further development of flexible and cheap bio- and gas sensors with reduced graphene oxide active layer. For spin coated

IOP Conf. Series: Materials Science and Engineering **934** (2020) 012040 doi:10.1088/1757-899X/934/1/012040

films we used high opacity PET film without surface treatment. Substrates with 20 x 20 mm dimensions were precleared with 2-propanol and water.

Deposition of the suspension was performed by Spin-NXG spin-coater (Apex Instruments) with maximum 2000 rpm.

Laser treatment was carried out by two facilities:

1) Commercially available Mercury III laser engraver (GCC Laser Pro, Taiwan) with 1060 nm CO2 laser with 25 W maximum power controlled by pulse width modulation (PWM) at 5 MHz frequency (Synard 48 Series, USA) and pulse length 1 μs (also can work as continuous wave mode). Output power was ranged from 2% to 6% of maximum power (25 W), stage movement velocity was varied from 25% to 100% of maximum possible.

2) own-designed system with 445 nm solid-state laser facility with motorized table [35]. Output laser power was varied in 10-400 mW range by pulse width modulation (PWM) at 30 MHz frequency. Pulses width ranged from 1 to 20 μs. Treatment time in a spot ($\sim 1{,}5 \cdot 10^3$ μm2) was constant 30 μs.

Raman spectroscopy was used for reduction process survey by estimating relative intensities of specific GO and rGO Raman peaks I_D (1350 cm^{-1}), I_G (1582 cm^{-1}), and I_{2D} (~ 2700 cm^{-1}). To obtain Raman spectra and optical images we used Centaur U HR spectrometer (Nano Scan Technologies Ltd, Russia) with Mitutoyo lens with x50 magnification and aperture NA 0.55.

We also used FTIR Nicolet iS10 spectrometer (ThermoFisher scientific, USA) to investigate presence of different functional groups in reduced films.

Resistance of reduced films was measured by two-contact method with IPS 16 (CJSC EKSIS, Russian Federation).

3. Results and discussion

3.1 Laser reduction by near IR laser

As it mentioned before, we used Mercury III laser engraver to reduce graphene oxide film with different parameters. The Mercury III laser engraver give possibility to vary movement speed of engraving head (i.e. laser) and power of the laser in percent to the maximum power (about 40 W laser) with increments of 1%. As we find out power about 7% of maximum power and higher leads to full destruction of the graphene oxide film and PET substrate. Reduction in this case could be done only in a range from 2 to 6 % with the same increment. So we formed a set of structures with 25, 40, 60, 80 and 100 % movement speed for each of 2, 3, 5, 6 % of laser power.

As a result, we observed differences in surface morphology, resistance and Raman spectra of these structures. By this results we identified 5 groups or reduced films with specific parameters. First group consists of almost nonreduced films with surface morphology similar to pure graphene oxide (figure 1a). These structures have resistance about 6 MOhm or more and can be obtained in case of low laser power (2-3%) and low movement speed (25-40%) (see blue cells in table 1). So we can mention that under low laser power and movement speed reduction threshold can't be exceeded.

Second morphology type have locally reduced graphene oxide conductive pathways with up to 0,5 mm length (200-300 μm average) with clearly seen graphene oxide film between reduced areas (figure 1 b). As we can mention in this case reduction threshold is exceeded but in some local areas due to local nonuniformities of a graphene oxide films. So in this case reduction take place only in top layers of graphene oxide film and it doesn't form continuous pathways on the macrosize. Due to the fact that long conductive channels are not formed, such structures have a high resistance from 1,5 to ~5 MOhm (yellow cells in table 1). Nevertheless, reduced areas in this morphology have Raman spectra close to the fourth type of morphology with relatively high intensity of 2D peak and low D peak (figure 2 c).

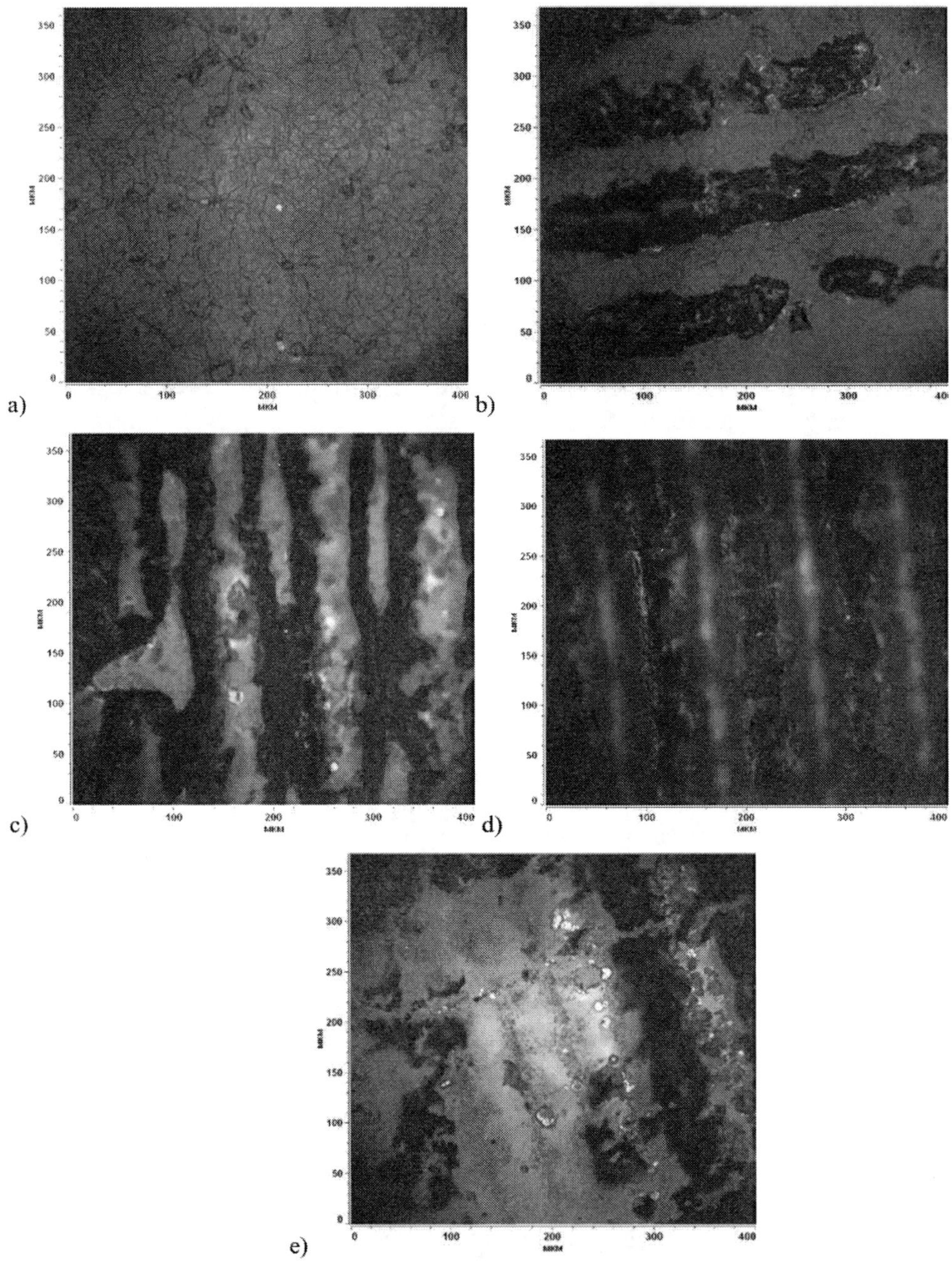

Figure 1. Optical images of surface morphology of near IR laser reduced graphene oxide films.

Third morphology type is reduced graphene oxide with conductive channels with caverns inside laser pathway (figure 1 c). According to Raman spectroscopy (figure 2 b) this is the most preferable type of reduction that can be done on laser engraver. This morphology type has resistance in range from ~70 kOhm to ~900 kOhm. Raman spectra in this case indicate the presence of more defects in the structure, and according to [36] have the largest amount of functional groups. These structures marked green in table 1. We can also note that in this case only small wrinkled graphene oxide areas are observed between reduced pathways. We attribute this fact to the higher heating of the film on larger areas that lead to formation of longer pathways independent of local GO film nonuniformity. But heating in this case is not enough to obtain form highly structured multilayer graphene grains like in fourth type of morphology. This type of reduced film has best perspective for nanocomposite formation due to the maximal number of functional groups. Among possible applications of such type of reduced graphene oxide film chemical or biological sensors are preferable.

Fourth morphology type have clearly seen conductive pathways without caverns (figure 1 d) and with relatively high graphitization level according to Raman spectra (figure 2 c). Such structures have resistance from ~27 kOhm to ~700 kOhm (orange cells in table 1). From table 1 it is clearly seen that this type of morphology шы usually obtained with 5-6 % of laser power with all movement speed except 60 and 100 % (in the last case the destroyed morphology is like an extremal type of the same morphology). In this case we can observe low Raman intensity of D peak and relatively high intensity of 2D peak. We can attribute this to the formation of highly structured grains of a few-layer graphene.

Fifth type of morphology looks similar to fourth but with local destruction of the whole film. Such structures are obtained only in case of high movement speed. This fact can be explained by the excess of ablation threshold due to high laser fluence in one point.

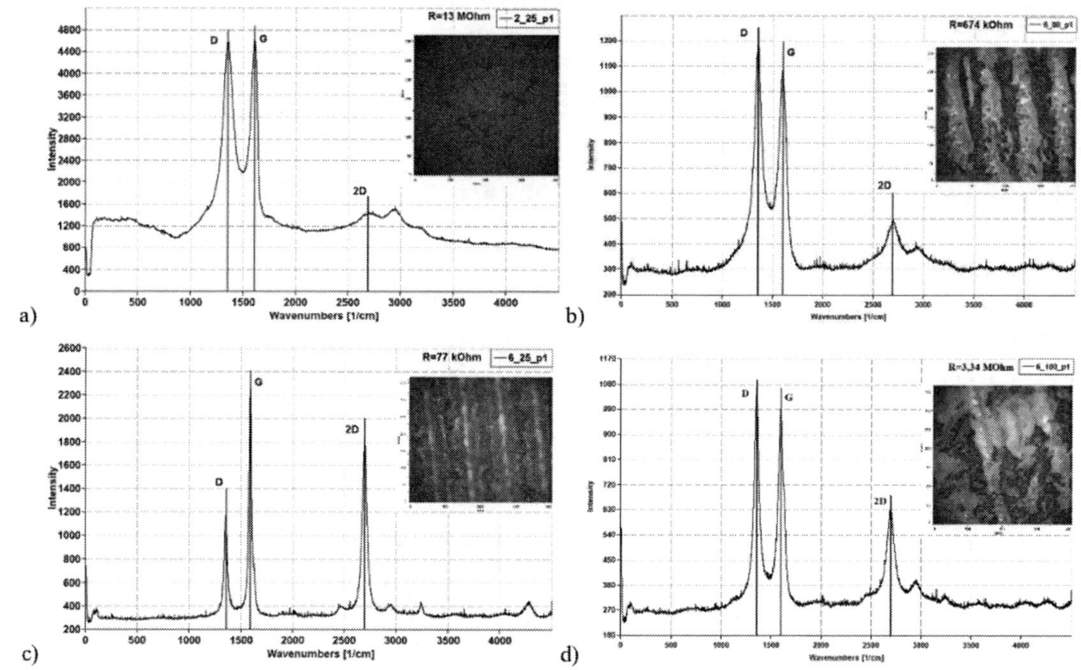

Figure 2. Raman spectra and morphology of near IR laser reduced GO films.

Figure 3 with of I_D/I_G and I_{2D}/I_G ratios dependence of the laser power and laser movement speed have very good coincidence with table 1. We can mention that nonreduced structures have I_D/I_G ratio about 1,0 and I_{2D}/I_G ratio ~0,35. Locally reduced films have variations in peaks intensity ratios from 0,75 to more than 1,1 in case of I_D/I_G and from 0,45 to 0,68 in case of I_{2D}/I_G. Structures that have best conductivity to functional groups presence relation shows have more than 1 I_D/I_G ratio and 0,48-0,55 I_{2D}/I_G ratio. Highly structured and most conductive structures have about 0,5-0,6 I_D/I_G ratio and 0,7-0,8 I_{2D}/I_G ratio. In case of film destruction, we once again observe variations in relative peak intensity with from 0,8 to more than 1 I_D/I_G ratio and from 0,35 to 0,6 I_{2D}/I_G ratio.

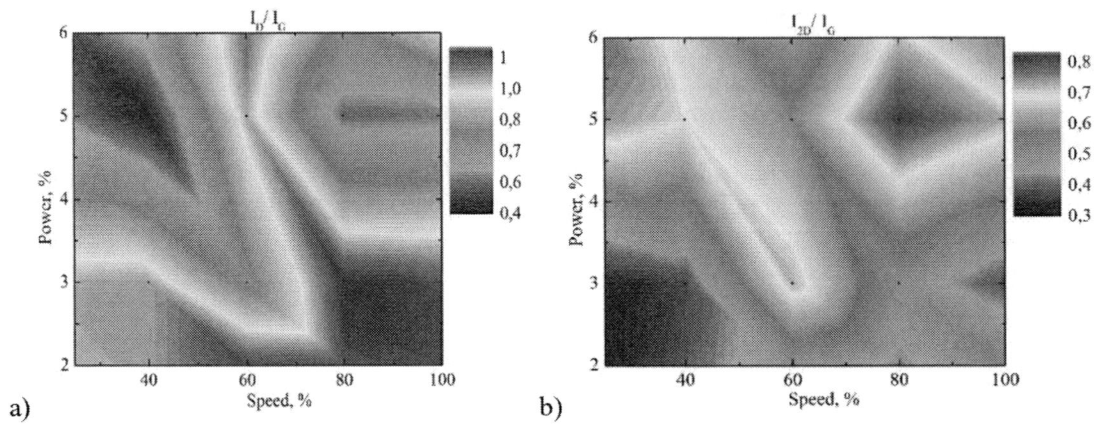

Figure 3. I_D/I_G (a) and I_{2D}/I_G (b) ratios vs near IR laser power and laser movement speed.

As a result we find near IR laser parameters for formation of relatively high conductive reduced graphene oxide films that can be used in different electronic applications (flexible electronics for example). Another relatively narrow group of parameters is perspective for functional nanocomposite formation that can further be used in different applications, for example chemical or biosensors.

Table 1. Resistance of spin-coated reduced graphene oxide structures vs laser power and movement speed (blue cells – nonreduced GO; yellow cells – partly locally reduced GO; green cells – slightly reduced GO with low I_{2D}, orange cells – highly reduced GO (high I_{2D}, low I_D), red – GO reduced with film destruction).

Laser Power, % to max / Movement speed, % to max	25	40	60	80	100
6	77 kOhm	40,3 kOhm	548 kOhm	674 kOhm	3.34 MOhm
5	27,5 kOhm	41,5 kOhm	198 kOhm	108,1 kOhm	755 kOhm
3	10 MOhm	8,2 MOhm	4,78 MOhm	891 kOhm	1.4 MOhm
2	13 MOhm	6 MOhm	1,50 MOhm	69,8 kOhm	10,4 MOhm

3.2 Laser reduction by microsecond pulsed laser

We made series structures modified with blue (445 nm) 600 mW laser power and pulse length 2,5 μs (laser fluence 1,5 μJ) and other modified with same laser power, but pulse length was 150 μs (fluence 93.4 μJ) laser spot size was 40 × 15 um in all cases.
Surface morphology of areas with low and high reduction level is on figure 4 (a) and (b) respectively. In case of low reduction level one can clearly see that structure consist of slightly overlapping laser traces on the whole area of the frame. In case of high reduction level overlapped laser traces forms line-shaped conductive pathways with some distance between them.
Comparing graphene oxide films resistance reduced with near IR laser and blue laser we can notice significant difference in resistance. In case of graphene oxide, reduced with near IR laser resistance of successfully reduced areas varied from 27 kOhm to ~900 kOhm, whereas in case of pulsed lased we obtained resistance about 6 kOhm average (with range from 4 to 12 kOhm). We assume this effect to

the laser wavelength possibility of more precise laser parameters control. Another important fact is that possible step in control software of laser engraver is too big to work with relatively thin films.

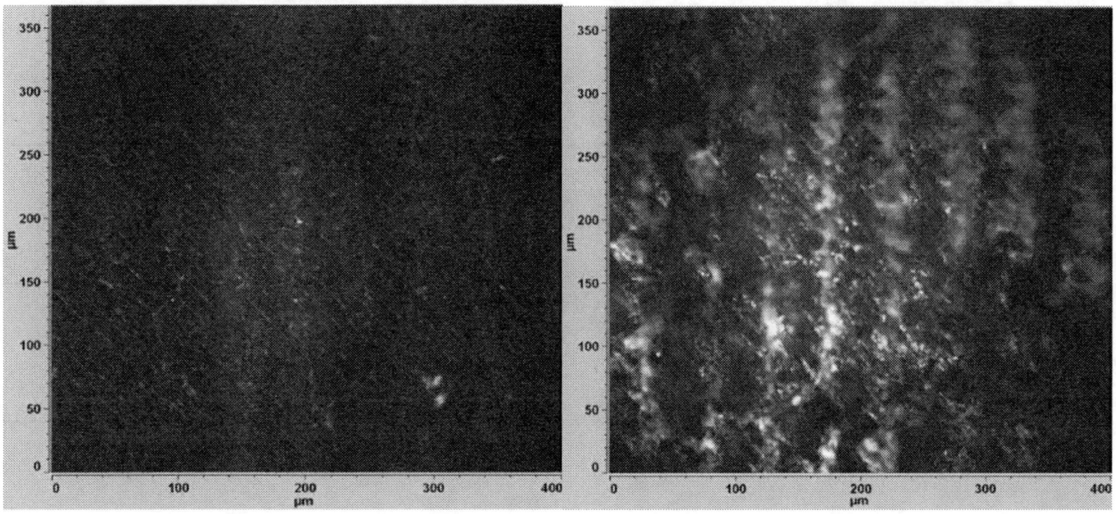

a) b)

Figure 4. Optical image of surface morphology of reduced graphene film in sensitive (a) and contact (b) area of sensor device.

So in general it's possible to reduce graphene oxide with "ready-to-use" laser systems but it's preferable to have more precise control of laser parameters. Another possible factor that provided a large variation in graphene oxide structures reduced with near IR laser is local nonuniformity of the initial film. We also observed poor adhesion of graphene oxide films reduced with near IR laser to the substrate, whereas in case of blue laser reduction adhesion is good enough.

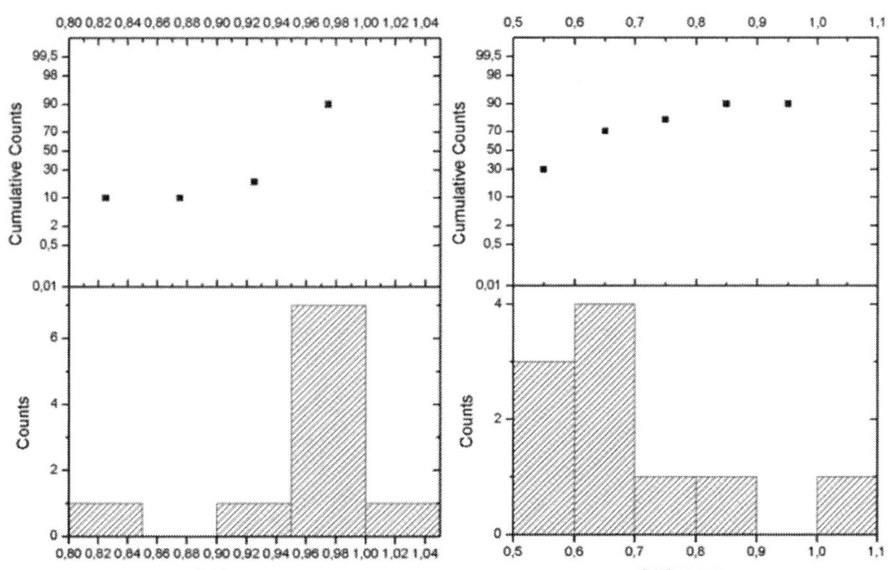

a) b)

Figure 5. I_D/I_G ratio for low fluence area (a) and high fluence reduced GO film (b).

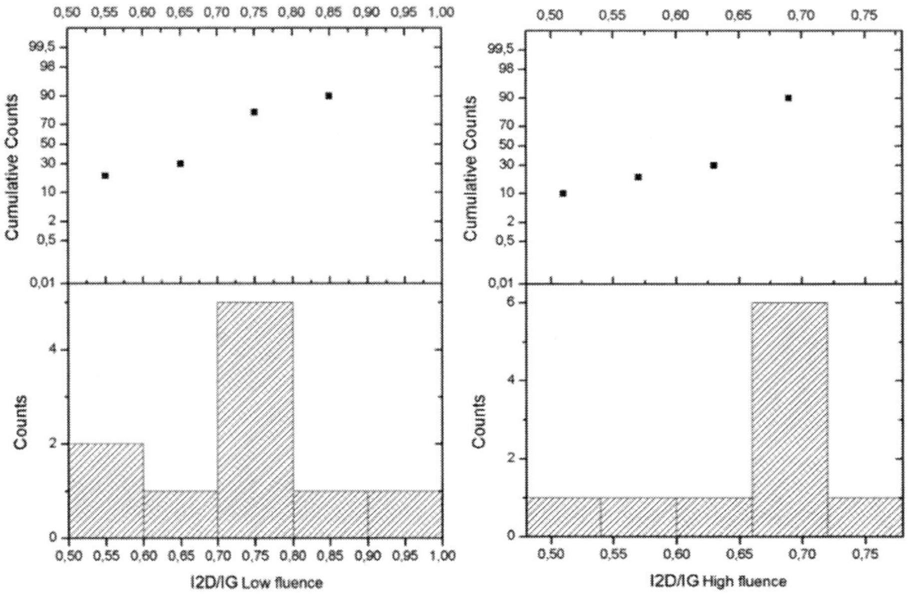

a) b)

Figure 6. I_{2D}/I_G ratio for low fluence area (a) and high fluence reduced GO film (b).

From figures 5 and 6 we can notice, that in case of reduction with low fluence ID/IG ratio lies in the range from 0,80 to 1,04 with statistics maximum at 0,98 to 1,00 whereas for high fluence it is in range from 0,5 to 1,1 with maximum at 0,5-0,7. Intensity of I2D/IG lies in range from 0,50 to 1,00 for low fluence, whereas for high fluence it's form 0,45 to 0,8. Statistic maximums in case of I2D/IG are 0,70 to 0,80 for low fluence and 0,65 to 0,70 for high fluence. According to the previous results with near IR laser we can see that ID/IG and I2D/IG ratios of areas with high conductivity and with high amount of functional groups are similar for both laser types. Near IR laser gives ID/IG ratio about 1 or more for areas with high amount of functional groups and blue laser gives 0,98 average, I2D/IG ratios for same cases are 0,48-0,55 (near-IR laser) and 0,6 (blue laser). For higher reduction these parameters are about 0,5-0,6 ID/IG ratio and 0,7-0,8 I2D/IG ratio (near IR laser) and average 0,6 ID/IG and 0,7 I2D/IG (blue laser).

We obtained FTIR spectroscopy data from both areas obtained by blue laser irradiation (Figure 6). From Figure 6 we can mention significant difference from spectra with low and high reduction level. Main difference is in clearly seen peak at ~1720 cm-1 in case of low reduction level. This peak should be attributed to C=O carboxyl and carbonyl groups according to [37]. I.e. in case of low laser fluence we observe conductive area with some residual carboxyl functional groups that is the main goal for such application of reduced graphene oxide films as different sensors (especially biosensors) due to possibility of covalent coupling of sensitive agents through these groups. We can also observe slight broad peak at ~ 3400 cm-1 that could be attributed to O-H stretching vibrations, i.e. hydroxyl groups presence [38]. Minor peaks at 2847 cm-1 (symmetric) and 2917 cm-1 (asymmetric) are –CH2 absorption band that was split into two sub-bands and can be assigned to alkyl moieties (as a result of the reduction of COOH to -CH3 segment) [39]. Wide band at about ~ 1080 cm-1 appears due to the presence of hydroxyls [40].

It is important to sign that we don't see any significant peaks of the area reduced with high fluence, that mean that this area have much higher graphitization level and its IR spectra up to 3000 cm-1 is very similar to those obtained by Mingru Su et al. from Si/Graphite@Graphene composite [41]. According to results of Raman spectroscopy this area have much lower level of defects with quite noticeable 2D peak which means that graphitization level of this area is higher than in case of low fluence.

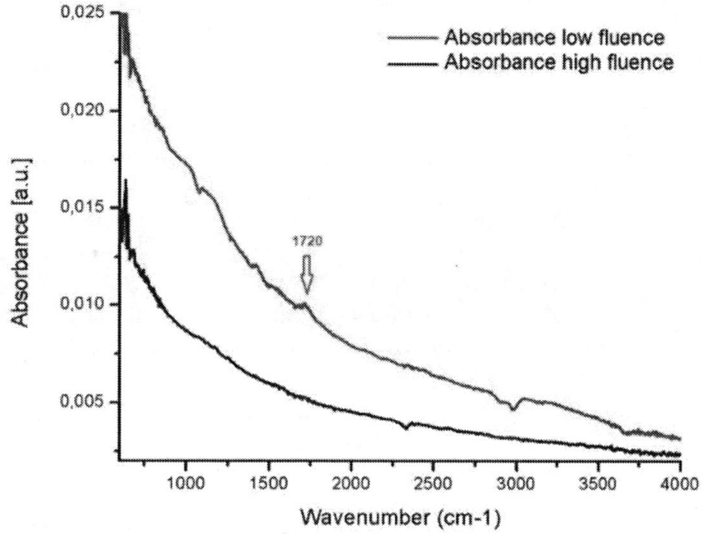

Figure 7. FTIR spectra of the graphene oxide film reduced with high and low fluence.

From FTIR spectra we can make a conclusion that after the graphene oxide reduction with low fluence a number of functional groups still remain in the film. Due to this fact low reduction level is good for the formation of functional composites, where there is a need to couple some nanoparticles, biomolecules or other additives. On the other hand, high reduction level can be used for different applications where a good conductivity is need, for example in flexible electronics, energy storage etc. We can also mention that we showed possibility of manageable graphene oxide reduction for different applications and future formation of functional composite materials.

4. Conclusions

In this work we empirically showed possibility of successful reduction of graphene oxide with relatively low-cost lasers (near IR and blue) and reviewed difference between "ready-to-use" device (commercial laser engraver) with powerful near IR laser and laboratory device, with low power blue laser. Both devices showed possibility of manageable reduction of graphene oxide but blue laser showed significantly less variation and better repeatability, of resistance, that was between 4 to 12 kOhm. This result is from one to 3 order of magnitude lower than in case of laser engraver. We also showed presence of different functional groups, in particular carboxyl, that is important for such application as chemical and biosensor development and functional composites formation.

Acknowledgement

This work is financially supported by Ministry of Science and Higher Education of Russian Federation grant number 14.574.21.0160 (grant unique I.D. RFMEFI57417X0160).

References

[1] Xiang L, Zhang H, Hu Y and Peng L-M 2018 Carbon nanotube-based flexible electronics. *J. Mater. Chem.* **6**, pp 7714-27.
[2] Perinot A, Passarella B, Giorgio M, Caironi M 2019 Walking the Route to GHz Solution-Processed Organic Electronics: An HEROIC Exploration. *Adv. Funct. Mater.* 1907641.

[3] Liu Y, Zhu Y and Cui Y 2019 Challenges and opportunities towards fast-charging battery materials. *Nature Energy* **4** pp 540–550.
[4] Piro B, Mattana G, Noël V 2019 Recent Advances in Skin Chemical Sensors. *Sensors 2019* **19** 4376.
[5] Luan C, Movva S, Wang K, Yao X, Zhang C and Wang B 2019 Towards next-generation fiber-reinforced polymer composites: a perspective on Multifunctionality. *Funct. Compos. Struct.* **1** 042002.
[6] Nelyub V A, Borodulin A S, Kobets L P, Malysheva G V 2016 A Study of Structure Formation in a Binder Depending on the Surface Microrelief of Carbon Fiber. *Polym. Sci. Ser.* **9** pp 286–89.
[7] Shahrubudin N, Leea T C, Ramlan R 2019 An Overview on 3D Printing Technology: Technological, Materials, and Applications. *Procedia Manufacturing* **3** pp 1286-96.
[8] Luo W, Hu T, Zhang C 2019 Digital twin for CNC machine tool: modeling and using strategy. *J Ambient Intell Human Comput* **10** pp 1129-1140.
[9] Deng T, Li J and Zheng Z 2020 Fundamental aspects and recent developments in metal surface polishing with energy beam irradiation. *International Journal of Machine Tools and Manufacture* **148** 103472.
[10] Borodulin A A, Kalinnikov A A, Kharaev A B, Shcherbin S 2019 Aromatic polysulfone to create polymer materials with high resistance to frost. *IOP Conference Series: Earth and Environmental Science* **302** 012062.
[11] Nelyub, V.A. Methods for developing carbon composites with functional properties 2020 *IOP Conference Series: Materials Science and Engineering* **709** 022037.
[12] Platonova E O, Vlasov E, Pavlov A A, Kireynov A, Nelyub V A, Polezhaev A V 2019 Self-healing polyurethane based on a difuranic monomer from biorenewable source. *Journal of Applied Polymer Science* **136** 47869.
[13] Popov A A, Tselikov G, Dumas N, Berard C, Metwally K, Jones N, Al-Kattan A, Larrat B, Braguer B, Mensah S, Da Silva A, Estève M-A and Andrei V 2019 Kabashin. Laser- synthesized TiN nanoparticles as promising plasmonic alternative for biomedical applications. *Scientific Reports* **9** 1194.
[14] Van H H, Badura K, Liang R, Okoli O and Zhang M 2017 Laser-induced graphitic healing of carbon nanotubes aligned in a sheet. *J. Laser Appl.* **29** 0022010.
[15] Bobrinetskiy I, Emelianov A, Nasibulin A, Komarov I, Otero N and Romero P M 2016 Photophysical and photochemical effects in ultrafast laser patterning of CVD graphene. *Journal of Physics D: Applied Physics* **49** 41.
[16] Li G 2020 Direct laser writing of graphene electrodes. *Journal of Applied Physics* **127** 010901.
[17] Rabchinskii M K, Dideikin A T, Kirilenko D A, Baidakova M V, Shnitov V V, Roth F, Konyakhin S V, Besedina N A, Pavlov S I, Kuricyn R A, Lebedeva N M, Brunkov P N and Vul A Y 2018 Facile reduction of graphene oxide suspensions and films using glass wafers. *Scientific Reports* **8** 14154.
[18] Marchena M, Wagner F and Arliguie T 2018 Dry transfer of graphene to dielectrics and flexible substrates using polyimide as a transparent and stable intermediate layer. *2D Materials* **5**, 035022.
[19] Jebaranjitham J N, Mageshwari C, Saravanan R and Mu T 2019 Fabrication of amine functionalized graphene oxide – AgNPs nanocomposite with improved dispersibility for reduction of 4-nitrophenol. *Composites Part B: Engineering* **171** pp 302-9.
[20] Nekrasov N, Kireev D, Emelianov A and Bobrinetskiy I 2019 Graphene-based sensing platform for on-chip ochratoxin a detection. *Toxins* **11** p 550.
[21] Kurapati D R et al. 2018 Covalent chemical functionalization enhances the biodegradation of graphene oxide. *2D Mater.* **5** 015020.
[22] Zhaoa C, Xinga L and Xiang J 2014 Formation of uniform reduced graphene oxide films on modified PET substrates using drop-casting method. *Particuology* **17** pp 66–73
[23] Wang J, Liang M and Fang Y 2012 Rod-Coating: Towards Large-Area Fabrication of Uniform Reduced Graphene Oxide Films for Flexible Touch Screens. *Adv. Mater.* **24** pp 2874-78.

[24] Holm A, Wrasman C J, Kao K-C 2018 Langmuir–Blodgett Deposition of Graphene Oxide—Identifying Marangoni Flow as a Process that Fundamentally Limits Deposition Control. *Langmuir* **34** pp 9683–91.
[25] Yadavab S and Kaur I 2016 Low temperature processed graphene thin film transparent electrodes for supercapacitor applications. *RSC Adv.* **6** 82.
[26] Struchkov, N S, Alexandrov E V, Romashkin A V, Silakov GO, Rabchinskii M K 2020 Uniform graphene oxide films fabrication via spray-coating for sensing application. *Fullerenes Nanotubes and Carbon Nanostructures* **28** pp 214-20.
[27] Pachauri N, Verma S, Singh P and Singh S P 2017 Few layered graphene oxide thin films: A potential matrix for immunosensors. *Integrated Ferroelectrics* **184** pp 85-91. DOI: 10.1080/10584587.2017.1368790
[28] Stankovich S 2007 Synthesis of graphene-based nanosheets via chemical reduction of exfoliated graphite oxide. *Carbon* **45** pp 1558-65.
[29] McAllister M J 2007 Single Sheet Functionalized Graphene by Oxidation and Thermal Expansion of Graphite. *Chem. Matter* **19** pp 4396-04.
[30] Zhang Y-L, Guo L, Xia H, Chen Q-D, Feng J and Sun H-B 2014 Photoreduction of Graphene Oxides: Methods, Properties, and Applications. *Adv. Optical Mater.* **2** pp 10–28.
[31] Gengler R Y N, Badali D S and Zhang D 2013 Revealing the ultrafast process behind the photoreduction of graphene oxide. *Nature Communications* **4** pp 2560-65.
[32] Sokolov D A, Shepperd K R and Orlando T M 2010 Formation of Graphene Features from Direct Laser-Induced Reduction of Graphite Oxide. J. *Phys. Chem. Lett.* **1** pp 2633 – 36.
[33] Bobrinetskiy I I, Emelianov A V, Smagulova S A, Komarov I A, Otero N and Romero P M 2017 Laser direct 3D patterning and reduction of graphene oxide film on polymer substrate. *Materials Letters* **187** pp 20-23.
[34] Alexandrov G N, Smagulova S A, Kapitonov A N, Vasil'eva F D, Kurkina I I, Vinokurov P V, Timofeev V B and Antonova I V 2014 *Nanotechnologies in Russia* **9** pp 5-6.
[35] Struchkov N S, Kondrashov V A, Rozanov R Y and Nevolin V K 2017 *Journal of Physics: Conference Series* **816** 012014.
[36] Kasischke M, Maragkaki S, Volz S, Ostendorf A and Gurevich E L 2018 Simultaneous nanopatterning and reduction of graphene oxide by femtosecond laser pulses. *Applied Surface Science* **445** pp 197–203.
[37] Eluyemi M S, Eleruja M A, Adedeji A V, Olofinjana B, Fasakin O, Akinwunmi O O, Ilori O O, Famojuro A T, Ayinde S A and Ajayi E O 2016 Synthesis and Characterization of Graphene Oxide and Reduced Graphene Oxide Thin Films Deposited by Spray Pyrolysis Method. *Graphene* **5** pp 143–154.
[38] Lee D, Cho H, Han D, Chand R, Yoon T, Kim Y 2017 Highly Selective Organic Transistor Biosensor with Inkjet Printed Graphene Oxide Support System. *J. Mater. Chem.* **5** pp 3580–85.
[39] Gupta B, Kumar N, Panda K, Kanan V, Joshi S and Fisher V I 2017 Role of Oxygen Functional Groups in Reduced Graphene Oxide for Lubrication. *Sci. Rep.* **7** 45030.
[40] Bagri A, Mattevi C, Acik M, Chabal Y J, Chhowalla M and Shenoy V B 2010 Structural Evolution during the Reduction of Chemically Derived Graphene Oxide. *Nature Chem.* **2** pp 581–87.
[41] Su M, Wang Z, Guo H, Li X, Huang S, Xiao W and Gan L 2014 Enhancement of the Cyclability of a Si/Graphite@Graphene composite as anode for Lithium-ion batteries. *Electrochimica Acta* **116** pp 230-36.

Estimation of the durability of polymer composites on a fabric basis, taking into account the influence of non-force factors

R A Kayumov[1,2], A M Suleymanov[1], I Z Muhamedova[1]

[1] Kazan State University of Architecture and Engineering, 1 Zelenaya st., Kazan, 420043, Russia

[2] Kazan National Research Technical University named after A.N.Tupolev, 68 K.Marks str., Kazan, 420015 Russia

E-mail: kayumov@rambler.ru, muhamedova-inzilija@mail.ru

Abstract. We considered the problem of evaluating the durability of film-fabric material (FFCM). The models of phase deformation of a film-fabric composite material was developed which took into account nonlinear viscoelasticity, temperature changes, material aging, microdamage accumulation processes and photodestruction. We presented the composite under consideration as an inhomogeneous body. When choosing the structure of the defining relations, we used the experimental data obtained by the authors for the phases of FFCM. Based on the modification of the second Fick's law, the kinetic law of propagation of the photodestruction parameter deep into the representative cell of FFCM was constructed. To conduct a qualitative analysis of the behavior of the composite material under study, we formulated the problem with some model constitutive relations and numerical experiments were performed.

1. Introduction

When designing elements of the structures and buildings made of polymer composite materials (PCM), the most important were the tasks of assessing their strength, stiffness and durability, taking into account the destruction of the material, since under real operating conditions the material of structures was subjected to the complex effects of aggressive environments, temperature, mechanical loads and other energy impacts, various combinations of which caused various mechanisms of destruction.

The aggressive media, penetrating into the volume of a structural element of FFCM, sometimes led to significant changes in its physical and mechanical characteristics, which also caused a change in the stress-strain state of structures. The researches, for example, [1-3] were devoted to improving the effectiveness of FFCM in the aggressive environments we had in service. In the work [1] we investigated the issues of improving the quality and efficiency of FFCM application for using in chemically and biologically aggressive environments by the methods of physical-and-chemical and structural modifications with small additives. In the work [2] was analyzed the effect of the content and grain size of the filler particles on the strength and durability of FFCM, technological and operational factors. We revealed dependencies that made it possible to predict the properties of PCM. In the work [3] we suggested using complex modifiers that significantly stiffen resistance of PCM to different kinds of fungi as well as improve their physical and mechanical properties.

In the works [4-6] we explored the subject of aging and PCM degradation under the influence of climatic factors and operational impacts. For example, in the work [5] we studied aging of PCM of aviation purpose, we also showed a collectively accelerating synergistic effect of climatic factors in the formation of material damage. In the work [6] we showed the test results of PCM compositions based on epoxy binders, carried out under the influence of natural climatic factors of a temperate climate. The analysis of changes in the total solar radiation and ultraviolet radiation of various ranges was carried out and the influence of the duration and intensity of irradiation on the change in the elastic-strength characteristics of polymer composites was determined. In the work [7] we submited the data on the change in the tensile strength, elastic modulus, and elongation under tension of the materials after several cycles of a rapid temperature change. In the work [8] we studied a strain-stress state (SSS) of fiberglass plastic under conditions of climatic aging. It was revealed that the destructive stresses of fiberglass plastic were significantly reduced during the first two years of its using, what should be taken into account when choosing fiberglass plastic with the declared characteristics. Ultraviolet did not significantly affect the stress-related properties of fiberglass plastic. In the work [9] we presented an assessment of the use of the composites based on a polyalkanimide matrix under thermal cycling. We also presented the data on the change in tensile strength, elastic modulus, and elongation in tension of materials after several cycles of a rapid temperature change. Vacuum ultraviolet radiation created by the sun in the outer space could lead to degradation of thermoregulatory coatings, what would lead to a change in their optical, mechanical and chemical properties. In the work [10] we represented the researches of high-strength thermal control coatings with the radiological protection function, carried out with the aim of reducing degradation of the optical properties of coatings under the influence of aggressive factors of the outer space and protecting on-board electronic equipment from electromagnetic interference. The effect of vacuum ultraviolet radiation on polymer composites led to a mass loss, a change and a deterioration in their optical properties [11]. We revealed the effect of smoothing the surface of composites. In the work [12] we made an experimental assessment of the effect of temperature aging on the mechanical behavior, residual strength, and rigidity of PCM. It turned out that, as a result of temperature aging, the mechanical properties of the composite deteriorated, defective structures grew, matrix embrittlement and fiber breakdown occurred. In the work [13] we offered a model which described degradation of the elastic moduli of PCM upon heating to the high temperatures. Using the model, we obtained analytical dependences between the elastic moduli of textile composites and the elastic characteristics of their matrices and fibers, as well as the geometric structural parameters of composites at the high temperatures. The results showed a good correlation with experimental data for carbon and glass phenolic composites.

2. Experimental part

To determine the parameters of the behavior patterns of the phases of a composite, we carried out the experiments to identify the mechanism of aging and destruction of the film-fabric composite material (FFCM) under the operating conditions. To ensure strict regulation of the simulation of FFCM operating conditions, the tests were carried out in the laboratory conditions under influence of the artificial factors that are adequate to the full-scale energy values according to the power-producing values [14].

The experimental studies revealed a unique effect of the FFCM aging and fracture mechanism under influence of the climatic factors in a stress-strain state. The facts cited in the literature [15–18] were explained, according to which the local gaps in the materials of fencing in many types of structures from soft shells that had been in operation for a certain period of time, as a rule, did not coincide with any of the places of the greatest tensile forces found as a result of calculation.

The climatic aging and fracture mechanism of stress-strain FFCM was as follows. Due to the high degree of initial crimping of the reinforcing threads, deformability of the material in the direction of the large tensile loads was very high and could reach 10-20% (the threads straighten in this direction). Moreover, in the orthogonal direction, if the tensile load was less than in the perpendicular, the threads of the reinforcing base were bent and created local stress zones in the polymer matrix of the composite. Additional influence of the climatic factors, mainly UV radiation of the sun, led to formation of microcracks in the above zones. Next, those microcracks grew into main through cracks, being oriented perpendicular to the direction of maximum deformation of the coating. It opened up an access of UV radiation to the less loaded threads of the reinforcing base, what led to an increase in the aging and fracture rate in the less loaded direction of the soft shell.

3. Structure of the defining (physical) relations for the FFCM components, connecting the static, kinematic and structural parameters of the material

To assess the durability of FFCM, a model of deformation of its representative cell was developed, taking into account nonlinear viscoelasticity, temperature changes, aging of the material, the processes of microdamage accumulation, and photodestruction. The considered cell was presented as an inhomogeneous body. When choosing the structure of the determining relations, the known experimental facts regarding the FFCM phases were taken into account. Based on the modification of the second Fick's law on diffusion the kinetic law of the propagation of the photodestruction parameter deep into the representative FFCM cell was constructed. To conduct a qualitative analysis of the behavior of the studied composite material, the problem was formulated for some model defining relations.

The defining relations for the FFCM phases were constructed taking into account the viscoelastic properties of polymeric materials. For the structural components of the polymer composite, nonlinear relations of the theory of hereditary elasticity were used [19]. The theory of damage accumulation developed by Yu. N. Rabotnov [20] was used to formulate criteria relations of the delayed fracture of the material in question.

The following parameters of the FFCM aging process were introduced as the main ones. First, this was a damage parameter ω, which described the accumulation of defects in the material, such as microcracks, micropores For it, we could use defining relations both in differential form and of a hereditary type [20-23].

In the numerical examples, the following kinetic equation is adopted:

$$\frac{d\omega}{dt} = B(T)\cdot\eta(\omega,\sigma_i), \tag{1}$$

where $B(T)$ is some fugacity of temperature, $\eta(\omega,\sigma_i)$ is a function depending on stress intensity σ_i and on a parameter of accumulated defects ω in the FFCM cell.

Under influence of the climatic factors, in particular ultraviolet irradiation, there was aging of the material, which covered the tissue. After crack opening, phase transformations and changes in the mechanical properties of the fabric material began. We called that process a photodestruction of the material. We introduced a scalar parameter W, which we called a photodestruction parameter. Diffusion of destruction into the thickness of the material in a certain layer of height h proceeded from the side of the surface exposed to ultraviolet radiation. Microcracks appeared on the surface, which also grew with time, which, in turn, again led to an increase in the layer thickness h. The photodestruction parameter was considered changeable in proportion to the irradiation intensity γ.

It is proposed to use the following evolution equation for the photodestruction parameter W_0 on the surface of a representative FFCM cell, subjected to irradiation:

$$\frac{dW_0}{dt} = \gamma\cdot\zeta(T)\cdot f(\sigma_{i0},W_0), \tag{2}$$

where $\zeta(T)$ is a nonlinear temperature function T, and $f(\sigma_{i0},W_0)$ is a function of two arguments: stress intensity σ_{i0} and degree of photodestruction W_0 on the surface of the material.

The kinetic law of the propagation of the photodestruction parameter W deep into the representative FFCM cell can be written on the basis of Fick's law in the following form:

$$\frac{\partial W}{\partial t}=\gamma\cdot\zeta(T)\cdot\left(\frac{\partial}{\partial x}(\mu\frac{\partial W}{\partial x})+\frac{\partial}{\partial y}(\mu\frac{\partial W}{\partial y})\right), \tag{3}$$

where $\mu=\mu(x,y)$ is a diffusion coefficient of photodestruction parameter W. We assume that, in the case of completion of irradiation, the destruction process should end, therefore the coefficient γ is introduced in (3).

The disadvantage of the equation (3) was that it was hyperbolic, therefore, in the very first moment after the appearance of destruction on the irradiated surface this destruction instantly spread over the entire FFCM depth, (although in infinitely small values) gradually increasing with time. Since this was not natural for the material degradation process, we introduced the hypothesis that the penetration depth of photodegradation h was determined by some dependence on the state parameters of the material. The irradiated surface of the representative FFCM element, was assumed to be flat, just for simplicity, therefore, the process of increasing the depth of penetration of photodestruction into the material is described by the ratio:

$$\frac{dh}{dt}=\gamma\cdot\zeta(T)\cdot g(\sigma_i,W_0,h). \tag{4}$$

Then for W the boundary conditions take the following form:

$$W(\sigma,\omega,T,x,y,t)\big|_{y=0}=W_0(x,t) \tag{5}$$

$$W(\sigma,\omega,T,x,y,t)\big|_{y=h}=0 \tag{6}$$

4. Problem solving with non-stationary temperature changes

This section presents the results of the qualitative studies of the influence pattern of temperature differences on the FFCM durability. The law of a temperature difference *T* during the year can be approximated as:

$$T=T_{cp}+\Delta T\cdot Sin(\omega t) \tag{8}$$

where T_{cp} is an average operating temperature, t is time, ΔT is a temperature difference.

The principle of linear summation of damage is taken as a basis of calculations of durability. Let us introduce as an example the following function f_{1i}, that determines the law of change in the mechanical characteristics of elasticity and creep of a component with the number *i* depending on the temperature *T*:

$$f_{1i}=(f_{0i}\cdot(Arctg[(T-T_{cp}+\frac{tg[[\frac{1}{f_{0i}}-1]\cdot\frac{\pi}{2}]}{\theta_{1i}})\cdot\theta_{1i}]+\frac{\pi}{2})/\frac{\pi}{2})^2, \tag{9}$$

here the constants f_{0i}, θ_{1i} determine the nature of the change in mechanical characteristics, which are increased with a raise of the temperature.

The Figures 1, 2 show the diagrams of durability dependency on the temperature difference for various deformations ε at $\theta_{11}=0.045$, $\psi_{11}=0.045$ and at $\theta_{11}=0.18$, $\psi_{11}=0.18$.

The following qualitative regularities of the change in durability due to temperature difference were revealed. If the temperature difference leads to small changes in the elastic characteristics and creep parameters, then with increase in $\Delta T>0$ durability t^* also decreases (Fig.1). However, at sufficiently large θ_{11} and ψ_{11} durability starts increasing from a certain value of ΔT, as can be seen in Fig. 2. This can be explained by the fact that at the large parameters of θ_{11} and ψ_{11}, and, at large , the material softens so quickly that the damages in the material start being accumulated more slowly.

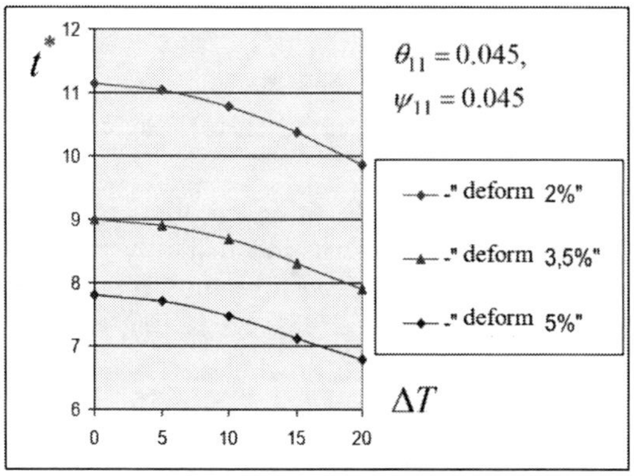

Fig. 1. Durability dependency on temperature difference at $\theta_{11} = 0.045,\ \psi_{11} = 0.045$

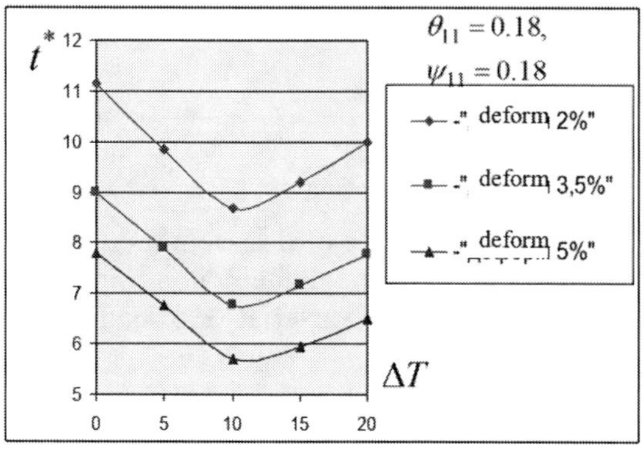

Fig. 2. Durability dependency on temperature difference at $\theta_{11} = 0.18,\ \psi_{11} = 0.18$

Acknowledgments
This work was supported by the Russian Federal Property Fund (project No.19-08-00349, sections 1,3) and the Russian Science Foundation (project No.19-19-00059, section 2)

References

[1] Ogrel LYu 2006 Modified epoxy binder for the manufacture of fiberglass pipes, gas ducts and gas outlet trunks of thermal power plants *Building materials* **5** Pp 57-59.

[2] Bobryshev A N, Tumanova N N, Pertsev V T, Zubarev P A, Kuvshinov P I 2012 Chemical resistance of mixed polymer composites in aggressive environments *Scientific Bulletin of the Voronezh State University of Architecture and Civil Engineering. Series: Physico-chemical problems and high technologies of building materials science* **5**. Pp 26-32

[3] Vetrova Yu V 2014 Improving the resistance of polymer building materials and composites in biologically aggressive environments *International Scientific-Research Journal*. 2014 **6-1 (25)** Pp 36-37.

[4] Kirillov V N, Startsev O V and Efimov V A 2012 Climatic resistance and damage to polymer composite materials, problems and ways of solution *Aviation materials and technologies: annivers., scientific and techn. compos.book (append. to magaz. «Aviation materials and technologies»)*. M.VIAM Pp 412-423.

[5] Kablov E N, Startsev O V, Krotov A S and Kirillov V N 2011 Climatic aging of composite materials for aviation purposes. III. Significant factors of aging *Deformation, destruction of materials* **1** Pp 34–40.

[6] Nizina T A, Selyaev V P, Nizin D R and Artamonov D A 2015 Climatic resistance of composite building materials in natural conditions of operation *Fundamental and applied researches of corrosion and aging of materials in climatic conditions: problems and prospects: book of reports. II All-Russian. Scientific and technical Conf.* (VIAM. M.)**1**

[7] Nizina T A, Artamonov D A, Nizin D R, Chernov A N and Andronychev D O 2017 Analysis of the effect of hardener on the climatic resistance of epoxy composite materials *News of Higher Education Institutions. Building* **1 (697)** Pp 55-64

[8] Martynov G V, Monastyreva D E, Morina E A and Makarov A I 2018 Stress-strain state of fiberglass plasticin conditions of climatic aging *Vestnik MGSU* **13** 12 (123). Pp 1509-1523
[9] Cherkashina N I and Pavlenko A V 2018 Effect of SiO2 crystal structure on the thermal cycling of polymer composites *Building Materials and Products* **1** 4 Pp 21-29
[10] Cherkashina N I, Pavlenko V I, Demchenko O V and Ivanitsky D A 2015 Use of polyhedral oligomeric silsesquioxanes for the synthesis of radiation-resistant polymer composites for thermoregulation purpose *Successes in modern natural science* **9-3** Pp 524-527
[11] Pavlenko V I and Cherkashina N I 2019 Effect of SiO2 crystal structure on the stability of polymer composites exposed to vacuum ultraviolet radiation *Acta Astronautica* **155** Pp 1-9
[12] Lobanov D S and Zubova E M 2019 Research of temperature aging effects on mechanical behaviour and properties of composite material by tensile tests with used system of registration acoustic emission signal *Procedia Structural Integrity* **18** pp 347-352
[13] Dimitrienko Yu 1998 I Modelling of mechanical properties of composite materials at high temperatures. Part 3. Textile composites *Applied Composite Materials* **5** 4 Pp 257-272.
[14] Suleymanov A M 2005 Installation for modeling the impact of operational factors on the materials of soft shells *Factory Laboratory. Diagnostics of materials.* (Moscow) **71** 12 pp 44-46.
[15] Suleymanov A M 2005 Work, aging and destruction of soft shell fencing materials *Building materials* (Moscow) **11** Pp 62- 65
[16] Ogrel L Yu 2006 Increase in efficiency of building polymer composites operated in aggressive environments *Dissertation for the degree of Doctor of Technical Sciences* (Belgorod)
[17] Bobryshev A N, Tumanova N N, Pertsev V T, Zubarev P A and Kuvshinov P I 2012 Chemical resistance of mixed polymer composites in aggressive environments *Scientific Bulletin of the Voronezh State University of Architecture and Civil Engineering. Series: Physico-chemical problems and high technologies of building materials science* **5** Pp 26-32
[18] Vetrova Yu V 2014 Improving the resistance of polymer building materials and composites in biologically aggressive environments *International Scientific-Research Journal* **6-1 (25)** Pp 36-37
[19] Kayumov R A, Tazyukov B F and Mukhamedova I Z 2019 Identification of the mechanical characteristics of a nonlinear-viscoelastic composite by the results of testing the shells of rotary *Mechanics of composite materials* **55** 2 Pp 247-258
[20] Rabotnov Yu N 1965 Creep of elements of constructions (M .: Science) p 752
[21] Ilyushin A A Mechanics of continua 1971 (M: Publishing house of Moscow University) p 248
[22] Demishkevich E B and Gavriushin S S 2020 A viscoelastic model of the long-term orthodontic tooth movement *Advances in Intelligent Systems and Computing* **902** Pp 315-322
[23] Dumansky A M and Liu Hao 2019 Analysis of anisotropy of time-dependent and nonlinear properties of unidirectional CFRP *Advances in Composite Science and Technology IOP Publishing IOP Conference Series: Materials Science and Engineering* **683** 012093 DOI:10.1088/1757-899X/683/1/012093

IOP Conf. Series: Materials Science and Engineering **934** (2020) 012042 doi:10.1088/1757-899X/934/1/012042

Estimation of the residual strength of supporting composite structural elements

R A Kayumov[1,2], D E Strakhov[1], A M Sulejmanov[1] and E B Tuysina[1]

[1]Kazan State University of Architecture and Engineering, 1 Zelenaya st., Kazan, 420043, Russia
[2]Kazan National Research Technical University named after A.N.Tupolev, 68 K.Marks str., Kazan, 420015 Russia

E-mail: strachovde@mail.ru (D E Strakhov)

Abstract. The proposed study will allow us to develop a calculation method to predict the service life of fiberglass structures providing reliability and resistance; a design method for polymer composite structures taking into account damage accumulation and aggressive factors that affect the decrease in the strength of the studied material. The method of aging and failure of polymer composite materials was developed. The significance of the results for the construction industry lies in the development of a methodology for assessing the service life of fiberglass structures of the investigated building structures.

1. Introduction

Currently, the need for the use of load-bearing structural elements made of polymer composite materials has significantly increased. This necessitates studying the mechanisms of damage and its development that affect the load bearing capacity and durability of elements and compounds.

The disadvantage of the methods for calculating such elements and compounds is the lack of accurate experimental data, as well as the variety of compositions of modified composites with different operational properties.

The calculation of polymer composite materials for resistance and durability, a description of their rheological properties, assessment of the characteristics of long-term creep-rupture is an undeniable task today.

Pultrusion process is the most versatile and technological in production of various fiberglass elements of load-bearing structures. The manufactured products have fiberglass reinforcement in one direction in which the mass is pulled.

The full range of effects on the polymer composite material during operation is temperature, aqueous, alkaline or acidic environment, the influence of the atmosphere, various biodeteriorations leading to climatic and natural aging [1-10].

The working capacity of the material is characterized by three qualitative characteristics - heat resistance, strength and durability, each of which is determined by a set of physical constants, where a change in one working parameter can be achieved by changing the other two [11]. Thus, experimentally determining the values of physical constants, it is possible to predict the durability of the material.

IOP Conf. Series: Materials Science and Engineering **934** (2020) 012042 doi:10.1088/1757-899X/934/1/012042

2. The study of flat elements reinforced with fiberglass

In this work, we investigate a polymer composite material in the form of flat extended elements reinforced with fiberglass. Research media were selected: air, distilled water and aqueous alkali solutions. It is known that when moisture penetrates PCM, weakening of bonds at the fiber-binder boundary occurs, microcracks appear [12-15], which leads to a decrease in strength. The first aqueous solution of alkali, modeling the liquid phase of concrete, adopted of NaOH – 8 g. and KOH – 22,4 g. per 1 liter of distilled water (GOST 31938-2012), the pH of the alkaline solution was in the range from 12,6 to 13. To exclude interaction of air with CO2 and evaporation, the alkaline solution was in a closed container. The concentration of the second aqueous solution was doubled. With the penetration of moisture into the PCM, weakening of bonds at the fiber-binder boundary occurs, microcracks appear [4–7], which leads to a decrease in strength.

The studied samples were deformed in the form of a three-hinged fastening (Figure 1), mechanical deformation was assumed to be 20, 30, and 60% of the destructive deformation. When conducting intermediate measurements the equipment was disassembled. After control measurements the equipment was reassembled. Test temperature accepted 18 ± 3 degrees. Samples were aged in selected media for 540 days with intermediate measurements. Measurements were carried out with KM-8 cathetometers with an accuracy of 0,001 mm.

The dependences of residual deformations on time are obtained (Figs. 2-5).

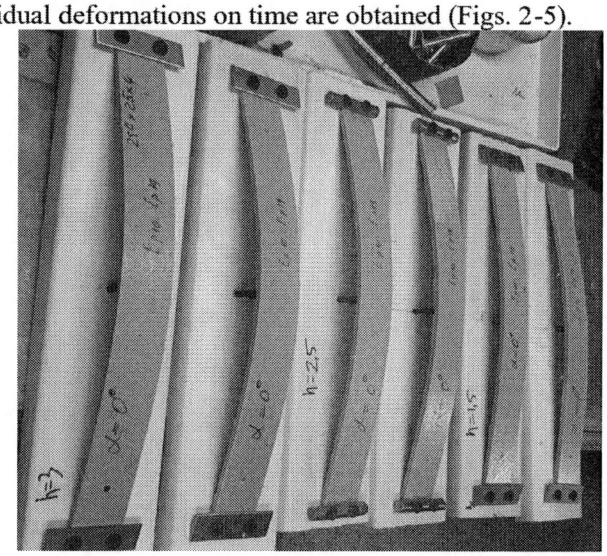

Figure 1. General view of the studied samples, the magnitude of the deflection of 30 mm, 25 mm and 15 mm (authors' illustration). General view of the studied samples, the magnitude of the deflection

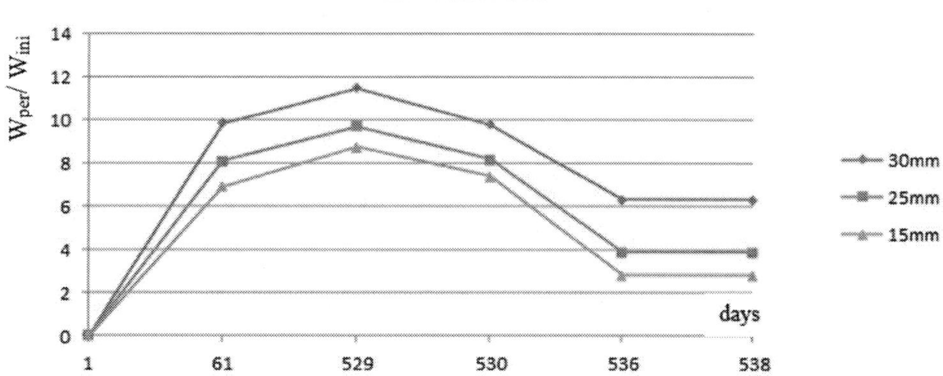

Figure 2. Air environment, deformation height 30 mm, 25 mm and 15 mm, 538 days (authors' illustration).

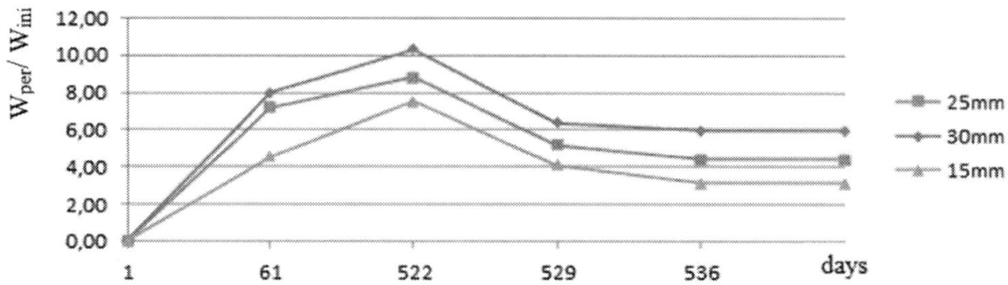

Figure 3. Aquatic environment, deformation height 30 mm, 25 mm and 15 mm, 538 days (authors' illustration).

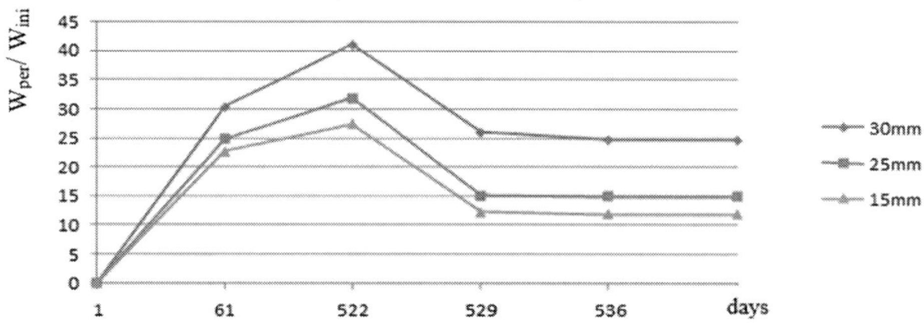

Figure 4. Alkaline composition, deformation height 30 mm, 25 mm and 15 mm, 538 days (authors' illustration).

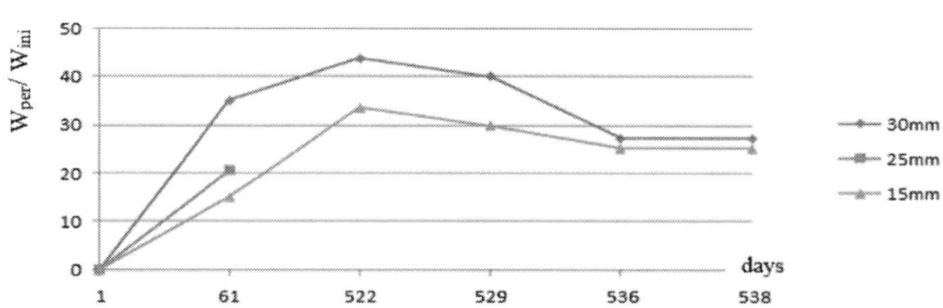

Figure 5. Doubled alkaline composition, deformation height 30 mm, 25 mm and 15 mm, 538 days. A sample 25 mm high collapsed during the test after 61 days (authors' illustration).

3. Prediction of long-term strength of composite depending on temperature and stress level.

This section presents the results of studies of round rods. The tests were carried out at various levels of long-term static load and without it. The load was applied by the method of longitudinal bending [16-22]. The temperature was also used as a factor that naturally accelerates the aging of the material. In total, two types of tests were carried out:

Type 1: samples were kept in an alkali solution (GOST 31938-2012) at various temperatures of 20 °C, 40 °C, 50° C, 60 °C, excluding Stress-Strain state (SSS).

Type 2: the samples were aged in an alkali solution (GOST 31938-2012) at a temperature of 50 °C, the tests were carried out in SSS at three levels of deformation, equivalent to 0,2, 0,4, 0,6 of the strength of the PCA with longitudinal bending (hereinafter R).

The dependences of changes in residual strength at different stages of exposure of the samples in unstressed (Figure 6) and stressed (Figure 7) states were obtained.

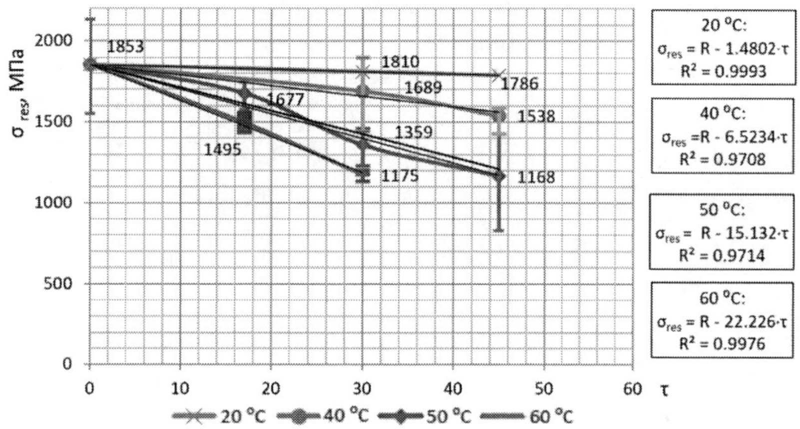

Figure 6. Experiment results for type 1
(conditions: different temperature levels and alkali solution)
with approximation by linear functions (authors' illustration).

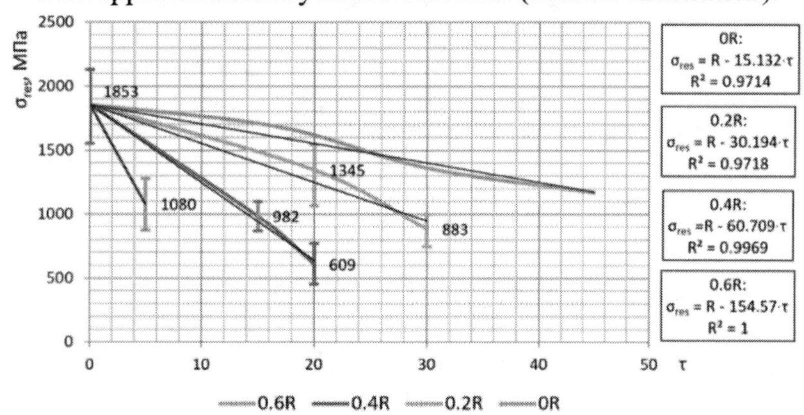

Figure 7. Experiment results of type 2
(conditions: various levels of mechanical load, temperature 50°C, alkali solution)
with approximation by linear functions (authors' illustration).

As can be seen from the analysis of the test results, the pattern of strength change with a high level of correlation is described by linear functions. Therefore, in the time interval corresponding to the time of accelerated testing, the residual strength σres can be determined by a function of the form:

$$\sigma_{res} = R\,(1 - K_{con} \cdot K_T \cdot \tau) \qquad (1)$$

where R is the initial strength of the material, mcon = σ / R is the level of long-term load in fractions of the initial strength; Kcon = f (mcon) is a function of the level of continuous load mcon, which determines the effect of a given SSS on the change in the initial strength of the material; KT = g (T) is a function of the exposure temperature T, which determines the effect of temperature on the change in the initial strength of the material; T is the exposure temperature of the samples, ° C; τ is the exposure time of the samples.

Thus, the task of determining the law of change in residual strength is reduced to determining the functions Kcon and KT. We assume that at the level of continuous load mcon = 0, the function Kcon = 1. Then, to determine the function KT, we consider its values obtained during the type 1 test:
KT (T = 20 ° C) = 1.4802; KT (T = 50 ° C) = 15.132;
KT (T = 40 ° C) = 6.5234; KT (T = 60 ° C) = 22.226.
Approximating the obtained experimental values of KT, we determine the general form function for the test material (Figure 8 a) in the temperature range at which studies were carried out. The scope can be expanded without additional research only if there are appropriate justifications.

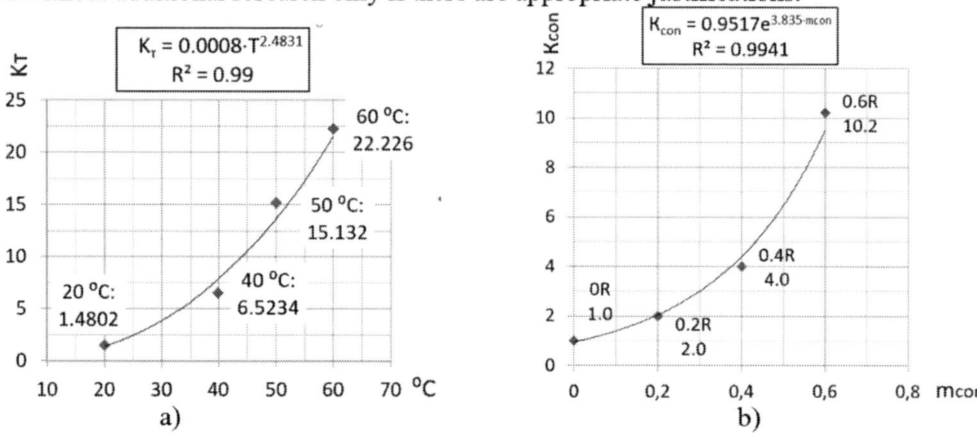

Figure 8. a) Definition of K_T function; b) Definition of the K_{con} function. The dots indicate the values of the coefficients obtained by approximating the experimental strength values (authors' illustration).

For a test of type 1, carried out at a temperature of T = 50 ° C, the coefficient KT = 15,132. Then, according to the values (Kcon • KT) obtained by approximating the test data of type 2, for various load levels we find:
K_{con} (m_{con})= (K_{con} • K_T) / K_T (T = 50°C);
K_{con} (m_{con} = 0) = 1;
K_{con} (m_{con} = 0,2) = 2;
K_{lt} (m_{lt} = 0,4) = 4,01;
K_{con} (m_{con} = 0,6) = 10,02.
Using the same approach as in determining the function for the coefficient KT, the empirical values of the coefficient Kcon (mcon) are approximated (Figure 8 b).
Thus, the prediction of the residual strength of PCR samples, tests that are carried out under conditions of an alkaline environment exposure at temperatures from 20 ° C to 60 ° C at loads from 0 to 0,6R, can be carried out according to the formula (1), where:
KT = 0.0008 • T2,4831; Kcon= 0,9517 • e3.835 • mcon.
It should be noted that for a relatively small interval, the coefficients KT and Kcon can be considered as functions of one variable: temperature or continuous level respectively.
However, the possibility of applying the described approach for a time interval in which the law of change in residual strength for different types or test conditions occurs according to different laws and differs from a linear one should be determined individually.

4. Assessment of continuous strength by the level of deformation.

The maximum elastic deformation on the surface of a flat sample is calculated by the formula:

$$\varepsilon_{\max} = \frac{(w_{\max} - w_{res})h}{4L^2} \tag{2}$$

Here w_{max} is the deflection of the central point of the sample, wres is the residual deflection of this point, h is the thickness of the sample, L is its length. We denote by R0 the initial tensile strength, and

F0 the value of the breaking load obtained in the experiment under "instant" loading. For each variant of the aggressive environment, one can obtain the dependence of the tensile strength on deformation after holding for a time t* from the condition:

$$\frac{R}{R_0} = \frac{F}{F_0} \tag{3}$$

The expression for R can be approximated by a function in the form:

$$R = R_0\left(1 + \varepsilon \cdot k_1 + \varepsilon(\varepsilon - \varepsilon_1) \cdot k_2 + \varepsilon(\varepsilon - \varepsilon_1)(\varepsilon - \varepsilon_2) \cdot k_3\right) \tag{4}$$

The coefficients k1, k2, k3 can be obtained from condition (3) for different values of F. This gives a system of equations, the solution of which can be written as:

$$k_1 = \frac{F_1 - F_0}{F_0 \varepsilon_1},\ k_2 = \frac{F_2 - F_0(1 + \varepsilon_2 k_1)}{F_0 \varepsilon_2(\varepsilon_2 - \varepsilon_1)},\ k_3 = \frac{F_3 - F_0\left(1 + \varepsilon_3 k_1 + \varepsilon_3(\varepsilon_3 - \varepsilon_1)k_2\right)}{F_0 \varepsilon_3(\varepsilon_3 - \varepsilon_1)(\varepsilon_3 - \varepsilon_2)} \tag{5}$$

Here, F1, F2, F3 are the values of the breaking load obtained in the first, second and third experiments, ε1, ε2, ε3 are the deformations corresponding to these loads, found by formula (2). Instead of deformation ε, we can approximately substitute stresses, expressing them in terms of ε according to Hooke's law. By calculating k1, k2, k3 for various sample exposure times, which were then brought to failure, we can obtain their dependence on time. Then we obtain relation (4), which allows predicting the residual strength at any operating time of composite structural elements.

5. Conclusion

A study of a polymer composite material in the form of flat and round elongated fiberglass reinforced elements in various environments and at different temperatures was made. Relations are proposed that can be obtained on the basis of data from a number of experimental studies on transverse or longitudinal bending. They allow to predict the level of permissible continuous load at a given structure life cycle. The advantage of this approach is that the number of test types is significantly reduced, as a result, the number of samples is reduced, less time and equipment are required to obtain the necessary data on the material.

Acknowledgments

This work was supported by the Russian Foundation for Basic Research (project № 19-08-00349, sections 1,3) and the Russian Science Foundation (project № 19-19-00059, section 2).

References

[1] Blaznov A N, Savin V F, Atyasova E V, Babenko F I and Fedorov Yu Yu 2014 Influence of temperature on the strength of basalt and fiberglass plastics *Polzunovskiy vestnik..* **2(4)** P. 154–159.

[2] Fedorov Yu Yu, Babenko F I, Gerasimov A A and Lapiy G P 2012 Influence of cold climate factors on the mechanical properties of composite rods made of glass and basalt plastics. *Metal processing (technology, equipment, tools)* **2** P. 56–61.

[3] Blaznov A N, Savin V F and Krasnov A A 2014 Prediction of the strength of composite rods under alkaline conditions South Siberian Scientific Bulletin **4 (8)** P. 12-14.

[4] Sen R, Mullins G and Salem T. (2002). Durability of E-Glass / Vinylester Reinforcement in Alkaline Solution *ACI Structural Journal* **99(3)** pp. 369-375.

[5] Cabral-Fonseka S, Correia J R, Rodrigues M P and Branco F A 2012 Artificial accelerated aging of GFRP pultruded profiles made of polyester and vinylester resins: characterization of physical – chemical and mechanical damage *Strain* **48** P. 162–173.

[6] Koretskaya L and Aleksandrova T 2011 Influence of water on the properties of fiberglass technology and materials **1 (31)** P. 38–41.

[7] Nikolaev E V, Barbotko S L, Andreeva N P, Pavlov M R and Grashchenkov D V 2016 Comprehensive study of the impact of climatic and operational factors on the new generation

of epoxy binder and polymer composite materials based on it Part 4. Natural climate tests of polymer composite materials on basis of epoxy matrix *VIAM Proceedings* **6 (42)**

[8] Kablov E N, Startsev O V, Krotov A S and Kirillov V N 2011 Climatic aging of composite materials for aviation purposes. III. Significant Aging Factors *Deformation and Failure of Materials* № 1 Electronic Version: www.viam.ru/public

[9] Efimov V A, Shvedkova A K, Korenkova T G and Kirillov V N 2013 Study of polymer structural materials under the influence of climatic factors and loads in laboratory and natural conditions *Aviation materials and technologies.*. P. 68–73.

[10] Karapetyan K A 2014 Resistance to deformation and destruction of reinforced plastics Ep. : *"Gitutiun" NAS RA*,. 164 p, Tab. 9, ill. 35, lit. 150 titles.

[11] Ratner S B and Yartsev V P 1992 Physical mechanics of plastics. As predicted performance. M.: Chemistry 320 p.

[12] Efimov V A, Kirillov V N, Dobryanskaya O A, Nikolaev E V, Shvedkova A K, Korenkova T G and Deev I S 2010 On the issue of the methodology for conducting full-scale climatic tests of polymer composite materials: collection of articles. *Reports of the eighth scientific conference on hydroaviation "Hydroaviation 2010"* / VIAM. Moscow,. P. 102–106.

[13] Aniskevich K K, Kurzemnieks A X and Yanson Yu O 1985 Study of the effect of prolonged exposure to temperature and moisture on the elastic properties and structure of organoplastics // *Mechanics of Composite Materials* **4** P. 620–623.

[14] Statnikov R B, Gavriushin S S, Dang M and Statnikov A 2014 Multicriteria design of composite pressure vessels //International Journal of Multicriteria Decision Making. T. 4. **3** C. 252-278.

[15] Vapirov Yu M, Krivonos V V and Startsev O V 1994 Interpretation of the anomalous change in the properties of the KMU-1u carbon fiber during aging in different climatic zones *Mechanics of composite materials* **30(2)** P. 266–273.

[16] Arnautov A K and Tarnopolsky Yu M 2004 Longitudinal bending as a method for determining the bending strength of composite materials. *Mechanics of composite materials* **40(1)** P. 25–42.

[17] Savin V F and Blaznov A N 2001 Method for determining the continuous strength of fiberglass reinforcement *Measuring, automation and modeling in industry and scientific research: Interuniversity collection* P. 214–219.

[18] Savin V F, Volkov Yu P, Lugovoi A N, Blaznov A N and Xe A I 2002 Longitudinal bending as a means of controlling the mechanical characteristics of composite structural materials Measurements, automation and modeling in industry and scientific research: Interuniversity digest P. 167–172 .

[19] Kiselev N M, Savin V F and Blaznov A N 2008 Fatigue strength of composite material rods under longitudinal bending and tension: *reports of the 7th All-Russian Scientific and Practical Conference May 21-23, 2008 "Technique and technology for the production of heat-insulating materials from mineral raw materials"* / BTI AltGTU Biysk P. 137–142.

[20] Blaznov A N, Volkov U P, Lugovoi A N, Savin V F and Khe A I 2003 Investigation of the deformation of fiberglass rods in longitudinal bending *Measurements, automation and modeling in industry and scientific research: Interuniversity digest* pp. 180–185.

[21] Petrov M G 2003 Analysis of the strength and durability of unidirectional fiberglass from the standpoint of the kinetic concept of fracture *Mechanics of composite materials and structures* Vol. 9 № 3 P. 376–397.

[22] Demishkevich E B and Gavriushin S S 2020 A viscoelastic model of the long-term orthodontic tooth movement // Advances in Intelligent Systems and Computing.. - T. 902. - C. 315-322.

[23] Dumansky A M and Liu Hao 2019 Analysis of anisotropy of time-dependent and nonlinear properties of unidirectional CFRP // Advances in Composite Science and Technology IOP Publishing IOP Conference Series: Materials Science and Engineering 683 012093 DOI:10.1088/1757-899X/683/1/012093.

IOP Conf. Series: Materials Science and Engineering **934** (2020) 012043 doi:10.1088/1757-899X/934/1/012043

Structure, properties and heat treatment of aluminum alloy BAC1 synthesized by 3D printing

R S Fachyrtdinov, P E Kuznetsova and I D Savichev

FSUE All Russian scientific-research institute of aviation materials, 105005 Russia, Moscow, Radio str. 17

E-mail: ravel@bmstu.ru, kuznetsovapolinae@yandex.ru, ilix77@yandex.ru

Abstract. Article is devoted to studying of influence of the process of selective laser melting and subsequent various heat treatment on the structure of aluminum alloy. On the basis of scientific and technical literature describes the need for and relevance of the application of modern technologies of 3D printing, the possibility of its application. The authors investigate the microstructure and determine the porosity of the synthesized samples of aluminum alloy of the Al-Si-Mg (Cu) system. The dependence of the porosity of the samples on the energy density is established. Also, the author investigated the influence of annealing and hardening modes with aging on the hardness of the samples of aluminum alloy. On the basis of the conducted researches characteristic properties and features of the synthesized material and preferable heat treatment were determined.

Keywords: selective laser melting, aluminum alloys, additive technologies, metal-powder composition, heat treatment.

1. Introduction

Currently, a wide range of different technologies is used for the manufacture of body parts of aircraft units. Mainly for metals used casting followed by machining. Silumins-aluminum alloys of Al-Si system are used for casting of cases. They have high technology in casting. However, the yield of suitable in the manufacture of products is reduced due to a variety of casting defects associated with the characteristics of crystallization of alloys.

For a number of complex profile elements of the design of units in the casting process reduced yield due to casting defects associated with the characteristics of crystallization of alloys, while the design and manufacture of equipment, as well as the development of technology require significant time and resources. This approach imposes restrictions on the choice of technological solutions of a node, due to the possibility of its manufacture. As a result, the weight of the unit increases, reliability decreases, the coefficient of use of the material falls. An alternative way to obtain complex shape parts are actively developing in the last few decades additive 3D technologies [1]. With the use of additive technologies, it is possible to implement step-by-step topological optimization of nodes by means of rapid changes in the structural appearance according to the results of tests and, accordingly, to increase the weight efficiency of the structure. It can be achieved by increasing the CMM by obtaining blanks of parts with the most approximate dimensions and reducing the cost of subsequent processing, and by reducing the marriage and increasing the yield of suitable in comparison with casting technologies. The advantages of additive manufacturing technologies can be applied in the aerospace industry, automotive manufacturing and other areas of mechanical engineering.

Interest in additive technologies arose in aviation, space industry and power engineering. It is in these industries that production is measured by tens or hundreds of products of complex shapes made of special materials. Often, the use of such methods in these industries is economically feasible. In many cases, additive technologies are even less expensive than traditional technologies. Direct cultivation of finished metal products is an alternative to traditional technological methods for the production of marketable products.

The essence of additive technologies is the layer-by-layer synthesis or layer-by-layer "growing" of a model or a finished product from a digital model, without the use of technological equipment. Thus, the construction of the part occurs by adding material in contrast to traditional technologies, where the creation of the part occurs by removing the "excess" material [2].

Layer-by-layer synthesis to obtain details of complex configuration was used in the late 19th century. Thin wax plates, paper sheets, wooden veneer were used as materials for cutting out elements, created structures, three-dimensional maps of the Earth's surface, landscape, technological equipment (forms for printing, castings). At the moment, the technology allows you to create structures and parts not manually, but on automated equipment and by 3D printing [3].

Cultivation methods involve the use of materials of different nature. Polymer materials, ceramics, metals and alloys, as well as titanium and aluminum alloys are used [4]. Machines in which, for example, metals are used as a material are divided according to the principle of feeding the material into: jet release, mixing the powder into a paste, surfacing with a jet feed and the most studied selective fusion of a metal powder with a powerful electron beam or selective laser fusion.

Selective laser fusion (SLS) technology is the process of manufacturing a product by layer-by-layer fusion of a powder, the particles of which are joined by melting them with the help of a laser beam that moves along a set trajectory. Today it is one of the most promising areas in the production of parts for the aviation and aerospace industries. Compared with traditional methods of creating products, SLS has a number of advantages [1; 5].

All SLS installations work on the same principle. First, a 3D model of the part is generated using CAD or scanning. The program divides the model into layers. A special device distributes a layer of powder on the surface of the substrate. The laser beam then melts the powder according to a specific algorithm, scanning the cross-section generated by the 3D model. After scanning each layer, the substrate on which the product is synthesized is lowered by a certain amount. The process is repeated, all layers of the product are made [6]. At the end of the printing process, the platform with the product is removed and the product is cleaned of the remaining powder. The thickness of the melted layer is usually about 30 microns, this is due to the quality of the resulting material and the possibility of complete further penetration without the formation of pores and discontinuities.

Aluminum alloys are widely used in the aircraft industry, and the modern technology of selective laser fusion will save time for manufacturing parts, but at the same time preserve the strength properties [7;8]. All mechanical characteristics and structure of the obtained parts will directly depend on the parameters of the SLS and at the same time differ from the properties of the part, which will be made by traditional casting methods with subsequent machining. In works [9-10] composite materials (CM) on the basis of aluminum matrix obtained by SLS technology are considered. These materials have a composite structure, that is, the dispersed phase is evenly distributed in the matrix, they are separated by a boundary and are not dissolved in each other. To create modern technology requires materials with a complex of properties that do not have standard metal alloys.

A rather narrow range of materials is used for SLS. The main requirement for the material for SLS-good casting properties to ensure alloying and continuity. For this reason, silumins are widely used, they are the most technologically advanced in casting. Al-Si alloys are used abroad:AlSi10Mg (ak9ch analogue), AlSi12 (AK12), AlSi7Mg. The most studied of them is the alloy AlSi10Mg [11-28].

2. Materials and methods.

The object of the study was a metal-powder composition of silumin system Al-Si-Mg-si, produced by gas atomization in FSUE "VIAM" with a given chemical composition, which is presented in table 1. The chemical composition of the powder was determined on the optical emission spectrometer MagellanQ8 according to GOST 24231.

Table 1. Chemical composition of metal-powder composition of silumin [11]

Alloy grade	Mass fraction of elements, %						
	Core components						
	Al	Si	Mg	Cu	Zr	Ce	Fe
BAC1	Base	10,5	0,63	0,7	0,25	0,23	0,15

Samples were made from powder of 10-63 microns fraction according to the selected modes on 1 cube with a side of 20 mm for each mode, also samples were subdivided in the directions of printing. Printing took place along the z and x axes.
Production of metallographic samples of the alloy was carried out according to the traditional method. Etching was carried out in 0.5% HF. The hardness of the samples was measured using the EMCO-TESTDuraJet instrument on the HRB scale.

3. Results and discussion.

The microstructure of the alloy after synthesis has a characteristic "track" structure, this is shown in figure 2 (a, b). It consists of a supersaturated solid solution of aluminum and dispersed inclusions of silicon and intermetallides. On the surface in the longitudinal section traces of laser passage in the form of "scales" are visible, and in the transverse melt baths are ellipses of irregular shape.

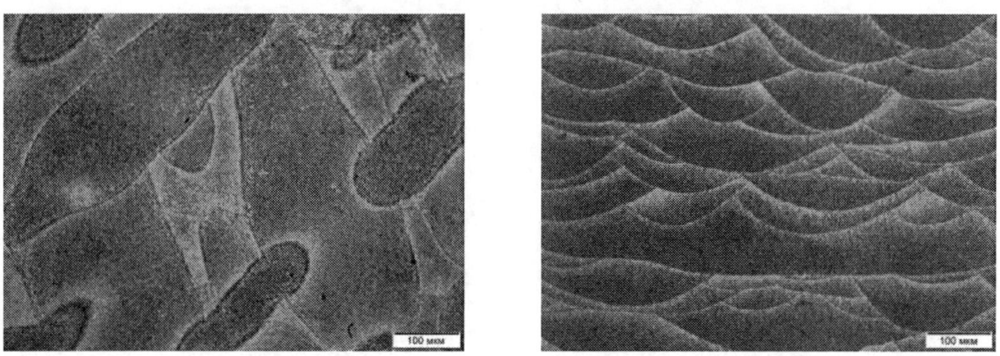

a b

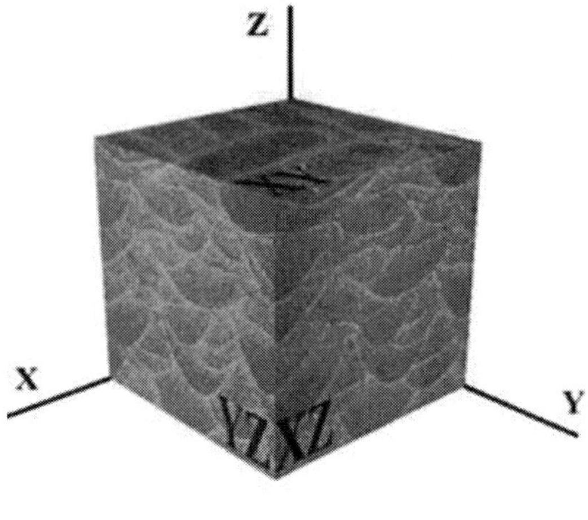

c

Figure. 1. The microstructure of the synthesized material:a) cross-section in the XY plane;b) longitudinal section in the xz plane;c)in general form.

When comparing the cast structure of the aluminum alloy Ak9ch and the synthesized sample from the powder composition (figure 2, b), it is seen that the structure of the synthesized sample is uniformly distributed particles of secondary fine phases in a solid solution of aluminum, while the cast structure shown in figure 3 has a pronounced dendritic structure.

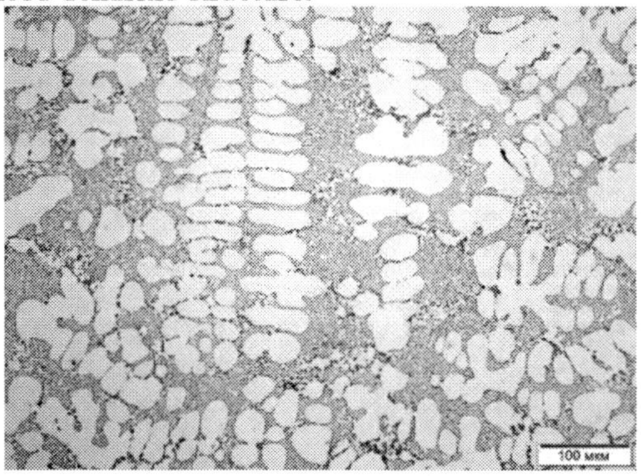

Figure. 2. The cast structure of the alloy AK9

The microstructure of the synthesized alloy depends on the parameters of the SLS. These parameters are intertrack distance, laser power, laser scanning speed. All parameters in the aggregate reflects the formula, which was given in [6].

$$E = \frac{P}{v * s * t}$$

where E is the laser energy density (J/mm3), P is the laser power (W), v is the scanning speed (mm / s), s is the distance between tracks (mm), t is the layer thickness (mm).
With different combinations of SLS parameters, samples with different porosity are obtained. As the energy density increases, the porosity of the synthesized material decreases.

IOP Conf. Series: Materials Science and Engineering **934** (2020) 012043 doi:10.1088/1757-899X/934/1/012043

a b

Figure. 3. The porosity of the samples: a) with an energy density of 56 j / mm3; b) with an energy density of 28 j/mm3

To study the effect of heat treatment on the structure and properties of the material, a batch of samples from the metal-powder composition of the alloy under study was synthesized. The hardness of the synthesized sample without heat treatment is 76 HRB. The hardness of cast silumin is in the range of 52.6-57.1 HRB. In the process of synthesis, a high concentration of silicon in a solid solution of aluminum is achieved - up to 6-8%. The formation of a supersaturated solid solution, the presence of dispersed inclusions of silicon and intermetallides explains the high hardness values.
Annealing of AlSi10Mg alloy was carried out at temperatures from 300 C to 450 C with a duration of exposure from 1 to 9 hours. Annealing at a temperature of 300 degrees, is taken for the lower threshold because at lower temperatures the relaxation of internal stresses does not occur completely [12-13]. The microstructures of annealed alloy samples are shown in figure 5.

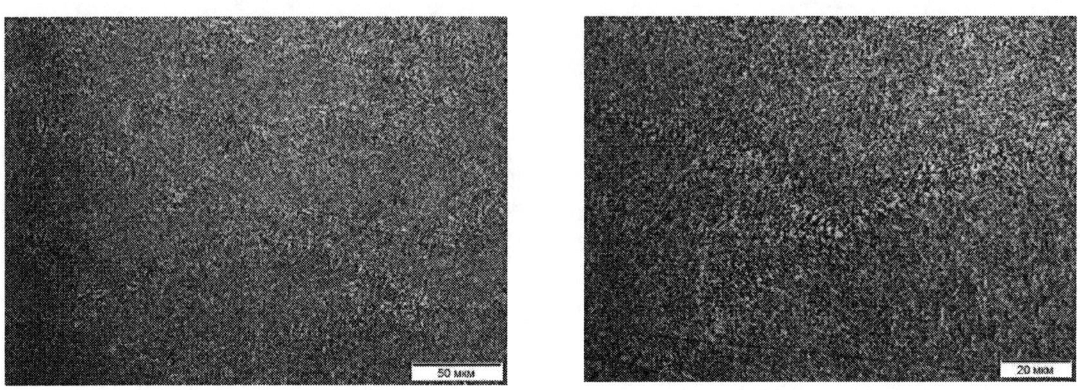

a b

IOP Conf. Series: Materials Science and Engineering **934** (2020) 012043 doi:10.1088/1757-899X/934/1/012043

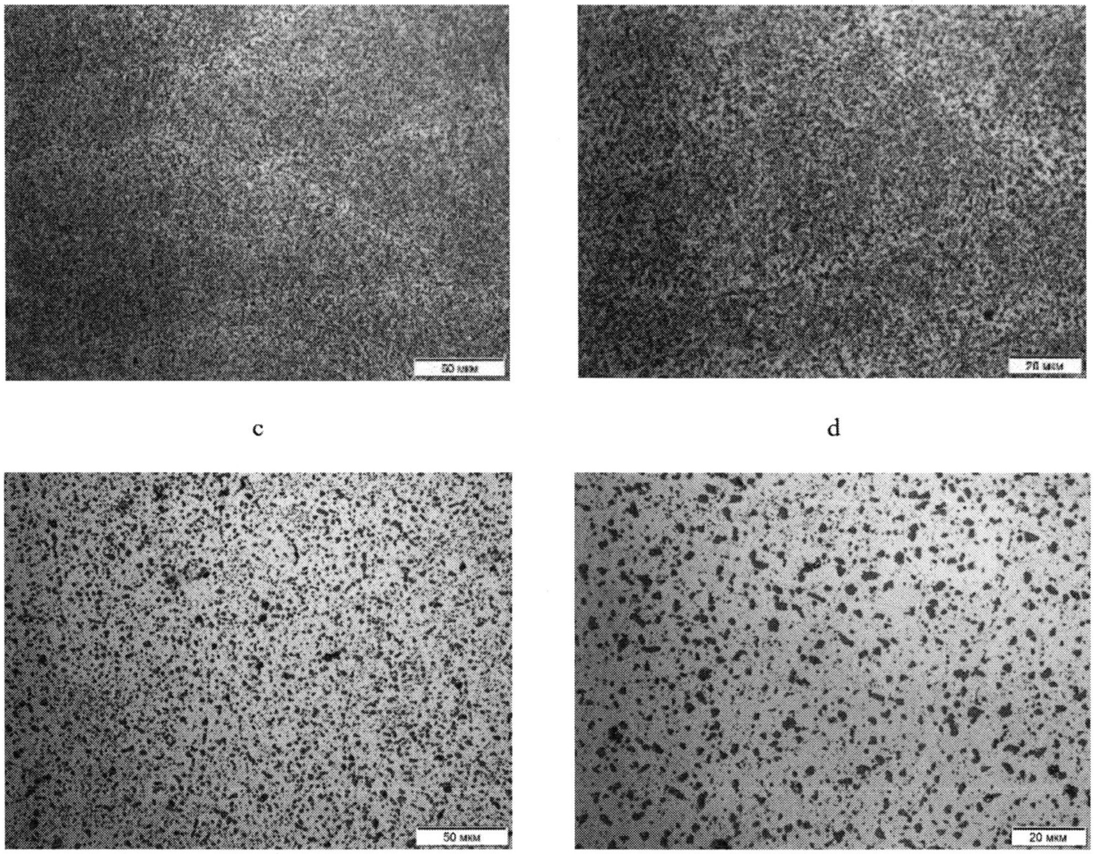

c d

e f

Figure. 4. Microstructures of synthesized alloy samples after annealing:
a), b) - at an exposure time of 1 hour and T=300S and magnification ×500 and ×1000
c), d) - at an exposure time of 3 hours and T=300S and an increase of ×500 and ×1000
e), f) - with a holding time of 9 hours and T=450S and an increase of ×500 and ×1000

The study of the microstructure showed that the annealing process is accompanied by a change in the structure. The hardness of the alloy also changes (figure 6). In the annealing process, eutectic silicon particles fall out of the solid solution of aluminum and coagulate, the material tends to an equilibrium state. The track structure is eliminated, the anisotropy of the structure is reduced, and therefore the anisotropy of the properties is reduced. This state is quite stable and devoid of flaws synthesized state and is quite applicable to aircraft parts.

The hardness of the alloy depends on the temperature-time parameters of the annealing mode (see figure 6). The graphs reflect the expected sharp drop in hardness as the temperature and annealing time increase. At 300 °C annealing and 3 hour exposure, a high level of hardening is maintained, but the anisotropy of the structure is not completely eliminated. Increasing the annealing temperature reduces the hardness and reduces the anisotropy of the structure and properties.

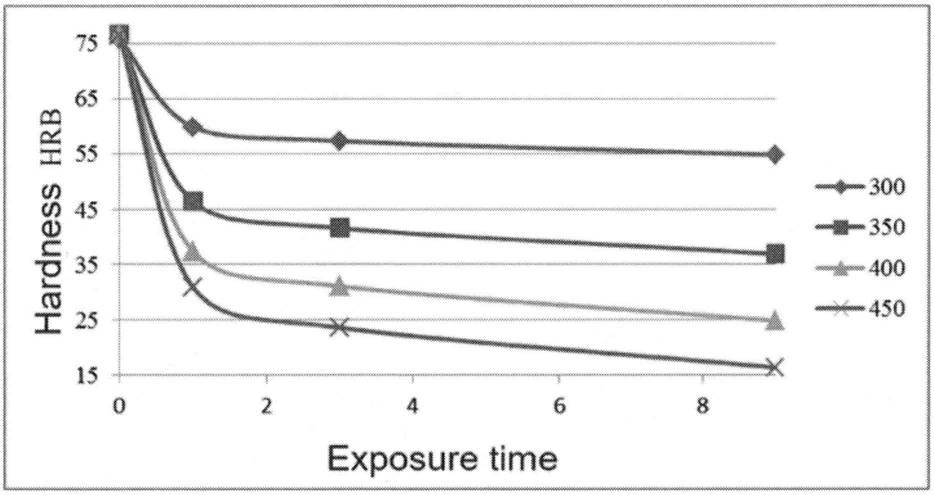

Figure. 5. Graph of dependence of hardness values on temperature and holding time after annealing

4. Conclusion

Methods of metallographic analysis was carried out to estimate the porosity of the samples grown on various regimes of selective laser melting. The dependence of the porosity and hardness of the samples on the laser energy density is established. With an increase in energy density of more than 50 j/mm3, the number of pores becomes less than 0.3%. As the energy density increases, the hardness of the synthesized material decreases. These data can be used in the further development of technologies for manufacturing products from alloys of the Al-Si-Mg system by the SLS method.

The structure of the synthesized samples has a track structure different from that of the cast sample having a C dendritic structure.

The effect of annealing on the structure and properties of al-Si-Mg-si silumin synthesized by 3D printing was studied.

It was found that during the annealing process, the material is softened, which is associated with the release and coagulation of silicon particles. The higher the temperature and duration of exposure, the higher the degree of softening. It was found that during the annealing process at 300 °C for 3 hours a relatively high level of hardness (57.5 HRB) is provided while maintaining the anisotropy of the structure. An additional increase in strength can be achieved by subsequent hardening treatment.

References

[1] Aleshin N P, Grigoriev M V, Shchipakov N A, Murashov V V, Evgenov A G, Vasilenko S A and Krasnov I S 2016 The classification of flaws of metal materials synthesized by the selective laser melting method and the capabilities of nondestructive testing methods for their detection //Russian Journal of Nondestructive Testing.**52(1)** C. 38-43. .

[2] Dovbysh V M, Zabednov P V, Zlenko M A Additivnyye tekhnologii i izdeliya iz metalla // sayt GNTS RF FGUP NAMI URL: http://nami.ru/55a62fc89524bAT_metall.pdf (data obrashcheniya: 03.06.2019).

[3] Dmitriev A M, Grigoriev S N, Korobova N V and Stupnikov V P 2012 A study of the quality of preforms from iron-base powders produced by forming combined with sintering by electric current pulses // *Metal Science and Heat Treatment*. **54(1-2)** C. 17-21.

[4] Kablov Ye N 2015 Additivnyye tekhnologii - dominanta natsional'noy tekhnologicheskoy initsiativy // *Intellekt i tekhnologii*,**2(11)** S.52-55.

[5] Additivnyye metody sozdaniya detaley – eto perekhod k shestomu tekhnologicheskomu ukladu.. // VIAM URL: https://viam.ru/interview/2521.

[6] Fakhurtdinov R S, Ryzhova M Y and Pakhomova S A 2017 Advantages and commercial application problems of vacuum carburization. // Polymer Science. Series D.**10(1)** C. 79-83.

[7] Gerasimov S A, Gress M A, Mukhin G G, Bayazitova V V and Lapteva V G 2008 Wear resistance of gasobaric nitrided layers on steel 12KH18N10T // *Metal Science and Heat Treatment.***50(1-2)** C. 73-76.

[8] Belov N A, Bazlova T A, Alekseeva E V and Naumova E A 2016 *STructure, phase composition, and strengthening of cast AL–CA–MG–SC alloys* // The Physics of Metals and Metallography **117(2)** C. 188-194

[9] Zaitsev A I, Zaitseva N E, Shimko R Yu, Arutyunyan N A, Dunaev S F, Kraposhin V S and Lam H T 2008 *Thermodynamic properties of al-mn, al-cu, and al-fe-cu melts and their relations to liquid and quasicrystal structure* // Journal of Physics: Condensed Matter 20(11) C. 114121..

[10] Kablov Ye N, Shchetanov B V, Grashchenkov D V, Shavnev A A and Nyafkin A N 2012 Metallomatrichnyye kompozitsionnyye materialy na osnove Al–SiC //Aviatsionnyye materialy i tekhnologii, **S. S. 373-380**

[11] Ivanova A O, Zavodov A V, Dynin N V and Fomina M A 2017 *Effektivnost' vvedeniya malykh dobavok perekhodnykh metallov v splavy tipa AlSi10Mg, prednaznachennyye dlya polucheniya izdeliy po tekhnologii selektivnogo lazernogo splavleniya* // Trudy VIAM: elektron. nauch.-tekhnich. zhurn. **7(55)** S.1.URL: http://www.viam-works.ru(data obrashcheniya 04.06.2019). DOI: 10.18577/2307-6046-2017-0-7-1-1.

[12] Fousová M, Dvorský D and Michalcová A 2018 *Changes in the microstructure and mechanical properties of additively manufactured AlSi10Mg alloy after exposure to elevated temperatures* // Materials Characterization. **137** C. 119-126.

[13] Uzan N E, Shneck R and Frage N 2018 *High-temperature mechanical properties of AlSi10Mg specimens fabricated by additive manufacturing using selective laser melting technologies (AM-SLM)* / // Additive Manufacturing. **24** C. 257-263.

[14] Zhou L, Mehta E and Schoulz B 2018 *Microstructure, precipitates and hardness of selectively laser melted AlSi10Mg alloy before and after heat treatment*// Materials Characterization. **143** C. 5-17.

[15] Weingarten C, Buchbinder D and Pirch N 2015 *Formation and reduction of hydrogen porosity during selective laser melting of AlSi10Mg*// Journal of Materials Processing Technology. **221** C. 112-120.

[16] Read N, Wang W and Essa K 2015 Selective laser melting of AlSi10Mg alloy: Process optimisation and mechanical properties development// Materials & Design. **65**.C. 417-424.

[17] Abele E, Stoffregen H A and Kniepkamp M 2015 *Selective laser melting for manufacturing of thin-walled porous elements* // Journal of Materials Processing Technology. **215**.C. 114-122.

[18] Wang L, Chen T and Wang S 2017 *Microstructural characteristics and mechanical properties of carbon nanotube reinforced AlSi10Mg composites fabricated by selective laser melting* / // Optik. **143** C. 173-179.

[19] Zaretsky E, Stern A and Frage N 2017 *Dynamic response of AlSi10Mg alloy fabricated by selective laser melting* // Materials Science and Engineering. **688** 364-370.

[20] Fiocchi J, Tuissi A and Bassani P 2017 *Low temperature annealing dedicated to AlSi10Mg selective laser melting products*// Journal of Alloys and Compounds. **695** C. 3402-3409

[21] Biffi, C A, Fiocchi J and Paolino A 2017 *Microstructure and preliminary fatigue analysis on AlSi10Mg samples manufactured by SLM*// Procedia Structural Integrity. **2017** C. 50-57.

[22] Asgari H, Baxter C and Hosseinkhani K 2017 *On microstructure and mechanical properties of additively manufactured AlSi10Mg_200C using recycled powder*// Materials Science and Engineering: A,**707** C. 148-158.

[23] Aversa A, Lorusso M and Cattano G 2017 *A study of the microstructure and the mechanical properties of an AlSiNi alloy produced via selective laser melting*// Journal of Alloys and Compounds. **695** C.1470-1478.

[24] Olakanmi E O 2013 *Selective laser sintering/melting (SLS/SLM) of pure Al, Al–Mg, and Al–Si powders: Effect of processing conditions and powder properties*// Journal of Materials Processing Technology. **213** C. 1387-1405.

[25] Li Y and Gu D 2014 *Parametric analysis of thermal behavior during selective laser melting additive manufacturing of aluminum alloy powder* // Materials & Design. **63** C. 856-867.

[26] EOS develops new AM machine // Metal Powder Report.2014.T. 1, **69** C. 43.

[27] Aboulkhair, N, Maskery I and Tuck C 2016 The microstructure and mechanical properties of selectively laser melted AlSi10Mg: The effect of a conventional T6-like heat treatment // Materials Science and Engineering: A, **667** C. 139-146.

[28] Hezrong D, Seyda V and Wycisk E 2016 *Additive manufacturing of metals* //ActaMaterialia. **117** C. 371-392.

[29] Aleshin N P, Grigor'ev M V, Shchipakov N A, Prilutskii M A and Murashov V V 2016 *Using nondestructive testing methods for in-production quality control of additive manufactured parts* // Russian Journal of Nondestructive Testing.**52(9)** C. 532-537.

[30] Gerov M V 2017 *Issledovaniye vliyaniya rezhima termoobrabotki na mekhanicheskiye kharakteristiki obraztsov iz alyuminiyevogo splava sistemy Al-Si-Mg, poluchennykh metodom selektivnogo lazernogo plavleniya* // Sbornik XIV Rossiyskaya yezhegodnaya konferentsiya molodykh nauchnykh sotrudnikov i aspirantov "Fizikokhimiya i tekhnologiya neorganicheskikh materialov" Sbornik trudov. S. 24–25.

[31] Kablov Ye N 2015 *Innovatsionnyye razrabotki FGUP «VIAM» GNTS RF po realizatsii «Strategicheskikh napravleniy razvitiya materialov i tekhnologiy ikh pererabotki na period do 2030 goda»* //Aviatsionnyye materialy i tekhnologii. **1(34)** S. 3–33. DOI: 10.18577/2071-9140-2015-0-1-3-33.

Design and analysis of the strength and durability of metal composite assemblies of aircraft structures.

A A Larionova[1] and A A Dudchenko[1]

[1]Moscow Aviation Institute (National Research University), Moscow, Russia

Abstract. This report is represented the decision of the optimization problem to determine the rational zone in attachment of helicopter's blade that based on calculations of the static strength and fatigue life of bearing screw from the loads arising in a typical flight.

Key words: fatigue life, fatigue strength, resource, polymer composite materials.

1. Introduction

Fatigue destruction of the blade during helicopter flight is caused by large variable loads acting on it with a significant number of change cycles. To increase the resource of the blade, use polymer composite materials.

2. Main part

The purpose of this work is to increase the fatigue life of the blade by optimizing the attachment of the butt of the composite blade to the metal node of the adapter.
In the course of the work, the following tasks were solved:
- optimization of the area of the bolt connection of the rotor blade of the helicopter (geometric modeling);
- calculation of the static structural strength by the FE method;
- calculation of the fatigue life of the materials of the blade and mounting bolts.

The analysis of literary sources was carried out [1,2,3].
The drawing of the butt part and the attachment point of the blade to the metal adapter is shown in Fig. 1,2.

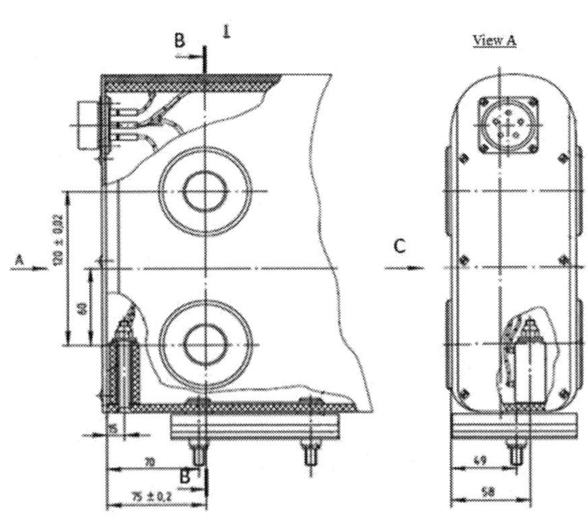

Fig.1 The butt part and the attachment point of the blade.

IOP Conf. Series: Materials Science and Engineering **934** (2020) 012044 doi:10.1088/1757-899X/934/1/012044

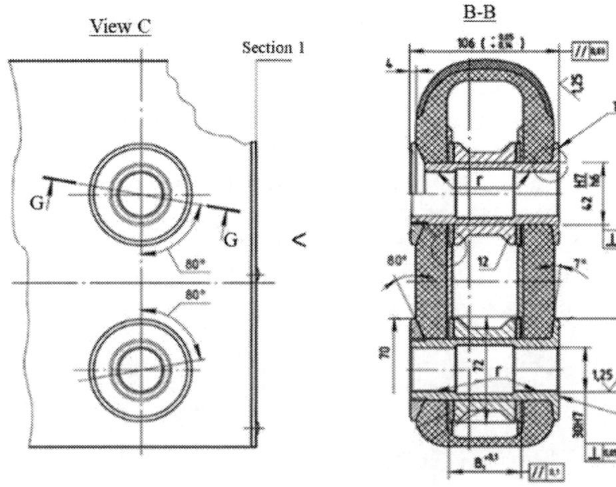

Fig.2 The butt part and the attachment point of the blade.

In the process, formulas were obtained that optimize the distribution of power load in the blade attachment site and thereby increase the service life.
To calculate the effective length of the zone of transfer of forces from polymer composite materials to metal, the technique described in [4, 5] was used.
The calculation was carried out taking into account the diameters of the bolts. The most effective diameter has been determined. A comparison was made of the characteristics of bolts with diameters of 6, 8 and 30 mm. 30 mm bolts are considered, as they are in a real design. It was also taken into account that single-bolt bolts are used for bolt diameters of 6 and 8 mm, and a double-bolt bolt for diameter of 30 mm. The calculation was carried out according to the method described in [6].
Since, according to the design calculation, for the case of fastening the structure with bolts with a diameter of 6 mm, the jumpers were shorter than the permissible length due to the larger number of bolts, in the further calculation only cases of fastening with bolts of 8 and 30 mm were considered [7].
The blade attachment node was also simulated in the MSC.Software Patran / Nastran software package and a static calculation of the structural strength by the finite element analysis method was performed. Static analysis showed that bolts with diameters of 8 and 30 mm satisfy the margin of safety.
Further, fatigue life calculations were additionally performed.
There are various models for evaluating the fatigue life of elements. One of them is the Mendell equation:

$$\sigma_{\max} = a + b \cdot \lg N, \qquad (1)$$

where $\sigma_{\max}$is the maximum value of cyclic tensile stresses of "smooth", that is, without stress stress concentrators of the composite;
N is the number of loading cycles to failure;
a and b are the constants of the equation, the values of which are determined by the approximation of the data by the dependence.
According to the data of [8], the fatigue curve constructed according to the Mendell equation in the form of equation (1) shows the highest level of accuracy of approximation of the experimental σ-N data.
The service life of the blade is determined by the formula:

$$T = \frac{N}{60 \cdot n_{ve} \cdot i \cdot \eta'_{\sigma} \cdot \eta_{\Sigma} \cdot \eta_N} \cdot \left(\frac{\sigma_W}{\Delta\sigma_{ekv}^{ef}}\right)^m, \qquad (2)$$

where n_{ve}- the number of revolutions of the screw per minute;
i is the number that determines the harmonics of the alternating voltages
(for the rotor blades i = 1);
η'_{σ} - voltage safety factor calculated in the reserve by the number of cycles (η'_{σ}= 3);
η_N - safety factor for the dispersion of endurance characteristics;
η_{Σ}- safety factor for the summation hypothesis
Using formula (2), we determine the resource of the blade when fastening with bolts d = 8mm:

$$T = 8627ч$$

and using bolts d = 30 mm:

$$T = 2371 \text{ ч}$$

Thus, the resource of the blade with bolts with a diameter of 8 mm is higher than the resource of the blade with bolts with a diameter of 30 mm. A smaller diameter is preferable, since it can significantly reduce the size of the cut fibers, and, therefore, increase the strength of the material of structures and reduce the stress concentration in the region of the holes. A bolt connection of 8 mm is also more preferable for structural reasons from the point of view of fail-safety.

Next, the fatigue life of shear bolts of metal composite compounds was calculated.

To calculate the fatigue life of the bolts, we select the fatigue curve for bolt material (Fig.3).

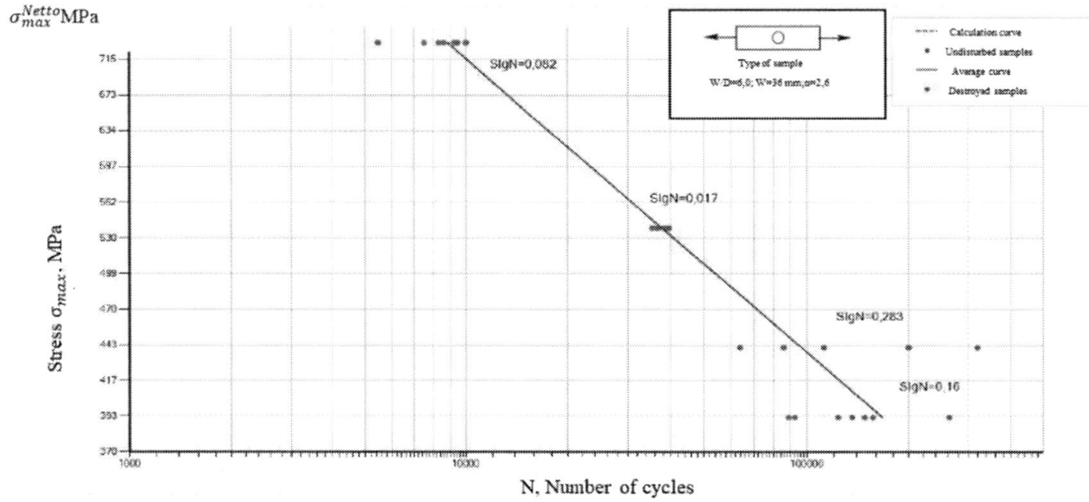

Fig. 3. Fatigue curve for bolt material

σ_{max}^{Netto}Approximating the fatigue curve, we obtain the values for the constants of the equation:

$$N \cdot \sigma^m = 10^C ,$$

where C = 17.363; m = 4,681 - constants,

N is the number of cycles.

Further, using stress diagrams, the bolts for fatigue life were calculated.

Based on the calculation results, we can conclude that structures with bolts of 8 and 30 mm have the fatigue strength necessary to work out the project resource.

3. Conclusion

In this paper, we describe a method for calculating fatigue life, due to which it is possible to increase the fatigue strength of the rotor blade of a helicopter in connection with a decrease in stress concentrators and increase the resource.

References

[1] Mikhailovsky K V and Baranovsky S V 2016 Method of designing a wing made of polymer composite materials based on parametric modeling. Part 2. Design of the power structure *News of higher educational institutions. Engineering* **12** pp 106-116

[2] Dumansky A M and Liu Hao 2019 Analysis of anisotropy of time-dependent and nonlinear properties of unidirectional CFRP *IOP Conference Series: Materials Science and Engineering* **683** 012093 DOI:10.1088/1757-899X/683/1/012093.

[3] Hao Liu, Dumansky A M and Alimov M A Development of a nonlinear defining relation of the hereditary type for the shear description of unidirectional carbon fiber reinforced plastic IM7-8552 *AIP Conference Proceedings* **2171** Is 1 DOI: 10.1063/1.5133175

[4] Dudchenko A A 2013 Calculation of the stress-strain state of the elements of aircraft, missile and engineering structures from composite materials (M .:NPU Federal Information System) p 232

[5] Dudchenko A A 2007 Strength and design of elements of aircraft structures made of composite material (M .: Publishing House of the Moscow Aviation Institute) p 200

[6] Larionova A A, Dudchenko A A and Sergeev V N 2016 Design "Proceedings of the Moscow Aviation Institute" **90** http://trudymai.ru/upload/iblock/e2e/larionova_dudchenko_sergeev_rus.pdf.
[7] Michael Chun-Yung Niu 1992 Composite airframe structures (Hong Kong: Conmilit Press Ltd) p 664
[8] Strizius V E 2015 Methods for assessing the fatigue strength of composite aircraft structural elements: a reference guide (M.: Mechanical Engineering. Innovative engineering) p270

IOP Conf. Series: Materials Science and Engineering **934** (2020) 012045 doi:10.1088/1757-899X/934/1/012045

Thinking development and aesthetic education of students in the process of teaching mathematics by example solutions for one problem

A B Zhanys[1], A O Dautov2, A U Aktayeva[2] and A Zh Askarova[3]

[1]Russian Academy of Kazakh Agro-Technical University named after S. Seifullin Natural Sciences № 7524 Kazakhstan, Nur-Sultan

[2]Mathematics KSU after named Sh. Ualikhanov, Kokshetau, Kazakhstan

[3]Associate Professor Higher Mathematics chair Astana, Pobeda Avenue 62a, building №2, Office 2516

E-mail: aray.zhanys@gmail.com,d.abeke@mail.ru,aaktaewa@list.ru, alfiya_askarova@mail.ru ,igibaevna@bk.ru

Abstract: The article discusses one of the leading goals of teaching mathematics - the development of students' thinking by means of mathematics, it is noted that this requires the creation of special situations, the resolution of which leads to the development of thinking. The possibility of developing logical and figurative thinking of students of secondary schools in the process of aesthetic education using the subject of mathematics is investigated. The beauty of problem solving allows you to increase interest in the study of mathematics. The development of intellectual and creative capabilities, their imagination, as well as broadening the horizons in the application of methodological programs in teaching and evaluating students' knowledge and skills. The present work is presented on the example of solving one problem by several mathematical methods. Using a mathematical model to solve various kinds of problems with the greatest number of solutions contributes to aesthetic education, allowing you to develop a culture and logical thinking, to form in students the optimal analysis of the task in solving it. The work is presented in an experiment on organizing aesthetic education by knowing the beauty of problem solving, which will fully reveal the criteria for applying the individual approach to learning through the development of thinking in solving various kinds of mathematical problems. The organization of the intensification of the process of mastering knowledge and applying aesthetic beauty is shown by the example of solving problems in mathematics of grades 7-9. We carried out a number of experimental works on the development of thinking and aesthetic education, relying on the activity-based approaches of educational and developmental teaching of mathematics.

Key words: development of students' thinking, aesthetic education, mathematical education, assessment, knowledge, skills.

1. Introduction

Among the main directions of modernization of education, one can single out the personal orientation of the content of education, the active nature of education, the focus of education on the formation of generalized methods of various types of activities, the formation of key competencies, the focus of education on the development of the student's personality, his cognitive and creative abilities.

The development of a person's personality involves the development of his thinking, and in this regard, mathematics has great potential.

2. Main part

V.M. Tikhomirov writes about this: "The goal of mathematical education, in my humble opinion, should be, first of all, development. The development of skills in operating with numbers and figures, spatial imagination, logical thinking - in a word, the development of intelligence. Nothing can teach this better than mathematics, - the whole experience of mankind speaks of this" [1].

A.V. Borovsky and N.Kh. Rozov [2], sharing the point of view expressed above, note that academic subjects should be viewed as means, tools of training, education and development.

The same idea was expressed in a resolution of the Congress of Mathematics Teachers, which was convened again in October 2010 after a hundred-year break: "... mathematical education is the most important and necessary component of personality development ..." [3].

Note that the development of intelligence refers mainly to the development of thinking. It is necessary to search for the basics of the development of thinking of students not only in the content and technologies of teaching mathematics, but also in psychology. Thinking should be developed through various types of educational activities performed by students in learning mathematics, and this development should go through such stages as visual-effective thinking, visual-figurative thinking, and verbal-logical thinking.

We, following N.N. Pospelov, we will mean by the development of thinking, students in the learning process "the formation and improvement of all types, forms and operations of thinking, the development of skills for applying the laws of thinking in cognitive and educational activities, as well as the ability to carry out the transfer of methods of mental activity from one area of knowledge to another" [4].

Here are examples of situations that are created in order to organize the educational and cognitive activities of students with the aim of developing their thinking.

1. The challenge. Point M moves along the sides of the square (Figure 1).

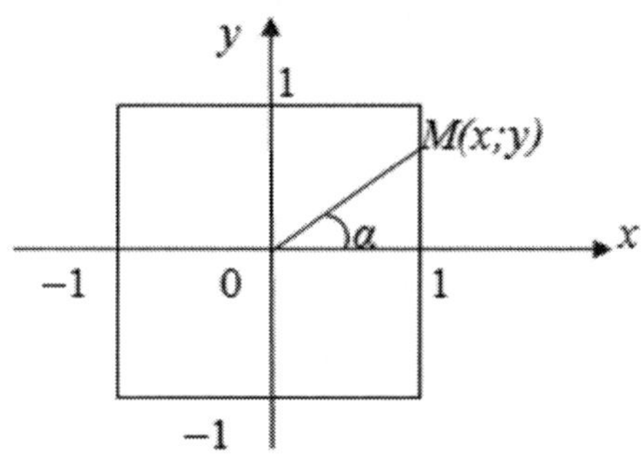

Figure 1. Location of a material point in the square boundaries

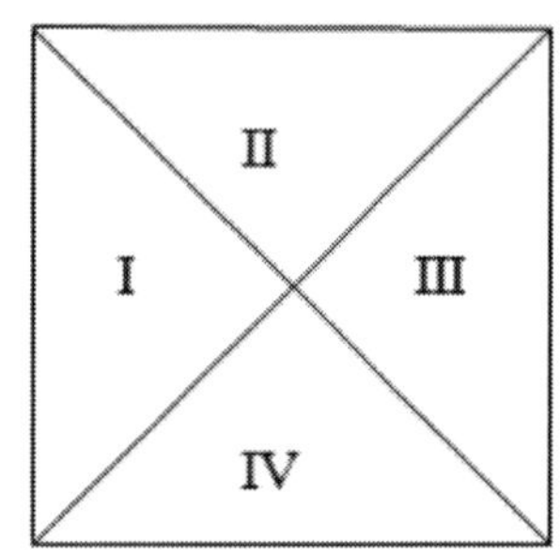

Figure 2. Comparative square quarters of trigonometric functions with a single circle

By analogy with the known trigonometric functions, we introduce new functions: sanα = x – y (the sine of the angle α); casα = x + y (the casanus of the angle α); tigα = x2 – y2 (tigent of angle α). Complete the following tasks:

establish links between these functions;

plot the graphs of these functions;

set the set of values of each of the square trigonometric functions;

define the signs of the square trigonometric functions in the quarters marked is shown in Figure 2;

establish a connection between classical and square trigonometric functions;

set the algorithm for solving square trigonometric equations: sant = α, cast = α, tigt = α, etc.

The situation described above is artificially created, but according to A.S. Krygovskoy: "The significance of creative activity itself is not what it has created" [5].

2. In the school course of mathematics, only arithmetic progressions with constant differences d = const are considered. The concept of arithmetic progression allows a completely natural generalization, if we assume that the difference of the arithmetic progression itself will be a function of the natural argument, that is, dn = d (n). In this case, we will deal with arithmetic progressions with variable differences. The study of such progressions should be carried out in the same way as the study of ordinary arithmetic progressions: the formula of the general term, the formula of the sum of the n first terms of the progression, a characteristic property, etc.

3. It is known that inside any triangle ABC (see Figure 3) there exists a point P (and there are two) such that.

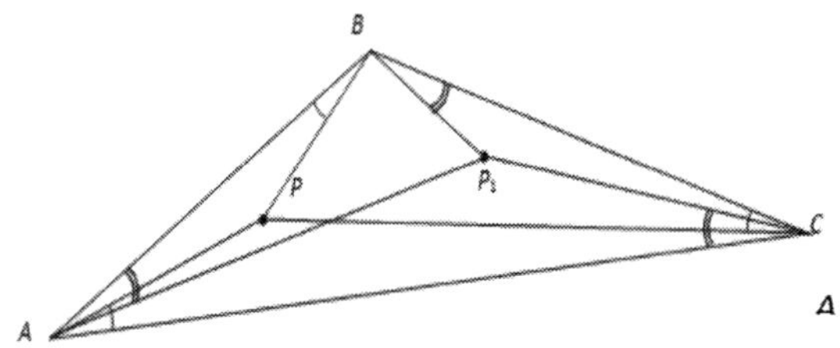

Figure 3. Krell-Brokar's triangle point composition

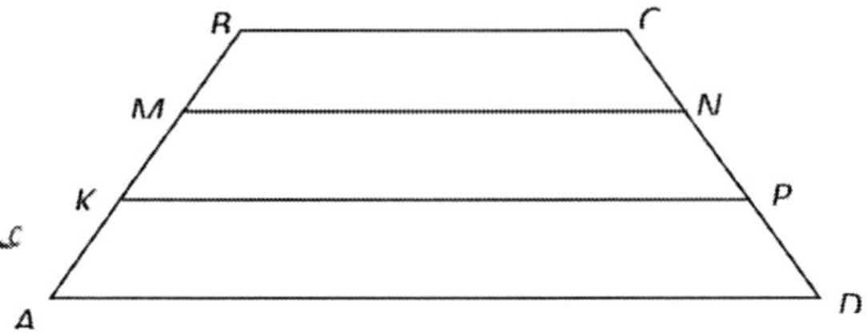

Figure 4. Trapezoid with two midlines

Points P and P1 are called Krell-Brokar points.
a) Perform the following research task: Find ways to build Krell-Brokar points using a computer. (We give suggestive hints: similar triangles are built on the sides of the original triangle or formulas are used for the coordinates of the Krell-Brokar points.)
4. Solve the problem: "A trapezoid is given (see Figure 4) with bases of 2 cm and 5 cm. The lateral side of the trapezoid is divided into three equal parts. Through the points of division drawn straight parallel to the bases of the trapezoid. Find the lengths of the obtained segments. "
The lengths of the segments MN and KP can be found by solving the following system of equations:

$$\begin{cases} MN = \dfrac{2+KP}{2}, \\ KP = \dfrac{5+MN}{2}. \end{cases}$$

Solve the same problem for the case when the side is divided into 6 equal parts. It is clear that this problem should not be solved in the same way as the previous one (we will have a system of five equations with five unknowns). This problem should be solved (through the points D1, D2, D3, D4, D5, C, segments parallel to the side of the trapezoid AB are drawn) is shown in Figure 5.

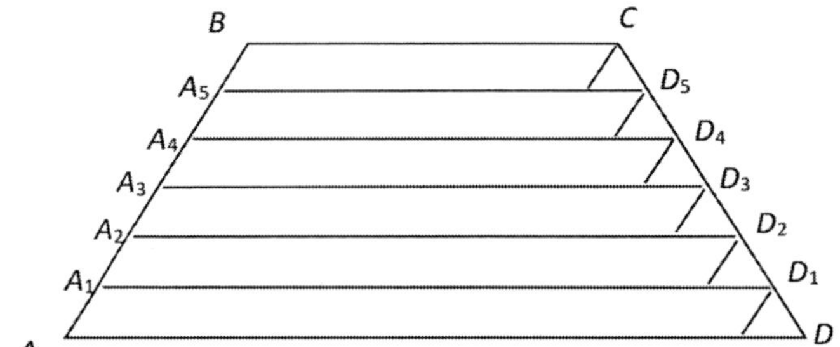

Figure 5. Trapezoid with five's midlines

Summarize these two tasks in the case when the bases of the trapezoid are equal, a and b, and the side is divided into n equal parts.
In the works of V.A. Dalinger [6-8], the reader will find many tasks of this nature.
We examined one aspect of the problem raised in the article, but another aspect is stated in the title of the article - the aesthetic education of students in the process of teaching mathematics. Let us dwell on this issue more systematically.
In the process of creating a harmoniously developed personality, the aesthetic education of students is of paramount importance. The role of mathematics as one of the important subjects of the school is difficult not to notice in aesthetic education, the potential of mathematics in this regard is huge. Mathematics is very rich in beautiful formulas, proofs, multi-type methods for solving problems and you can specify entire sections

where the multifaceted elements of beauty and magnificence are hidden, such as "Trigonometry", "Golden Section", "Symmetry", "Algebra and Number Theory", "Geometry in space", excellent for aesthetic education.

The effective development of the wide potential of mathematics presupposes a full-fledged perception of mathematical literacy, the cultivation of aesthetic feelings, taste and ideal through imaginative thinking and logical culture, the formation of a person's value orientation in his quest for the beauty of original problem solving.

In various pedagogical literature, aesthetic education is considered as a system of measures for the development of good artistic tastes in a person, the ability to correctly and truly judge the beautiful in art.

According to scientists autors, "... education through beauty is, on the one hand, an important means of developing educational motivation and, on the other hand, a source of emotionality of a person as one of the leading components of aesthetic culture" [9-23].

One of the founders of the development of the educational process problem is the outstanding teacher Yu.K. Babansky, he had and has a theory in conjunction with the educational process. The author notes that the aforementioned components of the aesthetic culture at the same time act as criteria for the aesthetic education of students. They determine the tasks and content of the aesthetic education of schoolchildren [9].

In the late 80s of the twentieth century V.P. Bespalko [10, p.31] introduced the learning process by the formula:

$$D_{pr}=M+P_D+Y$$

where D_{pr}– is the didactic process; M - motivation student for learning; P_D– his cognitive activity; Y – student activity management.

In the training manual P.I. Obraztsov and V.M. Kosukhin refers to the statement of V.D. Dyachenko: "... the content of training is always some kind of activity that, to one degree or another, is owned by the teacher and is not fully or partially owned by the student" [11].

3. Research methods

Teacher G.P. Bursa notes that aesthetic education in a comprehensive school is to instill in students good tastes, correct concepts, views and opinions in the field of music, painting, literature, etc. [12].

In the dictionary of pedagogy G.M. Kojaspirova, A.Yu. Kojaspirova pointed out that aesthetic education is the development and improvement in a person of the ability to perceive, correctly understand, value and create beauty in life and art, actively participate in creativity, creation according to the laws of beauty [13].

In addition, I.F. Kharlamov emphasizes that aesthetic education is organically linked to the term "aesthetics", which means the science of beauty. The very word aesthetics comes from the Greek aisthesis, which in translation into Russian means sensation, feeling. Therefore, in general terms, aesthetic education denotes the process of forming feelings in the field of beauty [14].

N.I. Firstova emphasizes that aesthetic education should be considered as an integral part of the comprehensive development of the individual. Through aesthetic education, it is possible to expand and deepen the knowledge and ideas of schoolchildren about reality, the formation of their views [15].

G.I. Sarantsev notes that much has been written about the beauty of mathematics. Authors see the beauty of mathematics in:

- harmony of numbers and forms;
- geometric expressiveness;
- harmony of mathematical formulas;
- the grace of mathematical evidence;
- order;
- a wealth of applications;
- universality of mathematical methods.

According to G.I. Sarantseva, the peculiarity of mathematics lies not only in that, as in art, it has a huge aesthetic potential, but also in the fact that mathematical activity obeys the laws of beauty [16].

Therefore, mathematics, along with art, is considered the most important means of introducing students to beauty, the formation of their aesthetic taste. Opinion, W.U. Sawyer, speaking of the importance of mathematical theory, calls beauty, harmony, "so attractive to the mind" as one of its indicators [17].

So, D. von Neumann noted that mathematics, like art, is driven almost exclusively by aesthetic motives. J. Hadamard argued that a scientist, seeing a structurally imperfect, asymmetric, "crooked" mathematical construction, begins to feel the need for active work to harmoniously supplement it [18].
According to V.G. Boltyansky, the beauty of a mathematical object can be expressed through isomorphism between the object and its visual model, the simplicity of the model and the unexpectedness of its appearance. This statement can be supported by the formula of "mathematical aesthetics" from his article "Mathematical Culture and Aesthetics" [19]:

BEAUTY = VISIBILITY + UNEXPECTED = ISOMORPHISM + SIMPLICITY + UNEXPECTED.

The beauty of objects reveals their property, which exists independently of consciousness, says A.V. Shcherva: "The sense of beauty is interpreted by the author as a product of the reflection in the human mind of the really existing aesthetic properties of the surrounding world" [20]. The psychological basis of this interpretation is presented in the intuitive attraction of the human psyche to grace and harmony, comprehended by feelings.
From here, from all the theses examined, it is possible to derive methodological recommendations on the formation of the aesthetic education of the subject of mathematics among schoolchildren.
One of the most important tasks in the formation of a person's worldview is the problem of the formation of an aesthetic attitude to mathematics, as is clear from the above quotes. When teaching mathematics, school students can and must learn to perceive, feel the beauty of mathematical expressions, theoretical constructions, evaluate the wide possibilities of a mathematical culture from an aesthetic point of view. To reveal to students the beauty of the content of mathematics, which will use special methods to develop creativity in the classroom, use these methods to assess students' knowledge, prepare them for life in modern conditions - an important factor in aesthetic education.
The development of modern textbooks did not eradicate the need for creativity, but, on the contrary, demanded an ever higher level of general cultural development, education, creativity and activity from a person. Modern educational methods using technology open up new didactic opportunities in realizing the goals of aesthetic education in mathematics, which should be used to familiarize oneself with beauty, nurturing aesthetic tastes and experiences, including through methods related to computer graphics and animation, and the development of multimedia tools etc. [21].
The use of technological tools in training in solving various types of problems can have a positive impact on the formation of aesthetic features, the growth of interest in the study of mathematics, as well as the increase in the level of fundamental knowledge using digital technologies, while improving traditional methods of various levels of education and social development of students 7- 9 classes.
To improve the knowledge and effectiveness of teaching a subject of mathematics, we decided to create an electronic textbook intended for students as well as mathematics teachers using innovative information and communication technologies and its main goal is to provide methodological assistance in a better use of technologies that allow teaching on The various stages are interesting, accessible and specific [22].
Research hypothesis: the formation of logical thinking in students by means of aesthetic education in the learning process in mathematics will be effective if:
1. provide a solution to problems with the aim of developing creative activity, striving for the beauty of the originality of tasks
2. integrate various tasks, including exercises, tasks that require several methods in their solution to ensure the integration of the process of teaching mathematics, this will purposefully provide educational and developmental functions, then in the final stage, the cognitive interest of students will be formed by ensuring conscious assimilation and development of personally significant qualities aimed at aesthetic development.
The goals and objectives of the electronic textbook:
1. identification of the relationship of mathematics with various fields of human activity and phenomena occurring in nature.
2. broadening the horizons in the field of application of the subject of mathematics for secondary schools.
3. the formation of a general and mathematical culture of personality.
4. aesthetic development of personality.
5. the development of logical and imaginative thinking among students in schools.
6. development of skills in working with electronic technologies.
7. Development of modern electronic textbooks in the field of natural sciences.

The assimilation of the material was not limited to the properties and relationships of the basic concepts of mathematics. For the assimilation of the theory, the formation of the skills and its application to solving problems was important to us.

In our opinion, it is most advisable to use the above goals in the process of illustrative demonstration work, integration in mathematics lessons, as well as in the disciplines of the natural sciences. In modern science, an interdisciplinary approach is one of the priority areas [23]. According to Kazakh and Russian scientists, the mathematical apparatus and mathematical methods can be used to study qualitatively different fragments of reality, they contribute to the disclosure of their unity and thereby indicate new ways of integrating new knowledge [20; 22-23]. For example, students will receive information on the use of these methods in formative assessment in mathematics using digital technologies.

This approach will allow students to fully understand the subject-matter relationship, learn how to use technological capabilities, and experience the aesthetic appeal of mathematics problems.

Strengthening the role of the electronic textbook is aimed at fulfilling practical tasks as a determining component in studying the theoretical information of a mathematics course, helping to bring students to understanding internal logic, deploying their ability to experiment and conduct modeling experiments with various options for solving problems, as well as analyze and synthesize any kind of information . This makes it possible to determine quick and complete information about the work done by students, which makes feedback more effective and productive in assessment.

Each practical and experimental task is accompanied by a lecture (using presentation and full text) on the topic of the lesson. For example, when studying a specific topic before performing practical tasks, students are invited to get acquainted with the concepts introduced in the lesson, the history of the occurrence of these concepts and brief explanations for them, using them in various fields of science to compare, analyze and reveal their beauty in nature and life.

Then it is proposed to perform several practical exercises on the topic with various solution methods, build various images, as well as their compositions, design the image to formulate a character that meets the requirements of the problem. Thus, when performing these tasks, one can comprehensively consider the possibilities of students of any link defined at the present stage and including an important component as a process of aesthetic education through the beauty of illustrativeness.

Thus, a number of measures were taken to identify the logical development of aesthetic potential in teaching mathematics. The problem of aesthetic forms of learning as a means of educating the educational process at school yielded its results in experimental work - to reveal the potential of the essence, to determine the significance of introducing aesthetics into the process of mastering the subject "Mathematics", to discover the beauty of solving problems, and to identify ways to implement practical work in secondary schools.

By the example of solving one math problem carried out within the walls of the profile school-gymnasium of multilingual education No. 3 named after M. Gabdullin, secondary school No. 2, the regional boarding school for gifted children "Daryn", as well as the AES NIS and KTL schools (school data the cities of Kokshetau, Akmola Oblast, Kazakhstan assisted in carrying out experimental work with students in grades 7–9 in mathematics) we were able to determine the criteria by aesthetic education of students in the process of general disciplinary education.

By forming aesthetic forms of learning, a link can be constructed to generalize and consolidate knowledge, skills and abilities using the example of one task. Thus, according to the practical activities of students, it is possible to identify certain links in understanding mathematical simplicity, that is, the aesthetic potential in solving one problem in various ways.

4. Practical part

Let us show, as an example, the task for independent work, where the student applies various methods to solve and, justifying, forms the essence of the same answers of the same task.

Task. Two cars left one city in one direction. The first was at a speed of 60 km / h, and the second - 90 km / h. The second car came out at 2h. later than the first. After how many hours and at what distance from the city will the car with the highest speed catch up with the car with the lowest speed?

Solution: 1-way. Arithmetic method. We give a brief record of the problem is shown in Figure 6.

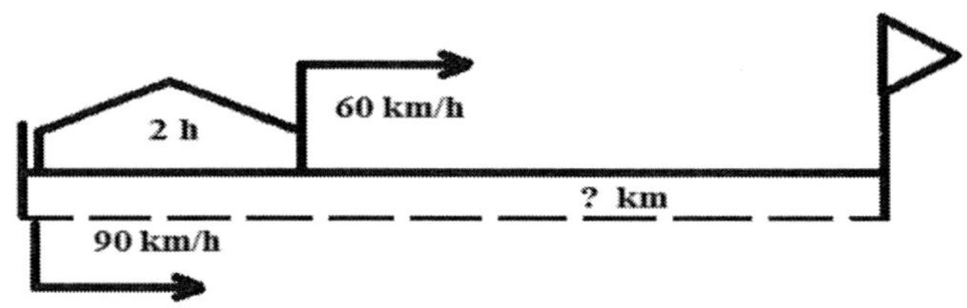

Figure 6. The scheme of movement of special vehicles

1) - the first car will pass in 2 hours;
2) - the difference in speed;
3) - after so many hours after its release, the second will catch up with the first car;
4) - at such a distance from the city, the second one will catch up with the first car.

2-way. The geometric method. Consider the solution to the problem on a schedule, that is, we will characterize a visual schedule of the movement of fire engines.

When solving geometrically, we will arbitrarily take the length of one segment vertically for 30 km, and the length of one segment horizontally for 1 hour. We set aside from 0 to 6 hours on a horizontal line, vertically we will lay off the segments of the path traveled by each vehicle in 1 hour, 2 hours, 3 hours, etc. (Figure 7).

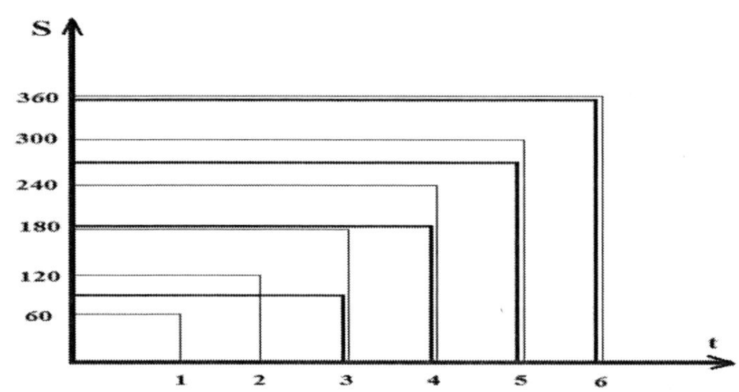

Figure 7. Scale of movement of cars

First, we postpone the segments of the path traveled by the first to the exit of the second car, and then the segments of the path that characterize the distances traveled by the first and second car (table 1):

Table 1. Algorithm of actions covered distances in accordance with time

Avto-1	$t_1 \backslash t_2$	Avto -2
180 km	3 h; 1 h	90 km
240 km	4 h; 2 h	180 km
300 km	5 h; 3 h	270 km
360 km	6 h; 4 h	360 km

The segments of the found paths turned out to be equal, which means that the second will catch up with the first after 4 hours, that is, after the exit of the first car. As you can see, this will happen at a distance of 360 km from the city is shown in Figure 7.

3-way. The method of geometric construction of the problem consists in dividing into intervals in accordance with the difference in speeds of two cars. Assuming that the length of one segment is 30 km, we will postpone segments of the path traveled by each of the vehicles. The interval of the path corresponds to 360 km from the beginning of their movement (Figure 8).

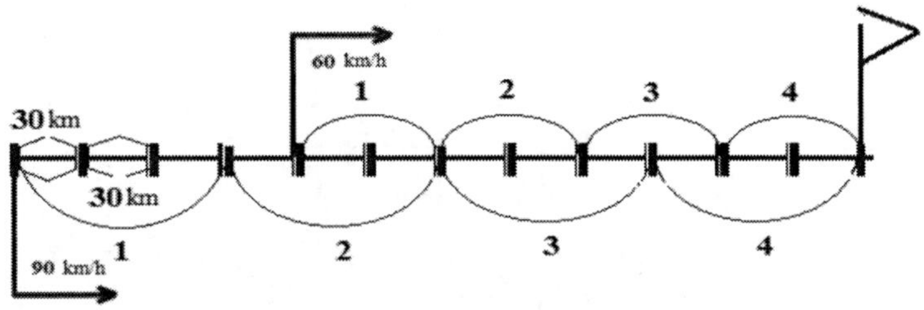

Figure 8. Interval path segments

4 way. Algebraic method. Let's make a linear equation:

$$\frac{S}{60} - \frac{S}{90} = 2 \quad (1)$$

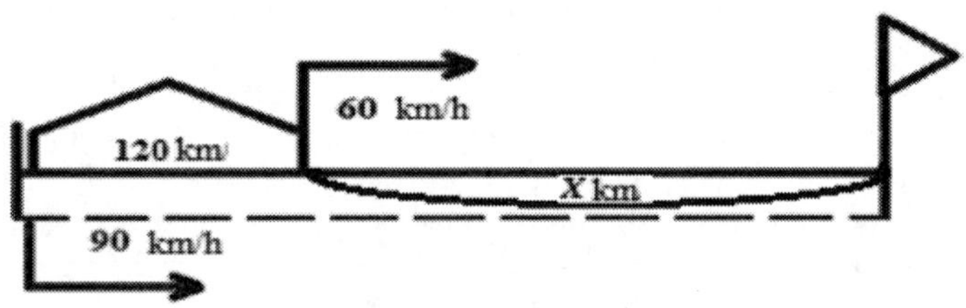

Figure 9. Unknown leg

And so where x is the path to go. Substituting into equation (1) we obtain the following equality by shown in Figure 9:

$$\frac{120+x}{60} - \frac{120+x}{90} = 2 \quad (2)$$

From equality (2), it is easy to calculate that x = 240 km, this is the distance of the distance traveled by the first car 4 hours after 120 km.
5 way. Algebraic method. Finding time using NOC speeds of two cars:
1) 2 * NOC (60; 90) = 2 * 180 = 360 km - from the beginning of their movement;
2) 360: 90 = 4h. - after so many hours after its release, the second will catch up with the first car;
In each of the proposed solution methods, the same result was obtained.
And also considered some of the practical problems of composites:
Continuous reinforced composite strength fibers
In the general case, the tension diagram of a unidirectional fiber composite (Figure 8) should consist of three main sections:
I - matrix and fibers are deformed elastically;
II - the matrix goes into an elastic-plastic state, fibers continue to deform elastically;
III - both components of the system are in a plastic state deformation.
Depending on the properties of the components of the composite, sections II and III on the curve may be absent.

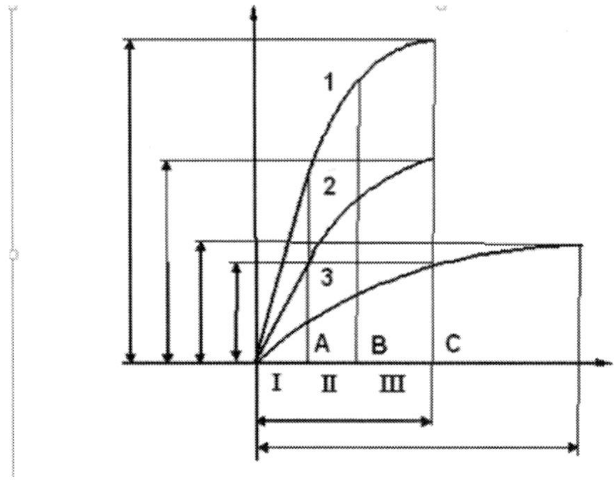

Figure 10. Diagrams of tensile fibers (1), unidirectional composite (2) and matrix (3)

The external load is equal to the sum of the loads on the matrix and the fibers, provided that the bond strength at the fiber-matrix interface is sufficient to ensure joint deformation of the components up to failure, i.e. Then, the tensile strength of the composite along the fibers, depending on the volume fraction of VB fibers for a typical composite reinforced with continuous unidirectional fibers, varies in direct proportion to the volume fraction of fibers:
where the average value of the tensile strength of the fibers in tension (compression, shear); - voltage in the matrix at the time of rupture of the fibers.
The value depends on the complex relationships between the properties of the matrix and the fibers and for engineering purposes is usually determined by the measured strength of the composite material with a known volume fraction of fibers.
The strains of the fibers and the matrix are rarely equal, so for brittle fibers in an elastic matrix, for example, for a ceramic-ceramic composite material, the value is given by the ratio:, where EM, EB are the elastic moduli of the matrix and fibers. For a matrix capable of being plastically deformed, for example, metal, a more suitable parameter in the calculation is the yield strength.
When compressed along the fibers, the destruction of the composite material occurs due to the loss of stability of the fibers, similar to destruction during the longitudinal bending of the rod.
In the case of the fracture mechanism, when the main type of matrix deformation is tensile directed perpendicular to the fiber axis, the compressive strength is determined by the expression

$$(\sigma_f)_s = 2V_f\left[V_f E_M E_f 3(1-V_f)\right]$$

In the case of failure by type of shear, the tensile strength is determined expression

$$(\sigma_f)_{sh} = (\sigma_f)_M (1-V_f)$$

Minimum and critical fiber concentration
The concept of the minimum concentration (volume fraction) of Vmin fibers was introduced in relation to composites in which the matrix is more plastic than fibers. At values of VB <Vmin, the destruction of the fibers does not lead to the immediate destruction of the entire composite, since the undamaged cross-section of the matrix has the ability to bear a higher load than the total cross-section of the destroyed fibers. The growing load will lead to the destruction of the fibers into smaller parts - multiple destruction. If VB <Vmin, then at an external load corresponding to the strength of the fibers, they are destroyed. The stress that acted on the fibers is redistributed to the matrix, the strength of which is insufficient to withstand this stress - there is a catastrophic destruction of the entire composite.

The value of the minimum fiber concentration V min is determined from
Conditions

$$(\sigma_f)_f V_{\min} + \sigma_M (1 - V_{\min}) = (\sigma_f)_c = (\sigma_f)_M (1 - V_{\min})$$

where from

$$V_{\min} = [(\sigma_f)_M - \sigma_M]/[(\sigma_f)_f - (\sigma_f)_M - \sigma_M]$$

The critical volume fraction of fibers Vcr is called such, at where the strength of the composite becomes equal to the strength of unreinforced matrices. The value is calculated from the condition

$$(\sigma_f)_f V_{cr} + \sigma_M^* (1 - V_{cr}) = (\sigma_f)_c$$

from here

$$V_{cr} = [(\sigma_f)_M - \sigma_M^*]/[(\sigma_f)_f - \sigma_M^*]$$

The concentration dependence of the strength of composites in which the matrix is more plastic than fibers. If the fibers have a greater plasticity margin than the matrix, then the transition from single to multiple fracture occurs at a fiber concentration Cv, determined from the condition

$$(\sigma_f)_f V_f' = (\sigma_f)_M (1 - V_f') + \sigma_f' V_f'$$

where from

$$V_f' = (\sigma_f)_M / [(\sigma_f)_f + (\sigma_f)_M - \sigma_f']$$

where - is the stress in the fibers during deformation of the destruction of the matrix.
The dependence of the strength of the composite on the volume fraction of fibers for this case is presented in is shown in Figure 11.
From Fig. Figure 11 shows that composites with a plastic matrix are destroyed by a single destruction mechanism at high fiber concentrations, and with a brittle matrix, at low fiber concentrations.

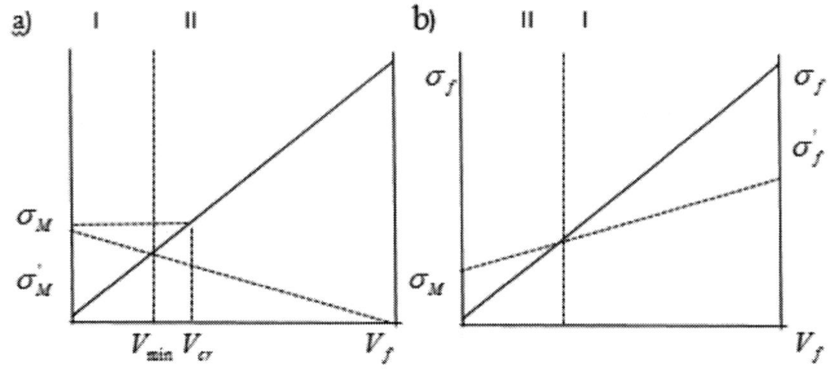

Figure 11. A typical view of the dependence of the tensile strength of the composite on the volume

fraction of fibers: a - for a unidirectional composite with a plastic matrix and brittle fibers; b - s Fragile matrix and plastic fibers (I-region of multiple fracture, II - region of single fracture)

The influence of fiber orientation on the destruction of the composite.
A description of the influence of fiber orientation is based on the theory of maximum principal stresses. Under the action of tensile stresses a applied at an angle ϴ to the direction of laying of the fibers (Figure 12), depending on the value of три, three types of destruction are possible.

Figure 12. Scheme of loading unidirectional composites at an angle ϴ to the wave axis

5. Approbation

When solving a problem in several ways, the knowledge of students in mathematics, the ability to apply them to problem solving, and the ability to build mathematical models were tested.
After instilling skills in solving one problem in several ways, we conducted an experiment. During the experiment, in the process of checking the assimilation of knowledge in the choice of methods for solving one problem, two stages were outlined: initial and final.
The experiment involved a student in grades 7–9 of the specialized school-gymnasium for multilingual education No. 3 named after M. Gabdullin, secondary school No. 2, the regional boarding school for gifted children "Daryn", NIS (Nazarbayev Intellectual School) and KTL (Kazakh-Turkish Lyceum).
Students were assigned to a control group (CG) and an experimental group (EG). Students in these groups were asked to solve one problem in five ways. On the basis of monitoring and summarizing the results of the experiment, the quality of knowledge was determined for students finding several ways to solve one problem, which made it possible to identify the flexibility of thinking, capable of multi-level cognition and comprehensive understanding when performing certain types of tasks.
Testing of knowledge, skills of students in grades 7-9 was carried out on the basis of the methodology proposed by A.A. Kyveryalg and is defined by the following formula [20]:

$$C_{ak} = \frac{n}{N}$$

C_{ak}– coefficient of assimilation of knowledge.
n – is the number of students who correctly used the methods of solving one problem.
N – is the number of all students participating.

Table 2. The levels of students' knowledge acquisition were divided into a control group (CG) and an experimental group (EG) in the choice of ways to solve one problem at the initial stage (school-gymnasium of multilingual education No. 3 named after M. Gabdullin, Grade 7 - 126 students, Grade 8 - 98 students, Grade 9 - 84 students)

Problem solving by methods	C_{ak}					
	CG			EG		
	7 Gr	8 Gr	9 Gr	7 Gr	8 Gr	9 Gr

1	0,86	0,88	0,86	0,87	0,84	0,81
2	0,78	0,84	0,90	0,75	0,80	0,88
3	0,67	0,80	0,83	0,63	0,76	0,79
4	0,83	0,76	0,76	0,84	0,63	0,71
5	0,95	0,82	0,79	0,92	0,86	0,79

The results of the experiment conducted with students in grades 7-9 of the above school from this table will be shown in the form of diagrams (Figure 13):

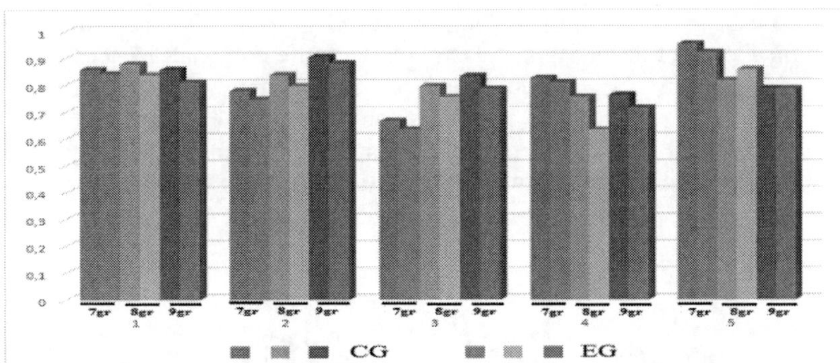

Figure 13. The skill levels of students in the CG and the EG in choosing ways to solve one problem at the initial stage of the pedagogical experiment (based on grades 7-9 of the school-gymnasium of multilingual education No. 3 named after M. Gabdullin)

Based on the table and the corresponding diagrams (Figure 13), we will draw some conclusions: it is necessary to conduct consistent work with EG students to choose various methods for solving one problem, since at the initial stage the knowledge, skills of students in grades 7-9 are still insufficient. Students need to be brought to an understanding of the essence of aesthetic education by developing their logical thinking in solving one problem in several ways. It should be noted that in this case, you can use an electronic textbook, which stimulates the activities of students and is an additional tool in the course of students learning.
After focused work with students of the experimental group, the results of mastering the knowledge in choosing ways to solve one problem have changed for the better. These results are given in table 3 and in the corresponding diagrams (Figure 14).

Table 3. Learning levels of students in choosing ways to solve one problem at the final stage (school-gymnasium of multilingual education No. 3 named after M. Gabdullin, Grade 7 - 126 students, Grade 8 - 98 students, Grade 9 - 84 students)

Problem solving by methods	C_{ak}					
	CG			EG		
	7 Gr	8 Gr	7 Gr	8 Gr	7 Gr	8 Gr
1	0,84	0,90	0,81	0,84	0,92	0,83
2	0,81	0,88	0,88	0,86	0,90	0,93
3	0,75	0,76	0,74	0,81	0,84	0,81
4	0,79	0,80	0,90	0,92	0,90	0,93
5	0,92	0,79	0,71	0,98	0,92	0,98

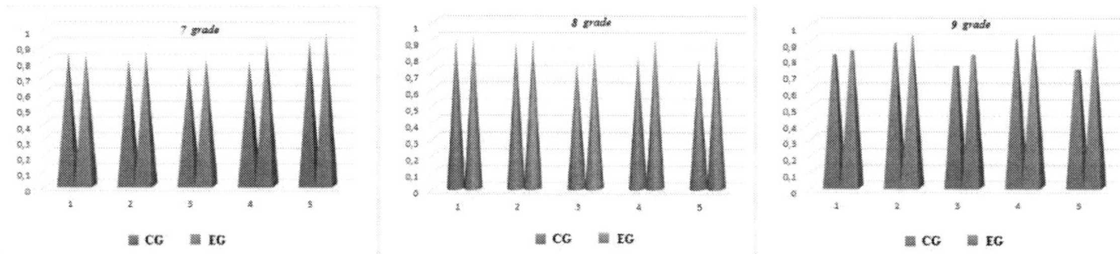

Figure 14. Learning levels of students in the CG and the EG in the choice of ways to solve one problem at the final stage of the pedagogical experiment (based on grades 7-9 of the school-gymnasium of multilingual education No. 3 named after M. Gabdullin)

A similar experiment was conducted on the basis of schools: secondary school No. 2, a regional boarding school for gifted children "Daryn", NIS and KTL (Figures. 15 and 16).

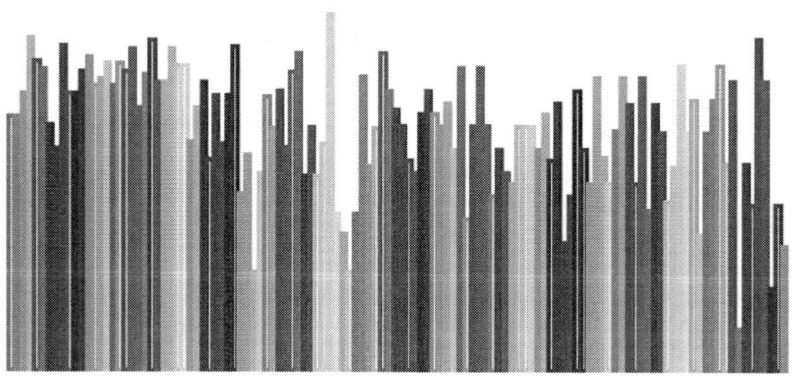

Figure 15. The levels of assimilation of students in the CG and EG in the choice of ways to solve one problem at the initial stage of a pedagogical experiment

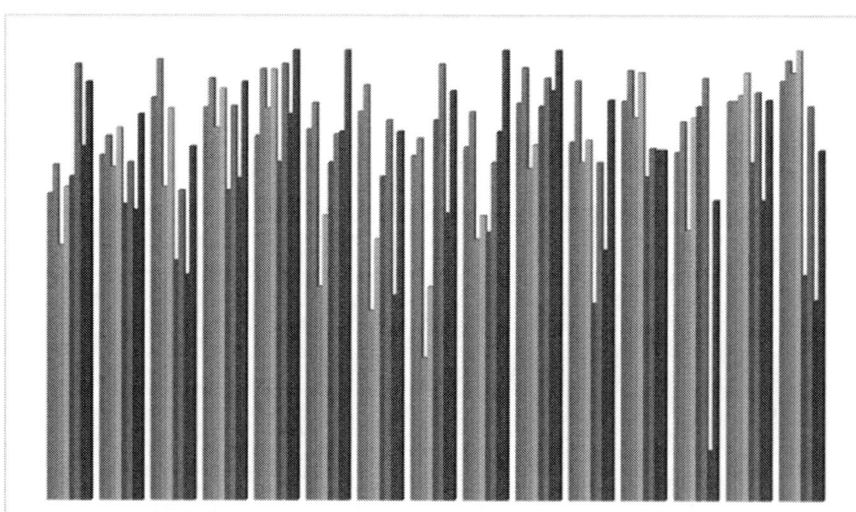

Figure 16. The levels of assimilation of students in the CG and EG in the choice of ways to solve one problem at the final stage of the pedagogical experiment

6. Conclusion

The results of the experiment confirmed the hypothesis of the study and made it possible to draw some conclusions:
1. The essence and content of the development of thinking in the aesthetic education of students are determined by choosing several methods for solving one problem in several ways.
2. The analysis of the compiled electronic textbook on the example of various tasks with numerous ways and methods of solving it.
Thus, a significant difference in the levels of formation of students' ability to solve problems of the same type by several methods based on mathematical calculation makes it possible to develop aesthetic education and logical thinking at a professional level. The assimilation of knowledge increases interest, makes it possible to see aesthetic beauty in solving one problem. This means that the upbringing of a mathematics lesson is just as important as the upbringing of the mind, thought and independence of the choice of a solution path, thereby ensuring the conscious assimilation and development of personally significant qualities.
Based on the aforementioned studies of scientists, in particular, our research, we can state that aesthetic education occupies an important place in the process of personality formation. No less significant is the fact that the knowledge of the world from the standpoint of its beauty, aesthetic value contributes to the formation of the concepts of student culture, which is one of the most important components in the educational process of students in secondary schools.
Using an electronic textbook, as an additional tool in learning, you can show the beauty of solving mathematical problems, the harmony of their solutions not only of one example, but also of many other tasks in mastering quite fundamental knowledge. About even greater manifestations of aesthetics in mathematical education, in revealing all the intellectual and creative possibilities of a person, in developing imagination, as well as broadening the horizons of students at each stage, according to Academician B.V. Rauschenbakh, "... penetrate the essence of the object ...".
So, solving the same problem in different ways shows that the essential component of the forms of aesthetic teaching as a means of educating the educational process in mathematics is the mastery of schoolchildren by knowledge related to understanding the art of a creative approach.
In conclusion of this article, I would like to note that solving the same problem in different ways (if necessary, using an electronic textbook) increases the level of general development of students, contributes to their intellectual growth, that is, affects the formation of personal and mental actions in the formation of personality.

References

[1] Tikhomirov V M 1999 *The genius who lives among us* // Extreme phenomenon. The book is about Kolmogorov. - M .: FAZIS, MIROS,p.256

[2] Borovsky A V and Rozov N Kh 2010 Activity principles in pedagogy and pedagogical logic - M .: MAKS Press, p.80

[3] Tsvyrko O L and Tsvyrko N I 2010 *Chromathematical approach in school mathematics* // Materials of the regional scientific-practical conference of the Tyumen State University,P.8

[4] Pospelov N N and Pospelov I.N. The formation of mental operations in high school students. - M .: Pedagogy, 1989 .-- 151 p

[5] Krygovskaya A S 1966 The development of students' mathematical activity and the role of tasks in this development // Mathematics in school. **6** S. 19-30

[6] Dalinger V A 2019 Methods of teaching mathematics. Search and research activities of students: a textbook and workshop for open source software /2nd ed., Rev. and add. - Moscow: Yurayt Publishing House,460 p. - (Series: Vocational Education). - ISBN 978-5-534-01288-0

[7] Dalinger V A 2007 Educational and research activities of students in the process of studying fractions and actions on them: a training manual. - Omsk: Publishing House of OmGPU,191 p

[8] Dalinger V A and Tolpekina N V 2004 Organization and content of search and research activities of students in mathematics: a manual. - Omsk: Publishing House of OmGPU, 263 p

[9] Babansky Yu K 1983 Pedagogy: Textbook. manual for students of ped. in-to /. - M.: Education,608 p

[10] Breytigam E K 2004 The activity-semantic approach in the context of developing teaching senior pupils the principles of mathematical analysis: Dis. ... doctors ped. sciences. - Barnaul: OmGPU, 433 p

[11] Samples P I and Kosukhin V M 2004 Didactics of the higher military school. Tutorial. –Orel:

Academy of Special Communications of Russia, p –317

[12] Reasonable V A 1969 Aesthetic education: Essence. Forms. Methods - M .: publishing house "Thought",81 p

[13] Kodzhaspirova G M and Kodjaspirov A Yu 2005 Dictionary of pedagogy. - M .: ICC "Mart"; Rostov n / a: Publishing center "Mart",448 p

[14] Kharlamov I F 1990 Pedagogy: Textbook. 2nd ed., Revised. and add. - M .: Higher. school,576 p

[15] Firstova N I 2002 Introduction of elements of aesthetic education in the context of school textbooks in mathematics// Humanitarianization of secondary and higher mathematical education: methodology, theory and practice: Materials of the All-Russian Scientific Conference. Saransk, September 18-20, 2002. Part 1 / Mordov. state ped institute - Saransk,p. 141-143

[16] Sarantsev G I 2003 *Aesthetic motivation in teaching mathematics*. - PO RAO, Mordov. ped institute - Saransk, p.136

[17] Sawyer W U 1991 *Intuitive understanding of mathematical proof* // Mathematics at school. **2**. - p. 75-77

[18] Sarantsev G I *Beauty is in mathematics, mathematics is in beauty* // Electronic resource URL: http://portalus.ru/CatalogId1193230519

[19] Boltyanskiy V G 1982 *Mathematical culture and aesthetics* // Mathematics at school. **2** p. 40-43

[20] Dalinger V A Simonzhenkov S D and Galyukshov B S 2019 *Probability theory and mathematical statistics using mathcad*: a textbook and workshop for applied undergraduate /2nd ed., Rev. and add. - Moscow: Yurayt Publishing House,145 p. (Series: Bachelor. Applied Course). - ISBN 978-5-534-10080-8

[21] Smirnova E I 2007 *Visual modeling in teaching mathematics*: theory and practice:Textbook/ Yaroslavl: IPC Indigo, p –454

[22] Zhokhov A L 2008 Cognition of mathematics and the fundamentals of the scientific worldview: worldview directed teaching of mathematics: a training manual. - Yaroslavl: Publishing House of YAGPU,p.–183

[23] Dalinger V A, Shatova N D, Kalt E A and Filonenko L A 2019 *Methods of developing teaching mathematics*: textbook. manual for STR /under the general. ed. V.A.Dalinger. - 2nd ed., Rev. and add. - Moscow: Yurayt Publishing House,297 p. - (Series: Vocational Education). - ISBN 978-5-534-07408-6

[24] Dalinger V A 2019 *Methods of teaching mathematics*. Search and research activities of students: textbook and workshop for universities /2nd ed., Rev. and add. - Moscow: Yurayt Publishing House,460 p. - (Series: Educational Process). - ISBN 978-5-534-09597-5

[25] Rozin V M 2005 Methodology: formation and current state. Tutorial. - M.: Moscow Psychological and Social Institute, p.- 414

[26] Zhokhov A L 1995 How to help shape the worldview of schoolchildren - Samara: Samara GPU Publishing House,p.- 288

IOP Conf. Series: Materials Science and Engineering **934** (2020) 012046 doi:10.1088/1757-899X/934/1/012046

High-speed induction heating during tensile strength testing of a composite material based on an inorganic binder at temperatures up to 1500 ° C

K V Klemazov[1], P Yu Yakushkin[1], M O Zabezhailov[1] and P A Stepanov[1,2]

[1]JSC «ORPE «Technologiya» named after A. G. Romashin, Obninsk, Kaluga Region, Russia

[2]Bauman Moscow State Technical University, Moscow, Russia

E-mail: klemazov_kirill@mail.ru

Abstract. High-speed induction heating was used during strength tests in an oxidizing atmosphere of a composite material based on an aluminum-chromophosphate binder. Induction heating allows heating modes close to those of aircraft operating modes, which avoids the appearance of crystalline phases such as Al(PO3)3, Cr2O3, which arise under conditions of slow heating under classical methods of strength testing at elevated temperatures.

Keywords: alumochromophosphate binder, composite material, tensile strength, high temperature tests, induction heating

1. Introduction

The development of promising high-speed aircraft is associated with the problem of choosing structural materials that are resistant to high temperatures [1], in a wide range of heating rates, coupled with power loading. The solution to this problem necessitates the conduct of heat-strength tests of materials in order to determine the physical and mechanical characteristics. Classical methods of mechanical testing at elevated temperatures involve quasi-stationary heating. Under such heating conditions, the mechanical properties of the material may differ from the properties that are observed under actual operating conditions. Induction heating allows high-speed high-temperature heating of materials to high temperatures.

Induction heating is widely used as a method of heating conductive materials [2, 3]. The advantages of induction heating include low inertia and easy access to the sample, which facilitates the measurement. For non-conductive materials, induction heating can be used only in the presence of an intermediate heating conductive element heated by the inductor, from which the test sample is heated, [4, 5]. Heating through an intermediate heater leads to a certain increase in inertia (the smaller, the thinner the intermediate heater). The choice of material for the intermediate heater depends on the temperature and the medium in which the heating is carried out.

Composite materials based on inorganic binders are promising materials that can be used under conditions of high-temperature high-speed heating [6]. One of the representatives of this class of materials based on an aluminochromophosphate binder (ACPB) [7]. The study of the mechanical properties of this class of materials under high-speed high-temperature heating is an important task, since it allows one to obtain properties in conditions close to real operating conditions.

2. Materials and research methods

The object of the study was a composite material made by contact molding from an alumochromophosphate binder with the addition of Al2O3 powder, which is applied to quartz fabric (ACPB composite material).

The calculation of the temperature distribution in the sample during stationary and non-stationary heating was performed using ANSYS Fluent. The following characteristics were calculated:

1. - conditions for achieving a stationary heating mode, the required exposure time;
2. - temperature distribution in the sample.

Tensile tests were carried out on the tensile setup IR5047-50 at temperatures of 1100, 1200, 1300, 1500 ° C. Loading speed 5 mm / min. 5 samples were tested for each temperature point. Temperature control was carried out with a Modline 5 partial radiation pyrometer. An HfB2 heater was placed between the grips (Figure 1) in thermal protection from porous Al2O3. The heating rate was 10 ° C/s. Upon reaching the required temperature was carried out isothermal exposure for 180 seconds. After isothermal exposure, a tensile test was carried out. A schematic representation of a laboratory setup is shown in Figure 2.

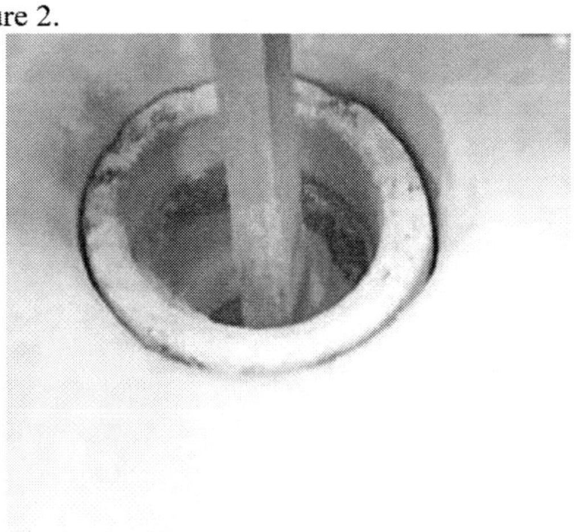

Fig. 1. HfB_2 based heater

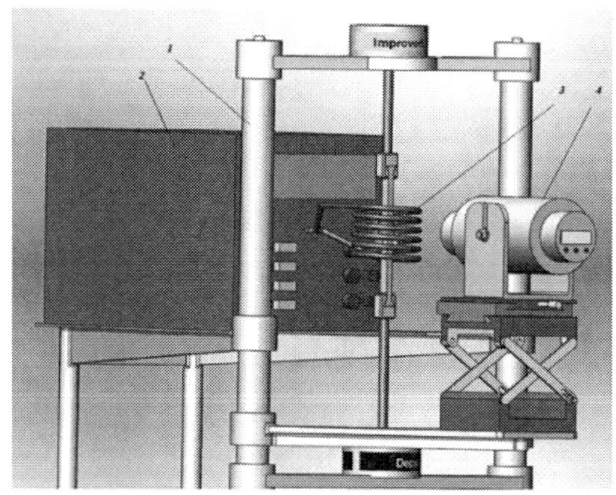

Fig. 2. Schematic representation of a laboratory setup

1) Tensile machine
2) Inductor
3) Induction coil
4) Pyrometer

Tensile strength σ, MPa is determined by the formula:

$$\sigma = \frac{P}{S}$$

where P is the maximum load during tensile testing, N;
S is the cross-sectional area of the working area of the sample, mm2.

X-ray phase analysis (XRD) was carried out on a PANalytical Empyrian Series 2 diffractometer in CuKα radiation with a nickel filter using the Data Collector, High Score Plus software package. To identify xrd patterns, the database "PDF2-2015" was used.

3. Thermophysical Calculation

IOP Conf. Series: Materials Science and Engineering **934** (2020) 012046 doi:10.1088/1757-899X/934/1/012046

The geometry and the designation of the elements are shown in Figure 3. The following characteristics were subject to a calculated assessment: (1) - conditions for achieving a stationary heating mode, the required exposure time (Figure 4); (2) - temperature distribution in the sample (Figure 5).

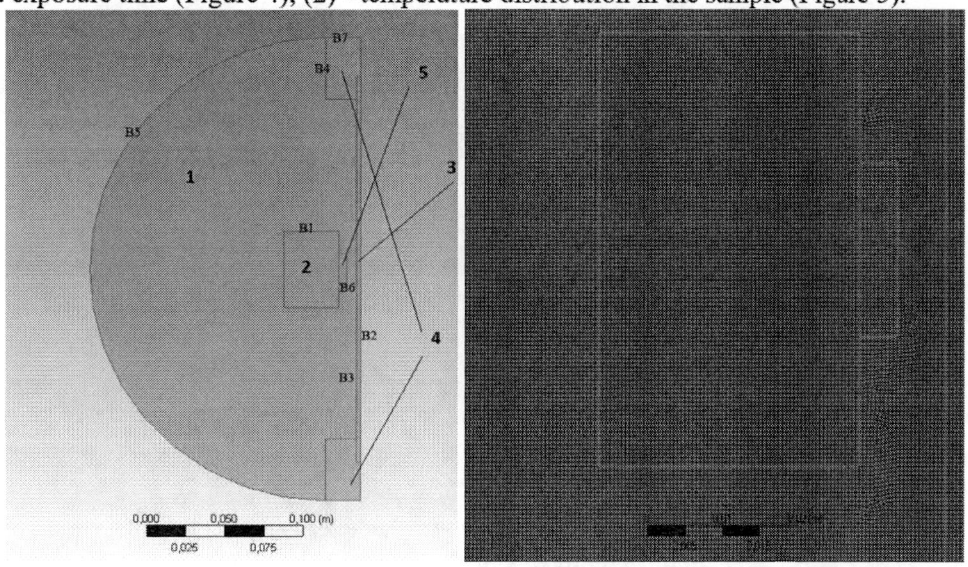

Fig. 3. The computational area and computational domains around the sample: 1 – air, 2 – thermal insulation, 3 – the sample, 4 – metallic uncooled grips, 5 –heater. The following notation is introduced for the boundaries of computational domains: B1 –thermal insulation-air boundary, B2 – axis of symmetry, B3 – the sample–air, B4 – the grips-air, B5 – outer border of calculating domain for air, B6 – the heater – air, B7 – outer border of calculating domain for the grips.

The calculation showed that the exposure time of 180 seconds is sufficient to achieve uniform heating of the working area of the sample. It should also be noted that in the experiments, the heating rate was ~ 10 ° C / s, while in the calculation it was 67 ° C/s. Thus, the estimated time for reaching stationary heating is an upper bound. The calculation of the temperature distribution in the sample showed that the temperature difference across the layer thickness when entering the test mode does not exceed 10 degrees up to a sample surface temperature of ~ 1700 ° C.

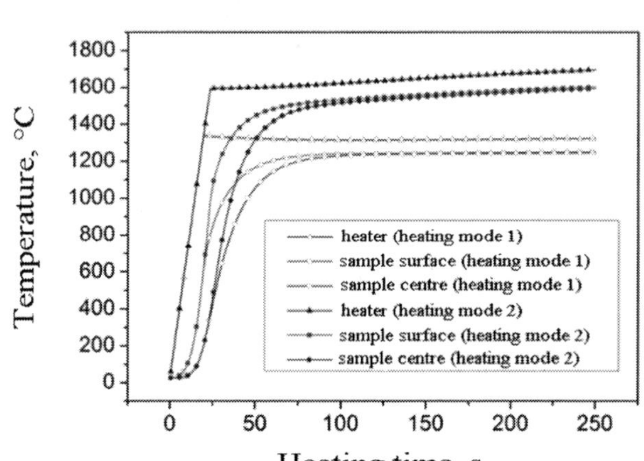

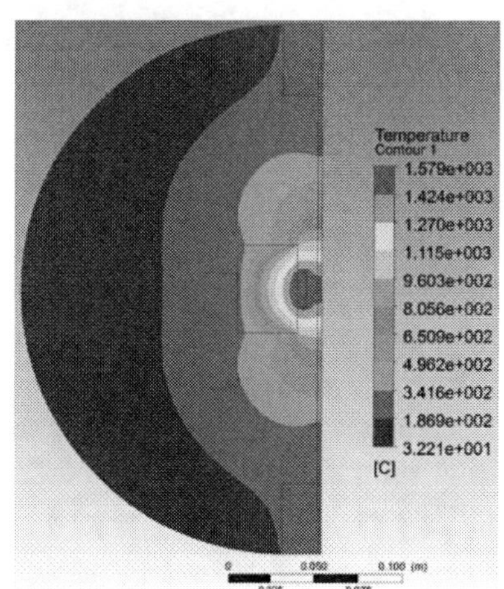

IOP Conf. Series: Materials Science and Engineering **934** (2020) 012046 doi:10.1088/1757-899X/934/1/012046

Fig. 4. Calculation of heating time.

Fig. 5 Temperature distribution in the sample

4. Mechanical tensile tests

Table 1 presents the results of determining the tensile strength of ACPB composite material at various temperatures.

Table1. Tensile strength of ACPB composite material samples after induction heating

Temperature (°C)	Average (MPa)	STD	Coefficient of variation (%)
1100	15,0	1,5	10
1200	14,3	0,7	5
1300	13,4	1,2	9
1400	7,5	0,8	10
1500	8,1	1,9	23

Figure 6 shows the samples after the test. The results are characterized by a low dispersion of strength values - up to 10% at test temperatures up to 1500 ° C.

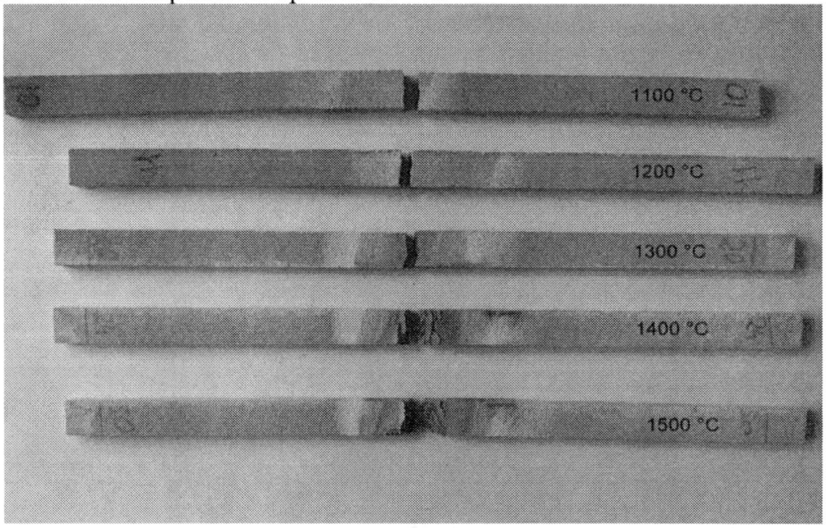

Fig. 6. The samples after induction heating tensile tests at various temperatures

At a temperature of 1400 ° C, a marked drop in strength occurs. This fact can be explained by the fact that at these temperatures, quartz fabric ceases to play a reinforcing role due to softening. In general, the nature of destruction changes (Figure 7). At a temperature of 1400 and 1500 ° C, cracks are observed perpendicular to the length of the sample with a thickness of ≈ 300 μm. A change in the surface colour of the sample can be explained by a change in the valency of chromium, which is contained in the aluminochromophosphate binder.

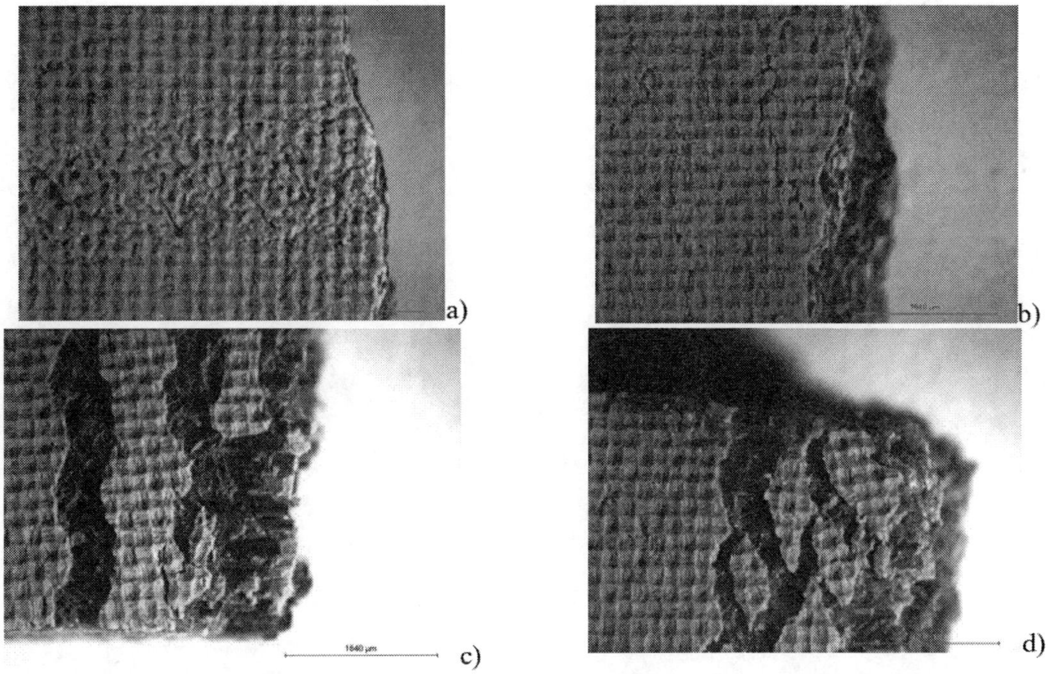

Fig. 7. Optical microscopy of the samples after induction heating tensile tests at various temperatures. a) 1100 °C, b) 1300 °C, c) 1400 °C, d) 1500 °C

5. X-ray phase analysis

The results of X-ray phase analysis of ACPB composite material after tensile tests are presented in Figure 8. According to the X-ray diffraction patterns, it is seen that after heating to a temperature of 1100 ° C, the formation of the crystalline phase AlPO4 occurs. The shift in the position of the maximum of the main AlPO4 peak with increasing temperature is associated with the ordering of the structure (Figure 9). However, with high-speed induction heating, additional crystalline phases, such as Al(PO3)3, Cr2O3, do not appear, which arise under conditions of slow heating under classical methods of strength testing at elevated temperatures [7].

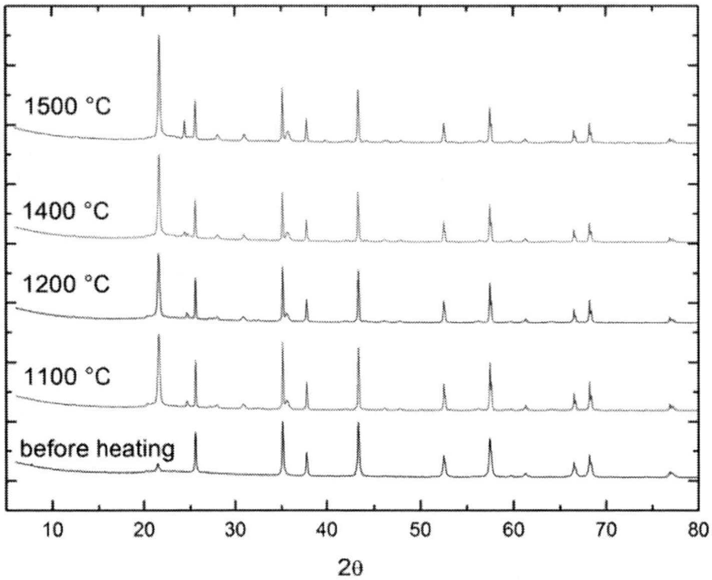

Fig. 8. XRD patterns of the samples.

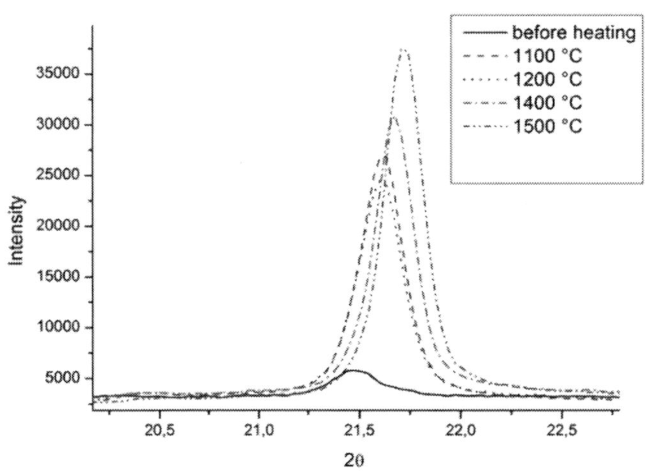

Fig. 9. Growing of $AlPO_4$ phase during induction heating

6. Conclusion

The developed prototype laboratory setup allows tensile tests to be carried out under conditions of high-temperature, high-speed heating in an air atmosphere to temperatures of about 1500 ° C.

During mechanical tests, it was revealed that at a temperature of 1400 ° C, a change in the nature of failure occurs, which is accompanied by a decrease in strength by about a factor of two relative to tests at 1300 ° C. This fact can be explained by the fact that at these temperatures, quartz fabric ceases to play a reinforcing role due to softening.

According to x-ray phase analysis, the appearance of crystalline phases in the area of induction heating is revealed. The crystalline AlPO4 phase is already formed at a temperature of 1100 ° C and its amount increases with increasing temperature. The shift in the position of the maximum of the main AlPO4 peak with increasing temperature is related to the ordering of the structure.
With high-speed induction heating, additional crystalline phases, such as Al(PO3)3, Cr2O3, do not appear, which arise under conditions of quasi-stationary heating with classical methods of strength testing at elevated temperatures.

References

[1] Fahrenholtz W G and Hilmas G E 2017 *Ultra-high temperature ceramics: Materials for extreme environments. Scripta Materialia,* **129** 94–99. doi:10.1016/j.scriptamat.2016.10.018
[2] Hyde C J, Sun W and Leen S B 2010 Cyclic thermo-mechanical material modelling and testing of 316 stainless steel. International Journal of Pressure Vessels and Piping, **87-6** 365–372. doi:10.1016/j.ijpvp.2010.03.007
[3] Markovskya P E 011 Tailoring of microstructure and mechanical properties of Ti–6Al–4V withlocal rapid (induction) heat treatment // *Materials Science and Engineering A*, **2** 3079–3089.
[4] Neuman E W, Hilmas G E and Fahrenholtz W G 2016 Ultra-high temperature mechanical properties of a zirconium diboride–zirconium carbide ceramic. *Journal of the American Ceramic Society*, **99-2** 597-603.
[5] Yakushkin P Yu and other 2016 High-intensity induction heating as a way of approximating test conditions to operating conditions // New materials, S. 268-269 [in Russian]
[6] Chen N. et al. 2014 Studies on high-temperature thermal transformation and dielectric property of aluminum–chromium phosphates //*Journal of Thermal Analysis and Calorimetry*. **116-2** C. 875-879.
[7] Stepanov P A, Atroshchenko I G, Starodubtseva N I, Shutkina O V and Melnikov D A 2014 *Development of high-temperature composite materials for heat-shielding and radio engineering purposes*. // Promising materials, **10** S. 17-21 [in Russian]

Solution of experimental tasks in the study of physics

S N Nurkasymova[1], U S Nurgaliyeva[1], A B Zhanys[2], A S Baigizova[1] and A S Bayegizova[1]

[1]Eurasian National University. L.N.Gumilev 010000 Republic of Kazakhstan, Nur-Sultan

[2]Kazakh Agrotechnical University named after Saken Seifyllin, Department of Information Systems, Republic of Kazakhstan Nur-Sultan city

E-mail: SauleNurkasim@mail.ru, aray.zhanys@gmail.com

Abstract: The solution of experimental problems has features that reveal wide opportunities for the development of students' creative abilities in the learning process. The task of developing the creative abilities of students is an integral part of the problem of developing their thinking.All problems are solved with the help of an expert. An experiment is a set of sciences about the laws and forms of thinking, about the mathematical-logical laws of calculus, about the most general laws of thinking. Experimental apparatus allows you to build the necessary judgment or inference, to highlight a certain concept with its essential features.

Keywords: Problem solving in physics, methods of composing tasks, methods of problem solving, the development of thinking, the development of logic, problem solving

1. Introduction

In the Message of the President to the people of Kazakhstan, the development and implementation ofinnovative technologies, increasing computer literacy contribute to the qualitative growth of human capital. In the context of globalization, informational opportunities are significantly expanding. The Internet creates access to global information space and Internet resources [1].

Physics is the fundamental basis for the theoretical preparation of students, without which its successful activity is impossible.

At all stages of learning, the practical application of theoretical knowledge is of great importance, one of which is to solve problems when studying a physics course. This is especially important when improving various forms of independent work of students.

Problem solving is one of the important means of repeating the consolidation and examination of students.

2. Methods of experimental problems

Systematic problem solving contributes to the development of students' thinking, their preparation for participation in creative searches, brings up diligence, perseverance, dedication and is a good means of controlling knowledge and skills.

Teaching a student to solve physical problems is one of the most difficult pedagogical problems. Therefore, it is very important what method of teaching the teacher uses.

Material submission methods must meet the following requirements [2]:

- a clear selection of the fundamental concepts of the topic and section;
- a clear formulation of laws explaining the phenomena of nature and the laws under which they obey;
- highlighting the functional relationship between the quantities using formulas and graphs;
- theoretical generalization of the material.

Problems in physics are very diverse in various ways, reflecting the goals. They can be classified according to various characteristics, reflecting the most typical features for many tasks of different content.

Based on the classification method of solution, we can distinguish the following types of tasks:
- quality tasks or tasks - questions;
- quantitative;
- experimental;
- brain teaser;
- graphic;
- tasks with production and technical content.

3. Basic elements of experimental problems

Experimental tasks — Experimental tasks are those in which an experiment is used to obtain initial data or to verify the correctness of solutions.

Qualitative experimental problems do not require quantitative data and mathematical calculations for their solution. Let's give an example:

"On the table there is an electrical circuit shown in the picture. How will the ammeter and voltmeter readings change when the rheostat slider moves in one direction or another? Why?"

Quantitative experimental problems in their formulation require obtaining quantitative data and mathematical calculations for their processing.

Qualitative experimental tasks with the production content of students make measurements using technical instruments and tools: calipers, micrometers, technical manometer, etc. For the preparation of tasks the teacher uses the production material of those enterprises, which work students.

Experimental tasks should be planned in advance when approaching a topic, when studying a new material and when it is fixed.

Experimental tasks can be used during the homework of students. The following can be attributed to the number of home experimental tasks: "Calculate the cost of electricity consumed in a flat per week using the readings of an electric meter"; "To determine what will be the image of the flame of a candle, if you look at it through a horizontally positioned and filled with water bottle", etc.

Problems in physics are very diverse in various ways, reflecting the goals. They can be classified according to various characteristics, reflecting the most typical features for many tasks of different content.

Taking the solution method as the basis for the classification, the following types of tasks can be shown (Fig-1) [4].

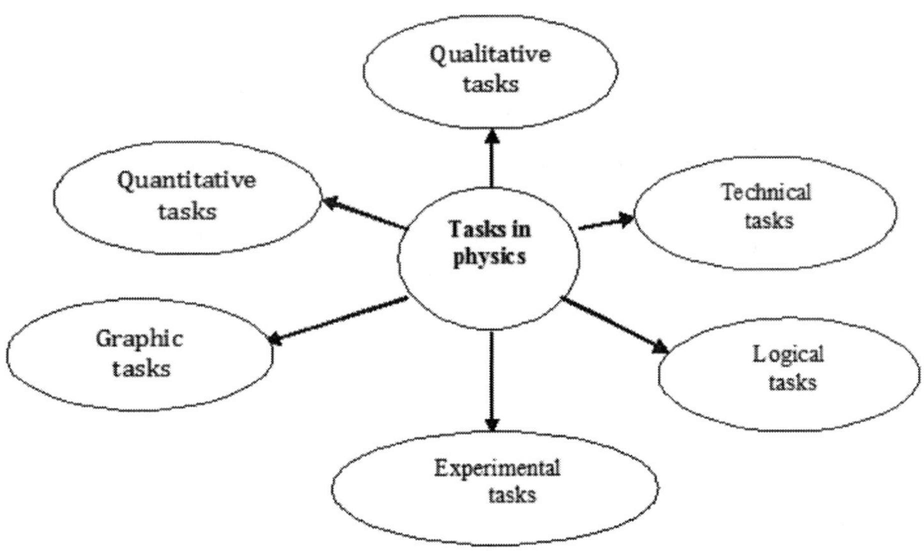

Fig. 1: Types of physical tasks.

4. The following types of logical problems exist:

Experimental tasks that do not require complex equipment and a lot of time to perform them are included in the content of tests and tests, for example, to measure the temperature with a thermometer, read out the atmospheric pressure using an aneroid, etc.

The formulation of experimental tasks can be different:
a) The experimental task is set as a demonstration experiment. In this case, the necessary equipment is put on the demonstration table, and the teacher receives the experimental data for the task. In this case, all requirements for the demonstration experiment should be fulfilled:
- clear target setting,
- providing visibility
- short duration
- convincing experience, etc.

b) The experimental task is set as a frontal laboratory task. In this case, the equipment is provided for each table and students receive experimental data and solve the problem on their own.
c) The experimental task has a research character. In this case, each table is given a different equipment, depending on the tasks that students receive. They independently assemble the installations and carry out the necessary experiment. The solution of these tasks should be distinguished by a great independence of students in the selection of equipment and analysis of physical phenomena.
Experimental research problems can be a type of extracurricular work in physics in the case when their solution requires a long time.
The solution and design of the experimental problem of a calculated nature consists of the following elements:
- formulation of the problem;
- condition analysis;
- measurements;
- calculations;
- experienced response verification.

5. Examples of experimental tasks.

5.1 Problem 1.

On the table there is a rectangular can, scales, weights, scale ruler, a vessel with water, sand. To ensure the vertical position of the banks when swimming in water, it is slightly loaded with sand. Determine the depth of the cans when it is immersed in water.
In this case, the condition of the problem can be expressed in a picture
with the signature of the question under it. Then proceed to the analysis, find out what measurements need to be performed to solve the problem.
Task analysis:
The can will be submerged in water until the force of gravity acting on it and the sand is balanced by the buoyancy force of water acting on the can from below. In this case:

$$F_{Arch} = F$$

But since the Archimedean force Farch is equal to the weight of the fluid displaced by the body, then:

$$F_{Arch} = 9{,}8\frac{N}{kg}\cdot\rho_B\cdot V_B$$

where V_B — is the volume of the submerged part of the can, ρ_B — is the density of water. The volume of the submerged part is equal to the product of the area of the base S and the depth of immersion in water h. Consequently,

$$F_{Arch} = 9{,}8\frac{H}{\kappa г}\cdot\rho_B\cdot V_B = 9{,}8\frac{N}{kg}\cdot\rho_B h S$$

from where,

$$h = \frac{F}{9{,}8\frac{N}{kg}\cdot\rho_B S}$$

The correctness of the found solution is checked by operations with the names of the quantities included in the formula.
From the formula (1) it is clear that to solve the problem it is necessary to
know the weight of the can of sand, the density of water and the area of the base of the can.
Measurements. Determine the weight of the P banks with sand using a dynamometer.
Measure the length l and width of the base. Determine the area of the base

$$S = l\, a\, .$$

Density of water $\rho_B = 1g/sm^3$
Experienced check on empirically. On the vertical wall of the can, a color line indicates the depth of immersion found from experience and subsequent calculations, and put the can in a vessel with water. Experience shows that the depth of immersion coincides with the value found.
In connection with the solution of the problem, the principle of determining the draft of the ship is explained.
In experimental qualitative problems, experience is set at the moment when a need arises in it. For example, to solve the problem of a baroscopic on a demonstration table, an installation is prepared in advance with a glass cap, inside which the baroscopic is placed. The air tray is connected to the Komovsky pump. The teacher explains the device of the baroscopic and, drawing attention to the balance of the lever, puts the question to the class: "What changes will occur with the baroscopic if air is pumped out?" The students offer an explanation. In this case, correct and incorrect explanations may be offered. Doubt in their correctness
is allowed by demonstration of experience (Figure - 2):

Fig.2: Bell air pump

However, there may be another way to find the answer to this problem. After the question is posed, the tasks immediately proceed to the execution of experience Watch the occurrence of the phenomenon. Then he is explained theoretically.
Some experimental tasks in physics in a classroom can be posed frontally. Examples of such tasks: "Measure the water pressure at the bottom of the glass using a ruler"; "Determine the power of the current consumed by the light bulb." In this case, they perform the role of frontal experiments.

5.2 Problem 2.

Question: How with the help of a ruler can you find the height of a tree on a sunny day without climbing into it? [9].

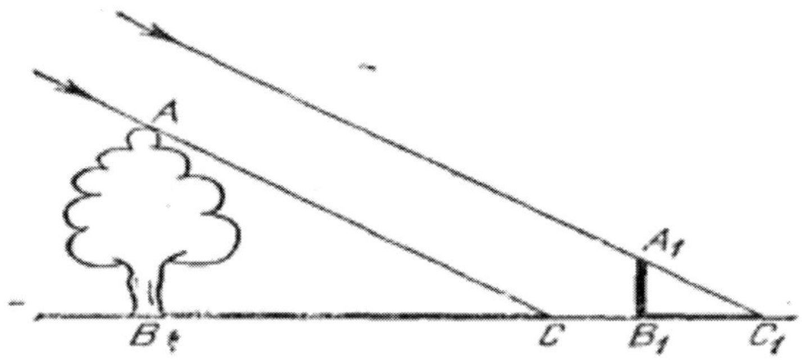

Fig. 3: Drawn tree to determine the height

Answer: Putting the ruler vertically, mark on the ground the length of the shadow B_1C_1, and also the length of the shadow BC from the tree. From the similarity of the triangle ABC and $A_1B_1C_1$ find the tree height equal to,

$$AB = \frac{A_1B_1}{B_1C_1} \cdot BC$$

where - A_1B_1 is the known ruler length.

5.3 Here Are Examples of Logical Problems. Problem 3.

There is a plank on both banks, but each one is somewhat shorter than the width of the stream. How can an adult and a child run from one bank to another? An adult and a child need to go through speeches: one from the left bank to the right bank, the second in the opposite direction.

Fig.4: There is a plank on both banks, but each one is somewhat shorter than the width of the stream

5.4 Problem 4.

The Fall of the pressure in pluming is demonstrated on the following model (the rice. 5). Narrow (for increase of friction) pipe A and her (its) branch B, supplied man metric tube, can be closed tap but and b.

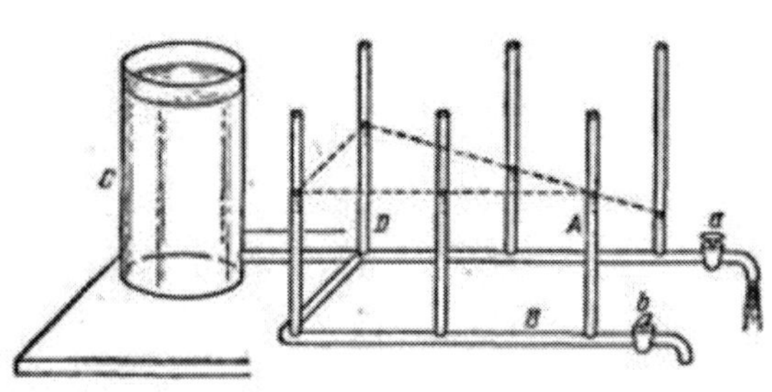

Fig. 5: Water tank and metric tube for demonstrating pressure drop

If fill water in container C and close the taps that we observe what spreads the height to liquids in turban a, b and C and as so much for possible explain?

5.5 Problem 5.

Why should high-speed trains (Fig. 6) slow down at a meeting, otherwise the windows in the cars will break? Which way does the glass fall out: inside the cars or out? Can this happen if trains move in one direction? Will you be attracted to the train or push away from it if you find yourself too close to the fast train?

Fig. 6: As an example, consider the movement of a railroad car at the rounding of a horizontal track

5.6 Problem 6.

In rainy windy weather, each of us noticed that the opened umbrellas sometimes "turn inside out" (Fig.7). Why is this happening?

Fig. 7: Open umbrellas "turned inside out"

5.7 Problem 7.

For the experience, we will manufacture a cylinder of thick, but not thick paper with a diameter of 5 cm, 25-30 cm long. A sharp movement along the horizontal surface of the table will inform the cylinder of a complex movement
(translational and rotational (Fig. 8). At high speed, the cylinder rises and describes a small vertical loop. Explain why this is happening.

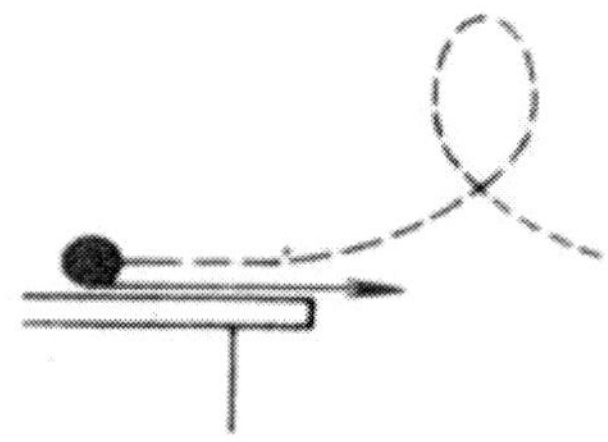

Fig. 8: Describes a small vertical loop

5.8 *Problem 8.*

That occurs if blow on sheet of paper along his(its) surfaces that occurs and as this possible explain (Fig - 9)?

Fig. 9: Blow on a sheet of paper along its surface

6. Conclusion

The solution of experimental problems begins with the formulation of the problem, then a brief record of the conditions and requirements is carried out, a hypothesis is formulated, the test of which is planned. Then, the implementation of the planned plan is carried out by various means: experimental, logical and mathematical. Then the result is checked.

The results of solving experimental problems are refined model ideas about the didactic material of the educational process and the creation of experimental tasks of a new generation. The tasks set the level of complexity and the scope of the task, due to the fact that the activity of the teacher is aimed at building a model of the educational process with predetermined quality objectives.

The solution of experimental problems with students in the professional competence of the teacher, who mediates between theory and practice.

With the mastery of this skill, a new technological thinking of the teacher begins: clarity, structure, clarity of the methodological language, the emergence of a reasonable standard in learning.

So, only through the pedagogical technology, the practice of its implementation and implementation, the volume and level of complexity of the teaching content are derived. The object of the activity of the methodologist is the design of the educational process on technology. The activity of the teacher - becomes the reorientation of the project - a sample of the educational process to the characteristics of the teacher, and his methodological enrichment.

References

[1] http //www.bnews.kz/
[2] Nurkasymova S N 2004 Methods of teaching physics - Semey, p.110-114
[3] Apatova N V 1994 Information technology in school education. M. p.165
[4] Nurkasymova S.N. The use of information technology in educational the process - Astana, 2010 p.109 - 112
[5] 2002 Journal "Informatics and Education",p.128

[6] Usova AV and Tulkibaeva N N 2001 Workshop on the solution of physical problems "Education" Moscow
[7] Tarasov L V 1990 Modern physics in high school. - M .: Enlightenment
[8] Bugaev A I 1990 Methods of teaching physics in high school. - M .: Enlightenment.
[9] Lange V N 1986 Experimental physical tasks for ingenuity. - M .: Enlightenment.
[10] Theory and methods of teaching physics in school. Private issues. Ed. S.E. Kamenetsky. // M .: Academy, 2000.
[11] Yastrebtseva E N 1998 Project "Harmony" // Computer at school. **3**
[12] LabVIEW computer modeling tools, 8. Measurement Studio, etc. Internet address: http://labview.nm.ru
[13] Khutorskoy A V 1998 The development of schoolchildren endowment: Methods of productive learning. - M., x. p. 66. http://www.corbina.net/~snark/
[14] Sevchenko N E 2000 Problems in physics with the analysis of their solutions. M .: Enlightenment -
[15] Saveliev I V 2003 Collection of questions and problems in general physics. M .: Science
[16] 2002 Journal "Informatics and Education".p.128
[17] Kotkin G L and Cherkassky V S 2001 Computer simulation of physical processes using MATLAB. Novosibirsk,169 b.
[18] Porshnev S V 2003Computer simulation of physical processes in the MATLAB package. –M,. p. 592
[19] Martynov N N and Ivanov A P 2000 MATLAB 5.X. Calculations, visualization, programming. M, KUDITS-IMAGE, p.332
[20] Evans, J.ST.B.T., 1972. On the problems of interpreting reasoning data: Logical and psychological approaches. Cognition, 1(4): 373-384.
[21] Hutorskoj A V 2000. The development of gifted students : Methods of Productive Teaching. pp: 66.
[22] Marco De Boni 2007 Using logical relevance for question answering. Journal of Applied Logic,5(1): 92-103.
[23] Physics at school - 2003
[24] Quant - 2004
[25] Physics at school - 2005
[26] http: //l-micro.ru/
[27] http://school56.spb.ru/lego/lego.index.html

IOP Conf. Series: Materials Science and Engineering **934** (2020) 012048 doi:10.1088/1757-899X/934/1/012048

Developing a New Type of Jointing the Metal Plates of a Heat Exchanger

D O Onishchenko[1,5], A Yu Korotchenko[2,6], Yu O Fokin[3,7], M V Tverskoy[4,8]

[1] D. Sc, Prof, E-2, Bauman Moscow State Technical University,105005, Russia

[2] D. Sc, MT-5, Bauman Moscow State Technical University, 105005, Russia

[3] Engineer, Bauman Moscow State Technical University, Moscow, 105005, Russia

[4] Engineer, Bauman Moscow State Technical University, Moscow, 105005, Russia

E-mail:[5]doctor@baumanracing.ru, [6]kor_15@mail.ru, [7]fokin_5689@mail.ru, [8]tversk@mail.ru

Abstract. This paper proposes using a new composition of the metal polymer compound to joint heat exchanger plates. This composition makes it possible to improve the factors of plate junction, such as the cost, safety, efficiency, and rate of the junction procedure.We consider the benefits and drawbacks of the two main ways to joint metal plates of brazed steel plate heat exchangers (BPHE), namely, copper brazing and laser beam welding; propose to joint plates using a metal polymer compound consisting of the metal powder of steel 316L and binding agents; show the metal polymer compound production process; provide the results of the component mixing process depending on the parameters, such as duration and rate of mixing; as well as present the resulting density of the specimens made of the metal polymer compound, depending on the mixing conditions

Keywords: Plate heat exchangers, metal-polymer composite, copper brazing, laser beam welding

1. Introduction

Plate heat exchangers (PHE) are currently one of the most promising devices in the modern chemical, oil refining, engineering, and other industries. The main PHE application (about 50%) is their use in heat supply (heating and hot water supply systems) cooling systems [1, 2].

The most demanded and promising heat exchangers are the brazed steel plate ones (BPHE) [3,4]. Due to their design and ease of manufacturing, BPHEs are much cheaper, more efficient and reliable than shell and tube or knock-down PHEs. Figure 1 shows the appearance of a brazed heat-exchange apparatus.

Figure 1. Brazed heat-exchange apparatus

BPHEs should comply with the following requirements:
transfer the required amount of heat from one medium to another with the required end temperatures at the best possible heat exchange capacity;
have a heat exchange surface area and other design elements of the apparatus exposed to the working fluid that are sufficiently resilient to heat conductor exposure;
work reliably at the given thermodynamic parameters of the working media (pressure, temperature, flow) and various physical state;
allow for regular cleaning of the heat exchange surfaces and ensure their maintainability in order to preserve continuous serviceability when operated with media that cause deposition on the heat exchanger walls;
be durable enough to ensure safe condition at tensions that occur as a result of the working medium pressure as well as due to thermally-induced displacements of various parts of the heat exchanger.
In order to ensure the given technical requirements at the stage of BPHE design and development, one should pay particular attention to the technology of plate junction into a heat exchange matrix, as well as the brazing alloy material.
Currently, there are two types of metal BPHE plate junction: copper brazing and laser beam welding.
Copper brazing is an efficient and cost-effective way to produce plate heat exchangers (Figure 2) [5]. This method uses a copper plate as the junction material to joint stainless-steel plates with each other. This technological process for plate junction is performed in a vacuum furnace in inert gas atmosphere. The brazing process should be closely controlled as copper can penetrate stainless steel causing metal embrittlement while it is liquid, affecting the strength of the manufactured product. Sometimes, it turns out to be difficult to apply or even determine the most suitable welding filler. Also, when this BPHE is used, copper particles can penetrate the working medium, which is not allowed in certain cases.

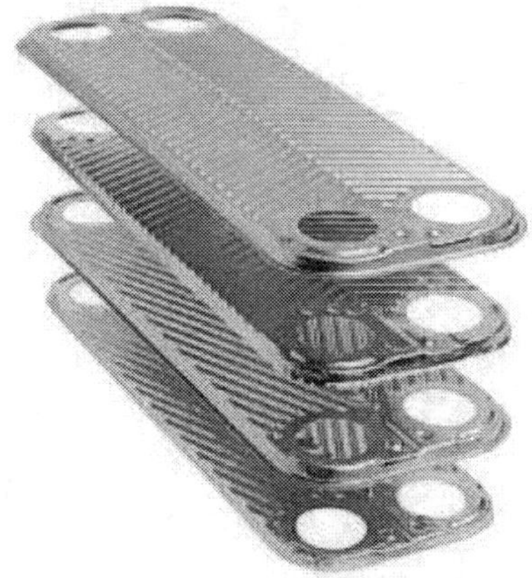

Figure 2. The process of plate junction by copper brazing

In laser beam welding, corrugated stainless steel plates are positioned opposite one another and the material at the contact points is melted with laser (Figure 3) [6]. As steel hardens, metal diffusion occurs on the surfaces of the plates. This method has several drawbacks. Sometimes, it is necessary to adapt the product design to the limitations of this welding technology. In addition, this manufacturing method is expensive. The process must take place in a chemically inert environment, otherwise materials react with oxygen contained in the air, worsening the quality of weld seams. The equipment required for such welding is also quite expensive.

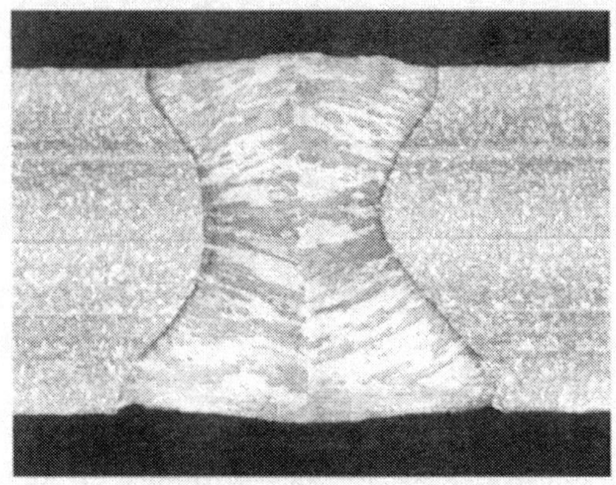

Figure 3. Laser beam welding process

When choosing a particular jointing method, one should consider many factors, such as cost, safety, productivity, and rate of the plate jointing process. The above methods of BPHE plate junction have certain advantages, but there is still a need for new junction methods that will improve these factors [7, 8, 9].

The purpose of the article is to develop a composition of metal-polymer composites for BPHE plate junction.

2. Source data

The metal-polymer composite consists of metal powder and a polymer binder. To develop the composition of metal-polymer composites, we chose steel 316L as the main powder.

The chemical composition of the powder is shown in Table 1 [10].

Table 1. Chemical composition

C %	Cr %	Ni %	Mn %	Mo %	Si %	Fe %
$\geq$ 0.03	15–17	14–16	$\leq$ 0.8	2.5–3	$\leq$ 0.6	Balance

The characteristics of the selected powder mixture's particle are provided below:

Bulk density: 7.6 g/cm3

Grain size analysis (Figure 4): D10 — 27 µm, D50 — 38 µm, D90 — 57 µm

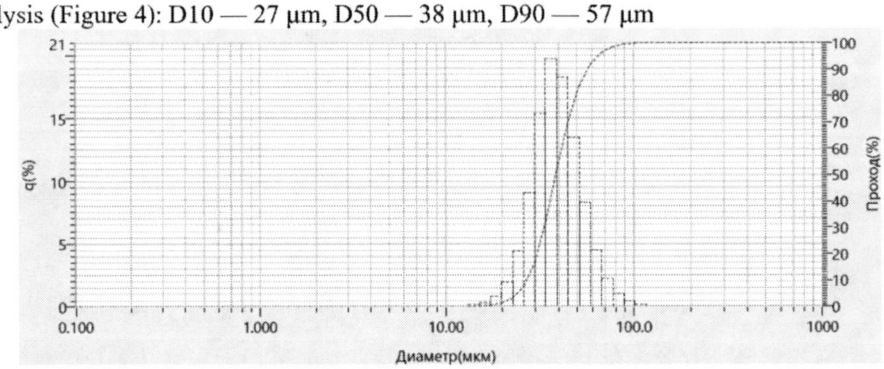

Figure 4. Metal powder grain size analysis

The binder system consists of three components:

a) the main component of the binder;

b) the auxiliary binder;

c) the multifunctional surfactant.

Paraffin oil wax YaV-1 was chosen as the main binder. Low density polyethylene (LDPE) was used as the auxiliary binder to prevent the metal-polymer composite from spreading over the plate during the sintering process. Stearic acid (and its derivatives) was selected as the immediately available and most suitable surfactant additive for the following reasons:

1) hydrophobically compatible with the wax binder;

2) decreases metal-polymer composite viscosity;

3) removed by distillation;

4) removed without carbon residues.

IOP Conf. Series: Materials Science and Engineering **934** (2020) 012048 doi:10.1088/1757-899X/934/1/012048

There are many publications, in which [11, 12, 13] the information on choosing the optimal ratio of metal powder and binder in a metal-polymer composite is provided. All the researchers give approximately the same powder-to-binder ratio at the level of 60% to 40% by volume [14]. In terms of the weight , this ratio gives about 90% or more powder in the metal-polymer composite.
When manufacturing the metal-polymer composite, we mixed the binder and powder components with a laboratory twin-screw mixer (see Figure 5).

Figure 5. Laboratory twin-screw mixer

After the components have been mixed, the finished composite is poured from the mixer onto a special metal rig and then solidifies (Figure 6).

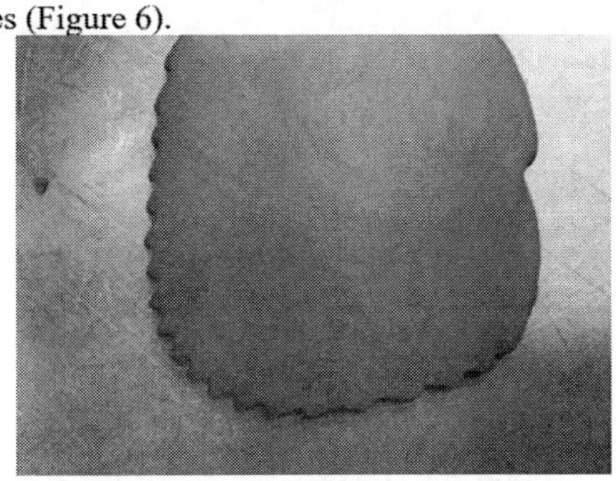

Figure 6. Metal-polymer composite

3. The experiment results. Discussion

Mixing the metal-polymer composite components is an important step, which determines its homogeneity and the quality of the final junction of the plates. Deficiencies in the metal-polymer composite quality cannot be rectified at the subsequent stages of PHE plate junction (debinding and high-temperature sintering). Failure to uniformly distribute the metal powder in the metal-polymer composite inevitably leads to defects in the final product.
The following parameters have an impact on the final properties of the metal-polymer composite: mixing rate and duration, design and geometry of the mixer blades, mixing temperature, particle size distribution of the metal powder, viscosity.
The mixing process was carried out for 30 to 50 minutes at 10 to 30 rpm. At the first stage, we loaded powder and paraffin, and then the remaining fractions of the binder. The maximum percentage of powder loading was 63.5% of the volume (maximum torque). See Table 2.

Table 2. Mixing parameters.

Sample	Powder Infeed (%)	Mixing rate (rpm)	Mixing temp (°C)
1	61%	10	130
2	61%	10	150
3	61%	30	130
4	61%	30	150

The torque of the mixing process at different rates and temperatures changes during the mixing process, and the metal-polymer composite becomes homogeneous at the final stage, while the torque remains the same for each sample. Increasing the mixing rate from 10 to 30 rpm reduces the mixing time required to achieve uniformity at a temperature of 130 °C. When mixing at a temperature of 150 °C, the time needed to achieve homogeneity of the metal-polymer composite reduced even more. In addition, increasing the mixing temperature reduced the time needed to achieve homogeneity. The amount of energy supplied, which was transferred through the mixer blades to the mixing process, increases as the rate grows.

Table 3. Raw material density.

Sample	1	2	3	4
Density, g/cm^3	5.01 ± 0.02	5.05 ± 0.02	5.12 ± 0.02	5.19 ± 0.02

We measured the density of the obtained metal-polymer composite and the raw material homogeneity. Table 3 shows that each sample has a small standard deviation and that the material achieved homogeneity. At a mixing temperature of 150 °C, an increase in mixing rate leads to a higher density of the metal-polymer composite. However, the density remains stable at a mixing temperature of 130 °C. Specimen 4 has the highest density, which is due to the combined increase in the mixing rate and temperature as a result of the viscosity effect of polyethylene, which has a melting point of 125 °C. Polyethylene viscosity decreased at higher temperatures, which contributed to good mixing quality.

Agglomeration of powder particles is a common phenomenon for nanometer-sized powders. For this reason, the share of nanoscale particles should not exceed 1% of the total volume. Agglomeration generally decreases as the mixing rate and temperature increase. It can be concluded that an increase in the mixing rate has a significant effect on the final quality of the metal-polymer composite. However, a high mixing rate leads to penetration of air into the metal-polymer composite. In turn, increasing the mixing temperature reduces the metal-polymer mixture viscosity, which facilitates the process of achieving homogeneity of the metal-polymer composite.

4. Conclusion

The objective of the research was to analyze the effect of mixing time, rate, and temperature on the density and homogeneity of the 316L metal-polymer composite. The experimental results showed that increasing the rate and temperature of mixing the components of the metal-polymer composite, as a rule, reduces the mixing time required to achieve homogeneity of the metal-polymer composite. In addition, an analysis of the metal-polymer composite density showed that the rate of 30 rpm and the temperature of 150 °C ensure the optimal mixing time and maximum density.

References

[1] Alexandrov A A, Arkharov A M and Afanasiev V N 2018. Thermotechnics (Federal state budgetary institution of higher professional education Bauman Moscow State Technical university (National research university of technology) p 876

[2] Gortyshov Yu F, Popov I A, Olympiev V V, Shchelchkov AV and Kaskov S I 2009 Thermohydrau lic efficiency of perspective methods of heat transfer intensification in heat exchange equipment channels (Center for innovative technologies) p 531

[3] Leontiev A I, Gortyshov Yu F, Olympiev V V, Dilevskaya E V, Popov I A, Kaskov S I and Shchelchkov AV 2006 Development of the fundamental principles of prototyping energy-effi cient

heat exchangers with surface heat exchange intensification (In Proceedings of the RNCT-4 conference) pp 253-257

[4] Gortyshov Yu F, Popov I A, Olympiev V V, Shchelchkov AV, Kaskov S I. and Gureev V M 2010 Yesterday, today and tomorrow heat transfer intensification (Proceedings of the fifth Russian na tional conference on heat-exchange) pp 37-40

[5] Baranovsky N V, Kovalenko L M and Yastrebenetsky A R 1973 Plate and spiral heat exchangers. Moscow (Engineering) p 81

[6] Kurganova Yu A and Kolmakov A G 2015 Structural metal matrix composite materials (Federal state budgetary institution of higher professional education Bauman Moscow State Tech nical university (National research university of technology))

[7] Bogonosov K A 2015 High-speed crystallization of copper in low-temperature laser plasma (Physcal institute P. N. Lebedeva of the Russian Academy of Sciences) p 23

[8] Belonosov K A and Frank S N 2013 High-speed crystallization of copper on its own liquid substrate (Brief reports on physics of the Physical Institute P. N. Lebedeva of the Russian Academy of Sciences) pp 18-22

[9] Bogomolov K A, Dmitriev V F, Maximova S N, Stavtsev A Yu (2011) High-speed nanocrystalliza tion in laser plasma (Scientific and technical journal Science and technology in industry) pp. 111-114

[10] Chemical composition 2019 (URL:http://metallicheckiy-portal.ru/marki_metallov/stn/aisi316l)

[11] Amin A M, Ibrahim M H I, Hashim M Y, Marwah O M F, Othman M H, Johar M A and Ng C H 2017 Green density optimization of stainless steel powder via metal injection molding by Taguchi-method (In MATEC Web of Conferences) p 38.

[12] Kong X, Barriere T and Gelin J C 2012. Determination of critical and optimal powder loadingsor 316L fine stainless steel feedstocks for micro-powder injection molding (Journal of Materials Processing Technology) pp 2173-2182

[13] Raza M R, Ahmad F, Muhamad N, Sulong A B, Omar M A, Akhtar M N and Aslam M 2016 Effects of solid loading and cooling rate on the mechanical properties and corrosion behavior of powder injection molded 316 L stainless steel (Powder technology) pp 135-142

[14] Korotchenko A A, Khilkov D E, Tverskoy M Vand Khilkova A A 2020 Use of additive technologiesfor metal injection molding (Engineering Solid Mechanics) pp 143-150

[15] Abashin M I, Barzov A A, Bochkarev S V, Galinovsky A L, Maslov B G 2014 Application of ultrasound diagnostics for assessing the quality of welds Welding production. **9** pp 26-29.

[16] Bochkarev S V, Tsaplin A I, Galinovskii A L, Abashin M I, Barzov A A 2017 Ultra-jet diagnosis of heat treated material microstructure Metal Science and Heat Treatment **59(5-6)** pp 384-388

[17] Abashin M I, Barzov A A, Bochkarev S V, Galinovsky A L, Maslov B G 2015 Using utrastream diagnostics for evaluating the quality of welded joints Welding International **29(9)** pp730-733

[18] Abashin M I, Bochkarev S V, Tsaplin A I, Kobernik N V 2015 Ultra-jet diagnostics of the quality of welds *News of higher educational institutions. Engineering* **12 (669)** pp52-61

[19] Galinovsky A L, Abashin M I, Petrishchev N D, Shtykov A A 2017 Comparative analysis of the meteoroid and laser impact on the ceramic-metal plate. In the collection: Perspective approaches and technologies for the design and production of parts and products of aerospace engineering. Proceedings of the International Youth Scientific and Technical Conference. Under the general editorship of A.L. Galinovsky pp38-41

[20] Barzov A A, Galinovsky A L, Abashin M I 2012 Rapid determination of the surface layer parameters of the material using the ultra-jet method Vestnik of the Rybinsk State Aviation Technological Academy named after P.A. Solovyov **(23)** pp57-59

IOP Conf. Series: Materials Science and Engineering **934** (2020) 012049 doi:10.1088/1757-899X/934/1/012049

Development of equipment for composite 3D printing of structural elements for aerospace applications

A V Azarov[1,2], V A Kolesnikov[1,3] and A R Khaziev[1,2]

[1]«Anisoprint» LLC, 143026, Moscow, The Skolkovo innovation center, Bolshoy, bulvar, 42, b. 1, info@anisoprint.com
[2] Bauman Moscow state technical university, 105005, Moscow, 2-nd Baumanskaya str. 5, b. 1, bauman@bmstu.ru
[3]«Karfidov Lab» LLC, 119049, Moscow, Leninskiy avenue. 2a, office 69, as@karfidovlab.com

E-mail: kolesnikov@anisoprint.com

Abstract. Development of an industrial equipment prototype based on a multi-axial robotic arm KUKA KR3 R540 for composite 3D printing with continuous fiber is presented. Composite Filament Co-extrusion Anisoprint technology is taken as a basis of the equipment. The following advantages of the manufacturing equipment are presented in the paper: ability to produce parts of complex spatial shapes with complex spatial reinforcement; mold fabrication in a single process with a part manufacturing. Complex spatial composite structures manufacturing is demonstrated.

Keywords: structural elements, aerospace applications, bi-matrix composite, industrial equipment prototype, reinforcement with continuous fiber, lattice structures, 3D printing.

1. Introduction

Moving to digital manufacturing based on three-dimensional printing (3D printing) is one of the key focus of modern technologies. [1-5]. Already nowadays there exist parts of aircrafts and rocket and space structures made by 3D printing. The following examples can be mentioned: rocket engine nozzles (Lockheed Martin) [6], a fuel tank (RedEye and Lockheed Martin) [7], parts of F35 fighter engine (Northrop Grumman) [6], etc. Boeing Company uses numerically controlled machines that use the layer-by-layer fabrication method to produce more than 200 parts for 10 different aircrafts [8].

Evolution of Fused Filament Fabrication (FFF) is three-dimensional printing of composite parts reinforced with continuous fibers [9-14]. Specific mechanical characteristics of such materials significantly exceed the characteristics of materials obtained by other methods of 3D printing. Composite 3D printing has all advantages of FFF method. Main of them is capabilities of print head movement that provide flexible control of part reinforcement. This feature is not available for conventional manufacturing of composites. Only three-dimensional printing in space allows maximum utilization of advantages of reinforced composites and obtaining parts with high weight efficiency and high mechanical properties.

The first results in composite 3D printing reinforced with continuous fiber appeared in 2013. Today both in the US and in Europe, many research centers and private companies are engaged in the development

of this branch of additive manufacturing [12-18]. However, current application of composites in industrial 3D printing is mostly limited by plastics filled with chopped fibers. Such materials possess much lower physical and mechanical properties then reinforced composites do [19]. There are additive manufacturing methods based on semi-extractive process with layer-by-layer placement of continuous fiber thermoplastic prepregs combined with machining [20]. Anisoprint suggests own method based on FFF process which is called Composite Filament Co-Extrusion (CFC) where reinforcing composite fiber and thermoplastic filament are simultaneously supplied into hot end [21].
Using thermoplastic polymers as a matrix eliminates the need for subsequent curing of a part after its printing. It enables fabricating parts of complex spatial shape due to solidification of the matrix during the printing process. This approach makes it possible to optimize reinforcement paths in accordance with acting loads. That allows obtaining parts of complex shapes, without resorting to post-processing that causes destruction and damage of reinforcing fibers [22]. The machining essentially reduces the strength and stiffness of composites. Examples of machining detrimental effects are shown in Figure 1 [23].
Despite the fact that 3D printing with continuous fibers is actively developing, currently there are no industrial solutions capable to fabricate composite parts of complex shape and complex spatial internal structure. This work is dedicated to development of multiaxial 3D printing equipment for manufacturing of composite parts reinforced with continuous fibers.

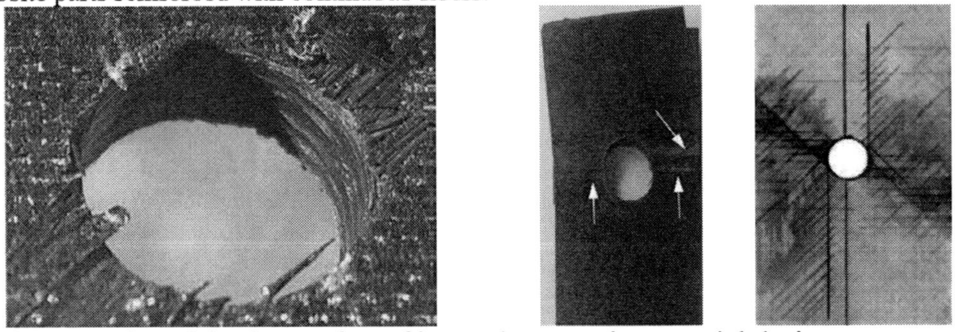

Figure 1. Examples of the destruction of layered composite material during post-processing.

2. Prototype of Equipment for Composite 3D Printing

This article describes the development of the industrial equipment prototype for 3D printing of composite load-carrying structural elements reinforced with continuous fibers. A reinforcing fiber is used in the fabrication process of composite elements. During the printing process this fiber is fed directly into the printing head together with a thermoplastic polymer. The fiber is pre-impregnated with a special polymer composition [24]. Thus, the output is a bi-matrix composite. One matrix provides a reliable impregnation of the reinforcing fiber and its adhesion to the used thermoplastic. The other ensures coupling between such composite fibers.
Within the concept of additive automated digital manufacturing the developed equipment allows to fabricate aerospace and aviation structural elements of complex shapes and internal structure with high specific mechanical characteristics.
Main elements of the equipment prototype for composite 3D printing include: a non-removable load-bearing frame decorated with a protective casing; a multi-coordinate kinematic system based on an industrial robot; a print head manufactured and operated by Anisoprint technology.
The frame was designed in the Siemens NX / NX Nastran computer-aided design and engineering system. An analysis of acting loads to the prototype during the manufacturing process was made. A finite element (FE) analysis of the prototype frame with further optimization of the structure was also conducted.
The load was defined as a concentrated force applied at the cantilever end of the outstretched robotic arm and acting vertically down. The magnitude of the applied load was equal to 800N. It corresponded to the triple weight of the entire industrial robot with printhead payload. Multiplier 3 (three) was applied

as a safety factor, taking into account dynamic loading and possible accident loads. Several iterations of design refinement were performed until the final design was obtained. Goal of the design was minimum weight of the frame with satisfaction of the stiffness restriction. Final weight of the frame was 254 kg and maximum deflection did not exceed 0.1 mm under applied loading. The results of FE analysis for the chosen design are presented in Figure 2.

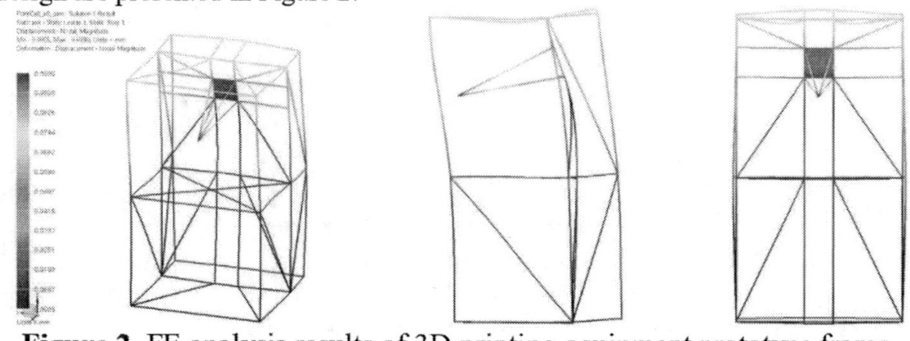

Figure 2. FE analysis results of 3D printing equipment prototype frame.

The industrial robot KUKA KR3 R540 with a specialized software and hardware complex of movement control KR C4 Compact RP DC4 XI1 X55 10 was used as a multi-axial kinematic system. It was dictated by the highest accuracy of positioning and repeatability in its class of equipment, the absence of the hardware redundancy and the possibility of deep industrial integration into manufacturing processes.

The printhead was installed on the interface of the robot. It worked according to the Anisoprint CFC process. The printhead had several features. The first one is a compact size. The second feature is an ability of cooling the print head input channels for stable supply of plastic filaments. The printing head has a special extruder nozzle to ensure the consolidation of the composite.

The printhead had the following characteristics: maximum extruder heating temperature up to 400 °C; supported binder materials: PLA, ABS, PA (nylon), PET-G, PC, PEI.

The 3D model and photo of the print head installed on the KUKA KR3 R540 robot is shown in Figure 3.

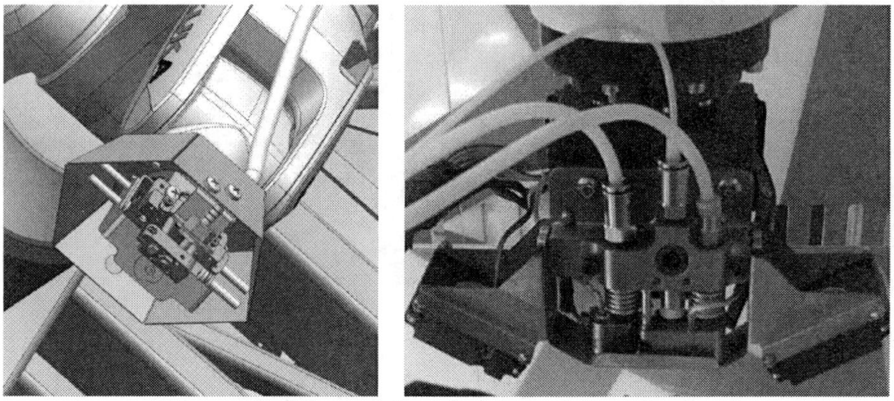

Figure 3. 3D model and photo of the print head

To achieve the maximum printing area of the multi-coordinate moving system the position of the industrial robot was chosen as "On the wall".

A simplified functional block diagram of the equipment and placing of control system for the material supply are shown in Figure 4. All units are located on the support frame.

IOP Conf. Series: Materials Science and Engineering **934** (2020) 012049 doi:10.1088/1757-899X/934/1/012049

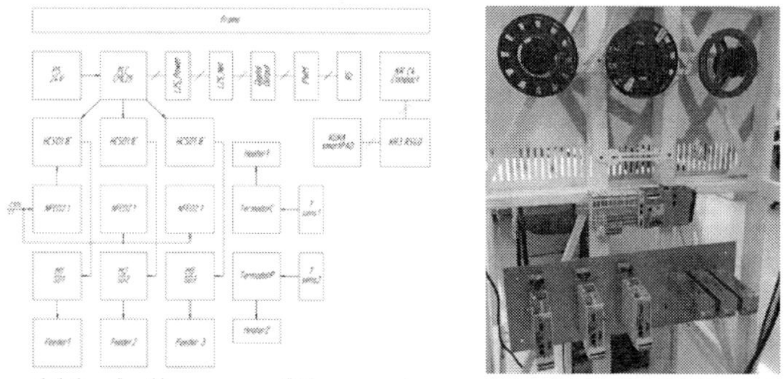

Figure 4. Functional block diagram of the equipment and placing of control system for the material supply on the frame.

The control system for the material supply consisted of the industrial controller Bosch Rexroth CML25 with I/o modules, the module for heating control via RS-485 Protocol and two-channel pulse-width modulation signal generation module. NFE02.1 module was used for line filters. HCS01.1E modules with brand-name MSM019B servo motor connection cables are used as drivers. At the same time, the controller drove the Thermodat modules intended to heat the nozzles of the printhead. Main advantages of the layout are compliance with the requirements of industrial operating standards, scalable hardware platform, standardized communication interfaces, the ability to connect high-grade human-machine interface components and I/O modules, etc.

Figure 5 shows the equipment prototype and CFC printing a branch pipe. The branch pipe was manufactured in one operation cycle.

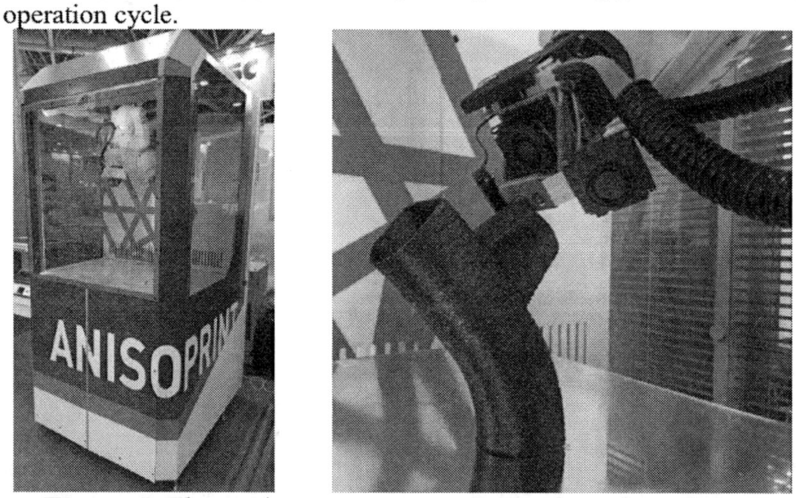

Figure 5. The equipment prototype.Manufacturing of Parts

3. Parts Manufacturing

For numerical control code generation, simulation of the printhead movement and reinforcement paths modelling were conducted in Rhinoceros 3D. Figure 6 shows the developed 3D model and the printing process of the part.

It should be noted that reinforcement of the part was made only within predefined zone. Such reinforcement can be realized either by applying trimming of reinforcing fibers, or by serpentine turns of the reinforcement paths. The second method was chosen to demonstrate capabilities of the technology. In this case, the serpentine turns are high curvature non-geodesic lines. Such layup cannot be carried out by conventional automated processes of composite manufacturing such as winding or automated fiber placement.

4. Discussion

Currently composite lattice structures are widely used in rocket and space industry. Such structures allow to fully utilize advantages of reinforced composites and reduce their basic drawbacks. It ensures their high weight efficiency in comparison with traditional composites and metal structures [25-28].

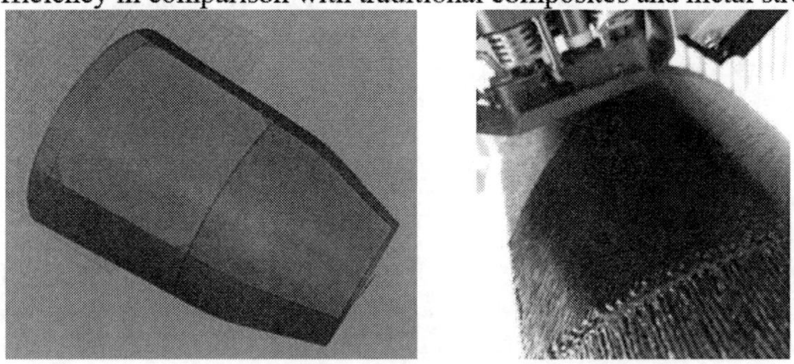

Figure 6. 3D model and Part printing.

Filament winding is mainly applied process for manufacturing of lattice structures. Ribs of the lattices corresponds to geodesic or quasi-geodesic lines. Shapes of the lattice structure as a rule are convex or close to convex. Figure 7 shows typical rocker and space structures. Irregular grids of complex structure and configuration can be evolution of anisogrid structures. Shapes of ribs of such structures are not restricted by geodesic and quasi-geodesic lines. The developed equipment for composite 3D printing allows manufacturing irregular composite lattice structures with complex shapes. It is not required application of expensive molds and tooling. Thus, the suggested technology can be considered as promising for manufacturing of lightweight aerospace structural elements of new generation [29-31].

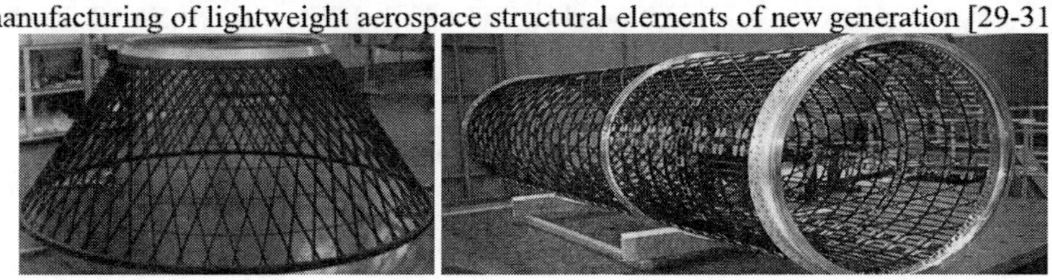

Figure 7. Composite lattice payload adapter and spacecraft frame (CRISM).

5. Conclusions

The developed equipment makes possible manufacturing of composite structures with complex shape and internal structure. It allows effective distribution of acting loads through structural elements. That provides production of structural elements of high weight efficiency.

Anisoprint Composite Filament Co-Extrusion process is basis of the developed technology. It provides flexible control of fiber volume fraction in composite. Application of thermoplastic polymers as a matrix can be considered as one of the technology advantage in terms of increasing impact and crack resistance of manufactured composite parts.

Modern hardware components provide full automation for manufacturing of composite parts.

References

[1] Fo F G A and others 1970 *Fiber Composite Materials* Ed. D M Frantsevich and D M Karpinos (Kiev: Naukova Dumka) p 403

[2] Kobets L P and Gunyaev G M 1974 *Structural plastics* Ed. E B Trostyanskaya (Moscow: Khimiya) p 304

[3] Konkin A A 1974 *Carbon and other heat-resistant fibrous materials* (Moscow: Khimiya) p 304

[4] Sayfulin R S 1983 *Inorganic Composite Materials* (Moscow) p 304

[5] Lubin D ed. 1988 *Composite Materials Reference* translation from English book I, 2. (Moscow: Mashinostroyeniye) p 448
[6] Aizinson I L and others 1988 *The main directions of the development of composite thermoplastic materials* (Moscow: Khimiya) p 48
[7] Kelly A and Rabotnov J ed. by 1983 *Handbook of composites* 1-4 (Amsterdam: North Holland)
[8] Bourell D, Leu M and Rosen D Roadmap for Additive Manufacturing - Identifying the Future of Freeform Processing 2009 *The University of Texas at Austin Laboratory for Freeform Fabrication Advanced Manufacturing Center* **TX**
[9] Guo N and Leu M 2013 Additive Manufacturing: Technology, Applications and Research Needs *Department of Mechanical and Aerospace Engineering, Missouri University of Science and Technology Rolla* **8** 3 215-43
[10] Jayakumar A 2014 Why the aerospace industry is investing in 3-D printing *The Washington Post*
[11] Cohen D, Sargeant M and Somers K 3-D printing takes shape 2014 https://www.mckinsey.com/usinessfunctions/operations/our-insights/3-d-printing-takes-shape access: nov. 2019
[12] Orbital Composites Electronic resource https://www.orbitalcomposites.com/ access: nov. 2019
[13] Matsuzaki R, Ueda M, Namiki M and other 2016 Three-dimensional printing of continuous-fiber composites by in-nozzle impregnation *Scientific Reports*
[14] 3D Printer Using Continuous Carbon Fiber Composite Materials Electronic document http://www.jscm.gr.jp/3Dprinting/images/introduction_CFRP3Dprinter.pdf access: nov. 2019
[15] Composites 2.0 Electronic document https://www.rs.tus.ac.jp/composites2/ access: nov. 2019
[16] Frank van der Klift, Koga Y, Todoroki A, Ueda M and other 2016 3D Printing of Continuous Carbon Fibre Reinforced Thermo-Plastic (CFRTP) Tensile Test Specimens Open Journal of Composite Materials 6 18-27
[17] +LAB Electronic resource http://www.piulab.it/projects/continuos-fiber-composites access: nov. 2019
[18] Invernizzi M, Natale G, Levi M, Turri S and Griffini G 2016 UV-Assisted 3D Printing of Glass and Carbon Fiber-Reinforced Dual-Cure Polymer Composites Materials *Materials* **9(7)** 583
[19] Zelensky E S, Kuperman A M, Gorbatkina Y A, Ivanova-Mumzhieva V G and Berlin A A 2001 Reinforced plastics - modern structural materials *Journal of the Russian Chemical Society D I Mendeleev* XLV **2**
[20] Zhiyo A, Schwil H and Frolov M 2015 Thermoplastic carbon plastics: Special unidirectional materials for aircraft engineering *II international technological forum "Innovations. Technologies. Production"* (Rybinsk)
[21] Demkovich N A, Volkov I A and Yablochnikov E I 2016 The use of numerical modeling systems when introducing new production technologies *Engineering and mechanical engineering* **18** 4(3) 459-63
[22] Chvalun S N, Novokshonova L A, Korobko A P and Brevnov P N 2008 Polymer-silicate nanocomposites: physicochemical aspects of in situ polymerization synthesis *Journal of the Russian Chemical Society D I Mendeleev* **LII** 5 52-7
[23] Wisnom M R The role of delamination in failure of fibre-reinforced composites. *Philos Trans A Math Phys Eng Sci.* 2012 Apr 28;370(1965):1850-70
[24] Adumitroaie A, Antonov F, Khaziev A, Azarov A, Golubev and Vasiliev V 2019 Novel Continuous Fiber Bi-Matrix Composite 3-DPrinting Technology *Materials* **12(18)** 3011
[25] Vasiliev V 2009 *Composite Pressure Vessels: Design, Analysis, and Manufacturing* (Bull Ridge Publishing) p 704
[26] Kondakov I 2016 Development and validation of methods for calculating the strength analysis of mesh composite fuselage structures *Civil Aviation High TECHNOLOGIE* **19** 06 137-46
[27] Azarov A V, Antonov F K, Vasiliev V V, Golubev M V, Krasovskii D S, Razin A F, Salov V A, Stupnikov V V and Khaziev A R Development of a two-matrix composite material fabricated by 3D printing. *Polym Sci* **D** 2017;10(1):87–90
[28] Azarov A, Antonov F, Golubev M, Khaziev A and Ushanov S 2019 Composite 3D printing for

the small size unmanned aerial vehicle structure *Composites* Part B **169** 157–63

[29] Reznik S V, Prosuntsov P V and Azarov A V 2015 Substantiation of the structural-layout scheme of the mirror-space-antenna reflector with a high shape stability and a low density per unit length // J. Eng. Phys. Thermophy. **88(3)** pp. 699-705. DOI:10.1007/s10891-015-1239-x

[30] Reznik S V, Prosuntsov P V and Azarov A V 2015 Modeling of the temperature and stressed-strained states of the reflector of a mirror space antenna // J. Eng. Phys. Thermophy **88(4)** pp. 978-983 DOI:10.1007/s10891-015-1273-8

[31] Reznik S V, Prosuntsov P V, Mikhailovsky K V and Shafikova I R 2016 Material science problems of building space antennas with a transformable reflector 100 m in diameter // IOP Conference Series: Materials Science and Engineering **153(1)** 012012. – 10 p. DOI: 10.1088/1757-899X/153/1/012001

Microstructure and erosion resistance of Mo-Si-B, Mo-Zr-Si-B, and Mo-Hf-Si-B coatings

Ph V Kiryukhantsev-Korneev[1*], A D Sytchenko[1] and N Ulan[1]

[1] «National University of Science and Technology «MISiS», Leninsky prosp., 4, 119049 Moscow, Russia

E-mail:*kiruhancev-korneev@yandex.ru

Annotation. Mo-Si-B, Mo-Zr-Si-B, and Mo-Hf-Si-B coatings were obtained by direct current magnetron sputtering. Microstructure and composition of coatings were studied using scanning electron microscopy, energy-dispersive analysis, and glow discharge optical emission spectroscopy. Erosion resistance were estimated in the liquid under the ultrasonic vibration using abrasive particles of SiO2, WC, Si3N4, Al2O3, and B4C. The results obtained showed that all coatings have a homogeneous and uniform structure with no pronounced columnar elements and contain hexagonal h-MoSi2 and h-MoB2 phases. It was found that the most pronounced erosive effect on the coating is silicon oxide powder, which was selected for the main experiments. The highest mass loss of 3.4 g was fixed for Mo-Zr-Si-B sample, while the minimal value of 2.1 mg was found for hafnium-contained coatings.

Key words: Zr- and Hf-doped Mo-Si-B coatings; magnetron sputtering; multicomponent cathodes; microstructure; erosion resistance

1. Introduction

Molybdenum and Mo-based alloys have high melting points, low coefficient of thermal expansion and high tensile strength at elevated temperatures [1,2] and are considered to be among the most attractive materials for high-temperature applications in the aviation, space and metallurgical industries. However, such alloys have low oxidation resistance at temperatures above 600 °C. This problem can be solved by the application of protective coatings deposited onto the surface of Mo-based alloys. Mo-Si-B coatings are considered to be one of the most effective due to their ability to form a continuous and dense Si:B:O layer on the surface as a result of exposure to high temperatures [3-7]. Various methods can be used for Mo-Si-B coatings deposition such as: spark plasma sintering [4], plasma transferred arc [5], pack cementation [6], magnetron sputtering [7], electro-spark alloying [8], thermal spraying [9], molten salts synthesis [10]. The Mo-Si-B coatings obtained by electro-spark alloying showed oxidation resistance up to a temperature of 700 °C [8]. While the sample obtained by Mo deposition and next pack cementation showed nice resistance up to 1300 °C [10]. The Mo-Si-B coatings produced by magnetron sputtering are able to withstand a temperature of 1700 °C for a short time [7]. The strength characteristics of Mo-Si-B ceramic can be increased by complex alloying [11-17]. The investigation of Mo-Si-B-based coatings is often limited to the determination of structure, mechanical and tribological properties, corrosion/oxidation resistance and thermal stability [18-21]. However, from a practical point of view, it is important to study the erosion resistance of coatings. For example, in the aviation industry, parts such as the blades of a gas turbine engine lose their performance due to erosion caused by exposure to sand, salt and ice crystals or volcanic ash. It is worth noting that the abrasive resistance of Mo-Si-B-based coatings on has not been previously investigated. The aim of present work is to study the erosion resistance of Mo-Si-B, Mo-Zr-Si-B and Mo-Hf-Si-B coatings obtained by magnetron sputtering.

2. Experimental Part

The coatings were obtained by magnetron sputtering in direct current mode. The MoSiB (37.5 at.% Mo, 50.0 at.% Si, 12.5 at.% B), MoZrSiB (26.1 at.% Mo, 44.8 at.% Si, 20.6 at.% B, 8.5 at.% Zr), and MoHfSiB (29.4 at.% Mo, 50.4 at.% Si, 14.9 at.% B, 5.3 at.% Hf) cathodes with a diameter of 120 mm and a thickness of 10 mm were made by hot pressing using milled products of the self-propagation high-temperature synthesis. Sputtering was carried out at a current of 2 A, voltage of 500 V, pressure 0.1-0.2 Pa in Ar (99.9995%) using a vacuum installation based on the UVN-2M pumping system. Mo plates 20×20×1 mm in size were used as a substrate material. Samples were mounted on a 2-axis rotating table with the possibility of applying a bias voltage and heating. The substrates were subjected to ultrasonic cleaning in C3H8O for 5 minutes and etching by Ar+ ions with energy of 2 keV for 40 minutes in a vacuum chamber using a slot ion source.
The structure and elemental composition were determined by scanning electron microscopy (SEM) and energy dispersion spectroscopy (EDS) using Hitachi s-3400N microscope with Noran 7 Thermo prefix. X-ray diffraction (XRD) phase analysis was performed on D8 ADVANCE, Bruker unit. Elemental profiles of the coatings were obtained using the glow discharge optical emission spectrometer PROFILER-2 (Horiba Jobin Yvon). The study of erosion resistance of coatings was carried out on UZDN-2T ultrasonic machine (Fig. 1a), which consisted of the ultrasonic wave generator, two vibrators with different lengths, working chamber with protective doors, dashboard, and cooling system for the power supply. Detail technical parameters of machine is given in [22].

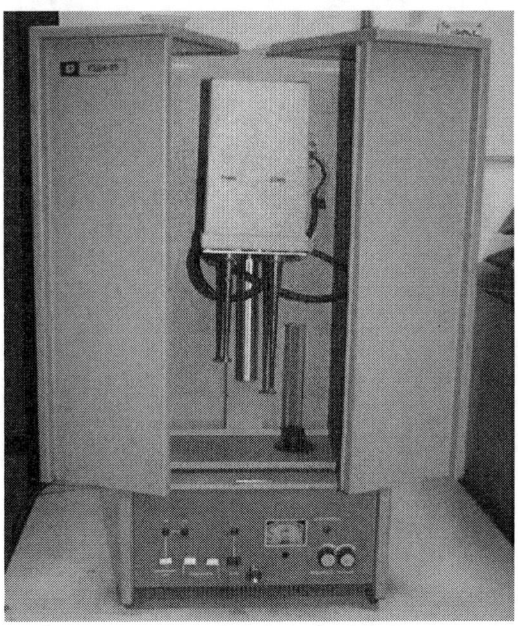

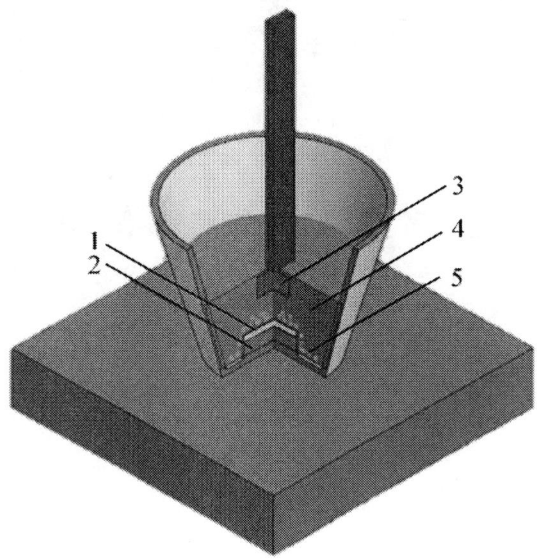

a b

Fig. 1. Photo of UZDN-2T machine (a) and scheme of erosion test (b): 1 – coating, 2 – substrate, 3 – vibrator, 4 – liquid, 5 – abrasive particles

In the working chamber there is a cell, which is a glass cylinder with a diameter close to the diameter of the measured samples (Fig. 1b). The sample was placed in a cell, then the working liquid (distilled H2O) and abrasive material (SiO2, WC, Si3N4, Al2O3 or B4C) with fixed mass were added. The sand (SiO2) with fractions of <0.46, 0.56-0.63, and >1.00 when the particle size was 290-500 microns, 570-

IOP Conf. Series: Materials Science and Engineering **934** (2020) 012050 doi:10.1088/1757-899X/934/1/012050

1070 microns and 1000-1500 microns, respectively, WC with a particle size of 5-15 microns, Si3N4 (5-25 microns), Al2O3 (7-14 microns) and B4C (7-21 microns) were used as abrasive. Mass of abrasive was changed in range 1-20 g. Water volume was 25 ml. At the next step, the vibrator was fixed at the distance of 10 mm over the surface of the sample and tested at a given frequency of 22 kHz with step expositions of 5 and 10 minutes. Samples were weighed on AND GR-202 analytical scales with a measurement accuracy of 0.0001 mg before and after the erosion tests

3. Results and discussion

Mo-Si-B, Mo-Zr-Si-B, and Mo-Hf-Si-B coatings obtained by the magnetron sputtering method had a dense and homogeneous structure with no pronounced columnar elements. Typical data typical data on the example of Mo-Zr-Si-B coatings are presented on Fig. 2. According to SEM micrographs of the fracture and GDOES elemental depth profiles the thickness of the coatings was in the range of 10-12 μm. The Mo-Zr-Si-B coating contained a homogeneous layer on the surface of which there was an increased content of Si (68 at. %) and reduced concentration of the other elements of the coating. At a depth of 0.45 microns and throughout the thickness the concentration of elements does not change. Concentration of elements can be recorded as, at. %: 42.5 Si, 31.5 B, 21.4 Mo, and 4.6 Zr. The carbon and oxygen concentration throughout the coating was <1.5 at. %. According to XRD, all coatings contained the hexagonal phases of MoSi2 and MoB2.

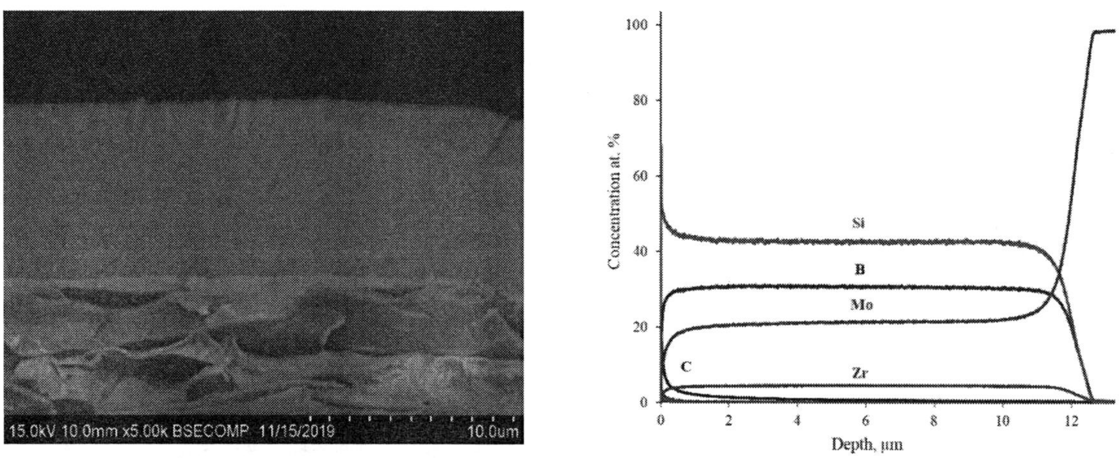

Fig. 2. Cross-section SEM image (a) and elemental GDOES profile (b) of Mo-Zr-Si-B coating

Mo-Zr-Si-B coating was studied in detail for erosion wear in contact with various time expositions, different mass and composition of abrasive particles. The results of the studies are shown in figure 3. The sample with Mo-Zr-Si-B in contact with 1 and 20 g of SiO2 (0.56-0.63) showed similar values, for the 25 min test the difference in mass loss was 0.1 mg. However, 5 g of silicon oxide led to a noticeable change in the mass of the coating (Δm=1.7 mg). The same wear is observed when using SiO2 with fraction smaller than 0.46. The use of powder with fraction >1.0 resulted in a 10% increase in sample mass loss. The wear rate of the coating by treatment of particles of SiO2, WC, Si3N4, Al2O3 was linear, throughout the test erosion of samples increased. When using particles of abrasives WC, Si3N4 and Al2O3 were obtained similar results, Δm was 0.4, 0.5 and 0.3, respectively. In the case of B4C, the mass of the Mo-Zr-Si-B coating reduced during first 15 minutes of testing, after it increased perhaps due to powder sticking to the sample. The coating showed good resistance to abrasive particles like Al2O3, WC, and Si3N4. Greater and more visible wear was observed when sand SiO2 with different weights and fractions were used. Therefore, the uncoated substrate (Mo), Mo-Si-B, Mo-Zr-Si-B, and Mo-Hf-Si-B coatings were further tested for erosion wear with SiO2 (0.56-0.63) powder with mass of 5 g. Results are shown on Fig. 4.

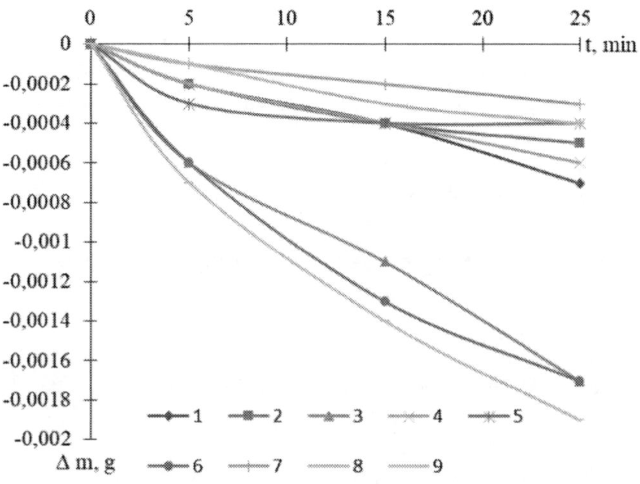

Fig. 3. Mass change of the Mo-Zr-Si-B coating as a result of erosion wear by abrasive particles: 1 – SiO_2 (0.56-0.63) 20g, 2 – Si_3N_4 20g, 3– SiO_2 (0.56-0.63) 5g, 4 – SiO_2 (0.56-0.63) 1g, 5 – B_4C 5g, 6 – SiO_2 (0.46) 5g, 7 – Al_2O_3 5g, 8 – WC 5g, 9 – SiO_2 (1.00) 5g.

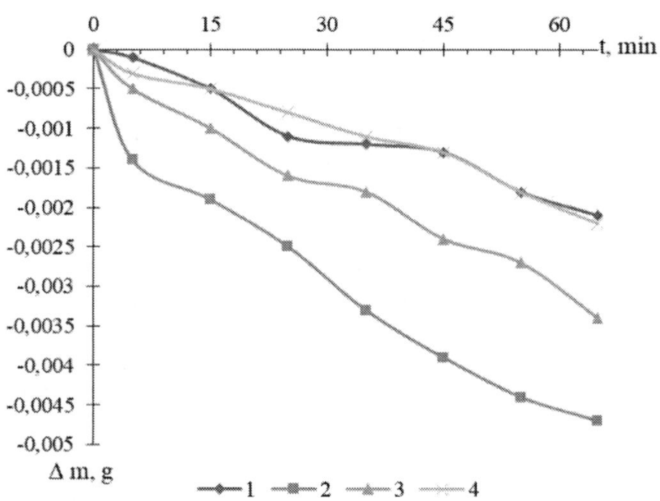

Fig. 4 Mass change of the coatings as a result of erosion wear by abrasive particles of SiO_2; 1 – Mo substrate, 2 – Mo-Si-B coating, 3 – Mo-Zr-Si-B coating, 4 – Mo-Hf-Si-B coating.

The Mo-Si-B coating exhibited the greatest mass loss (Δm=4.7 mg), the wear rate of the sample was 0.07 mg/min. The introduction of Zr into the base coating reduced the rate of erosion wear by at least 25%, which may be due to the denser structure of the Mo-Zr-Si-B coating. The Mo-Hf-Si-B coating had the best erosion resistance and demonstrated Δm=2.1 mg. Note that uncoated Mo substrate possess smaller mass reduction Δm=2.2 mg than that for samples with Mo-Si-B and Mo-Zr-Si-B coatings. This phenomenon may be explained by intensive plastic deformation of metal substrate during interaction with the abrasive.

4. Conclusion

Mo-Si-B, Mo-Zr-Si-B, and Mo-Hf-Si-B coatings were obtained by DC magnetron sputtering of multicomponent ceramics targets. The samples were tested for erosion resistance using SiO2 powders of different fractions, WC, Si3N4, Al2O3, and B4C. Coatings had a homogeneous and uniform structure with no pronounced columnar elements. The Mo-Zr-Si-B sample was resistant to erosion by WC, Si3N4, Al2O3 and B4C particles with a mass loss of 0.3-0.5 mg. When SiO2 was used as an abrasive, the mass loss increased to 1.9 mg during the 25 min test. The using of SiO2 sand particles allowed to obtain quite visual results and to identify differences between materials at short time exposures. The erosion resistance of coating estimated by 65 min test increased in direction Mo-Si-B → Mo-Zr-Si-B → Mo-Hf-Si-B.

Acknowledgements

The work was supported by the Russian Science Foundation (Agreement No. 19-19-00117).

References

[1] Liu G, Zhang G J, Jiang F, Ding X D, Sun Y J, Sun J and Ma E 2013 *Nat. Mater.* **12** 344–350
[2] Cheng P, Zhang Z, Zhang G, Zhang J, Wu K, Liu G, Fu W and Sun J *Mater. Sci. Eng. A* 2017 **707** 295–305
[3] Thom A J, Summers E, Akinc M *Intermetallics* 2002 **10** 555–570
[4] Wen S H, Zhou C G and Sha J B *Surf. Coat. Techn.* 2018 **352** 320–329
[5] Deng X, Zhang G, Wang T, Ren S, Shi Y, Bai Z and Cao Q *Ceram. Int.* 2019 **45** 415–423
[6] Perepezko J H *Int. J. Refract. Met. H* 2018 **71** 246–254
[7] Kiryukhantsev-Korneev Ph V, Iatsyuk I V, Shvindina N V, Levashov E A and Shtansky D V 2017 *Corros. Sci.* **123** 319–327
[8] Kudryashov A E, Lebedev D N, Potanin A Yu and Levashov E A 2018 *Surf. Coat. Technol.* **335** 104–117
[9] Totemeier T C, Wright R N and Swank W D 2018 *Intermetallics* ч **12** 1335–44
[10] Kuznetsov S A, Rebrov E V, Mies M J M, de Croon M H J M and Schouten J C 2018 *Surf. Coat. Technol.* ч **201** 971–978
[11] Sakidja R, Perepezko J H 2007 *J. Nucl. Mater.* **366** 407–416
[12] Jéhanno P, Heilmaier M, Saage H, Böning M, Kestler H, Freudenberger and Drawin S 2007 *Mat. Sci. Eng. A* **463** 216–223
[13] Mitra R, Srivastava A K, Prasad N E and Kumari S 2006 *Intermetallics* **14** 1461–71
[14] Schneibel J H, Ritchie R O, Kruzic J J and Tortorelli P F 2005 *Metall.Mater. trans. A* **36a** 525–531
[15] Li Z K, Yu J L, Zheng X, Zhang J J, Liu H, Bai R and Wang H, Wang D H and Wang W S 2011 *Powder Technol* **214** 54–56
[16] Yu J L, Li Z K, Zheng X, Zhang J J, Liu H, Bai R and Wang H 2012 *Mat. Sci. Eng. A* **53** 392–395
[17] Majumdar S, Kumar A, Schliephake D, Christ H-J, Jiang X and Heilmaier M 2013 *Mat. Sci. Eng. A* **573** 257–263
[18] Kiryukhantsev-Korneev Ph V and Potanin A Yu 2018 *Russ. J. of Non-Ferr. Met.* **59** 698–708
[19] Kiryukhantsev-Korneev Ph V, Bondarev A V, Shtansky D V and Levashov E A 2018 *Prot. Met. Phys. Chem.*+ ч **51** 794–802
[20] Ritt P, Sakidja R, Perepezko J H 2012 *Surf. Coat. Technol.* **206** 4166–72
[21] Riedl H, Vieweg A, Limbeck A, Kalaš J, Arndt M, Polcik P, Euchner H, Bartosik M and Mayrhofer P H 2015 *Surf. Coat. Technol.* **280** 282–290
[22] Kiryukhantsev-Korneev Ph V, Phiri J, Gladkov V I, Ratnikov S N, Yakovlev M G and Levashov E A 2019 *Prot. Met. Phys. Chem. Surf.* **55** 913–923

IOP Conf. Series: Materials Science and Engineering **934** (2020) 012051 doi:10.1088/1757-899X/934/1/012051

Methods of teaching computer science in the system pedagogical knowledge

F K Nadyrova[1], A B Zhanys[2] and A M Mubarakov[1]

[1]Eurasian National University named L.N. Gumilev, Kazakhstan, Nur-Sultan

[2]Kazakh Agrotechnical University named after Saken Seifyllin, Department of Information Systems, Republic of Kazakhstan, Nur-Sultan city

E-mail: aray.zhanys@gmail.com

Abstract: Modernization of education began to wear a permanent, permanent character in the modern world, since the dynamics of changes in the needs of individuals, society and the state for education is steadily growing. For quite a long time, the main subject of the modernization of school education was its content. However, in recent years, it has become increasingly clear that the center of gravity of the renewal and development of education should be shifted towards methods, innovative forms and means of education. The system-activity approach as a methodology for constructing state standards of general education closely links educational results with certain types of schoolchildren's learning activities. It is clear that the use of new types of educational activities will force to look for new innovative forms of their implementation.The subject of the methodology of teaching informatics and a place in the vocational training system for an informatics teacherComputer science as a science and a subject in school.The connection of teaching methods of computer science with pedagogy, psychology and computer science.Methodical system of teaching computer science in a secondary school. General characteristics of its main components (goals, content of training, methods, forms and means of training).

1. Introduction

The subject of informatics teaching methods has a special place in the system of vocational training for an informatics teacher as a whole. Computer science as a science and a subject in school is bearing fruit at this time. The connection of the methods of teaching computer science with pedagogy, psychology and computer science is this part of the methodological system of teaching computer science in a secondary school. General characteristics of its main components (goals, content of training, methods, forms and means of training).

The introduction to the school of the general subject "Fundamentals of Informatics and Computer Engineering" began the formation of a new field of pedagogical science - the methodology of teaching computer science, the object of which is computer science training. A computer science teaching methodology course appeared in the country's universities in 1985. In 1986, the publication of the methodological journal "Computer Science and Education" began. According to the classification of scientific specialties, this section of pedagogy, which studies the laws of teaching computer science at the present stage of its development in accordance with the goals set by society, has received a new name - "Theory and Methods of Teaching and Education (computer science; by educational level)". In accordance with the State educational standard of higher vocational education, specialty 030100 "Informatics" (2000), the MPI course became known as "Theory and Methodology of Teaching Informatics". However, the established name "Methodology of teaching computer science" continues to be used in the names of training courses, textbooks, regulatory documents. An important role in the development of computer science teaching methods was played by didactic studies of the goals and contents of general cybernetic education, accumulated by the domestic school before the introduction of the subject of computer science, the practical experience of teaching students the elements of cybernetics, algorithmization and programming, elements of logic, computational and discrete mathematics, etc. In accordance with the general goals of teaching, the methodology of teaching computer science sets itself the following main tasks: to determine the specific goals of studying computer science, as well as the content of the corresponding general educational subject, and its place in the curriculum of high school; to develop and offer the school and the teacher-practitioner the most rational methods and organizational forms of training aimed at achieving the goals; consider the totality of computer science teaching aids (textbooks, software, hardware, etc.) and develop recommendations for their use in the practice of the teacher. A number of publications rightly noted that over a very long period the content of the methodological training of a future computer science teacher was the weakest part (and the most poorly provided part) of his professional training.

The content of the subject of MPI defines its two main sections: the general methodology, which examines the general theoretical foundations of the methodology of teaching computer science, the combination of basic software and hardware, and the private (specific) methodology - methods for studying specific topics of the school course in computer science on propaedeutic, basic and core stages of training. The methodology of teaching computer science is a young science, but it is not being formed from scratch. Being an independent scientific discipline, in the process of formation it absorbed the knowledge of other sciences, and in its development it relies on the results they obtained. These sciences are philosophy, pedagogy, psychology, age-related physiology, computer science, as well as generalized practical experience in the methods of other secondary schools. As N.V. notes Sofronova, "the teaching of computer science at the modern level is based on information from various fields of scientific knowledge: biology (self-governing biological systems such as humans, other living organisms), history and social studies (social systems), the Russian language (grammar, syntax, semantics and etc.), logic (thinking, formal operations, truth, falsehood), mathematics (numbers, variables, functions, sets, signs, actions), psychology (perception, thinking, communication). " In the context of global informatization of all branches of human activity and the penetration of informatics into all other sciences, we can safely say that the teaching methods of informatics are connected with almost any science. This connection was especially strengthened in connection with the transition of the system of general secondary education to specialized education: no doubt, elective computer science courses will be in demand in all profiles and school disciplines.

Goals and objectives of teaching computer science at school

The main goals of teaching computer science at school are formulated in normative documents. However, it should be noted that both the subject and content of the computer science course, as well as its objectives, are still widely discussed and debated. The first draft of the state educational standard in computer science (1997) notes three aspects of the general educational significance of the course and, accordingly, three areas in teaching computer science:

- "worldview aspect" related to the formation of ideas about a system-information approach to the analysis of the world, the role of information in management, the specifics of self-governing systems, the general laws of information processes in systems of various nature;
- "Algorithmic (programmer's) aspect", which is now associated more with the development of schoolchildren's thinking;
- "user" aspect associated with the formation of computer literacy, the preparation of students for practical activities in the context of widespread use of information technology. "

Pedagogical functions of the computer science course

The pedagogical functions of the computer science course, like any educational field, as well as the school subject reflecting it, are determined by the contribution of the educational field to solving the main tasks of general education:

• formation of the foundations of a modern scientific worldview;
• development of thinking;
• preparing schoolchildren for practical activities, work and continuing education.

2. Practical part

In recent years, robotics has been introduced into the computer science course, which has become of great interest to current technical developed children.

Robotics is an applied science engaged in the development of automated technical systems. Robotics relies on disciplines such as electronics, mechanics, and programming.

Robotics is one of the most important scientific and technological achievements in which there are problems with artificial intelligence. At the present stage, problems of robotics are found in the school. Lego robots are built into the learning process.

Fig. 1. Robotic device

In modern society, robots are being introduced into our lives; many processes are being replaced by robots. Fields of application of robots are different: medicine, construction, geodesy, meteorology, etc. There are a lot of processes in life, a person can't even imagine without robotic devices (mobile robots): a robot for all kinds of children's and adult toys, a robot - a nurse, a robot - a nanny, a robot - a housekeeper, etc. Specialists with knowledge in this area are in great demand. And the question of introducing robotics into the educational process since elementary school is relevant. If a child is interested in this area from a very young age, he can discover so many interesting things. Therefore, the introduction of robotics in the educational process and extracurricular hours are becoming increasingly important and relevant. The main equipment used in teaching children robotics in schools is LEGO Mindstorm constructors. In our area, methodological recommendations have been developed for embedding robotics in the educational process. LEGO Mindstorms is a constructor (a set of mating parts and electronic components) for creating a programmable robot. First introduced by LEGO in 1998.

All school kits based on the LEGO® constructor of the FirstRobot NXT are designed for students to work mainly in groups. Therefore, students simultaneously acquire cooperation skills, and the ability to cope with individual tasks that are part of the overall task. In the design process, to ensure that the created models work, and meet the tasks that are set for him. Students get the opportunity to learn from their own experience, to be creative in solving the problem. Students master the tasks of varying difficulty in stages. A key step-by-step principle for LEGO® education provides students with the opportunity to work at their own pace.

The constructor of PervoRobot NXT allows the teacher to improve, take new ideas that allow you to attract and retain the attention of students, organize learning activities using various subjects and conduct integrated classes. Additional elements contained in each set of designers allow students to create models of their own invention, to design robots that are used in life.

These constructors show students the relationship between different areas of knowledge. At computer science lessons, solve problems in physics, mathematics, etc. Models of the NXT FirstRobot Constructor give an idea of the work of mechanical structures, of force, motion and speed, to perform mathematical calculations. These sets help to study sections of computer science - this is modeling and programming.

The purpose of using Lego - design in the system of additional education is to master the skills of initial technical design, the development of fine motor skills, the study of the concepts of design and basic properties (rigidity, strength, stability), the skill of interaction in a group. At the disposal of children are designers equipped with a microprocessor and sensor sets. With their help, a student can program a robot - a smart machine to perform certain functions.

Fig. 2: Models of the NXT FirstRobot Constructor give an idea of the work of mechanical structures

The main task of modern education is to create an environment that makes it easier for a child to discover his own potential. This will allow him to act freely, knowing this environment, and through it the surrounding world. The new role of the teacher is to organize and equip the appropriate educational environment and encourage the child to cognition and activity.
The new FSES require the development of the basics of design and development, and robotics programs fully satisfy these requirements. The educational environment of LEGO combines the LEGO sets specially arranged for classes in the group, a carefully thought out system of tasks for children and a clearly formulated educational concept.
What is LEGO design? LEGO-construction is one of the most famous and widespread pedagogical systems nowadays, which widely uses three-dimensional models of the real world and a subject-game environment for teaching and developing a child. Lego translated from Danish means “smart game”. The LEGO designer encourages to work, equally, both the head and hands of the student. The designer helps children to realize their ideas, build and fantasize, enthusiastically working and seeing the end result. It is LEGO that allows you to study while playing and study in the game.
In this, I see the relevance of introducing the “Fundamentals of Robotics” course at school.
The main goal of the course is to educate a creative, technically competent, harmoniously developed personality with logical thinking, able to analyze and solve problems associated with programming and algorithmization.
Studying the “Fundamentals of Robotics” creates the prerequisites for socializing the personality of students and provides the opportunity for continuous technical education, and mastering computer technologies using lego kits and other robot designers is the way for schoolchildren to modern promising professions and a successful life in the information society. Of course, occupations with work technology will not lead to the fact that all children will want to become programmers and robot builders, engineers, researchers. First of all, classes are designed for the general scientific preparation of schoolchildren, the development of their thinking, logic, mathematical abilities, research skills. The robot does not give ratings and does not give homework, but it makes you work mentally and constantly.
In the last academic year, the school received equipment for organizing classes in robotics in the framework of the republican project "Profile engineering classes." This course helps us solve the following educational problems:
• Development of creative abilities of children.
• Formation of communication skills.
• Formation of an active “I concept”.

Simplicity in building a model, combined with the great constructive capabilities of LEGO, allow children to see a do-it-yourself model at the end of the lesson that fulfills their own task.
The course program can be conditionally divided into two large parts:
• Design
• Programming
Being engaged in designing, children learn simple mechanisms, learn to work with their hands, they develop elementary design thinking, fantasy, and study the working principles of many mechanisms.

Children are tireless designers, their creative capabilities and technical solutions are witty, original. Schoolchildren learn to construct "step by step". Such training allows them to move forward at their own pace, stimulates the desire to learn and solve new, more complex problems. Any recognized and appreciated success leads to the fact that the child becomes more self-confident.

During the lessons, the communicative activity of each child increases, the ability to work in pairs, in a group is formed, and creative abilities develop.

At the programming stage, students move to a higher level: the gaming component begins to give way to a serious thoughtful study of the LEGO environment, which requires thoughtfulness and patience.

Lego is always a new discovery, a new idea! A new impetus to the development of innovative thinking ...

Robotics is fun! Thanks to robotics, my students became active, observant, quick-witted. The world does not stand still, it is always developing, and who knows, maybe it is my students who will create a nanotechnology device or a new robot of the 21st century.

I hope that my students after mastering the skills of robotics will quickly move on to solving complex technical problems and become glorious successors of engineering professions.

3. Conclusion:

Given the above problems, at the moment the robotics program at school is still not available everywhere. However, even without the use of special equipment, designers, and real robots in school programs in computer science and ICT, it is worth starting a study of the introduction to robotics. This will make it possible to familiarize students with the subject, and will also help in further steps in this field of knowledge. In this case, it is enough to conduct only two classes, after which the children will be able to independently engage in robotics.

The basics of robotics for children in elementary school will allow students to understand what a robot is and how it works. Also, children will be interested to know that the concept of "robot" was invented by science fiction writer Karel Chapek in the distant 1920. These are the basics of robotics, allowing you to plunge into a world full of amazing inventions and high technologies that instantly excite a huge interest in this science in children.

In addition, the basics of robotics will help children who choose to study robots in further education.

Technologies do not stand still, they are constantly evolving, and it is entirely possible that it is your child or student who will construct a nanorobot that can treat complex diseases. The program of robotics at school is a huge step towards the technologies of the future, towards the development and perfection of technologies.

References

[1] E Tanenbaum 2002 Computer networks (St. Petersburg: Peter)

[2] Background information on computer networks and telecommunications www.index.com

[3] K Zaker 2002 Computer networks. Upgrading and troubleshooting. (St. Petersburg: "BHV-Petersburg")

[4] M V Kuzmina, S I Melekhina, A A Pivovarov, Yu A Skurikhina, N I Chuprakov 2016 Educational robotics *Educational-methodical allowance for educators on educational development Robotics in the conditions of implementation of the requirements of the Federal state educational standards* (Kirov: LLC Printing house "Old Vyatka") p210

[5] M V Kuzmina 2016 All-Russian Festival of Scientific and Technical Creativity "Technopark of the Young" - "School of Multimedia Technologies" // Media. Information. Communication. [Electronic resource]: International electronic scientific and educational journal **17** http://mic.org.ru/sem-konf/560-shkola-multimedijnykh-tekhnologij-vovserossijskom-detskom-tsentre-smena-12-25-maya-2016

[6] M V Kuzmina 2016 MediaRoboVyatka Initiative - experience and prospects scientific and experimental activities on the integration of media education and Robotics // Media. Information. Communication. [Electronic resource]: *International electronic scientific and educational journal* **18**

[7] M V Kuzmina 2017 Multimedia Journalism and "JuniorSkills" at WorldSkills Championship - competency, media creation, professional test // *Media. Information. Communication.* [Electronic resource]: International electronic scientific and educational journal **20**

[8] M V Kuzmina, T V Masharova 2016 Meta-subject matter in education and development media education *Education in the Kirov region Scientific and methodical journal* **3 (39)** (Kirov: KOGOAU DPO IRO Kirovskaya area) p. 36-41.

[9] A B Zhanys, A F Tursinbaeva Internet technology in electronic commerce. International Conference Information Technologies in Business and Industry 2018 IOP Publishing

IOP Conf. Series: JournalofPhysics: Conf. Series 1015 (2018) 032154 doi :10.1088/1742-6596/1015/3/032154

Developing of the autoclave-free composite manufacturing technology

A D Novikov, A O Lubyanskiy and P I Ignatov

Engineering center "Composites of Russia", Bauman Moscow State Technical University, Russian Federation, Moscow, Lefortovskaya emb., 1.

Abstract. Nowadays there are numerous ways of making parts of polymer composite materials. The main controlled properties of a part depending on a used technology are porosity and binder percentage, which influence on quality and strength of a part, and also manufacturing cycle time and costs of the used equipment. The main aim of the research was to increase the efficiency of the PCM moulding comparing to the resin infusion method. For reaching the aim there was developed the moulding technology with using elastic diaphragm inflation, as well as there were conducted series of tests by using the dynamo-mechanic analyzer. The way of making the mould and pattern of reinforcing plies layout were offered. For technology approbation there were made a serie of carbon fiber reinforced plastic sport helmets, and got results of time properties of manufacturing. It was discovered that the number of parts made for the time unit by using the inflation method twice exceeds the number of ones made by vacuum infusion method. Specific properties of constructions made by the offered technology also exceed properties of parts made by traditional methods. Inflation technology was successfully put in production and proved its efficiency, reliability and profitability.

Key words: polymer composite materials, composite technologies, autoclave-free moulding, inflating technology, composite materials manufacturing, sport helmet, manufacturing optimization

1. Introduction

Composite materials are widely used in different fields of technics due to their unique properties. At the present moment there exist a lot of ways of making parts of these materials [1-5]. In composite industry there is a trinity of material, technology and construction [6], which means that every composite part requires the unique combination of producing method and the plies layout. Many factors influence on the final quality of the part: the technology of moulding, personnel skills, raw materials and their pre-preparing, environment factors (e.g. temperature, humidity, etc.), facility's equipment [7-9]. Thus while creating a new manufacturing technology negative factors are to be significantly reduced or eliminated. Cycle time of producing the part and the price of the used equipment are also important [10]. Nowadays RTM, autoclave moulding, Compression RTM and experimental technologies such as VARTM with incomplete vacuum are the technologies providing the most optimal ratio of the matrix and filler as well as pores percentage [11-17].

2. Elastic diaphragm inflation technology

As the basis of the method there was taken the process of making composite helmets with thermoplastic binder (figure 1).

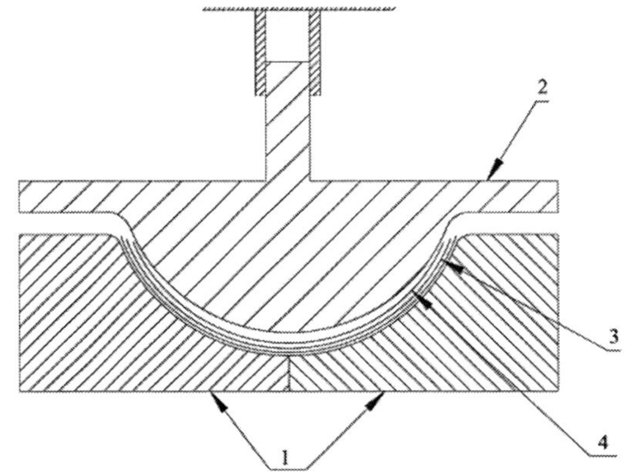

Figure 1. Schematic of motorcycle helmet making:
1 – separating mould, 2 – punch on a hydrocylinder, 3 – layer of the thermoplastic binder, 4 – filler layer

While manufacturing the helmets the separating aluminium mould, divided on the face of symmetry is used. Filler (4) and thermoplastic layers (3) are put in the assembled mould. After that the punch with the inner shape of the helmet (2) is pressed by using the hydraulic cylinder. After connecting the punch and the mould thermoplastic binder is softened by heating. Then, after compressing of all the layers and cooling the part down, the punch removes from the mould and the part ejects.
Due to the relatively high complicity of the technology realization, there was offered the following technology, whose schematic is shown on the figure 2.

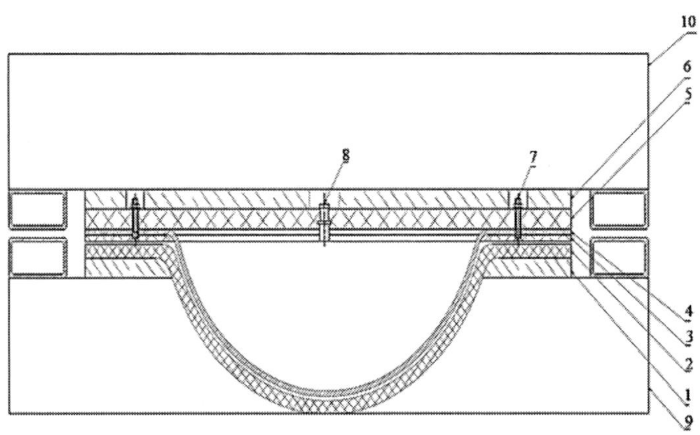

Figure 2. Schematic of the assembled mould: 1 – 15 mm plywood, 2 – fiberglass mould, 3 – lower cover, 4 – silicon bag, 5 – upper plate, 6 – plywood support, 7 – M5 bolts, 8 – air inlet, 9 – lower frame half, 10 – upper frame half

Before being moulded the stack of dry layers lays out on the mould (2) just the same way as for vacuum infusion. For binder spreading and stack compression the silicon elastic diaphragm (4), inflated by the excessing pressure from the air pipeline, is used. Hermetic area is created by the upper plate (5) and silicon diaphragm (4), pressed to each other and held by the lower plate (3). Plates and silicone membrane are pulled together by 16 bolts (7). To avoid displacements of the hermetic area relatively to the mould there are used two halves of the frame: upper (10) and lower (9), which are pulled together by 4 M16x90 bolts. To fit both of the frames parts tight to the mould and the upper cover, 15 mm plywood plates (1), (6) are adhered.

3. Part moulding process

Part making starts from preparing the preform:

Mould surface is covered with Loctite 770NC release agent. Then, filler plies (carbon biaxial tape BT70-20, hybrid carbon-aramid cloth) are laid on the preform mould in reverse order, due to that the first layer would become inner after locating the preform in the mould (table 1).

Table 1. Order of preform's plies layout

Ply no.	Filler material
1	Carbon biaxial tape
2	Carbon biaxial tape
3	Carbo-aramid cloth

During first ply laying out on the preform's mould, the minimal amount of adhesive spray should be used. Moreover, it's better to spray only on the flanges so not have fabric surface glued to the main mould surface, otherwise there may be a risk of strong preform deformation.

Elantas Elantech 152 (80g of resin and 24g of hardener per each helmet) was chosen as the binder. Needed amount of the compound pours in the mould and the helmet's preform locates into the right position (3).

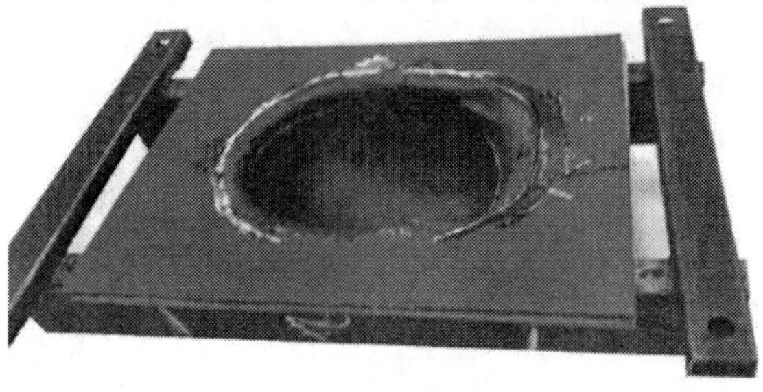

Figure 3. Mould with the poured binder

Figure 4. Assembled mould

For silicon bag's protecting from damaging interacting with epoxy resin, the silicon lubricant is used. When a preform is located in the mould, it's covers with the silicon bag and then these two parts are pulled together by the frame. After both parts of the frame are strongly screwed, thermo-resistant pipe connects to air inlet and pressure applies. Fully assembled mould is shown on the figure 4.

After binder hardening the inlet closes, air releases, the frame unscrews and the silicon bag takes off the mould. To avoid temperature deformations, the helmet cools down to the room temperature, and then ejects. Further, the helmet is cut on the CNC milling machine.

4. Measurements of parts' elastic modules

To compare elastic modules of parts there were cut 5 specimens 5x20 mm off the zones with maximal radius of curvature. For tests the were used one helmet made by each of technology. Cutting out zones' locations are shown on the figure 5.

IOP Conf. Series: Materials Science and Engineering **934** (2020) 012052 doi:10.1088/1757-899X/934/1/012052

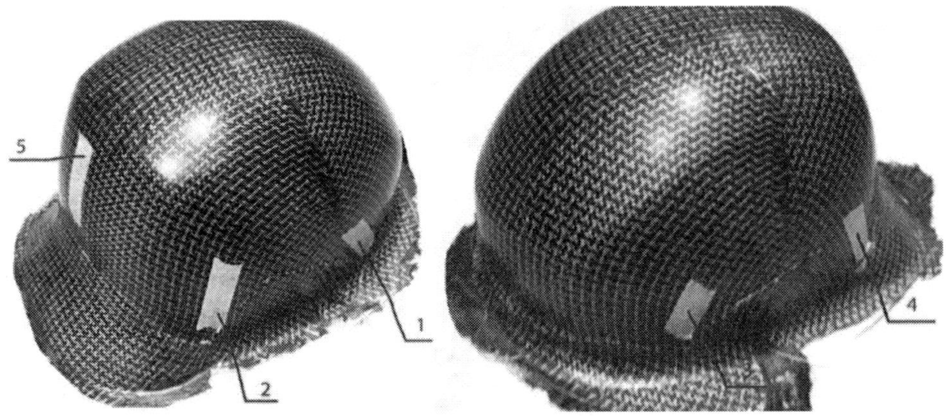

Figure 5. Specimens' cut-out zones on the helmet

Tests were conducted on dynamo-mechanical analyzer (DMA) NETZSCH DMA 242 E [17]. Three-point bending with sinusoidal specimen loading was used as the deformation mode [19],[20]. Results of tests are provided in table 2.

Table 2. Specimen's elastic modules' values

	Silicon bag inflation	Vacuum infusion
Cutting out zone no.	*E*, GPa	*E*, GPa
1	11,22	10,83
2	12,24	11,95
3	11,27	10,57
4	12,16	11,96
5	13,77	12,13

Figure 6 shows the values of specimens' made by vacuum infusion and elastic diaphragm technologies elastic modules.

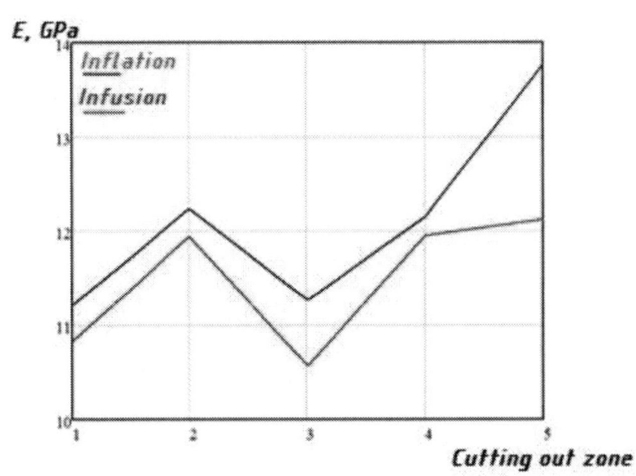

Figure 6. Dependence of parts' elastic modulus on cutting out zone. Blue line – elastic diaphragm inflation, red line – vacuum infusion

IOP Conf. Series: Materials Science and Engineering **934** (2020) 012052 doi:10.1088/1757-899X/934/1/012052

For mass comparison there were chosen 15 helmets from each batch made by different methods. Weighing was conducted on the laboratory scales OHAUS SJX3201/E [21], which value of division is 0,1 g and accuracy is ±0,05 g.
Computing the mean values of helmets' mass, equal 148,52 g for vacuum infusion and 138,32 g for membrane inflation, it can be concluded that using the new technology the mass decreased for 6,8%.

Conclusion
1. The new technological process for making composite parachute helmets was developed with efficiency higher than vacuum infusion.
2. The new technological mould for the process was developed.
3. The technology was worked out for helmet manufacturing. During work there were found and eliminated the main defects appeared on parts
4. Trial parts were created. The mechanical properties of specimens and their mass were compared. Based on the received data it was concluded that elastic modules of parts made by using the new technology is 9% higher, filler percentage increased for 4,4% due to binder amount decreasing.

References

[1] V A Lukasik, V P Medvedeev 2005 *Manufacturing technology and properties evaluation of thermoset and thermoplastic plastics* pp 59-61
[2] A V Postnov , M V Postnova , V I Postnov 2018 *new technologies in the manufactures from metal-polymeric materials* News of Samara sci. center of Russian Academy of Sciences. **#4**
[3] K S Gabriels, O A Karaeva, A M Kudrin, E V Kulakova 2015 *Physical and mechanical properties of fiberglass produced with polyester resin by vacuum infusion* Voronezh State Technical University news. **#1.**
[4] Lubin G 1998 *Handbook of composites* Van Nostrand Reinhold Company p 219
[5] Yu A Kurganova, Yijin Chen and S P Shcherbakov Evaluation of the effectiveness of the introduction of a discrete filler in the aluminum matrix melt //Published under licence by IOP Publishing LtdIOP Conference Series: Materials Science and Engineering, Volume 683, Advances in Composite Science and Technology 5–8 December 2018, Bauman Moscow State Technical University
[6] Whelan A 1994 *Polymer Technology Dictionary* (Amsterdam, Ne: Springer Netherlands), p. 523
[7] I M Bulanov, V V Vorobey 1998 *Technologies of aerospace composite constructions* (Moscow, Bauman Moscow State Technical University) c 121-124
[8] N Popham 2019 *Resin infusion for the manufacture of large composite structures* (Dorset, UK: Sunseeker Poole Limited) p 213
[9] M I Dushin, A V Khrulkov, R Yu Karavayev 2015 *Parameters that influence the formation of porosity in the products made of polymer composite materials (PCM) produced by out-of-autoclave methods* All-Russian Scientific Research Institute of Aviation Materials
[10] V E Basin 1981 *Adhesion strength* (Moscow: Chemistry) p 59-66
[11] A N Babin 2015 *New generation of binders for autoclave-free technologies* (Moscow: Works of All-Russian Scientific Research Institute of Aviation Materials) – **#8 (44)** p
[12] M Poorzeinolabedin, L Parnas and S H Dashtan 2014 Resin infusion under flexible tooling process and structural design optimization of the complex composite part *Materials and Design* **#64**
[13] N P Timoshkov, A V Khrulkov, L N Yazvenko 2018 *Composite materials for non-autoclave technology technologies* (Moscow: Works of All-Russian Scientific Research Institute of Aviation Materials) **#3 (63)**
[14] Justin A and Advani S G 2007 Simulation and experimental validation of flow flooding chamber method of resin delivery in liquid composite molding *Composites Pt. A: Applied Science and Manufacturing* **№38**
[15] V N Krysin, M V Krysin 1989 *Technological processes of molding, winding and gluing of constructions* (Moscow: Machinery) p 35-37
[16] I V Leschukova 2018 *Principal technologies of aviation composite constructions producing* Innovational Science **#1**

IOP Conf. Series: Materials Science and Engineering **934** (2020) 012052 doi:10.1088/1757-899X/934/1/012052

[17] F L Matthews R D Rawlings 1999 *Composite materials: Engineering and* Woodhead Publishing Ltd, Abington Cambridge p 211-214
[18] Analyzer DMA 242 E Artemis [Electronic resource] / NETZSCH. – link: https://www.netzsch-thermal-analysis.com/ru/produkty-reshenija/dinamicheskii-mekhanicheskii-termicheskii-analiz/dma-242-e-artemis/
[19] E V Chernikova, A A Efimova, V V Spiridonov 2013 *Physic-chemical and physic-mechanical properties of polymers* (Moscow: Machinery), p 43-44
[20] V V Vasiliev 1988 *Composite constructions mechanics* (Moscow: Machinery) p 197-200
[21] Laboratory scalesOHAUS SJX3201/E [Electronic resource] / OHAUS Corporation. – Link: http://www.ohaus-cis.ru/products/portable-scales/scout-sjx/

Optimization Model for Dairy Farms of Northern Kazakhstan

A H Sukoot, R Z Suleimenova and A B Zhanys

Department of Information Systems S.Seifullin Kazakh Agro-technical University

Abstract: Dairy cattle breeding is one of the most important branches of animal husbandry. It serves as a source of such valuable food products as milk, meat, as well as a source of raw materials for industry. Milk is an almost indispensable basis for nutrition in childhood, both humans and animals. It contains all the necessary nutrients. By its diverse composition, not one of the known foods can compete with it. In milk, there are almost all currently known vitamins.

1. Introduction

The features that characterize dairy cattle breeding are: the ubiquity of milk and dairy products for their uninterrupted supply to the population, the need for organic combination of dairy cattle with other sectors of agriculture, the considerable labor intensity and a large share of the products of this industry in the entire volume of agricultural production in most regions of the country . Dairy farming has a great influence on the economy of all agriculture, therefore milk production is of great economic importance. Suffice it to say that this is one of the few industries that generate daily income. The share of dairy products in terms of price in total livestock products is more than 35%.

The average milk yield per cow in 2017 was 4,024 kg of milk. As a result of the implementation of the priority national project for the development of agriculture, the foundation has been laid for the creation of a new base for dairy cattle breeding. Over the years, many new facilities for 168.6 thousand cows were put into operation; more than 1,150 dairy complexes and farms were modernized and reconstructed using the most advanced technologies.

We must recognize that the majority of subjects in the past and this year, faced with the problematic issues of organizational and economic problems that hinder further growth of milk production.

Among the main reasons for the decrease in the growth rate of milk production and the reduction of its volumes are the following:

Firstly, in a number of regions a low growth rate of milk production of cows was noted, and in 14 territories its unreasonable decrease.

According to the average annual milk yield per cow, the main indicator in assessing the state of dairy cattle breeding, Kazakhstan lags noticeably behind countries with developed livestock breeding. In 32 territories, the average milk yield per cow is less than 3,500 kg.

With such indicators in modern conditions, conducting intensive dairy cattle breeding is simply impossible.

The increase in the average milk yield per cow in the country over the past three years by 8-10% per year and the achieved productivity of cows are clearly insufficient to ensure an increase in milk production.

Taking into account the presence of the number of cows and the prevailing trend of its annual decline, the achievement of the planned production volumes is possible only due to higher rates of increase in the milk productivity of cows, provided that the breeding stock is stabilized.

There are opportunities for this. A sufficient number of dairy and combined breeds, which are characterized by high productive qualities, are bred in the country.

The intensification of dairy cattle breeding requires addressing such fundamental issues as the creation of an appropriate feed base, changing the structure of feed in the direction of increasing feed protein.

2. Organizational and economic characteristics of Olzha-Sadchikovskoe LLP

Olzha-Sadchikovskoye LLP is located in the Kostanay region. According to the climatic and soil conditions, the farm is located in the northern zone. The average annual air temperature is 1.8 - 2.3, the annual rainfall is 378 - 441 mm, the moisture supply of spring wheat is 70 - 75%, the number of dry days is 32, the probability of droughts: medium and weak - 100%, very intense - 25%. There are no lands damaged by erosion in the farm. Heat resources are sufficient for the ripening of grain and vegetable crops.

Let us consider in more detail the production sizes of Olzha-Sadchikovskoye LLP for 3 years based on the data in table 1.

Table 1 Analysis of the size of the enterprise

Indicator	2015 y.	2016 y.	2017 y.	Growth rate, %
Area of agricultural land, ha	5170	5170	5170	100,0
Including arable land	3947	3947	3947	100,0
Hayfields	223	223	223	100,0
Pastures	1000	1000	1000	100,0
The average annual cost of fixed assets of production, thousand. tenges.	44349	45918	49282	111,12
The amount of energy, horsepower	12045	11190	11186	92,87
The annual average number of employees, people.	245	230	215	87,76
The cost of gross output (at cost price), thousand.	40147	42122	48267	120,22

During the study period, the area of agricultural land remains unchanged. The average annual number of employees decreased by 30 people, the number of energy resources - by 7.13%. Because of the above changes, the value of fixed assets increased by 11.12%, and the value of gross output increased by 20.22% or by 8120 thousand tenges.

The main factor determining the volume of gross output of this farm is the number of cattle. Growing cattle is a process that requires a lot of time and money.

One of the key factors of production of any type of goods and services is labour resources. In modern conditions, labour resources are considered as the most valuable capital the company has.

The value of labour resources is of great importance. So, sufficient security of the enterprise with the necessary labour resources, their rational use, high level of labour productivity is of great importance for increasing products volume and boosting production efficiency. In particular, the volume and timeliness of all work, the efficiency of using equipment, machines, mechanisms and, as a result, the volume of production, its cost, profit and a number of other economic indicators depend on the security of the enterprise with labour resources and the efficiency of their use.

The indicator of the effectiveness of the use of labour resources is its performance. Improving labour productivity is of great economic and social importance. [3]

Let us see the composition and structure of employees of LLP Olzha-Sadchikovskoe using the data in Table 2.

Table 2 The composition and structure of employees of the organization

Name	2015 y.	2016 y.	2017 y.	Deviation (+, -)
Person	Specific Share, %	Person	Specific Share, %	Person
Workers engaged in agricultural production - total, including	109	44,49	110	47,83
Permanent workers:	82	33,47	79	34,35
tractor drivers - machinists	34	13,88	31	13,48
machine milking operators	32	13,06	32	13,91

Cattle farm workers	16	6,53	16	6,96
Workers	27	11,02	31	13,48
Executives	10	4,08	11	4,78
Specialists	17	6,94	20	8,70
Workers employed in utility industries	51	20,82	46	20,00
Workers employed in other types of activities	85	34,69	74	32,17
By organization - total	245	100	230	100

The table shows that the largest share in the structure of employees of the enterprise is employees employed in the main production. In 2017, their share was 80%. For the analyzed period, the total number of employees in the organization decreased by 30 people.

Based on the theoretical basis of economic efficiency, it can be concluded that the key to increasing it is reducing the cost of production.

The cost of production, as an economic category, combines all the expenses of the farm in monetary form, the reimbursement of which is necessary for carrying out the process of simple production. It shows what it costs to each agricultural enterprise production and marketing of products.

Distinguish between the cost of production, which includes all the expenses associated with receiving and transporting products to the place of storage, and the total cost, which is calculated taking into account the expenses of marketing products. The cost of production consists of several types of expenses that have different production purposes. The ratio of their total costs represents the cost structure. It depends on the specialization of the industry, technical equipment and organization of production. [8]

Let us analyze the change in the cost of milk and other indicators of economic efficiency of dairy cattle using Table 3.

The table shows that for the analyzing period, the average annual livestock decreased by 5%, gross milk production also decreased - by 9.89%. The total cost of milk production and the cost of milk increased by 48%. Milk sales decreased by 31.1%. During the analyzing period, there was a significant decrease in profits from the sale of milk - by 96.37% and, as a result of a decrease in profits, the level of profitability of milk production significantly decreased.

Table 3 Economic efficiency of milk production

Indicators	2015 y.	2016 y.	2017 y.
The average annual livestock, head	620	612	589
Gross milk production, quintal	20526	19170	18495
Average annual milk production per cow, quintal	33,11	31,32	31,40
Total costs for milk production, thousand. tenges.	13899	14809	20700
Milk net cost - total, thousand. tenges.	12372	13113	18347
The cost of 1 quintal of milk, tenges.	603	684	992
Produced milk, quintal	20526	18017	14142
The total cost of milk, thousand tenges	11220	11939	14644
Proceeds from the sale of milk, thousand tenges	13922	13891	14742
The average selling price of 1 quintal of milk, tenges	678,26	770,99	1042,43
Profit from the sale of milk, thousand tenges	2802	1952	98
Profitability level,%	24,08	16,35	0,66

In order to find out what expenses affect the cost, it is necessary to consider the composition and cost structure of 1 quintal of milk. Table 4 presents data on the composition and cost structure of 1 quintal of milk.

Table 4 shows that the largest share in the cost structure of 1 quintal of milk is taken up by material expenses, which include the cost of feed, electricity, petroleum products, etc. At the end of the analyzed period, their share in the total cost amounted to 58.71%.

Next, we consider table 5, which will give us data on the factors affecting the cost of milk.

Table 4 Composition and cost structure of 1 quintal of milk

Expenditures	2015 y.	2016 y.	2017 y.	Deviation (+,-)
Amount, tenge	Specific Share, %	Amount, tenge	Specific Share, %	Amount, tenge
Labor costs, including allocations for social needs	112	16,54	161	20,83
Material costs	377	55,69	151	19,53
Maintenance of main equipment	188	27,77	461	59,64
Total expenses	677	100	773	100

Table 5 Factors affecting the cost of milk

Indicators	2015 y.	2016 y.	2017 y.
Production expenses per cow, thousand tenge.	22,42	24,20	35,14
Milk yield from one cow, quintal	33,11	31,32	31,40
Labor input, man-hours/ 1 quintal of milk	5,26	4,90	5,79
Labor input, man-hours / 1 cow	174,19	153,59	181,66

Considering the data in the table, we can conclude that for the period under review, production costs per cow increased by 56.74%, milk yield per cow decreased by 5.16%, labour costs by 1 quintal of milk and labour costs by 1 cow relatively increased by 10, 08% and 4.29%.

It can be concluded that the cost of production is increasing every year. This suggests that the efficiency of milk production in LLP Olzha-Sadchikovskoye is decreasing.

Along with the strengthening of the feed base, agricultural enterprises can recommend the introduction of a flow-shop technology for milk production, in which animals are serviced according to their physiological state in groups of cows, heifers and milk production.

Before the introduction of the flow-shop system, it is necessary to determine how many livestock places should be in each workshop. An approximate calculation can be made by the following formula:

$$C = (D * Q) / 365$$

Where C (Count) is the number of livestock places; D - the duration of stay of animals in the workshop in days; 365 - the number of days; Q (quantity) - the number of animals on the farm [11].

Let us give an example of calculating the required number of cattle units in workshops, for a complex with a population of 100 heads (Table 6).

Table 6 Calculation of cattle units in workshops for 100 heads of cattle

Name of workshop	required units, %	Cattle, heads
Dry cows	14	70
Calving	11	55
Insemination	25	125
Milk production	50	250

Based on these calculations, we can determine that in which barns it is better to place groups of animals, which barns should we reconstruct. Schedule of animals in a particular workshop is compiled by the zoo-technical service, according to the plan of mating and admission of offspring.

For effective use of continuous-shop technology, a number of conditions are needed: a solid feed base, the availability of feed in an assortment for different physiological groups of cows, uniform calving of cows on a farm, a method of tagging animals, convenient for reading by livestock breeders. The advantages of the flow-shop system for milk production are presented in table 7.

Table 7 The advantages of the flow-shop system for milk production [11]

Technical operations	Indicators	Actual level	With flow-shop system
Dry cow maintenance	Fatness	Average (450 kg)	Higher (over 550 kg)
	Safety of Pregnancy	82%	100%
	Condition of the mammary gland	Possible diseases	Absence of disease
Cow service in the calving workshop	The normal birth process time	The possibility of disease, death of newborn calves	Getting a healthy calf
Service of cows in the workshop of milking and insemination	Milking of cows	Increasing daily milk yield by 10-12%	Increasing daily milk yield by 25-30%
	The quality of the milk	Probability of deviation from the required level	In accordance with the required level
Insemination of cows	Pregnancy of cows	Not fully controlled	Matches technological parameters

This technology makes it possible to reduce the rawness of cows, to increase the yield and safety of calves, to increase the efficiency of feed use, since they are spent in accordance with the physiological state of cows and their and productivity. In result, increases product quality and decreases the cost.

3. The main directions of increasing the production and sale of milk

The main ways to increase the economic efficiency of production can be the introduction of the latest technology in milking and feeding, this technology saves manual labor and, accordingly, wages, feed and energy. During the work of milking rooms with new generation, only one operator is needed to serve the entire herd of cattle. The following measures are recommended to increase milk production:
1. Specialization and concentration of production.
2. Creating a strong feed base that can meet the nutritional needs of livestock for the full realization of the genetic potential of productive qualities.
3. Mastering a set of measures to improve the technology of keeping and feeding livestock.
4. Rationalization of rearing stock rearing systems, providing an increase in milk production with a significant improvement in the payment of feed products.

Measures should be taken to fundamentally improve natural pastures, attention should be paid to the creation of long-term cultivated, seeded hayfields and pastures, and autumn-winter pastures should be created and used at the expense of sorghum, cereal grain mixtures and other crops of the green conveyor. The creation of grass stands with a high content of legumes is the main direction of intensification of fodder production. Lucerne-grass and clover-grass stands with yield, productivity and protein content are 8-12 times higher than natural ones.

The development of more advanced and modern methods of ensiling and haying reduces the losses during harvesting and during storage and obtains high-quality feed. The possibility of making haylage and silage using preservatives from various raw materials is widely used. This creates a lesser dependence when creating a solid feed base for cattle breeding on weather conditions and provides for obtaining and using high-grade feed throughout the year, which is especially important when shifting the production of dairy cattle products to an industrial basis.

Formulations for premix and feed additives should be improved. At the same time, it is necessary to use such a compounding formula in order to replace the grain part of the feed mixtures with cheaper and effective ingredients - beet pulp, molasses, crumbs, grass meal, etc.

Development of resource-saving technologies based on the biological features of various age and sex groups of livestock. The use of rational, low-cost methods dramatically increases the economy of milk production.

Intensification of dairy cattle breeding is based on increased culling of cows and a more rapid replacement of the main herd of young, highly productive animals. This is accompanied by the creation of such conditions so that each cow must give a viable calf for a year, which is intensively grown and subsequently goes to repair the herd.

The most important factor in the intensification of dairy farming is the use of highly productive, intense cattle breeds.

Thus, the use of all factors of intensification of milk production in cattle breeding allows us to raise the industry to a higher level of development, to make it highly profitable. [11]

4. Conclusions and recommendation

Based on the study of the economic status and efficiency of milk production in Olzha-Sadchikovskoye LLP, the following conclusions can be made:

1. For the study period, the area of agricultural land remains unchanged. The average annual number of employees decreased by 30 people, the number of energy resources - by 7.13%. As a result of the above changes, the value of fixed assets increased by 11.12%, and the value of gross output increased by 20.22% or 8120 thousand tenges.
2. In 2017, the share of marketable crop production increased compared to 2015 by 7.02% or by 3339 thousand tenge. As for livestock products, the share in 2017 decreased by 7.02% or by 2261 thousand tenge compared with 2015. Changes in crop production occurred due to an increase in grain production by 7.54% or by 3443 thousand tenge. In total, the number of sales increased by 2,032.7 thousand tenge.
3. During the analyzed period, the average annual livestock decreased by 5%, gross milk production also decreased - by 9.89%. The total cost of milk production and the cost of milk increased by 48%. Milk sales decreased by 31.1%. During the analyzed period, there was a significant decrease in profits from the sale of milk - by 96.37% and, as a result of a decrease in profits, the level of profitability of milk production significantly decreased.
4. The largest share in the cost structure of 1 quintal of milk is material costs, which include the cost of feed, electricity, petroleum products, etc. At the end of the analyzed period, their share in the total cost amounted to 58.71%.
5. During the period under review the production costs per cow increased by 56.74%, milk yield per cow decreased by 5.16%, labour costs for 1 quintal of milk and 1 cow increased by 10.08% and 4.29% accordingly. The cost of production is increasing every year. It shows that the efficiency of milk production in LLP Olzha-Sadchikovskoye is decreasing.
6. The main ways to increase the economic efficiency of production can be the introduction of the latest technology in milking and feeding, this technology allows you to save manual labour, and therefore wages, feed and energy. During the operation of the milking rooms of the new generation, only one operator is needed to serve the entire herd of cattle.
7. One of the measures to improve the economic efficiency of milk production is the creation of a solid feed base capable of meeting the nutritional needs of livestock in order to fully realize the genetic potential of productive qualities. For the successful development of animal husbandry it is necessary to create a strong forage base. Equally important in this is the process of harvesting feed.

References

[1] Afanasyev V N 2003 Agriculture statistics / V.N. Afanasyev // *Finance and Statistics*. M. 272 p.
[2] Bakanov M I 1997 Theory of economic analysis: Textbook. / M.I. Bakanov, Sheremet A.D. // *Finance and Statistics* .- 4th edition –M 354 p.
[3] Vodyanikov V T 2008 Organization and production management at agricultural enterprises / V.T. Vodyannikov, A.I. Lysyuk, N.Ye.Zimin, and others // - M .: *Publishing house "KolossS"*, Publishing House of the State Agrarian University "AGRUS" 506s.
[4] Volkova N A 2007 *Economics of agriculture and processing enterprises* / Volkova N A, Stolyarova O A, Kosterin E M.; edited by. Volkovoy N A // - Moscow: KolosS, 426s.
[5] Dobrynin V A 1990 *Economics of Agriculture: Textbook.* / Dobrynin V A// - Moscow: Kolos 476s.
[6] Dontsova L V and Nikiforova N A 2004 Analysis of financial statements: textbook. - 2nd ed. -

Moscow: *Business and Service Publishing House* - 336 p.
[7] Kovalenko N Y 2008 *Economics of agriculture: a textbook for universities* / Kovalenko N Ya // - Moscow: YURKNIGA
[8] Kovanov S I and Svobodin V A 2000 *Economic performance of agricultural enterprises* - M .; Agropromizdat,158s.
[9] Minakov I A 2000 *Economics of agriculture* / Sabetova L A, Kulikov N I and others. Edited by Minakov A A - M .: Kolos.
[10] Savitskaya G V 2001 *Analysis of the economic activity of the enterprise: study guide* / G.V. Savitskaya. - 6th ed., Pererab. and add. - Minsk: New Knowledge 704 p.
[11] Internet resources: lakta.ru/pages/19.html
[12] Smirnov E R 2015 The World Dairy Market in 2011–2013 // *Dairy industry*. –2013. 2 25-27.
[13] Main indicators of agriculture in Russia in 2014: statistical bulletin. -M .: IIC *"Statistics of Russia"*.
[14] Agriculture, hunting and hunting, forestry in Russia. 2013: Statsb / Rosstat -M., 2013. –462 p.
[15] The socio-economic situation of the Oryol region in 2014: a brief stat. Sat Territorial body of the Federal State Statistics Service for the Oryol region. Eagle. 2015. 79 p.
[16] Grudkin A A and Babenkova S N 2014 Directions for improving the mechanisms of state support for dairy cattle breeding // *Economics of agricultural and processing enterprises*. –. –No 4. – C. 33-37.
[17] The state program for the development of agriculture and regulation of the markets of agricultural products, raw materials and food for 2013-2020. –Ministry of Agriculture of the Russian Federation, 2011.
[18] Labinov V V 2013: Russian dairy market. *State support measures* // Dairy industry. –2014. **3** 5–10.
[19] Grudkina T I et al. *2012 Strategic management of competitiveness of the agrarian sector of the regional economy: monograph* / Grudkina T. I., Agoshkova N. E., Agoshkova N. N. and others-Orel: Orel Pub SAU,396 s.
[20] Stavtsev A N and Polshakova N V 2014 *The state and prospects of development of dairy cattle breeding of the Oryol region* // Agrarian Russia. **2** . –C. 25-27
[21] Veklenko E V, Prusov N S, Kabluchkov E Yu and Tkacheva N I 2013 Rationale for improving the organization of the fodder base for dairy cattle // *Bulletin of the Kursk State Agricultural Academy*. **2**. –C. 37-40.
[22] Morozova N I, Musaev F A, Podol S R and Ulkina M A 2012 *Flow-shop technology of milk production* // Dairy industry. **8**. C.8-9
[23] Agoshkova N Ye 2014 *Organization and planning of production activities of the main and auxiliary-service units in agricultural enterprises: recommendations for production*. Orel: Publishing house OrelGAU. 105 sec.
[24] Agoshkova N E 2013 *Improving the management of the reproduction of fixed assets in the regional agro-industrial complex* /Agoshkova N E and Agoshkova N N// Regional Economics: Theory and Practice.**12**. –C. 31-39

Possibilities of Obtaining Diffused Layers upon Saturation of Multilayer Materials with Nitrogen and Carbon.

K B Polikevich[1], A I Plokhikh[1] and L P Fomina[1]

[1]Bauman Moscow State Technical University named after N.E. Bauman (National Research University); (Bauman MSTU, 2nd Baumanskaya Str., bldg. 1, 105005, Moscow, Russian Federation

Abstract. The paper analyzes the processes of saturation of multilayer metal materials with interstitial elements: nitrogen and carbon. The formation of diffusion layers is presented. It is found that the thickness of the laminar layer and the degree of structuring of the sample affect the penetration depth of the diffusant.

1. Introduction

Currently, it is relevant to create promising materials that could ensure a reduction in the weight and overall dimensions of parts and structures with an increase in their operational characteristics. These materials include a new class of anisotropic structural metal materials with a gradient macro- or microstructure, consisting of hundreds and thousands of layers separated by large angular boundaries.

The technological process for obtaining such materials includes the following steps: preparation of cards from sheets of the necessary materials (A and B), their assembly into the initial package, compaction, vacuumization and further plastic deformation by hot pack rolling at a temperature of 1000°C. The cycle is repeated a predetermined number of times until the required layer thicknesses and workpiece sizes are obtained [1-6].

The theory and practice of diffusion saturation of metals and alloys show that imperfections in the crystal structure of the material have a significant effect on the penetration of alloying elements [7]. It is also known that the magnitude of grain boundary diffusion significantly exceeds the values of the coefficient of volume diffusion, therefore, it may be quite promising to perform surface saturation processes as applied to materials with a layered structure, the structure of which could increase the penetration depth of alloying elements and reduce the duration of the saturation process [8 - 11].

2. Materials and Research Methods

A multilayer composition, consisting of rotary alternating stainless steels of ferritic (AISI403) and austenitic (AISI304) classes, was chosen as the objects of study. After the first technological cycle, the number of layers in the multilayer workpiece was 100 (50 of each grade) with a laminar layer thickness of 100 μm and a sheet thickness of 10 mm. After the second cycle, the number of layers in the material increased to 2000, with a single layer thickness of 1 μm.

Surface carbon saturation was carried out in a gaseous medium at a temperature of 950oC for 5 hours. Nitrogen saturation was also carried out in a gaseous medium at a temperature of 540oC with a degree of ammonia dissociation from 20 to 40% for 45 hours.

The samples cut along the direction of the diffusion flux and perpendicular to the operating surface were studied using an OLYMPUS optical microscope, as well as using a VEGATS5130 scanning electron microscope in the secondary electron mode at an accelerating voltage of 5 to 20 kV.

3. Results and Discussion

The study of the microstructure after carbon saturation shows a difference in the formation of diffusion layers in austenitic and ferritic steels. A characteristic feature, detected by metallographic etching, is the boundary of diffusion penetration of carbon, which is fixed in the layers of AISI403 steel at a distance of about 650 μm from the initial surface (Figure 1).

IOP Conf. Series: Materials Science and Engineering **934** (2020) 012054 doi:10.1088/1757-899X/934/1/012054

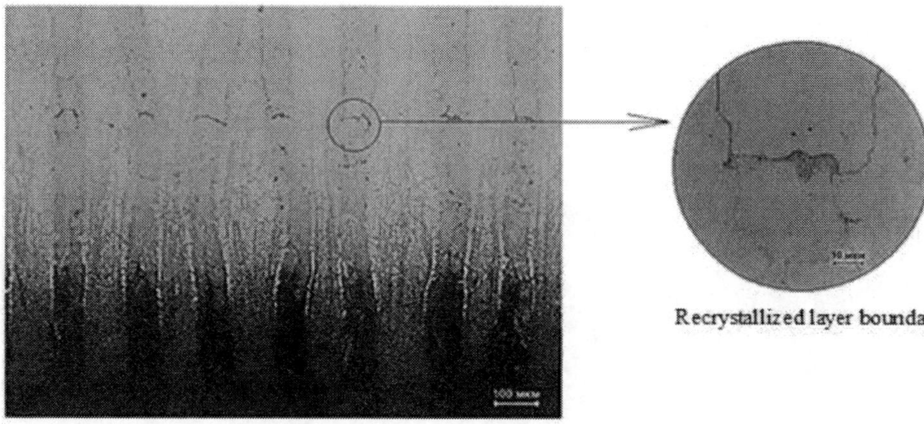

Figure 1. The microstructure of the cemented layer in the multilayer material of the composition of AISI403+ AISI304 (the thickness of the laminar layer is 100 μm)

At the same time, the formation of globular-shaped carbides is observed in the AISI304 steel layers, and a lamellar structure similar to that of perlite is observed in the AISI403 steel layers (Figure 2).

The formation of various types of carbide inclusions and the recrystallization zone between the layers of the formerly AISI403 and AISI304 steel can be explained as follows. The steels involved in the formation of the composition have a different ability to dissolve carbon. So steel AISI304, having an FCC lattice, at a saturation temperature is capable of dissolving a significant amount of carbon, which is also facilitated by the presence of chromium. The diffusion restraining factor is nickel, which increases the thermodynamic activity of carbon at the layer boundary (Figure 3).

The layers of AISI403 steel have a BCC lattice in which the carbon solubility is limited. Therefore, at the beginning of diffusion saturation, carbon actively diffuses along the interlayer boundary. With an increase in its concentration in the interlayer boundary, diffusion into the AISI304 steel layers begins with the formation of globular-shaped carbides of the Cr23C6 type, which in turn is facilitated by nickel, which reduces the solubility of carbon [12].

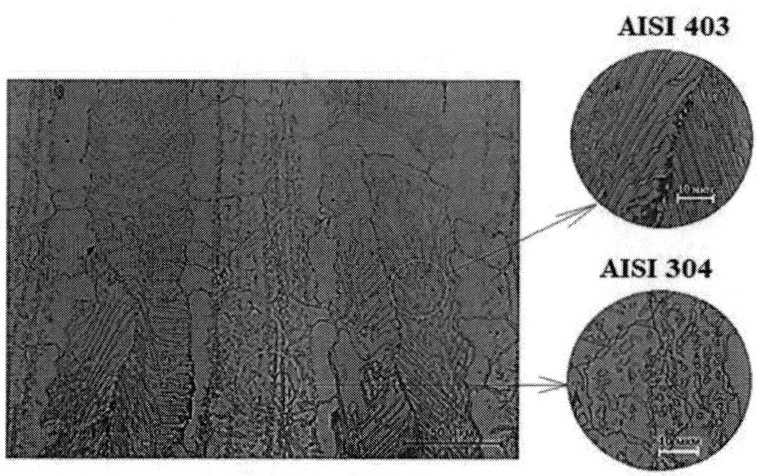

Figure 2. The form of carbide inclusions in the layers of steels AISI403 and AISI304

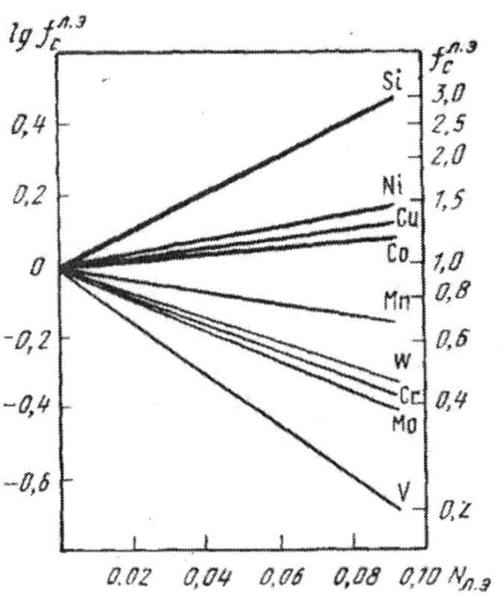

Figure 3. The influence of the concentration of the molar fraction of alloying elements (Na.e.) on the relative coefficient of thermodynamic activity of carbon (fc) in austenite [13]

According to the isothermal section of the triple diagram of the state of Fe-C-Cr, it can be seen that cementite type carbides Kc (Fe, Cr)$_3$C, which are observed in the microstructure of AISI403 steel layers, can be formed at a significant carbon concentration and low chromium content (Figure 4). Therefore, with an increase in concentration during diffusion saturation with carbon, carbides are formed at the boundary from the side of the AISI403 steel layer. This, in turn, leads to a depletion of the solid solution of this steel and recrystallization with the formation of the γ phase at the interlayer boundary, ready to dissolve a new portion of carbon diffusing along the interface, with the repetition of the described cycle. It is also important that recrystallization, with the formation of new light grains, occurs mainly due to layers that originally were AISI304 steel.

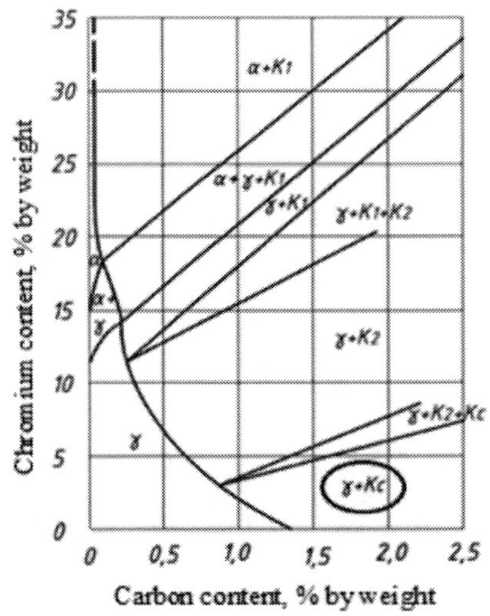

Figure 4. Isothermal cross-section at 1000 °C of the triple diagram of the state of the Fe-C-Cr system [14]

Similarly, a study was conducted on the diffusion saturation with nitrogen according to the regime mentioned above. For the study of structure formation, we also selected samples of multilayer materials with a layer thickness of 100 μm and a workpiece thickness of 10 mm after the first production cycle (Figure 5).

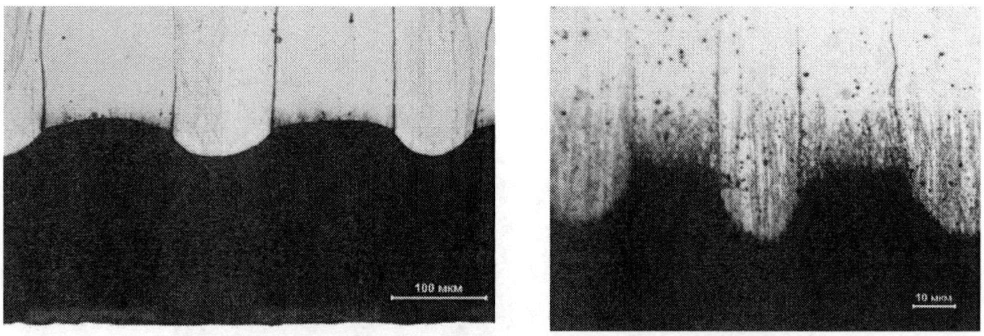

Figure 5. The microstructure of the nitrated layer in the multilayer material of the composition of AISI403+ AISI304 (the thickness of the laminar layer is 100 μm)

The study of the microstructure has shown that, in contrast to samples saturated with carbon, the boundary of the diffusion front of nitrogen (Figure 5), which moves ahead along the boundaries of the laminar layers, as well as along the defects in the crystal structure, manifests quite clearly. You can also see that the boundary of the diffusion front has different curvature in the AISI304 steel layers and in the AISI403 steel layers. As the results of X-ray microanalysis show, a convex meniscus is observed in the AISI403 steel layers, and a concave meniscus is observed in the AISI304 steel layers (Figure 6, Table 1).

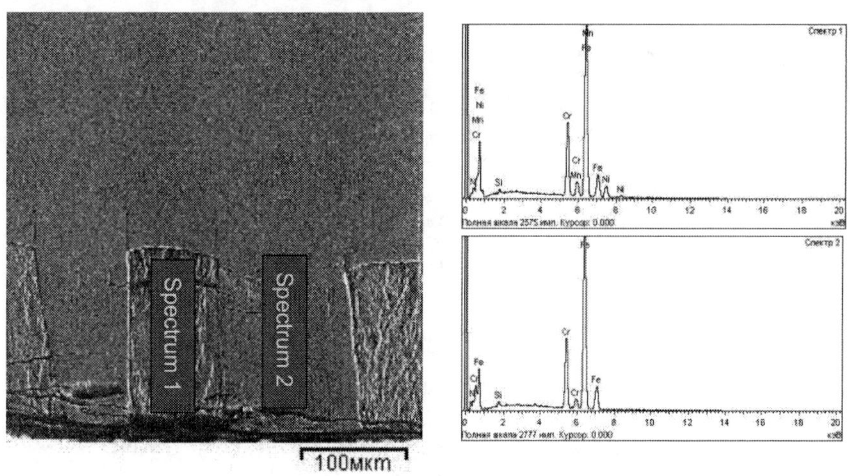

Figure 6. The electronic image of the structure of the multilayer material AISI304+AISI403, spectrum 1 is AISI304 steel layer, spectrum 2 is AISI403 steel layer

Table 1. The results of layer-by-layer X-ray microanalysis of the multilayer material AISI304+ AISI403

Element	Spectrum 1 (AISI304)	Spectrum 2 (AISI403)
Si K	0.33	0.69
Cr K	18.24	18.53
Mn K	1.69	-
Ni K	7.02	-
Fe K	72.72	80.78
Total:	100.00	

By the nature of the influence of alloying elements on thermodynamic activity, nitrogen has much in common with carbon, which determines its active diffusion along the interlayer boundaries of the studied multilayer material. Chromium reduces the thermodynamic activity of nitrogen, so the photomicrographs show the formation of a concave meniscus in the AISI403 steel layers, while in the AISI304 steel layers a convex meniscus is observed due to the inhibitory effect of nickel.

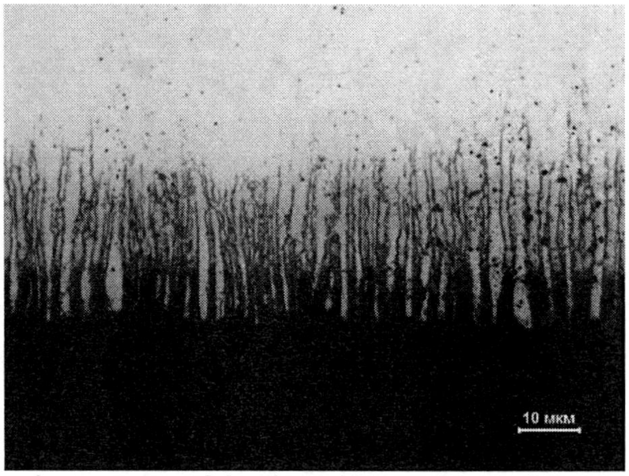

Figure 7. The microstructure of the nitrated layer of the multilayer material of the composition of AISI403+ AISI304 (the thickness of the laminar layer is 2 μm)

4. Conclusion

Studies have shown the effect of the thickness of the laminar layer of a multilayer material on the depth of nitrogen penetration. It is established that in the multilayer material AISI304+AISI403 with the number of layers of 2000 and with a unit layer thickness of 2 μm, there is a significant advance in the diffusion penetration of nitrogen along the interlayer boundaries (Figure 7).

References

[1] Kolesnikov A G, Plokhikh A I, Komisarchuk Yu S and Mikhaltsevich I Yu 2010 A study of special features of formation of submicro- and nanosize structure in multilayer materials by the method of hot rolling//Metal Science and Heat Treatment. **52(5–6)** p. 273–278

[2] Tabatchikova T I, Plokhikh A I, Yakovlev I L and Klyueva S Yu 2013 Structure and properties of a steel-based multilayer material produced by hot pack rolling//The Physics of Metals and Metallography. **114** – Issue 7 – p. 580-592.

[3] Tabatchikova T I, Yakovlev I L, Plokhikh A I and Del'gado Reina S Yu 2014 Studying a multilayer material based on stainless steels and produced by hot pack rolling//The Physics of Metals and Metallography **115** – Issue 4 – p. 403-412.

[4] Chernyshov G G.,Panichenko S A and Chernyshova T A 2003 Welding of metal composites // Welding International..**17(6)** C. 487-492.

[5] Schmiedt A, Luecker L, Kolesnikov A, Plokhikh A I and Walther F 2017 Production- and microstructure-based fatigue assessment of metallic aisi 304/430 multilayer materials produced by hot pack rolling //Materialpruefung **59(2)** C. 123 -129.

[6] Xia C Z, Li Y J, Wang J and Puchkov U A 2010 Microstructure of welding zone of super-ni laminated composite and 18-8 steel tig joint // Zhongguo Youse Jinshu Xuebao **20(6)** C. 1149-1154.

[7] Petelin A L and Plokhikh A I 2013 Diffusion along grain boundaries in multilayer materials//Steel in translation/ **11** p.710-713

[8] Gerasimov S A, Golikov V A, Gress M A, Mukhin G G and Snop V I 2004 High-pressure gas nitriding of steels // Metal Science and Heat Treatment.. **46(5-6)** C. 227-229.

[9] Ryzhov N M, SmirnovA E and Fakhurtdinov R S 2004 Control of carbon saturation of the diffusion layer in vacuum carburizing of heat-resistant steels Metal Science and Heat Treatment **46(7-8)** C. 340-344.

[10] Gerasimov S A, Zhikharev A V, Berezina E V, Zubarev G I and Pryanichnikov V A 2004 New ideas on the mechanism of structure formation in nitrided steels // Metal Science and Heat Treatment **46(1-2)** C. 13-17.

[11] KraposhinV S, Kondrat'ev S Y, Anastasiadi G P and TalisA L 2017 Experimental investigation of in-situ transformations of the M7C3 carbide in the cast FE–CR–NI alloy // The Physics of Metals and Metallography **118(3)** C. 227-232.

[12] Plokhikh A I 2013 On the Possibility of Using Multilayer Metal Materials for Machine Parts Hardened by Thermochemical Treatment // News of Volgograd State Technical University. Series Problems of Materials Science, Welding and Strength in Mechanical Engineering **7(6)** p. 13-17.

[13] Special Steels Gol'dshtejn M I, Grachev S V and Veksler YU G 1985 M.: Metallurgiya. 408 p.

[14] Special Steels: trans. from German./Gudremon E; Edited by cor. memb. AN SSSR Zajmovskogy A S [et al.] 1966 Moscow: Metallurgiya **1** p. 736

IOP Conf. Series: Materials Science and Engineering **934** (2020) 012055 doi:10.1088/1757-899X/934/1/012055

Overview of 4G, 5G radio spectrum spectrum in the world and Kazakhstan

A S Tolegenova[1], A B Zhanys[1], S N Nurkasymova[2] and L A Soboleva[1]

[1]Kazakh Agrotechnical University named after Saken Seifyllin, Republic of Kazakhstan

[2]Eurasian National University. L.N.Gumilev 010000 Republic of Kazakhstan, Nur-Sultan

E-mail: aray.zhanys@gmail.com, SauleNurkasim@mail.ru

Annotation: Mobile operators support data transfer on 2G (GRPS, EDGE), 3G (EV-DO, HSPA) networks. In some regions, data services are provided on WiMax networks. In 2012 ALTEL launched 4GLTE data distribution network. GSM standard in Kazakhstan has started to decline in mobile communication revenue since its first launch in 2013. By the end of 2015, they totaled 19% to 249 bln. tenge. In February 2016, income from voice communications was less than the revenue from Internet access services. Mobile communication in Kazakhstan is carried out in GSM (900, 1800) and UMTS (900, 2100) formats. In 2012 the mobile accessibility index was 120%. Mobile accessibility in Kazakhstan in 2015 will increase by 185%, to 31 mln. SIM card is already registered. More than 35% of these SIM cards (11 million) are Internet access. GSM and UMTS mobile operators are: Kcell and Activ trademarks are owned by Kcell (GSM Kazakhstan OJSC JSC Kazakhtelecom OJSC until 2012) (controlling interest is owned by TeliaSonera, acquired by Kazakhtelecom JSC,% Scandinavian-Turkish holding of FinTur 47); Altel JSC (owned by Kazakhtelecom JSC), ALTEL 4G GSM + LTE (the share of the company in the mobile market is 3.6% (December 2012). Altel JSC - First national mobile operator of the Republic of Kazakhstan [1]."Kar-Tel" LLP (subsidiary of Russian holding Vympel-Communications) and "Beeline Business" (previously K-Mobile)); Mobile Telecom-Service LLP (former Neo) with Tele2 trademark. Kcell (KCEL) is a mobile operator GSM900, GSM1800, UMTS / WCDMA (2100 MHz), LTE / 4G, LTE Advanced (700/800 MHz and 1700/1800 MHz). "Activ" and "Kcell" brands. In 2018, KazakhTlecom acquired the shares.

1. Introdusion

Telephones - Basic Routes: 4.928 million (2018). Cell Phones - Cell Phones: 14,995 million (2019). Phone code: +7.

Inside: between cities through landlines, radio relay connections and satellite communications. The number of registered connections has gradually increased and reached 40 per 100 people. Mobile (mobile) connections are rapidly increasing and their number is about 88 per 100 people.

International: international communication network with the post-soviet republics and China and microwave radio-relay communication. Via Trans-Asia-Europe (TAE) satellite and fiber-optic cable. Satellite ground stations: Intelsat. KazSAT-1, KazSAT-2, KazSAT-3, Kazakhstan series of satellites KazSAT-4.

In 2019, the World Radiocommunication Conference (WRC-19) will be held every four years. The most probable radio frequency bands for communication are:

- 24.25-27.5 GHz, 37-43.5 GHz;
- The range 70/80 GHz may be agreed on RAD-19;
- For RAD-19 5G, the band 31.8-33.4 GHz is virtually impossible to agree on since it is problematic to ensure compatibility with existing services.

With regard to low frequencies, there is no final solution for bandwidth and bandwidth for 5G. In Kazakhstan, the majority of the 3,4-3,8 GHz band is received by military personnel.

The LTE Union has proposed the use of the non-standard 4.4-4.99 GHz frequency band for the 5G European zone. This will allow increasing the frequency spectrum for deploying new generation cellular networks. However, there is no commercial equipment for this range of frequencies.
Modern mobile communication is dense within distributed frequencies and 5G networks receive additional frequencies - now in a short, millimeters range of radio waves from 30 to 300 GHz.
Their length (1-10 mm) is smaller than the usual centimeter wave for cellular communication. They give a lot of data, but they are far away and easily protected. It is necessary to deploy a dense network of small cellular communication to help them.
In low-density environments, this process may take years, but engineers work with technologies that let the short wave be transmitted to longer distances up to several kilometers [2].

2. Overview of telecommunication operators in Kazakhstan

Delivery stations: AM-60, FM-24, Shortwave-14 (2018). Radio: 11.47 million (2018). Television stations: 26 (plus 14 repetitions) (2018). Internet service providers: 18 (with their own international channels) (2018). Internet host: 180,217 (2018)
Internet users: 400,000 (2009); 1,247,000 (20011); 5 300 000 (2016). Internet domain:. <Url>.
The company was founded in 1998 and is mainly focused on the corporate segment, Kcell, focused on the mass market, and offers mobile services under the brand Activ. As of December 31, 2016 the subscriber base of Kcell was 9 986 thousand people.
Kcell was established on September 30, 1998. February 7, 1999 - Kcell brand was launched. On September 9, 1999 the brand ActiV was launched. In 2004, the company's subscriber base amounted to $ 1.5 million. Man. In December 2007, the 6 millionth subscriber was added. That year more than 5,000 settlements were covered. In 2009 Kcell corporate brand was restored. In 2010, Kcell hosted the OSCE Summit and VII Winter Asian Winter Games
In 2016, Kcell launched a 4G / LTE commercial network. Together with Beeline Kazakhstan, the agreement on the joint 4G / LTE network was significantly accelerated. This agreement has allowed us to provide high-quality services while reducing capital costs and accelerating network deployment. By the end of December 2016, coverage of 4G / LTE network from Kcell reached 35% of Kazakhstan's population. In the first half of 2017, residents of Ridder, Taraz, Turkestan, Kyzylorda, Petropavlovsk, Shemonaikha and Ekibastuz had access to 4G / LTE high-speed mobile Internet. From May 12, 2017, the company has developed 4G + (LTE Advanced) communication.
In the beginning of 2016 the development of entertainment services in the network of mobility in mob (mobi-tv, mob-music, mobi-press, mobi-cinema, book) started [3].

3. Analysis of the current telecommunication network of Nur-Sultan

Number of mobile subscribers in the Republic of Kazakhstan is 31.2 mln. The subscriber base increased by 7.1% per annum (by the end of March 2018), the density of cellular communication numbers per 100 people increased from 167 to 176.
According to the Committee for Communication, Informatization and Information of the Ministry of Internal Affairs of the Republic of Kazakhstan, by mid-February, Kcell is one of the leading telecommunication operators, representing 31% of all base stations in the country (9384 units), Tele2 26% 7 712 Unit), Beeline share is estimated at 24% (7193 base stations); Altel has 19% of 5,678 base stations.
In the regional context, the base stations are clearly distributed among operators. Thus, Kcell dominates in 16 to 10 areas. The Company is located in Atyrau and Mangistau oblasts (54% of market and 45% respectively) in South Kazakhstan region (43%) and Astana agglomeration (38% in Astana and 45% in Akmola region). In Almaty, Kcell has priority over the number of base stations with Altel - 29% of online accounts for each of the operators. Altel also trusts Astana, which owns 29% of the base network. The share of Tele2 in the Karaganda region is 37%, Zhambyl and Kostanai oblasts (34%), and 30% of base stations in EKR. Beeline's position is stronger in Western Kazakhstan (39% of total base stations) and Pavlodar region (31%).
Out of the total number of base stations, 17.5 thousand (58% of the network) support 2G standard; There are 10 thousand (34%) 3G subscribers, and only 4 out of 292 stations (8%). So far, all 4G stations have been owned by Altel, but in 2018 all Kazakhstani operators have the right to operate at LTE frequency and serve as standard 4G, and experts predict the increase in competition in this sector (Figure 1).

IOP Conf. Series: Materials Science and Engineering **934** (2020) 012055 doi:10.1088/1757-899X/934/1/012055

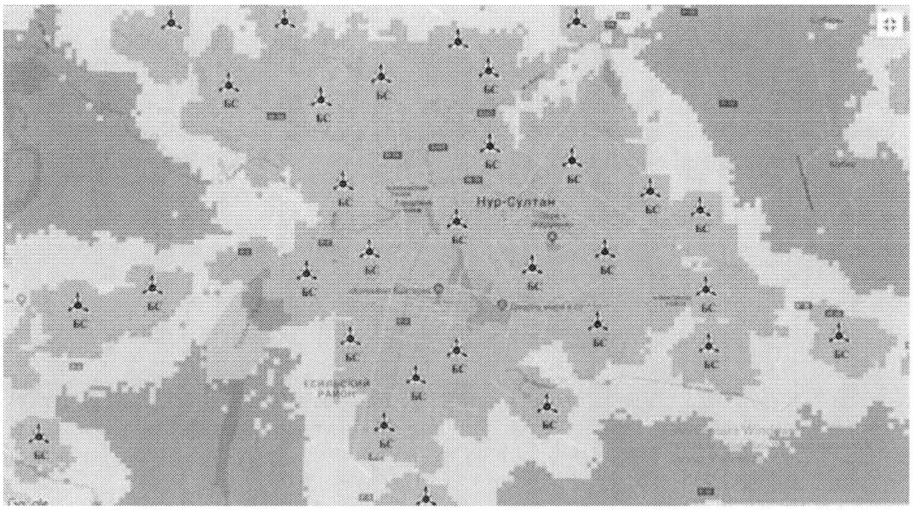

Fig. 1: Wireless 4G era in Nur City

The number of base stations affects the coverage area and, consequently, ensures a high level of cellular communication. The highest density of base stations in population density is observed in Astana and Atyrau regions where one station serves an average of 1 thousand people. Each station in Almaty, Mangystau and North Kazakhstan oblasts has about 1.5 thousand people. The largest truck belonging to the base stations of the Karaganda region - 4,3 thousand, in the West Kazakhstan region - 3,1 thousand people [4].

4. Selecting base stations and antennas

The optimal solution for choosing equipment is Huawei DBS 3900 base station.

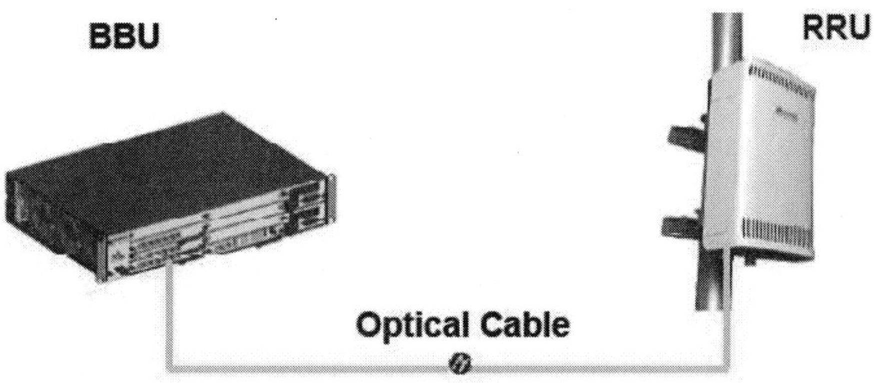

Fig. 2: Optical communication between the base station and the antenna

Huawei DBS3900 base station DBS3900 is a distributed base station that uses the HuaweiBTS platform. Being the base station of the mobile communication system, the DBS3900 consists of the Base Frequency Processing Unit (BBU) and Remote Radio Frequency Units (RRU).

The DBS3900 uses an external radio frequency module that meets the requirements of construction networks in terms of capacities, flexibility, and upgrades.

The Huawei-RRU3929 remote radio station base station is selected. Installed near the antenna. RRU3929 is a portable radio frequency unit. Provides basic and radio frequency signals processing. One RRU3929 is a function of two transceivers. If two RRU3929 modules are installed at the RRU3929 substation, they perform the function of the four transceivers. RRU3929 is easy and easy to install. Substrate RRU3929 can be installed on steel fittings, wall or concrete base. The RRU3929 and BBU3900 units use the CPRI interface, which provides the use of two modular optical cables. This will allow you to significantly reduce the cost of building a room, installing equipment and operating costs. Introducing the DBS3900 with the integrated BTS will speed up the development of cellular networks, ensure their compatibility with other networks, and exploit broadband technologies [5].

BBU3900 is a basic unit processing unit for centralized management and maintenance, as well as centralized equipment for providing base station signal processing and providing standard reference signal.
The device has physical interfaces for communicating with BSC and RRU3004.
The BBU3900 is mounted on a 2 U cabinet with a height of 47.5 cm and can be installed in a 19 "cabinet or wall mounted.
BBU3900 installs additional Abis interface controls and GPRS synchronization signals. BBU3900 - compact equipment, easy installation. It consumes a small amount of energy and offers a full range of services.
The DBS3900 DBS3900 features have the following advantages in providing coverage:

- RRU3929 supports cascade connection of three RRU modules. One RRU module is installed within 40 km of CWU;
- static sensitivity of the TCH / FS channel -113 dBm (normal temperature at normal temperature);
- RRU3929 maximum output power is 40 W (900 M) or 30 W (1800 m);
- maximum configurations up to 12 hundred and multi-purpose network support;
- Supported broadcast distribution and "Antenna upgrade".

BBU3929 supports 72 transmitters.
Network features. E1 / T1, optical FE, radio relay and satellite broadcast support. Supports structure (topology) - star, tree, chain, ring and mixed topology. GSMb and UMTSBBU3900. Optimized messages using the Abis interface. Support for BTS and BSC packages detection and recovery. When the synchronization system operates in an internal vibration mode, the system can operate continuously for 7 days.
Support for different synchronization modes: Support for synchronization synchronization from Abis interface, support for GPS synchronization, 2MHz BITS external sync.
RRU3929 operates in the bands 1800 MHz and 900 MHz. For sites with GSM / UMTS / LTE configuration, 6 RRU3929 is required. For sites with GSM / LTE configuration, 3 RRU3929 is required.
BBU is characterized by high environmental adaptability:

- Operating Temperature Range: -20 C ~ 55 C
- BBU can work with a wide range of operating voltage: -
- 38.4 V DC ~ -57 V DC (rated voltage -48 V DC).

The used power supply converts 220V AC to -48V DC for BBU operation. RRU - Environmentally friendly equipment. RRU has a closed integral structure. Water resistant to IP65 standard. Measures against moisture, saline and saline fog are in accordance with Class 1 specification.RRU Operating Temperature Range: -40 C ~ +50 CRRU can operate with a wide range of operating voltage ranges: -36 V DC ~ -57 V DC (rated voltage -48 V DC). The used RRU power source is 220V AC - 48V AC. Remote Electronic Grinding (RET) antenna support. Using the RET antennas enables you to adjust the coverage of the network by adjusting the angle of antennas in the car hall. At the same time, the cost of maintenance and maintenance can be saved [6].
Support for double polar antennas reduces the number of antennas in the cell. AISG1.1 protocol support.
To compare the results obtained with the simplified model calculations for Macro-Cells by Okumura-Hat, we obtain the next radius of coverage (1.06 +1.47) / 2 = 1.27.
Calculating the number of base stations (BTS) After determining the value of the cell radius (P 1270 m), we place the base stations on the Taldykorgan map. At the same time, we take into consideration the effectiveness of the sector allocation. That is, to the non-residential areas of the city, to the enterprises and factories.
As a result, the required number of 4G stations will depend on the area of Nur-Sultan about 100 square meters. km, no settlement.
Let's sum up the results:

- coverage area - 100 km2;
- Total network accessibility - 1088 000 subscribers;
- Subscriber density - 10880ab / km2.

Mobile device antennas should be short enough and light enough for space design to create smartphone designs. F-shaped antennas (planar inverted-F antenna - PIFA) are a good choice for mobile communication devices, as they meet the small size, power and efficiency requirements. These antennas support multi-channel frequencies for connecting mobile devices, WiFi, and Bluetooth technologies, making them the best candidates for IoT-compliant objects and devices.

IOP Conf. Series: Materials Science and Engineering **934** (2020) 012055 doi:10.1088/1757-899X/934/1/012055

In this example, the antenna is part of a 5G mobile device and is a PIFF construction in the ABS case (Acrylonitrile butadiene stirrine (ABS) - PTFE - PTFE - PTFE - based fluoroplast) mounted on thermoplastic FR4 panels. and a composite silicon substrate is covered with glass. The antenna itself consists of a thin copper layer with high permeability from the PTFE block, which has a concentrated port between the excellent conductivity (PEC) screen and the sensor contact (for these specialists and the antenna signal is transmitted by signal) and also located near the feeder, the impedance used to match the PEC screen to another contact. Provides design antenna compatibility to fit the antenna to a 50 Ω reference value.
Antenna modeling can be performed on the boundary conditions of the PEC for its electrodes due to the low operating frequency in the communication network. The loss of metal may not be due to the high conductivity of the copper layer. PIFA radiation is modeled in a spherical area surrounded by a perfectly coincident layer (PML) that imitates the irradiation in infinite space. Availability of boundary conditions ensures the absence of airborne energy sources. A port with an impedance center of 50 OM for excitation of PIAA and for estimating its incoming resistance.
By using simulation, PIFA can calculate and visualize the field share. The results indicate that the strongest field is located at the top of the specimen in one of the remote ends of the electrode from the remote sensor contact. These results are similar to data for a quarter-wave monopoly antenna, which is actually PIFA precursors [7].
The simple alignment does not meet the requirements for using the network. In addition, operators assume that space for the installation of antennas for Massive MIMO technology is supposed to provide smooth evolution. As a result, the only antenna that combines all tapes with a bandwidth of less than 3 GHz and a massive MIMO space should have a standard antenna of 5G technology (Figure 3).

Fig. 3: The solution for the installation of a 5G technology-oriented antenna for Telefonica Deutschland

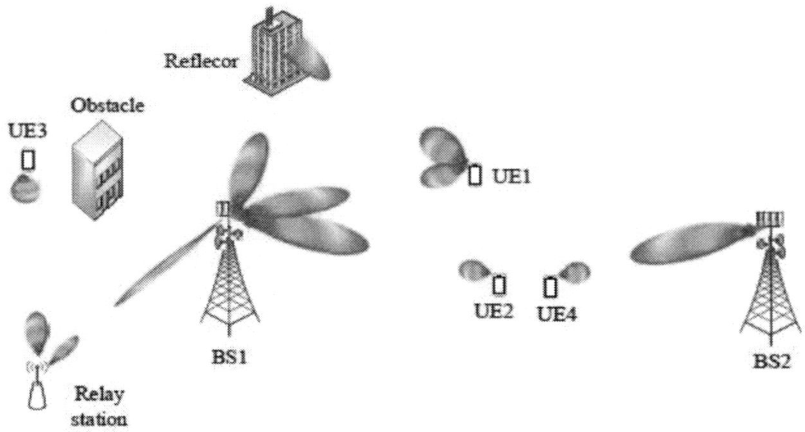

Fig. 4: The principle of working Massive MIMO antenna

In the process of transition from 4G to 4.5G / 5G, there was a lack of space to install the antennas when placing the lines. As part of the allocation of 700 MHz, 1400 MHz and other new bands, the 4T4R technology has become a standard configuration.

We consider a serious problem in the prospective evolution of broadband networks that make the antenna installation a challenge to the network. Huawei and Telefonica Deutschland offer antenna installation solutions that include 5G technology and include a 14-port multi-band antenna and an MGIM antenna of 5 GHz.

The installation of the antenna set up by Telephoneica Deutschland and Huawei 5G will help solve the problem of spatial deficiencies, "said Jaime Lluch, Director of the Telephony Deutschland radio network [8].

5. Application of 5G additional technologies

It is important to note the additional technologies used in the new generation standard:

Massive MIMO. Technology allows one subscriber to send up to 8 data streams. It is a complex of several antennas, which forms an express diagram of the route. The use of multiple radiation levels improves the level of acceptance of the signal and eliminates the interference of other subscribers with a positive effect on network bandwidth and frequency spectrum.

M2M. M2M technology is needed to interact with devices without the direct involvement of the person, ie to automate processes. The scope of M2M is very broad. For example, in payment terminals, security systems, and vehicle management systems. Technology minimizes processes and also reduces their dependence on human factor, allowing rapid response to system failures.

Flexible Duplex is a technology that can be used to streamline the traffic, for example, in the upstream channel you can send information to the downstream channel.

FBMC / UFMC (Filter Bank Multicarrier / Universal Filter Multi-Carrier). The technology allows to achieve spectral efficiency and improve channel selectivity.

Ultra-dense networking - this technology allows you to organize dense networks due to virtualization, thanks to which you can serve a large number of subscribers in a dedicated area, which in turn allows to create a complex network hierarchy.

One of the obstacles to launch 5G standard networks is a shortage of frequency resources. The resources in future networks, including millimeters, should be expanded. The primary band for the 5G network creates a band that is less than 6 GHz. It is planned to use 6Ghz for further universal access and organization of backbone communication.

Test equipment for 5G network was tested at the end of 2016 in a number of countries, including Kazakhstan. During the testing of the equipment, indicators that are close to the required values have been obtained from next generation networks [9].

IOP Conf. Series: Materials Science and Engineering **934** (2020) 012055 doi:10.1088/1757-899X/934/1/012055

6. Design of 5G network in Nur-Sultan city model

When switching to modern standards of the fourth and fifth generation, the stations should only be connected to fiber optics to meet their requirements. In modern BS design, optical fibers are an integral part of providing information between BS components and blocks. For example, the following illustration shows the construction of a modern base station using a fiber optic cable (Figure 5) to transfer the data from the RRU (remote control) antenna (orange).

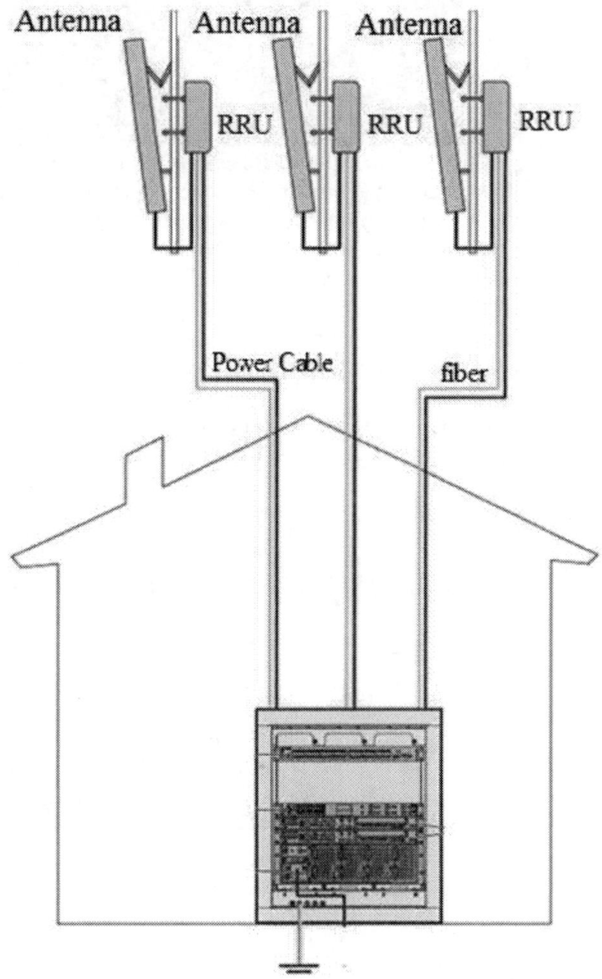

Fig. 5: An example of the base station and antenna accommodation.

Equipment of the base station is located in non-residential areas of the building or is installed in special containers (installed on walls or poles) as modern equipment is very compact and easily accessible to a server computer's system unit. The radio module is often installed near the antenna, which reduces the power loss and dissipation of the antenna. Here are three modules installed directly on Flexi Multiradio base station equipment (Fig. 6).

Fig. 6: Base station Flexi Multiradio

Another important technology to achieve greater spectrum effect is the mass media. The MIMO array is sometimes called large-scale MIMO; the number of antennas on the base station is much more than the number of MIMO formats available for the signal source. A large number of base station antennas with respect to user devices are responsible for channel-side, which gives greater benefit to quasi-orthogonal and spectral effects [10].

Designers face difficulties in scaling up the number of antennas hundreds of scales:

- The simulation speed of the traditional antenna design tools for large antennas arrays is slow.
- It is difficult to model the antenna jack.
- Hybrid radiation is required to optimize the number of radio frequency circuits (Figure 7).

Fig. 7: Masive MIMO antenna

Qualcomm has introduced antennas and transmitters to smartphones designed to work in the 5G promising standards. The first mobile devices with such modules are expected to appear in the first half of 2019.

5G data technology is still described by the IMT-2020 standard of the International Telecommunication Union. It is assumed that 5G networks maintain data rates at a much higher level than previous generations, and also support a large number of devices connected to the base station. The standard features of the standard are not yet approved, but 5G networks will use 4G networks and two main frequency ranges from 600 MHz to 6 GHz, to be used from 24 to 86 GHz. The greatest increase in data transmission speeds and other features is associated with the second band, called mmWave because it will be transmitted by millimeter radio waves.

Qualcomm has introduced a compact antenna module QTM052 intended for smartphones and other mobile devices. It works with the Snapdragon X50 modem introduced in 2017, which was successfully tested in 2017. The module is made up of four antennas forming an array of phased antennas. In this regard, it controls the direction of radiation and thus improves the signal quality. The module supports three frequency bands - 26.5-29.5 gigahertz (n257), 27.5-28.35 gigahertz (261) and 37-40 gigahertz (n260). The devices with the Snapdragon X50 modem can use up to four antennas, which will allow the hands to weaken the signal depending on the hands or other factors. The company has already started producing the module serial production from the manufacturer, and its serial devices should be available in the market in the first half of 2019 [11].

Qualcomm also introduced the QPM56xx module for operation with 3.3-4.4 GHz (n77), 3.3-3.8 GHz (n78) and 4.4-5.0 GHz (6 GHz) n79). The company has begun manufacturing production modules for the manufacturer of this device.

While some features of the 5G standard have not yet been approved, some telecom operators have begun implementing this technology. For example, the two largest US mobile operators, AT & T, recently announced that it will use 5G communication networks in 12 cities across the country by the end of 2018.

In order to evaluate the performance of new 5G algorithms and architectures, engineers must prove the concept prototype and prepare new trials for field trials. Typically, prototypes are made using FPGA hardware with built-in processors for some parts of the design. These platforms are often known as hardware testing and they provide fast testing and testing of new technologies and design changes in this area.

It may be difficult for the usual R & D team to implement FPGA-based radio prototypes and test sites without external assistance. R & D engineers have expertise in strong signal processing and communication algorithm development, but hardware experience is minimal. This experience increases the difference between discrepancy and workflow. While R & D engineers typically use high-end languages such as MatLab, hardware engineers use their own design tools and hardware descriptive languages (HDLs).

Leading companies have made model designs with MATLAB and Simulink to eliminate these deficiencies and allow R & D engineers to quickly switch to new 5G algorithms and make changes to their FPGA-based test platforms.

There were network equipment suppliers and wireless operators. Public testing for 5G technology is described in the real world. In field tests, many parameters are measured and dynamically monitored, such as signal signal (RSRP) and received signal quality (RSRQ). The performance of the system is recorded by a wide range of signals and a wide range of parameters.

After accepting the measurements, designers can get useful and effective results from the data. Compared to 4G systems, 5G field trials collect data far beyond data collection and analysis at high speeds and mass MMIM radiation specimens [12].

If implemented properly, this visualization architecture will allow engineers to comprehensively depict system performance in real world scenarios and display results for network planning decisions reporting.

The field test system requires a test and analysis system that supports extensive data capture, data processing, analysis, and sharing of results (Figure 8).

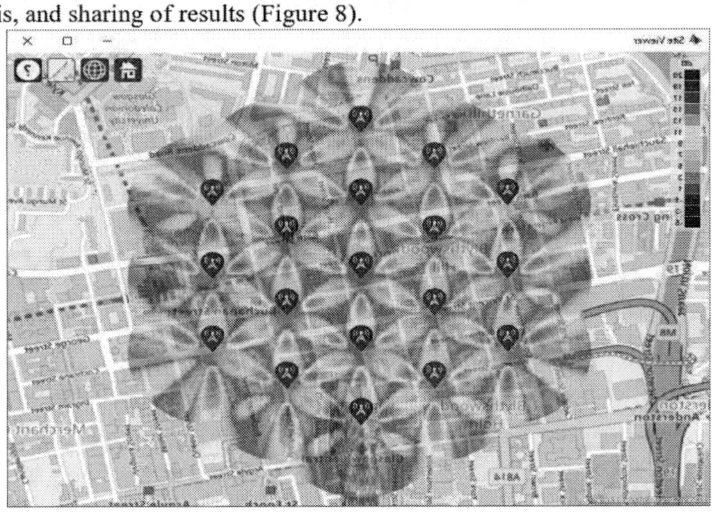

Fig. 8: Base stations location in wireless 5G era

It was decided to cover the issue with subscriber-focused network coverage and accessibility, ie the coverage of the network's network will be adapted to the needs of subscribers rather than the network of the previous generation. It is planned to use automatic antenna arrays that dynamically change the antenna structure of the antenna systems. It is also planned to use the range of all available frequencies, particularly within a short range of millimeters [13].

The 5G network architecture is divided into three internal systems (clouds), and cloud technologies that do not exist in the modern world:

Access cloud (Access) means the addition of distributed and centralized access technologies and systems. Also backward compatibility with 4G and 3G networks is planned;

IOP Conf. Series: Materials Science and Engineering **934** (2020) 012055 doi:10.1088/1757-899X/934/1/012055

Cloud Management - session management, mobility and quality of services;
Transport cloud (Forward) - transmission of physical data to the network by high reliability, speed and load balancing.
Small nucleus with small radius. Unlike the nest and the macro cell. Small honey and equipment for their production. The concept includes femtones and pecio cells in the cellular network, as well as other solutions to address the problem of wireless access to data networks by forming areas with very minimal radius of networks in the network. high density data, high position at home or abroad. Small nucleus with small radius. Unlike the nest and the macro cell. Small honey and equipment for their production. The concept includes femtones and pecio cells in the cellular network, as well as other solutions to address the problem of wireless access to data networks by forming areas with very minimal radius of networks in the network. high density data, high position at home or abroad.
Setting up a small unit for bussing, KPN and JCDecaux outdoor advertising operator was conducted in Amsterdam, the Netherlands. The Ericsson Radio 2203 (C-RAN) bus station was located on the Rembrandt Square. The device supports the aggregation of the carrier b3 + b20 (800 MHz and 1800 MHz).
For the rest of the world, including Russia, the theme of installing small urban infrastructure is not new, for example, several times in the MTS 2014 to install small open cells at bus stations within the pilot project. The company has told about Moscow, St. Petersburg, Novosibirsk, Chelyabinsk, Yekaterinburg, Nizhny Novgorod, Kazan. The equipment was tested differently - NEC, Alcatel-Lucent and Nokia. Installation of kiosks, lighting poles, and advertising structures [14].
This month, a pilot project was launched in Bern, Switzerland. The small boxes of the Ericsson Vault remote radio transceiver protected by the IP68 standard are installed on the ventilated hatches located on the fiber-optic infrastructure of the main network. The surface of the Antenna system is located just across the Katrein Street Connect (Figure 9).

Fig. 9: Placement of SW in city center

I've heard about it for the first time, but the idea of using locals for the deployment of railway base stations was discussed at Motorola at the end of the 1990s. But I have not heard about practical actions yet [15].
RadioPlanner software is intended for calculating and optimizing radio coverage when designing mobile networks, terrestrial radio and television broadcasting from 30 MHz to 3000 MHz. (figure 10)

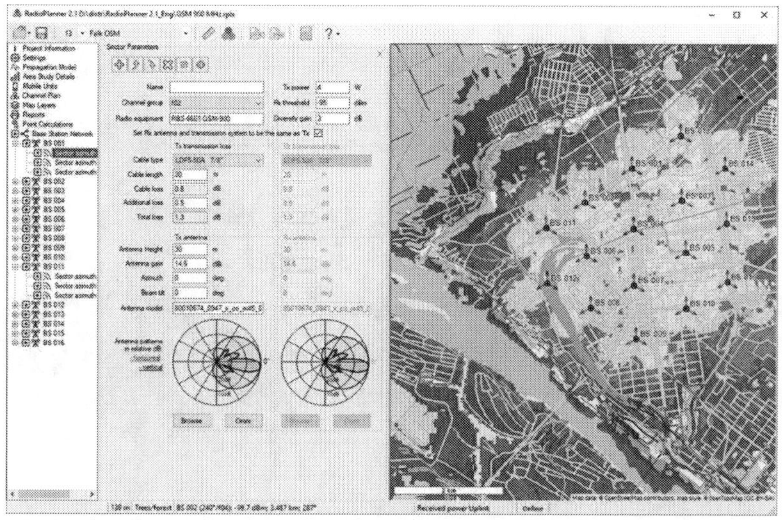

Fig. 10: Setting the angle of the directional diagram of base stations located through RadioPlanner 2.1.

7. Conclusion

In any case, 5G networks in the city should be condensed and it is clear that smaller cells are used. The choice of installation is a challenging issue, where technical capabilities are important, as well as important conditions, as well as the conditions for leasing electric and transportation. All the issues are being resolved, but at the same time remain economically viable for the operator - this task is not trivial. Promoting creativity is unlikely that operators will be able to live without dwindling due to the increase in the number of users of mobile broadband access and the increase in the consumption of mobile traffic.

Main features of RadioPlanner. Method for forecasting a predetermined method for a particular route for land-to-ground services in UU and VHF groups The calculation of the radiofrequency zones using the GIS-propagation method according to the recommendation of ITU-R P.1812-3 (09/2013), as well as the GOST- R 55897-2013 Mobile Radio Networks. Areas of activity. Methods of calculation. Quantitative Height (CMV) Shuttle Radar With TopographyMission (SRTM) 3. Using a Resilient Barrier Barrier (Urban Construction, Suburban, Forests, Water Surface), Resistant to OpenStreetMap and Global Forest Change Projects [16].

Use maps that are automatically downloaded from TOPO CTT, Open Street Map, Google, Bing, and any other tile servers as a mapping substrate. The mappings of the Radio Planner 2.1 software are illustrated in figures A1 - A5 (Figure 11-15).

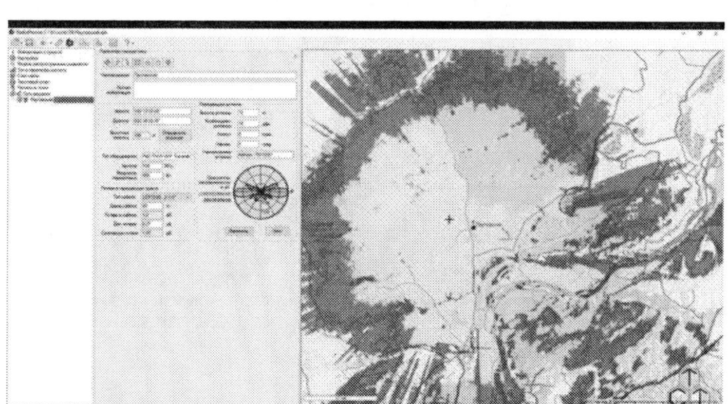

Fig. 11: RadioPlanner 2.1 will open a map of the city using Google Maps

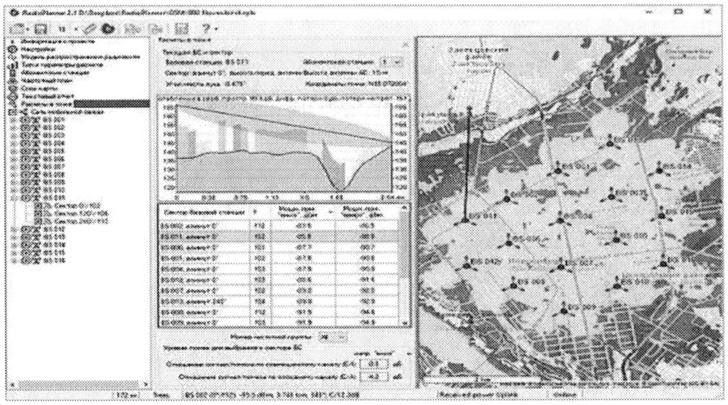

Fig. 12: Placing base stations on the map in RadioPlanner 2.1

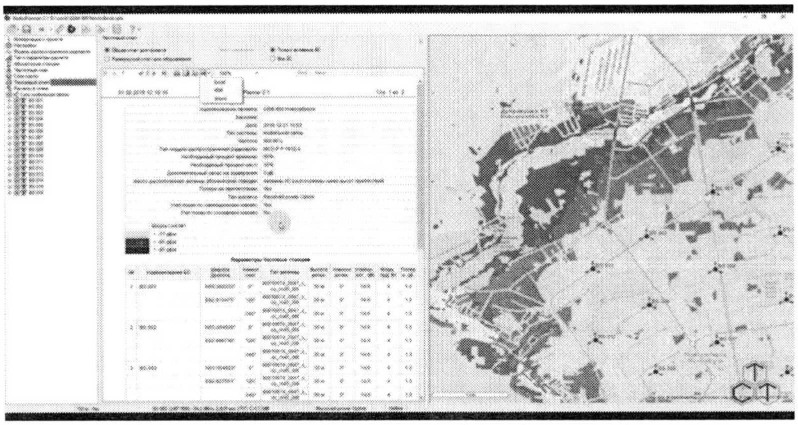

Fig. 13: Defining the optimal angle of base stations in RadioPlanner 2.1

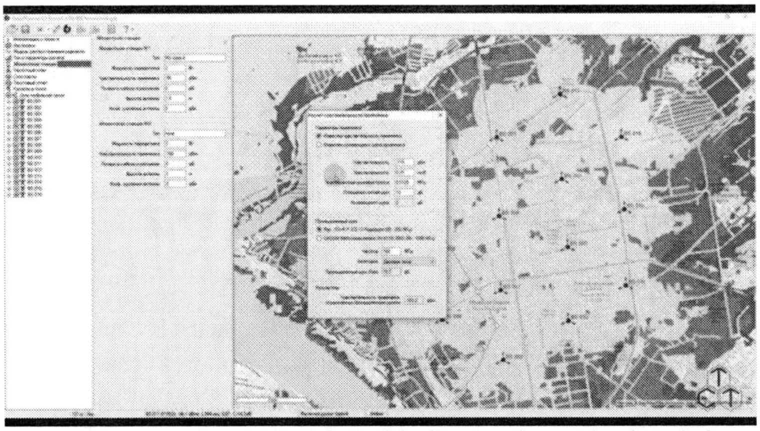

Fig. 14: Determine the duration of the base station optimal angle in RadioPlanner 2.1

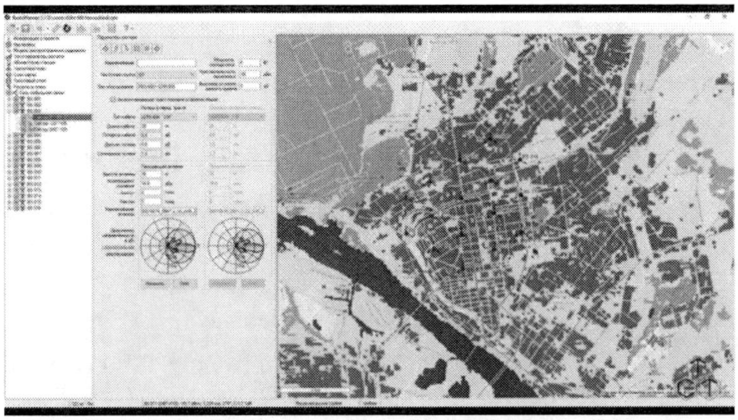

Fig. 15: Result of settings configured in RadioPlanner 2.1

References

[1] Masashi Kishimota "5G Mobile Communications". 2015 16 S 49-56
[2] Steputin A N and Nikolayev A D 2016 Kniga dvukhtomnik «Mobil'naya svyaz' na puti k 6G». **8**. S. 9-12.
[3] Denisov D 2013 Tekhnologii 5G setey. **3** S. 56-58
[4] 3GPP TS 36 104: «E-UTRA Base Station (BS) radio transmission and reception» (Release 9). April 2011.
[5] Kim A V and Tikhvinskiy V O 2013 Novyy mobil'nyy gorizont: itogi MWC-13. 3 S. 15-18.
[6] Tikhvinskiy V O, Arkhipkin V YA 2013 LTE World Summit-2013: na puti k 5G. Zhurnal Elektrosvyaz'. 7 S 19-21.
[7] Sporyshev K A,Tkachev A M and Postolnyi A S 2008 Analiz podhodov k ocenke prognoziruemyh zon obsluzivania bazovymi stanciami v sistemah podvijnoy radiosvazi.Harkovskii universitet Wozdushnyh Sil im. I.Kojeduba, Harkov..
[8] Niri S G 2013 Towards 5G//LTE World Summit-2013.Hardouin E.5G:an operator's perspective// LTE World Summit.
[9] Osseiran A 2013 The 5G Mobile and Wireless Communications: Challenges and Scenarios//LTE World Summit-.
[10] Azar Y, Wong G N, Wang K, Mayzus R, Schulz J K, Zhao H, Gutierrez Hwang D and Rappaport T S 2013 28 GHz Propagation Measurements for Outdoor Cellular Communications Using Steerable Beam Antennas in New York City: 2013 IEEE International Conference on Communications (ICC). — Budapest, Hungary. June 9-13.
[11] Likontsev D N 2008 Spetsial'nyye voprosy kursa rasprostraneniya radiovoln Tashkent.
[12] Babkov B Yu 2011 Obshie podhodi k zadacham planirovania i optimizacii 2G – 4G setey podvizhnoy svyazi. Sankt-Peterburg.
[13] Tikhvinskiy V O, Yurchuk A B and Terent'yev S V 2010 Seti mobil'noy svyazi LTE. Tekhnologii i arkhitektura. – M: Eko-Trendz 284 s.
[14] Harri Holma and Antti Toskala 2009 LTE for UMTS. OFDMA and CS-FDMA Based Radio Access. – John Wiley Ltd.
[15] Stefania Sesia 2009 LTE. The UMTS Long Term Evilution. From Theory to Practice. – John Wiley Ltd.

Research of thermal control system for nanosatellite with carbon fiber reinforced plastic body

Wang Yu, Zarni Soe-Moe, O V Denisov*, L V Denisova

Bauman Moscow State Technical University, 2-nd Baumanskaya, 5, b.1, 105005, Moscow, Russia

E-mail: denisov.sm13@mail.ru

Abstract. One of the key problems in the design of nanosatellites is to provide a specified temperature range for the operation of electronic equipment. In structures of modern nanosatellites mainly used the elemental base of smartphones. To process a large amount of information, more advanced processors with high thermal power are required. The specified thermal mode of the on-board computer can be achieved using a remote heat removal system or by direct contact of the processor cover with the carbon fiber reinforced plastics (CFRP) body of the nanosatellite. Using a model of nanosatellite as an example, a thermal control system with miniature loop heat pipes and highly heat-conducting refrigerators-emitters is designed. The influence of the thermal conductivity coefficient in the reinforcement plane of the nanosatellite CFRP body on its temperature state in low Earth orbit is studied.

Key words: carbon fiber reinforced plastic, nanosatellites, thermal conductivity coefficient, loop heat pipes, thermal control system.

1. Introduction

Currently, the space technology, associated with small spacecraft – micro-nanosatellites, is actively developing around the world. The great demand for the application of micro-nanosatellites is due to the relatively low cost of operation and great potential in the global space goods market. Over the past 5 years, the number of launches of micro-nanosatellites has increased more than twice. According to forecasts, by 2023, more than four hundred devices would be launched [1, 2]. The cost of launching spacecrafts largely depends on their mass. Therefore, composite materials with high specific characteristics – CFRP are promising for the manufacture of nanoatellites. High heat-conducting CFRP are perspective [3].

One of the key problems in the design of nanosatellites is the guarantee of a specified temperature range for the operation of electronic equipment (Figure 1) [4]. For the normal functioning of most processors, the allowable temperature should not exceed 90 °C [5]. The constantly increasing information load leads to the need to use more advanced processors with high thermal design power (TDP) in the on-board computers of nanosatellites. The thermal regime of nanosatellites is determined by external thermal loads and heat generated by airborne equipment [6-8]. In the conditions of orbital flight, the use of the usual thermal control system for processer is impossible due to the lack of convective heat transfer. The small overall dimensions and dense layout make it difficult to discharge heat by radiation into the internal volume of nanosatellites. This can lead to overheating of the processor and failure of the on-board computer.

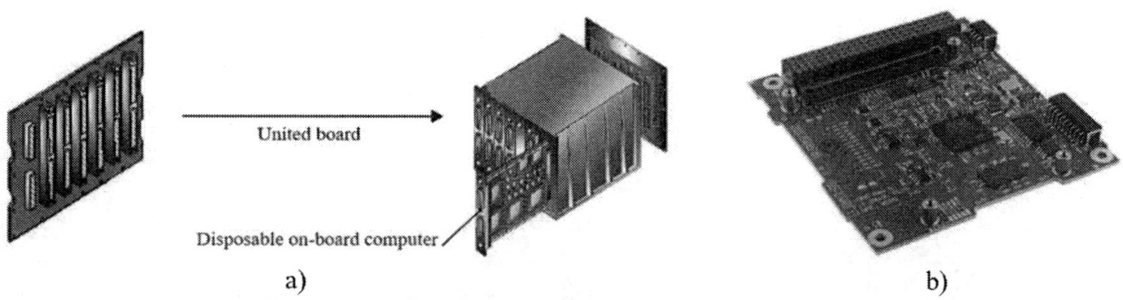

Figure 1. Radio-electronic equipment of a nanosatellite:
a) equipment in a unified format; b) ARM9-based ISIS computer

To solve this problem, it is necessary to ensure the removal of excess heat from the on-board computer processor to the CFRP body of the nanosatellite and its discharge into the surrounding space. This can be implemented in two ways. The first – using remote heat removal systems – miniature loop heat pipes (LHP), which have high effective thermal conductivity and are easily adapted to operating conditions [9-11]. They are distinguished by relative simplicity of design, complete autonomy and transmission a large amount of heat to the required distance. The second way to remove excess heat is through direct contact of the processor cover with the nanosatellite body. At the same time, to increase the cooling area of the surface, it is advisable to use heat-conducting refrigerators-emitters.

Purpose of the work – theoretical justification of the possibility of using LHP and heat-conducting refrigerators-emitters to ensure the operability of the on-board computer processor of nanosatellites with CFRP body.

2. Objects and materials

The nanosatellite has a frameless structure in the form of a thin-walled parallelepiped, in the center of which is installed a motherboard 6 with a processor 3 and six RAM elements 1 (figure 2). Between the processor 3 and the processor cover 4, a thin layer of thermal grease such as Evercool Nano Diamond is applied, which ensures perfect thermal contact between them. The processor and RAM elements are the only heat sources with the TDP of 15 W and 1 W, respectively. The processor Intelcorei7-8650U for light laptops can serve as an analogue.

The material of nanosatellite is orthotropic CFRP with thickness 2 mm (table 1). Assumed that λ_y and λ_z are the thermal conductivity coefficients, respectively, in the Y and Z directions of the reinforcement plane of each face of the nanosatellite body, and λ_x is thermal conductivity coefficient in the normal direction X of each face of the nanosatellite body.

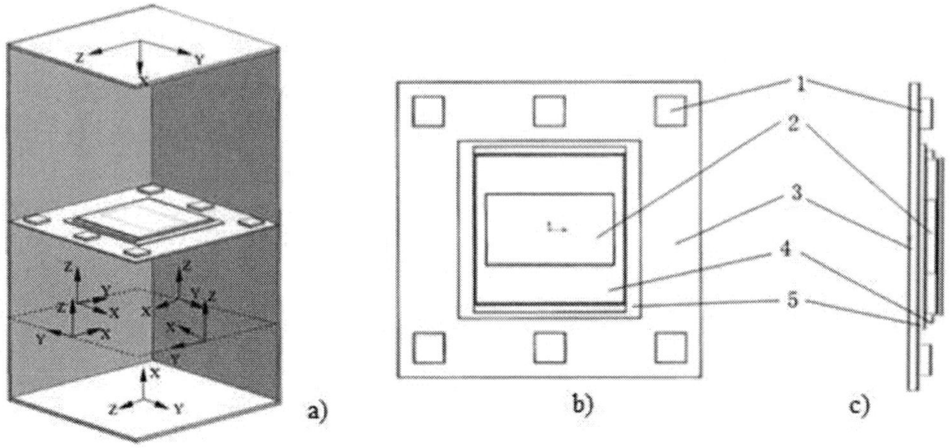

Figure 2. Structural diagram of a nanosatellite: a) general view, b) top and side views
1 – RAM; 2 – processor; 3 – motherboard; 4 – processor cover; 5 – processor board

Table 1. Geometrical dimensions and materials of nanosatellite construction

Name	Material	Coefficient of thermal conductivity λ, W/(m·K)	Size, mm
Motherboard	fiberglass	0.244	100x100x1
Processor board	fiberglass	0.244	60x60x0.8
Processor	silicon	148.0	24x42x1.3
RAM (6 pieces)	silicon	148.0	10x10x1.5
Thermal grease Evercool Nano Diamond	diamond micro particles	8.0	24x42x0.1
Processor cover	Aluminium alloy	144.0	53x53x0.5
Satellite body	CFRP	λ_x= 0.5; λ_y = var; λ_z=var	100x100x200

3. Thermal control systems

3.1 Loop heat pipes.

When the processor is cooled with LHP, the excess heat is transferred through the aluminum cover 4 to the copper interface 2, inside of which locates an evaporator 1 with a capillary-porous structure (figure 3). The evaporator 1 and the condenser 7 are connected by pipelines 11 to separate the circulation movement of the vapor (steam line) and liquid (condensate line) phases of the coolant. To increase the intensity of heat discharge into the surrounding space and to increase the contact area between the heat pipe capacitor 7 and the inner surface of the wall of the nanosatellite body 9, thermal grease 8 is used.

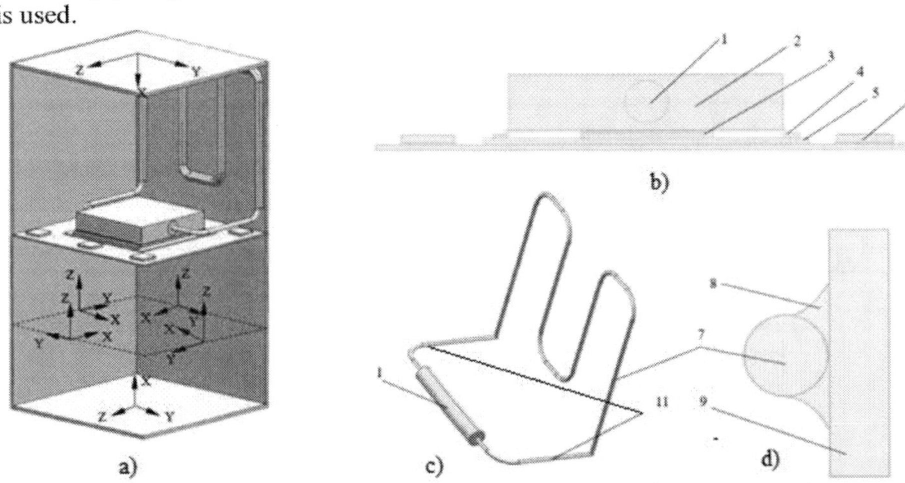

Figure 3. Thermal control system with the help of a loop heat pipe:
a) – general view; b) – assembly of the evaporator interface; c) – elements of LHP;
d) –diagram of the contact of the capacitor with the surface of the nanosatellite shell;
1– evaporator; 2 – evaporator interface; 3 – processor; 4– processor cover; 5 – processor board; 6 – RAM; 7 – capacitor; 8 – thermal grease; 9 – nanosatellite body;
10 – motherboard; 11 – pipelines (steam line and condenser line)

One of the main characteristics of LHP is its effective thermal conductivity coefficient λef. According to published data [12–16], the value of λef, can vary over a wide range and is higher the thermal conductivity of aluminum more than 100 times.

The temperature state of the processor is analysed in the module – Steady-State Thermal of ANSYS, which showed that for the given geometric dimensions of the LHP (table 2) with λef values from 5000

W/(m·K) to 20000 W/(m·K), the processor temperature changes less than 10%. In further calculations, it was assumed that the LHP is a monolithic body with effective thermal conductivity coefficient λ_{ef} = 20000 W/(m·K).

Table 2. The geometric dimensions of the loop heat pipe

Component	Characteristics	Value, mm
Evaporator	diameter / length	8.0/50.0
Pipelines	diameter / length	3.0/68.5
Capacitor	diameter / length	3.0/400.0
Vaporizer interface	length / width / thickness	50.0/50.0/10.0

3.2 Refrigerators-emitters.

When the processor is cooled with the help of refrigerators-emitters, the aluminum cover of processer is in contact with the nanosatellite body (figure 4). To ensure perfect thermal contact between them, a thin layer of thermal grease is applied. The processor board is connected to the motherboard with the information cable 4. The refrigerators-emitters are screens made of high-conductivity CFRP, that will unfold after the nanosatellite is put into orbit of the Earth.

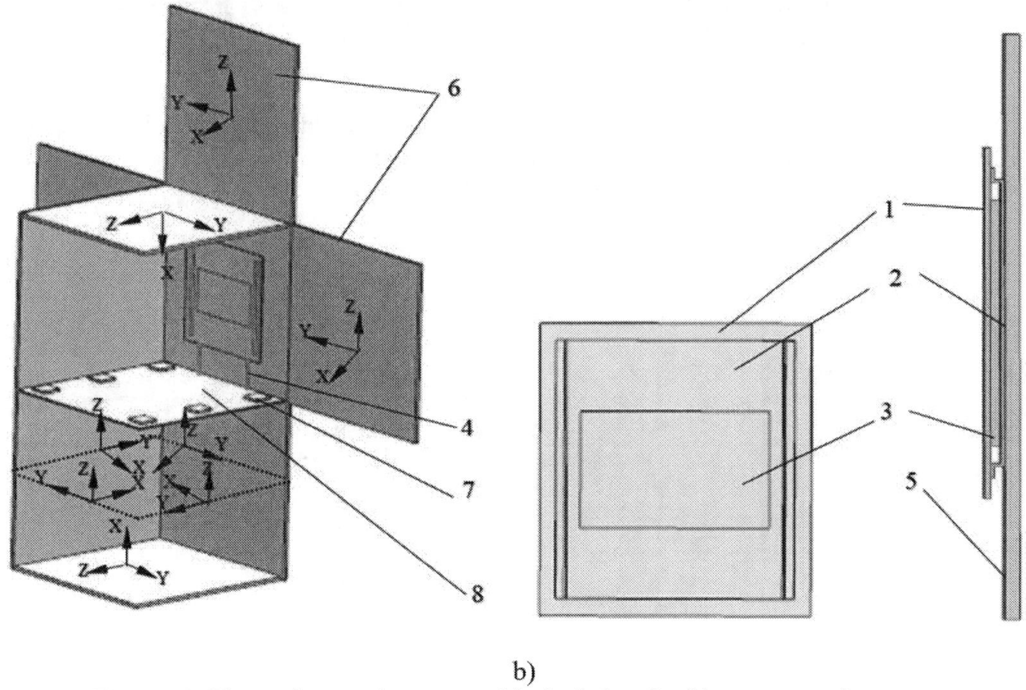

Figure 4. Thermal control system with the help of refrigerators-emitters:
a) – general view; b) – contact diagram of the processor with the surface of the nanosatellite body
1 – processor board; 2 – processor cover; 3 – processor; 4 – information cable; 5 – nanosatellite body; 6 – refrigerators-emitters; 7 – RAM; 8 – motherboard

4. Mathematical simulation

The simulation was carried out in the Siemens NX program in elliptical orbit conditions with an angle of inclination 95° and anoperating period 5980 s. The following values served as initial data: thermal conductivity coefficients in Y and Z directions of the reinforcement plane nanosatellite body λy = 6.0

W/(m·K) and λz= 4.0 W/(m·K); the emissivity coefficient of the nanosatellite surface was assumed to be ε= 0.9; the solar absorption coefficient As = 0.3. The geometric dimensions of the nanosatellite elements and the thermophysical characteristics of the materials were taken from tables 1, 2.

When the processor is cooled with LHP, the temperature of the processor on the illuminated side of the orbit exceeds itspermissible value and amounts to 132°C (figure 5). The maximum processor temperature on the shadow side of the orbit is only 4 ° C lower. In this case, the temperature of the nanosatellite body varies from minus 30°C to plus 50°C (figure 6). RAM elements are heated unevenly. The RAM element located between the interface and the LHP capacitor near the illuminated face of the nanosatellite body has the maximum temperature of 174°C. Therefore, at further stages of nanosatellite design, it is necessary to determine a more rational location of RAM elements on the circuit board or to change the structure of the LHP interface in order to remove the additional heat.

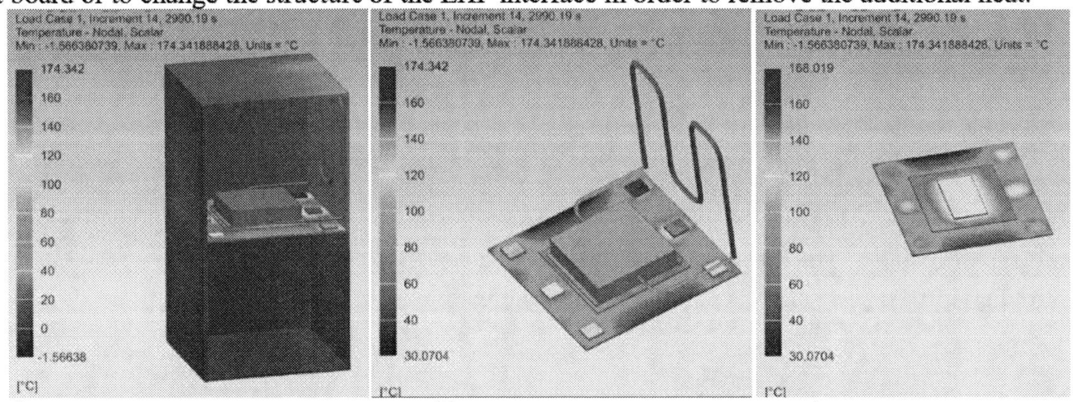

a) b) c)

Figure 5. Temperature state of the nanosatellite on the illuminated side of the orbit when cooling with LHP (t = 2990 s):
a) general view; b) interior fittings; c) processor

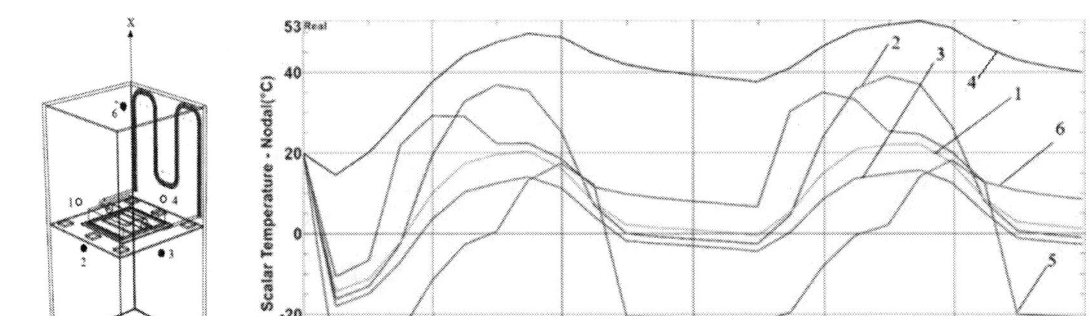

Figure 6. The temperature in the center of each face of the nanosatellite body during orbital flight when the processor is cooled with LHP

Similar calculations were carried out for scheme 2, when the processor is cooled with refrigerators-emitters. The maximum processor temperature on the illuminated side of the orbit also exceeds its permissible value and amounts to 147 ° C. At the same time, all RAM elements remain relatively cold and their temperature does not exceed 67°C (figure 7).

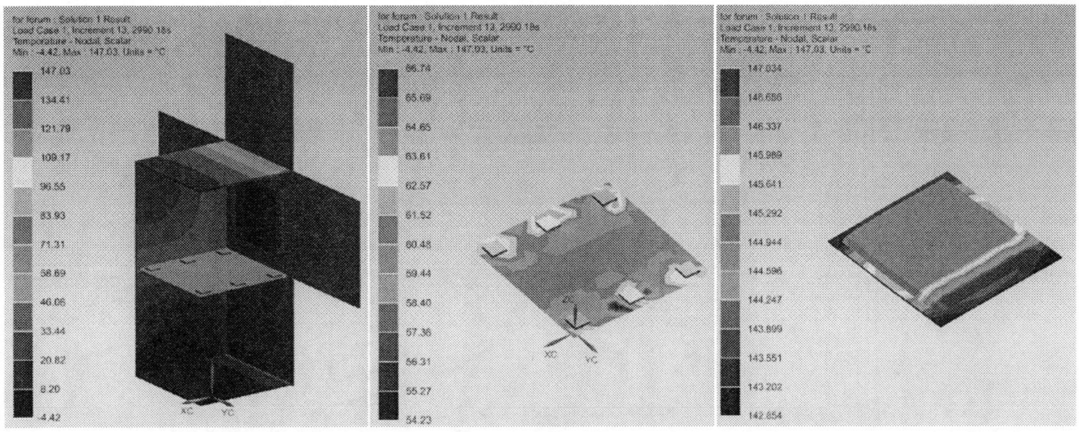

Figure 7. Temperature state of the nanosatellite on the illuminated side of the orbit ($t = 2990$ s) when the processor is cooled with the refrigerators-emitters: a) general view; b) motherboard and memory; c) processor

The anisotropy of the thermal conductivity coefficient in the reinforcement plane [17, 18] of each face of the CFRP nanosatellite body has a different degree of influence on the temperature state of the processor depending on the selected thermal control system. For example, when cooling the processor with LHP, increasing the thermal conductivity in the Y and Z directions of each face of the CFRP body from 4.0 W/(m·K) to 100.0 W/(m·K)reduces the temperature of the processor by 10 ... 16°C, and when cooling the processor with the help of refrigerators-emitters – by 55 ... 60°C (figures 8, 9). Thermal control system based on LHP is preferable for composite materials with low values of λ_y and λ_z,. When using highly heat-conducting CFRP, it is more efficient to remove heat from the processor using refrigerators-emitters.

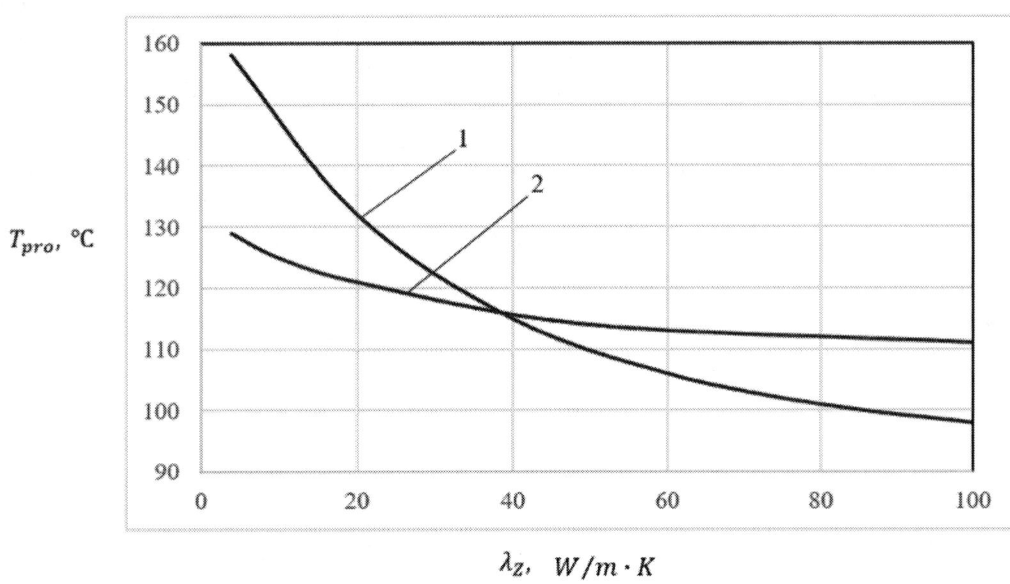

Figure 8. The influence of the thermal conductivity coefficient λ_z on the processor temperature on the illuminated side of the orbit for different thermal control systems ($\lambda_y = 4.0$ W/(m·K)): 1 – refrigerators-emitters; 2 – LHP

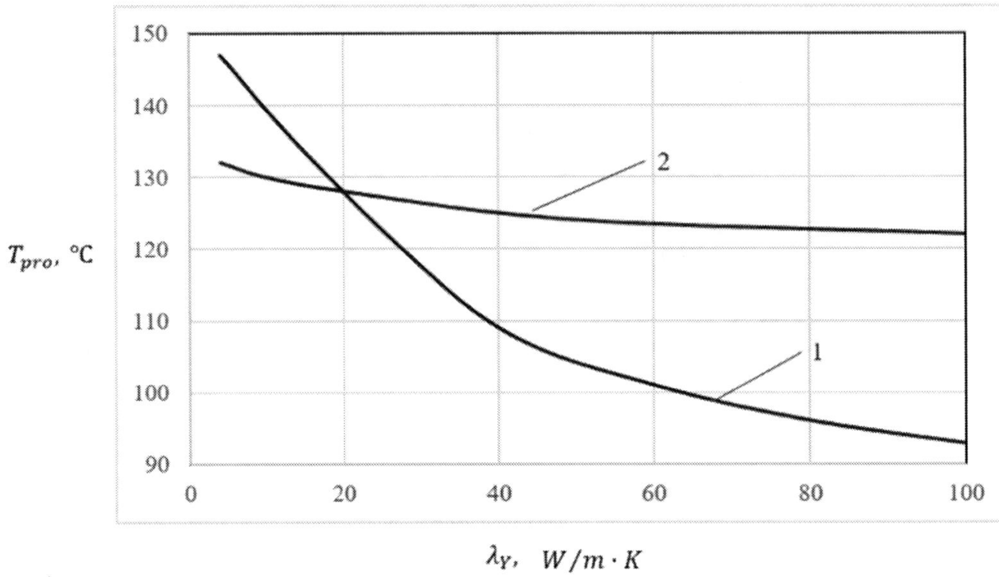

Figure 9. The influence of the thermal conductivity coefficient λ_y on the processor temperature on the illuminated side of the orbit for different thermal control systems (λ_z = 4.0 W/(m·K)): 1 – refrigerators-emitters; 2 – LHP

With the values of the thermal conductivity coefficients of the CFRP $\lambda_y = \lambda_z$ = 100.0 W/(m·K), the processor temperature will be respectively 90°Cand 73°C for cooling systems with loop heat pipes and highly heat-conducting refrigerators-emitters.

5. Conclusions

The possibility of using loop heat pipes and highly conductive refrigerator-emitters to cool a processor with the thermal power of 15 W (class of light notebooks) in nanosatellites with a CFRP body has been studied. It has been established that at values of thermal conductivity coefficients in the reinforcement plane of the CFRP body up to 20 ... 40 W/(m·K), the heat is more effectively removed by LHP, and at higher values of λ_y and λ_z – by refrigerators-emitters. For normal processor operation, the nanosatellite body under consideration must be made of CFRP with a thermal conductivity coefficient of 100 W/(m · K) and higher.

Reference

[1] Makridenko L A, Boyarchuk K A 2005 Mikrosputniki. Microsatellites. Development trend. Market features and social significance *Voprosyelektromekhaniki* **102** pp12-27

[2] grandviewresearch.com [Internet]. San Francisco: Nanosatellite and Microsatellite Market Analysis Report By Mass, By End Use (Defense& Security), By Application (Communication & Navigation, Scientific Research), And Segment Forecasts, 2019 – 2025, [updated 2019 May]. Available from: www.grandviewresearch.com/industry-analysis/nanosatellite-microsatellite-market

[3] Mikhailovskii K V, Prosuntsov P V and Reznik S V 2012 Development of high-conductivity polymer composite materials for space structures *Vestnik MGTU im. N.E. Baumana. Seriya «Mashinostroenie»* **9** pp 98-106

[4] Ivanov M A and Kirillov K Yu 2014 On board control system for moon artificial satellite *Spacecrafts and technologies* pp. 4-11

[5] Modern mobile technology. URL:https://www.notebook-center.ru/processor_tdp.html

[6] Nikolskii V V 2012 Designing of ultra-small spacecrafts: training manual (Saint-Petersburg: Baltic State Technical University)

[7] Chebotarev V E and Kosenko V E 2011 The basics of designing a spacecraft information support: tutorial (Krasnoyarsk:Siberian State Aerospace University)

[8] Tsaplin S V and Bolychev S A 2019 Investigation of the thermal conditions of the optoelectronic telescopic module of a nanosatellite in operating conditions. *Vestnik of Samara University. Aerospace and Mechanical Engineering* **2** pp96-111

[9] Dunn P D and Reay D A 1976 Heat Pipes (Oxford - New York - Toornto - Sydney - Paris – Braunschweig: Pergamon Press) p318

[10] Maydanik Yu F, Vershinin S V and Chernysheva M A 2016 Development and Research of Ammonia Miniature Loop Heat Pipe under Various External Conditions *Journal of Thermal Processes in Engineering* **7** pp 312-320

[11] Maydanik Yu F, Vershinin S V and Chernysheva M A 2016 Development and Research of Ammonia Miniature Loop Heat Pipe under Various External Conditions *Journal of Thermal Processes in Engineering* **7** pp312-320

[12] Ren Chuan, Wu Qing-Song, Hu Mao-Bin 2007 Heat transfer with flow and evaporation in loop heat pipe's wick at low or moderate heat fluxes *International Journal of Heat and Mass Transfer* **50**(11-12) pp2296-2308

[13] Afanasev V N, Nedaivozov A V and Yakomaskin A A 2014 Experimental Study of Processes in Loop Heat Pipes *Vestnik MGTU im. N.E. Baumana. Seriya «Mashinostroenie»* **2** pp. 44-61

[14] Pozhilov A A, Zaitsev D K, Smirnov E M and Smirnovsky A A 2017 Numerical simulation of heat and mass transfer in a 3D model of a loop heat pipe evaporator *St. Petersburg Polytechnical State University Journal. Physics and Mathematics* **10 (3)** pp 52–61 DOI: 10.18721/JPM.10305

[15] Chernysheva M A, Vershinin S V and Maydanik Y F 2007 Operating Temperature and Distribution of a Working Fluid in LHP *International. Journal of Heat and MassTransfer* **50**(13-14) pp 2704-2713

[16] Derevyanko V A, Nesterov D A, Kosenko V E, Zvonar V D, Chebotarev V E, Fatkulin R F and Suntsov S B 2013 Flat heat pipes for heat removal from electronic equipment in space vehicles *Siberian Journal of Science and Technology* **6 (52)** pp111-116

[15] Reznik S V, Prosuntsov P V, Denisov O V, Petrov N M, Shulyakovsky A V and Denisova L V 2016 Numerical and experimental estimation of heat conductivity of carbon plastic in a reinforcement plane on the basis of contactless measurement of temperature *Journal of Thermal Processes in Engineering* **12** pp557-563

[18] Reznik S V, Prosuntsov P V, Denisov O V, Petrov N M and Vonheong Lee 2017 Nano-satellite body composite material thermal conductivity determination computational and theoretical method *RUDN Journal of Engineering Researches* **18(3)** pp345-352 DOI 10.22363/2312-8143-2017-18-3-345-352

IOP Conf. Series: Materials Science and Engineering **934** (2020) 012057 doi:10.1088/1757-899X/934/1/012057

Increase of the resistance to high-temperature effects of carbon composite materials

K S Panina[1,2], E A Danilov[1], Yu A Kurganova[2]

[1]JSC "Research Institute of Structural Materials based on Graphite" NIIgrafit "

[2] Dynamics and flight control of motion of rockets and spacecraft, Bauman Moscow State Technical University, 2-nd Baumanskaya, 5, b.1, 105005, Moscow, Russia

Abstract. The article considers the possibility of increasing resistance to high-temperature effects of carbon composite polymeric materials by combining the components of the binder and its modification by silicon carbide powder. The technology for producing a single-layer composite is described. According to the results of the experiments, the composition and amount of the binder and SiC additives leading to a significant reduction in mass loss during high-temperature tests were revealed.

1. Introduction

Modern aircraft designs are aimed at the transition to high power and flight speeds, which in turn leads to an increase in operating temperatures and aggressiveness of working units and parts of aerospace engineering. Accordingly, the requirements for materials used are toughened: for example, temperature resistance, resistance to thermal shock, and resistance ablation, erosion, oxidation, strength characteristics, weight and size parameters, etc. Therefore, obtaining and improving high-temperature materials used as heat-protective coatings (HPC) in aircraft structures is an urgent task.

Carbon fiber-based polymer composite materials (PCMs) have a high density and greater processability compared to high-temperature ceramics.

Some groups of polymers such as phenol-aldehyde, organosilicon, polyimide and their compositions can be considered in role of matrices.

Phenol-formaldehyde resins and organosilicon compounds are of the greatest interest for the creation of thermoelectric heaters operating at temperatures above 1500 ° C.

Resol phenol-formaldehyde resins (PF) have high thermal stability and are able to withstand high temperatures in the range of 1000-1500 ° C for several seconds [1,2]. PF are widely used to produce carbon-carbon materials through high-temperature pyrolysis. For example, a matrix based on carbon fibers is impregnated with a phenol-formaldehyde resin and then pyrolyzed at 2000 ° C to form a strong coke residue, which remains until the carbon sublimation temperature [1].

Thus, FPcan be a promising material as a PCM matrix developed for products subject to short-term high-temperature effects in th e range of 1300-3600K.

Organosilicon polymers have elasticity, heat resistance, heat resistance, weather resistance, resistance to gases at elevated temperatures.

One of the most important properties is the formation during high-temperature degradation of silicon and its compounds (carbides, nitrides, oxides, etc.) [3,4], which have higher melting and sublimation temperatures. This property is widely used to obtain ceramic matrices in composite materials that effectively protect against external influences. For example, under the influence of high temperatures, the healing properties of SiO2 are used, which is formed as a result of oxidation of the SiC matrix or Si and restricts the access of oxygen to the fiber surface; such a matrix works for up to1650-1700 ° C for several hours. [4] Organosilicon is widely used for heat-shielding materials in China and Japan [4], active development is underway in the Russian Federation: for example, a quasi-plastic composite material reinforced with carbon fibers UKN-5000 LU-24P and organoelement polymers with a temperature resistance of up to 1500 ° C; the Steklarm is a material with a glass-ceramic matrix obtained from polymer precursors (polysilosanes, polycarbosilanes, polyborethoxysilosanes), it have high-strength heat-resistant at 1500 ° C[6,7].

Such properties of organosilicon polymers as the elasticity in its original form and the formation of a ceramic component under the influence of heating of the working medium under operating conditions can be successfully implemented to protect products that must be flexible, withstand harsh high-temperature environmental conditions.

Fillers and catalysts, modifying additives, flame retardants, such as metals and their compounds, silicon-based compounds, carbon, calcium fluorides, phosphorus-based flame retardants, ceramic powders, and other various sizes and quantities are actively used to increase operational characteristics [8-11].

For example, it is possible to use finely dispersed fillers, including finely dispersed silicon carbides, and thermally expanding graphite as a high-temperature additive in elastomeric matrices [10].

The introduction of flame retardants into a polymer binder can have a significant effect on increasing the heat resistance of polymer composites used for parts operating in oxidizing environments.

In this work we consider layered composite materials based on carbon fibers with a matrix obtained from a mixture of phenol-formaldehyde resin and an organosilicon polymer with inorganic additives to create thin-walled HPC.

2. Materials and methods

The studied materials are single-layer composite materials based on carbon fibers. The polymer matrix consists of phenol-formaldehyde resin in the form of a powder, silicone and hardener. Micron powder of silicon carbide was used as a modifying additive.

Composite materials are made by traditional technological methods of liquid-phase combination of filler and matrix.

The PF powder or a solution of PF in isopropyl alcohol is mixed with silicone and then subjected to intensive mixing using the Ultraturex apparatus for 2-3 minutes. After that, silicon carbide powder and condensation agent are added to the mixture.

The samples are dried at 60 ° C for 4 hours and cured at 120 ° C for 4 hours after uniformly application a binder to the carbon fabrics layer.

To determine the optimal composition of the samples, fire resistance tests are carried out in the flame of an acetylene burner for 60 s. After the test, the relative weight loss is determined.

3. Results and discussions

To determine the ratio of the main components, a series of samples was prepared with matrices containing parts by weight of components: silicone without PF, silicone: PF (3: 1), silicone: PF (2: 1), silicone: PF (1: 1), silicone : PF (1: 2), silicone: PF (1: 3). PF was dissolved in isopropyl alcohol. The test results in the burner flame are shown in Figure 1.

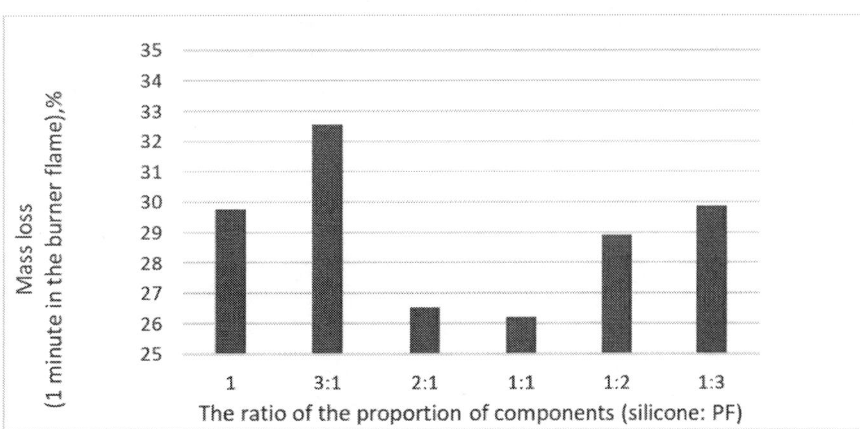

Figure 1.Dependence of mass loss after exposure to flame for a minute on the ratio of the proportions of polymer binders.

The smallest weight loss is exemplified by the compositions: silicone: FFS (2: 1) - 26.52%, silicone: FFS (1: 1) - 26.21%. However, the latter has a higher viscosity, therefore, the polymer composition silicone: FFS (2: 1) was chosen as the basis for further tests.

To assess the effect of silicon carbide particles on the fire resistance of polymer composites, samples were prepared based on silicone and phenol-formaldehyde resin powder (2: 1) and silicon carbide additives in amounts of 1, 3.5% of the mass.
Photographs of the original structure are shown in Figures 2-5.

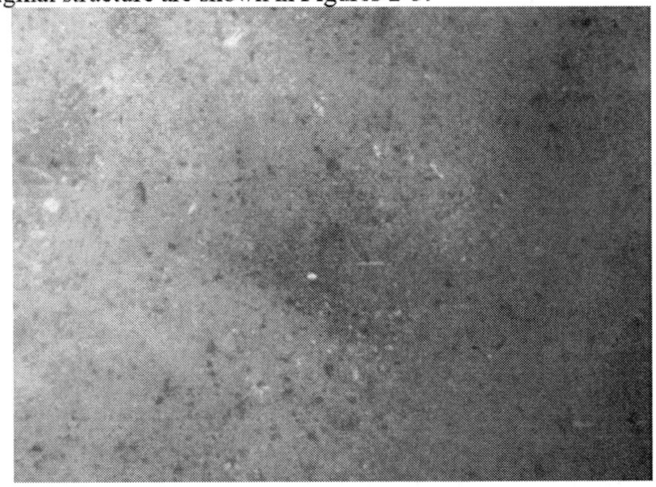

Figure 2.Microstructure of samples based on silicone and phenol-formaldehyde resin powder (2: 1) without silicon carbide additives (x100).

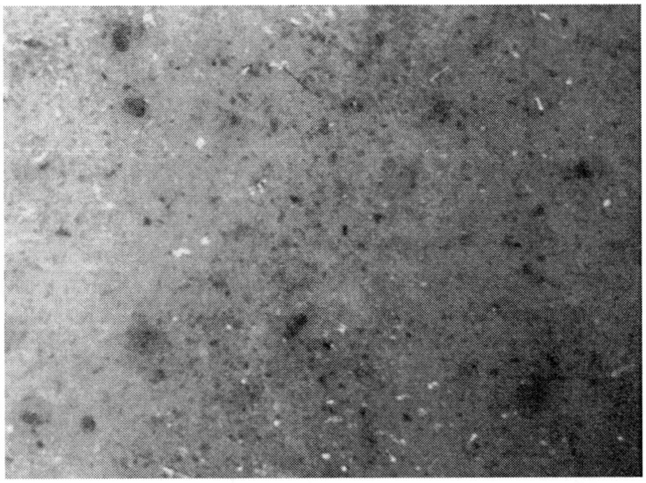

Figure 3.Microstructure of samples based on silicone and phenol-formaldehyde resin powder (2: 1) with the addition of silicon carbide in an amount of 1% mass (x100).

Figure 4. The microstructure of samples based on silicone and phenol-formaldehyde resin powder (2: 1) with the addition of silicon carbide in an amount of 3% by mass (x100).

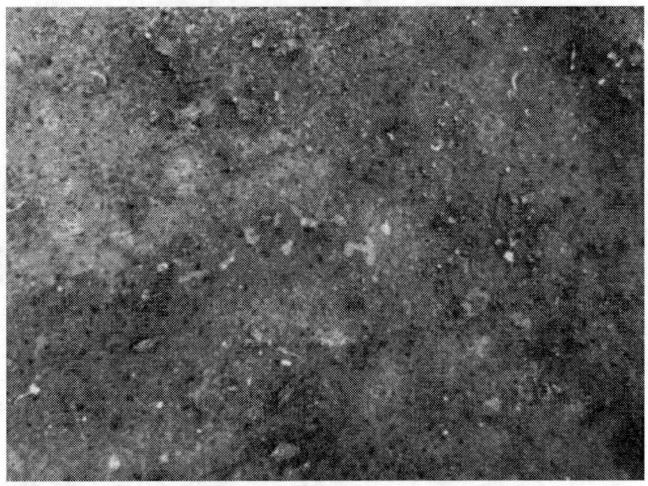

Figure 5. Microstructure of samples based on silicone and phenol-formaldehyde resin powder (2: 1) with the addition of silicon carbide in an amount of 5% by mass (x100).

The microstructure of the polymer matrix shows continuity, uniformity, the added silicon carbide particles are distributed evenly throughout the binder.
Burner flame test results are shown in Figure 6.

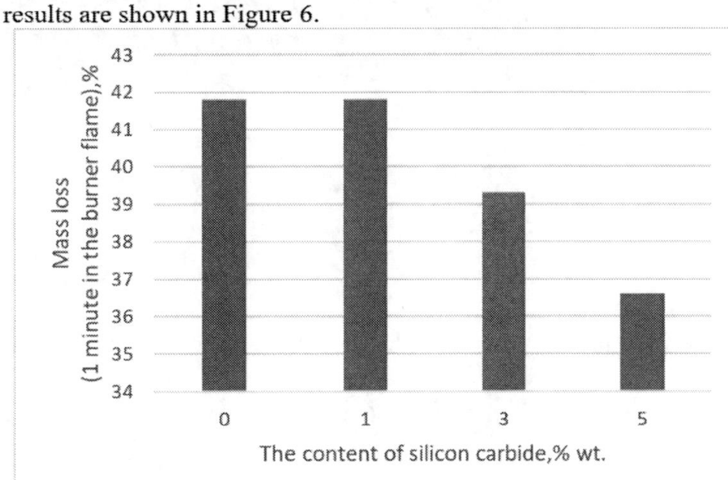

Figure 6. Effect of the addition of silicon carbide on weight loss after exposure to flame for a minute

According to the results of the introduction of the additive, there is a tendency to reduce weight loss during combustion. The best result is shown by a sample with the addition of 5% SiC. The relative change in sample weight decreased from 41.8% for samples without additives to 36.6% for samples containing 5% SiC. A 15.2% reduction in losses indicates the effectiveness of the modification of the binder with a ceramic filler.
Samples after testing are shown in Figures 7-10.

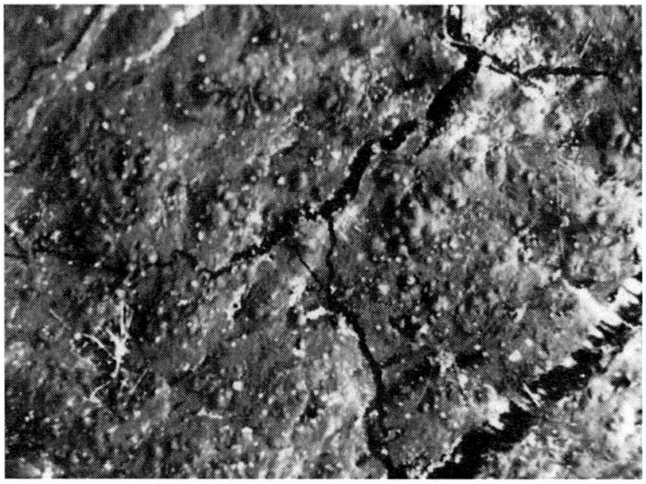

Figure 7. The surface after testing in a flame of a burner of samples based on silicone and phenol-formaldehyde resin powder (2: 1) without additives of silicon carbide.

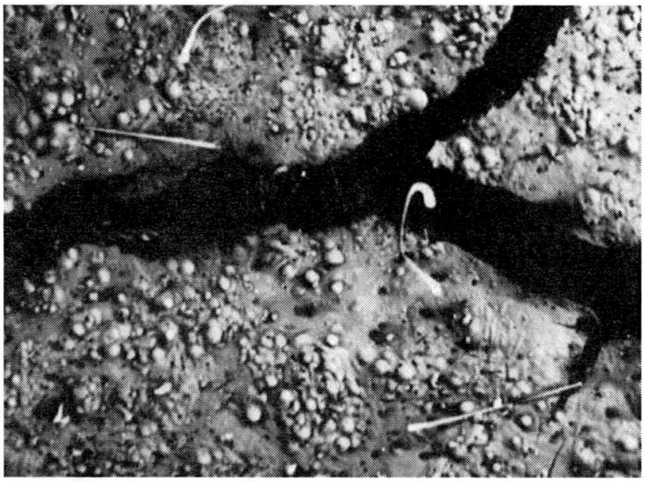

Figure 8. The surface after testing in a flame of a burner of samples based on silicone and phenol-formaldehyde resin powder (2: 1) with the addition of silicon carbide in an amount of 1% mass (x100).

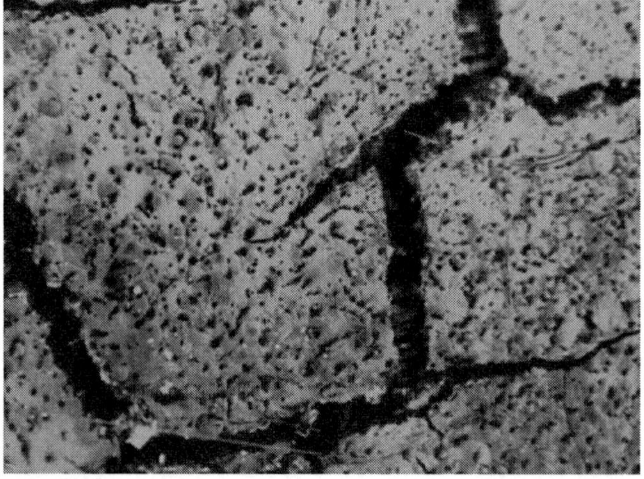

Figure 9. The surface after testing in a flame of a burner of samples based on silicone and phenol-formaldehyde resin powder (2: 1) with the addition of silicon carbide in an amount of 3% by mass (x100).

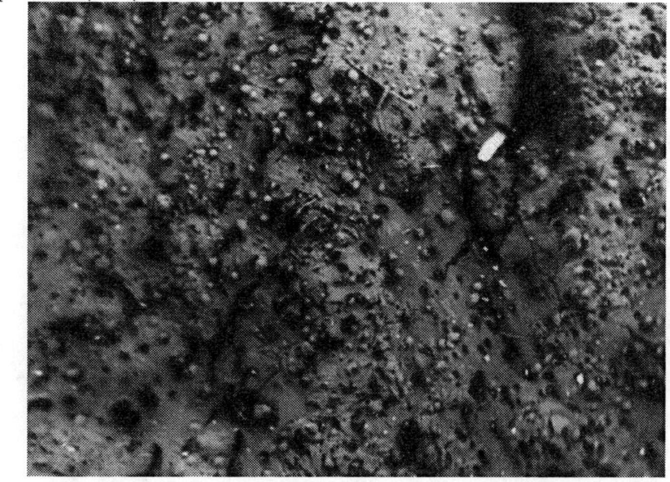

Figure 10. The surface after testing in a flame of a burner of samples based on silicone and phenol-formaldehyde resin powder (2: 1) with the addition of silicon carbide in an amount of 5% by mass (x100)

In the process of flame exposure, a carbon residue forms on the surface of the samples, which protects the carbon fibers from oxidation. The presence of spherical oxides is observed.

4. Conclusions

Thus, a silicone-phenol-formaldehyde resin-silicon carbide system composite has been proposed as heat-protective thin-walled composite materials. Based on the results of the experimental data, the composition of the binder silicone was selected as the basis: phenol-formaldehyde resin with respect to parts 2: 1 by weight. The results of the experiments revealed the effect of inorganic additives of silicon carbide on reducing weight loss. The most effective is the addition of 5% SiC. Samples with this content show a loss reduction of 15.2% relative to unfilled binder. Thus, the combination of the considered polymers and the modification of the mixture with the addition of silicon carbide leads to an increase in the resistance to high temperatures of the material in an oxidizing environment. This composite can be recommended as a heat-shielding material for parts and components of aviation and space technology.

References

[1] Sabadaha E N, Prokopchuk N P, Schutova A L and Globa A I 2017 Thermostable composites//*Works of BGTU. Series 2 Chemical technologies, biotechnology, geoecology*. **2(2)** Page 108-115.

[2] Pakhomova SA and Povalyayev A I 2019 Silicon nitride-based ceramic composite materials for corrosion-resistant rolling bearings // IOP Conf. Series: Materials Science and Engineering 683 012040. DOI:10.1088/1757-899X/683/1/01204

[3] Rudakova T A, Perov N S and Ozerin A N 2011 Fire protection elastic coatings on organosilicon binder//News of UFU. Technical sciences.. Page 215-219.

[4] Garshin A P and Kulik V I 2017 Oznaya directions of increasing corrosion and heat resistance of refractory fibre-reinforced ceramic-matrix composites//Scientific research and development.. T. 12. Page 49-59.

[5] Kraev I D and other. 2017 Prospects for the use of organosilicon elastomers in the development of modern polymer materials and coatings for various purposes // Proc. VIAM. **12** C. 5–5.

[6] Suntsev S S 2010High-temperature composites and coatings based on glass and ceramics for aerospace engineering//Ros. chemical. G. (J. Ros. chemical. D.I. Mendeleeva).. T. 1. Page 25-33.

[7] Gunyaev G M and Gofin M J 2013 Carbon-carbon compositions//*Aviation materials and technologies*.. Page 62-90.

[8] Polonik V D and Prokopchuk N R 2013 Properties of elastomeric compositions with fluorine-containing additive//Works of BGTU. Series 2 Chemical technologies, biotechnology, geoecology **4** Page 141-143.

IOP Conf. Series: Materials Science and Engineering **934** (2020) 012057 doi:10.1088/1757-899X/934/1/012057

[9] Kablov V F, Novopoltseva O M, Kochetkov V G, Lapina A G and Zvada S S 2015 Study of the influence of fine swelling systems of flame retardants on fire resistance of elastomeric compositions//*Journal of Technological University*. **18(14)** Page 1-2.

[10] Patent RU № 2594014. Method of producing binder based on phenol-formaldehyde resin of rubber type for laminar material, binder and laminate based on binder and reinforcing fibrous base. Federal State Budget Educational Institution of Higher Education "Kazan National Research Technology

[11] Shvets N I et al. 2013 Phenol formaldehyde binder of reduced combustion//Fire and explosion safety **22(5)** Page 26-32.

IOP Conf. Series: Materials Science and Engineering **934** (2020) 012058 doi:10.1088/1757-899X/934/1/012058

Ceramic Matrix Composite Materials for Structural Elements of Aviation and Rocket-Space Equipment

P A Stepanov[1,2], I G Atroshchenko[1] and O V Nikulina[1]

[1] JSC ORPE Technologiya named after A.G. Romashin, 15 Kievskoye Shosse, Obninsk, Russia
[2] Bauman Moscow Technical State University, ul. Baumanskaya 2-ya, 5/1, Moscow, Russia

E-mail:lunya23@yandex.ru

Abstract. Ceramic matrix composite materials based on inorganic binders (based on aluminum-chromium-phosphate binder and quartz glass slurry) and textured quartz filler are obtained by various molding methods at JSC ORPE Technologiya named after A.G. Romashin. The materials were studied in the high-temperature region (at the ambient temperature up to 1200°C during short-term exposure). The main specifications of the materials obtained in comparison with those of quartz ceramic material are shown.

Key words: ceramic matrix materials, inorganic binder, quartz fabric, slurry, ceramics.

1. Introduction

Modern aircraft represent sophisticated technical complex. At present, in international practice of aerospace industry there is a trend of increasing the speed and maneuvering ability of aircraft based on severe competition among manufacturers on the global market of civil and military equipment. This involves constant severization of requirements for radiotransparent structural elements. One of the major challenges in modern aviation and rocket equipment is the development of radiotransparent aircraft elements.

There is a system approach to the development of aircraft element structures which requires solution of a number of problems:

-selection of the material of the radiotransparent structure considering the required radio-technical characteristics and aircraft operation conditions;
-development of methods to provide the specified radio-technical characteristics;
-selection of the materials of the connection parts and development of their structure;
-selection of moisture-protective coating in case of using porous aircraft materials;
-development of methods for simulating the specified operation conditions.

The development of aircraft elements at all production stages is performed considering severe requirements for the level and stability of dielectric properties of radiotransparent materials and for the structure of aircraft elements in general. In addition to stable dielectric properties, materials of aircraft elements should have appropriate thermal insulating properties, the main ones being low thermal conductivity and relatively high thermal capacity.

Due to the operation of the aircraft elements under conditions of dust, rain, wind, and high or low temperature which adversely affect the composite material when getting into pores or destroying the wall, it is necessary to use moisture-protective and anti-erosion coatings. Physical and mechanical properties of the coatings should be selected through careful coordination of their dielectric properties, thermal conductivity, thermal stability and thermal coefficients of linear expansion.
When analyzing the whole set of requirements for modern aircraft elements for aviation and rocket equipment, and underlying their contradictory nature, the research should be done individually for each specific aircraft element during the development of its structure and manufacturing technology.
The relevance of the work is caused by the need to develop new materials for thermally-loaded radiotransparent aircraft elements for aviation and rocket equipment.
Among many structural materials the most promising for radomes are: polycrystalline glass, oxide ceramics, glass ceramics, nitride ceramics, and composite materials (including glass-fiber reinforced plastics).
Ceramic materials and polycrystalline glass have high thermal stability and radio transparency up to1600°C and higher, therefore they are widely used in manufacture of radomes for high-speed rockets. However, these materials, unlike composites, have low impact strength (capability of a material to absorb energy fast) and crack growth resistance (a property of materials to resist crack growth under mechanical and other effects). Ceramic materials and articles made thereof are also difficult to manufacture and require expensive equipment and tooling to be used.
Among radiotransparent composite materials one should distinguish glass-fiber reinforced plastics, polymer composite materials based on inorganic oriented quartz, glass and silica fabric fillers, which are widely applied in the structures of aviation and rocket equipment units. Applying glass-fiber reinforced plastics for aircraft structural elements enables one to drastically reduce the weight of articles, and therefore to improve fuel efficiency of airplanes and increase payload.
However, polymer composite materials have significant restrictions in the operation temperature range (< 450°C).
A promising option for developing reinforced heat-resistant composite materials is to develop structural composite materials based on silica or quartz fabrics and inorganic binders. The binders can be composed of various inert fillers by means of which mechanical, thermophysical, dielectric and other properties can be controlled.

2. Development of the material

Composite materials based on inorganic oriented quartz, glass and silica fabric fillers are widely applied in the structures of aviation and rocket equipment units. For new generation airplanes, rockets and engines, the use of glass-fiber reinforced plastics enables one to drastically reduce the weight of articles, and therefore to improve fuel efficiency of airplanes and increase payload.
The use of combinations, reinforcing and binding components makes it possible to produce materials with a wide range of controllable properties, which predetermines a great variety of applications of composite materials.
To maintain high physical and technical properties and reliability of the functioning of thermally-loaded structure elements during operation, it became necessary to use composite materials serviceable in the high-temperature region. Advanced polymer composite materials allow long-term operation at a temperature not above 400-450°C. One of the solutions of this problem is the use of inorganic binder, e.g. binders based on slurries, phosphates and borides.
Phosphate binders are aqueous solutions of hydrophosphates. These binders can be composed of various inert fillers which allow mechanical, thermophysical, electrophysical and other properties to be controlled. Aluminum-chromium-phosphate binder stands out among various phosphate binders. Due to the high aluminum phosphate melting temperature, phosphate binder-based articles which are serviceable up to 1500°C can be obtained.

JSC ORPE Technologiya uses vacuum and contact molding methods to produce the ceramic matrix composite material KhAFSkv based on inorganic (aluminum-chromium-phosphate) binder and textured filler.

To protect the filler from corrosive action of acid medium of aluminum-chromium-phosphate binder, the optimal finishing composition was selected. As a finishing composition 5-15% alcohol-acetone solution of silicone resin was selected. The thin film of the solution was applied to the filler, and the amount of the film did not deteriorate dielectric properties of the material under high temperatures and maintained strength properties of the filler (e.g. quartz or silica fabric). The use of more concentrated solution led to the deterioration of dielectric properties of the material under high temperatures due to the fact that when the resin is degraded without the oxidizer carbon compounds which significantly affect radio-technical characteristics of the material can be formed [2].

Microground electrocorundum and aerosol powders were used as modifiers of the aluminum-chromium-phosphate binder. The use of fine modifiers is caused by the fact that applying coarser fractions resulted in strength degradation of the filler during molding, since these modifiers are abrasive materials and destroy the fiber under pressure.

The optimal binder-powder ratio was selected: 55-65% and 35-45% respectively. Additional introduction of microground powder as a modifier improves thermophysical properties of the material though deteriorating mechanical ones (porosity increases).

Ceramic matrix composite material is obtained by vacuum and contact molding methods. For details refer to [2, 3].

One more area of the development at JSC ORPE Technologiya is ceramic matrix composite material KhAFS-Sh from quartz slurry-based binder and textured filler. Let us observe this material in detail.

The binder is composed of: aluminum-chromium-phosphate binder, silica sol and quartz slurry. Aqueous slurry based on quartz glass is obtained by bead milling with corundum lining and alundum milling media with subsequent stabilization by mechanical mixing. To achieve general resistance to chemicals (aluminum-chromium-phosphate binder having pH=1-2), the binder with quartz slurry was modified by silica sol KZ-TM-30 with a density of 1.207 g/cm3 and pH=10-12. Silica sol is colloidal solution composed of dispersed medium which is constituted by water, and dispersed phase representing amorphous silica micelles. Silica sol micelles saturated with water molecules have polymeric nature, highly developed surface and large number of functional (silanol) groups, which provide high reaction capacity and possibility of modifying the surface of particles by adsorbing different ions. When acid-alkali solution is changed towards acid medium, and if sol was alkaline, gelation process takes place, in which case the joining of nanoparticles into large combinations, which are not capable of further milling, occurs. The introduction of silica sol does not only neutralize pH of the binder medium but also reinforces the binder and reduces the curing temperature. [3]

In terms of processability, ecology and level of impact on the fiber and quartz filler, the optimal compositions are those which are close to neutral ones with pH=6-7 units. These compositions contained: 4-8% of aluminum-chromium-phosphate binder, 20-40% of silica sol, and the rest was quartz glass slurry. [4]

To accomplish the objectives, three methods for manufacturing ceramic matrix material KhAFSkv-Sh were tested: contact, autoclave, and vacuum ones. To determine the optimal heat treatment temperature, specimens were provided to perform a series of investigations on the impact of heat treatment temperature on material characteristics (structural and mechanical (ultimate strength at maximal bending). The heat treatment was applied at the following peak temperatures: 350, 600, 800, 900, 1000, 1100, and 1200 0C.

From the data on properties obtained, the most optimal method for manufacturing ceramic matrix material was vacuum molding method, and it should also be noted that higher strength characteristics obtained from the specimens of this molding method were observed after the heat treatment mode at 600°C with maximal exposure time of two hours.

3. Results and discussion

After the optimization of the process for obtaining the material, specimens for the investigation of a set of properties were fabricated. The investigation results showed high service reliability of the ceramic matrix composite materials obtained. The investigations were conducted using the methods developed at JSC ORPE Technologiya named after A.G. Romashin.

The microstructure of the specimens of ceramic matrix materials and (quartz) ceramics was investigated using the Nikon microscope with magnification of 1.00 KX. The analysis of the results of the microstructure (Figure 1) showed that the specimens of ceramic matrix composite material have amorphous structure typical of polymer composite materials, whereas ceramic material, according to the microstructural analysis data, has inclusions in the form of the crystalline phase which can lead to the material destruction.

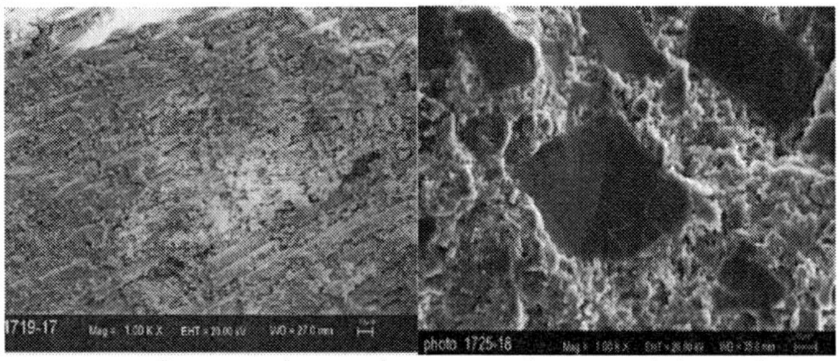

Figure 1. Microstructure of a ceramic matrix composite material specimen (left) and quartz ceramics specimen (right).

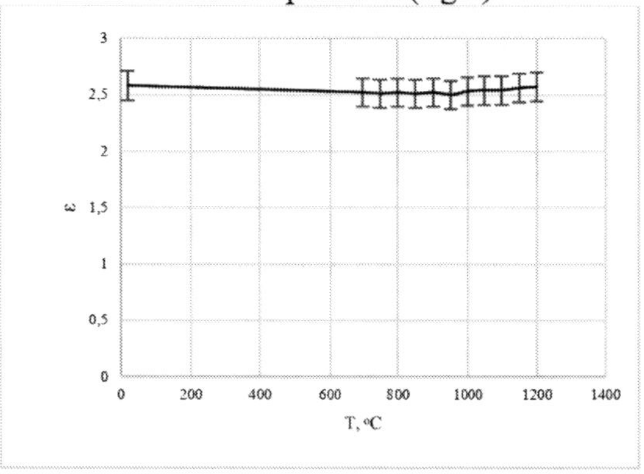

Figure 2. Dielectric constant of ceramic matrix composite material as a function of temperature.

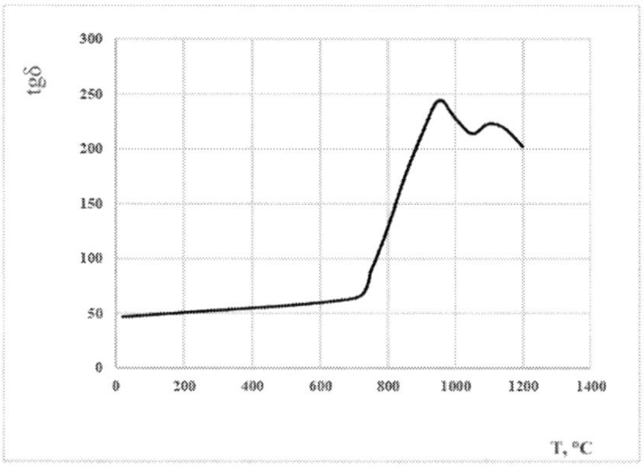

Figure 3. Dielectric loss tangent of ceramic matrix composite material as a function of temperature.

Dielectric constant and dielectric loss tangent were measured by a composite resonator using the method developed at JSC ORPE Technologiya. The method for determination of dielectric constant (ε) is based on measuring the difference of resonant wavelengths of the resonator before and after the placement of the dielectric specimen into the resonator at the fixed measuring frequency. The method for determination of dielectric loss tangent (tgδ) is based on measuring the variation of intensity of electromagnetic oscillations when the dielectric specimen is placed into the resonator. The research results are shown in Figures 2 and 3.

The increase of dielectric loss tangent after 700°C is most likely to be caused by phase transition which occurs in the specimen when it is heated. However, this fact requires further investigations.

Thermophysical properties of ceramic matrix composite material were studied using the Netzsch LFA 457 apparatus according to the methods of the Enterprise. The ceramic matrix material showed the stability of thermophysical properties (thermal conductivity, temperature conductivity, and specific thermal capacity coefficients) in the wide ambient temperature range (from 20°C to 1200°C). The research results are shown in Figures 4-6.

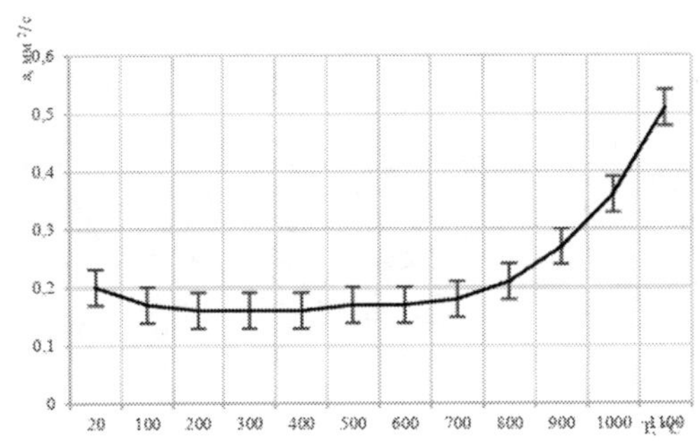

Figure 4. Temperature conductivity coefficient of ceramic matrix composite material as a function of temperature.

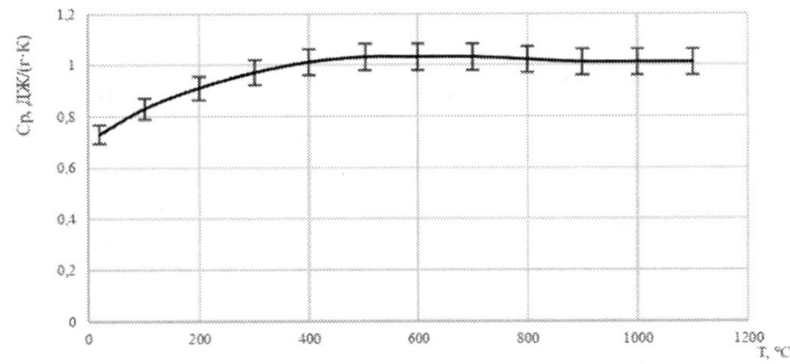

Figure 5. Specific thermal capacity of ceramic matrix composite material as a function of temperature

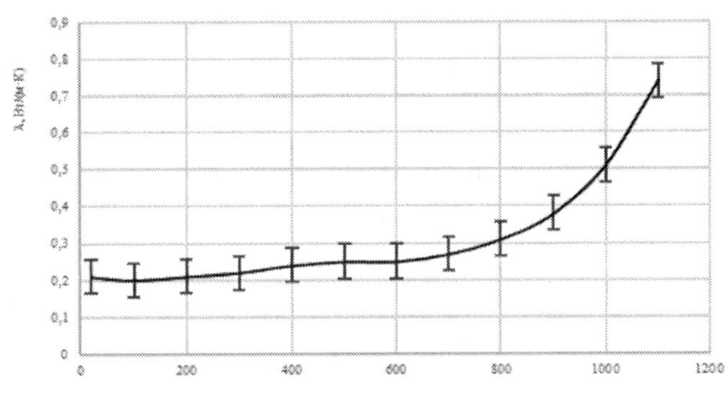

Figure 6 . Thermal conductivity coefficient of ceramic matrix composite material as a function of temperature.

The research results can form the basis for the development of ceramic matrix composite materials for special challenges of aviation and rocket engineering industries, e.g. for the development of aircraft nose radomes.

Table 1 provides the comparison of physical and technical characteristics of ceramic matrix materials and quartz ceramics developed at JSC ORPE Technologiya.

Table 1. Comparison of physical and technical characteristics of materials

	Ceramic matrix composite material (KhAFSkv-Sh)	Ceramic matrix composite material (KhAFSkv)	Quartz ceramics
Structural characteristics			
Packed density (ρ), g/cm³	1.429	1.58-1,64	1.94-2.05
Porosity P, %	32-35	31-35	7-12
Water absorption W, %	20-24	19-22	4-5
Strength characteristics at 20°C			

Impact strength a, kJ/m^2	6.6±0.7	10±0.8	0.78±0.05
Dielectric characteristics			
Dielectric constant (Ɛ)			
20°C	2.58	3.10-3.30	3.30 – 3.53
1100°C	2.54	3.30	-
Dielectric loss tangent at a frequency of 10^{10} Hz			
20°C	0.004	0.009	0.004
1200°C	0.020	0.13	0.01
Thermophysical characteristics at 20 - 1100°C			
Temperature conductivity coefficient a, mm^2/s			
20°C	0.20	0.36	0.57
1100°C	0.51	0.29	0.46
Thermal conductivity coefficient λ, W/(m·K)			
20°C			
1100°C	0.83	0.43	0.21
	1.23	0.49	0.74
Specific thermal capacity, kJ/(kg K)	0.73 – 1.01	0.74-1.19	0.74 – 1.29

4. Conclusion

JSC ORPE Technologiya developed the manufacturing process for heat-resistant ceramic matrix composite material KhAFS-Shkv based on quartz slurry and quartz fabric filler with stable thermophysical and dielectric characteristics in the range of T= 20-1200°C. Physical and technical characteristics of the new material were studied.

It was shown that water absorption of the materials KhAFS-Shkv and KhAFSkv was ~ 24%, in comparison with quartz ceramics with water absorption ~ 5%. This requires additional application of moisture-protective coating to ceramic matrix composite material and articles made thereof.

The material developed has significantly higher impact strength a=6.6 kJ/m2 in comparison with quartz ceramics (a=0.78 kJ/m2). It allows KhAFS-Shkv to be considered not only as thermal protective, but also impact-resistant structural material.

The ceramic matrix composite material obtained has low thermal conductivity, coefficient of thermal expansion, and dielectric loss tangent in the temperature range of 20°-1100°C.

Highly heat-resistant ceramic composite material KhAFS-Shkv based on quartz slurry and quartz fabric filler with stable thermophysical and dielectric characteristics is promising for the development of thermally-loaded structures of special aircraft for rocket and aerospace engineering.

References

[1] Gurtovnik I, Sokolov V, Trofimov N and Shalgunov S 2002 Glass-Fiber Reinforced Plastic Radiotransparent Articles, Moscow, Mir Publ. 368

[2] Stepanov P, Atroshchenko I, Starodubtseva N, Shutkina O and Melnikov D 2014 *Advanced Materials* **10** 17

[3] Borodai F, Nepovinnykh L, Stepanov P; Roletskaya N and Shutkina O 2016 Method for Manufacturing Highly Heat-Resistant Radiotransparent Material (Article) Based on Phosphate Binder and Quartz Fabric. The RF Patent no. 2596619, C04B35/80, JSC ORPE Technologiya named after A.G. Romashin, Obninsk

[4] Rusin M, Vasilenko V, Stepanov P, Atroshchenko I, Shutkina O and Romashin V 2015 *Refr. and* Ind. Ceramics **58** 239,.

[5] Sudnik LV 2014 Modernization of technology for assessing operational dynamic properties of composite structural ceramics by using hydroabrasive ultrajet / L.V. Sudnik, A.L.

Galinovsky, V.I. Kolpakov, S.G. Mulyar, M.I. Abashin, A.S. Provators // Science and education: scientific publication of MSTU. N.E. Bauman 3

[6] Kolpakov V I and Vasilieva T V 2015 Numerical modeling of the introduction of a compact drummer into a composite barrier made of composite materials: guidelines for laboratory work. - M.: Publishing House of NUK IU MSTU. N.E. Bauman, 24 p., Ill.

[7] Mulyar S G 2015 Calibration of mathematical models of shock interaction of a bullet with oxide ceramics taking into account changes in the concentration of nanoparticles of boehmite powder / S.G. Mulyar, V.I. Kolpakov, A.L. Galinovsky, M.I. Abashin, E.V. Vinokurova // Reports of the international conf *XVII Kharitonov thematic scientific readings* (Sarov: FSUE RFNC-VNIIEF) 546-550

[8] Kolpakov V I 2016 Mechanisms for breaking concrete barriers with compact high-speed impactors. Experiment and numerical modeling / V.I. Kolpakov, G.G. Savenkov, V.A. Morozov, A.A. Lukin *Bulletin of the Tambov University. Series Natural and technical sciences.* **21**. Is.3 1055-1057.

[9] Galinovsky A L, Kolpakov V I, Semashko V V, Sudnik L V and Chaykin S V 2017 Numerical modeling of the explosive compaction of copper cylindrical billets // *Engineering Journal: Science and Innovation* **10(70)**1-16URL: http://engjournal.ru/catalog/mesc/tempp/1690.html. DOI 10.18698 / 2308-6033-2017-10-1690

[10] Kolpakov V I 2017 Simulation of the impact interaction of high-speed particles with the structural elements of the screen protection of a spacecraft: guidelines for laboratory work / V.I. Kolpakov, T.V. Vasilieva (Moscow: Publishing house of MGTU im. N.E. Bauman) 46 [2] c., Ill.

[11] Kolpakov V I and Ilyukhina A A 2019 Features of mathematical modeling of the destruction of structures from different materials under the action of a high-speed waterjet jet *Journal of Engineering: Science and Innovation* **9 (93)** 1-8 DOI: 10.18698 / 2308-6033-2019-9-1913 URL: http://engjournal.ru/articles/1913/1913.pdf

[12] Kolpakov V I and Iliukhina A A 2019 Specifics of mathematical modeling of erosion of structures made out of different materials by high-speed hydro-abrasive water jet // AIP Conference Proceedings 2171, 170002 (2019); https://doi.org/10.1063/1.5133313. Published Online: 15 November. Pp.170002-1—170002-6.

[13] Sudnik L V, Galinovsky A L, Kolpakov V I, Khafizov M V, Mulyar S G and Sayfutdinov R R 2013 Formation and ultrastructure diagnostics of sintered material from nanosized bemit powder Nanoengineering **1** (19) 26-31

[14] Kolpakov V I, Malkin A I and Galinovsky A L 2019 Justification for the use of metal composites as materials for protective screens of spacecraft // Collection of materials of the International scientific-practical conference: "Key trends in composites: science and technology." (Moscow) 355-360.

[15] Kruglov P V and Kolpakov V I 2019 Use of composite materials to increase the effectiveness of variable-geometry aircraft // Conf. Series: Materials Science and Engineering **683** 012033 () 1-5. doi: 10.1088 / 1757-899X / 683/1/012033. URL: https://doi.org/10.1088/1757-899X/683/1/012033.

IOP Conf. Series: Materials Science and Engineering **934** (2020) 012059 doi:10.1088/1757-899X/934/1/012059

Evaluation the complex of thermal properties for epoxy-based GFRP used in wing of tourist class reusable space vehicle

E R Ashikhmina*, N M Petrov and P V Prosuntsov

Bauman Moscow State Technical University, 2nd Baumanskaya str., b. 5, Moscow, 105005, Russia

E-mail: ashikhmina.ekaterin@gmail.com

Abstract. In general, the prediction the effective properties of fiber reinforced composites is a complicated task due to its heterogeneous properties. Therefore, new approaches can make possible obtaining of the reliable thermal properties of composites with least time and financial costs. This work presents the approach for determination of thermal properties of epoxy-based glass fiber reinforced composites. Firstly, to evaluate the thermal behavior of considered material the numerical modeling was conducted via finite element analysis in commercial software MSC Digimat. To validate the model, two-stage experimental study was carried out. As the result of the experimental research thermal conductivity through-plane was obtained by laser flash method as well as thermal conductivity in-plane – by specialized heat equipment developed in Bauman University with subsequent solution of the inverse heat conduction problem. Summing up, the analytical and experimental data showed good agreement that demonstrates the feasibility of the approach.

1. Introduction

Over the past decades the interest to the business-driven space transportation systems has considerably increased. Human spaceflight and space exploration have become a domain of not only government, but also private companies, entrepreneurs and research groups [1]. The prominent branch of space industry that have advantageous potential is space tourism [2]. The space tourism is expected to progress and to become the catalyst for less expensive, more frequent access to space [3] and point-to-point suborbital space travel routes [4]. Nowadays more than 40 private companies and research groups all over the world have been engaged with the development of transportation systems into space – tourist class reusable space vehicle. Today, the most known space tourism vehicles are SpaceShipTwo (Virgin Galactic, USA), New Shepard (Blue Origin, USA), Dragon (SpaceX, USA), Spica (Copenhagen Suborbitals, Denmark), IAR 111 (ARCA Space Corporation, Romania) and others. In Russia, several projects are also under development – suborbital reusable complex (KosmoKurs), Oduvanchik (BMSTU), Fanstream (MAI), and others [4-6]. There are a few types of space routes: orbital, suborbital that proposes reach the Karman line and then falling back to Earth, and the spaceflight to celestial bodies (Moon, Mars, etc.). Suborbital spaceflight is more attractive for business caused by comparatively less tour cost, passenger health requirements, significantly lower level of thermal and force loads results from the atmospheric re-entry, etc. [7-9].

In Bauman University [7, 8, 10] during some years the winged spacecraft Oduvanchik is being developed. The general scheme of RSV design is presented in Figure 1 [10, 11].

Figure 1. Suborbital tourist class reusable space vehicle, Bauman University, 2008

Weight and cost efficiency are the key issues in spacecraft design. Hybrid polymer matrix composites (herein after *hybrid composites*) that contains in microstructure two or more types of reinforcing components have proved their effectiveness for structural application in industrial fields, aerospace as well, for decades. Combination of expensive carbon fibers that demonstrate high strength-to-weight performance with less expensive but lower in terms of mechanical and thermal properties glass fibers would provide appropriate design solution from technical and economical positions [12]. By changing the types of reinforcing fibers, their ratios, orientation angles, it is possible to vary properties over a wide range, as well as to create materials with the required anisotropy of properties and adjust the required characteristics of such compositions [13, 14].

In this study particular attention is focused on the essential structural element of the suborbital spacecraft that significantly determines its design appearance and weight efficiency – the wing. For wing design the reliable information on the materials properties used in its structure is required. However, in the reference literature such information is practically absent due to the specialness of the composites [15]. Therefore, to conduct an in-depth analysis of the structure, it is necessary to create an approach for obtaining the materials properties for the principal structural element of the wing, notably with the least time and financial costs [16-18].

According to flight profile [19] the spacecraft operates in a wide range of velocities (its maximum speed is 3.6 M at an altitude of about 40 km) and is exposed to intensive aerodynamic heating [20, 21]. Therefore, the spacecraft design requires data on the thermal properties of hybrid composites applied in structure (thermal conductivity coefficients in different directions, heat capacity and density) in dependence on temperature. If the last two properties can be determined with a sufficient accuracy level using the mixture rule [22], then it is proposed to use the numerical analysis to find the thermal conductivity coefficients followed by validation of its results based on the experimental data. The properties of the used materials are mostly defined by experimental testing, numerical simulation can help to eliminate the excessive number of trials and thus to reduce time, labor and financial costs [9, 10, 12, 23].

For technological and economic reasons, in the wing made of hybrid composites, glass fibre fabric with harness-satin weave and FibArm Tape 230 unidirectional carbon fibre tape with AKSA / Carbon fiber A-49 24K4 fiber, as well as an epoxy resin Noapox 7510 are used.

Consequently, current work aims to develop an approach for obtaining of thermal properties of epoxy-based glass fibre fabric (herein after *GFRP*) for the spacecraft wing [9] by numerical simulation validated by experimental data. GFRP was chosen to test the developing approach as this material is well-studied and can be additionally validated by known data.

2. Numerical modeling of GFRP thermal properties in MSC Digimat for the wing skin

Traditionally the process of obtaining of composite materials properties with a complex microstructure has been carried out by only experimental methods that is an expensive and time consuming process. Nowadays, the situation has being changed and there are a number of software for determination of the thermal behavior of materials via multi-scale modeling tools [24]. Through this tools realized in such software one can find the effective properties of considered materials with complex woven microstructure, inclusions, coatings, defects, etc., and use it as an input to in-depth analysis of structures. In this case the composite constituent material is considered based on homogenization procedure ignoring of internal boundaries and generation of representative volume element (RVE) model of a composite. The behavior of RVE can be considered as the basis for macro-level analysis by taking the effective properties obtained using homogenization process [25, 26].

Figure 2. Geometric model of RVE of GFRP generated using Digimat-FE (matrix is hidden for clarity)

In this research a special attention was paid to commercial software MSC Digimat. The approach of obtaining GFRP properties includes the generation of a RVE, the application of a heat load to its faces, and the subsequent calculation the values of the thermal conductivity coefficients via finite element analysis. The geometric model RVE of GFRP is depicted in Figure 2. Herewith the glass fiber and epoxy matrix were assumed as isotropic, and perfect thermal contact on their coupled boundaries was performed. The porosity factor was equal to 0. Due to these assumptions in the result of numerical modeling the upper estimate of thermal conductivity was obtained.

The initial data for modeling the thermal properties were the coefficients of thermal conductivity, density and heat capacity of individual components [8, 11]. It was believed that monofilaments have the shape of an infinitely elongated cylinder, and their volume content in the matrix is 50%. The results of modeling the coefficients of thermal conductivity in different directions of GFRP are given in Table 1.

Table 1. Results of numerical modeling of GFRP mechanical properties

Property	Value
Thermal conductivity in-plane (x direction), W/(m·K)	0.554
Thermal conductivity in-plane (y direction), W/(m·K)	0.547
Thermal conductivity through plane, W/(m·K)	0.513

It is shown that the thermal conductivity of GFRP, especially in-plane, weakly depends on the direction. The results show a mismatch value of no more than 0.5%.

3. Experimental study for determination GFRP thermal properties for the wing skin

(a) (b)

Figure 3. GFRP samples for experimental

research of (a) $\lambda_\perp$; (b) λ_{II}

To confirm the reliability of the numerical modeling results, its validation by comparison with the data obtained by experimental studies was conducted. For this purpose, two types of samples were produced (Figure 3). The first sample type was cylindrical with a diameter of ∅12.5 mm and a thickness of 2 mm and used for investigation of thermal conductivity through plane ($\lambda_\perp$). The second type was produced in a form of plate and was applied to get data about thermal conductivity in-plane (λ_{II}).

3.1 Experimental measurement of the $\lambda_\perp$ via Laser Flash method

An array of methods has been developed for determination of $\lambda_\perp$ [27]. Among the most advanced and reliable is the laser flash method, which is implemented in the Laser Flash Apparatus (NETZSCH, Germany) [28] (Figure 4). In current study this apparatus was used to measure $\lambda_\perp$.

(a)

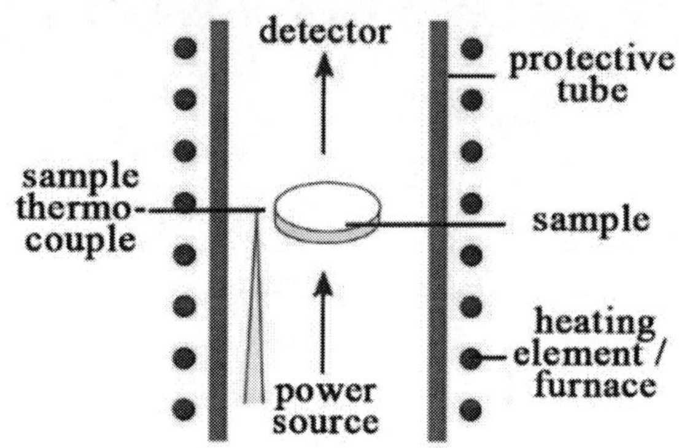

(b)

Figure 4. (a) Laser Flash Apparatus LFA 457 MicroFlash (NETZSCH, Germany); (b) measuring methodology for determination $\lambda_\perp$ [28]

As the result of measurement the temperature dependence $\lambda_\perp$ of the GFRP in the range from 25°C to 150°C (Figure 5) was revealed that showed satisfactory compliance with the modeling data.

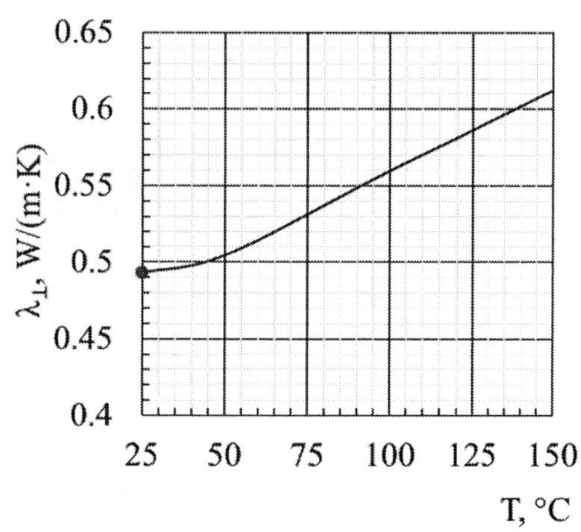

Figure 5. Temperature dependence of $\lambda_\perp$ (the red point related to $\lambda_\perp$ calculated in MSC Digimat)

3.2 Experimental measurement of the λ_{II}

The obtaining of λ_{II} is a complicated scientific and technical task. Nowadays, there are no standard methods for determination of λ_{II} [27, 29]. However, for this purpose in the Bauman University developed a

IOP Conf. Series: Materials Science and Engineering **934** (2020) 012059 doi:10.1088/1757-899X/934/1/012059

specialized experimental heating equipment [30]. The equipment (Figure 6) includes two heaters (2) that are covered with thermal insulation (5) and are fixed between copper plates for temperature equalization across the width of the sample (3). The sample is made in the form of a plate with dimensions 25x150 mm and thickness of 2 mm (3). To measure the temperature on the sample surface at test points, a Fluke Ti-400 infrared camera was used [31]. It was assumed that GFRP is quasi-isotropic and the experiment was aimed only on one parameter estimation.

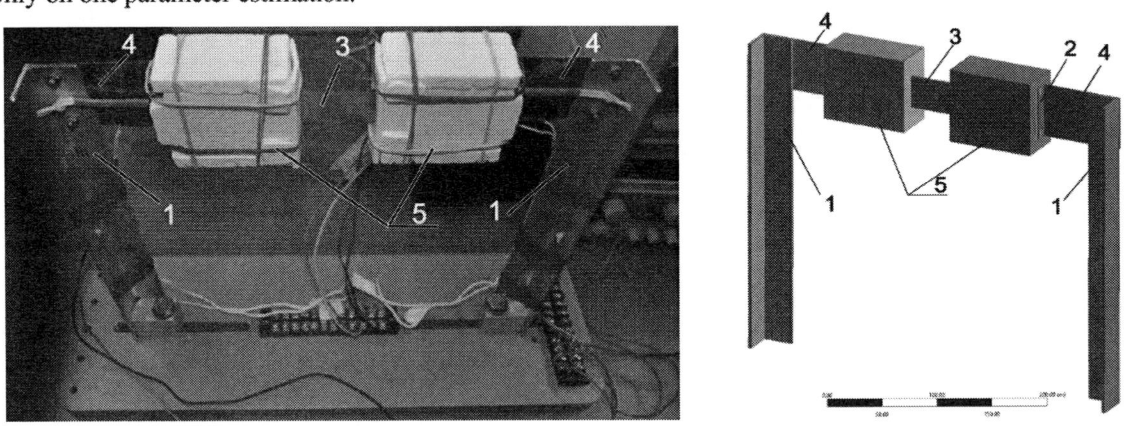

Figure 6. Equipment for λ_{II} measurement: 1 – steel console; 2 – heaters with copper plates; 3 – sample; 4 – aluminum plates; 5 – heat insulation

The measured by infrared camera time dependences of surface temperature of the GFRP sample center and at a distance of 5, 10 and 15 mm (Figure 6b) from the center were used as source data to solve the inverse heat conduction problem and determine λ_{II} (Figure 6a). For this purpose, a three-dimensional geometric model of the heating equipment was developed for subsequent simulation of the transient process of radiation-conductive heat transfer. The heaters temperature was set constant in its volume and in accordance with experimentally determined dependencies. It was assumed that free convection on the sample surface as well as release of radiation energy from the surface sample were occurred. Parameter estimation problem was assumed the determination of λ_{II} that provides a minimum of the residual functional of the experimental and calculated temperatures at the test points was considered. Based on the inverse problem solution only temperature dependence of λ_{II} has been found, the other thermal properties were determined by the rule of the mixture as well as $\lambda_{\perp}$ were appropriated from an experimental study via the laser flash method described in paragraph 3.1.

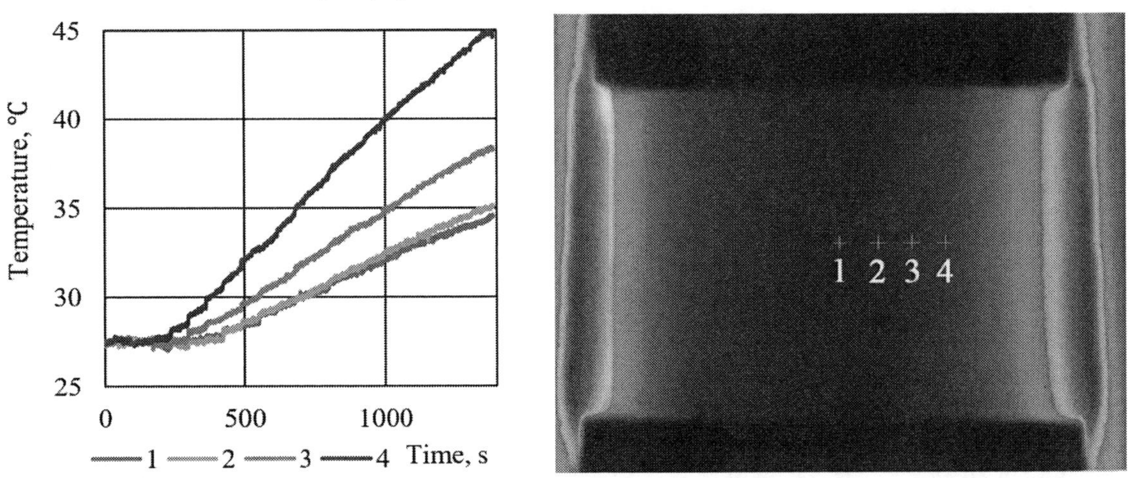

(a) (b)

Figure 7. (a) experimental temperature dependencies for GFRP sample; (b) images made by infrared camera (1 – temperature in the sample center; 2, 3, 4 – at a distance of 5, 10 and 15 mm from the center respectively)

As the result of the inverse problem solution, λ_{II} was obtained for GFRP sample. Results analysis showed a good agreement between the experimental data and numerical modeling obtained in MSC Digimat (Table 2).

Table 2. Results comparison of the numerical modeling of GFRP mechanical properties and experimental data on λ_{II} measurement

Property	Value
λ_{II} results of numerical modeling in MSC Digimat, W/(m·K):	0.51
λ_{II} obtained through solution of inverse heat conduction problem, W/(m·K)	0.49

4. Conclusion

Summing up, the comprehensive approach for determination of thermal properties of GFRP used in the hybrid composite wing skin of the tourist class reusable space vehicle was developed. The numerical modeling of epoxy-based GFRP thermal properties demonstrated good agreement with the experimental results. Also, manufactured two types of experimental samples of GFRP allowed to obtain the temperature dependence of $\lambda_{\perp}$ via laser flash method and λ_{II} using heating equipment developed in Bauman University and subsequent inverse problem solution. In both experiments on thermal conductivity determination the difference between the experimental and analytical calculations was less than 4% that validates the numerical modeling results.

In future this approach can be applied for determination thermal properties of other materials that used in the wing skin of the tourist class reusable space vehicle.

References

[1] Iacomino C 2019 Springerbr Appl Sci (Cham: Springer)
[2] Seedhouse E 2014 Suborbital. Industry at the Edge of Space (Chichester: Springer International Publishing)
[3] Depasquale D, Charania A C and Olds J 2006 The Emerging Orbital Space Tourism Industry: New Insight into Demand and Prospects for Success
[4] Webber D 2010 Acta Astronaut vol 66 pp 1645-51
[5] Guerster M and Crawley E F 2018 Architectural options and optimization of suborbital space tourism vehicles. In: 2018 IEEE Aerospace Conference, pp 1-18
[6] Pelt M V 2005 Space Tourism. Adventures in Earth Orbit and Beyond (New York: Copernicus Books)
[7] Ashikhmina E R and Prosuntsov P V 2019 AIP Conference Proceedings vol 2135 p 020005
[8] Ashikhmina E R and Prosuntsov P V 2019 AIP Conference Proceedings vol 2171 p 030014
[9] Ageyeva T G, Ashikhmina E R and Prosuntsov P V 2018 Vestn. Mosk. Gos. Tekh. Univ. im. N.E. Baumana, Mashinostr. [Herald of the Bauman Moscow State Tech. Univ., Mech. Eng.] vol 1 pp 4-19
[10] Reznik S, Ageyeva T and Shafikova I 2012 ACMTAA 2012: Proceedings of the Second International Conference on Advanced Composite Materials and Technologies for Aerospace Applications pp 7-13
[11] Ageyeva T G, Ashikhmina E R and Prosuntsov P V 2017 Engineering Journal: Science and Innovation vol 12 pp 4-19
[12] Swolfs Y, Verpoest I and Gorbatikh L 2019 International Materials Reviews vol 64 pp 181-215
[13] Zhang J, Chaisombat K, He S and Wang C 2012 Materials and Design vol 36 pp 75-80
[14] Botelho E C, Silva R A, Pardini L C and Rezende M C 2006 Materials Research vol 9 pp 247-256

[15] Smerdov A A 2018 Mechanics of Composite Materials vol 54 pp 269-80
[16] Reznik S V, Prosuntsov P V and Mikhailovskii K V 2019 J Eng Phys Thermoph+ vol 92 pp 89-94
[17] Reznik S V, Mikhailovskii K V and Prosuntsov P V 2017 J Eng Phys Thermoph+ vol 90 pp 291-300
[18] Reznik S V and Novikov A D 2016 Matec Web Conf vol 92
[19] Ageeva T G, Dudar E N and Reznik S V 2010 Aviakosmicheskaya tekhnika i tekhnologiya. [Aerospace Technic and Technology] vol 2 pp 3-8
[20] Anuchin S A, Lanin A V, Prosuntsov P V, Kordo M N, Zabezhailov M O and Rusin M Y 2018 J Eng Phys Thermoph+ vol 91 pp 585-93
[21] Maiorova I A, Prosuntsov P V and Zuev A V 2016 J Eng Phys Thermoph+ vol 89 pp 528-33
[22] Zinovev P A and Smerdov A A 2006 Optimalnoe proektirovanie kompozitnykh materialov: uchebnoe posobie po kursu «Proektirovanie kompozitnykh konstruktsiy: Izd-vo MGTU im. N.E. Baumana)
[23] Reznik S V, Prosuntsov P V and Mikhailovskii K V 2015 Journal of Engineering Physics and Thermophysics vol 88 pp 594-601
[24] Jagath Narayana K and Burela R G 2019 Journal of Physics: Conference Series vol 1240 p 012103
[25] Mori T and Tanaka K 1973 Acta Metallurgica vol 21 pp 571-74
[26] Nemat-Nasser S and Hori M 2013 Micromechanics: overall properties of heterogeneous materials vol 37 (Elsevier)
[27] Kerschbaumer R C, Stieger S, Gschwandl M, Hutterer T, Fasching M, Lechner B, Meinhart L, Hildenbrandt J, Schrittesser B, Fuchs P F, Berger G R and Friesenbichler W 2019 Polymer Testing vol 80 p 106121
[28] NETZSCH. URL: www.netzsch-thermal-analysis.com
[29] Arumugam V, Kumar C and Joshi G 2016 Composite Interfaces vol 23 pp 1-26
[30] Denisov O V, Minakov D S and Kirbay A A 2015 Science and Education: Scientific Publication vol 7 pp 171-84
[31] Fluke Ti480 PRO Infrared Camera. URL: www.fluke.com

IOP Conf. Series: Materials Science and Engineering **934** (2020) 012060 doi:10.1088/1757-899X/934/1/012060

Influence of modes of washing-out of the polyvinyl acetate-based substrate on the properties of carbon fiber reinforced plastics obtained by TFP

V A Nelyub[1], A S Borodulin[1], M A Orlov[1], I A Polikarpova[1], A N Kalinnikov[1], A I Soloviev[1] and S I Dzhafarova[1]

[1]Bauman MSTU, Moscow, Russia

Abstract. The article shows the advantage of using water-soluble polyvinyl acetate-based non-woven fabric as a substrate in the manufacture of tfp preforms of products made from polymer composite materials, including gas turbine engine blades. As compared with the applied woven materials, elastomers and films, water-soluble non-woven fabric is easily removed from the surface of the semi-finished product without disturbing its structure and without causing damage to the fibers. It has been established that high-quality washing-out of the substrate does not affect the properties of carbon fiber-reinforced plastics.

1. Introduction

Currently, carbon fiber-reinforced plastics based on volume-reinforced preforms created by TFP (Tailored fiber placement) technology [1-4] are more and more common in the manufacture of products from polymer composite materials (PCM), as they can withstand multiaxial loads, unlike traditional textile materials.

The TFP process is about the directional laying of the roving on the base layer (substrate) using a piercing zigzag stitch consisting of a system of upper and lower threads (Fig. 1) [5]. The three-dimensional structure of the semi-finished products is created by introducing a third direction of reinforcement: each subsequent layer of fibers is attached to each other with piercing threads (polyamide, aramid, etc.) in the process of creating the preform.

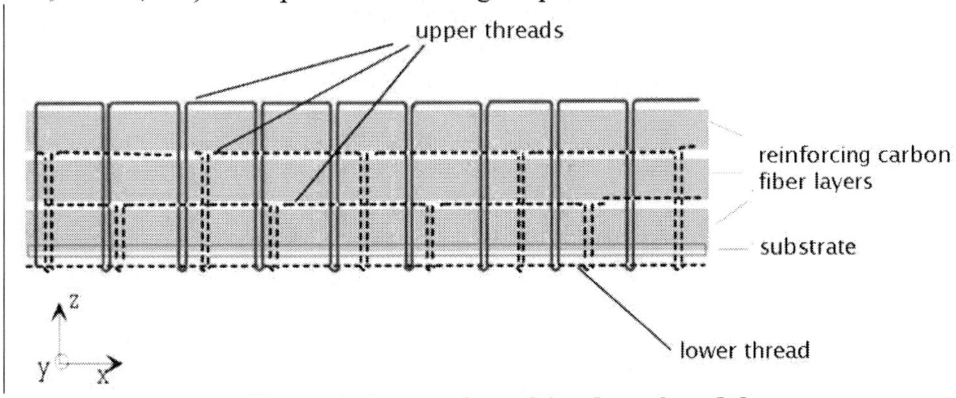

Figure 1. Layout view of the tfp preform [5]

As a rule, the base carrier layer is an auxiliary material to be removed from the finished composite part at the stage of production of dry preforms or at the impregnation stage. At present, flexible elastomer sheets (in particular, rubber sheets) [6], woven reinforcing materials (carbon and glass fabrics), and separation thermoplastic films [7-10] are used as such a base for roving. The use of reinforcing fabric as the main layer leads to loss of bearing capacity and deterioration of the mechanical properties of the finished product. This is due to the partial destruction of the reinforcing fibers in connection with the penetration of the needle into the tissue structure [11]. Moreover, when separating any sewn base, there is a major risk of damage and configuration changes of the fibrous semi-finished product.

Earlier [12, 13], we used a water-soluble polyvinyl acetate-based non-woven fabric as a substrate in the manufacture of preforms. However, we did not study the effect of the substrate on the properties of carbon fiber-reinforced plastics in detail. Despite the advantages of water-soluble non-woven fabric, there are no indications of its widespread use for the manufacture of preforms in the literature. Thus, this study is a development of the previous studies [12, 13], and is devoted to the influence of technological modes of washing-out of water-soluble non-woven fabric on the microstructure, weight and strength properties of samples of sewed carbon fiber-reinforced plastics.

2. Materials and methods

Preform samples were obtained on a JCW 0100-500 numerically controlled embroidery machine (ZSK Stickmaschinen GmbH, Germany).

Preforms of planar sample of sewed carbon fiber-reinforced plastics with dimensions of 40x20x4 mm with reinforcement scheme [0°, + 45°, 90°, -45°] N and model samples of gas turbine engine (GTE) compressor blades were made on the basis of HTS 45 carbon fiber (TohoTenax®, Japan) and Rusar-S piercing fixing thread (Thermotex LLC, Mytishchi, Russia). A water-soluble polyvinyl acetate-based non-woven fabric (Aurora, China), 40 μm in thickness, was used as a support layer (substrate). Planar samples consisting of 8 layers were sewn onto a substrate in one technological operation. The preform samples of the GTE compressor blades had identical characteristics (size, weight, laying, sewing parameters), and their component blocks were cut separately from the common substrate with the same outsize to ensure the same substrate material content in all tested samples. Each preform of the experimental blades consisted of 12 separate blocks sewn onto the substrate. These blocks were subsequently laid out on top of each other to obtain the final design of the semi-finished product. After manufacturing the preforms, their weight together with the substrate was determined. Next, the preforms were put in the water for a period of 1 to 15 minutes in the temperature range from 20 to 100 °C to remove the substrate. The washing-out quality of the substrate was evaluated visually. Then the preforms were dried and their weight was re-determined.

After drying, the preforms were used to produce samples of planar carbon fiber-reinforced plastics and GTE blades using vacuum molding technique.

Araldite LY-556 epoxy resin (Huntsman, Switzerland) was used as a binder. Iso-methyltetrahydrophthalic anhydride (i-MTHFA) (90 mph) (Triune Chemicals Materials (TCM), China) served as a hardener, and 2-methylimidazole (0.2 mph) was an accelerator (Modifikator LLC, Tomsk, Russia).

The preforms were impregnated at a room temperature of 23 °C and a discharge of about -1 atm for 30 minutes. After impregnation, the samples were hardened for 4 hours at 110 °C without removing the discharge.

The samples with a length of 40 mm, a width of 20 mm and a thickness of 4 mm were cut from the hardened planar samples and GTE blades. For each type of material, the number of samples was at least 5.

The interlaminar shear strength τM in the reinforcement plane was determined using the short-beam method [14] on a Zwick Z100 universal testing machine (Zwick Roell, Switzerland). The loading speed was 10 mm/min, the distance L between the supports was 20 mm. The ultimate shear stress was determined by the formula:

$$\tau_M = \frac{3}{4} \times \frac{F_{max}}{bh},$$

where Fmax – load corresponding to destruction of the sample, b – sample width, h – sample thickness.
To study the microstructure of GTE blades based on tfp preforms with a substrate in the product and without a substrate (after it was washed out), we used a SkyScan 1172 tomograph (Bruker, Germany).

3. Results and their discussion

The influence of modes of washing out of the substrate from the fibrous structure of tfp preforms of GTE blades on their weight and quality are presented in Table 1.

Table 1. The influence of modes of washing out of the substrate on the properties of preforms obtained by stripping a carbon tow.

Parameter \ Modes	1	2	3	4	5
	T = 20 °C τ = 15 min	T = 40 °C τ = 10 min	T = 60 °C τ = 5 min	T = 80 °C τ = 3 min	T = 100 °C τ = 1 min
Substrate dissolution time, s	43	36	21	10	4
Substrate dissolution quality	+	+	+	+	-
Washing-out quality	-	+	+	+	-
Preform weight without a substrate, g	33±0.1	33±0.1	33±0.1	33±0.1	33±0.1
Preform weight after washing-out and drying, g	33.65	33.15	33.13	33.10	33.67

Note: «+» - good-quality performance of operation, «-» – low-quality.

As the Table shows, the substrate dissolution time decline substantially with increasing temperature. Moreover, in almost all cases, we can obtain high-quality washing-out of the base layer. In this case, there is no change in the fibrous structure of the obtained preform, which remains flexible, indicating the absence of traces of polyvinyl acetate. It should be noted the substrate is not washed out in a qualitative manner at room temperature and short exposure times at 100 °C. The preform remains rigid, i.e. dissolved polyvinyl acetate was not completely removed. Additional washing with running water also did not completely remove the polymer. Thus, based on results of testing of tfp preforms of the compressor blades, we can say that modes 2, 3, and 4 are suitable for washing-out of the water-soluble substrate with a water temperature range of 40 ÷ 80 °C and an exposure time of 3 ÷ 10 min. Mode 4 was chosen as the optimal mode of operation for washing-out of non-woven fabric from the blade preforms (or preforms of similar sizes), since less time is spent on washing-out, which is a significant advantage in the series production, ensuring considerable time saving.
Figure 2 shows the microstructure of the material of the GTE blades with the substrate washed out from the preform and without washing-out of the substrate, obtained on an X-ray tomograph. As Fig.2b) shows, those blades that are made on the basis of preforms without washing-out of the substrates demonstrate an increased content of the matrix between the layers, caused by the presence of the substrate, which is expressed by the increased distance between the reinforcing layers. Apparently, this kind of layer may be considered as a defect. In the case of materials with the washed out substrates, no layers enriched in the matrix are observed.

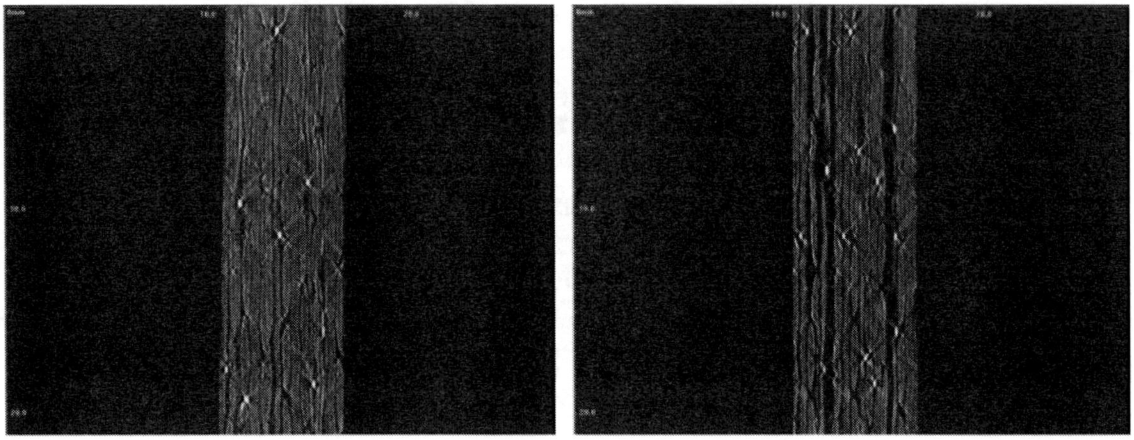

a) b)

Figure 2. Tomogram of material of GTE blades based on tfp preforms without a substrate (a) and with a substrate (b)

Table 2 shows what influence the presence of the substrate in the structure of the material has on the strength characteristics of planar samples and samples of GTE blades, the preforms of which consist of several blocks of sewn semi-finished products, stacked on top of each other to form the desired thickness of the product.

When the substrate was washed out, the shear strength was 10–15% higher in the case of planar samples and 25–30% higher in the case of compressor blades.

Table 2. The influence of the method of producing carbon fiber-reinforced plastics on the shear strength.

Samples / Parameter	Planar samples		Compressor blades	
Production of a preform	without washing-out of the substrate	with washing-out of the substrate	without washing-out of the substrate	with washing-out of the substrate
Internal delaminations	None	None	Present	None
Interlaminar shear strength τ_M in the reinforcement plane, MPa	105	118	110	142

In the production of prefabricated preform structures in one technological operation, the substrate was below the preform, and was incommensurably small compared with the thickness of the entire preform. The presence of a substrate in the composition of planar semi-finished products had no effect on carbon fiber-reinforced plastics, since it completely dissolved in the PCM during polymerization. Therefore, it is not necessary to wash the substrate out of such preforms. In the production of compressor blade preforms consisting of separate sewn parts and subsequently stacked on top of each other, the number of non-woven fabric layers corresponded to the number of blocks, which led to an increase in the weight of the semi-finished product and the final product, as well as to internal delamination and reduction in the shear strength of the finished products. Therefore, washing out of the substrate is needed in the production of complex composite preforms.

4. Conclusion

It has been established that despite the fact that the polyvinyl acetate-based substrate dissolves in the volume of carbon fiber-reinforced plastics during their shaping in the mounting and does not have any function in the future, it increases the weight of the preforms and the weight of finished products. In addition, the substrate presence in the composite preform leads to such structural defects of the finished products as delamination, which reduces the interlayer strength in the reinforcement plane. Thus, it is recommended to remove non-woven fabric (wash out) from the preforms.

Acknowledgments
This study was supported financially by the Ministry of Science and Higher Education of the Russian Federation, subsidy agreement No.14.574.21.0160 of September 26, 2017, unique agreement identifier RFMEFI57417X0160.

References

[1] Carosella S 2012 Tailored Fibre Placement – an option for automated high volume Preform Production *Proc. Int. Conf. SAMPE Europe Symposium (SEMAT 12)* (Munich, Germany) 52-58
[2] Khaliulin V I, Khilov P A and Toroptsova D M 2015 Prospects of Applying the Tailored Fiber Placement (TFP) Technology for Manufacture of Composite Aircraft Parts *Russian Aeronautics (Iz.VUZ)* **58(4)** 495–500
[3] Mattheij P, Gliesche K and Feltin D 2000 3D reinforced stitched carbon/epoxy laminates made by tailored fibre placement *Composites Part A: Applied Science and Manufacturing* **31(6)** 571-581
[4] Bezzametnova D M 2015 Primenenie TFP-tekhnologii pri izgotovlenii konstrukcij iz kompozicionnyh materialov *Proc. Int. Conf. "ANTE-2015" (Problemy i perspektivy razvitiya aviacii, nazemnogo transporta i energetiki: stat'ya v sbornike trudov konferencii. Kazan)* 117-122 [in Russian]
[5] Mattheij P, Glische K and Feltin D 1998 Tailored Fiber Placement-Mechanical Properties and Applications *Journal of Reinforced Plastics and Composites* **17(9)** 774-786
[6] Joern P and Eberth U 2008 Method for producing one-or multi-layer layer fibre preforms according to a tfp method. Pat. EP 1915244 (A1)
[7] Joern P and Eberth U 2009 Method for producing single- or multi-layer fiber preforms by the tfp process as well as a backing layer. Pat. US 2009050259 (A1)
[8] Lo Faro Karmelo, Karter Dzheffri Tomas, Makgrehjl Patrik Terens and Maskell Robin Kelvin 2006 Elastic polymeric element, substrate for curable composition, curable composition, methods for preparation thereof, curing method, products. Pat. RU 2272051 (C2)
[9] Gessler A 2002 Method for producing multilayer Tailored Fiber Placement (TFP) preforms using meltable fixing fibers. Pat. WO 0245932 (A1)
[10] Grelen H 2013 Method for manufacturing a preform and its use in the manufacture of a composite part. Pat. EP 2599614 (A1)
[11] Temment H, Degenhardt R and Raible T 2006 Tailored Fiber Placement Optimization Tool *Proc. Conf. of the 25th Int. Congress of the Aeronautical Sciences (2006, Hamburg, ICAS)* 576–583
[12] Orlov M A, Kalinnikov A N and Polikarpova I A 2018 Razrabotka tekhnologii sozdaniya predvaritel'nyh zagotovok elementov gazoturbinnogo dvigatelya iz polimernyh kompozicionnyh materialov metodom avtomatizirovannoj nashivki uglerodnogo volokna *Proc. Int. Conf. «Novye tekhnologii, materialy i oborudovanie rossijskoj aviakosmicheskoj otrasli:sbornik materialov Vserossijskoj nauchno-prakticheskoj konferencii s mezhdunarodnym uchastiem» (Kazan)* **1** 270-275 [in Russian]
[13] Orlov M A and Polikarpova I A 2018 Razrabotka tekhnologii izgotovleniya preform elementov gazoturbinnogo dvigatelya iz polimernyh kompozicionnyh materialov s primeneniem avtomatizirovannoj nashivki rovinga *Proc. Int. Conf. «Nauka – Obshchestvo – Tekhnologii – 2018* (Moscow, Moscow Polytechnic University) 256-260 [in Russian]
[14] ISO 14130:1997. Fibre-reinforced plastic composites – Determination of apparent interlaminar shear strength by short-beam method

IOP Conf. Series: Materials Science and Engineering **934** (2020) 012061 doi:10.1088/1757-899X/934/1/012061

Analytical and experimental research of epoxy-based glass/carbon hybrid polymer composites

E R Ashikhmina, T G Ageyeva, S Yu Fedorov and Yu V Klimovich

Bauman Moscow State Technical University, 2nd Baumanskaya str., b. 5, Moscow, 105005, Russia

E-mail: ashikhmina.ekaterin@gmail.com

Abstract. Nowadays, the significant potential of hybrid polymer composites in terms of weight and cost has contributed to their application in different branches of industry. However, it is known that determination of unique properties of hybrid composites remains significantly challenging as laborious experimental work is usually required. This study aims to develop a numerical material model for prediction of mechanical properties of glass/carbon epoxy-based composites. To validate the material model the subsequent experimental research is proposed. It shows good agreement between numerical simulation results and experimental one.

1. Introduction

Fiber reinforced polymer composites are a well-known class of engineering materials that have proved their excellent performance in heavy-loaded aerospace structures through decades. Besides their high strength-to-weight ratio together with some other exceptional properties, fiber reinforced polymers offer a certain design freedom to developers. Thus, the fibers and matrix choice, fiber orientation and stacking sequence are the variable parameters during the composite material development. One of the possible ways to extend the design space is to use two types of fibers instead of just one, thus combining benefits of the both types of fibers within one material. Composites materials, that consisted of at least two dissimilar reinforcements embedded either in thermoplastic or thermoset matrix, are called hybrid polymer composites [1-4].

Three levels of hybridization of fiber reinforced composites can be distinguished, i.e. intrayarn (fibre-by-fibre), intralayer (yarn-by-yarn) and interlayer (layer-by-layer) [5]. One of the best known types of hybrid composites is fiber-metal laminates (FMLs), that consist of thin metal sheets (usually aluminum) bonded into a laminate with intermediate thin fiber reinforced composite layers [6]. Although FMLs are widely used in aviation since 1980s and today a number of grades are commercially available [7], a bunch of challenges arises during FML manufacture and usage [7]. One of the main problems of FMLs is residual thermal stresses, that occur due to the significant difference in the thermal properties of metals and typical composite reinforcement fibers (glass, carbon, etc.). Hybrid composites reinforced with two different types of synthetic fibers are almost free from the mentioned drawback of FML, have lower density and can even exhibit so-called “hybrid effect” [1, 8]. However, fiber reinforced hybrid composites are much less studied than FMLs. Moreover, numerous combinations of the variable parameters give birth to a considerable number of hybrids with different properties. Therefore, a problem of optimum material properties combination arises during the design stage [9-11]. This problem is multilevel and requires application of different techniques to solve it in each level (optimization methods, modelling, testing, etc.). Although, the properties of the developed material are mostly defined by experimental testing, modelling can help to eliminate the excessive number of trials and thus to reduce time, labor and financial costs.

Modern software allows to simulate material microstructure and define the material physical and mechanical properties. In this study we aimed to model the mechanical properties of hybrid polymer composite reinforced with glass and carbon fibers by means of combination of numerical and analytical methods, and validate the obtained results with the experimental ones [2, 12, 13].

In this paper to study the mechanical properties of composites, the lay-up consists of totally 16 glass and carbon monolayers that stacked with angles of orientations: [0°; ±45°; 90 °] were considered. The properties of the following glass/carbon ratios in the lay-up were used: 0/100; 50/50; 100/0 in order to obtain methodology for mechanical properties determination of composites, also to investigate dependence of properties upon components volume fraction. Woven glass fibre fabric T10 [14], carbon FibArm Tape 230 [15] and epoxy resin Noapox 7510 were used for analysis and experimental research.

2. Numerical simulation of the hybrid polymer composite material properties in MSC Digimat

Over the past decades several approaches as well as software have been developed for the prediction of the behaviour of multi-component materials, especially composites. The Digimat [16] software is of particular interest for modelling a complex material microstructure by generation a representative volume element (RVE). Digimat includes several algorithms for microstructure homogenization and properties determination: the FEA-based homogenization approach (Digimat-FE) and the mean-field approach (Digimat-MF). In Digimat-MF only ellipsoidal reinforcements for homogenization can be used, also the semi-analytical Mori–Tanaka methods that is correlated with the Eshelby approach, interpolative double-inclusion scheme (Lielens' model) and others are realized. [17, 18]. Due to such constraints in this study the mechanical properties of the hybrid composites are obtained via FEA-based homogenization approach that performed in Digimat-FE that includes different types of reinforcements are implemented.

To determine the properties, the RVEs of a carbon tape and a glass fabric are generated. The initial characteristics of the carbon and glass monofilaments and the epoxy matrix are presented in Table 1.

Table 1. Parameters of the numerical models for determination of materials mechanical properties (CFRP – carbon fiber reinforced plastic; GFRP – glass fiber reinforced plastic)

Parameter	CFRP	GFRP
Reinforcement architecture / type	UD tape/ AKSA A-49 24K4	Satin fabric 8/3 /T10
Geometric model of representative volume element (RVE) generated using Digimat-FE (matrix is hidden for clarity)		
RVE dimensions (mm)	0.04 x 0.04 x 0.4	4 x 2.2 x 0.24
Fiber diameter (μm)	10	8
Reinforcement volume fraction (vol.%)	60	60
Modulus of elasticity in longitudinal / transverse directions (GPa)	240 / 50	80 / 80
Modulus of rigidity (GPa)	42	–
Density (kg/m^3): reinforcement /epoxy matrix	1 790/1 300	2 520/1 300

As the result of the simulation the mechanical properties of the GFRP and CFRP monolayers are obtained. In order to conduct more accurate calculation, the properties are calculated for various orientation angles: 0°, 45°, -45°, and 90° in MSC Digimat (table 2).

Table 2. Mechanical properties of GFRP and CFRP monolayers for various orientations obtained in MSC Digimat and user for subsequent analytical calculation.

Property	GFRP (0°)	GFRP (±45°)	GFRP (90°)	CFRP (0°)	CFRP (±45°)	CFRP (90°)
Modulus of elasticity in longitudinal direction E_1 (GPa)	29.8	23.5	23.6	122.2	12.4	8.5
Modulus of elasticity in transverse direction E_2 (GPa)	24.5	25.2	29.4	9.4	12.4	116.9
Modulus of rigidity G_{12} (GPa)	5.8	12.2	6.0	3.8	9.4	3.4
Poison's ratio ν_{12}	0.19	0.29	0.20	0.27	0.40	0.02
Poison's ratio ν_{21}	0.16	0.30	0.15	0.02	0.40	0.27

Unfortunately, the RVE size of a glass fabric is considerably larger than the RVE of a carbon tape. For savings in computational time the calculation of the mechanical properties of composites are defined in analytical form using the expressions of composite mechanics. Based on the known values of the technical elastic constants, the coefficients of the stiffness matrix for each layer in its natural coordinate system are obtained [19-22] using following expressions:

$$g_{11}^{i} = \frac{E_1^i}{1-\nu_{12}^i \cdot \nu_{21}^i} \tag{1}$$

$$g_{22}^{i} = \frac{E_2^i}{1-\nu_{12}^i \cdot \nu_{21}^i} \tag{2}$$

$$g_{12}^{i} = \nu_{12} \cdot g_{22}^{i} = \nu_{21} \cdot g_{11}^{i} \tag{3}$$

$$g_{66}^{i} = G_{12}^{i}, \tag{4}$$

where g_{11}^i, g_{22}^i, g_{12}^i, g_{66}^i are the coefficients of the stiffness matrix of each layer; E_1^i, E_2^i are the modulus of elasticity in longitudinal and transverse directions respectively, Pa; ν_{12}^i, ν_{21}^i are the Poisson's ratio; G_{12}^i is the modulus of rigidity, Pa.

For the determination of the coefficients of the stiffness matrices of each of 16 layers in the coordinate system of multilayer material we have used the following formulas [19-22]:

$$g_{xx}^{i} = g_{11}^{i} \cdot \cos^4 \varphi_i + g_{22}^{i} \cdot \sin^4 \varphi_i + (2 \cdot g_{12}^{i} + 4 \cdot g_{66}^{i}) \cdot \sin^2 \varphi_i \cdot \cos^2 \varphi_i \tag{5}$$

$$g_{yy}^{i} = g_{11}^{i} \cdot \sin^4 \varphi_i + g_{22}^{i} \cdot \cos^4 \varphi_i + (2 \cdot g_{12}^{i} + 4 \cdot g_{66}^{i}) \cdot \sin^2 \varphi_i \cdot \cos^2 \varphi_i \tag{6}$$

$$g_{xy}^{i} = (g_{11}^{i} + g_{22}^{i} - 4 \cdot g_{66}^{i}) \cdot \sin^2 \varphi_i \cdot \cos^2 \varphi_i + g_{12}^{i} (\sin^4 \varphi_i + \cos^4 \varphi_i) \tag{7}$$

$$g_{ss}^{i} = (g_{11}^{i} + g_{22}^{i} - 2 \cdot g_{12}^{i}) \cdot \sin^2 \varphi_i \cdot \cos^2 \varphi_i + g_{66}^{i} (\sin^4 \varphi_i - \cos^4 \varphi_i)^2, \tag{8}$$

where g_{xx}^i, g_{yy}^i, g_{xy}^i, g_{ss}^i are the coefficients of the stiffness matrix of monolayers in the coordinate system of multilayer material.

The coefficients of the stiffness matrix for laminate structure are defined:

$$g_{xx} = \sum_{i=1}^{n} g_{xx}^{i} \cdot \tilde{h}_i ; \tag{9}$$

$$g_{yy} = \sum_{i=1}^{n} g_{yy}^{i} \cdot \tilde{h}_i ; \tag{10}$$

$$g_{xy} = \sum_{i=1}^{n} g_{xy}^{i} \cdot \tilde{h}_i ; \tag{11}$$

$$g_{ss} = \sum_{i=1}^{n} g_{ss}^{i} \cdot \tilde{h}_i , \tag{12}$$

where $\tilde{h}$ is the relative thickness of a monolayer.
The resulting technical constants of the laminate structure of hybrid composite are obtained:

$$E_x = \frac{g_{xx} - g_{xy}^2}{g_{yy}} ; \tag{13}$$

$$E_y = \frac{g_{yy} - g_{xy}^2}{g_{xy}} ; \tag{14}$$

$$G_{xy} = g_{ss} , \tag{15}$$

where E_x and E_y are the modulus of elasticity of the longitudinal and transverse directions respectively, Pa; G_{xy} is the modulus of rigidity, Pa.

3. Validation of the numerical results via experimental research

Several samples of hybrid composite are manufactured and have the following glass/carbon ratios: 0/100; 50/50; 100/0. For statistical processing, 5 samples of each type of material are made. Photos of the used materials are presented in Figure 1.

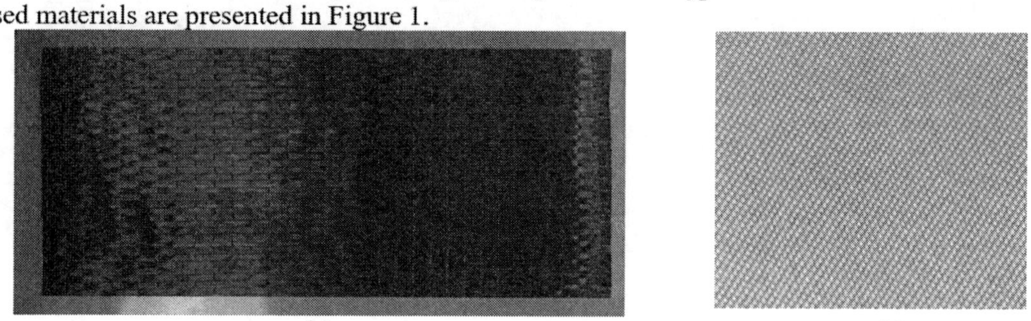

(a) (b)

Figure 1. Photos of the (a) UD tape/ AKSA A-49 24K4 (b) satin fabric 8/3 /T10 used for samples fabrication

To manufacture the composite samples, the vacuum infusion is applied. The scheme of the manufacturing process is presented in Figure 2 [23, 24]. For the production of the flat samples the dry preforms are placed into the aluminum mold that stowed in the vacuum bag [24].

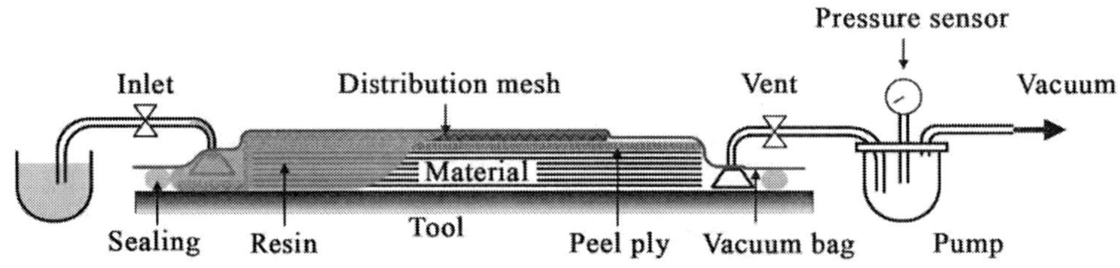

(a)

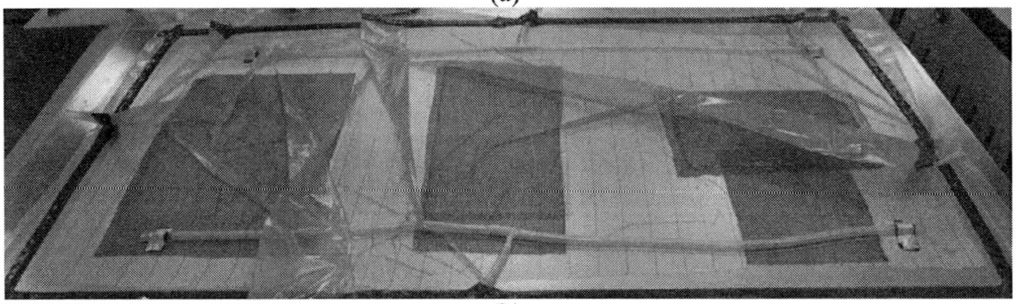

(b)

Figure 2. (a) Scheme of the vacuum infusion technology [24]; (b) vacuum bag for samples manufacturing

Samples with the rectangular cross section with the end tabs made of orthogonally-reinforced GFRP are used for tensile tests (Figure 3a). Five samples of GFRP, CFRP and hybrid composite are applied for each trial. As an example, the GFRP sample is presented in Figure 3b.

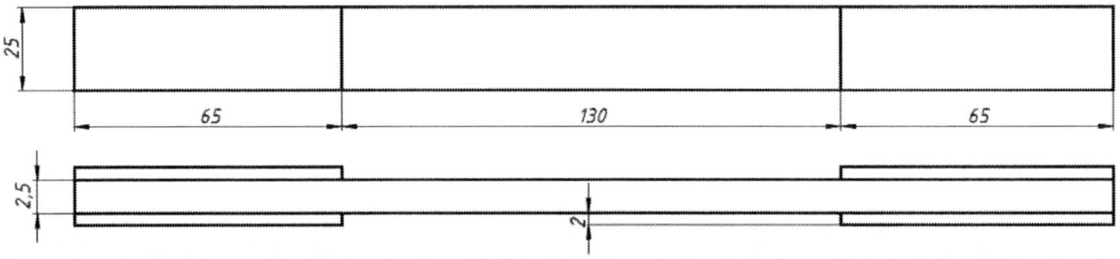

Figure 3. (a) Schematic representation of the samples for tensile tests (according to GOST (State standard) 56785-2015); (b) GFRP sample

For the proposed lay-up the hybrid FRP samples (GFRP/CFRP = 50/50) as well as neat CFRP and GFRP samples are manufactured. The experiments are conducted at room temperature.
As the results of the tensile testing the relation between elastic modulus and GFRP/CFGP ration and the relation between ultimate tensile strength and GFRP/CFRP ration are obtained (Figure 4). The experimental results are compared with the numerical ones.

IOP Conf. Series: Materials Science and Engineering **934** (2020) 012061 doi:10.1088/1757-899X/934/1/012061

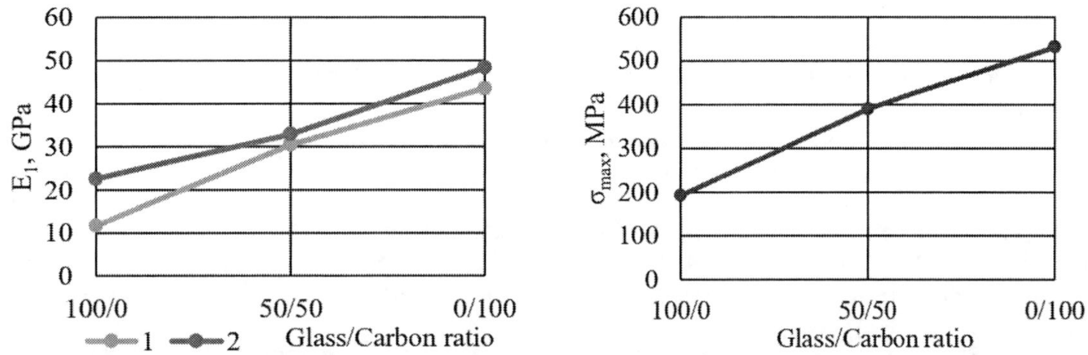

(a) (b)

Figure 4. (a) Comparison of the mechanical properties obtained in MSC Digimat (1) with experimental ones (2) in dependence on components ratios; (b) Ultimate tensile strength in dependence on components ratios

The experimental results of the dependencies of the elastic modulus on the volume content of the components have good agreement with the numerical results. The numerical results are slightly overestimated than the experimental, that is due to the fact that the structural defects were not considered during the modelling. However, this can be taken into account in the further research, as Digimat provides such a possibility. The dependence of the ultimate tensile strength on the GFRP/CFRP ration illustrates the hybrid effect, as when 50% of the CFRP are added to GFRP the strength of the composite is almost doubled.

4. Conclusion

In this study an approach to determination of the mechanical properties of composites, including hybrids, is implemented. As the result of numerical simulation by means of MSC Digimat the mechanical properties of composites with different reinforcing ratios are obtained. To validate the numerical analysis results the experiment is conducted. The simulation results are compared with experimental data and have shown good agreement that confirms the feasibility of the approach [21-22, 24].

References

[1] Swolfs Y, Verpoest I and Gorbatikh L 2019 *International Materials Reviews* **64** pp 181-215
[2] Reznik S V, Prosuntsov P V and Mikhailovskii K V 2019 *J Eng Phys Thermoph+* **92** pp 89-94
[3] Dimitrienko Y I and Dimitrienko I D 2017 *Int J Multiscale Com* **15** pp 219-37
[4] Prosuntsov P V, Reznik S V, Mikhailovsky K V, Novikov A D and Aung Z Y 2016 *IOP Conf. Series: Materials Science and Engineering* **153** p 012012
[5] Longana M L, Yu H, Hamerton I and Potter K D 2018 *Advanced Manufacturing: Polymer & Composites Science* **4** pp 48-55
[6] Botelho E C, Silva R A, Pardini L C and Rezende M C 2006 *Materials Research* **9** pp 247-256
[7] Wanhill R 2017 GLARE®: A Versatile Fibre Metal Laminate (FML) *Concept Aerospace Materials and Material Technologies (Aerospace Materials* vol 1*)* ed N Eswara Prasad and R J H Wanhill pp 291-307
[8] Zhang J, Chaisombat K, He S and Wang C 2012 *Materials and Design* **36** pp 75-80
[9] Ashikhmina E R and Prosuntsov P V 2019 *AIP Conference Proceedings* **2135** p 020005
[10] Ashikhmina E R and Prosuntsov P V 2019 *AIP Conference Proceedings* **2171** p 030014
[11] Ageyeva T G, Ashikhmina E R and Prosuntsov P V 2018 *Herald of the Bauman Moscow State Technical University. Series Mechanical Engineering* **1** pp 4-19
[12] Prosuntsov P V and Taraskin N Y 2016 *Matec Web Conf* **92**
[13] Anuchin S A, Lanin A V, Prosuntsov P V, Kordo M N, Zabezhailov M O and Rusin M Y 2018 *J*

Eng Phys Thermoph+ **91** pp 585-93

[14] Fibre glass. Constructive woven fabrics. Specifications. 2002 In: *Fibre glass. Constructive woven fabrics. Specifications (State Standard),* (Minsk: JSC NPO Stekloplastic)

[15] Kuzina E and Rimshin V 2019 *E3S Web Conf.* **97**

[16] Digimat. The Nonlinear Multi-scale Material and Structure Modeling Platform. 2019

[17] Trzepieciński T, Ryzińska G, Gromada M and Biglar M 2018 *Journal of the European Ceramic Society* **38** pp 2911-19

[18] Hori M and Nemat-Nasser S 1993 *Mechanics of Materials* **14** pp 189-206

[19] Zinovev P A and Smerdov A A 2006 *Optimalnoe proektirovanie kompozitnykh materialov: uchebnoe posobie po kursu «Proektirovanie kompozitnykh konstruktsiy*: Izd-vo MGTU im. N.E. Baumana)

[20] Smerdov A A 2018 Mechanics of Composite Materials **54** pp 269-80

[21] Shelofast V V, Kulish G G, Smerdov A A and Rostovtcev M Y 2017 *International Journal of Applied Engineering Research* **12** pp 728-37

[22] Zinoviev P A and Smerdov A A 1999 Compos Sci Technol **59** pp 625-34

[23] Nelyub V A and Borodulin A S 2018 Polymer Science – Series D **11** pp 147-153

[24] Abusrea M and Arakawa K 2016 Proc 17th European Conference on Composite Materials (ECCM17) (Munich)

Optimal design of advanced 3D printed composite parts of rocket and space structures

A V Azarov[1,3], T A Latysheva[2], A R Khaziev[1,3]

[1]Anisoprint LLC, 143026, Moscow, Skolkovo Innovation center, Bolshoi Bulvar, 42, str. 1, info@anisoprint.ru

[2]Progresstech, Ltd, 107023, Moscow, Electrozavodskaya str., d. 14, str. 1, engineering@progresstech.ru

[3]Bauman MSTU, 105005, Moscow, 2nd Baumanskaya str., d. 5, str. 1, bauman@bmstu.ru

E-mail: fml1580@gmail.com

Abstract. In this article the method of structural and topology optimization for 3D printed composite parts reinforced with continuous fibers is presented. Within this method part shape is changed simultaneously with material orientation in order to minimize the mass of a part under its strength constraints. The presented method is demonstrated on example of cantilever plate. Quadratic stress criteria is used as strength constraint. Algorithm is implemented as python code interacted with Ansys software. The obtained optimized design is shown. Finite element analysis of the optimized plate is performed.

Key words: topology optimization, fiber steering, composite 3D printing, optimal design of composites

1. Introduction

Mass minimization of advanced parts of rocket and space structures is an important problem solved by designers. The reduction of mass enables to use less fuel which is a great economic and ecologic benefit. Continuous fiber reinforced composite materials have high mechanical properties along fibers and are lightweight. To realize these benefits of composites to the fullest extent composite parts should be designed so that the load is taken along the fibers. The proper use of composites provides obtaining lightweight parts with high mechanical properties. Lattice structures are example of such use of composites [1]. Composite Payload Lattice Adapter is shown in Figure 1.

Continuous fiber reinforced composite 3D printing technology has developed rapidly during last years [2-7]. It enables high flexible fiber steering so that parts with almost any topology and internal structure can be produced. Anisoprint company has developed 3D printers which embed reinforcing fiber pre impregnated with thermoset polymer into molten thermoplastic polymer and produce the so-called bi-matrix composites [8-10]. Continuous fiber 3D printed Side Panel Element produced by Anisoprint is shown in Figure 2.

Figure 1. Lattice Payload Adapter (CRISM).

Figure 2. Side Panel Element (Anisoprint).

In order to design composite part in the best way structural and topology optimization is needed. Topology optimization is the process of finding the best material distribution within the given design space, for given loads and boundary conditions. Structural optimization is the process of finding the best orientation of fibers within the given topology. To get the optimal design for anisotropic part these optimizations should be performed simultaneously.

Mass minimization problem with strength constraints has been successfully solved for isotropic materials [11-21]. One of the algorithms for solving this problem is Proportional Topology Optimization method (hereinafter referred to as PTO) [20]. This is a non-gradient algorithm which implies easy implementation and small computational cost.

As we deal with anisotropic material we need to optimize material orientation as well. For this purpose, a fiber steering technique was proposed [22]. Within this approach reinforcing fibers are oriented along the principal stress direction. This orientation implies minimum local compliance of the structure and assures that the load is directed along the fiber. Another approach was suggested by Kentaro Sugiyama et al. [23].

In this article a new method of mass minimization with strength constraints for 3D printed composite parts is presented. It combines PTO method with fiber steering technique. The method is demonstrated on a rectangular cantilever plate with applied concentrated force. Optimal shape and layout are obtained for the plate. Strength constraint is satisfied. Finite element analysis is performed on the optimized model. Stresses appear to be uniformly distributed.

2. Design variables

Design variables for PTO method are chosen using Solid Isotropic Material with Penalization (SIMP) technique. Design space is supposed to be divided into finite elements (FE). The so-called density ρ, varying from 0 to 1, is assigned to each element. It states for the presence of material in particular element. So, $\rho = 0$ means that the element is empty i.e. does not contain material, while $\rho = 1$ means that the element is full, i.e. it is completely filled with material. Only 0 and 1 values are physically essential. But we allow intermediate densities as well in order to have continuous problem. Within SIMP method elastic moduli E and G of material in element with density ρ are calculated through the real moduli E_{real} and G_{real} of material in the following way:

$$E = \rho^p E_{real}$$
$$G = \rho^p G_{real}$$

$p > 1$ is a penalty factor. The increase of p makes it ineffective to have intermediate densities.

Optimization variables for fiber steering technique are angles of material orientation θ. So, besides the density, the angle is assigned to each finite element. Thus, the number of optimization variables is twice as big as the number of elements.

3. Problem formulation.

Using chosen optimization variables, the problem of mass minimization with strength constraints can be written as follows:

$$\min_{\rho} \sum_{i=1}^{N_{el}} d_i v_i \rho_i$$

$$\sigma^* \leq 1$$
$$0 \leq \rho_i \leq 1, \quad i = 1, \dots, N_{el}$$
$$0 \leq \theta_i \leq 180^\circ, \quad i = 1, \dots, N_{el}$$

Here N_{el} is a total number of elements in the model, d_i is the density of material in the ith element, v_i is the volume of the ith element, σ^* is some function of stresses representing strength constraint.

4. The PTO algorithm.

Algorithm of PTO is described in [20]. Slightly modified version used in this work is shown in Figure 3. To start some initial densities need to be chosen, say, all densities are equal to the same value. Sum of all densities in the current iteration is called target material amount. Before the first iteration it is calculated using initial densities.

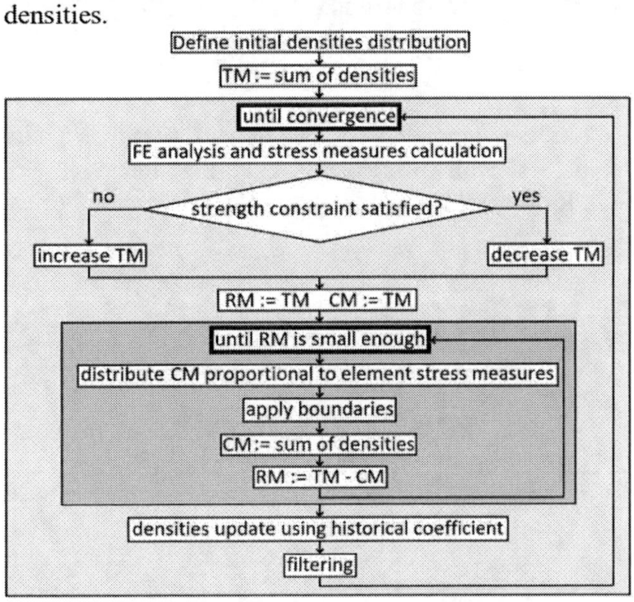

TM - target material amount, CM - current material amount, RM - remaining material amount

Figure 3. Proportional Topology Optimization algorithm

Then a cycle begins and continues until chosen convergence criteria is satisfied. Each loop of the cycle begins with FE analysis of the current model. Stresses are obtained and stress measures of elements are calculated. In case of isotropic material von Mises stresses were taken [20]. Then a strength constraint is checked. If it is satisfied the target material amount is increased. Otherwise it is decreased. Target material amount is then distributed using an inner cycle. On each iteration of this cycle some part of target material amount remains. This part is called remaining material amount. At first it is made equal to target material. The inner cycle continues until remaining material amount is small enough so that almost all target material is distributed on elements. Current material amount is used to keep information about material amount left to be distributed. At first it is equal to target material amount. On each inner iteration current material amount is distributed proportional to element stress measures. After this distribution some densities can appear to exceed 1. Such densities are made equal to 1 and material deleted by this procedure is acquired to remaining material. After the inner cycle densities are updated using historical coefficient α and densities from previous iteration ρ^{prev} in the following way:

$$\rho := \alpha \rho^{prev} + (1 - \alpha)\rho$$

Historical coefficient α represents inertia of densities changing. In case of $\alpha = 0$ densities do not depend on the previous iteration at all. In case of $\alpha = 1$ densities are not changed at all. Finally filtering technique is applied to densities. Instead of element density ρ_i weighted average of all densities inside the circle with the center in this element and radius r are taken:

$$\rho_i^{filter} = \frac{\sum_{j=1,dist(i,j)<r}^{N} w_{ij}\rho_j}{\sum_{j=1,dist(i,j)<r}^{N} w_{ij}}$$

w_{ij} – are weights inversely proportional to the distance between the ith and the jth elements. After densities are filtered one outer iteration ends and another one begins.

5. Fiber steering technique.

Within fiber steering technique angles are changed iteratively in the following way. On each iteration principal stress directions $\boldsymbol{n}_\alpha$ are found for all elements. In case of 2D problem there are two in-plane principal stress directions $\boldsymbol{n}_1$ and $\boldsymbol{n}_2$ for an element. Local compliance corresponding to principal stress direction $\boldsymbol{n}_\alpha$ is calculated by the following formula:

$$c = \frac{1}{2}\{\sigma_{xy}\}^T C^{-1}(\boldsymbol{n}_\alpha)\{\sigma_{xy}\}$$

Here $\{\sigma_{xy}\}$ is the vector of stress components for the element. $C(\boldsymbol{n}_\alpha)$ is a stiffness matrix for material which reinforcement coincides with principal stress direction $\boldsymbol{n}_\alpha$. Principal stress direction corresponding to minimum local compliance is chosen. Current material orientation is moved towards the chosen direction.

6. Presented algorithm

The algorithm presented in this work combines PTO method with fiber steering technique and is described in Figure 4.

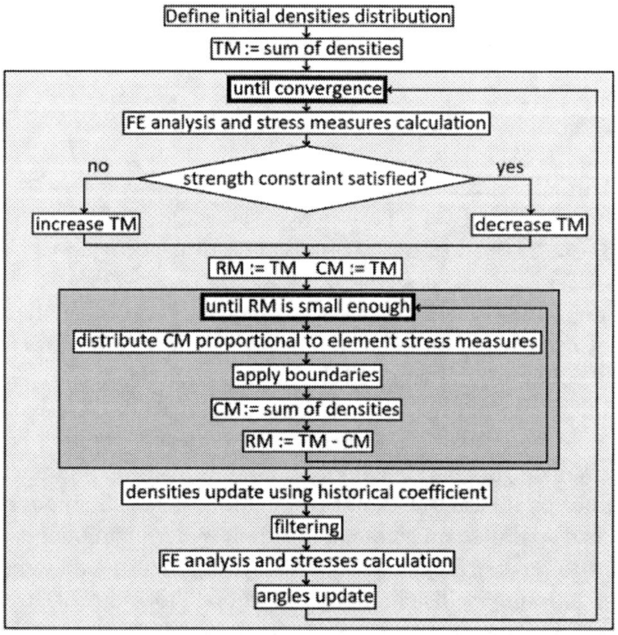

TM - target material amount, CM - current material amount, RM - remaining material amount

Figure 4. Combination of PTO method and fiber steering technique

The usual procedure of PTO takes place and fiber steering technique is added at the end of each outer loop. Finite element analysis is conducted after filtering and stresses are calculated in all elements. Using this information principal stress directions with minimum compliance are chosen and angles are moved towards them.

Quadratic failure criterion containing the difference between tensile and compressive strengths is applied to strength evaluation in this work:

$$\sigma_1\left(\frac{1}{\bar{\sigma}_1^{\,+}}-\frac{1}{\bar{\sigma}_1^{\,-}}\right)+\sigma_2\left(\frac{1}{\bar{\sigma}_2^{\,+}}-\frac{1}{\bar{\sigma}_2^{\,-}}\right)+\frac{\sigma_1^2}{\bar{\sigma}_1^{\,+}\bar{\sigma}_1^{\,-}}+\frac{\sigma_2^2}{\bar{\sigma}_2^{\,+}\bar{\sigma}_2^{\,-}}+\left(\frac{\tau_{12}}{\bar{\tau}_{12}}\right)^2\leq 1$$

Reserve factors are calculated using the equation:

$$n=\frac{2}{s+\sqrt{s^2+4r^2}}$$

where

$$s=\sigma_1\left(\frac{1}{\bar{\sigma}_1^{\,+}}-\frac{1}{\bar{\sigma}_1^{\,-}}\right)+\sigma_2\left(\frac{1}{\bar{\sigma}_2^{\,+}}-\frac{1}{\bar{\sigma}_2^{\,-}}\right)\qquad r^2=\frac{\sigma_1^2}{\bar{\sigma}_1^{\,+}\bar{\sigma}_1^{\,-}}+\frac{\sigma_2^2}{\bar{\sigma}_2^{\,+}\bar{\sigma}_2^{\,-}}+\left(\frac{\tau_{12}}{\bar{\tau}_{12}}\right)^2$$

Inverse reserve factors are taken as element stress measures. In the work of Emre Biyikli and Albert C. To strength constraint bounds maximal stress measure to be less or equal to 1 [20]. In the current work p-norm of stress measures is used instead of maximum in order to ignore local stress peaks. P-norm is used by Erik Holmberg, Bo Torstenfelt and Anders Klarbring in the work [19]. P-norm of some values $a_1, a_2, \dots, a_n$ looks like the following:

$$\|a_1,a_2,\dots,a_n\|_P=\left(\frac{\sum_i (a_i)^P}{n}\right)^{\frac{1}{P}}$$

P-norm of some set of values tends to the maximal value when p tends to infinity:

$$\lim_{P\to\infty}\|a_1,a_2,\dots,a_n\|_P=\max_i a_i$$

In this work $P = 20$ in which case only view element stress measures exceeded the value of P-norm.

7. Test problem formulation.

In order to demonstrate described algorithm of optimization a rectangular plate (see Figure 5) is considered.

Figure 5. Cantilever plate with applied concentrated force *F*

The left edge of the plate is clamped. A concentrated force is applied at the center of the right edge as shown in Figure 5. and directed down. Plate width is 0.2 m. Plate height is 0.1 m. Plate thickness is 0.005 m. The magnitude of the force is 100 N. The following material properties are used:

$$E_1 = 64100000000 \qquad E_2 = 4100000000 \qquad G_{12} = 1170000000 \qquad \upsilon_{12} = 0.023$$

Such properties correspond to material obtained using Composite Filament Co-Extrusion (CFC) Anisoprint technology. The plate was meshed into 2048 square elements.

8. Results.

To solve the problem Python code interacting with Ansys software package is written. Initial mass of the part is 141 g. The result of optimization is a frame shown in Figure 6.

IOP Conf. Series: Materials Science and Engineering **934** (2020) 012062 doi:10.1088/1757-899X/934/1/012062

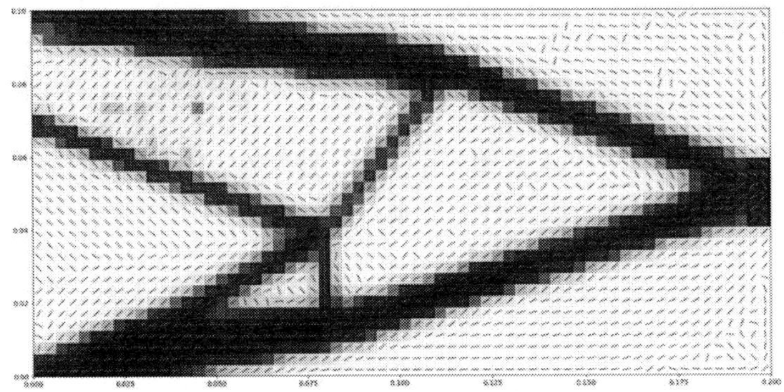

Figure 6. Result of optimization

The lower ribs of the frame are thicker than the upper one. It follows from the fact that compressive strength is lower than tension strength. In the central part of the frame the strut elements can be seen. These elements prevent distortion of the frame shape under shear load. Material is oriented along the ribs of the frame. Weight of the obtained part is 35.9 g which is 25.5% of the fully filled space. FE analysis is performed. Stresses σ_1, σ_2 and τ_{12} in material coordinate system are shown in Figures 7(a) – 7(c). Stresses are close to uniform in the most part of the plate. On the left side of the plate near rigid support stresses achieve their maximum. Reserve factors are shown in Figure 7(d). In the most part of the plate they exceed 1 quite a lot and are close to 1 near the rigid support. Strength criterion is satisfied but the method still needs to be improved so that all reserve factors are close to 1.

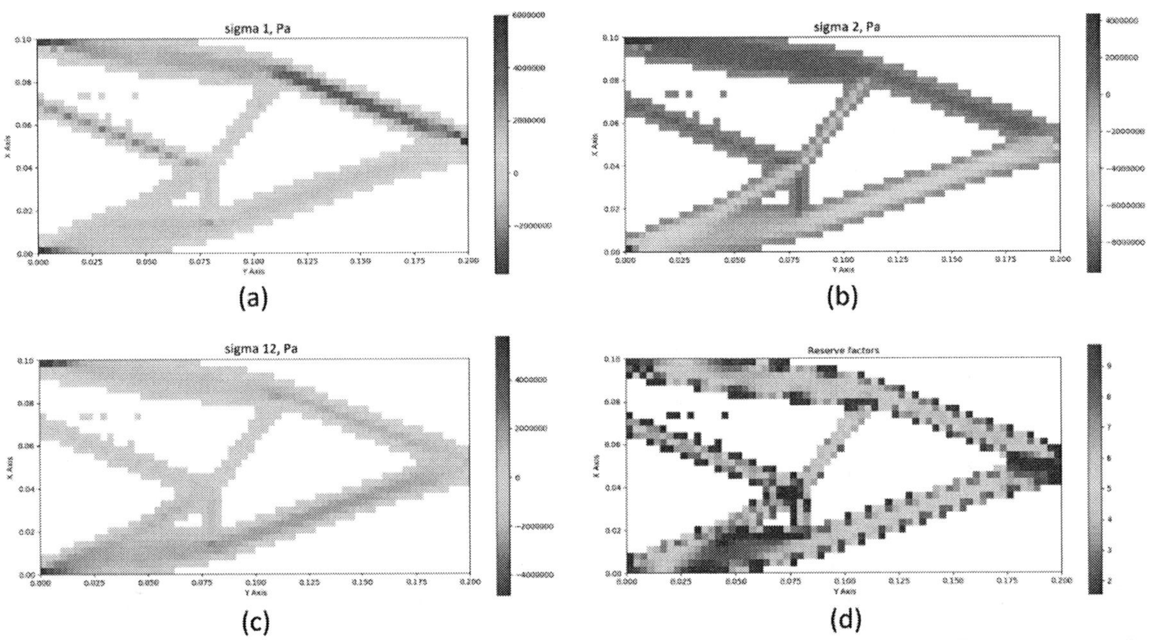

Figure 7. FE analysis of optimized plate: (a) elemental stresses σ_1 in material coordinate system; (b) elemental stresses σ_2 in material coordinate system; (c) elemental stresses σ_{12} in material coordinate system; (d) elemental reserve factors.

9. Conclusion.

New algorithm for mass minimization with strength constraints for Anisoprint CFC technology is proposed. This algorithm combines PTO method and fiber steering technique. Optimization problem for cantilever plate with concentrated force is solved using described algorithm. Quadratic failure criterion containing the difference between tensile and compressive strengths is used as strength constraint. The optimal design of the plate is obtained. Strength constraint is satisfied. Finite element analysis is conducted for the plate. Stresses are shown to be close to uniform which proves optimality of the solution. The proposed method enables to get lightweight strong parts and can be applied to design of rocket and space structure where mass is of prime importance [24-26].

References

[1] V.V. Vasiliev, Morozov E.V. Advanced Mechanics of Composite Materials and Structural Elements. Third edition. Amsterdam: Elsevier, 2013, 818p.
[2] Guo N and Leu M 2013 Additive Manufacturing: Technology, Applications and Research Needs Department of Mechanical and Aerospace Engineering, Missouri University of Science and Technology Rolla 8 3 215-43
[3] Jayakumar A 2014 Why the aerospace industry is investing in 3-D printing The Washington Post
[4] Cohen D, Sargeant M and Somers K 3-D printing takes shape 2014 https://www.mckinsey.com/usinessfunctions/operations/our-insights/3-d-printing-takes-shape access: nov. 2019
[5] Orbital Composites Electronic resource https://www.orbitalcomposites.com/ access: nov. 2019
[6] Matsuzaki R, Ueda M, Namiki M and other 2016 Three-dimensional printing of continuous-fiber composites by in-nozzle impregnation Scientific Reports
[7] 3D Printer Using Continuous Carbon Fiber Composite Materials Electronic document http://www.jscm.gr.jp/3Dprinting/images/introduction_CFRP3Dprinter.pdf access: november 2019
[8] Azarov A V, Antonov F K, Vasiliev V V, Golubev M V, Krasovskii D S, Razin A F, Salov V A, Stupnikov V V, Khaziev A R Development of a two-matrix composite material fabricated by 3D printing. Polym Sci D 2017;10(1):87–90
[9] Azarov A, Antonov F, Golubev M, Khaziev A and Ushanov S 2019 Composite 3D printing for the small size unmanned aerial vehicle structure *Composites* Part B **169** 157–63
[10] Adumitroaie A, Antonov F, Khaziev A, Azarov A, Golubev and Vasiliev V 2019 Novel Continuous Fiber Bi-Matrix Composite 3-DPrinting Technology *Materials* **12(18)** 3011
[11] Lee, E., James, K., & Martins, J. R. R. A. (2012). Stress-constrained topology optimization with design-dependent loading. Structural and Multidisciplinary Optimization, 46(5), 647-661. https://doi.org/10.1007/s00158-012-0780-x
[12] AM Mirzendehdel, B Rankouhi, K Suresh. (2018) Strength-Based topology optimization. Additive Manufacturing 19, 104-113
[13] de Leon, D. M., Alexandersen, J., Jun, J. S., & Sigmund, O. (2015). Stress-constrained topology optimization for compliant mechanism design. Structural and Multidisciplinary Optimization, 52(5), 929-943. https://doi.org/10.1007/s00158-015-1279-z
[14] Bourdin B (2001) Filters in topology optimization. International journal for numerical methods in engineering 50:2143–2158
[15] Cheng G, Jiang Z (1992) Study on topology optimization with stress constraint. Engineering Optimization 20:2:129–148
[16] Duysinx P, Bendsøe MP (1998) Topology optimization of continuum structures with local stress constraints. International journal for numerical methods in engineering 43:1453–1478
[17] Le C, Norato J, Bruns T, Ha C, Tortorelli D (2010) Stress-based topology optimization for continua. Structural and Multidisciplinary Optimization 41:605–620
[18] Meneghelli LR, Cardoso EL (2013) Optimization of structures and components, Springer International Publishing Switzerland, chap Design of compliant mechanisms with stress constraints using topology optimization, pp 35–48
[19] Holmberg E., Tornstenfelt Bo, Klarbring A., (2013) Stress constrained topology optimization. Structural and multidisciplinary optimization (Print), ISSN 1615-147X, E-ISSN 1615-1488, Vol. 48, no 1, p. 33-47
[20] Biyikli E, To AC (2015) Proportional Topology Optimization: A New Non-Sensitivity Method for Solving Stress Constrained and Minimum Compliance Problems and Its Implementation in MATLAB. PLoS ONE 10(12): e0145041. https://doi.org/10.1371/journal.pone.0145041
[21] Rozvany, George. (2000). The SIMP method in topology optimization - Theoretical background, advantages and new applications. 8th Symposium on Multidisciplinary Analysis and Optimization. 10.2514/6.2000-4738.

[22] Safonov, A. A., 3D topology optimization of continuous fiber-reinforced structures via natural evolution method. Composite Structures (2019), doi: https://doi.org/10.1016/j.compstruct.2019.02.063
[23] Kentaro Sugiyama , Ryosuke Matsuzaki, Andrei V. Malakhov, Alexander N. Polilov, Masahito Ueda, Akira Todoroki, Yoshiyasu Hirano 3D printing of optimized composites with variable fiber volume fraction and stiffness using continuous fiber // Composites Science and Technology 186 (2020) 107905
[24] Reznik S.V., Prosuntsov P.V., Azarov A.V. Substantiation of the structural-layout scheme of the mirror-space-antenna reflector with a high shape stability and a low density per unit length // J. Eng. Phys. Thermophy. – 2015. – Vol. 88, No. 3. – pp. 699-705. DOI:10.1007/s10891-015-1239-x
[25] Reznik S.V., Prosuntsov P.V., Azarov A.V. Modeling of the temperature and stressed-strained states of the reflector of a mirror space antenna // J. Eng. Phys. Thermophy. – 2015. – Vol. 88, No. 4. – pp. 978-983 DOI:10.1007/s10891-015-1273-8
[26] Reznik S.V., Prosuntsov P.V., Mikhailovsky K.V., Shafikova I.R. Material science problems of building space antennas with a transformable reflector 100 m in diameter // IOP Conference Series: Materials Science and Engineering. 2016. Vol. 153, No 1. 012012. – 10 p. DOI: 10.1088/1757-899X/153/1/012001

IOP Conf. Series: Materials Science and Engineering **934** (2020) 012063 doi:10.1088/1757-899X/934/1/012063

Modeling of volumetric deformation during compaction of Al-SiC composite materials based on mechanically activated batch

S N Sergeenko

Federal State Budget Educational Institution of Higher Education Platov South-Russian State Polytechnic University (NPI), Novocherkassk, Russia

E-mail: sergeenko@gmail.com

Annotation. A 3D Spline model of the effect of the silicon carbide content C_{SiC} and the cold compaction pressure p on the bulk deformation of the powder molding ε_v during the cold compaction of the Al-SiC charge was constructed. The nonlinear character of the dependence $\varepsilon_v(p)$ is established for the studied values of the silicon carbide content. The two-stage nature of the cold compaction of the Al-SiC powder mixture was revealed, which is characterized by intensive compaction at the initial stage. A sigmoidal model of cold compaction of a powder mechanically activated Al-SiC charge, which takes into account the initial stage of lightweight structural compaction, is constructed. The dependences of the influence of the SiC powder content on the coefficients of the sigmoidal cold compaction equation are established. A logistic model of the influence of C_{SiC} on the coefficients of the sigmoidal equation of cold compaction and the average size of agglomerates formed during the mechanical activation of a mixture of Al-SiC composite powder material is proposed. The energy equation of hot compaction is obtained, the value of the compressibility coefficient is determined, and a linear equation is constructed that relates the relative densities before and after hot compaction of the Al-SiC composite material.

1. Introduction

To solve the urgent problem of manufacturing valve seats of internal combustion engines, promising technologies for producing powder and cast Al-SiC composite materials have been proposed [1]. The physicomechanical properties of Al-SiC materials are determined by the content and average particle size of silicon carbide [2, 3]. The possibility of producing hot-compacted Al-SiC composite materials based on mechanically activated powder mixtures in an aqueous H3BO3 solution was studied at SRSPU (NPI). The increased efficiency of mechanical activation is shown, in comparison with the mixing technology for the production of hot-compacted materials. It was established that an increase in the relative density of the cold compacted molding and a decrease in the SiC content in the charge reduces the residual porosity of the hot compacted Al-SiC powder composite material [4]. As a result of previous studies, the hereditary effect of mechanical activation on the processes of compaction and formation of the mechanical properties of sintered and hot-compacted powder materials was established [4-25].

2. Purpose of work

Construction of models of cold and hot compaction of mechanically activated powder mixture of Al-SiC composite material.

3. Research Methodology

IOP Conf. Series: Materials Science and Engineering **934** (2020) 012063 doi:10.1088/1757-899X/934/1/012063

The manufacturing technology of the samples included mechanical activation of the Al-SiC powder mixture, its cold compaction with a pressure of 47-140 MPa, heating (t = 823 K; τ = 0.12 ks) in an air atmosphere, followed by hot compaction on a copra with reduced (specific) work (w), equal to 130 MJ/m3. The powder mixture was mechanically activated (3.24 ks) in a SAND-1 planetary ball mill with a rotational speed of 4.8 s-1 and a mass ratio of balls (db = 10 mm) and a charge of 20 in a saturated aqueous solution of boric acid (20% of the mass of the charge). Moreover, the formed protective layer on the surface of the powder particles prevents oxidation during short-term heating of the molding in an air atmosphere. Powders of aluminum PA-4 and silicon carbide (black) SiC (5-20 wt.%) With an irregular, fragmented form of particles were used as starting materials [4]. The construction of a 3D Spline model of cold compaction of a powder mixture of Al-SiC composite material was carried out in a Statistica medium.

4. The results of experimental studies

Based on the results of experimental studies [4], the values of volumetric strain ε_v (1) were determined taking into account the relative density θ_0 of the charge in the bulk state ($p = 0$) and the cold-compacted molding θ at pressure p

$$\varepsilon_v = ln\frac{\theta_0}{\theta}. \tag{1}$$

Analysis of the 3D Spline model $\varepsilon_v(p,\ C_{SiC})$ of the effect of the silicon carbide content C_{SiC} and pressure p on the volumetric strain ε_v of the powder molding during cold compaction of the Al-SiC charge (Figure 1) showed that the dependences $\varepsilon_v(p)$ are nonlinear for all studied SiC values. The dependence of the cold compaction $p(\varepsilon_v)$ of an Al-SiC powder mixture (10 wt%) can be described by the nonlinear model (SDSCum) (2) ($a = -146.39$; $b = 146.03$; $\boldsymbol{c} = -0.357$; $d = 0.216$; $e = 0.0002$), built in the Table Curve 2D environment, with an increased value of the coefficient of determination (r2 = 0.996)

$$p = a + \frac{b}{2 \cdot d}\left\{2 \cdot e \cdot \left[ln\left(exp\frac{\varepsilon_v+0,5d}{e} + exp\frac{c}{e}\right) - ln\left(exp\frac{c+0,5d}{e} + exp\frac{\varepsilon_v+0,5d}{e}\right)\right] + d\right\} \tag{2}$$

As a result of a joint analysis of the 3D Spline model (Figure 1) and the resulting equation (2), the two-stage nature of the cold compaction of the Al-SiC powder mixture was revealed. At the first stage (/p/<47 MPa), intense compaction is observed, associated with structural deformation, characterized by facilitated displacement of powder particles and their redistribution. With a further increase in the cold compaction pressure (/p/>47 MPa) (in the second stage), the dependence $p(\varepsilon_v)$ can be described by linear equation (3) by analogy with the rheological equation for a compressible medium

$$p = K \cdot \varepsilon_v, \tag{3}$$

where p is the pressure of cold compaction, MPa;
K-compressibility coefficient during cold compaction, MPa;
ε_v-is the logarithmic volumetric deformation calculated taking into account the calculated value of the relative density of the powder material θ_c characterizing the completion of the stage of lightweight structural compaction ($p\sim0$)and taken as the initial one (at $\theta_0 = \theta_c$; $\varepsilon_v\sim0$).
Transforming equation (3) taking into account dependence (1), it was found that θ affects the pressure p for the second stage of cold compaction (4) when the relative density changes from the initial values θ_c ($p\sim0$) to the current θ

$$p = K \cdot ln\theta_c - K \cdot ln\theta = K \cdot \varepsilon_{pr} - K \cdot ln\theta = p_{max} - K \cdot ln\theta, \tag{4}$$

where p_{max}, $\varepsilon_{pr} = ln\theta_c$ are the calculated values of the cold compaction pressure and the ultimate volumetric deformation, which ensure the production of a non-porous (compact) powder composite material ($\theta = 1$), respectively.

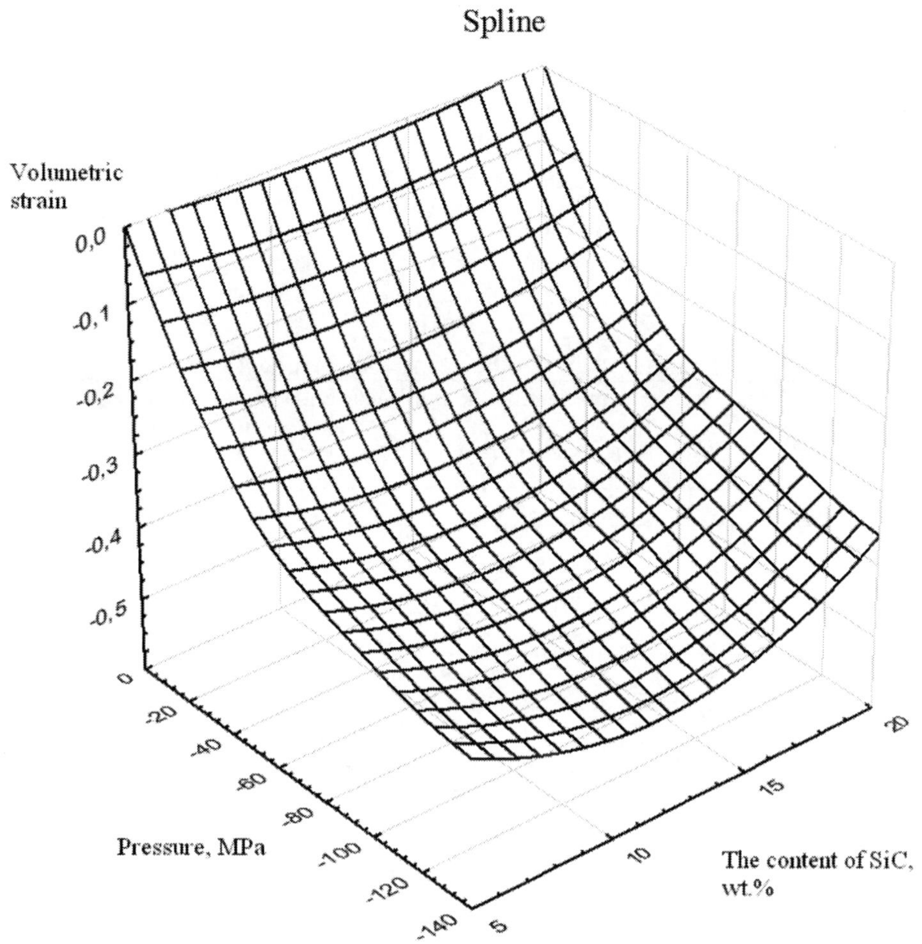

Figure 1. 3D Spline model $\varepsilon_v(C_{SiC},\ p)$ of the effect of SiC content and cold compaction pressure p on the volumetric strain ε_v of the Al-SiC composite material

In the general case, during compaction of a compressible porous body, the dependence $p(\varepsilon_v)$ is nonlinear. To establish the relationship between the pressure of unilateral compression p and volumetric strain ε_v, Professor Reiner M. proposed nonlinear equations $p(\varepsilon_v)$ (5) and (6)

$$p = K \cdot \varepsilon_v \cdot exp(\varepsilon_v), \quad (5)$$

$$p = K \cdot \frac{1-\psi}{exp(\varepsilon_v)-\psi} \varepsilon_v \cdot exp(\varepsilon_v), \quad (6)$$

where ψ is the coefficient of the limiting relative volume.

When studying the cold compaction of Al-SiC composite material for the entire compaction stage, taking into account the initial stage of lightweight structural compaction, the sigmoid equation $p(\varepsilon_v)$ was proposed (7)

$$p(\varepsilon_v) = a + \frac{b}{1+exp\left(-\frac{\varepsilon_v - c}{d}\right)}. \quad (7)$$

The values of volumetric strain ε_v of the compressible powder material during cold compaction were determined taking into account the relative molding density θ_0 in the bulk state ($p = 0$) according to formula (1). Cold compaction of the charge of the Al-Si composite material is characterized by a continuously sequential compaction – decompression – re-compaction process, depending on the Si content [4]. Based on the results of experimental studies (table 1) for the stage of continuous compaction (0-140 MPa), we determined (table 2) the parameters (a, b, c, d) of the sigmoid equation (7) and the determination coefficients r^2 using the standard package Table Curve 2D.

Table 1. The values of volumetric strain ε_v.

C_{SiC}, wt.%	The values of ε_v at the investigated pressure p, MPa			
	$p = 0$	$p = -47$	$p = -94$	$p = -140$
5	0	-0,268	-0,347	-0,396
10	0	-0,320	-0,389	-0,447
15	0	-0,295	-0,373	-0,481
20	0	-0,257	-0,297	-0,350

Table 2. Parameter values of the sigmoidal (7) model $p(\varepsilon_v)$.

C_{SiC}, wt.%	Parameters of the nonlinear equation $p(\varepsilon_v) = a + b/\left(1 + exp\left(-\frac{\varepsilon_v}{c}\right)^d\right)$				
	a	b	c	d	r^2
5	-674,1	678,6	-0,536	0,107	0,999
10	-222,5	223,1	-0,41	0,07	1,0
15	-156,5	157,1	-0,35	0,06	1,0
20	-158,6	158,6	-0,28	0,03	1,0

An analysis of the results (table 2) showed that the content of SiC powder affects the values of the coefficients (a, b, c, d) of the sigmoid equation (7) of cold compaction of the charge of the Al-SiC composite material. Dependences $c, d(C_{SiC})$ are described by linear equations (8), and $a, b(C_{SiC})$ logistic (LgstcDoseRsp) (9) with increased values of the determination coefficient r^2 (table 3)

$$c, d(C_{SiC}) = a_1 + b_1 \cdot C_{SiC}, \tag{8}$$

$$a, b(C_{SiC}) = a_2 + \frac{b_2}{1+\left(\frac{C_{SiC}}{c_2}\right)^{d_2}}. \tag{9}$$

Table 3. Parameter values of linear $c, d(C_{SiC})$ (8) and logistic $a, b(C_{SiC})$ (9) models.

$a, b,$ c, d	Parameters of linear models $c, d(C_{SiC}) = a_1 + b_1 \cdot C_{SiC}$			Logistic Model Parameters $a, b(C_{SiC}) = a_2 + b_2/\left(1 + \left(\frac{C_{SiC}}{c_2}\right)^{d_2}\right)$				
	a_1	b_1	r^2	a_2	b_2	c_2	d_2	r^2
a	-	-	-	-674,1	516,6	8,9	-16,9	0,999
b	-	-	-	157,8	520,9	8,85	15,82	0,999
c	0,13	0,018	0,951	-	-	-	-	-
d	0,13	-0,005	0,954	-	-	-	-	-

An analysis of the effect of C_SiC on the mean size d_0 of agglomerates formed during the mechanical activation of the Al-SiC mixture [4] showed that the dependence $d_0(C_{SiC})$ can be described by logistic equation (10) with an increased value of the determination coefficient (r^2 = 0.999)

$$d_0(C_{SiC}) = 128 + \frac{325,5}{1+\left(\frac{C_{SiC}}{10,5}\right)^{-27,4}}. \tag{10}$$

The values of the parameter c_2 (table 3, equation 10) represent the critical value of the silicon carbide content in the charge (C_{SiC} = 10 wt.%), Characterizing the transition to the formation of large agglomerates (~ 450 μm), as well as an abrupt change in the values of the parameters a, b (table 2) of the sigmoidal equation (7) of cold compaction of the Al-SiC composite material. It was shown in earlier work that when an optimal SiC content (10 wt.%) is introduced into an Al-based charge, extreme values of the parameters of the nonlinear model of particle size distribution of mechanically activated powder powders are observed [4, 25].

When describing the hot compaction of Al-SiC composite material by the linear model (3), we obtain the energy equation of compaction (11) taking into account the relative density of the cold-compacted molding θ and the hot-compacted workpiece θ_1

$$w = 0{,}5 \cdot K_1 \cdot \varepsilon_v^2 = 0{,}5 \cdot K_1 \cdot (ln\theta - ln\theta_1)^2, \qquad (11)$$

where w is the reduced (specific) work of hot compaction of a unit volume of a compact (non-porous) composite material, MJ/m3;

K_1 - compressibility coefficient during hot compaction of the composite material. Transforming equation (11) for a fixed value of the reduced work (w = 130 MJ/m3), we obtain a linear (12) dependence $\theta_1(\theta)$ connecting the relative density values before (θ) and after hot compaction (θ_1) of composite material Al-SiC

$$\theta_1 = \theta \cdot exp\left(\frac{2 \cdot w}{K_1}\right)^2. \qquad (12)$$

Based on the analysis of the results of studies [4] of hot ($t = 823\ K$; $\tau = 0{,}12\ ks$; $w =$ 130 MJ/m3) compacting of the Al-SiC composite material (15 wt.%), we obtained linear regression equation $\theta_1(\theta)$ (13) with an increased determination coefficient (r^2 = 0.984) and the value of the compressibility coefficient (K_1 = 2000 MPa) was calculated

$$\theta_1 = 1{,}433 \cdot \theta. \qquad (13)$$

5. Conclusions

Based on the studies, a 3D Spline model of cold compaction of a mechanically activated mixture of Al-SiC composite material was constructed. A nonlinear sigmoidal model of the effect of cold compaction pressure on volumetric deformation is proposed. Logistic models of the coefficients of the sigmoidal equation of cold compaction and the average size of agglomerates formed during mechanical activation, depending on the content of silicon carbide in the mixture, are constructed. An energy equation of compaction is obtained that describes the effect of volumetric deformation of a composite material on the reduced work of hot compaction. A linear model is constructed that relates the relative densities before and after hot compaction of the Al-SiC composite material for a fixed reduced (specific) work w.

References

[1] Patel N S, Patel A D and Ranjan R K 2018 Studies on Properties of Al–SiC Metal Matrix Composite Material for Making IC Engine Valves *Proceedings of the World Congress on Engineering* 2018 **II** WCE July 4-6 2018 London U K www.iaeng.org/publication/.../WCE2018_pp667-70.pdf

[2] Youssef Y and El-Sayed M 2016 Effect of Reinforcement Particle Size and Weight Fraction on the Mechanical Properties of SiC Particle Reinforced Al Metal Matrix Composites / *International Review of Mechanical Engineering* **10** No 4 pp 261-265

[3] El-Kady O and Fathy A 2014 Effect of SiC particle size on the physical and mechanical properties of extruded Al matrix nanocomposites *Materials & Design* **54** pp 348-53

[4] Sergeenko S N and, Alabid N S 2016 Hot-deformed powder materials based on mechanochemically activated charges Al – SiC *Tsvetnye Metally* **9** (885) pp 68-77

[5] Slabkii D V and Sergeenko S N 2016 Hot-deformed Al–Ni powder materials based on alloy D-16 mechanically-activated turnings. *Metallurgist* **59** 11 pp 1228-33

[6] Dyuzhechkin M K, Sergeenko S N and Popov Y V 2016 Features of structure and property formation for hot-deformed materials of the Al–Si and Al–Si–C systems based on mechanochemically activated charges *Metallurgist* **59** 9-10 pp 835-42

[7] Sergeenko S N, Juraev F T, Al Hilfi A X and Postnikov A A 2016 Features of mechanochemical activation and compaction of the powder mixture Al-Si-Ni-*Fizika i khimiya obrabotki materialov* [Physics and chemistry of materials treatment] **4** pp 63-72

[8] Fedoseeva M A and Sergeenko S N 2015 Structure and properties of Al–FeTi powder material based on mechanochemically activated alloy D-16 turnings *Metallurgist* **59** 5-6 pp 535-9

[9] Dorofeev Yu G, Bezborodov E N and Sergeenko S N 2003 Special features of formation of compacted material from mechanochemically activated fining of aluminum alloy D16 *Metal Science and Heat Treatment* **45** 1-2 pp 73-5

[10] Dorofeev Yu G, Bezborodov E N and Sergeenko S N 2003 Hot-strained powder materials on the base of mechanically alloyed Al-C mixture *Fizika i khimiya obrabotki materialov* [Physics and chemistry of materials treatment] **3** pp 64-72

[11] Dorofeev Yu G, Bezborodov E N and Sergeenko S N 2003 Hot wrought powders on the base of D16 chips mechanochemically activated *Tsvetnye Metally* **1** pp 81-5

[12] Dorofeev Yu G, Bezborodov E N and Sergeenko S N 2002 Development of technology for producing hot-deformed powder material based on mechanochemically activated chips of D-16 aluminum alloy *Materialovedeniye* **9** pp 40-5

[13] Dorofeev Yu G, Bezborodov E N and Sergeenko S N 2002 Kinetics of mechano-chemical activation of aluminium-based powder charge in saturated solution of orthoboric acid *Fizika i khimiya obrabotki materialov* [Physics and chemistry of materials treatment] **3** pp 51-4

[14] Dorofeev Yu G, Bezborodov E N and Sergeenko S N 2002 Effect of kinetics of mechanochemical activation of al powders on the processes of finish hot compacting *Fizika i khimiya obrabotki materialov* [Physics and chemistry of materials treatment] **4** pp 79-81

[15] Yu G, Bezborodov E N and Sergeenko S N 2002 Special features of formation of compacted material from mechanochemically activated chips of aluminum alloy D-16 *Tekhnologiya legkikh splavov* [The Technology of Light Alloys] **2** pp 23-8

[16] Dorofeyev Yu G, Bezborodov E N and Sergeenko S N 2002 Features of the formation on the dynamic hot compacted material from mechanochemically activated chips of aluminum alloy D-16 *Tekhnologiya legkikh splavov* [The Technology of Light Alloys] **3** pp 37-41

[17] Dorofeyev Yu G, Bezborodov Ye N and Sergeenko S N 2001 Features of high-density hot-deformed material based on "chip" powder of aluminum alloy D-16 *Izvestiya Vysshikh Uchebnykh Zavedenij. Tsvetnaya Metallurgiya* **10** pp 28-31

[18] Sergeenko S N 2012 Technologies for producing the powder materials on the base of mechanically alloyed mixture (review) *Tekhnologiya metallov* [Metals technology] **1** pp 46-56

[19] Sergeenko S N 2012 Technologies for producing the powder materials on the base of mechanically alloyed mixture (review) *Tekhnologiya metallov* [Metals technology] **5** pp 46-55

[20] Sergeenko S N 2012 Technologies for producing the powder materials on the base of mechanically alloyed mixture (review)]. *Tekhnologiya metallov* [Metals technology] **6** pp 47-56

[21] Dorofeyev Yu G, Sergeyenko S N and Kolomiyets R V 2007 Effect of the composition of a powder charges Ni-Fe-NaCl on the kinetics mechanical activation in liquid media *Fizika i khimiya obrabotki materialov* [Physics and chemistry of materials treatment] **4** pp 57-61

[22] Sergeenko S N 2019 Kinetics of dispersion-agglomeration processes during mechanical activation of the charge of 110g13 powder steel *Chernye Metally* **7(1051)** pp 47-52

[23] Sergeenko S N 2019 Compaction and consolidation of 110g13 steel powder based on mechanically activated charges *Chernye Metally* **8(1052)** pp 56-61

[24] Sergeenko S N 2019 Methods of Hot Compaction and Infiltration of Powder Materials Based on Recycled Non-Ferrous Alloy Chips. *Aspects Min Miner Sci.* **3(5)** AMMS.000572.2019. DOI: 10.31031/AMMS.2019.03.000572

[25] Sergeenko S N 2019 Simulation of particle size distribution of mechanochemically activated charges of Al-SiC composite materials *IOP Conf. Ser.: Mater. Sci. Eng.* **683** 012085

IOP Conf. Series: Materials Science and Engineering **934** (2020) 012064 doi:10.1088/1757-899X/934/1/012064

Optimization of the power structure of carbon-ceramic composite materials

N Yu Taraskin and V L Grigorieva

Bauman Moscow State Technical University, 2nd Baumanskaya str., b. 5, Moscow, 105005, Russia

E-mail: trzzz@mail.ru

Abstract. In this paper analyze of structures and materials for the heat shields of the descent vehicles is considered. In addition, the structure, manufacturing technology and characteristics of carbon-ceramic composite materials are analyzed. A basic finite-element model of a heat-shielding coating is based, and topological optimization is performed. Based on the results of optimization, the power structures of the heat-shielding coating are developed and their strength characteristics are evaluated. The general structure of the heat shield is developed including account the heat-shielding coating.

1. Introduction

Today, the leading space powers are working on the creation of a new generation of spacecraft, which are being developed as modular systems. The main module of the spacecraft is the descent vehicle (DV), in which the astronauts are during orbiting, during the flight and when returning to Earth. Heat shield is a necessary element of the spacecraft that protect the power structure of the DV from the action of a high-enthalpy gas flow. The main part of a heat shield of a traditional design is a heat-protective coating (HPC).

In the DV of the previous generation, heat protection from ablative materials is used, the scheme of which is shown in Figure 1. The principle of operation of the thermal current transformer of this type is that the bulk of the thermal energy is spent on decomposition of the material. Ablation materials are a reliable, proven solution, DV with thermal protection of this type to this day deliver people from orbit to the ground.

However, the features of the work of ablation materials do not allow to make spacecraft reusable and significantly limit the possibilities of their use. In addition, since the ablative materials burn out during the flight of the DV, they cannot be part of the power structure. There is a need to create an additional design that will bear all the power loads.

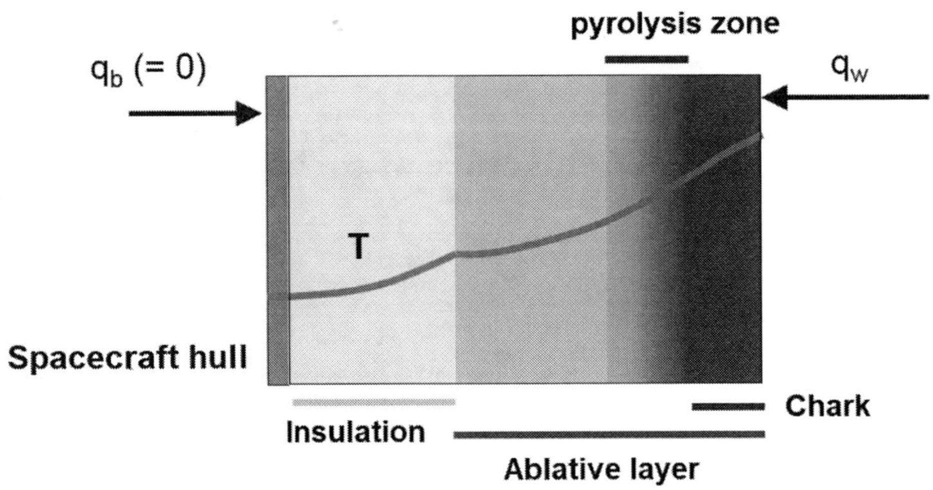

Figure 1. The operation scheme of the ablation TZP.

It is relevant to create new HPCs from high-temperature materials for reusable use so that there is the possibility of flights not only to the ISS, but also to the Moon and Mars. Such a material should be resistant to high temperatures (up to 2000 ° C), ability to work in an oxidized environment and good mechanical characteristics. These properties are possessed by carbon-ceramic composite materials (CCCM) and are suitable for use in HPC DV. The characteristics are presented in table 1. The high mechanical characteristics of CCCM are ensured by the use of a reinforcing filler of carbon fiber, protected from oxidation by a coating of silicon carbide. The effectiveness of CCCM as a HPC has already been substantiated [1]. Carbon fibers give the material good mechanical properties. The ceramic matrix protects the carbon skeleton from oxidation and burning at high temperatures [2,3]. Such a coating, due to its high heat resistance, high and stable frictional, mechanical and thermophysical properties, makes it possible to incorporate HPC into the power structure of DV, which will help to save weight and space in comparison with traditional schemes where HPC performs an exclusively heat-protective function.

2. Features of obtaining carbon-ceramic composite materials

One of the most common CCCMs is C / SiC. There are three main technologies for producing C / SiC composites: chemical gas saturation (CVI) [4], liquid polymer saturation and pyrolysis (LPI) [5], liquid silicon saturation (LSI) [6]. In Russia, well-developed technology is gas phase saturation [7]. Creating a material takes place in four main stages. At the first stage, a preform is created from PAN fibers. The second step is pyrolysis, which results in the formation of a rigid carbon frame. At the third stage, an interfacial layer is formed due to carbon deposition. At the final stage, a ceramic matrix is formed by gas deposition in the pores of the workpiece. Typically, methyltrichlorosilane ($CH3SiCl3$) and hydrogen as a catalyst are used as the process gas.

CCCM obtained by CVI technology has high mechanical characteristics, which are presented in table 1 [8]. Due to these characteristics, the C / SiC material is able to perform not only a heat-shielding function, but also bear power loads.

Table 1. Characteristics of UKKM.

Density	g/cm^3	1.7-2.1

IOP Conf. Series: Materials Science and Engineering **934** (2020) 012064 doi:10.1088/1757-899X/934/1/012064

Porosity	%	2-15
Young's modulus	GPa	90-100
Tensile strength	MPa	200-350
Compressive strength	MPa	450-700
Bending strength	MPa	160-700
Thermal conductivity	W/(m·K)	11.5-22.6
Heat capacity	J/(kg·K)	620-1600

3. The methodology of the calculations

To search for the most effective power design of the HPC, you can use topological optimization. Topology optimization refers to changes in design, including the creation of new body boundaries and the removal of existing ones. The goal of topological optimization is to increase or decrease a given design property when certain conditions are met.

At the first stage, the strength problem is formulated and solved by the finite element method (FEM) in ANSYS Mechanical. For this, the contour geometric region of the structure, its mechanical properties, fixing conditions, and acting loads were set. A basic geometric model was created, consisting of two layers (Figure 2). The first was the outer layer, it should remain monolithic, since it takes on the main heat flux. Over the second, topological optimization was carried out. At a given coating depth, there is no need to use dense UKCM from the point of view of thermal calculations [9,10], since a more porous material has lower thermal conductivity and interferes with the passage of heat deeper into the material. However, the porous UKKM no longer has sufficient mechanical characteristics; therefore, stiffeners must be created from a dense UKKM, into the niches between which a less dense CCCM can be placed.

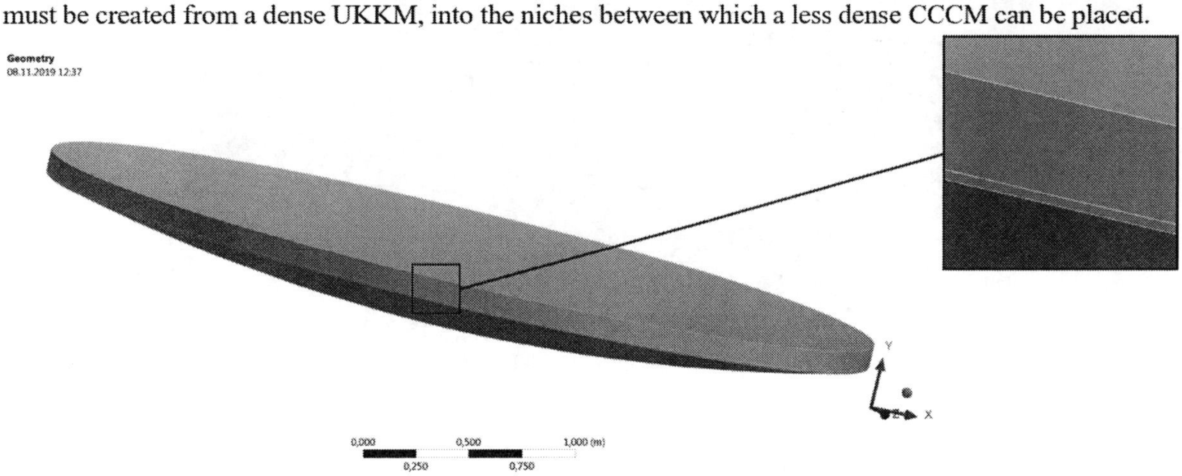

Figure 2. Basic geometric model of HPC

Sealing around the shield perimeter was used as the boundary conditions, there is no support in the center of the shield, since the bearing capacity of exclusively HPC without the auxiliary power structure of the apparatus was estimated. External pressure adopted 10 MPa.

4. Results

IOP Conf. Series: Materials Science and Engineering **934** (2020) 012064 doi:10.1088/1757-899X/934/1/012064

As a result of topological optimization, the following HPC geometry was obtained (Figure 3). In the zone of highest stresses, stiffeners are present, and in the center of the shield, where the voltage is lower, the thickness of the material is reduced. Considering the results obtained, several geometric models of the power structure were built. Based on them, finite element models were created and strength analysis was carried out.

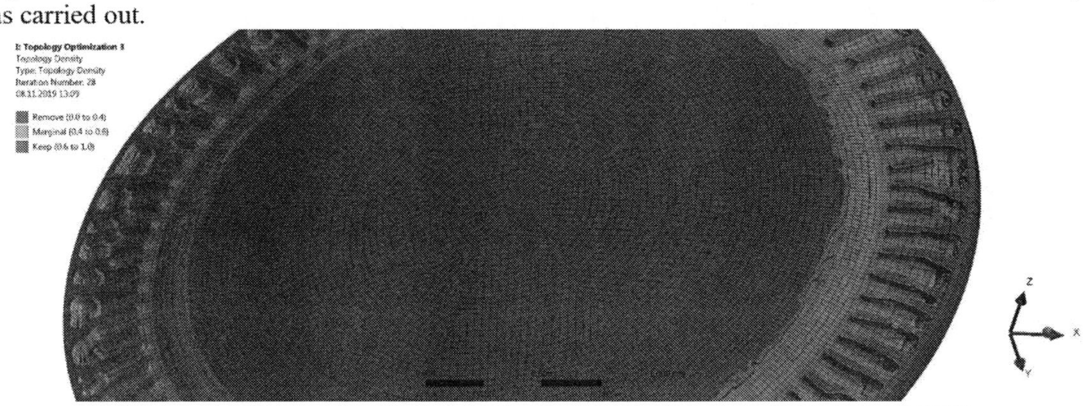

Figure 3. The results of the topological optimization of the model TZP.

Based on the results of topological optimization, radial ribs were used in model № 1. The height of the ribs increases to the edge of the shield, in the zone of maximum stresses. The strength of this force structure was insufficient, the maximum stresses significantly exceeded the characteristics of the material (Figure 4). In model № 2, material was added in the zone of maximum compressive loads, as well as lengthened stiffeners. The added changes significantly reduced the load on the shield along the edges, in the area not supported by stiffeners, the stresses remained high (Figure 5). In model № 3, the stiffeners were extended to the center of the shield, which significantly reduced stresses (Figure 6). However, the stress level remained above the capabilities of the material.

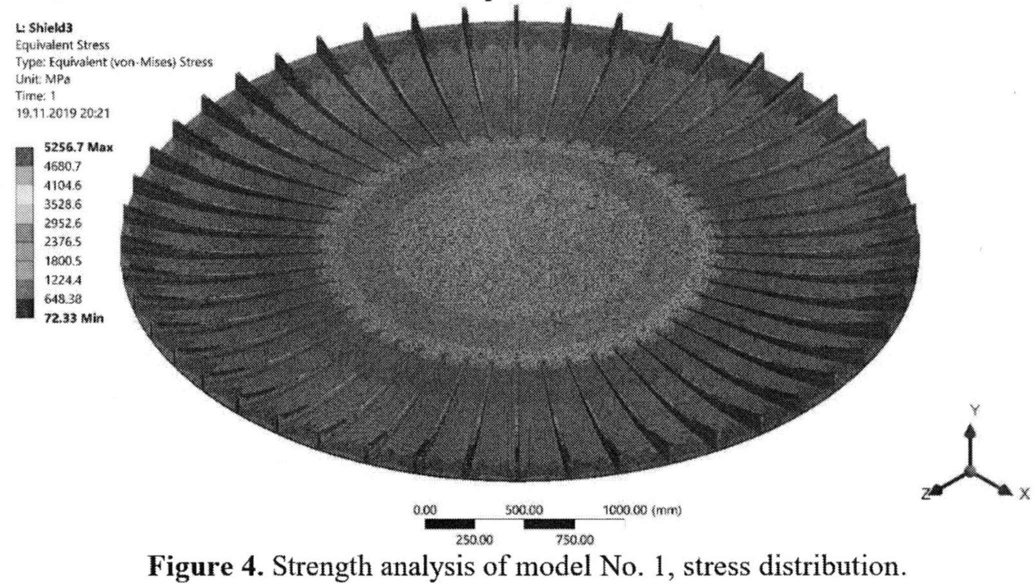

Figure 4. Strength analysis of model No. 1, stress distribution.

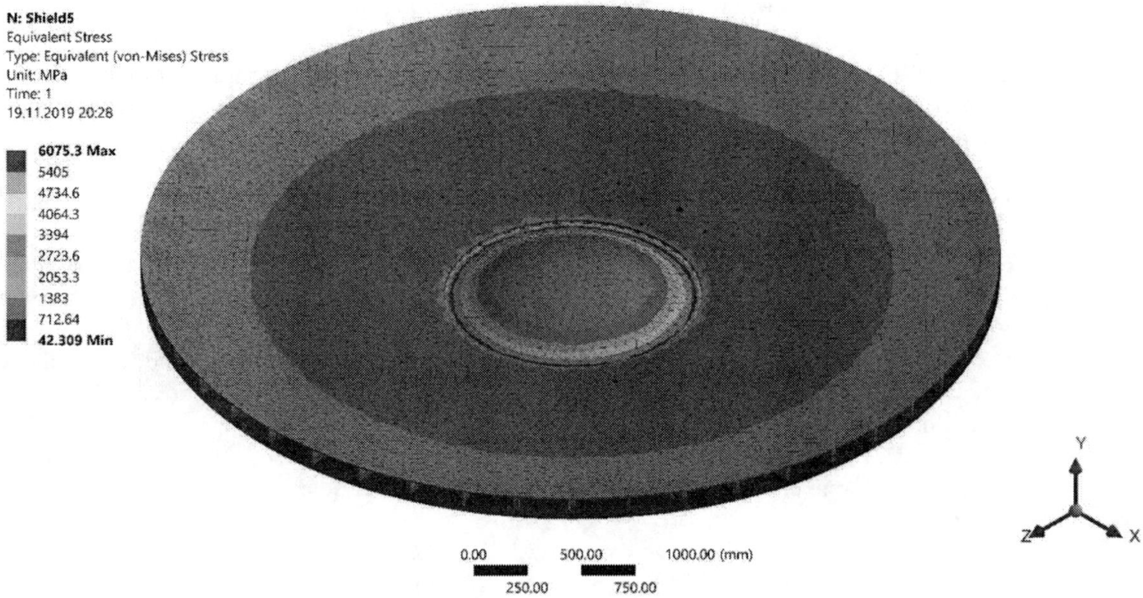

Figure 5. Strength analysis of model No. 2, stress distribution.

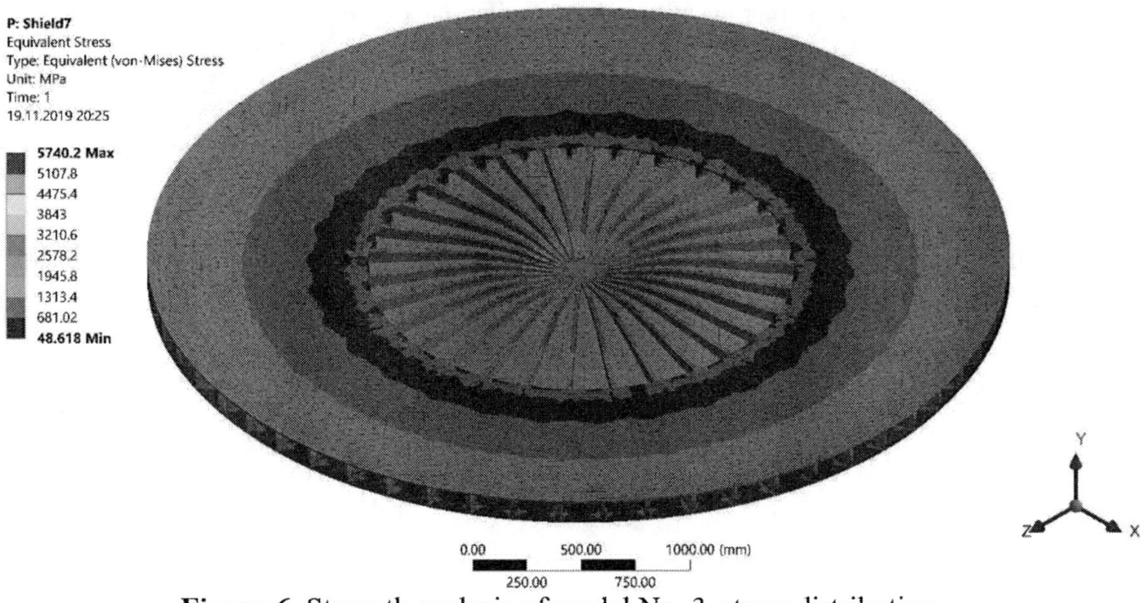

Figure 6. Strength analysis of model No. 3, stress distribution.

The results show that the HPC based on CCCM is capable of taking on an impressive part of the mechanical loads that the heat shield of the DV is subjected to. However, it will be necessary to use an additional power frame to impart structural strength, which will be less heat-loaded, which means that it can be made of more durable structural materials.

References

[1] Reznik S V, Prosuntsov P V and Mikhailovskii K V 2015 Prediction of thermophysical and thermomechanical characteristics of porous carbon–ceramic composite materials of the heat

shield of aerospace craft // J. Eng. Phys. Thermophy **88(3)** pp. 594-601. DOI:10.1007/s10891-015-1227-1.

[2] Reznik S V, Mikhailovskii K V and Prosuntsov P V 2017 Heat and mass transfer in the chemical vapor deposition of silicon carbide in a porous carbon–carbon composite material for a heat shield // J. Eng. Phys. Thermophy. **90(2)** P. 291-300. DOI 10.1007/s10891-017-1567-0

[3] Reznik S V, Prosuntsov P and Mikhaylovskiy K 2018 Development verification of coatings made from porous ceramic-matrix composite materials // MATEC Web of Conferences 224, 03019 5 p. DOI.ORG/10.1051/matecconf/201822403019 ICMTMTE 2018

[4] Leuchs M 2003 CVI-Verfahren zur Herstellung faserverstrkter Keramik – Herstellung, Eigenschaften, Anwendungen, Keramische Verbundwerkstoffe / M. Leuchs, A. Muhlratzer, W. Krenkel. – WILEY-VCH, Weinheim, Germany P. 95–121.

[5] Haug T 1992 Fiber Reinforced Ceramics for Aerospace Applications, Advanced Materials and Structures from Research to Application / T. Haug, R. Ostertag, W. Schfer, J. Brandt, H. Hornfeld, M. Neitzel // SAMPE European Chapter P. 163–173.

[6] Krenkel W 2001 Cost Effective Processing of CMC Composites by Melt Infiltration (LSI Process) / W. Krenkel. – Issue 3. – Ceramic Engineering and Science Proceedings. **22** P. 443–454.

[7] Sposob poluchenia kompozicionnogo materiala// Patent Rossii № 2687343. 2019 Byul. no. **14.** / Bogachev E A, Skuratovsky Yu E, Timofeev A N [i dr].

[8] Narottam P B 2005 Handbook of Ceramic Composites / P. B. Narottam // Kluwer academic publishers, Boston,. – 554 p.

[9] Prosuntsov P V and Taraskin N Y 2016 Theoretical and Numerical Characterization of the Thermal Physical Properties of Carbon Ceramic Material // Matec Web of Conferences. **72** 010929.

[10] Reznik S V, Mikhailovskiy K V and Prosuntsov P V 2016 Modeling the heat and mass transfer in the pores of the thermal protection carbon-carbon frame during the gas-phase deposition of silicon carbide // MATEC Web of Conferences 01075 (2017) Thermophysical Basis of Energy Technologies 6 p. DOI: 10.1051/matecconf/2017920107592.

IOP Conf. Series: Materials Science and Engineering **934** (2020) 012065 doi:10.1088/1757-899X/934/1/012065

Biharmonic Problems and their Application in Engineering and Medicine

H. A. Matevossian[1,2,3], **G. Nordo**[4], **T. Sako**[5]

[1] Federal Research Center "Computer Science and Control", Russian Academy of Sciences, Vavilov str., 40, Moscow 119333, Russia;
[2] Steklov Mathematical Institute, Russian Academy of Sciences, Gubkin str., 8, 119991, Moscow, Russia;
[3] Moscow Aviation Institute (National Research University), Volokolomskoe Shosse, 4, Moscow 125993, Russia;
[4] MIFT, Universita di Messina, Viale F. Stagno D'Alcontres, 31 – Contrada Papardo, salita Sperone, 31 - 98166 Sant'Agata, Messina, Italy;
[5] College of Science and Technology, Nihon University, 7-24-1, Narashinodai, Funabashi, Chiba 274-8501, Japan

E-mail: hmatevossian@graduate.org, giorgio.nordo@unime.it, sako.tokuei@nihon-u.ac.jp

Abstract. In the present paper we study some properties of solutions of the Steklov and Neumann boundary value problems for the biharmonic equation. For solving these biharmonic problems with application in engineering and medicine, we need to solve boundary value problems Dirichlet and Cauchy for the Poisson equation using the scattering model. In order to select suitable solutions, we solve the Poisson equation with the corresponding boundary conditions Dirichlet and Cauchy, that is, some criterion function is minimized in the Sobolev norms. Under appropriate smoothness assumptions, these problems may be reformulated as boundary value problems for the biharmonic equation.
KeyWords: Biharmonic operator, boundary value problems, scattering model, variational methods.
MSC: 35J15; 35J35; 35J40; 58J50; 58J90.

1. Introduction

Let $\Omega \subset \mathbb{R}^n$, $n \geq 2$, be a bounded Lipschitz domain with connected boundary $\partial\Omega$, and $\Omega \cup \partial\Omega = \overline{\Omega}$ is the closure of Ω. We consider the following boundary value problems for the biharmonic equation in Lipschitz domains:

$$\Delta^2 u = f, \qquad x \in \Omega \tag{1}$$

with the Steklov boundary conditions

$$\begin{cases} u = g_1 & \text{on} \quad \partial\Omega, \\ \Delta u + \tau \frac{\partial u}{\partial \nu} = g_2 & \text{on} \quad \partial\Omega, \end{cases} \tag{2}$$

or the Neumann boundary conditions

$$\begin{cases} Mu \equiv \sigma\Delta u + (1-\sigma)\frac{\partial^2 u}{\partial \nu^2} = h_1 & \text{on} \quad \partial\Omega, \\ Nu \equiv \frac{\partial \Delta u}{\partial \nu} + (1-\sigma)\frac{1}{2}\cdot\frac{\partial}{\partial t_{ij}}\left(\frac{\partial^2 u}{\partial\nu\partial t_{ij}}\right) = h_2 & \text{on} \quad \partial\Omega, \end{cases} \tag{3}$$

where ν is the outer unit normal vector to the domain with the Lipschitz boundary $\partial\Omega$, $\tau \geq 0$, $\tau \not\equiv 0$, and $\tau > 0$ on a set of positive $(n-1)$ – dimensional measure on $\partial\Omega$, $\overline{\Omega} = \Omega \cup \partial\Omega$ is the closure of Ω. The coefficient σ is a constant known as the Poisson ratio, $\frac{1}{n-1} < \sigma < 1$.

For $n = 2$, these problems and also the Neumann problem are related to the study of the transverse vibrations of a thin plate with a free edge and which occupies at rest a planar region of shape $\partial\Omega$. The coefficient σ represents the Poisson's ratio of the material that the plate is made of. For more details on the physical interpretation of the Neumann problem and on the Poisson's ratio σ, we refer, for example, to [3]. Note the paper [4], where the author studies the dependence of the vibrational modes of a plate subject to homogeneous boundary conditions upon the Poisson's ratio $0 < \sigma < \frac{1}{2}$, providing also a perturbation formula for the frequencies as functions of the Poisson's coefficient.

The standard elliptic regularity results are available in [7]. This monograph covers higher order linear and nonlinear elliptic boundary value problems, mainly with the biharmonic (polyharmonic) operator as leading principal part. Underlying models and, in particular, the role of different boundary conditions are explained in detail. As for linear problems, after a brief summary of the existence theory and L^p and Schauder estimates, the focus is on positivity. The required kernel estimates are also presented in detail.

In [6] and [7], the spectral and positivity preserving properties for the inverse of the biharmonic operator under Steklov and Steklov–type boundary conditions are studied. These are connected with the first Steklov eigenvalue. It is shown that the positivity preserving property is quite sensitive to the parameter involved in the boundary condition.

Elliptic problems with parameters in the boundary conditions are called Steklov problems from their first appearance in [23]. In the case of the biharmonic operator, these conditions were first considered in [2], [10], [21], who studied the isoperimetric properties of the first eigenvalue.

In [5], the boundary value problems for the biharmonic equation and the Stokes system are studied in a half space, and, using the Schwartz reflection principle in weighted L^q -space the uniqueness of solutions of the Stokes system or the biharmonic equation is proved.

Boundary value problems for a biharmonic (polyharmonic) equation and for the elasticity system in unbounded domains are studied in [11]– [20], in which the condition of the boundedness of the following weighted Dirichlet integral of solution is finite, namely

$$\int_\Omega |x|^a |\partial^\alpha u|^2\, dx < \infty, \quad a \in \mathbb{R},$$

where $a \in \mathbb{R}$ is a fixed number and $|\partial^\alpha u|^2$ denotes the Frobenius norm of the Hessian matrix of u. In particular, in [11]– [20] has been studied the dimension of the space of the solutions to the boundary value problems for a biharmonic (polyharmonic) equation and for the elasticity system, providing explicit formulas which depends on n and a.

In [8], [9], generalizations of the Hardy inequality were established for bounded and for a wide class of unbounded domains, and applied these to investigate boundary value problems for elliptic equations and systems. In particular, the problems of the existence, the uniqueness, the stability and the asymptotic expansions of solutions of boundary value problems were studied.

Notation: $C_0^\infty(\Omega)$ is the space of infinitely differentiable functions in Ω with compact support in Ω; $H^m(\Omega)$ is the Sobolev space obtained by the completion of $C^\infty(\overline{\Omega})$ with respect to the norm

$$\|u; H^m(\Omega)\| = \left(\int_\Omega \sum_{|\alpha| \le m} |\partial^\alpha u|^2 dx \right)^{1/2}, \; m = 1, 2,$$

where $\partial^\alpha \equiv \partial^{|\alpha|}/\partial x_1^{\alpha_1} \dots \partial x_n^{\alpha_n}$, $\alpha = (\alpha_1, \dots, \alpha_n)$ is a multi-index, $\alpha_i \ge 0$ are integers, and $|\alpha| = \alpha_1 + \dots + \alpha_n$; $\overset{\circ}{H}{}^m(\Omega)$ is the space obtained by the completion of $C_0^\infty(\Omega)$ with respect to the norm $\|u; H^m(\Omega)\|$. $\overset{\circ}{H}{}^m_{loc}(\Omega)$ is the space obtained by the completion of $C_0^\infty(\Omega)$ with respect to the family of semi-norms

$$\|u; H^m(\Omega \cap B_0(R))\| = \left(\int_{\Omega \cap B_0(R)} \sum_{|\alpha| \le m} |\partial^\alpha u|^2 \, dx \right)^{1/2}$$

for all open balls $B_0(R) := \{x : |x| < R\}$ in $\mathbb{R}^n$ for which $\Omega \cap B_0(R) \ne \emptyset$. Finally $H^{1/2}(\partial\Omega)$ is the usual trace space on the boundary and $H^{-1/2}(\partial\Omega)$ is its dual (see, for ex., [1]).

If we set $\sigma = 1$, for the biharmonic equation the problem with the Neumann boundary conditions reads

$$\begin{cases} \Delta u = h_1 & \text{on} \quad \partial\Omega, \\ \frac{\partial \Delta u}{\partial \nu} = h_2 & \text{on} \quad \partial\Omega. \end{cases} \tag{4}$$

Definition 1. *A solution of the biharmonic equation* (1) *in* Ω *is a function* $u \in H^2(\Omega)$ *such that, for every function* $\varphi \in C_0^\infty(\Omega)$*, the following integral identity holds:*

$$\int_\Omega \Delta u \, \Delta \varphi \, dx = \int_\Omega f \, \varphi \, dx, \quad f \in L^2(\Omega). \tag{5}$$

Definition 2. *A function* u *is a solution of the Steklov problem* (1),(2) *with* $g_1 = g_2 = 0$, *if* $u \in H^2(\Omega) \cap \overset{\circ}{H}{}^1(\Omega)$ *such that for every function* $\varphi \in H^2(\Omega) \cap \overset{\circ}{H}{}^1(\Omega)$*, the following integral identity holds*

$$\int_\Omega \Delta u \, \Delta \varphi \, dx + \int_{\partial\Omega} \tau \frac{\partial u}{\partial \nu} \frac{\partial \varphi}{\partial \nu} \, ds = 0. \tag{6}$$

Definition 3. *A function* u *is a solution of the Neumann problem* (1),(4) *with* $h_1 = h_2 = 0$, *if* $u \in H^2(\Omega)$ *such that the integral identity* (5) *holds for every function* $\varphi \in C_0^\infty(\Omega)$.

2. Scattering model

In the section we derive the mathematical model used for describing the radar process. In our parametrization the unknown is the height function H. As will be shown the height function is determined in two steps. In the first step $\mathfrak{L}(H)$, with $\mathfrak{L}$ a certain second-order differential operator, is determined. After retrieving H the equation $\mathfrak{L}(H) = f$ must be solved. To a good approximation the operator $\mathfrak{L}$ can be replaced by the Laplacian. So the second step simply consists of solving the Poisson equation over some smooth bounded domain, usually a rectangular region in the plane. The problem here is that no natural boundary conditions are available.

Here we will briefly discuss the mathematical inverse problem to be resolved in order to recover the ground topography height function from radar data. First cylindrical coordinates (r, φ, z) are introduced according to Fig. 1, where it is understood that the aircraft is flying at a constant speed along the z-axis. Further r denotes the distance from a point on the ground

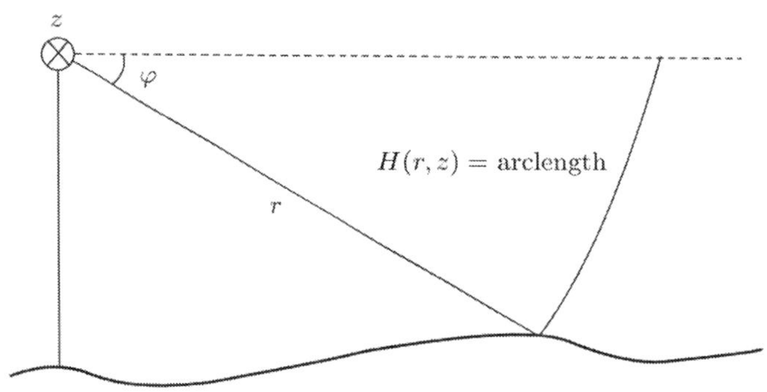

Figure 1. The ground surface measured at a fixed aircraft position.

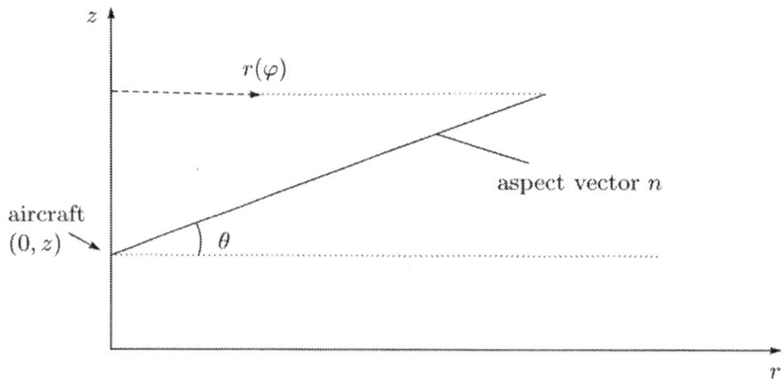

Figure 2. The measuring geometry as seen from above.

surface to the z-axis and φ is the angle between radius vector and a horizontal plane through the z-axis. Then the ground surface may be described by a function $H(r, z)$ through the equation

$$\frac{H(r,z)}{r} - \varphi = 0. \tag{7}$$

When r is large, $-H(r, z)$ is approximately a Cartesian height function. Fig. 2 shows a top view of the same scene. We have also indicated an aspect vector from the aircraft to some point on the ground, forming an angle θ with a vertical plane through the aircraft. Normalized to unit length, the aspect vector is denoted by $\hat{n}$. Accordingly

$$\hat{n} = \cos\theta\,\hat{r}(\varphi) + \sin\theta\,\hat{z}. \tag{8}$$

Here $\hat{r}(\varphi)$ denotes the cylindrical unit basis vector corresponding to the r-coordinate for the ground point as shown in the Fig. 2. For a point on the ground surface with coordinates (r, φ, z) we obtain, from Eq. (7), the following expression for the ground surface normal $\bar{m}$,

$$\bar{m} = grad\left(\frac{H(r,z)}{r} - \varphi\right) = \frac{\partial(H/r)}{\partial r}\hat{r} + \frac{1}{r}\frac{\partial H}{\partial z}\hat{z} - \frac{1}{r}\hat{\varphi}. \tag{9}$$

Let $\hat{m}$ denote the normalized normal. Then

$$\hat{m} \circ \hat{n} = \left(r\cos\theta\frac{\partial(H/r)}{\partial r} + \sin\theta\frac{\partial H}{\partial z}\right) : \sqrt{1 + \left(\frac{\partial(H/r)}{\partial r}\right)^2 + \left(\frac{\partial H}{\partial z}\right)^2}. \tag{10}$$

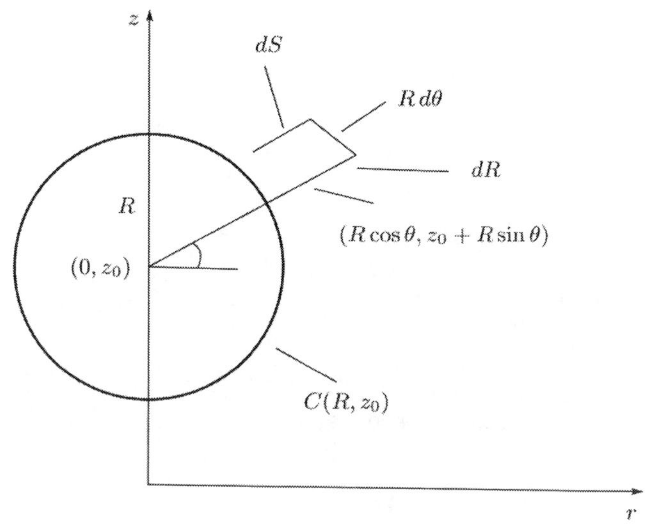

Figure 3. The coordinate system used to describe an infinitesimal surface element, dS.

Note that (r, φ, z) in Eq. (10) are related to the ground surface point and not to the position of the aircraft.

Let $(z_0, 0)$ be a position of the aircraft and R the distance to some point on the surface. According to Fig. 3 the coordinates (r, z) are then equal to $(z_0 + R\sin\theta, R\cos\theta)$. Next, to obtain a scattering model we will assume that the reflectivity from a ground surface element (see Fig. 4) is

$$\approx \frac{\hat{m} \circ \hat{n}}{R} dR\, d\theta. \tag{11}$$

From Fig. 4, where a vertical plane through $(z_0, 0)$ (the aircraft) and the ground point $(z_0 + R\sin\theta, R\cos\theta)$ is displayed, we conclude that the solid angle $d\Omega$ under which the surface element dS is seen from the antenna is approximately

$$\frac{dR\,\cos\alpha\, Rd\,d\theta}{R^2} = -\frac{\hat{m} \circ \hat{n}}{R} dR\, d\theta.$$

In expression (11) we are consequently assuming that the local reflectivity is proportional to the solid angle occupied by the infinitesimal surface element dS. The total reflected signal $G(R, z_0)$ from all points at a distance R from the antenna may now be obtained by integration over the circle $C(R, z_0) = \{(r, z) : (z - z_0)^2 + r^2 = R^2\}$ in Fig. 3.

$$G(R, z_0)dR = c \int_{-\pi}^{\pi} \frac{\hat{m} \circ \hat{n}(z_0 + R\cos\theta, R\sin\theta)}{R} d\theta dR$$

i.e.

$$RG(R, z_0) = c \int_{-\pi}^{\pi} \hat{m} \circ \hat{n}(z_0 + R\cos\theta, R\sin\theta) d\theta. \tag{12}$$

Assuming that $\hat{m} \circ \hat{n}$ is small Eq. (10) may be replaced by

$$\hat{m} \circ \hat{n} = r\cos\theta \frac{\partial (H/r)}{\partial r} + \sin\theta \frac{\partial H}{\partial z}.$$

IOP Conf. Series: Materials Science and Engineering **934** (2020) 012065 doi:10.1088/1757-899X/934/1/012065

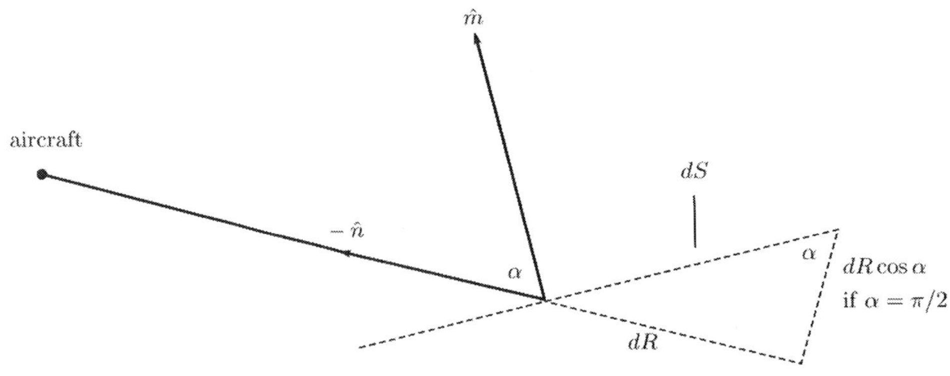

Figure 4. The infinitesimal surface element, dS, as it is seen from the aircraft.

By inserting this into Eq. (12) we get, after multiplying by R,

$$R^2G(R,z_0) = c\int_{-\pi}^{\pi}(rR\cos\theta\frac{\partial(H/r)}{\partial r} + R\sin\theta\frac{\partial H}{\partial z})d\theta.$$

Using the parametrization

$$z = z_0 + Rsin\theta, \quad r = Rcos\theta,$$

this may be rewritten as a curve integral over $C(R, z_0)$, with $dz = R\cos\theta d\theta$ and $dr = -R\sin\theta d\theta$,

$$R^2G(R,z_0) = c\int_{C(R,z_0)}(r\frac{\partial(H/r)}{\partial r}dz - \frac{\partial H}{\partial z}dr). \tag{13}$$

By applying Green's formula we get

$$R^2G(R,z_0) = c\iint_{D(R,z_0)}\mathfrak{L}(H)(r,z)dzdr, \tag{14}$$

where D is the disc,

$$D(R,z_0) = \{(r,z) : (z-z_0)^2 + r^2 \leq R^2\}$$

and

$$\mathfrak{L}(H) = \frac{\partial}{\partial r}(r\frac{\partial}{\partial r}(H/r)) + \frac{\partial^2}{\partial z^2}(H). \tag{15}$$

The problem of finding the height function H from radar data $G(r, z)$ may now be divided into two parts:

(a) First solve the integral equation (14) for $\mathfrak{L}(H)(r,z) = f(r,z)$.

(b) Next solve the partial differential equation

$$\mathfrak{L}(H) = f \tag{16}$$

for H. We note that if r is large and if $\hat{m} \circ \hat{n}$ is small it is reasonable to make the approximation

$$\mathfrak{L}(H) \approx \frac{\partial^2 H}{\partial r^2} + \frac{\partial^2 H}{\partial z^2} = \Delta H$$

IOP Conf. Series: Materials Science and Engineering **934** (2020) 012065 doi:10.1088/1757-899X/934/1/012065

so that Eq. (16) becomes Poisson's equation. To consider the first problem (a), both members in Eq. (14) are differentiated with respect to R. Then we get

$$\frac{1}{R}\frac{d}{dR}(R^2G(R,z_0)) = c\int_{-\pi}^{\pi}\mathfrak{L}(H)(z_0 + R\cos\nu, R\sin\nu)d\nu,$$

where the right-hand side is proportional to the average of $\mathfrak{L}(H)$ over the circle $C(R, z_0)$. In [2] an explicit solution is given for this problem of recovering the function $\mathfrak{L}(H)(r,z)$ when the average of $\mathfrak{L}(H)$ is known for all circles $C(R, z_0)$ with center on the z-axis and with arbitrary radius R. The solution formula is

$$\mathfrak{L}(H)^{(F,F)}(\sigma,\omega) \sim |\omega|[\frac{1}{R}\frac{d}{dR}\{R^2G(r,z)\}]^{(F,H_0)}(\sigma, \sqrt{\omega^2+\sigma^2}). \tag{17}$$

Here the notation (F, F) means that we have taken the Fourier transform with respect to both the variables and (F, H_0) means that we have taken Fourier transform with respect to the first variable and the Hankel-zero transform with respect to the second. After some calculations Eq. (17) may be rewritten

$$\mathfrak{L}(H)^{(F,F)}(\sigma,\omega) \sim |\omega|\sqrt{\omega^2+\sigma^2}[RG(r,z)]^{(F,H_1)}(\sigma, \sqrt{\omega^2+\sigma^2}). \tag{18}$$

Formula (18) may now be used in order to recover the function $\mathfrak{L}(H)$ in spatial coordinates. Of course, approximating $\mathfrak{L}(H)$ by ΔH we could rewrite Eq. (18) as

$$H^{(F,F)}(\sigma,\omega) \sim |\omega|\frac{1}{\sqrt{\omega^2+\sigma^2}}[RG(r,z)]^{(F,H_1)}(\sigma, \sqrt{\omega^2+\sigma^2}), \tag{19}$$

where H_1 denotes that we have taken the Hankel-one transform with respect to the second variable. Then we could obtain H directly by a two timensional Fourier transform. However, our solution might be expected to have errors caused by, e.g. noisy radar data and errors caused by the particular numerical implementation of the inversion formula (17) (or Eq. (18)) and therefore we would rather prefer to divide the solution procedure into the two steps described above and to use the second step, the solution of Poisson's equation, so that we perform some kind of regularization of the final solution. Note also that by using Eq.(19) as our solution formula we have tacitly assumed periodic boundary conditions for the Poisson equation.

3. Solution concepts for the Poisson equation

In this section we discuss different possibilities of defining a unique height function. Essentially our approach consists in minimizing some norm of the solution provided that it also satisfies the Poisson equation. In particular we consider the L^2- and H^1-norms. We also show how these two optimization problems may be reformulated as boundary value problems for the biharmonic equation. Note that the corresponding Poisson problem is well-posed unless $\sigma = 1$.

In the domain Ω for the Poisson equation we consider the following boundary value problems

$$\Delta u = f, \quad x \in \Omega \tag{20}$$

with the Dirichlet boundary condition

$$u = g \quad \text{on} \quad \partial\Omega, \tag{21}$$

or the Cauchy boundary conditions

$$u = \nabla u \cdot \nu = h \quad \text{on} \quad \partial\Omega, \tag{22}$$

where ν is the outer unit normal vector to $\partial\Omega$.

The boundary operators are independent of any particular choice of orientation for the rectangular coordinate systems. Finally, for Ω a rectangular region in, e.g., the plane

$$\Omega = \{(x, y) : a < x < b,\ c < y < d\}, \tag{23}$$

there may be the following boundary conditions

$$u(a, y) = u(b, y), \quad u(x, c) = u(x, d), \tag{24}$$

or the periodic boundary conditions

$$u_x(a, y) = u_x(b, y), \quad u_y(x, c) = u_y(x, d). \tag{25}$$

Provided g is smooth enough boundary conditions (21) define a unique solution of Eq. (20). For (22) and (25) the solution is determined up to a constant. It is also possible to use different mixtures of these three types of boundary conditions. Note that for cases (22) and (25) the following consistency conditions must hold, respectively:

$$\int_\Omega f\,dx = \int_{\partial\Omega} h\,ds \quad \text{and} \quad \int_\Omega f\,dx = 0.$$

We now consider a different way to select a solution to Eq. (20). Here we use a criterion function and optimize this criterion over the set of solutions to the Poisson equation. Scattering model of Section 2 shows the physical interpretation of function $u(x, y)$ is a surface function. We need to pick out the smoothest surface (in some sense) that fulfills Eq. (20), using the Sobolev space norms as criterion functions. Denote by $V_{f,i}$ the following set:

$$V_{f,i} = \{u \in H^i(\Omega) : \Delta u = f,\ f \in L^2(\Omega)\},\ i = 0, 1, 2, \tag{26}$$

where $H^0(\Omega) = L^2(\Omega)$. The equality $\Delta u = f$ is to be interpreted in the sense of distributions, i.e.,

Definition 4. *A solution of the Poisson equation* (20) *in* Ω *is a function* $u \in H^1(\Omega)$ *such that the following integral identity holds:*

$$\int_\Omega u\,\Delta\varphi\,dx = \int_\Omega f\varphi\,dx, \quad \forall\varphi \in C_0^\infty(\Omega).$$

Lemma 1. $V_{f,i}$ *is a closed, convex and nonempty set of* $H^i(\Omega)$.

Let α be a multi-index and $\beta_1 > 0$ a given parameter. We consider the following optimization problems:

$$I_0(u) \equiv \min_{u \in V_{f,0}} \int_\Omega |u|^2\,dx, \tag{27}$$

and

$$I_1(u) \equiv \min_{u \in V_{f,1}} \int_\Omega |u|^2\,dx + \beta_1 \int_\Omega \sum_{|\alpha|=1} |\partial^\alpha u|^2\,dx. \tag{28}$$

Theorem 1. *Problems* (27) *and* (28) *have unique solutions* u_0 *and* u_1*, respectively.*

From problems (27) and (28) we have the following results characterizing the solutions.

Theorem 2. *Let $u_o = \Delta v$. For the solution u_0 of the problems (27), where $v \in H^2(\Omega) \cap \overset{\circ}{H}{}^1(\Omega)$ is the unique solution of the following biharmonic problem*

$$\begin{cases} \Delta^2 v = f & on \quad \Omega, \\ v = \Delta v + \tau \frac{\partial v}{\partial \nu} = 0 & on \quad \partial\Omega. \end{cases} \tag{29}$$

Theorem 3. *Let $u_1 = \Delta v$. For the solution u_1 of the problems (28), where $v \in H^2(\Omega)$ is the unique solution in the class $\{\psi \in H^1(\Omega) : \Delta\psi \in H^1(\Omega)\}$ of the following biharmonic problem*

$$\begin{cases} \Delta^2 v = f & in \quad \Omega, \\ v = \beta_1 \Delta v, \ \nabla v \cdot \nu = 0 & on \quad \partial\Omega. \end{cases} \tag{30}$$

We conclude this section by a theorem relating the solution of problems (27) and (28). First we recall the following definition.

Definition 5. *$\Omega \subset \mathbb{R}^n$ is called star-shaped if there exists $x_0 \in \Omega$ such that for all $x \in \Omega$ the set $\{t \in \mathbb{R} : \quad x_0 + t(x - x_0) \in \Omega\}$ is an interval.*

Theorem 4. *Assume that $\Omega \subset \mathbb{R}^n$ is open, bounded and star-shaped. If $u_{1,\beta_1} \in H^1(\Omega)$ denotes the solution of problem (28) with the parameter $\beta_1 > 0$, and if $u_0 \in L^2(\Omega)$ denotes the solution of problem (27), then*

$$u_{1,\beta_1} \to u_0 \quad in \quad L^2(\Omega) \quad as \quad \beta_1 \to 0+.$$

Acknowledgement

The results of this article were presented at the II International Conference on Composites "Advances in Composite Science and Technologies" (Bauman Moscow State Technical University, Moscow, November 20–21, 2019).

References

[1] Adams R A 1975, *Sobolev spaces*, Academic Press, New York, 1975.
[2] Brock F 2001, "An isoperimetric inequality for eigenvalues of the Stekloff problem", *Z. Angew. Math. Mech. (ZAMM)*, **81**(1), 69–71.
[3] Courant R and Hilbert H 1953, *Methods of Mathematical Physics*, Vol. I. Interscience Publishers, Inc.: New York.
[4] Duffin R J 1969, "The influence of Poisson's ratio on the vibrational spectrum", *SIAM J. Appl. Math.*, **17**, 179–191.
[5] Farwig R 1994, "A note on the reflection principle for the biharmonic equation and the Stokes system", *Acta Appl. Math.*, **34** 41–51.
[6] Gazzola F and Sweers G 2008, "On positivity for the biharmonic operator under Steklov boundary conditions", *Arch. Rational Mech. Anal.*, **188**(3), 399–427.
[7] Gazzola F, Grunau H.-Ch and Sweers G 2010, *Polyharmonic Boundary Value Problems: Positivity Preserving and Nonlinear Higher Order Elliptic Equations in Bounded Domains.* Lecture Notes Math. **1991**, Springer-Verlag.
[8] Kondrat'ev V A and Oleinik O A 1988, "Boundary value problems for the system of elasticity theory in unbounded domains. Korn's inequalities", *Russ. Math. Surveys*, **43**(5), 65–119.
[9] Kondratiev V A and Oleinik O A 1990, "Hardy's and Korn's Inequality and their Application", *Rend. Mat. Appl.*, Serie VII. **10**(3), 641–666.
[10] Kuttler J R and Sigillito V G 1968, "Inequalities for membrane and Stekloff eigenvalues", *J. Math. Anal. Appl.*, **23**(1), 148–160.
[11] Matevosyan O A 1998, "On solutions of boundary value problems for a system in the theory of elasticity and for the biharmonic equation in a half–space", *Diff. Equations.*, **34**(6), 803–808.
[12] Matevosyan O A 2001, "The exterior Dirichlet problem for the biharmonic equation: Solutions with bounded Dirichlet integral", *Math. Notes*, **70**(3), 363–377.
[13] Matevosyan O A 2014, "Solutions of the Robin problem for the system of elastic theory in external domains", *J. Math. Sci. (N. Y.)*, **197**(3), 367–394.

[14] Matevossian O A 2015, "On solutions of the Neumann problem for the biharmonic equation in unbounded domains", *Math. Notes*, **98**, 990–994.

[15] Matevosyan O A 2016, "On solutions of the mixed Dirichlet–Navier problem for the polyharmonic equation in exterior domains", *Russ. J. Math. Phys.*, **23**(1), 135–138.

[16] Matevosyan O A 2016, "On solutions of one boundary value problem for the biharmonic equation", *Diff. Equations*, **52**(10), 1379–1383.

[17] Matevossian H A 2017, "On the biharmonic Steklov problem in weighted spaces", *Russ. J. Math. Phys.*, **24**(1), 134–138.

[18] Matevossian H A 2017, "On solutions of the mixed Dirichlet–Steklov problem for the biharmonic equation in exterior domains", *P-Adic Numbers, Ultrametric Analysis, and Appl.*, **9**(2), 151–157.

[19] Matevossian H A 2018, "On the Steklov–type biharmonic problem in unbounded domains", *Russ. J. Math. Phys.*, **25**:2, 271–276.

[20] Matevossian H A 2019, "On the polyharmonic Neumann problem in weighted spaces", *Complex Variables & Elliptic Equations*, **64**:1, 1–7.

[21] Payne L E 1970, "Some isoperimetric inequalities for harmonic functions", *SIAM J. Math. Anal.*, **1**(3), 354–359.

[22] Sobolev S L 1988, *Some Applications of Functional Analysis in Mathematical Physics*, 3th ed., Nauka, Moscow; *Applications of Functional Analysis in Mathematical Physics.* Amer. Math. Soc., Providence 1991.

[23] Stekloff W 1902, "Sur les problemes fondamentaux de la physique mathematique", *Ann. Sci. de l'E.N.S.*, 3^e serie, **19**, 191-259 et 455-490.

IOP Conf. Series: Materials Science and Engineering **934** (2020) 012066 doi:10.1088/1757-899X/934/1/012066

Research of physical and mechanical properties of reinforced carbon fiber plastics based on tfp-preforms

M A Orlov[1], I A Polikarpova[1], V I Solodilov[1,2]

[1]Bauman MSTU, Moscow, Russia
[2]Semenov Institute of Chemical Physics, RAS, Moscow, Russia

E-mail: ipolikarpova@emtc.ru

Abstract. The article provides for studies of physical and mechanical properties of samples of carbon fiber reinforced plastics on the basis of tfp-preforms produced using various stitching technological parameters. The effect of the carbon fiber laying density and the piercing step of the aramid yarn on the strength properties of composites is shown. Optimal technological parameters of the creation of tfp-preforms were chosen.

1. Introduction

Polymer composite materials (PCM) is one of the most perspective class of materials for modern challenges in design of different structures from planes to sport equipment etc. There are different approaches to improve useful properties of polymer composite materials and products based on them. Some of them including additional metallization [1-4], treatment the fibers [5, 6], modification of polymer matrix [7-9] and develop technologies for the composite preform formation. One of these technologies is the TFP (Tailored fiber placement) method, which is relatively new, but quite promising for the manufacture of PCM with increased elastic and strength properties.

The TFP technology is an automated process for the production of volume reinforced preforms of polymer composite material (PCM) products, which is the directional laying of the roving on a bearing base, whereby the reinforcing fiber is stitched to the base using an additional thread in the form of a zigzag stitch [10, 11]. This technology is widely used abroad – some companies have mastered this process and are actively introducing it into production [12-15]. TFP is the most promising additive technology among other methods for creating volume preforms due to its high accuracy, reproducibility, process automation, the possibility to use complex reinforcement schemes and to create local areas, as well as the production of complex profile preforms with the geometry of a finished product [16, 17]. During creating preforms by automated laying of the roving the physical and mechanical properties of finished products are affected not only by the reinforcement scheme and the properties of the materials used. To achieve optimal strength properties, the technological stitching parameters (the roving laying density ρ and the piercing step Δ) should be also taken into account. The laying density ρ_{mm} is determined by the distance between the axial lines of adjacent carbon fibers and is characterized by the distance S between them. It should be noted that in the control program for stitching preforms, this value has no dimension and is denoted by conventional units (CU). $\rho_{mm} = 0{,}05\rho_{CU}$. The piercing step Δ is determined by the stitch length of the piercing aramid yarn. When selecting the technological parameters it is necessary understand that when the preform is stitched, the reinforcing rope is inevitably damaged. At the same time, it is necessary to select the optimal process parameters so that the maximum transverse characteristics are achieved without significantly reducing the strength along the direction of reinforcement.

The aim of the study is to determine how the technological parameters of carbon fiber stitching affect the transversal properties of model flat samples of carbon fiber reinforced plastics (CFRP).

2. Materials and methods

Test samples of stitched carbon fiber reinforced plastics were made on the basis of the HTS 45 carbon fiber (TohoTenax®, Japan). The piercing thread that fixes the carbon roving to a water-soluble substrate (Aurora, China) was the Rusar-S high strength aramid yarn (Thermotex LLC, Mytishchi, Russia). Control samples of carbon fiber reinforced plastics were obtained by winding of the HTS 45 carbon fiber (TohoTenax®, Japan) on a plate in the direction of 0-90° with further pressing.

The Araldite LY 556 epoxy resin («Huntsman», Switzerland) was used as a binder. Isomethyltetrahydrophthalic anhydride (IMTHFA) (90 phr) (Triune Chemicals Materials (TCM), China) served as a hardener, and 2-methylimidazole (0.2 phr) as an accelerator (Modifikator LLC, Tomsk, Russia).

Preform stitching was carried out on the JCW 0100-500 automatic numerically controlled sewing machine (ZSK Stickmaschinen GmbH, Germany). The technological stitching parameters are presented in Table 1. The produced preforms consisted of 8 layers with the reinforcement scheme [0°; 90°].

All preforms were impregnated by vacuum infusion at a room temperature of 23 °C and a pressure of -1 atm for 30 minutes. Curing mode: 4 h at 110 °C.

The obtained plates were cut into samples of the required size to perform bending and torsion tests that are the most sensitive test methods to changes in the properties of piercing and stitched materials.

Table 1. Technological parameters of sample stitching modes.

Laying density ρ, mm/CU	Stitch step Δ, mm	Preform thickness, mm	Laying density ρ, mm/CU	Stitch step Δ, mm	Preform thickness, mm
2.35/47	4	4.4	3/60	4	4.2
2.35/47	5	4.0	3/60	5	3.5
2.35/47	7	4.4	3/60	7	3.6
2.35/47	9	4.0	3/60	9	3.3
2.35/47	10	3.9	3/60	10	3.2
2.85/57	4	3.9	3.5/70	4	3.2
2.85/57	5	3.6	3.5/70	5	3.1
2.85/57	7	3.6	3.5/70	7	3.2
2.85/57	9	3.2	3.5/70	9	3.1
2.85/57	10	3.3	3.5/70	10	3.0

The three-point bending tests under quasistatic loading, four-point bending tests and plate torsion tests of the samples were carried out using the universal Zwick Roell Z100 testing machine (Zwick GmbH & Co. KG, Germany); and the three-point bending tests of the samples under conditions of low-velocity impact were carried out on the KPS-2 impact spring machine [18].

The three-point and four-point bending tests were carried out in accordance with ISO 14125:1998 «Fibre-reinforced plastic composites. Determination of flexural properties». The torsion test was carried out in accordance with ISO 14130:1997 «Fibre-reinforced plastic composites. Determination of apparent interlaminar shear strength by short-beam method».

The statistical processing of the measurement results was performed for each type of research by calculating the mean value and the standard deviation.

3. Results and discussion

Figure 1 presents photos with typical types of destruction of the tested samples during the three-point bending. It is apparent that the destruction of the stitched samples (Figure 1a-1d) is localized in the central area. This type of cracking is typical of carbon fiber reinforced plastics that break when bent from normal stresses. However, the central area of the samples shows also longitudinal cracks formed upon destruction from shear stresses. In this case, the movement of cracks from shear stresses ends near the aramid fibers, which is explained by their greater resistance to crack growth. This type of destruction is typical of all steps and densities of carbon yarn stitching. The direction of reinforcement (0 or 90°) in which the test samples were cut also does not affect the nature of destruction.

As far as the control samples obtained by pressing are concerned, typical destructions, i.e. delaminations, are observed in the central area (Figure 1e). This type of destruction is caused by the predominance of shear stresses over normal ones due to the ratio of the sample thickness to the distance between the supports h/l = 8÷11 selected depending on the stitching density. Usually the fracture character was independent of the type of reinforcing fibers and of loading rate [19].

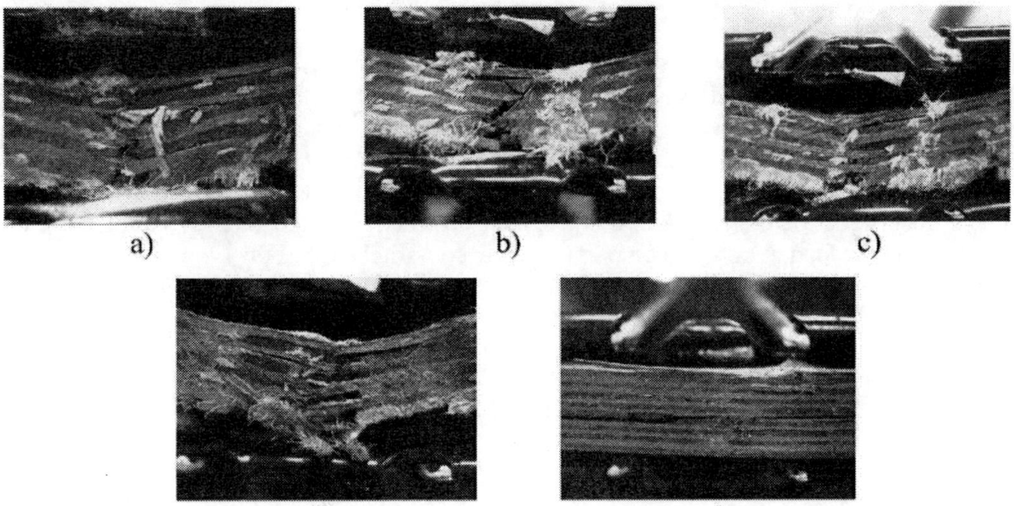

a) b) c)

d) e)

Figure 1. Photos with typical types of destruction of the tested samples during the three-point bending. The piercing step $\Delta = 7$ mm, the roving laying density ρ:
a) 47 CU, b) 57 CU, c) 60 CU, d) 70 CU, e) control samples

Figures 2 and 3 present typical loading diagrams of stitched carbon fiber reinforced plastics and the control samples in which a linear section corresponding to the elastic behavior of the sample at the initial stage of loading is observed. After the ultimate strength has been achieved, a series of peaks characterizing the further destruction of the sample can be seen on the diagrams.

A rather high level of peaks on the diagrams corresponding to secondary destruction is typical of carbon fiber reinforced plastics tested for three-point bending at a quasistatic loading rate (Figure 2). Cracks in non-stitched carbon fiber reinforced plastics (Figure 2b) propagate jerkily as evidenced by high secondary load peaks. For stitched materials (Figure 2a), load fluctuations are significantly less during the deformation growth. This may indicate that a crack in such materials propagates smoothly, without significant jumps. This mechanism of destruction is primarily attributed to the presence of aramid fibers in the structure of carbon fiber reinforced plastics. In this case, the aramid yarn slows down the movement of cracks and their further development. It should be noted that the type of loading diagrams practically remained the same when changing the step or the density of carbon thread stitching.

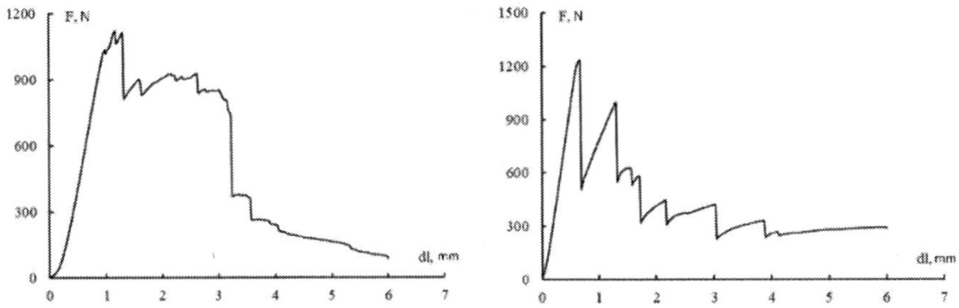

a) b)

Figure 2. Typical loading diagrams of stitched carbon fiber reinforced plastics (a) and the control samples (b), tested for three-point bending at a quasistatic loading rate.

The similar trend is observed when testing materials under conditions of low-velocity impact (Fig. 3). As under quasistatic loading, the area of secondary destruction of the stitched samples is characterized by less amplitude load jumps in comparison with non-stitched carbon fiber reinforced plastics. The nature of the change in the load during loading is practically independent of the stitching step and density as well as the direction of reinforcement.

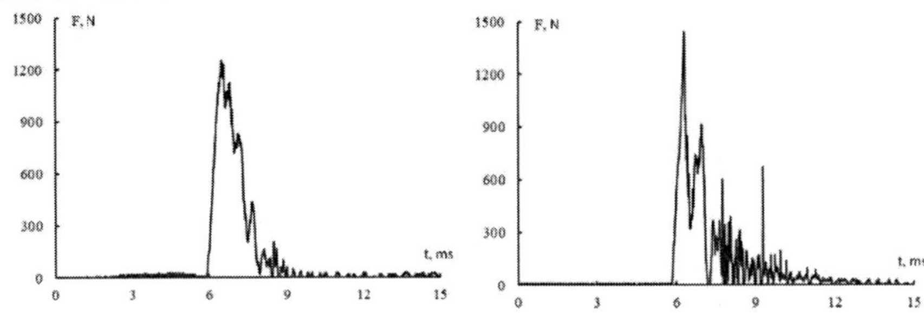

a) b)

Figure 3. Typical loading diagrams of stitched carbon fiber reinforced plastics (a) and the control samples (b), tested for three-point bending at a dynamic loading rate

Figures 4 and 5 present the calculation results for the total fracture energy of the stitched and control carbon fiber reinforced plastics under quasistatic and dynamic loading according to the three-point loading diagram depending on the stitching step and density. The energy level for the control sample is shown in the figure by a horizontal line, and the spread in the values of this parameter by a shaded area.

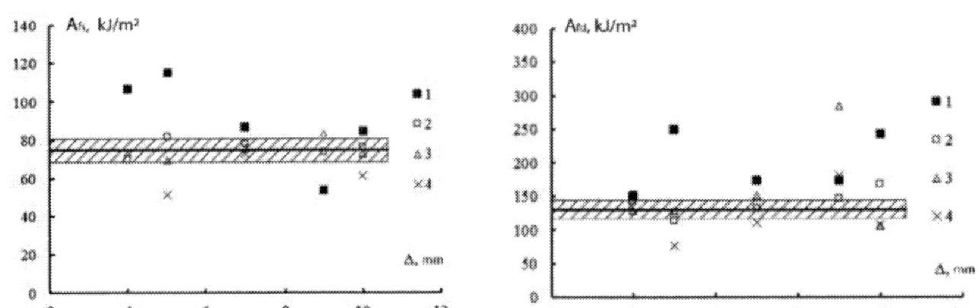

a) b)

Figure 4. Dependence of the fracture energy of the examined samples, tested for three-point bending at a quasistatic (a) and dynamic (b) loading rate on piercing step Δ_{mm}.
Roving laying densityρ: 1 – 47 CU, 2 – 57 CU, 3 – 60 CU, 4 – 70 CU

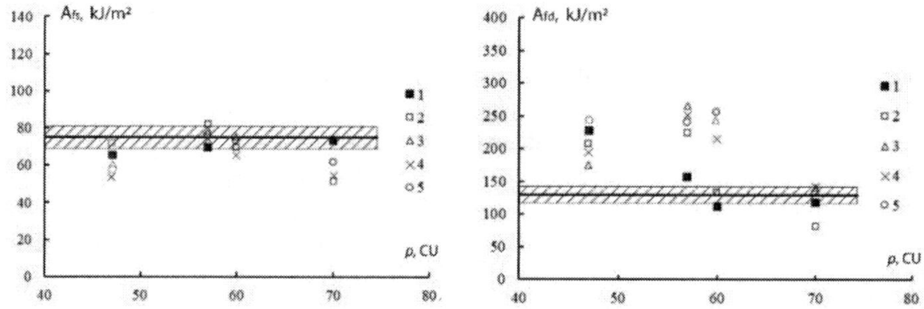

a) b)

Figure 5. Dependence of the fracture energy of the examined samples, tested for three-point bending at a quasistatic (a) and dynamic (b) loading rate on roving laying density ρ.
The piercing stepΔ: 1 – 4mm, 2 – 5mm, 3 – 7mm, 4 – 9mm, 5 – 10mm

No unambiguous dependence of the fracture energy on the examined parameter can be observed in both cases. However, it is apparent that the fracture energy is consistent with the energy for the control sample and even exceeds the control values by 50 % under conditions of low-velocity impact for materials with a stitching density of 57 and 60 CU and a step of 7 to 10 mm.

No clear dependences on the stitching step and density could be established for both the strength during the three-point bending (Figure 6) and the fracture energy. As in the case of calculating the fracture energy values, the most optimal technological parameters were a step of 7 to 10 mm and a density of 50 to 60 CU.

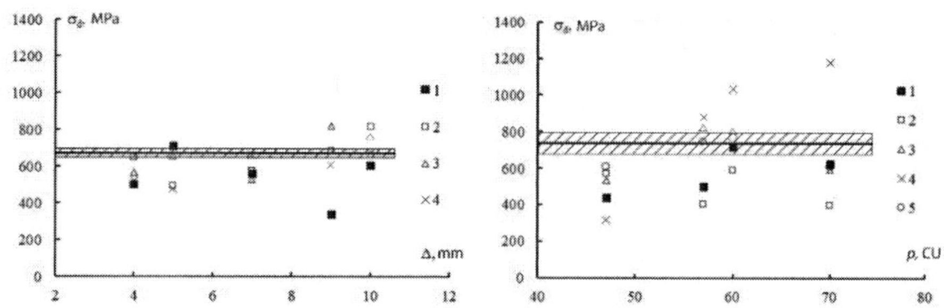

a) b)

Figure 6. Dependences of the strength during the three-point bending on the stitching step Δ (a) and the roving laying density ρ (b).
a) ρ: 1 – 47 CU, 2 – 57 CU, 3 – 60 CU, 4 – 70 CU
b) Δ: 1 – 4mm, 2 – 5mm, 3 – 7mm, 4 – 9mm, 5 – 10mm

A mixed mechanism of destruction of the stitched samples is observed during the four-point bending. The samples were fractured from both normal and shear stresses as evidenced by cracks that propagated along and across the carbon fiber reinforced plastics. At the same time, the samples are fractured from normal stresses to a greater extent, while the development of cracks from shear stresses is inhibited by aramid fiber piercing. The stitching density and step also have a little effect on the type of destruction and are of random nature. As far as the control carbon fiber reinforced plastics are concerned, a type of destruction typical of these materials, i.e. from normal stresses, is observed.

The behavior of the stitched and control carbon fiber reinforced plastics during the four-point bending is similar to the three-point bending. However, despite the fact that the stitching density and step as well as the direction of reinforcement do not have practically any effect on the nature of crack propagation, too frequent piercing with aramid yarns may significantly worsen the properties of the material acting as a concentrator of stress or defect created in the cross section of the sample.

The four-point bending strength is practically insensitive to changes in the technological parameters (Figure 7). At the same time, the absolute strength values are almost two times lower than the strength corresponding to the non-stitched carbon fiber reinforced plastic samples. The similar situation is observed for the elasticity and shear moduli. The comparison of the obtained values of these moduli for the stitched carbon fiber reinforced plastics with the values of the control material also showed an almost two-fold decrease in these parameters. No clearly expressed dependences of the E and G moduli on the stitching step or density could be obtained. However, a trend toward increasing the elasticity and shear moduli can be seen at a stitching step of 7 to 10 mm for any stitching density.

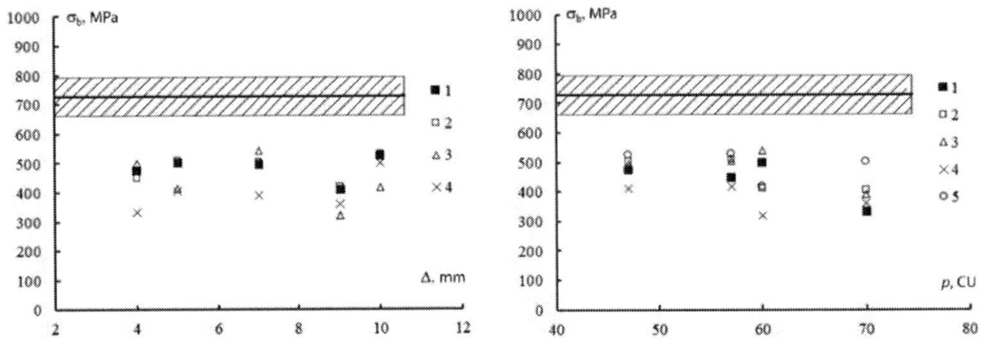

a) b)

Figure 7. Dependences of the strength during the four-point bending on the stitching stepΔ (a) and the roving laying density ρ (b).
a) ρ: 1 – 47 CU, 2 – 57 CU, 3 – 60 CU, 4 – 70 CU
b) Δ: 1 – 4mm, 2 – 5mm, 3 – 7mm, 4 – 9mm, 5 – 10mm

Table 2 presents the mean values of the fracture energies, strengths, and elasticity moduli of the stitched carbon fiber reinforced plastics obtained within the optimal range of the technological parameters and the control carbon fiber reinforced plastic. It is apparent that the manufacture of products using the TFP technology allows to approach the properties of highly oriented pressed composites with high strength characteristics.

Table 2. Comparison of physical and mechanical properties of the stitched and pressed CFRP.

Parameter	Stitched CFRP	Pressed CFRP
A_{fs}, kJ/m^2	~ 80	75±6
A_{ds}, kJ/m^2	~ 180	126±13
σ_s, MPa	~ 500	714±36
σ_d, MPa	~ 550	731±61
σ_b, MPa	~ 550	738±35
E_b, GPa	~ 45	79±7
G, GPa	~ 8	10.0±0.1

4. Conclusion

The studies of the effect of the technological stitching parameters on the physical and mechanical properties of CFRP suggest that the properties of stitched materials are consistent with the properties of pressed carbon fiber reinforced plastics with a highly oriented structure the bending strength of which differs by not more than 25 %. The most energy consuming structure of stitched carbon fiber reinforced plastics is achieved within the range of the stitching density of 50-60 CU and the stitching step of 7-10 mm. At the same time, carbon fiber reinforced plastics on the basis of TFP preforms have a higher fracture energy and shear strength in comparison with pressed ones. As soon as normal stresses prevail over the transverse ones during the short beam bending of the stitched materials, we can conclude about the increased shear characteristics of finished products and the possibility of their further improvement. It should be assumed that this is attributed to the presence of aramid yarns in the reinforcing structure that prevent the movement and development of cracks. We can hope that the crack growth in materials on the basis of TFP preforms will not be so disastrous even under impact loads in contrast to pressed carbon fiber reinforced plastics. An increased stitching step will entail the creation of a less defective structure with higher strength characteristics of carbon fiber reinforced plastics.

Acknowledgments
This study was supported financially by the Ministry of Science and Higher Education of the Russian Federation, subsidy agreement No.14.574.21.0160 of September 26, 2017, unique agreement identifier RFMEFI57417X0160.

References

[1] Nelyub V. A. An Impact Assessment of Metallization of a Carbon Tape Surface on the Mechanical Characteristics of the Filaments 2018 *Polymer Science, Series D* **11(4)** 423–425
[2] Nelyub V.A., Gorberg B.L., Grishin M.V., Sarvadii S.Y., Shub B.R., Berlin A.A. and Malysheva G.V. Properties and Technology of Applying Metal Coatings to Carbon Tape 2019 *Fibre Chemistry* **50(6)** 524-527
[3] Bocharov A., Vigovskiy V. and Nelyub V. Metal Coating of Carbon Fabric and Properties of the Carbon-Fiber-Reinforced Plastic 2019 *Materials Today: Proceedings* **11** 107-111
[4] Nelyub V.A. Technologies of Metallization of Carbon Fabric and the Properties of the Related Carbon Fiber Reinforced Plastics 2018 *Russian Metallurgy (Metally)* **13** 1199-1201
[5] Nelyub V. and Malysheva, G. Modern treatment technologies of carbon fibre for ensuring the high strength carbon fibre reinforced plastic production 2017 *MATEC Web of Conferences* **129** 02001
[6] Nelyub V.A., Borodulin A.S., Kobets L.P. et al. A study of structure formation in a binder depending on the surface microrelief of carbon fiber 2016 *Polymer Science, Series D* **9** 286–289
[7] Kopitsyna M. N., Bessonov I. V., Gusev S. A., Kireinov A. V., Kopitsyn D. S., Solodilov V. I. and Kotomin S. V. The Properties of Epoxy Binders Modified by Furan Resin and Polysulfone 2018 *Polymer Science. Series B* **60(3)** 299 – 306
[8] Malysheva G. V., Akhmetova E. Sh. and Marycheva A. N. Estimation of glass transition temperature of polysulfone-modified epoxy binders 2014 *Glass Physics and Chemistry* **40(5)** 543-548
[9] Alekseeva S.I., Fronya M.A., Viktorova I.V., Nogtev D.S. and Kononov D.M. Effect of modifiers on mechanical properties of polymer-based composite materials 2011 *Bulletin of the Lebedev Physics Institute* **38(12)** 371-373
[10] Mattheij P, Gliesche K and Feltin D. 3D reinforced stitched carbon/epoxy laminates made by tailored fibre placement 2000 *Composites Part A: Applied Science and Manufacturing* **31** 571-581.
[11] Crothers P, Drechsler K, Feltin D, Herszberg I and Kruckenberg T. Tailored fibre placement to minimize stress concentration 1997 *Composites Part A: Applied Science and Manufacturing* **28** 619-625.
[12] Ginger Gardiner. Tailored Fiber Placement: Besting metal in volume production 2013 *Plastic Technology* **2**
[13] Carosella S. Tailored Fibre Placement – Merging stitching and fibre placement technologies 2012 *Proc. Conf. ICS Europe, JEC Group* (Paris, France)
[14] Carosella S. Tailored Fibre Placement – an option for automated high volume Preform Production 2012 *Proc. Conf. SAMPE Europe Symposium (SEMAT 12)* (Munich, Germany) p. 52-58
[15] Temment H, Degenhardt R and Raible T. Tailored Fiber Placement Optimization Tool 2006

Proc. of the 25th Int. Congress of the Aeronautical Sciences (Hamburg) pp. 576–583.

[16] Spickenheuer A, Leipprand A, Bittrich L, Uhlig K, Richter E and Heinrich G. Process-dependent material properties for structural simulation of composites made by tailored fibre placement 22-26 June 2014 *Proc. of ECCM16-16th European conference on composite materials (*Seville, Spain)

[17] Mattheij P, Gliesche K and Feltin D. Tailored Fiber Placement-Mechanical Properties and Applications 1998 *Journal of Reinforced Plastics and Composites* **17** 774-786.

[18] Antonov A V, Zelenskii E S and Kuperman A. M. Impact Resistance of Reinforced Plastics Based on Polysulfone-Epoxy Blends 2014 *Journal of Reinforced Plastics and Composites* **22(4)** 361–372

[19] Solodilov V I, Korokhin R A, Gorbatkina Y A and Kuperman A M. Comparison of Fracture Energies of Epoxy-polysulfone Matrices and Unidirectional Composites Based on Them 2015 *Mechanics of Composite Materials* **51(2)** 177–190.

Thermoplastic poly ether sulfones for composite materials.

I P Storoshuk[1,2], V M Alekseev[1], N G Pavlukovich[2], A S Borodulin1, A N Kalinnikov[1] and A V Poleshaev[1,2]

[1] N.E.Bauman Moscow State Technical University (BMSTU), 105005, 2nd Bauman str., 5/1, Moscow, Russia

[2] A.N.Nesmeyanov Institute of Organoelement compounds RAS, 28 Vavilova str., 119991, Moscow, Russia

E-mail: storozhuk-ip@inbox.ru

Abstract. Results of the work on controlled synthesis of thermoplastic engineering poly ether sulfones with distinct molecular weights are presented. It is shown that control of time of the chemical reaction with other synthesis parameters being fixed allows for synthesis of poly ether sulfone with desired molecular weight and therefore with desired melt viscosity. The latter is an important parameter for creation of high-quality thermoplastic carbon fiber prepregs.

1. Introduction

Polyethersulfones (PES) are a family of high performance thermoplastics. They are characterized by high glass temperatures (from 220 to 270°C depending on the exact chemical structure), high tensile strength (from 500 to 800 MPa) and high elastic modulus (2,3-3,0 GPa), which makes them perspective materials for creating of high endurance thermoplastic carbon fiver composites [1-16]. However, polyethersulfones, in the same way as other high performance thermoplastics, suffer from 2 significant drawbacks, that hold back its wider application in industry: high cost and high melt viscosity. The cost of materials can be reduced by optimizing the manufacturing process of the initial reagents and the polymers themselves, and melt viscosity can be reduced by using polymers with lowest molecular weights at which the main polymer properties are still achieved [1,17,18].

The goal of our work was to find optimal synthesis conditions, to control molecular weights of the synthesized polymers, and to determine dependence of glass transition temperature on molecular weight of the polymer.

2. Receiving samples

The synthesis of polyethersulfone (named PES-1) using 4-(4-hydroxyphenyl)sulfonylphenol and 1-chloro-4-(4-chlorophenyl)sulfonylbenzene was conducted according to the following equation:

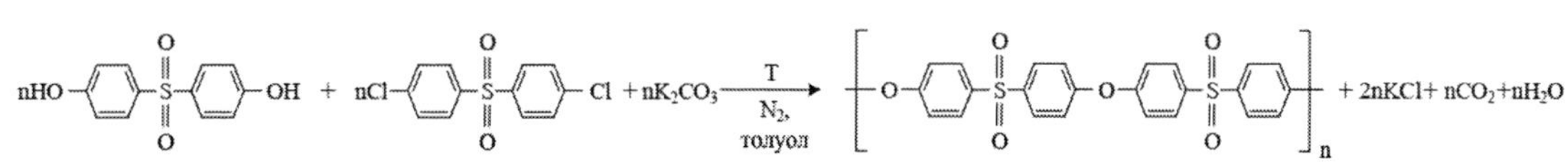

(1)

The reaction was conducted at 180°C in 22-24 hours. After completion of the reaction, the synthezed polymer wasprecipitated into constantly stirred distilled water, taken is 5-6-fold excess in relation to the organic solvent (dimethyl sulfoxide, dimethylacetamide or N-methylpyrrolidone). The precipitated polymer powder was filtrated and washed multiple times with demineralized water from KCl formed during the reaction. The purity degree was determined by a qualitative test for chlorine ions by

IOP Conf. Series: Materials Science and Engineering **934** (2020) 012067 doi:10.1088/1757-899X/934/1/012067

reaction with $AgNO_3$. Further, polymer was dried in air at a temperature of 120°C for 6 hours. The polymer yield was 97-98%.

3. Discussion of results

The molecular weight of the synthesized polyethersulfones was determined using gel permeation chromatography (GPC). The weight-average molecular weight (M_w) of the polymers was 74000-97000 g/mol, the number-average molecular weight (M_n) was 24000-31000 g/mol, and the molecular weight distribution (M_w/M_n) was 3,0-3,1. A general view of the GPC curve is shown in Fig.1, in which the small maxima to the right of the main one relate to the cyclic oligomers, the content of which is about 2-2,5% by mass and to the residual solvent.

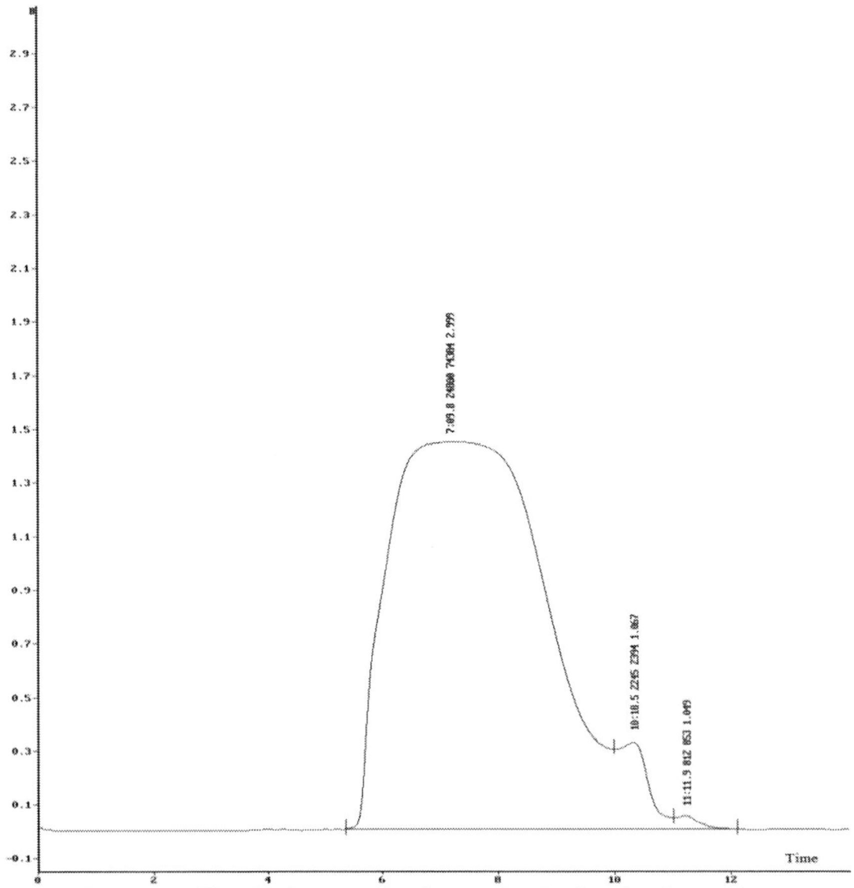

Figure 1. The GPC-curve for the synthesized polyethersulfone.

For the purpose of simplicity and efficiency the molecular weights of the synthesized polymers were estimated by determining the reduced viscosity of their 0,5% solutions in N-methylpyrrolidone on viscosimeter VPZh-2 at 25°C according to GOST 18249-72 «Plastics. Method for determining viscosity of dilute polymer solutions» (state standart). For samples taken during the synthesis the reduced viscosity was between 0,34 and 0,93 dL/g. For comparison, the reduced viscosity of polyethersulfones«Ultrason E 3010» и «Ultrason E 6020» («BASF»), measured under the same conditions, were 0,6 dl/g and 0,73 dl/g respectively.
Glass temperature of the synthesized polymers was determined according to GOST R 55135-2012 (ISO 11357-2:1999) «Plastics. Differential Scanning Calorimetry (DSC)» using the device «NETZH DSC 204 F1 Phoenix». The dependences of glass transition temperatures and reduced viscosity for the samples taken during PES-1 synthesis versus time of the synthesis are shown in Fig.2.

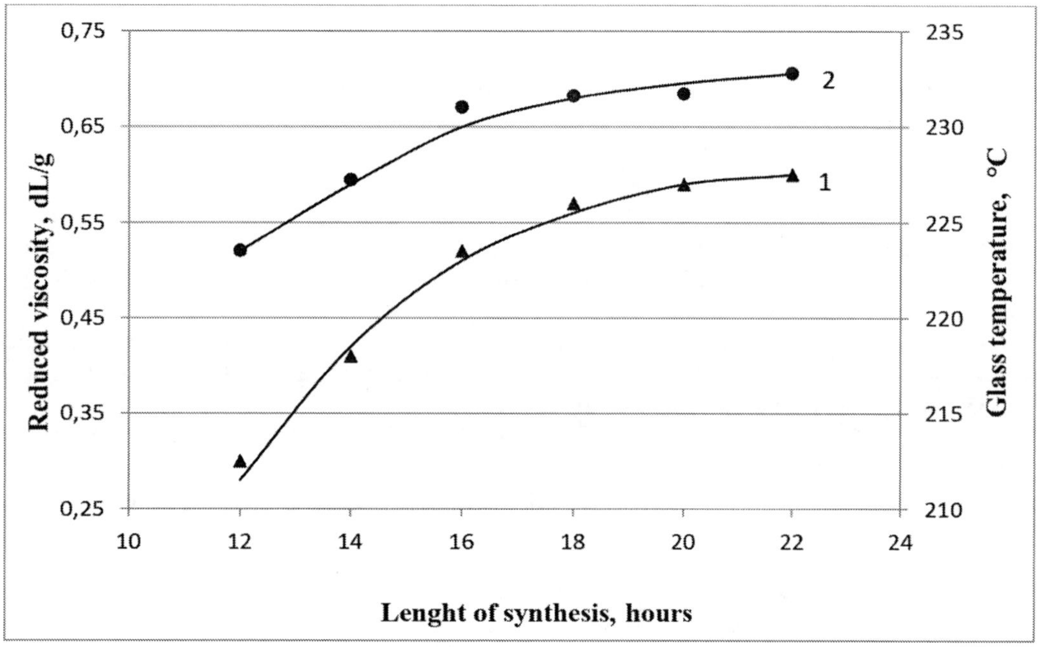

Figure 2. The dependencies of reduced viscosity (1) and glass temperature (2) PES-1 on time.

As can be seen from Fig.2, the very high glass temperature of PES-1 is achieved with the reduced viscosity of 0,45-0,5 dL/g. This suggests that upon reaching such a value of reduced viscosity for the taken samples it is possible to stop the synthesis and begin polymer precipitation into water. Suchapolymer, accordingto the theoretical concepts of reaching the maximum of most physical properties (such as glass temperature, tensile strength, elongation at break) with an optimal molecular weight will have an optimal melt viscosity in conjunction with good performance properties. An increase in molecular weight above the optimal value leads to a significant increase in the melt viscosity without a noticeable increase in physical properties, which complicates its processing from melt by common methods such as injection molding and extrusion, or makes processing impossible at all.

Thus, a method for synthesis of high performance polyethersulfones, allowing to synthesize polymers with a controlled optimal molecular weight that provides good performance properties of carbon composites and good technological properties when impregnating carbon fabrics with the polymer's melt, was developed.

Acknowledgments

This work was supported by the Ministry of Science and Higher Education of the Russian Federation.

References

[1] Beider E Y, Petrova G N 2015 Thermoplastic binders for polymer composite materials *Works of VIAM* №11 pp.40-49
[2] Kerber M L, Vinogradov V M, Golovkin G S 2011 Polymer composite materials: structure, properties, technology S-Pb: Profession 233
[3] Mihailin Yu A 2006 Heat-resistant polymers and polymer materials S-Pb: Profession 624
[4] Mihailin Yu A 2010 Structural polymer composite materials S-Pb: Scientific foundations and technologies 822
[5] Fitcer E 1988 Carbon fibers and carbon composites M.: Mir 336

[6] Kablov E N 2002 Aviation materials science: results and prospects *Bulletin of the Russian Academy of Sciences* **72** №1 pp.3-12
[7] Komarov G A 2009 State, prospects and problems of PCM application in engineering *Polymeric materials* №2 pp.5-9.
[8] Grashenkov D V, Chursova L V 2012. Strategy for the development of composite and functional materials *Aviation materials and technologies* №5 pp.231–242
[9] Petrova G N, Beider E Ya, Starostina I V 2016 Injection thermoplastics for aircraft products *Encyclopedic reference book* №6 pp.10-15
[10] Nelub V A, Berlin A A 2019 Electrical properties of carbon fiber *Himicheskaya promishlennost segodnya* **1** 38-43
[11] Nelub V A, Borodulin A S, Buyanov I A, Chudnov I V, Malisheva G V, Haraev A M, Luchkina L V, Bedanokov A Y, Bazheva R Ch, Beshtoev B Z, Kalinnikov A N 2017 Patent RF № 2653058 Ognestoykiy aromaticheskiy poliefirsulfon
[12] Borodulin A S, Kalinnikov A N and and oth. 2019. Patent RF № 2683268. Aromaticheskie poliefiry
[13] Borodulin A S, Kalinnikov A N, Bazheva R C at all. 2018. Synthesis and properties of aromatic polyethersulfones//International Journal of Mechanical Engineering and Technology (IJMET). – v. 9 (13), pp. 1109–1116.
[14] Borodulin A S, Kalinnikov A N, Bazheva R C at all. 2018. Receipt and investigation of performance characteristics of super constructions polyesters. - International Journal of Mechanical Engineering and Technology. - v.9 (13), pp. 1117-1127.
[15] Borodulin A S, Kalinnikov A N, Kharaev A M at all. 2019. New Polymeric Binders for the Production of Composit// International Conference on Modern Trends in Man-ufacturing Technologies and Equipment 2018. International Journal of Materials today: proceedings. –V. 11. Issue P1, pp. 3107–3111.
[16] Nelub V A, Borodulin A S and oth. 2018 Poliefirsul'fony s uluchshennymi teplofizicheskimi svoystvami *Klei, germetiki, tekhnologii* (Moscow:Nauka I tekhnologii) 7 15-20
[17] Van-Krevelen D V 1976 Properties and chemical structure of polymers M.Himiya – 416
[18] Bartenev G M, Frenkel C Ya 1990 Physics of polymers L.Himiya – 432

IOP Conf. Series: Materials Science and Engineering **934** (2020) 012068 doi:10.1088/1757-899X/934/1/012068

Evaluation construction made of polymer composite materials by molding using reusable flexible punches production profitability

M V Stoyanova[1], A D Novikov[2], A S Borodulin[3]

[1,2,3] Bauman Moscow State Technical University, 2nd Baumanskaya street, Moscow, Russia

E-mail: [1] MStoyanova@emtc.ru, [2] nad@emtc.ru, [3] asb@emtc.ru.

Abstracts. One of the most common methods for the production of composites is the vacuum infusion method, which consists in the use of a disposable flexible punch pressed to the surface of the mold by atmospheric pressure. A new technology is the use of reusable flexible punches also pressed by atmospheric pressure. The paper compares the profitability of the same construction by these methods production.

1. Introduction

Using vacuum infusion technology, a wide range of products can be manufactured. So, with this technology it is possible to produce both small products – car bumpers, helmets, and very large ones – aircraft wings, ship hulls [1]. The largest product size is limited only by the size of the workshop in which the production is located.

The essence of the technology is that the layers of the future product are laid in a sealed vacuum bag, from which air is subsequently removed by a vacuum pump and impregnated with a polymer binder [2]. Creating a vacuum inside the bag allows the use of atmospheric pressure as a press. By pressing the bag to the product surface, the layers of the reinforcing filler are compressed, which leads to an increase in the mechanical characteristics of the composite by reducing the amount of binder [3].

To carry out the impregnation of the product, a conducting mesh made of polymeric materials having volumetric weaving is laid, which creates a vacuum between the reinforcing filler and the bag. Due to the fact that the conducting mesh laid on reinforcement layers can become part of the future product, it is separated by a sacrificial layer, also known as Peel Ply. In addition, to install the package requires a sealing harness, ensuring the tightness of the vacuum bag. The sequence of laying the layers used to implement the vacuum infusion technology is shown in Fig. 1.

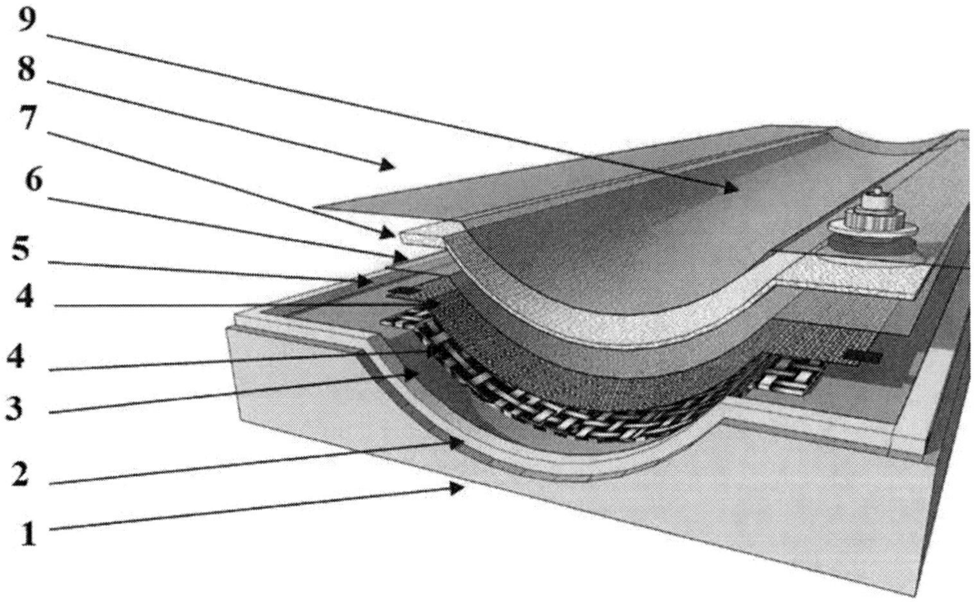

Figure 1. Layering procedure in infusion technology: 1 – mold; 2,5 – sealing tape; 3 – sealant agent; 4 – filler layer, 6 – peel ply, 7 – conducting mesh, 8 – vacuum film, 9 – pneumatic valve

Due to the nature of the technology, this method can produce constructions whose thickness does not exceed 15-20 mm. In other words, infusion is excellent for creating shell structures, while it is almost not applicable for thick-walled parts.

Vacuum infusion allows for relatively small financial costs to produce constructions with high surface quality and a more controlled percentage of binder in the composite in comparison with contact molding. A great advantage of the technology is also a low pore content compared to contact molding [4]. But first of all, the content of pores depends on the preparation of the binder immediately before molding. It is worth paying attention to the fact that for the implementation of infusion technology, a conventional rotary vane vacuum pump and a trap for a binder are sufficient. The cost of the pump, as a rule, is low in comparison with the equipment used to implement other technologies, such as presses, autoclaves, component accessories, etc.

The main disadvantage of infusion can be considered a large number of used technological materials that are disposed of after a single use. These materials include: peel ply, conducting mesh, sealing tape, vacuum and spiral tubes, as well as the vacuum film itself. Such a large number of purely technological materials significantly increases the cost of products manufactured by infusion technology in medium-series production. It is most successfully used in single or small batch production [5].

As a technology that allows you to take advantage of both vacuum infusion and press technology, you can use technology with a reusable flexible punch [6]. With their help, it is possible to produce composite products based on both thermoplastic and thermoplastic matrices [7]. Fig. 2 shows the design of the stand with a reusable flexible punch. This equipment also includes a vacuum system that cannot be shown in this figure.

Figure 2. Equipment with a reusable flexible punch: 1 – a countertop; 2 – a pressure cover with a punch; 3 – a control panel; 4 – a heating element (a block of infrared lamps)

In relation to composites, molding technology using such equipment can be classified as small-scale or single. This is due, first of all, to the fact that the bulk of the time, as well as in infusion technology, is occupied by the manual laying out of the layers of the future composite for tooling, preparation of a binder, impregnation, etc. The main difference from the traditional method of infusion is the use of a reusable flexible punch that acts as a vacuum bag, and the lack of the need to search for air leaks from under the sealing tow [8].

These features can reduce the complexity and time consumption compared to traditional molding methods, and the convenience of layout (pump, degasser and vacuum traps are located directly under the countertop) to reduce the size of the working space due to the fact that molds and material can be laid directly on the countertop, which is rather important both for industrial and engineering companies [9].

2. Objects and methods

A skateboard was chosen as a product for comparing the two technologies (Fig. 3). This product is suitable for molding both with vacuum infusion technology and with the silicone punch.

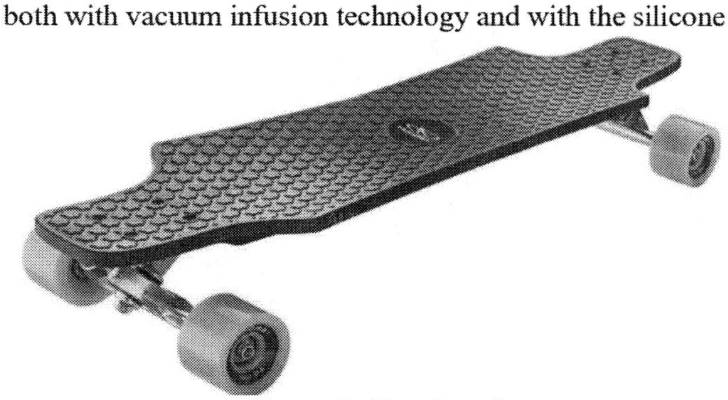

Figure 3. Skateboard

Mold of the product was designed in an Autodesk Inventor (Fig. 4) and was milled on a CNC machine from an MDF board 30 mm thick (Fig. 5). This material was chosen due to its low cost and ease of machining.

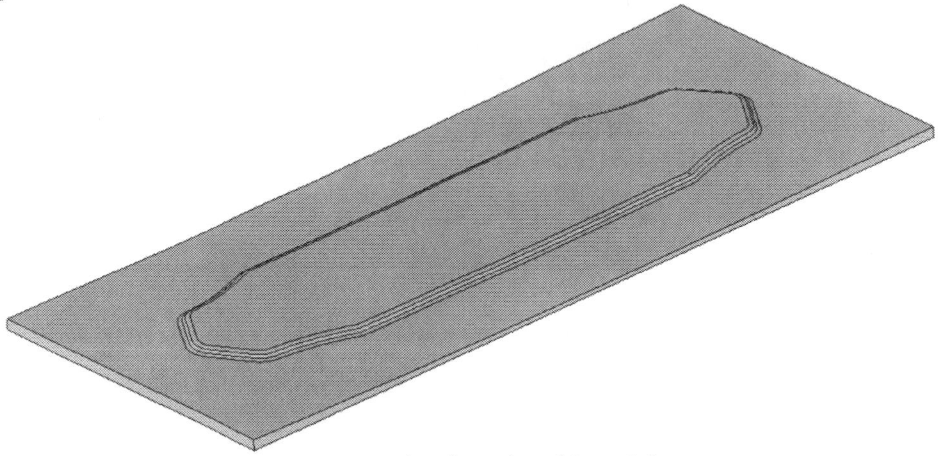

Figure 4. Skateboard mold model

Figure 5. Skateboard mold

Due to the fact that the mold material, as a whole, is a product of wood processing, it has a high porosity and surface roughness. When composites are molded, the surface of the product repeats the surface of the tooling, including microroughnesses, scratches, and other similar defects, which in the case of MDF can significantly reduce the quality of the product. In addition, these defects increase the level of adhesion, which, due to the low strength of MDF, leads to its destruction after several moldings. To increase durability, you need to get a glossy mold surface. This was achieved by surface treatment with Sealer Mikon 399MC pore-filling compound applied evenly on two-layer mold. This composition creates a thin solid polymer film, which is then easily processed by hand or mechanical grinding tools.

3. Results and Discussion

The production time of a typical construction for engineering company [10] was chosen as an indicator of the technologies effectiveness under consideration. The laying scheme is shown in Fig. 6.

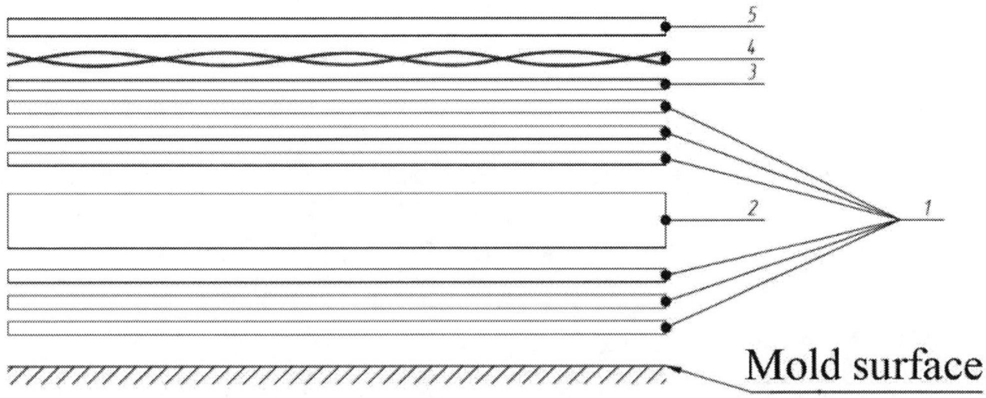

Figure 6. Layout of reinforcing and technological layers for the skateboard manufacture: 1 – carbon fabric; 2 – SORIC nonwoven fabric; 3 – peel ply; 4 –conducting mesh; 5 – vacuum film (silicone punch)

To compare the time spent, 3 moldings were made using each of the technologies described above. The time was measured by an independent expert using a stopwatch. The molder worked at the usual, measured pace. The average time of each of the stages of production is presented in table 1.
Not taken into account the time spent on connecting the tubes to the vacuum pump, as well as cutting and cutting tissue due to their insignificance.

Table 1.
Comparison of production time using infusion technology and a reusable punch

Infusion	**Silicon punch**	**Average stage time (min)**
Reinforcing Layers		40
Laying peel ply		5
Laying conducting mesh		5
Installation of spiral and vacuum tubes for supplying a binder and pumping air	Installation of spiral tubes for supplying a binder and pumping air	7
Bonding of sealing tape	–	10
Vacuum bag installation	–	20
Vacuum bag compression	–	8
Vacuum bag leak repair	–	10
Total time: 105	Total time: 57	

After this, impregnation and curing is performed under an infrared lamp for 1 hour [11, 12]. Removing the product from the mold takes 10 minutes.
As practice has shown, the time for forming a single product using a silicone punch is 33% faster than by infusion technology. Obviously, with an increase in the number of layers, the percentage will decrease, since the difference in technology is only the installation of the package and the sealing harness for it [13].
It should also be noted that due to the absence in the technology of disposable vacuum bags, usually made of polyethylene, polypropylene or nylon, the amount of garbage in production is reduced. This, in turn, helps to increase the environmental friendliness of production.

The calculation of the cost of materials necessary for the production of one product using vacuum infusion technology is presented in table 2. As a currency, conventional units are used, reduced to the cost of real materials.

Table 2.
Calculation of the cost of materials required for vacuum infusion

	Cost per square meter or meter of material, units	Number of layers	Area, m^2	Cost, units
Carbon fabric	160	5	0,2	160
SORIC nonwoven fabric	82	1	0,2	16.4
Peel ply	25	1	0,2	5
Conducting mesh	18	1	0,2	3.6
Vacuum film	8	1	1	8
Sealing tape	3.5	1	2,8	10
			Total	203

Since the price of a product made using a silicone punch does not include a vacuum bag and a sealing tape, therefore, it will amount to 185 units.
Considering that the working day at the production site is 8 hours, find the theoretical number of manufactured products per day. Accept that one person with one equipment works on the site.
Since in the calculation obtain non-integer values of the quantities of units per day, to approximate the conditions of real life, round them down (Table 3).

Table 3.
Calculation of the productivity of infusion technologies and reusable punch

	By infusion technology, min	By reusable punch technology, min
Molding	105	57
Curing	60	60
Product retrieval	10	10
Total time	175	127
Working hours	480	
Pieces per day	2	3

With this production method, each of the actions of the molder is sequential, that is, after molding, curing follows, in which, the worker is not engaged in production activities or rests. To increase the productivity of the worker, it makes sense to make a second mold. In this case, the molder is able to lay out the fabric in parallel with the curing process of the product in a polymerization furnace. It is worth noting that the existing vacuum-membrane press has dimensions that can accommodate only one product on its countertop. When using more dimensional installations, accordingly, it is possible to further increase the number of products manufactured in one cycle.
To compare labor productivity during molding using two rigs, both during molding using a membrane press and the method of vacuum infusion, Gantt charts were constructed (Fig. 7, 8). The numbers on the diagram indicate the processes of product forming.

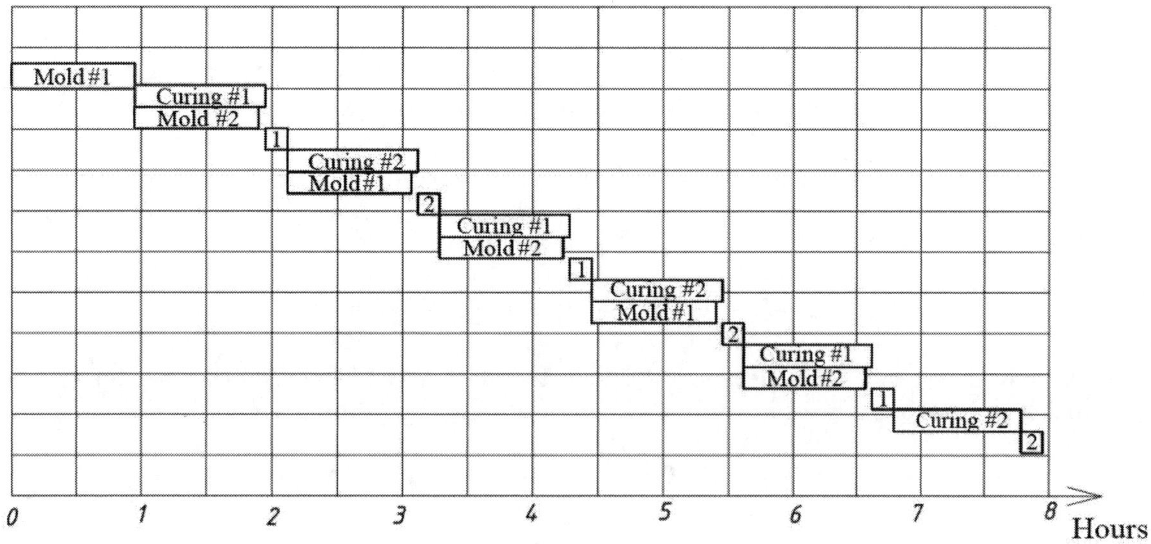

Figure 7. Gantt chart for forming using a reusable punch

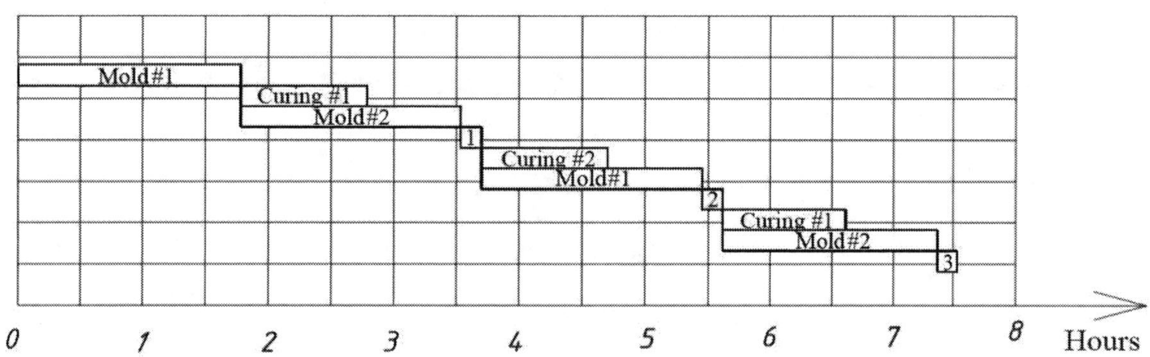

Figure 8. Gantt chart for molding by vacuum infusion technology

As can be seen from the above charts, the productivity when working with a reusable punch is 6 products per day, and during infusion – 3 products per day. It is worth noting that despite the fact that the last product produced per day does not have enough time for curing in the oven, curing can occur at room temperature in the absence of a worker using automatic shutdown equipment. Therefore, it makes sense to consider that the infusion rate is 4 pieces per day.

It should be noted that due to the fact that the vacuum punch becomes unusable over time, the cost of changing it is also worth considering. The average cost of one linear meter of a punch with a thickness of 3 millimeters and a maximum elongation of 600% is 795 units. Since the length of the table is 1.3 meters, the cost of the punch for the existing vacuum press will be m = 1033.5 units. According to the manufacturer, the punch is able to withstand t = 300-500 molding cycles. Take this value equal to t = 400.

Knowing the labor productivity p_i for both technologies, it is possible to find the number of manufactured products per month. Assume that in the month d = 21 working days.

$$n_i = p_i \cdot d \quad (1)$$

For infusion:

$$n_{ing} = p_{inf} \cdot d = 4 \cdot 21 = 84 \text{ pieces per month;} \quad (1.1)$$

IOP Conf. Series: Materials Science and Engineering **934** (2020) 012068 doi:10.1088/1757-899X/934/1/012068

For reusable punch:

$$n_{RM} = p_{RM} \cdot d = 6 \cdot 21 = 126 \text{ pieces per month;} \quad (1.2)$$

Take the worker's salary equal to 300 rubles per hour, therefore, for a month his salary will be:

$$z = 21 \cdot 8 \cdot 300 = 5040 \text{ units.} \quad (2)$$

The cost of buying a new punch will be per month:

$$M = m \cdot \frac{n_{RM}}{t} = 1033.5 \cdot \frac{126}{400} = 325.5 \text{ units.} \quad (3)$$

Since the cost of materials for production is 203 units for infusion technology and 185 units for a silicon punch, the total cost of materials per month will be:
for infusion:

$$T_{inf} = 2030 \cdot n_{inf} = 203 \cdot 84 = 17052 \text{ units;} \quad (4.1)$$

for reusable punch:

$$T_{RM} = 185 \cdot n_{RM} + M = 185 \cdot 126 + 325.5 = 23635.5 \text{ units;} \quad (4.2)$$

Total costs include both the cost of materials and the salary of the worker:
for infusion:

$$S_{inf} = T_{inf} + z = 17052 + 5040 = 22092 \text{ units;} \quad (5.1)$$

for reusable punch:

$$S_{RM} = T_{RM} + z = 23635.5 + 5040 = 28675.5 \text{ units;} \quad (5.2)$$

The resulting income is calculated as 1.3 of the cost of production:
for infusion:

$$D_{inf} = 1{,}3 \cdot 22092 = 26510.4 \text{ units;} \quad (6.1)$$

for reusable punch:

$$D_{RM} = 1{,}3 \cdot 28675.5 = 34410.6 \text{ units.} \quad (6.2)$$

Consequently, profits from sales of skateboards will amount to:
for infusion:

$$P_{inf} = D_{inf} - S_{inf} = 26510.4 - 22092 = 4418.4 \text{ units;} \quad (7.1)$$

for reusable punch:

$$P_{RM} = D_{RM} - S_{RM} = 34410.6 - 28675.5 = 5735.1 \text{ units.} \quad (7.2)$$

Since the profit is made from the sale of all skateboards produced in a month, the cost of one of them will be:
for infusion:

$$I_{inf} = \frac{D_{inf}}{n_{inf}} = \frac{26510.4}{84} = 315.6 \text{ units;} \quad (8.1)$$

for reusable punch:

$$I_{RM} = \frac{D_{RM}}{n_{RM}} = \frac{34410.6}{126} = 273.1 \text{ units.} \quad (8.2)$$

Thanks to the use of new technology, it is possible to reduce the final cost of the product by

$$\frac{315.6-273.1}{315.6}\cdot 100\% = 13{,}5\% \quad (9)$$

Since the profit on the new technology is higher by

$$P_{RM} - P_{inf} = 5735.1 - 4418.4 = 1316.7 \text{ units per month;} \quad (10)$$

All results of economic calculations are presented in table 4.

Table 4.
Economic calculation results

	Infusion	Reusable punch
Number of pieces per day	4	6
Pieces per month	84	126
Cost of materials	203	185
Cost of punch	-	1033.5
The number of molds that the punch can withstand	-	400
Worker's salary	5040	
Monthly Costs	22092	28675.5
Monthly income	26510.4	34410.6
Profit	4418.4	5735.1
The cost of one skateboard	315.6	273.1

4. Conclusion

The presented technology can be used to replace vacuum infusion technology without compromising product quality. Due to the vacuum bag installation process optimization, the production time of one product is reduced, and, as a result, the productivity of the method increases. At the same time, the cost of the product is reduced by 14%.

References

[1] Novikov A D, Reznik S V and Prosuntsov P V 2017 Comparison of space antennas mirror reflectors parameters made of composite materials *MATEC Web Conf* **92** p 5

[2] Borodulin A S 2013 Plasticizers for epoxy adhesives and binders *Polymer Science D* **6(1)** pp 59-62

[3] Borodulin A S 2013 Polyester resins for production of goods from polymer composite materials by pressing methods *Polymer Science Series D* **6(3)** pp 269-270

[4] Borodulin A S, Marycheva A N and Malysheva G V 2015 Simulation of impregnation kinetics of fabric fillers in the production of fiberglass *Glass Physics and Chemistry* **41(6)** pp 660-664

[5] Bulanov I M and Vorobey V V 1998 Technology of rocket and aerospace structures from composite materials (M.: BMSTU) p257

[6] Belyakov E V 2012 Development of the composite structures contact molding technology taking into account the individual properties of the feedstock: dissertation abstract (Moscow) p16

[7] Nelyub V A, Borodulin A S, Kobets L P and Malysheva G V 2017 Viscous hysteresis in filled siloxane binders *Polymer Science Series D* **10(1)** pp 19-20

[8] Gabriels K S, Karaeva O A, Kudrin A M, Kulakov E V 2015 Physico-mechanical properties of fiberglass obtained using a polyester resin by the method of vacuum infusion *Bulletin of Voronezh State Technical University* **11**(1) pp 52-55

[9] Stoyanova M V 2015 The phenomena of engineering in Russia and drivers of its growth *News of science: Proceedings of materials the international scientific conference* p 5

[10] Gorlacheva E N, Brom A E, Omelchenko I N and Stoyanova M V 2017 Knowledge management in engineering enterprises: the results of empiric research *Proceed. of International forum on knowledge asset dynamics: knowledge management in the 21st century: resilience, creativity and co-creation* pp 1785-1795

[11] Kobets L P, Malysheva G V and Borodulin A S 2016 Rheological properties of system «elastomer binder–discrete carbon fiber» *Inorganic Materials: Applied Research* **7(1)** pp 15-19

[12] Belov P A, Kobets L P and Borodulin A S 2014 Impregnation kinetics of fibers with liquids: Simulation within the generalization of Navier-Stokes equations *Inorganic Materials: Applied Research* **4(4)** pp 403-406

[13] Belov P A, Borodulin A S, Kobets L P and Malysheva G V 2016 Kinetics of fiber impregnation by a binder. Gradient generalization of Navier–Stokes–Darcy equations *Polymer Science - Series D* **9(2)** pp 205-208

IOP Conf. Series: Materials Science and Engineering **934** (2020) 012069 doi:10.1088/1757-899X/934/1/012069

Environmental effectiveness assessment of the technology for molding products made of polymer composite materials using a reusable flexible punch

A D Novikov[1], M V Stoyanova[2], A S Borodulin[3]

[1,2,3] Bauman Moscow State Technical University, 2nd Baumanskaya street, Moscow, Russia

E-mail: [1] nad@emtc.ru, [2] MStoyanova@emtc.ru, [3] asb@emtc.ru.

Abstracts. The use of composite materials in all industries has a negative impact on the environment in general and man in particular. Polymeric binders include hazardous substances, and the auxiliary components used in production are made of non-processed materials. It is possible to reduce the negative impact on the environment by using a new technology using reusable flexible punches.

1. Introduction

Among a wide range of materials that can be attributed to composite materials, polymers are most widely used due to unique characteristics and availability in production. Polymer composite materials (PCM), as the name implies, are based on polymeric materials, however this generalization is too common and includes a very wide range of materials, and the choice of a polymer for a composite depends on operating conditions, strength and economic parameters of the construction, production equipment, etc. Today epoxy and polyester resins are used as binders in most engineering industries [1], due to the range of mechanical, operational, rheological and other characteristics that meet the challenges facing rocket and aviation equipment, as well as in the manufacture of ships, cars and rail transport. Equally important is the direction of engineering companies that develop engineering, and for which it is necessary to ensure efficient and safe production of products from composites [2].

Polymer composite materials based on epoxy and polyester compounds are include components that can have a negative impact on human health and the environment as a whole. Epoxy binders are harmless to humans in the cured state, but in the uncured form are sufficiently strong poisons. The composition of epoxy resins includes phenols and bisphenols, as well as epichlorohydrin [3], which affects the kidneys. Epoxies are dangerous both in direct contact with the skin, and when inhaling fumes.

Polyester resin, in turn, is relatively safe in itself, but hazardous accelerators, thinners and hardeners may be included [4]. Styrene – a good solvent for polymers, is the most dangerous one. Styrene affects the central and peripheral nervous systems, digestive tract, metabolism. Polyester compounds are also dangerous when in direct contact with the skin and when inhaling vapors.

If we consider another component of polymer composite materials, a filler, then it is not completely harmless. Among fillers, glass and carbon fabric are common. Fiberglass is woven from glass fibers with a diameter of 10-15 microns. So thin glass fibers are flexible enough, but it can still break. The

IOP Conf. Series: Materials Science and Engineering **934** (2020) 012069 doi:10.1088/1757-899X/934/1/012069

fragility of this diameter glass fibers leads to the large number of debris, easily penetrating the pores of the skin and causing itching. Inhalation of air with glass fibers debris may cause lung irritation. Similar factors are observed when working with carbon fabrics, although carbon fibers are less brittle, but their diameter is less, from 5 to 10 microns. Fragments of fibers are formed during cutting and laying of fabrics in most methods used, and are associated with direct contact with humans.

2. Objects and methods

Among industrial methods for manufacturing PCM products, contact molding, vacuum infusion, autoclave molding, RTM and winding can be distinguished [5]. There are other methods, but they are variations of the methods listed and the advantages and disadvantages described below apply, including. Contact molding process is the almost manual. The point of this method is application of polymer onto a fibrous filler using brushes and rollers [6]. This method does not require any technological equipment, it is quite simple, but the most dangerous for the technologist, because the whole polymerization process takes place in the atmosphere. In addition, products obtained by this method do not differ in high values of mechanical and strength characteristics [7]. To improve the composite characteristics, as well as to reduce the harmful effects, sealed vacuum bags are used. Preform (a set of dry reinforcing fabric layers) is placed into this bag, and then the preform is impregnated with a polymer binder [8]. This method is called vacuum infusion, and combines relative safety, economic efficiency and product quality [9]. The disadvantages of the method include a large complexity and a large amount of waste.

The highest product quality can be achieved using autoclaves [10] – closed vessels in which the binder is polymerized at high pressure and temperature (more than 5 atmospheres and 250 °C) [11]. However, an autoclave is a very expensive device, which, among other things, according to safety rules. For example, it cannot be placed in educational institutions.

Least-laborious for the technologist is RTM method. When manufacturing products using the RTM method, the technologist only needs to put the cut fabric in a metal mold, then close half the hydraulic press and start the impregnation in automatic mode [12]. This method is also characterized by high environmental friendliness, because the technologist contacts the binder only at the stage of replenishment of sealed containers. The disadvantages of the method include the very high cost of equipment and the limited size of the resulting product.

The most technologically advanced method for the production of composites is the winding method, where continuous fiber is applied layer-by-layer to a rotating mandrel. This method combines the low cost of equipment, high productivity and low labor costs. However, this method has two critical drawbacks: low environmental friendliness and the ability to manufacture only rotation parts.

A new method for manufacturing composite products using reusable flexible punches stands combines the advantages of vacuum infusion and RTM. In this method, the technologist practically does not contact the polymer compound, while the low complexity and cost of the equipment increases the manufacturability of the production.

3. Results and Discussion

One of the ways to reduce the influence of harmful components on the human body and the environment is by impregnating the filler in closed systems [13]. For example, when using equipment with a reusable elastic punches (Fig. 1).

In such equipment, the components of the binder, resin and hardener, are poured into sealed vessels in advance, and until the curing process the operator does not come into contact with them, which significantly reduces the harm from working with the compound.

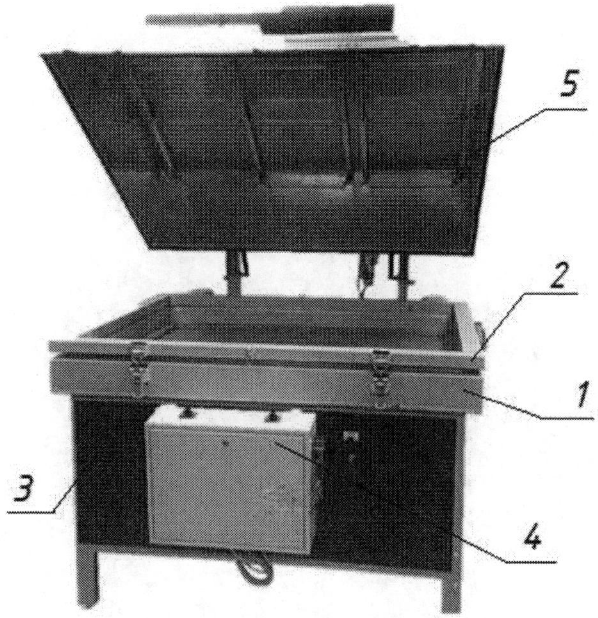

Figure 1. Equipment with reusable elastic punch: 1) countertop, 2) pressure cap with a silicon punch 3) control panel, 4) infrared lamp unit

Also, the use of such equipment reduces the amount of waste in production. In the classical technology for the production of composite structures, auxiliary materials (films, separating fabrics, plastic nets) are used, which cannot be recycled after use (Fig. 2) [14]. Recycling of these materials is impossible, which affects the environment, and with the increase in the use of composites in industry, the amount of such waste also grows, and exponentially, because in the production of one composite product, three types of waste are generated.

Figure 2. Assembled bag for impregnation of samples using vacuum infusion technology

In the manufacture of products using equipment with a reusable elastic punch, two types of auxiliary materials are excluded, and the punch itself is recyclable (Fig. 3).

Figure 3. Assembled bag for impregnation of samples using reusable elastic punch

An important advantage of the equipment with a reusable elastic punch use is the possibility of using thermoplastics as a binder. One of the main drawbacks of PCM with the use of thermosetting binders today is the inability to recycle. The solution is to use thermoplastic polymers, but this is not possible in the case of using traditional production technologies – vacuum infusion, RTM and Light-RTM or autoclave molding. However, equipment using a reusable elastic punches allows this, because the silicone from which the press membrane is made has a melting point of 240 °C, sintering of thermoplastics can be carried out on membrane-vacuum presses. Strictly speaking, membrane-vacuum presses were originally adapted for gluing polymer films onto products. The use of injection or impregnation with a molten thermoplastic binder will not allow full impregnation of the fibrous tissue filler, however, it is possible to use a film binder. The use of such binders will not only allow the processing of composite structures, but also virtually eliminate the contact of the technologist with harmful substances.

4. Conclusion

Despite the sufficient prevalence of PCM in many industries, the manufacturing process of structures made of these materials is dangerous for humans, and it is necessary to develop new technologies that reduce the negative effect not only on the technologist directly involved in the manufacturing process, but also on the environment as a whole. Table 1 shows a conditional comparison of common technologies from which it can be seen that RTM technology and the use of a reusable flexible punches have the most environmental friendliness, with comparative performance and product quality. However, the cost of the product obtained using equipment with a reusable elastic punches is much lower, because lower as the cost of equipment, and molds. This method is quiet useful for modern engineering companies and for education [15].

Table 1.
Comparison of common technologies for the production of PCM products

Technology	Environmental friendliness	Performance	Product quality	Product cost
Contact molding	Low	Low	Low	Low
Vacuum infusion	Average	Average	Average	Average
Autoclave molding	Average	Average	Высокое	High
RTM	High	High	Average	High
Winding	Low	Average	Average	Low
Reusable elastic punch	High	High	Average	Average

Compared with other methods, the technology using a reusable elastic punch has an average product quality, i.e. the ratio of binder and fiber filler, which directly affects the mechanical characteristics of the product. Yielding to autoclaving, the quality of the products obtained is comparable to the quality of products obtained by the technology of vacuum infusion, RTM and winding.
Thus, the use of technology using a reusable elastic punch can significantly reduce the harmful effects on humans and the environment, not inferior in other parameters to other technologies.

References
[1] Kobets L P, Borodulin A S and Malysheva G V 2016 Study of Microcapillary Impregnation of Carbon Fibres by Epoxy Binders *Fibre Chemistry* **48 (1)** pp 311-315
[2] Gorlacheva E N, Brom A E and Omelchenko I N and Stoyanova M V 2017 Knowledge management in engineering enterprises: the results of empiric research *Proceed. of International forum on knowledge asset dynamics: knowledge management in the 21st century: resilience, creativity and co-creation* pp 1785-1795
[3] Nelyub V A, Borodulin A S, Kobets L P and Malysheva G V 2017 Viscous hysteresis in filled siloxane binders *Polymer Science Series D* **10(1)** pp 19-20
[4] Reznik S V, Prosuntsov P V and Novikov A D 2017 Comparison of space antennas mirror reflectors parameters made of composite materials *MATEC Web Conf* **110** p 4
[5] Reznik S V and Novikov A D 2017Comparative analysis of the honeycomb and thin-shell space antenna reflectors *MATEC Web Conf* **92** p 5
[6] Kobets L P, Malysheva G V and Borodulin A S 2016 Rheological properties of system «elastomer binder–discrete carbon fiber» *Inorganic Materials: Applied Research* **7(1)** pp 15-19
[7] Bazhenov S L 2014 Mechanics and Technology of Composite Materials (Dolgoprudny) p 328
[8] Borodulin A S, Marycheva A N and Malysheva G V 2015 Simulation of impregnation kinetics of fabric fillers in the production of fiberglass *Glass Physics and Chemistry* **41(6)** pp 660-664
[9] Belov P A, Kobets L P and Borodulin A S 2014 Impregnation kinetics of fibers with liquids: Simulation within the generalization of Navier-Stokes equations *Inorganic Materials: Applied Research* **4(4)** pp 403-406
[10] Borodulin A S 2013 Polyester resins for production of goods from polymer composite materials by pressing methods *Polymer Science Series D* **6(3)** pp 269-270

[11] Borodulin A S 2013 Plasticizers for epoxy adhesives and binders *Polymer Science Series D* **6(1)** pp 59-62

[12] Khaliulin V I, Batrakov V V, Khilov P A, Konstantinov D U and Danilov Y S 2015 Experience in manufacturing integrated structures using RTM *Bulletin of Kazan State Technical University named after A.N. Tupolev* **4(2)** pp 45-49

[13] Khrulkov A V, Dushin M I, Popov U O and Kogan D I 2012 Research and development of autoclave and autoclave-free PCM molding technologies *Aviation materials and technology* **5** pp 292-301

[14] Vlasenko F S, Utkina T S, Sumenkova O D, Veshkin E A and Khlebnikov N V 2016 Technological testing of domestic auxiliary materials for autoclave molding processes *Composite world* **4(67)** pp 56-58

[15] Stoyanova M V 2015 The phenomena of engineering in Russia and drivers of its growth *News of science: Proceedings of materials the international scientific conference* p 5

IOP Conf. Series: Materials Science and Engineering **934** (2020) 012070 doi:10.1088/1757-899X/934/1/012070

Numerical simulation of the heat transfer process in the anti-icing heating system for Arctic marine applications

Yu I Dimitrienko, V Yu Chibisov, M N Koryakov.

Bauman Moscow State Technical University, Moscow, Russia Federation

Abstract In the Arctic, there is a serious problem of icing of structures, which can cause a catastrophic disruption of the functioning of the equipment necessary to maintain the life support of sea vessels and ground structures. For marine vessels, this problem is especially relevant in view of the limitations on heating systems and the mass of permissible heating equipment. The article is a continuation of work on the research and development of an electrical heating system for the air intake grilles of marine vessels. A numerical simulation of the heat transfer conditions for the design of the air intake grille with a built-in electric heating system was carried out. Some variant calculations of the numerical simulation of heat transfer processes in the system are presented — air, a streamlined structure — an air intake grill case — a heating system. A technique has been developed for selecting the optimal parameters of the electric heating system for various operating conditions of air intake grilles.

1. Introduction

One of these problems is the problem of choosing the optimal geometric dimensions of air intake grids, due to the special operating conditions (icing, freezing, thawing, etc.) of Arctic vessels [1-3]. Icing narrows the cross-section of the intake ducts, reduces the amount of incoming air, which can also lead to the failure of the ship's ventilation system and lead to death. To solve this problem, it is necessary to conduct research and numerical modeling of heat transfer processes of the de-icing system. To calculate the heat transfer in the design of blinds with internal heating from electric heaters, it is necessary to calculate the convective the heat flux removed from the surface of the blinds when exposed to cold air flow, as well as the calculation of the coefficient of convective heat-exchange's. To calculate these parameters, a General system of equations of dynamics of a linear-viscous heat-conducting gas is used, describing movement of cold air masses on the surface of the blinds.

The anti-icing heating system (AIHS) for the ship's application was developed at BMTU, the Simplex Research and Education Center, together with Bee Pitron (St. Petersburg) [4-7]. AIHS is designed to protect the intake grilles of marine ventilation systems from icing when exposed to negative temperatures up to -60 ° C. AIHS design fragment is shown in Figure 1.

Figure 1. General view of the developed AIHS module

2. The method of numerical simulation of the heat transfer process in the anti-icing heating system

The method of numerical simulation of the heat transfer process in the anti-icing heating system of the air intake grilles was presented in work 1. The method consists of 2 main stages: modeling the processes of aerodynamic flow around the air intake grilles with a stream of cold air (stage 1) and modeling the heat transfer process in the leaves of the air intake grilles taking into account electric heating (stage 2).

The optimization of the electric heating system and the design of the heat-insulating lattice was carried out by selecting the optimal values of the distance between the lattice flaps under given restrictions on the permissible temperature of the lattice surface. The electric power was minimized, which must be supplied to the heating system of the air intake grille under given external flow conditions (flow velocity and its temperature) and a given restriction on the external temperature of the grill flaps. The permissible temperature was chosen from the condition that there was no ice formation under conditions of low temperature and the presence of sea moisture.

3. The results of numerical simulation of the parameters of the air flow around the structure of the air intake grille

Several series of variant numerical modeling of heating of the shutter leaf structures with AIHS were carried out. The following parameters varied:

- cold air temperature in the range from minus 10 to minus 50;
- the velocity of the flowing stream of cold air in the range from 5 to 20 m / s;
- the power of the electric heater W per 1 element of the blinds of the AIHS lattice, in the range from 100 to 1000 W;

The value of the heat flux density during electric heating was calculated by the formula:, where is the thermal equivalent of the transfer of electrical energy to heat; S is the area of the heated surface of the shutters.

Figures 2 – 5 show some of the results of numerical variant modeling of the parameters of the air flow around the structure of the air intake grille.

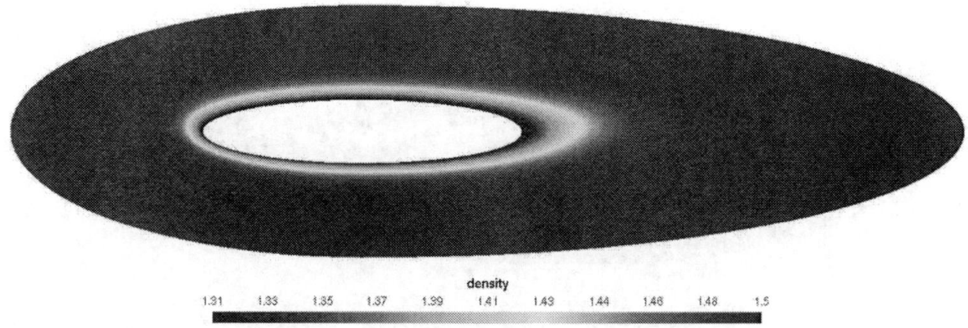

Figure 2: The calculated field of the density of the air flow flowing around the design of the blinds of the airspace grille at a temperature of cold air, a speed of 20 m/s, the heat flux density of the heater 210 W / m^2.

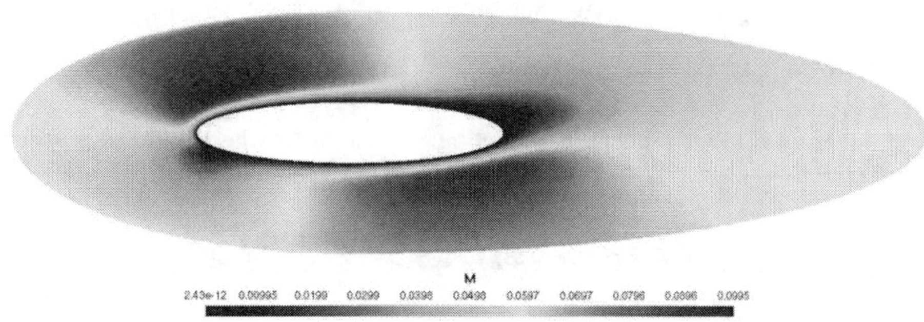

Figure 3: The calculated field of the Mach number of the air flow flowing around the design of the air curtain louvres at the cold air temperature, speed 20 m / s, heater heat flux density 210 W / m^2.

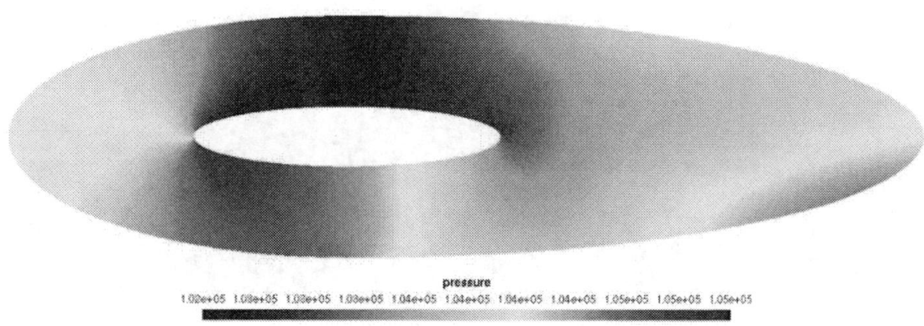

Figure 4: The calculated field of pressure of the air flow flowing around the design of the air curtain louvres at a cold air temperature, a speed of 20 m / s, heater heat flux density 210 W / m^2.

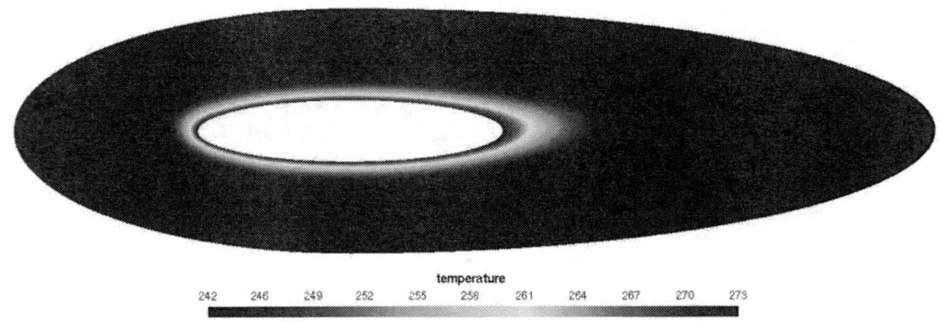

Figure 5: The calculated temperature field of the air flow flowing around the design of the air curtain louvres at a cold air temperature, a speed of 20 m / s, heater heat flux density 210 W / m^2.

As a result of the calculations, the values of the heat transfer coefficient of convective heat fluxes acting on the surface of the gratings were obtained 0.4 W / (m^2 K).

4. The results of the variant modeling of the heat transfer process in the design of the air intake grille

Several series of variant numerical modeling of heating of the shutter leaf structures with AIHS were carried out. Figures 6 - 7 show some of the results of numerical variant modeling

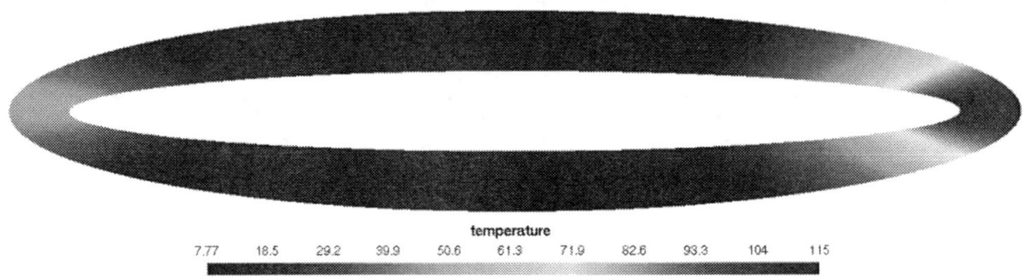

Figure 6: The calculated temperature field in the design of the shutter flap of the airspace shutter at a cold air temperature θ_e = - 50 °C, speed 5 m / s, heater heat flux density 770 W / m^2.

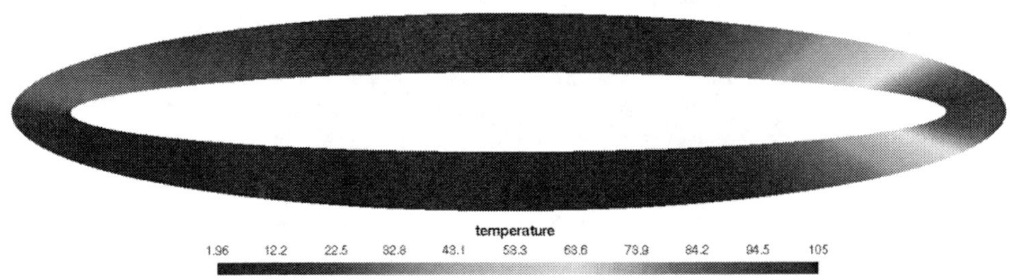

Figure 7: The calculated temperature field in the design of the shutter flap of the airspace grille at a temperature of cold air θ_e = - 50 °C, velocity 20 m / s, heater heat flux density 550 W / m^2.

The numerical values of the maximum and minimum temperatures on the housing of the louver shell for various calculation options are presented in tables 1.

Table 1. Numerical values of the maximum and minimum temperatures on the louvre casing at different air temperature and a flow velocity values.

	Temperature of the case of louvre, °C	Electric heating parameters (heat flux density, W/m²)			
		210	250	300	350
	air temperature (-10 ° C), flow velocity 5 m / s				
1	Maximum temperature	53,400	59,20	62,40	62,60
2	Minimum temperature	0,926	1,54	4,56	7,82
	air temperature (-10 ° C), flow velocity 20 m / s				
1	Maximum temperature	56,40	61,20	68,40	72,60
2	Minimum temperature	1,90	3,54	5,56	8,82
	air temperature (-50 ° C), flow velocity 5 m / s				
1	Maximum temperature	69,50	78,30	97,10	115,30
2	Minimum temperature	-12,90	-7,75	0,04	7,70

Figure 8 shows 3D calculations of the temperature field for an air intake grill at an air temperature of minus 50 ° C and a heat flux density of 670 W / m².

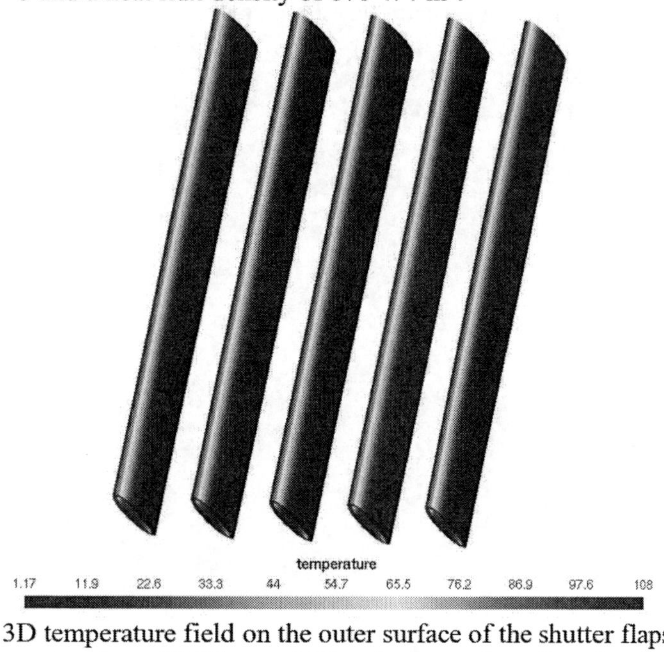

Figure 8. The calculated 3D temperature field on the outer surface of the shutter flaps of the air-outlet grille at a cold air temperature θ_e = -50 °C, speed 20 m / s, heater heat flux density 670 W / m².

5. Conclusion

A technique has been developed for selecting the optimal parameters of the electric heating system for various operating conditions of air intake grilles.

As a result of numerical simulation of heat transfer parameters for a typical air intake grille, it was found that with a decrease in the temperature of cold air from minus 10 to minus 50 ° C, the required power of the electric heater for heating the blinds increases from 210 W / m² to 570 W / m². Changes in wind speed from 5 to 20 m / s have little effect on the required power of electric heating.

The numerical simulation of the heat exchange processes of the heating system under the temperature conditions of the tests allowed us to verify the developed physical and mathematical model of heat transfer of the design parameters of the heating system.

References

[1] Lozowski E, Szilder K, and Makkonen L, 2000 Computer simulation of marine ice accretion. *Philosophical Trans of the Royal Society: Mathematical, Physical and Engineering Sciences*, Vol. 358, 2811–2845.

[2] Feher, L., and M. Thumm, Design of Avionic Microwave De-/Anti-Icing Systems//. in M.Willert-Porada(ed), Microwave devices. Springer 792 p., ISBN: 3540432523, pp. 695–702.- 2006.

[3] Petrenko, V.F., and C. Sullivan, 2003, Methods and Systems for Removing Ice from Surfaces. US Patent 6,653,598 B2, 25 November 2003.

[4] A. G. Kirchanov, Yu. I. Dimitrienko, V. Yu. Chibisov, I.K. Krasnov. 2014 Method of control of icing of louvers of air intake grating. *RF patent* RUS 2563710 23.05.2014

[5] Kirchanov A.G., Kokotkov V.V., Dimitrienko Yu.I., Ryizhkov A.V., Gavrilov A.Yu. Sposob zashchity vozdukhozabornykh reshetok s zhalyuzi ot obledeneniya i ustroystvo dlya yego osushchestvleniya [Way of protection of air-intake lattices with blinds from frosting and the device for its implementation]. *RF Patent* 2563715. Publ. August 26, 2015

[6] Yu I Dimitrienko*, V Yu Chibisov, M N Koryakov, S V Lebedev Simulation of thermal conditions of anti-icing systems IOP Conference Series: Material Science and Engeneering, 2019. volume 683 № 012004 doi:10.1088/1757-899X/683/1/012004 pp.1-6.

[7] Dimitrienko Yu I , Chibisov V Yu, Kirchanov A G, Voroshilov R Ya. 2015 Development and testing of anti-icing heating system for ship air-inlet grids *Engineering journal: science and innovation*, **12**(48), DOI: 10.18698/2308-6033-2015-12-1453

[8] Dimitrienko Yu. I., Kotenev V. P., Zakharov A. A. 2011 *The method of tape adaptive grids for numerical simulation in gas dynamics.* M.: Fizmatlit. 280 p.

AUTHOR INDEX

Institute of Physics
Dirac House, Temple Back
Bristol BS1 6BE UK

ISSN: 1757-8981
ISBN 978-1-7138-4770-0

9 781713 847700

CURRAN ASSOCIATES INC.
proceedings.com